Student Solutions M

to accompany

University Physics with Modern Physics

Second Edition

Wolfgang Bauer
Michigan State University

Gary D. Westfall
Michigan State University

Prepared by
diacriTech Inc.

The McGraw-Hill Companies

Student Solutions Manual to accompany
UNIVERSITY PHYSICS WITH MODERN PHYSICS, SECOND EDITION
WOLFGANG BAUER AND GARY D. WESTFALL

Published by McGraw-Hill Higher Education, an imprint of The McGraw-Hill Companies, Inc., 1221 Avenue of the Americas, New York, NY 10020.

This book is printed on acid-free paper.

1 2 3 4 5 6 7 8 9 0 QVS/QVS 1 0 9 8 7 6 5 4 3

ISBN: 978-0-07-740958-6
MHID: 0-07-740958-2

www.mhhe.com

Student Solutions Manual to accompany

***UNIVERSITY PHYSICS,* Second Edition**

Table of Contents

PART 8 RELATIVITY AND QUANTUM PHYSICS

Chapter 1: Overview

Exercises

1.35. (a) Three (b) Four (c) One (d) Six (e) One (f) Two (g) Three

1.39. Write "one ten-millionth of a centimeter" in scientific notation. One millionth is $1/10^6 = 1\cdot 10^{-6}$. Therefore, one ten-millionth is $1/\left[10\cdot 10^6\right] = 1/10^7 = 1\cdot 10^{-7}$ cm.

1.43. $1 \text{ km} = 1 \text{ km}\cdot\dfrac{1000 \text{ m}}{1 \text{ km}}\cdot\dfrac{1000 \text{ mm}}{1 \text{ m}} = 1{,}000{,}000 \text{ mm} = 1\cdot 10^6 \text{ mm}$

1.47. **THINK:** The cylinder has height h = 20.5 cm and radius r = 11.9 cm.
SKETCH:

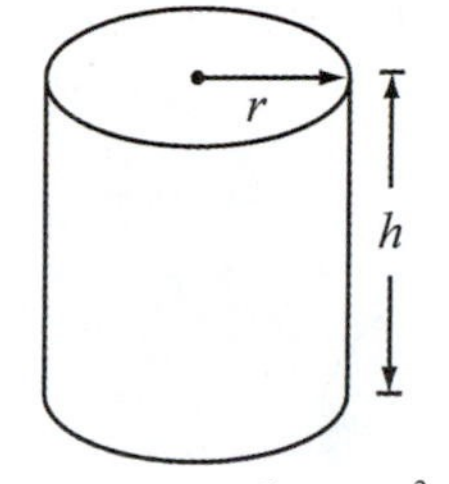

RESEARCH: The surface area of a cylinder is $A = 2\pi rh + 2\pi r^2$.
SIMPLIFY: $A = 2\pi r(h+r)$
CALCULATE: $A = 2\pi(11.9 \text{ cm})(20.5 \text{ cm} + 11.9 \text{ cm}) = 2422.545 \text{ cm}^2$
ROUND: Three significant figures: $A = 2.42\cdot 10^3 \text{ cm}^2$.
DOUBLE-CHECK: The units of area are a measure of distance squared so the answer is reasonable.

1.51. **THINK:** The given quantities, written in scientific notation and in units of meters, are: the starting position, $x_o = 7\cdot 10^{-3}$ m and the lengths of the flea's successive hops, $x_1 = 3.2\cdot 10^{-2}$ m, $x_2 = 6.5\cdot 10^{-2}$ m, $x_3 = 8.3\cdot 10^{-2}$ m, $x_4 = 10.0\cdot 10^{-2}$ m, $x_5 = 11.5\cdot 10^{-2}$ m and $x_6 = 15.5\cdot 10^{-2}$ m. The flea makes six jumps in total.
SKETCH:

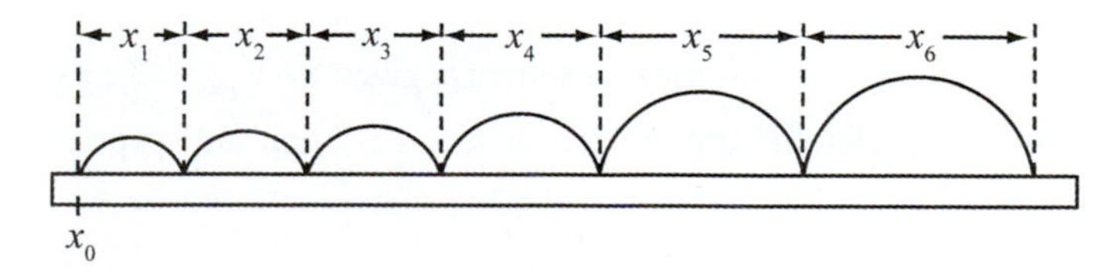

RESEARCH: The total distance jumped is $x_{\text{total}} = \sum_{n=1}^{6} x_n$. The average distance covered in a single hop is:

$$x_{\text{avg}} = \frac{1}{6}\sum_{n=1}^{6} x_n.$$

SIMPLIFY: $x_{\text{total}} = x_1 + x_2 + x_3 + x_4 + x_5 + x_6$, $x_{\text{avg}} = \dfrac{x_{\text{total}}}{6}$

CALCULATE: $x_{\text{total}} = (3.2 \text{ m} + 6.5 \text{ m} + 8.3 \text{ m} + 10.0 \text{ m} + 11.5 \text{ m} + 15.5 \text{ m})\cdot 10^{-2} = 55.0\cdot 10^{-2} \text{ m}$

$$x_{\text{avg}} = \frac{55.0\cdot 10^{-2} \text{ m}}{6} = 9.16666\cdot 10^{-2} \text{ m}$$

ROUND: Each of the hopping distances is measured to 1 mm precision. Therefore the total distance should also only be quoted to 1 mm precision: $x_{\text{total}} = 55.0\cdot 10^{-2}$ m. Rounding the average distance to the right number of significant digits, however, requires a few more words. As a general rule of thumb the average distance should be quoted to the same precision as the least precision of the individual distances, if there are only a few measurements contributing to the average. This is the case here, and so we state $x_{\text{avg}} = 9.17 \cdot 10^{-2}$ m. However, suppose we had 10,000 measurements contributing to an average. Surely we could then specify the average to a higher precision. The rule of thumb is that we can add one additional significant digit for every order of magnitude of the number of independent measurements contributing to an average. You see that the answer to this problem is yet another indication that specifying the correct number of significant figures can be complicated and sometimes outright tricky!

DOUBLE-CHECK: The flea made 6 hops, ranging from $3.2 \cdot 10^{-2}$ m to $15.5 \cdot 10^{-2}$ m, so the total distance covered is reasonable. The average distance per hop falls in the range between $3.2 \cdot 10^{-2}$ m and $1.55 \cdot 10^{-1}$ m, which is what is expected.

1.55. **THINK:** The radius of a planet, r_{p}, is 8.7 times greater than the Earth's radius, r_{E}. Determine how many times bigger the surface area of the planet is compared to the Earth's. Assume the planets are perfect spheres.

SKETCH:

RESEARCH: The surface area of a sphere is $A = 4\pi r^2$, so $A_{\text{E}} = 4\pi r_{\text{E}}^{\ 2}$, and $A_{\text{p}} = 4\pi r_{\text{p}}^{\ 2}$, and $r_p = 8.7 r_E$.

SIMPLIFY: $A_{\text{p}} = 4\pi (8.7 r_{\text{E}})^2$

CALCULATE: $A_{\text{p}} = (75.69) 4\pi r_{\text{E}}^{\ 2}$, and $A_{\text{E}} = 4\pi r_{\text{E}}^{\ 2}$. By comparison, $A_{\text{p}} = 75.69 A_{\text{E}}$.

ROUND: Rounding to two significant figures, the surface area of the planet is 76 times the surface area of Earth.

DOUBLE-CHECK: Since the area is proportional to the radius squared, it is expected that the surface area of the planet will be much larger than the surface area of the Earth, since its radius is 8.7 times Earth's radius.

1.61. **THINK:** The formula for the volume of a sphere is given by $V_{\text{sphere}} = (4/3)\pi r^3$. The formula for density is given by $\rho = m/V$. Refer to Appendix B in the text book and express the answers in SI units using scientific notation.

SKETCH:

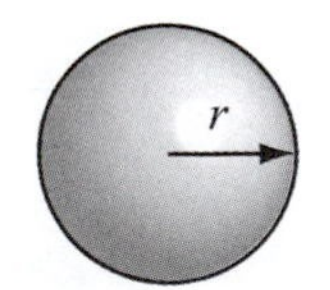

RESEARCH: The radius of the Sun is $r_{\text{S}} = 6.96 \cdot 10^8$ m, the mass of the Sun is $m_{\text{S}} = 1.99 \cdot 10^{30}$ kg, the radius of the Earth is $r_{\text{E}} = 6.37 \cdot 10^6$ m, and the mass of the Earth is $m_{\text{E}} = 5.98 \cdot 10^{24}$ kg.

SIMPLIFY: Not applicable.

CALCULATE:

(a) $V_{\text{S}} = \frac{4}{3}\pi r_{\text{S}}^{\ 3} = \frac{4}{3}\pi (6.96 \cdot 10^8)^3 = 1.412265 \cdot 10^{27}$ m^3

(b) $V_{\text{E}} = \frac{4}{3}\pi r_{\text{E}}^{\ 3} = \frac{4}{3}\pi (6.37 \cdot 10^6)^3 = 1.082696 \cdot 10^{21}$ m^3

(c) $\rho_S = \frac{m_S}{V_S} = \frac{1.99 \cdot 10^{30}}{1.412265 \cdot 10^{27}} = 1.40908 \cdot 10^3 \text{ kg/m}^3$

(d) $\rho_E = \frac{m_E}{V_E} = \frac{5.98 \cdot 10^{24}}{1.082696 \cdot 10^{21}} = 5.523249 \cdot 10^3 \text{ kg/m}^3$

ROUND: The given values have three significant figures, so the calculated values should be rounded as:

(a) $V_S = 1.41 \cdot 10^{27} \text{ m}^3$

(b) $V_E = 1.08 \cdot 10^{21} \text{ m}^3$

(c) $\rho_S = 1.41 \cdot 10^3 \text{kg/m}^3$

(d) $\rho_E = 5.52 \cdot 10^3 \text{ kg/m}^3$

DOUBLE-CHECK: The radius of the Sun is two orders of magnitude larger than the radius of the Earth. Because the volume of a sphere is proportional to the radius cubed, the volume of the Sun should be $(10^2)^3$ or 10^6 larger than the volume of the Earth, so the calculated volumes are reasonable. Because density depends on mass and volume, and the Sun is roughly 10^6 times larger and more massive than the Earth, it is not surprising that the density of the Sun is on the same order of magnitude as the density of the Earth (e.g. $10^6 / 10^6 = 1$). Earth is primarily solid, but the Sun is gaseous, therefore it is reasonable that the Earth is denser than the Sun.

1.65. **THINK:** Let $\vec{L}$ be the position vector. Then $|\vec{L}| = 40.0$ m and $\theta = 57.0°$ (above x-axis).

SKETCH:

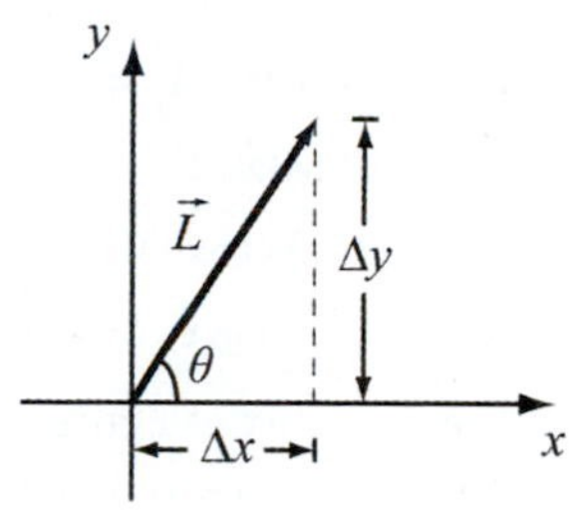

RESEARCH: From trigonometry, $\sin\theta = \Delta y / |\vec{L}|$ and $\cos\theta = \Delta x / |\vec{L}|$. The length of the vector $\vec{L}$ is given by the formula $|\vec{L}| = \sqrt{\Delta x^2 + \Delta y^2}$.

SIMPLIFY: $\Delta x = |\vec{L}|\cos\theta$, $\Delta y = |\vec{L}|\sin\theta$

CALCULATE: $\Delta x = (40.0 \text{ m})\cos(57.0°) = 21.786 \text{ m}$, $\Delta y = (40.0 \text{ m})\sin(57.0°) = 33.547 \text{ m}$

ROUND: $\Delta x = 21.8 \text{ m}$ and $\Delta y = 33.5 \text{ m}$.

DOUBLE-CHECK: $|\vec{L}| = \sqrt{\Delta x^2 + \Delta y^2} = \sqrt{(21.8 \text{ m})^2 + (33.5 \text{ m})^2} \approx 40.0 \text{ m}$, to three significant figures.

1.67. **THINK:** The lengths of the vectors are given as $|\vec{A}| = 75.0$, $|\vec{B}| = 60.0$, $|\vec{C}| = 25.0$ and $|\vec{D}| = 90.0$. The question asks for the vectors to be written in terms of unit vectors. Remember, when dealing with vectors, the x- and y-components must be treated separately.

SKETCH:

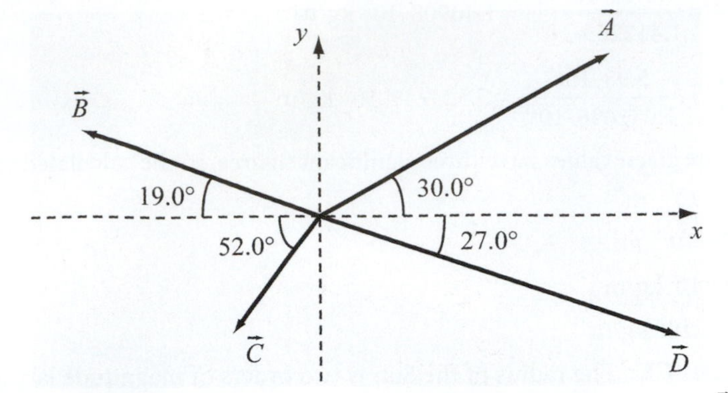

RESEARCH: The formula for a vector in terms of unit vectors is $\vec{V} = V_x\hat{x} + V_y\hat{y}$. Since $\sin\theta = \dfrac{\text{opposite}}{\text{hypotenuse}}$ and $\cos\theta = \dfrac{\text{adjacent}}{\text{hypotenuse}}$, $\sin\theta = \dfrac{A_y}{|\vec{A}|}$ and $\cos\theta = \dfrac{A_x}{|\vec{A}|}$.

$\theta_A = 30.0°$, $\theta_B = 19.0° = 161.0°$ (with respect to the positive x-axis),

$\theta_C = 52.0° = 232.0°$ (with respect to the positive x-axis),

$\theta_D = 27.0° = 333.0°$ (with respect to the positive x-axis).

SIMPLIFY: $A_x = |\vec{A}|\cos\theta_A$, $A_y = |\vec{A}|\sin\theta_A$, $B_x = |\vec{B}|\cos\theta_B$, $B_y = |\vec{B}|\sin\theta_B$, $C_x = |\vec{C}|\cos\theta_C$, $C_y = |\vec{C}|\sin\theta_C$, $D_x = |\vec{D}|\cos\theta_D$, and $D_y = |\vec{D}|\sin\theta_D$.

CALCULATE: $A_x = 75.0\cos 30.0° = 64.9519\hat{x}$, $A_y = 75.0\sin 30.0° = 37.5\hat{y}$

$B_x = 60.0\cos 161.0° = -56.73\hat{x}$, $B_y = 60.0\sin 161.0° = 19.534\hat{y}$

$C_x = 25.0\cos 232.0° = -15.3915\hat{x}$, $C_y = 25.0\sin 232.0° = -19.70027\hat{y}$

$D_x = 90.0\cos 333.0° = 80.19058\hat{x}$, $D_y = 90.0\sin 333.0° = -40.859144\hat{y}$

ROUND: The given values had three significant figures so the answers must be rounded to:

$\vec{A} = 65.0\hat{x} + 37.5\hat{y}$, $\hat{B} = -56.7\hat{x} + 19.5\hat{y}$, $\vec{C} = -15.4\hat{x} - 19.7\hat{y}$, $\vec{D} = 80.2\hat{x} - 40.9\hat{y}$.

DOUBLE-CHECK: Comparing the calculated components to the figure provided shows that this answer is reasonable.

1.71. **THINK:** The problem involves adding vectors, therefore break the vectors up into their components and add the components. NW is 45° north of west. $\vec{d}_1$ = 20 paces N, $\vec{d}_2$ = 30 paces NW, $\vec{d}_3$ = 12 paces N, $\vec{d}_4$ = 3 paces into ground ($\vec{d}_4$ implies 3 dimensions). Paces are counted to the nearest integer, so treat the number of paces as being precise.

SKETCH:

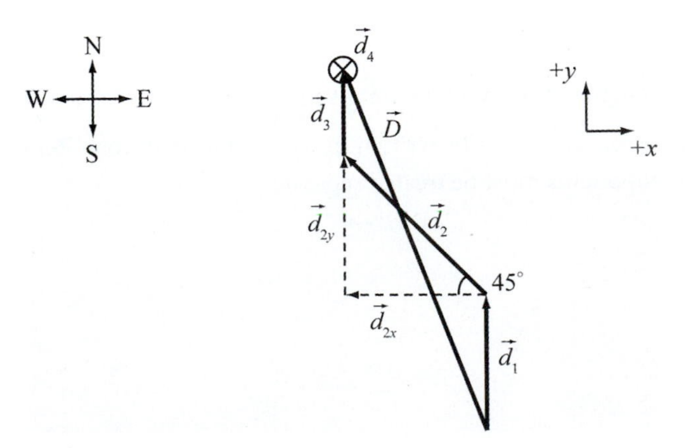

RESEARCH: $\vec{D}=\vec{d}_1+\vec{d}_2+\vec{d}_3+\vec{d}_4$, $\vec{d}_i=d_{ix}\hat{x}+d_{iy}\hat{y}+d_{iz}\hat{z}$, $|\vec{D}|=\sqrt{D_x^{\,2}+D_y^{\,2}+D_z^{\,2}}$, $\vec{d}_1=d_1\hat{y}$, $\vec{d}_2=-d_{2x}\hat{x}+d_{2y}\hat{y}=-d_2\cos(45°)\hat{x}+d_2\sin(45°)\hat{y}$, $\vec{d}_3=d_3\hat{y}$, and $\vec{d}_4=-d_4\hat{z}$.

SIMPLIFY: $\vec{D}=\vec{d}_1+\vec{d}_2+\vec{d}_3+\vec{d}_4=-d_2\cos(45°)\hat{x}+(d_1+d_3+d_2\sin(45°))\hat{y}-d_4\hat{z}$ and $|\vec{D}|=\sqrt{(-d_2\cos(45°))^2+(d_1+d_3+d_2\sin(45°))^2+(-d_4)^2}$.

CALCULATE: $\vec{D}=-30\frac{\sqrt{2}}{2}\hat{x}+\left(20+12+30\frac{\sqrt{2}}{2}\right)\hat{y}-3\hat{z}$

$|\vec{D}|=\sqrt{(-21.213)^2+(53.213)^2+(-3)^2}=57.36$ paces

ROUND: $\vec{D}=-15\sqrt{2}\hat{x}+\left(32+15\sqrt{2}\right)\hat{y}-3\hat{z}$ and round the number of paces to the nearest integer: $|\vec{D}|=57$ paces.

DOUBLE-CHECK: Distance should be less than the sum of the magnitudes of each vector, which is 65. Therefore, the calculated answer is reasonable.

1.75. $4{,}308{,}229\approx 4\cdot 10^6$; $44\approx 4\cdot 10^1$, $\left(4\cdot 10^6\right)\left(4\cdot 10^1\right)=16\cdot 10^7=2\cdot 10^8$

1.79. **THINK:** Sum the components of both vectors and find the magnitude and the angle from the positive x-axis of the resultant vector. $\vec{A}=(23.0,59.0)$ and $\vec{B}=(90.0,-150.0)$.

SKETCH:

(a)

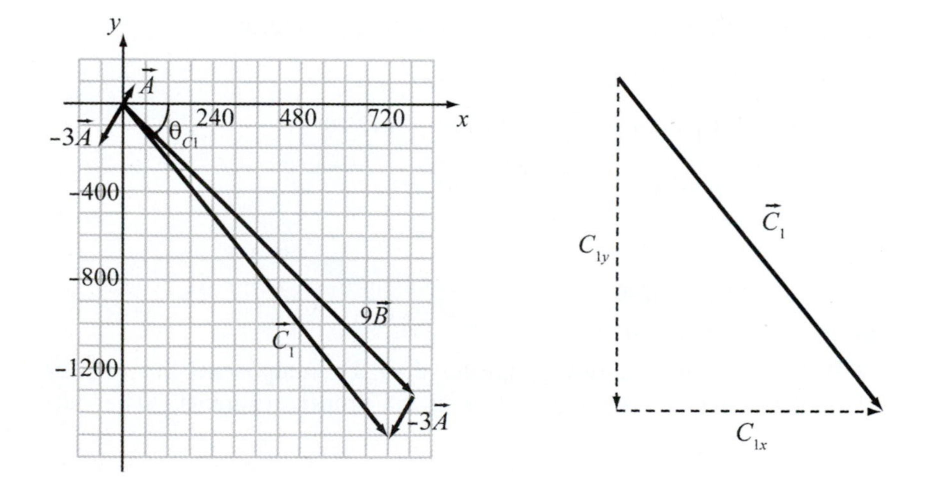

(b)

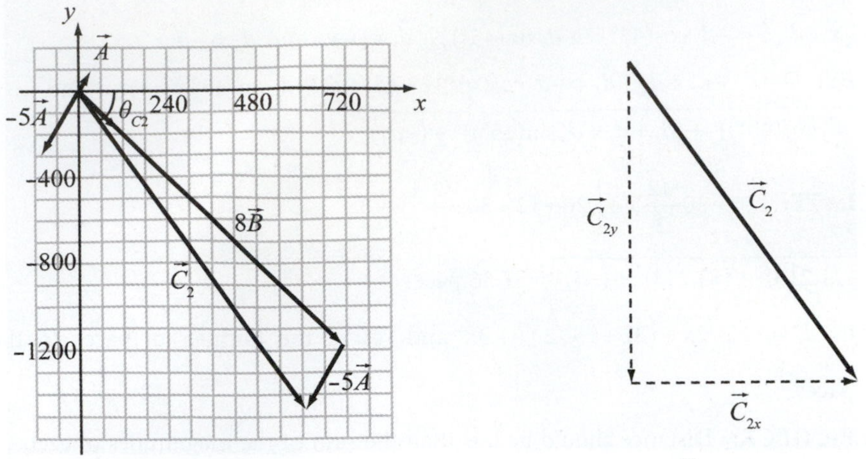

RESEARCH: $\vec{C}=(C_x,C_y)$, $C_i=nA_i+mB_i$, $\left|\vec{C}\right|=\sqrt{C_x^{\,2}+C_y^{\,2}}$, $\tan\theta_C=\dfrac{C_y}{C_x}$

SIMPLIFY:

(a) Since $n=-3$ and $m=9$, $C_x=-3A_x+9B_x$ and $C_y=-3A_y+9B_y$. Also, $\theta_C=\tan^{-1}\left(C_y/C_x\right)$.

(b) Since $n=-5$ and $m=8$, $C_x=-5A_x+8B_x$ and $C_y=-5A_y+8B_y$. Also, $\theta_C=\tan^{-1}\left(C_y/C_x\right)$.

CALCULATE:

(a) $C_x=-3(23.0)+9(90.0)=741.0,\quad C_y=-3(59.0)+9(-150)=-1527.0$

$\vec{A}=(A_x,A_y)=(-30.0\text{ m},-50.0\text{ m})$

(b) $=(30.0\text{ m},50.0\text{ m})$. $\left|\vec{C}\right|=\sqrt{(605.0)^2+(-1495.0)^2}=1612.78$

$$\theta_C=\tan^{-1}\left(\frac{-1495.0}{605.0}\right)=-67.97^\circ$$

ROUND:

(a) $\vec{C}=1.70\cdot 10^3$ at -64.1° or 296°

(b) $\vec{C}=1.61\cdot 10^3$ at -68.0° or 292°

DOUBLE-CHECK: Each magnitude is greater than the components but less than the sum of the components and the angles place the vectors in the proper quadrants. The calculated answers are reasonable.

1.83. **THINK:** Consider the 90° turns to be precise turns at right angles.

(a) The pilot initially flies N, then heads E, then finally heads S. Determine the vector $\vec{D}$ that points from the origin to the final point and find its magnitude. The vectors are $\vec{d}_1=155.3$ miles N, $\vec{d}_2=62.5$ miles E and $\vec{d}_3=47.5$ miles S.

(b) Now that the vector pointing to the final destination has been computed, $\vec{D}=d_2\hat{x}+(d_1-d_3)\hat{y}=(62.5\text{ miles})\hat{x}+(107.8\text{ miles})\hat{y}$, determine the angle the vector makes with the origin. The angle the pilot needs to travel is then 180° from this angle.

(c) Before the pilot turns S, he is farthest from the origin. This is because when he starts heading S, he is negating the distance travelled N. The only vectors of interest are $\vec{d}_1=155.3$ miles N and $\vec{d}_2=62.5$ miles E.

SKETCH:

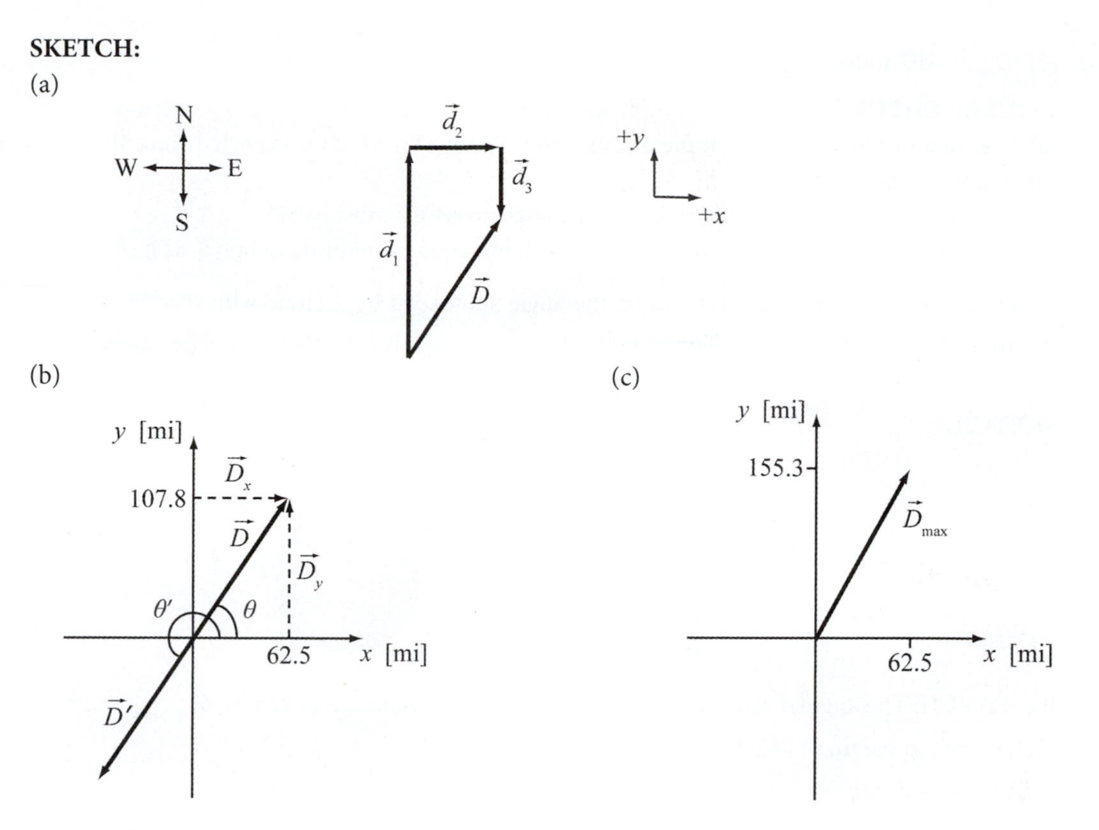

RESEARCH:

(a) $\vec{D}=\vec{d}_1+\vec{d}_2+\vec{d}_3=D_x\hat{x}+D_y\hat{y},\ \vec{d}_i=d_{ix}\hat{x}+d_{iy}\hat{y},\ \left|\vec{D}\right|=\sqrt{D_x^{\ 2}+D_y^{\ 2}}$

(b) $\tan\theta=\dfrac{D_y}{D_x},\ \theta'=\theta\pm180°$

(c) $\vec{D}_{\text{max}}=\vec{d}_1+\vec{d}_2,\ \vec{d}_i=d_{ix}\hat{x}+d_{iy}\hat{y},\ \left|\vec{D}_{\text{max}}\right|=\sqrt{D_x^{\ 2}+D_y^{\ 2}}$

SIMPLIFY:

(a) $\vec{d}_1=d_1\hat{y},\ \vec{d}_2=d_2\hat{x},\ \vec{d}_3=-d_3\hat{y}$

Therefore, $\vec{D}=d_2\hat{x}+(d_1-d_3)\hat{y}$ and $\left|\vec{D}\right|=\sqrt{d_2^{\ 2}+(d_1-d_3)^2}$.

(b) $\theta=\tan^{-1}\left(\dfrac{D_y}{D_x}\right)$ and $\theta'=\tan^{-1}\left(\dfrac{D_y}{D_x}\right)\pm180°$

(c) $\vec{d}_1=d_1\hat{y},\ \vec{d}_2=d_2\hat{x},\ \vec{D}_{\text{max}}=d_2\hat{x}+d_1\hat{y}\Rightarrow\left|\vec{D}_{\text{max}}\right|=\sqrt{d_2^{\ 2}+d_1^{\ 2}}$

CALCULATE:

(a) $\left|\vec{D}\right|=\sqrt{(62.5\text{ miles})^2+(155.3\text{ miles}-47.5\text{ miles})^2}$

$=124.608\text{ miles}$

(b) $\theta'=\tan^{-1}\left(\dfrac{107.8\text{ miles}}{62.5\text{ miles}}\right)\pm180°$

$=59.896°\pm180°=239.896°\text{ or }-120.104°$

(c) $\left|\vec{D}_{\text{max}}\right|=\sqrt{(62.5\text{ miles})^2+(155.3\text{ miles})^2}=167.405\text{ miles}$

ROUND:

(a) $\left|\vec{D}\right|=125\text{ miles}$

(b) $\theta'=240.°$ or $-120.°$ (from positive x-axis or E)

(c) $\left|\vec{D}_{\text{max}}\right| = 167$ miles

DOUBLE-CHECK:

(a) The total distance is less than the distance travelled north, which is expected since the pilot eventually turns around and heads south.

(b) The pilot is clearly NE of the origin and the angle to return must be SW.

(c) This distance is greater than the distance which included the pilot travelling S, as it should be.

1.87. **THINK:** 24.9 seconds of arc represents the angle subtended by a circle with diameter = $2r_{\text{M}}$ located a distance D_{EM} from Earth. This value must be converted to radians. The diameter of Mars is $2r_{\text{M}} = 6784$ km.

SKETCH:

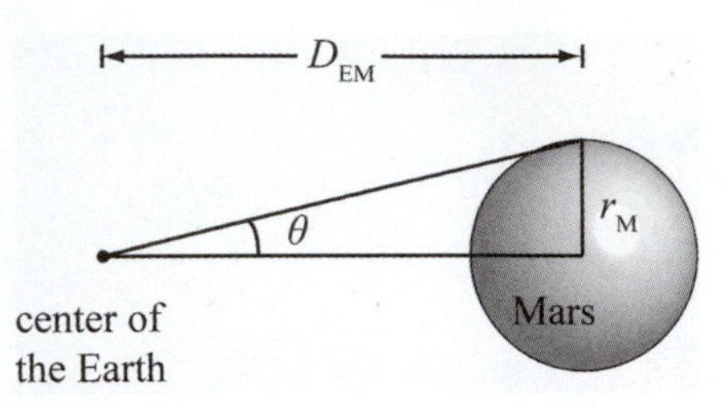

RESEARCH: The angular size is related to the angle θ shown in the sketch by $\theta_{\text{angular size}} = 2\theta$. From the sketch, we can see that

$$\tan\theta = \frac{r_{\text{M}}}{D_{\text{EM}}}.$$

Because Mars is a long distance from the Earth, even at closet approach, we can make the approximation $\tan\theta \approx \theta$.

SIMPLIFY: Putting our equations together gives us $\theta_{\text{angular size}} = 2\theta = \dfrac{2r_{\text{M}}}{D_{\text{EM}}}$.

CALCULATE: We first convert the observed angular size from seconds of arc to radians

$$24.9 \text{ arc seconds} \cdot \frac{1°}{3600 \text{ arc seconds}} \cdot \frac{2\pi \text{ radians}}{360°} = 1.207 \cdot 10^{-4} \text{ radians}.$$

The distance is then

$$D_{\text{EM}} = \frac{2r_{\text{M}}}{\theta_{\text{angular size}}} = \frac{6784 \text{ km}}{1.207 \cdot 10^{-4} \text{ radians}} = 5.6205 \cdot 10^{7} \text{ km}.$$

ROUND: We specify our answer to three significant figures, $D_{\text{EM}} = 5.62 \cdot 10^{7}$ km.

DOUBLE-CHECK: The mean distance from Earth to Mars is about $7 \cdot 10^{7}$ km. Because the distance calculated is for a close approach and the distance is less than the mean distance, the answer is reasonable.

1.91. **THINK:** Consider the Sun to be at the centre of a circle with Mercury on its circumference. This gives $r_{\text{M}} = 4.6 \cdot 10^{10}$ m as the radius of the circle. Earth is located a distance $r_{\text{E}} = 1.5 \cdot 10^{11}$ m from the Sun so that the three bodies form a triangle. The vector from Earth to the Sun is at $0°$. The vector from Earth to Mercury intersects Mercury's orbit once when Mercury is at a maximum angular separation from the Sun in the sky. This tangential vector is perpendicular to the radius vector of Mercury's orbit. The three bodies form a right angle triangle with r_{E} as the hypotenuse. Trigonometry can be used to solve for the angle and distance.

SKETCH:

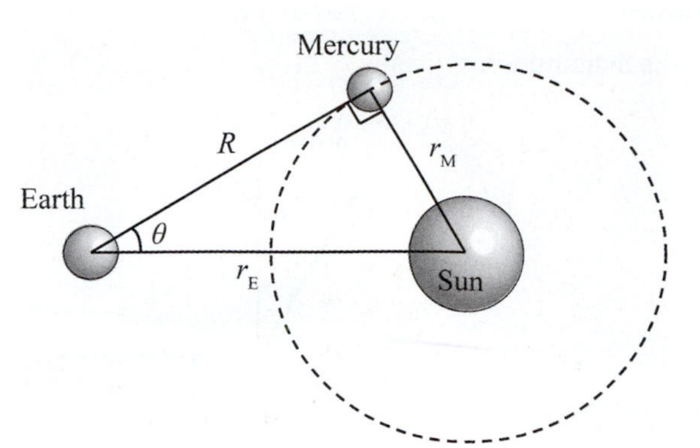

RESEARCH: $r_E^2 = r_M^2 + R^2, \;\; r_E \sin\theta = r_M$

SIMPLIFY: $R = \sqrt{r_E^2 - r_M^2}, \;\; \theta = \sin^{-1}\left(\frac{r_M}{r_E}\right)$

CALCULATE: $R = \sqrt{(1.5\cdot 10^{11})^2 - (4.6\cdot 10^{10})^2} = 1.4277\cdot 10^{11}$ m, $\theta = \sin^{-1}\left(\frac{4.6\cdot 10^{10}}{1.5\cdot 10^{11}}\right) = 17.858°$

ROUND: $R = 1.4\cdot 10^{11}$ m, $\theta = 18°$

DOUBLE-CHECK: If it had been assumed that the maximum angular separation occurred when the Earth to Sun to Mercury angle was 90°, $\theta = \tan^{-1}(r_M / r_E)$ would be about 17°. The maximum angle should be greater than this and it is, so the answer is reasonable.

Multi-Version Exercises

1.93. **THINK:** The lengths of the x and y components of the vectors can be read from the provided figure. Remember to decompose the vectors in terms of their x and y components.

SKETCH:

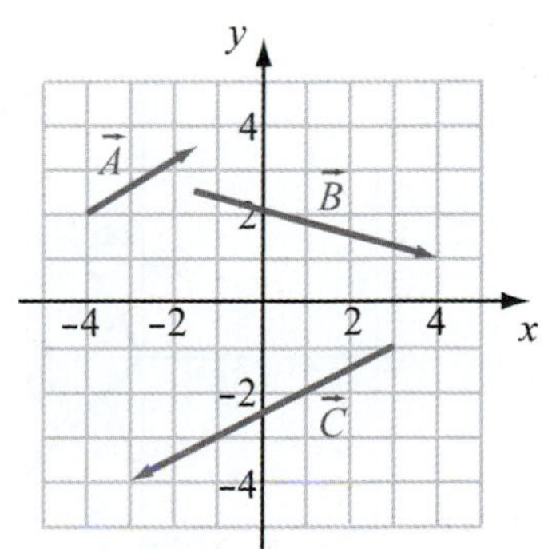

RESEARCH: A vector can be written as $\vec{V} = V_x\hat{x} + V_y\hat{y}$, where $V_x = x_f - x_i$ and $V_y = y_f - y_i$.

SIMPLIFY: Not applicable.

CALCULATE: $\vec{A} = (-1.5-(-4))\hat{x} + (3.5-2)\hat{y} = 2.5\hat{x} + 1.5\hat{y}$, $\vec{B} = (4-(-1.5))\hat{x} + (1-2.5)\hat{y} = 5.5\hat{x} - 1.5\hat{y}$

$\vec{C} = (-3-3)\hat{x} - (4-(-1))\hat{y} = -6\hat{x} - 3\hat{y}$

ROUND: Not applicable.

DOUBLE-CHECK: Comparing the signs of the x- and y-components of the vectors $\vec{A}$, $\vec{B}$ and $\vec{C}$ to the provided figure, the calculated components all point in the correct directions. The answer is therefore reasonable.

1.99. **THINK:** The two vectors are $\vec{A} = (A_x, A_y) = (30.0 \text{ m}, -50.0 \text{ m})$ and $\vec{B} = (B_x, B_y) = (-30.0 \text{ m}, 50.0 \text{ m})$. Sketch and find the magnitudes.

SKETCH:

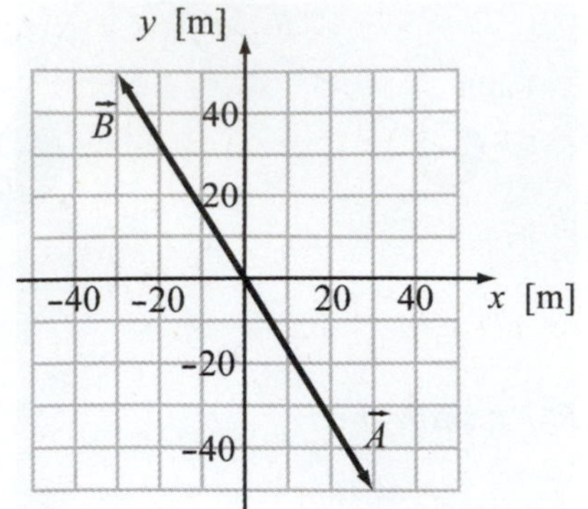

RESEARCH: The length of a vector $\vec{C} = C_x\hat{x} + C_y\hat{y}$ is $|\vec{C}| = \sqrt{C_x^2 + C_y^2}$.

SIMPLIFY: $|\vec{A}| = \sqrt{A_x^2 + A_y^2}$, $|\vec{B}| = \sqrt{B_x^2 + B_y^2}$

CALCULATE: $|\vec{A}| = \sqrt{(30)^2 + (-50)^2} = 58.3095 \text{ m}$, $|\vec{B}| = \sqrt{(-30)^2 + (50)^2} = 58.3095 \text{ m}$

ROUND: $|\vec{A}| = 58.3 \text{ m}$, $|\vec{B}| = 58.3 \text{ m}$

DOUBLE-CHECK: The calculated magnitudes are larger than the lengths of the component vectors, and are less than the sum of the lengths of the component vectors. Also, the vectors are opposites, so they should have the same length.

1.103. **THINK:** The two vectors are $\vec{A} = (23.0, 59.0)$ and $\vec{B} = (90.0, -150.0)$. Find the magnitude and angle with respect to the positive x-axis.

SKETCH:

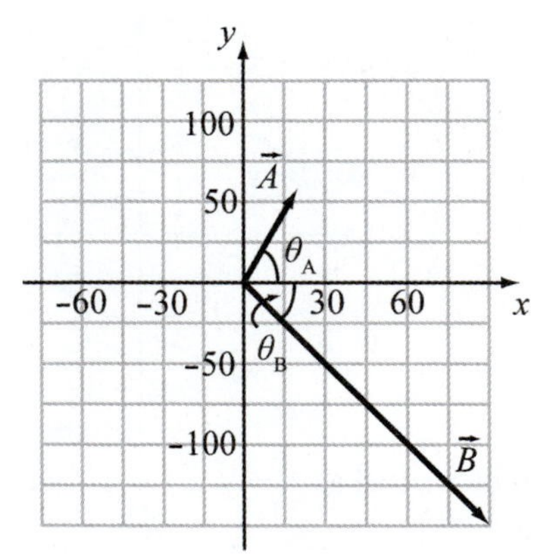

RESEARCH: For any vector $\vec{C} = C_x\hat{x} + C_y\hat{y}$, the magnitude is given by the formula $|\vec{C}| = \sqrt{C_x^2 + C_y^2}$, and the angle θ_C made with the x-axis is such that $\tan\theta_C = \dfrac{C_y}{C_x}$.

SIMPLIFY: $|\vec{A}| = \sqrt{A_x^2 + A_y^2}$, $|\vec{B}| = \sqrt{B_x^2 + B_y^2}$, $\theta_A = \tan^{-1}\left(\dfrac{A_y}{A_x}\right)$, $\theta_B = \tan^{-1}\left(\dfrac{B_y}{B_x}\right)$

CALCULATE: $|\vec{A}| = \sqrt{(23.0)^2 + (59.0)^2} = 63.3246$, $|\vec{B}| = \sqrt{(90.0)^2 + (-150.0)^2} = 174.9286$

$$\theta_A = \tan^{-1}\left(\frac{59.0}{23.0}\right) = 68.7026°, \ \theta_B = \tan^{-1}\left(\frac{-150.0}{90.0}\right) = -59.0362°$$

ROUND: Three significant figures: $\vec{A} = 63.3$ at $68.7°$, $\vec{B} = 175$ at $-59.0°$ or $301.0°$.

DOUBLE-CHECK: Each magnitude is greater than the components but less than the sum of the components, and the angles place the vectors in the proper quadrants.

1.107. **THINK:** The scalar product of two vectors equals the length of the two vectors times the cosine of the angle between them. Geometrically, think of the absolute value of the scalar product as the length of the projection of vector $\vec{B}$ onto vector $\vec{A}$ times the length of vector $\vec{A}$, or the area of a rectangle with one side the length of vector $\vec{A}$ and the other side the length of the projection of vector $\vec{B}$ onto vector $\vec{A}$. Algebraically, use the formula $\vec{A} \bullet \vec{B} = A_x B_x + A_y B_y$ to find the scalar product from the components, which can be read from the graphs.

SKETCH: Using the geometric interpretation, sketch the projection of vector $\vec{B}$ onto vector $\vec{A}$ and then draw the corresponding rectangular area, for instance for case (e):

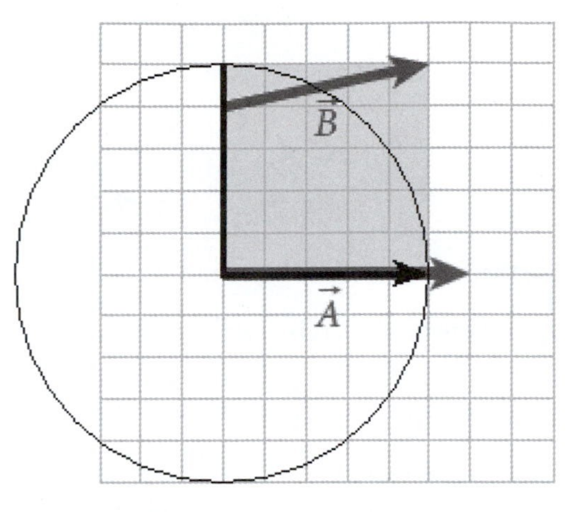

Note, however, that this method of finding the scalar product is cumbersome and does not readily produce exact results. The algebraic approach is much more efficient.

RESEARCH: Use the formula $\vec{A} \bullet \vec{B} = A_x B_x + A_y B_y$ to find the scalar product from the components, which can be read from the graphs. $(A_x, A_y) = (6,0)$ in all cases. In (a), $(B_x, B_y) = (1,5)$; in (b), $(B_x, B_y) = (0,3)$; in (c), $(B_x, B_y) = (2,2)$; in (d), $(B_x, B_y) = (-6,0)$; in (e), $(B_x, B_y) = (5,1)$; and in (f), $(B_x, B_y) = (1,4)$.

SIMPLIFY: Using the formula $\vec{A} \bullet \vec{B} = A_x B_x + A_y B_y$, find that in part (a), $\vec{A} \bullet \vec{B} = 6 \cdot 1 + 0 \cdot 5$. In part (b), $\vec{A} \bullet \vec{B} = 6 \cdot 0 + 0 \cdot 3$. In part (c), $\vec{A} \bullet \vec{B} = 6 \cdot 2 + 0 \cdot 2$. In part (d), $\vec{A} \bullet \vec{B} = 6 \cdot -6 + 0 \cdot 0$. In part (e), $\vec{A} \bullet \vec{B} = 6 \cdot 5 + 0 \cdot 1$. Finally, in part (f), $\vec{A} \bullet \vec{B} = 6 \cdot 1 + 0 \cdot 4$.

CALCULATE: Performing the multiplication and addition as shown above, the scalar product in (a) is 6 units, in (b) it is 0 units, and in part (c) the scalar product is 12 units. In parts (d), (e), and (f), the scalar products are –36 units, 30 units, and 6 units, respectively. The one with the largest absolute value is case (d), $|-36| = 36$.

ROUND: Rounding is not required in this problem.

DOUBLE-CHECK: Double-check by looking at the rectangles with sides the length of vector $\vec{A}$ and the length of the projection of vector $\vec{B}$ onto vector $\vec{A}$. The results agree with what was calculated using the formula.

1.113. **THINK:** We are given the change in the star's radius. So, if we can express the surface area, circumference, and volume in terms of the radius, we can find by what factors these change as the radius changes.

SKETCH: We can think of the star as a sphere in space with radius r.

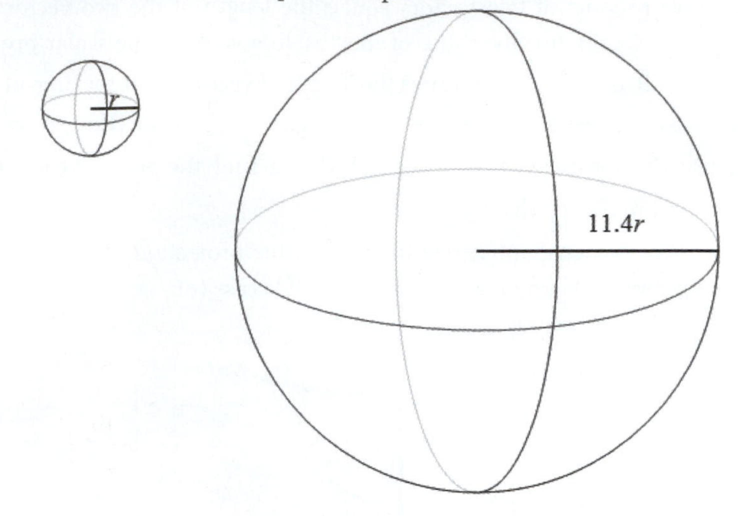

RESEARCH: We can use the formulas for volume and surface area of a sphere given in Appendix A. We find that the volume of the sphere on the left is $\frac{4}{3}\pi r^3$ and its surface area is $4\pi r^2$. Similarly, the volume of the sphere on the right is $\frac{4}{3}\pi(11.4r)^3$ and its surface area is $4\pi(11.4r)^2$. The circumference of a sphere is the same as the circumference of a great circle around it (shown in red in the sketch). Finding the radius of the circle will give us the radius of the sphere. Using this method, we find that the circumference of the sphere on the left is $2\pi r$, while the sphere on the right has a circumference of $2\pi(11.4r)$.

SIMPLIFY: We use algebra to find the volume, surface area, and circumference of the larger sphere in terms of the volume, surface area, and circumference of the smaller sphere. (a) We find the surface area of the sphere on the right is $4\pi(11.4r)^2 = 4\pi(11.4^2 r^2) = 11.4^2 \cdot 4\pi r^2$ (b) The circumference of the larger sphere is $2\pi(11.4r) = 11.4\cdot(2\pi r)$. (c) The volume of the larger sphere is $\frac{4}{3}\pi(11.4r)^3 = \frac{4}{3}\pi(11.4^3 r^3) = (11.4^3)\frac{4}{3}\pi r^3$.

CALCULATE: Since we don't know the star's original radius, we take the ratio of the new value divided by the old value to get the factor by which the surface area, volume, and circumference have increased. In part (a), we find that the ratio of the new surface area to the original surface area is $\frac{11.4^2 \cdot 4\pi r^2}{4\pi r^2} = 11.4^2 \frac{4\pi r^2}{4\pi r^2} = 11.4^2 \cdot 1 = 129.96$. (b) Similarly, we can divide the new circumference by the original one to get $\frac{11.4\cdot(2\pi r)}{2\pi r} = 11.4\frac{2\pi r}{2\pi r} = 11.4\cdot 1 = 11.4$. (c) The new volume divided by the original volume is $\frac{(11.4^3)\frac{4}{3}\pi r^3}{\frac{4}{3}\pi r^3} = (11.4^3)\frac{\frac{4}{3}\pi r^3}{\frac{4}{3}\pi r^3} = 11.4^3 \cdot 1 = 1481.544$.

ROUND: For all of these calculations, we round to three significant figures. This gives us that (a) the surface area has increased by a factor of130., (b) the circumference has increased by a factor of 11.4, and (c) the volume has increased by a factor of 1.48×10^3.

DOUBLE-CHECK: Think about what these values represent. The circumference is a one-dimensional quantity, with units such as km, which is proportional to r. The surface area is a two-dimensional quantity with units such as km^2, and is proportional to r^2. The volume is a three-dimensional quantity with units such as km^3 and is proportional to r^3. So it makes sense that, when we increase the radius by a given amount (11.4 in this case), the circumference increases in proportion to that amount, while the surface area increases by that amount squared, and the volume increases by the cube of that amount.

Chapter 2: Motion in a Straight Line

Exercises

2.29. **THINK:** What is the distance traveled, p, and the displacement d if $v_1 = 30.0$ m/s due north for $t_1 = 10.0$ min and $v_2 = 40.0$ m/s due south for $t_2 = 20.0$ min? Times should be in SI units: $t_1 = 10.0 \text{ min}(60 \text{ s/min}) = 6.00 \cdot 10^2$ s, $t_2 = 20.0 \text{ min}(60 \text{ s/min}) = 1.20 \cdot 10^3$ s.

SKETCH:

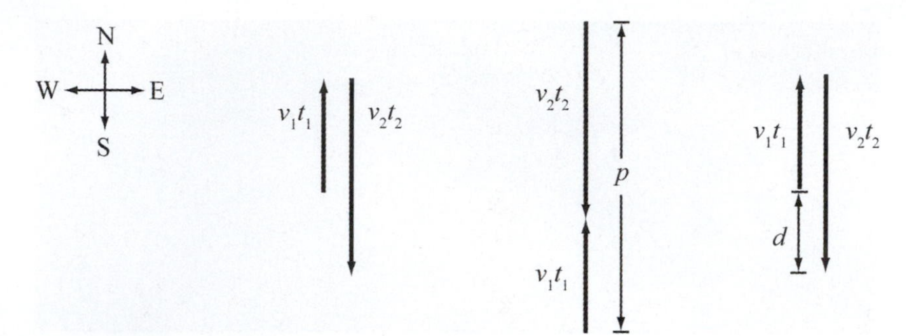

RESEARCH: The distance is equal to the product of velocity and time. The distance traveled is $p = v_1t_1 + v_2t_2$ and the displacement is the distance between where you start and where you finish, $d = v_1t_1 - v_2t_2$.

SIMPLIFY: There is no need to simplify.

CALCULATE: $p = v_1t_1 + v_2t_2 = (30. \text{ m/s})(6.00 \cdot 10^2 \text{ s}) + (40. \text{ m/s})(1.20 \cdot 10^3 \text{ s}) = 66{,}000. \text{ m}$

$d = v_1t_1 - v_2t_2 = (30. \text{ m/s})(6.00 \cdot 10^2 \text{ s}) - (40. \text{ m/s})(1.20 \cdot 10^3 \text{ s}) = -30{,}000. \text{ m}$

ROUND: The total distance traveled is 66.0 km, and the displacement is 30.0 km in southern direction.

DOUBLE-CHECK: The distance traveled is larger than the displacement as expected. Displacement is also expected to be towards the south since the second part of the trip going south is faster and has a longer duration.

2.33. **THINK:** The provided graph must be used to answer several questions about the speed and velocity of a particle. Questions about velocity are equivalent to questions about the slope of the position function.

SKETCH:

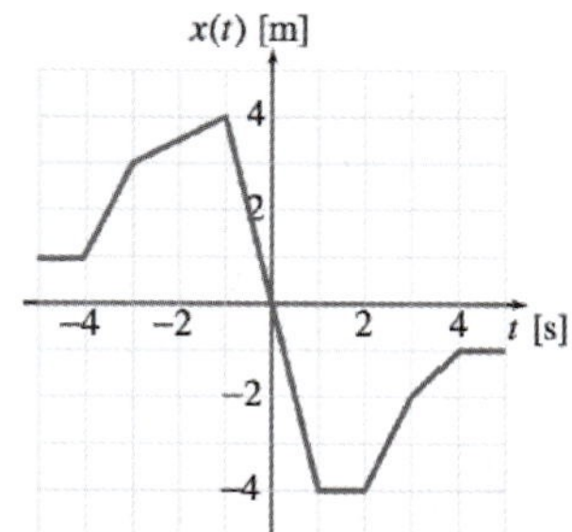

RESEARCH: The velocity is given by the slope on a distance versus time graph. A steeper slope means a greater speed.

$$\text{average velocity} = \frac{\text{final position} - \text{initial position}}{\text{time}}, \quad \text{speed} = \frac{\text{total distance traveled}}{\text{time}}$$

(a) The largest speed is where the slope is the steepest.

(b) The average velocity is the total displacement over the time interval.

(c) The average speed is the total distance traveled over the time interval.

(d) The ratio of the velocities is $v_1 : v_2$.

(e) A velocity of zero is indicated by a slope that is horizontal.

SIMPLIFY:

(a) The largest speed is given by the steepest slope occurring between –1 s and +1 s.

$$s = \frac{|x(t_2) - x(t_1)|}{t_2 - t_1}, \text{ with } t_2 = 1 \text{ s and } t_1 = -1 \text{ s.}$$

(b) The average velocity is given by the total displacement over the time interval.

$$\bar{v} = \frac{x(t_2) - x(t_1)}{t_2 - t_1}, \text{ with } t_2 = 5 \text{ s and } t_1 = -5 \text{ s.}$$

(c) In order to calculate the speed in the interval –5 s to 5 s, the path must first be determined. The path is given by starting at 1 m, going to 4 m, then turning around to move to –4 m and finishing at –1 m. So the total distance traveled is

$$\begin{aligned} p &= |(4 \text{ m} - 1 \text{ m})| + |((-4 \text{ m}) - 4 \text{ m})| + |(-1 \text{ m} - (-4 \text{ m}))| \\ &= 3 \text{ m} + 8 \text{ m} + 3 \text{ m} \\ &= 14 \text{ m} \end{aligned}$$

This path can be used to find the speed of the particle in this time interval.

$$s = \frac{p}{t_2 - t_1}, \text{ with } t_2 = 5 \text{ s and } t_1 = -5 \text{ s.}$$

(d) The first velocity is given by $v_1 = \frac{x(t_3) - x(t_2)}{t_3 - t_2}$ and the second by $v_2 = \frac{x(t_4) - x(t_3)}{t_4 - t_3}$,

(e) The velocity is zero in the regions 1 s to 2 s, – 5 s to – 4 s, and 4 s to 5 s.

CALCULATE:

(a) $s = \frac{|-4 \text{ m} - 4 \text{ m}|}{1 \text{ s} - (-1 \text{ s})} = 4.0 \text{ m/s}$

(b) $\bar{v} = \frac{-1 \text{ m} - 1 \text{ m}}{5 \text{ s} - (-5 \text{ s})} = -0.20 \text{ m/s}$

(c) $\bar{s} = \frac{14 \text{ m}}{5 \text{ s} - (-5 \text{ s})} = 1.4 \text{ m/s}$

(d) $v_1 = \frac{(-2 \text{ m}) - (-4 \text{ m})}{3 \text{ s} - 2 \text{ s}} = 2.0 \text{ m/s}$, $v_2 = \frac{(-1 \text{ m}) - (-2 \text{ m})}{4 \text{ s} - 3 \text{ s}} = 1.0 \text{ m/s}$, so $v_1 : v_2 = 2:1$.

(e) There is nothing to calculate.

ROUND: Rounding is not necessary in this case, because we can read the values of the positions and times off the graph to at least 2 digit precision.

DOUBLE-CHECK: The values are reasonable for a range of positions between –4 m and 4 m with times on the order of seconds. Each calculation has the expected units.

2.37. **THINK:**

(a) I want to find the velocity at t = 10.0 s of a particle whose position is given by the function $x(t) = At^3 + Bt^2 + Ct + D$, where A = 2.10 m/s^3, B = 1.00 m/s^2, C = –4.10 m/s, and D = 3.00 m. I can differentiate the position function to derive the velocity function.

(b) I want to find the time(s) when the object is at rest. The object is at rest when the velocity is zero. I'll solve the velocity function I obtain in (a) equal to zero.

(c) I want to find the acceleration of the object at t = 0.50 s. I can differentiate the velocity function found in part (a) to derive the acceleration function, and then calculate the acceleration at t = 0.50 s.

(d) I want to plot the function for the acceleration found in part (c) between the time range of –10.0 s to 10.0 s.

SKETCH:

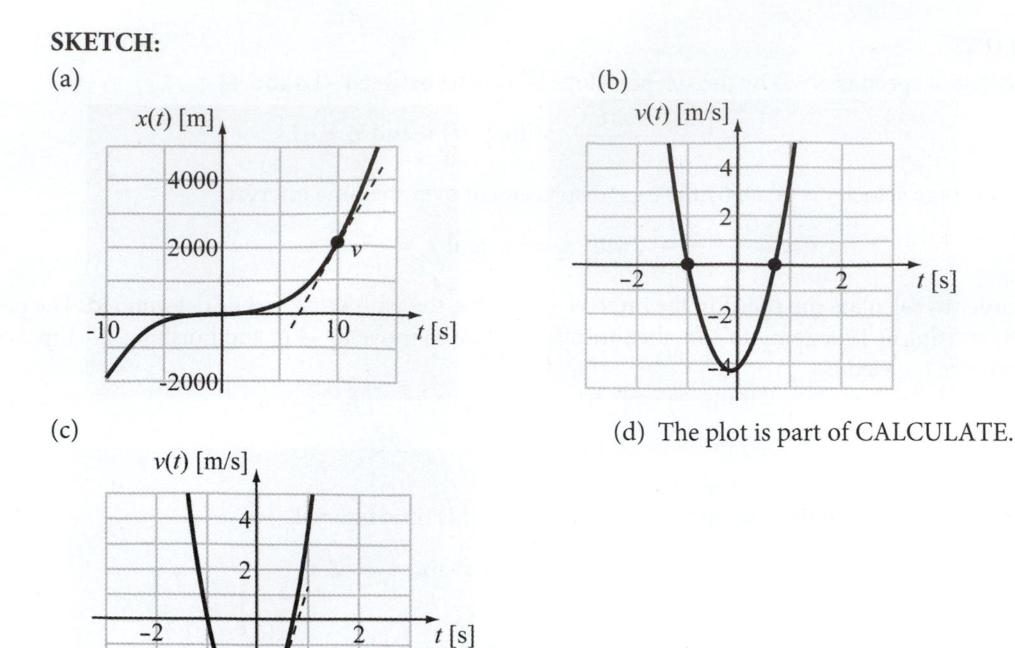

(d) The plot is part of CALCULATE.

RESEARCH:

(a) The velocity is given by the time derivative of the positive function $v(t)=\frac{d}{dt}x(t)$.

(b) To find the time when the object is at rest, set the velocity to zero, and solve for t. This is a quadratic equation of the form $ax^2+bx+c=0$, whose solution is $x=\dfrac{-b\pm\sqrt{b^2-4ac}}{2a}$.

(c) The acceleration is given by the time derivative of the velocity: $a(t)=\frac{d}{dt}v(t)$.

(d) The equation for acceleration found in part (c) can be used to plot the graph of the function.

SIMPLIFY:

(a) $v(t)=\frac{d}{dt}x(t)=\frac{d}{dt}(At^3+Bt^2+Ct+D)=3At^2+2Bt+C$

(b) Set the velocity equal to zero and solve for t using the quadratic formula:

$$t=\frac{-2B\pm\sqrt{4B^2-4(3A)(C)}}{2(3A)}=\frac{-2B\pm\sqrt{4B^2-12AC}}{6A}$$

(c) $a(t)=\frac{d}{dt}v(t)=\frac{d}{dt}(3At^2+2Bt+C)=6At+2B$

(d) There is no need to simplify this equation.

CALCULATE:

(a) $v(t=10.0\text{ s})=3(2.10\text{ m/s}^3)(10.0\text{ s})^2+2(1.00\text{ m/s}^2)(10.0\text{ s})-4.10\text{ m/s}=645.9\text{ m/s}$

(b) $t=\dfrac{-2(1.00\text{ m/s}^2)\pm\sqrt{4(1.00\text{ m/s}^2)^2-12(2.10\text{ m/s}^3)(-4.10\text{ m/s})}}{6(2.1\text{ 0m/s}^3)}$

$=0.6634553\text{ s},-0.9809156\text{ s}$

(c) $a(t=0.50\text{ s})=6(2.10\text{ m/s}^3)(0.50\text{ s})+2(1.00\text{ m/s}^2)=8.30\text{ m/s}^2$

(d) The acceleration function, $a(t) = 6At + 2B$, can be used to compute the acceleration for time steps of 2.5 s. For example:

$$a(t=-2.5\text{ s})=6(2.10\text{ m/s}^3)(-2.5\text{ s})+2(1.00\text{ m/s}^2)=-29.5\text{ m/s}^2$$

The result is given in the following table.

t [s]	−10.0	−7.5	−5.0	−2.5	0.0	2.5	5.0	7.5	10.0
a [m/s^2]	−124.0	−92.5	−61.0	−29.5	2.0	33.5	65.0	96.5	128.0

These values are used to plot the function.

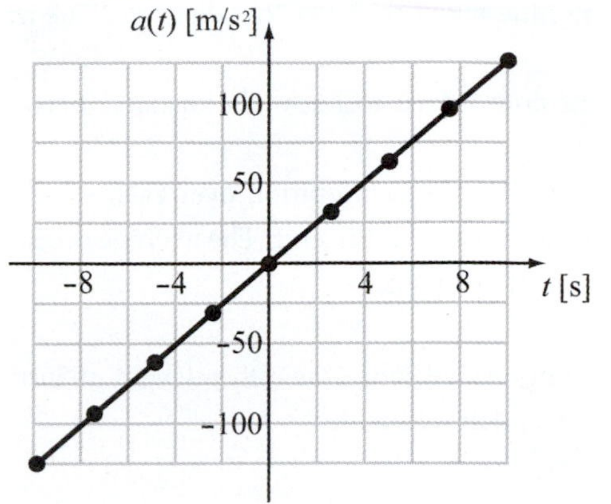

ROUND:

(a) The accuracy will be determined by the factor $3(2.10\text{ m/s}^3)(10.0\text{ s})^2$, which only has two significant digits. Thus the velocity at 10.0 s is 646 m/s.

(b) The parameters are accurate to two significant digits, thus the solutions will also have three significant digits: $t = 0.663$ s and −0.981 s

(c) The accuracy is limited by the values with the smallest number of significant figures. This requires three significant figures. The acceleration is then $a = 8.30\text{ m/s}^2$.

(d) No rounding is necessary.

DOUBLE-CHECK:

(a) This result is reasonable given the parameters. For example, $t^2=(10.0\text{ s})^2=100.\text{ s}$, so the velocity should be in the hundreds of meters per second.

(b) Since the function is quadratic, there should be two solutions. The negative solution means that the object was at rest 0.98 seconds before the time designated $t = 0$ s.

(c) These values are consistent with the parameters.

(d) The function for the acceleration is linear which the graph reflects.

2.41. **THINK:** I want to find the magnitude of the constant acceleration of a car that goes 0.500 km in 10.0 s: $d = 0.500$ km, $t = 10.0$ s.

SKETCH:

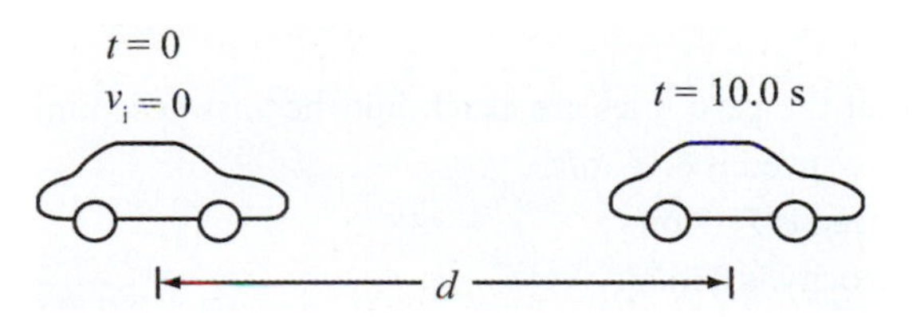

RESEARCH: The position of the car under constant acceleration is given by $d=\frac{1}{2}at^2$.

SIMPLIFY: Solving for acceleration gives $a=\frac{2d}{t^2}$.

CALCULATE: $a = \frac{2(0.500 \text{ km})}{(10.0 \text{ s})^2} = 0.0100 \text{ km/s}^2$

ROUND: The values all have three significant figures. Thus, the average acceleration is $a = 0.0100 \text{ km/s}^2$, which is 10.0 m/s^2.

DOUBLE-CHECK: This acceleration is on the order of a high performance car which can accelerate from 0 to 60 mph in 3 seconds, or 9 m/s^2.

2.45. **THINK:** (a) Since the motion is all in one direction, the average speed equals the distance covered divided by the time taken. I want to know the distance between the place where the ball was caught and midfield. I also want to know the time taken to cover this distance. The average speed will be the quotient of those two quantities.

(b) Same as in (a), but now I need to know the distance between midfield and the place where the run ended.

(c) I do not need to calculate the acceleration over each small time interval, since all that matters is the velocity at the start of the run and at the end. The average acceleration is the difference between those two quantities, divided by the time taken.

SKETCH:

In this case a sketch is not needed, since the only relevant quantities are those describing the runner at the start and end of the run, and at midfield.

RESEARCH:

The distance between two positions can be represented as $\Delta d = d_f - d_i$, where d_i is the initial position and d_f is the final one. The corresponding time difference is $\Delta t = t_f - t_i$. The average speed is $\Delta d / \Delta t$.

(a) Midfield is the 50-yard line, so $d_i = -1$ yd, $d_f = 50$ yd, $t_i = 0.00$ s, and $t_f = 5.73$ s.

(b) The end of the run is 1 yard past $d = 100$ yd, so $d_i = 50$ yd, $d_f = 101$ yd, $t_i = 5.73$ s, and $t_f = 12.01$ s.

(c) The average velocity is $\Delta v / \Delta t = (v_f - v_i)/(t_f - t_i)$. For this calculation, $t_i = 0$, $t_i = 12.01$ s, and $v_i = v_f = 0$ m/s, since the runner starts and finishes the run at a standstill.

SIMPLIFY:

(a), (b) $\frac{\Delta d}{\Delta t} = \frac{d_f - d_i}{t_f - t_i}$

(c) No simplification needed

CALCULATE:

(a) $\frac{\Delta d}{\Delta t} = \frac{(50 \text{ yd}) - (-1 \text{ yd})}{(5.73 \text{ s}) - (0.00 \text{ s})} = 8.900522356 \text{ yd/s} \cdot \frac{3 \text{ ft}}{1 \text{ yd}} \cdot \frac{0.3048 \text{ m}}{1 \text{ ft}} = 8.138638743 \text{ m/s}$

(b) $\frac{\Delta d}{\Delta t} = \frac{(101 \text{ yd}) - (50 \text{ yd})}{(12.01 \text{ s}) - (5.73 \text{ s})} = 8.121019108 \text{ yd/s} \cdot \frac{3 \text{ ft}}{1 \text{ yd}} \cdot \frac{0.3048 \text{ m}}{1 \text{ ft}} = 7.425859873 \text{ m/s}$

(c) $\frac{\Delta v}{\Delta t} = \frac{v_f - v_i}{t_f - t_i} = \frac{0 \text{ m/s} - 0 \text{ m/s}}{12.01 \text{ s} - 0.00 \text{ s}} = 0 \text{ m/s}^2$

ROUND:

(a) We assume that the yard lines are exact, but the answer is limited to 3 significant figures by the time data. So the average speed is 8.14 m/s.

(b) The average speed is 7.43 m/s.

(c) The average velocity is 0 m/s^2.

DOUBLE-CHECK:

The average speeds in parts (a) and (b) are reasonable speeds (8.9 ft/s is about 18 mph), and it makes sense that the average speed during the second half of the run would be slightly less than during the first half, due to fatigue. In part (c) it is logical that average acceleration would be zero, since the net change in velocity is zero.

2.49. **THINK:** I want to find the position of a car at t_f = 3.0 s if the velocity is given by the equation $v = At^2 + Bt$ with $A = 2.0\text{ m/s}^3$ and $B = 1.0\text{ m/s}^2$.

SKETCH:

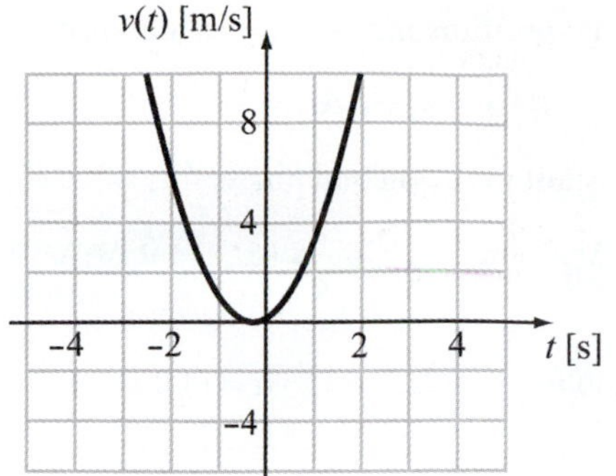

RESEARCH: The position is given by the integral of the velocity function: $x = x_0 + \int_0^{t_f} v(t)dt$.

SIMPLIFY: Since the car starts at the origin, $x_0 = 0$ m.

$$x = \int_0^{t_f} v(t)dt = \int_0^{t_f} \left(At^2 + Bt\right)dt = \frac{1}{3}At_f^3 + \frac{1}{2}Bt_f^2$$

CALCULATE: $x = \frac{1}{3}(2.0\text{ m/s}^3)(3.0\text{ s})^3 + \frac{1}{2}(1.0\text{ m/s}^2)(3.0\text{ s})^2 = 22.5\text{ m}$

ROUND: The parameters are given to two significant digits, and so the answer must also contain two significant digits: $x = 23$ m.

DOUBLE-CHECK: This is a reasonable distance for a car to travel in 3.0 s.

2.53. **THINK:** A motorcycle is accelerating at different rates as shown in the graph. I want to determine (a) its speed at t = 4.00 s and t = 14.0 s, and (b) its total displacement between t = 0 and t = 14.0 s.

SKETCH:

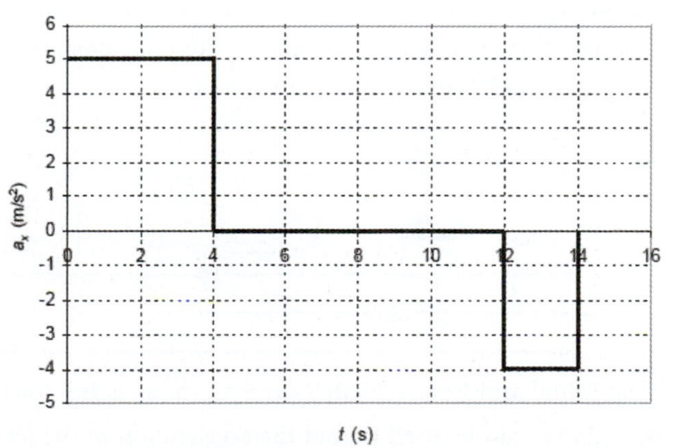

RESEARCH:

(a) The velocity of the motorcycle is defined by the area under the curve of the acceleration versus time graph. This area can be found by counting the blocks under the curve then multiply by the area of one block: 1 block = (2 s) 1 m/s² = 2 m/s.

(b) The displacement can be found by separating the acceleration into three parts: The first phase has an acceleration of $a_1 = 5\text{ m/s}^2$ for times between 0 to 4 seconds. The second phase has no acceleration, thus the motorcycle has a constant speed. The third phase has a constant acceleration of $a_3 = -4\text{ m/s}^2$. Recall the position and velocity of an object under constant acceleration is $x = x_0 + v_0t + (1/2)at^2$ and $v = v_0 + at$, respectively.

SIMPLIFY: At $t = 4.00$ s and 14.0 s, there are 10 blocks and 6 blocks respectively. Recall that blocks under the time axis are negative. In the first phase the position is given by $x=(1/2)a_1(\Delta t_1)^2$ where Δt is the duration of the phase. The velocity at the end of this phase is $v=a_1\Delta t_1$. The position and velocity of the first phase gives the initial position and velocity for the second phase.

$$x=x_0+v_0\Delta t_2=\frac{1}{2}a_1(\Delta t_1)^2+a_1\Delta t_1\Delta t_2$$

Since the velocity is constant in the second phase, this value is also the initial velocity of the third phase.

$$x=x_0+v_0\Delta t_3+\frac{1}{2}a_3(\Delta t_3)^2=\frac{1}{2}a_1(\Delta t_1)^2+a_1\Delta t_1\Delta t_2+a_1\Delta t_1\Delta t_3+\frac{1}{2}a_3(\Delta t_3)^2$$

CALCULATE:

(a) $v(t=4.00\text{ s})=10(2.00\text{ m/s})=20.0\text{ m/s}$, $v(t=14.0\text{ s})=6(2.00\text{ m/s})=12.0\text{ m/s}$

(b)

$$x=\frac{1}{2}(5.0\text{ m/s}^2)(4.00\text{ s}-0\text{ s})^2+(5.0\text{ m/s}^2)(4.00\text{ s}-0\text{ s})(12.0\text{ s}-4.0\text{ s})+(5.0\text{ m/s}^2)(4.00\text{ s}-0\text{ s})(14.0\text{ s}-12.0\text{ s})$$
$$+\frac{1}{2}(-4.0\text{ m/s}^2)(14.0\text{ s}-12.0\text{ s})^2$$
$$=232\text{ m}$$

ROUND:

(a) Rounding is not necessary in this case, because the values of the accelerations and times can be read off the graph to at least two digit precision.

(b) The motorcycle has traveled 232 m in 14.0 s.

DOUBLE-CHECK: The velocity of the motorcycle at t = 14 s is less than the speed at t = 4 s, which makes sense since the bike decelerated in the third phase. Since the bike was traveling at a maximum speed of 20 m/s, the most distance it could cover in 14 seconds would be 280 m. The calculated value is less than this, which makes sense since the bike decelerated in the third phase.

2.57. **THINK:** I am given $v_0=70.4\text{ m/s}$, $v=0$, $\Delta x=197.4\text{ m}$, and constant acceleration. I am asked to find the velocity v' when the jet is 44.2 m from its stopping position. This means the jet has traveled $\Delta x'=197.4\text{ m}-44.2\text{ m}=153.2\text{ m}$ on the aircraft carrier.

SKETCH:

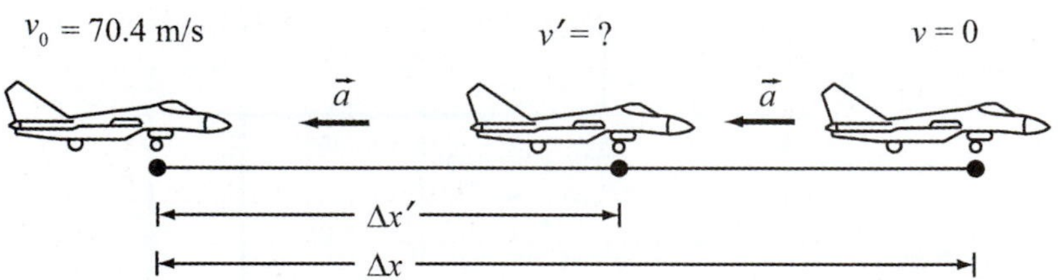

RESEARCH: The initial and final velocities are known, as is the total distance traveled. Therefore the equation $v^2=v_0^2+2a\Delta x$ can be used to find the acceleration of the jet. Once the acceleration is known, the intermediate velocity v' can be determined using $(v')^2=v_0^2+2a\Delta x'$.

SIMPLIFY: First find the constant acceleration using the total distance traveled, Δx, the initial velocity, v_0, and the final velocity, v: $a=\dfrac{v^2-v_0^2}{2\Delta x}=-\dfrac{v_0^2}{2\Delta x}$ (since $v=0$ m/s). Next, find the requested intermediate velocity, v':

$$\left((v')^2=v_0^2+2a\Delta x' \Rightarrow (v')^2=v_0^2+2\left(-\frac{v_0^2}{2\Delta x}\right)\Delta x' \Rightarrow v'=\sqrt{v_0^2-\frac{v_0^2}{\Delta x}\Delta x'}\right)$$

CALCULATE: $v'=\sqrt{(70.4\text{ m/s})^2-\frac{(70.4\text{ m/s})^2}{197.4\text{ m}}(153.2\text{ m})}=33.313\text{ m/s}$

ROUND: At $\Delta x'=153.2\text{ m}$, the velocity is $v'=33.3$ m/s.

DOUBLE-CHECK: This v' is less than v_0, but greater than v, and therefore makes sense.

2.61. **THINK:**

(a) I know that $v_0=0\text{ m/s}$, $v=5.00\text{ m/s}$, and a is constant. I want to find v_{avg}.

(b) t = 4.00 s is given. I want to find Δx.

SKETCH:

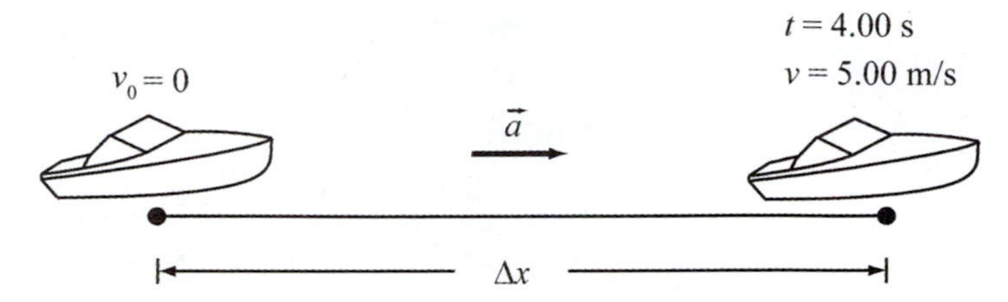

RESEARCH:

(a) $v_{\text{avg}}=\frac{v_0+v}{2}$

(b) a is unknown, so use $\Delta x=\frac{1}{2}(v_0+v)t$

SIMPLIFY: It is not necessary to simplify the equations above.

CALCULATE:

(a) $v_{\text{avg}}=\frac{5.00\text{ m/s}+0\text{ m/s}}{2}=2.50\text{ m/s}$

(b) $\Delta x=\frac{1}{2}(5.00\text{ m/s}+0\text{ m/s})(4.00\text{ s})=10.00\text{ m}$

ROUND:

(a) v is precise to three significant digits, so $v_{\text{avg}}=2.50$ m/s.

(b) Each v and t have three significant digits, so $\Delta x=10.0$ m.

DOUBLE-CHECK:

(a) This v_{avg} is between the given v_0 and v, and therefore makes sense.

(b) This is a reasonable distance to cover in 4.00 s when $v_{\text{avg}}=2.50$ m/s.

2.63. **THINK:**

(a) The girl is initially at rest, so $v_{1_0}=0$, and then she waits $t'=20\text{ s}$ before accelerating at $a_1=2.2\text{ m/s}^2$. Her friend has constant velocity $v_2=8.0\text{ m/s}$. I want to know the time required for the girl to catch up with her friend, t_1. Note that both people travel the same distance: $\Delta x_1=\Delta x_2$. The time the girls spends riding her bike is t_1. The friend, however, has a t' head-start; the friend travels for a total time of $t_2=t'+t_1$.

(b) The initial conditions of the girl have changed. Now $v_{1_0}=1.2\text{ m/s}$. The initial conditions of the friend are the same: $v_2=8.0\text{ m/s}$. Now there is no time delay between when the friend passes the girl and when the girl begins to accelerate. The time taken to catch up is that found in part a), $t=20\text{ s}$. I will use $t=16.2\text{ s}$ for my calculations, keeping in mind that t has only two significant figures. I want to know the acceleration of the girl, a_1, required to catch her friend in time t.

SKETCH:

(a)

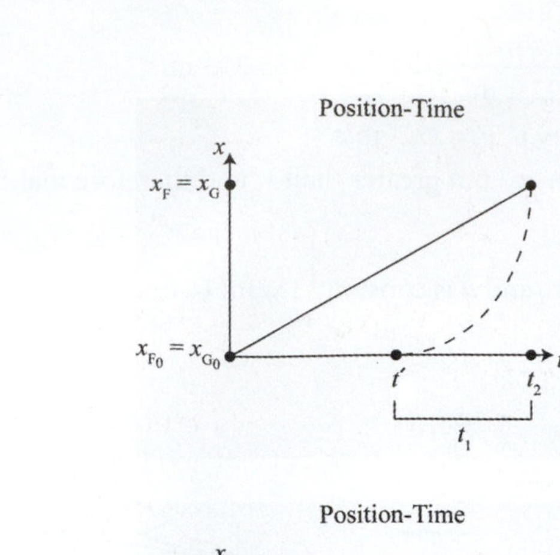

(b)

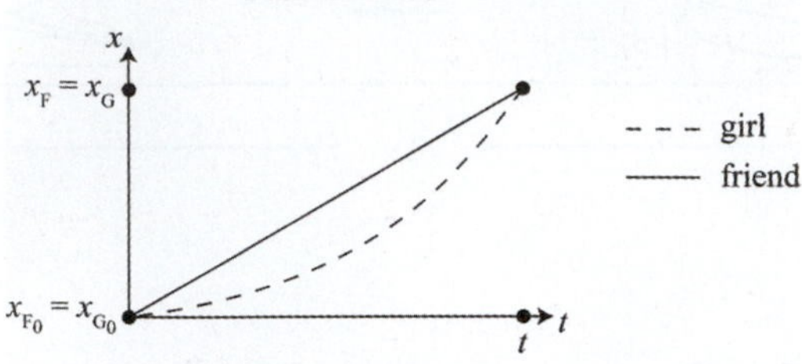

RESEARCH:

(a) The distance the girl travels is $\Delta x_1 = v_{1_0} t_1 + \frac{1}{2} a_1 t_1^2$. The distance her friend travels is $\Delta x_2 = v_2 t_2$.

(b) $\Delta x_1 = v_{1_0} t + \frac{1}{2} a_1 t^2$, $\Delta x_2 = v_2 t$

SIMPLIFY:

(a) Since $v_{1_0} = 0$, $\Delta x_1 = \frac{1}{2} a_1 t_1^2$. Also, since $t_2 = t' + t_1$, $\Delta x_2 = v_2 (t' + t_1)$. Recall that $\Delta x_1 = \Delta x_2$. This leads to $\frac{1}{2} a_1 t_1^2 = v_2 (t' + t_1)$. Now solve for t_1: $\frac{1}{2} a_1 t_1^2 = v_2 t' + v_2 t_1 \Rightarrow \frac{1}{2} a_1 t_1^2 - v_2 t_1 - v_2 t' = 0.$

The quadratic formula gives:

$$t_1 = \frac{v_2 \pm \sqrt{v_2^2 - 4\left(\frac{1}{2} a_1\right)(-v_2 t')}}{2\left(\frac{1}{2} a_1\right)} = \frac{v_2 \pm \sqrt{v_2^2 + 2 a_1 v_2 t'}}{a_1}$$

(b) As in part (a), $\Delta x_1 = \Delta x_2$, and so $v_{1_0} t + \frac{1}{2} a_1 t^2 = v_2 t$. Solving for a_1 gives:

$$\frac{1}{2} a_1 t^2 = v_2 t - v_{1_0} t \Rightarrow a_1 = \frac{2\left(v_2 - v_{1_0}\right)}{t}$$

CALCULATE:

(a) $$t_1 = \frac{(8.0 \text{ m/s}) \pm \sqrt{(-8.0 \text{ m/s})^2 + 2(2.2 \text{ m/s}^2)(8.0 \text{ m/s})(20 \text{ s})}}{2.2 \text{ m/s}^2} = \frac{(8.0 \text{ m/s}) \pm \sqrt{64 \text{ m}^2/\text{s}^2 + 704 \text{ m}^2/\text{s}^2}}{2.2 \text{ m/s}^2}$$

$$= \frac{(8.0 \text{ m/s}) \pm 27.7 \text{ m/s}}{2.2 \text{ m/s}^2} = 16.2272, -8.9545$$

(b) $$a_1 = \frac{2(8.0 \text{ m/s} - 1.2 \text{ m/s})}{16.2 \text{ s}} = 0.840 \text{ m/s}^2$$

ROUND:

(a) Time must be positive, so take the positive solution, $t_1 = 16$ s.

(b) $a_1 = 0.84 \text{ m/s}^2$

DOUBLE-CHECK:

(a) The units of the result are those of time. This is a reasonable amount of time to catch up to the friend who is traveling at $v_2 = 8.0$ m/s.

(b) This acceleration is less than that in part (a). Without the 20 s head-start, the friend does not travel as far, and so the acceleration of the girl should be less in part (b) than in part a), given the same time.

2.69. **THINK:** Take "downward" to be along the negative *y*-axis. I know that $v_0 = -10.0 \text{ m/s}$, $\Delta y = -50.0 \text{ m}$, and $a = -g = -9.81 \text{ m/s}^2$. I want to find *t*, the time when the ball reaches the ground.

SKETCH:

$v_0 = -10.0$ m/s $\quad y_0 = 50.0$ m

$\vec{a}$

$y = 0$

RESEARCH: $\Delta y = v_0 t + \frac{1}{2}at^2$

SIMPLIFY: $\frac{1}{2}at^2 + v_0 t - \Delta y = 0$. This is a quadratic equation. Solving for t:

$$t = \frac{-v_0 \pm \sqrt{v_0^2 - 4\left(\frac{1}{2}a\right)(-\Delta y)}}{2\left(\frac{1}{2}a\right)} = \frac{-v_0 \pm \sqrt{v_0^2 - 2g\Delta y}}{-g}$$

CALCULATE: $t = \dfrac{-(-10.0 \text{ m/s}) \pm \sqrt{(-10.0 \text{ m/s})^2 - 2(9.81 \text{ m/s}^2)(-50.0 \text{ m})}}{-9.81 \text{ m/s}^2}$

$= -4.3709 \text{ s}, \; 2.3322 \text{ s}$

The time interval has to be positive, so $t = 2.3322$ s.

ROUND: All original quantities are precise to three significant digits, therefore $t = 2.33$ s.

DOUBLE-CHECK: A negative v indicates that the stone is (still) falling downward. This makes sense, since the stone was thrown downward. The velocity is even more negative after 0.500 s than it was initially, which is consistent with the downward acceleration.

2.73. **THINK:** The bowling ball is released from rest. In such a case we have already studied the relationship between vertical distance fallen and time in Example 2.5, "Reaction Time", in the book. With this result in our arsenal, all we have to do here is to compute the time t_{total} it takes the ball to fall from Bill's apartment down to the ground and subtract from it the time t_1 it takes the ball to fall from Bill's apartment down to John's apartment.

SKETCH:

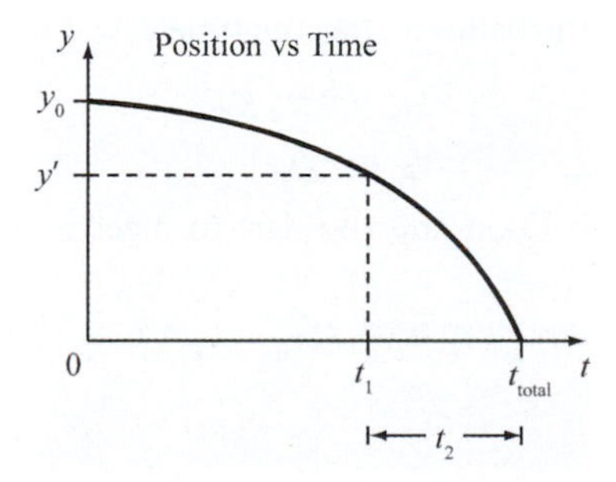

RESEARCH: We will use the formula $t = \sqrt{2h/g}$ from Example 2.5. If you look at the sketch, you see that $t_{\text{total}} = \sqrt{2h_{\text{total}}/g} = \sqrt{2y_0/g}$ and that $t_1 = \sqrt{2h_1/g} = \sqrt{2(y_0 - y')/g}$.

SIMPLIFY: Solving for the time difference gives:

$$t_2 = t_{\text{total}} - t_1 = \sqrt{2y_0/g} - \sqrt{2(y_0 - y')/g}$$

CALCULATE: $t_2 = \sqrt{2(63.17 \text{ m})/(9.81 \text{ m/s}^2)} - \sqrt{2(63.17 \text{ m} - 40.95 \text{ m})/(9.81 \text{ m/s}^2)}$
$= 1.4603 \text{ s}$

ROUND: We round to $t_2 = 1.46$ s., because g has three significant figures.

DOUBLE-CHECK: The units of the solution are those of time, which is already a good minimum requirement for a valid solution. But we can do better! If we compute the time it takes an object to fall 40.95 m from rest, we find from again using $t = \sqrt{2h/g}$ that this time is 2.89 s. In the problem here the bowling ball clearly already has a significant downward velocity as it passes the height of 40.95 m, and so we expect a time t_2 shorter than 2.89 s, which is clearly fulfilled for our solution.

2.77. **THINK:** It is probably a good idea to read through the solution of the "Melon Drop" problem, Solved Problem 2.5 in the textbook before getting started with the present problem. The present problem has the additional complication that the water balloon gets dropped some time before the dart get fired, whereas in the "Melon Drop" problem both projectiles get launched simultaneously. For the first 2 seconds, only our water balloon is in free fall, and we can calculate its position y_{b_0} and velocity v_{b_0} at the end of this time interval.

a) After the initial two seconds the dart also gets launched, and then both objects (water balloon and dart) are in free-fall. Their initial distance is y_{b_0}, and their relative velocity is the difference between the initial velocity of the dart and v_{b_0}. The time until the two objects meet is then simply the ratio of the initial distance and the relative velocity.

b) For this part we simply calculate the time it takes for the balloon to free-fall the entire height h and subtract our answer form part a).

SKETCH:

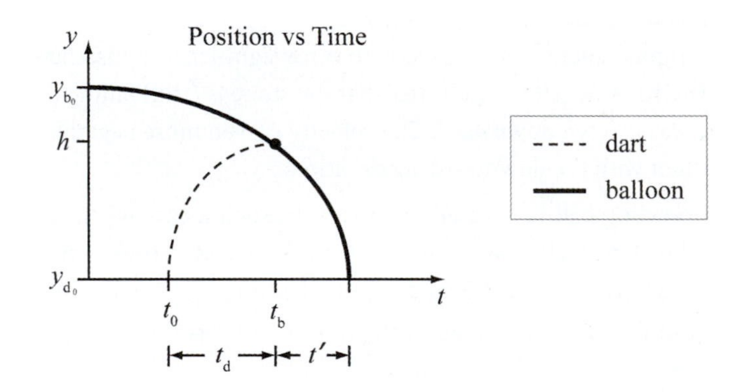

RESEARCH:

(a) The position and velocity of the balloon after the time $t_0 = 2$ s are

$$y_{b_0} = h - \tfrac{1}{2}gt_0^{\,2}$$
$$v_{b_0} = -gt_0$$

The time is takes then for the balloon and the dart to meet is the ratio of their initial distance to their initial relative velocity:

$$t_d = y_{b_0}/(v_{d_0} - v_{b_0})$$

Our answer for part a) is the sum of the time t_0, during which the balloon was in free-fall alone, and the time t_1, $t_b = t_d + t_0$.

(b) The total time it takes for the balloon to fall all the way to the ground is

$$t_{\text{total}} = \sqrt{2h/g}$$

We get our answer for part b) by subtracting the result of part a) from this total time:

$$t' = t_{\text{total}} - t_d$$

SIMPLIFY:

(a) If we insert the expressions for the initial distance and relative speed into $t_d = y_{b_0}/(v_{d_0} - v_{b_0})$, we find $t_d = y_{b_0}/(v_{d_0} - v_{b_0}) = (h - \frac{1}{2}gt_0^2)/(v_{d_0} + gt_0)$. Adding t_0 then gives us our final answer:

$$t_b = t_0 + (h - \tfrac{1}{2}gt_0^2)/(v_{d_0} + gt_0)$$

(b) For the time between the balloon being hit by the dart and the water reaching the ground we find by inserting $t_{\text{total}} = \sqrt{2h/g}$ into $t' = t_{\text{total}} - t_b$:

$$t' = \sqrt{2h/g} - t_b$$

CALCULATE:

(a) $t_b = 2.00 \text{ s} + \dfrac{80.0 \text{ m} - \frac{1}{2}(9.81 \text{ m/s}^2)(2.00 \text{ s})^2}{20.0 \text{ m/s} + (9.81 \text{ m/s}^2)(2.00 \text{ s})} = 3.524 \text{ s}$

(b) $t' = \sqrt{2(80.0 \text{ m})/(9.81 \text{ m/s}^2)} - 3.524 \text{ s} = 0.515 \text{ s}$.

ROUND:

(a) $t_b = 3.52s$

(b) $t' = 0.515$ s

DOUBLE-CHECK: The solution we showed in this problem is basically the double-check step in Solved Problem 2.5. Conversely, we can use the solution method of Solved Problem 2.5 as a double-check for what we have done here. This is left as an exercise for the reader.

2.83. **THINK:** The initial velocity is

$$v_0 = 60.0 \frac{\text{km}}{\text{h}} \cdot \frac{1 \text{ h}}{3600 \text{ s}} \cdot \frac{1000 \text{ m}}{1 \text{ km}} = 16.67 \text{ m/s}.$$

The final velocity is $v = 0$. The stop time is $t = 4.00$ s. The deceleration is uniform. Determine (a) the distance traveled while stopping, Δx and (b) the deceleration, a. I expect $a < 0$.

SKETCH:

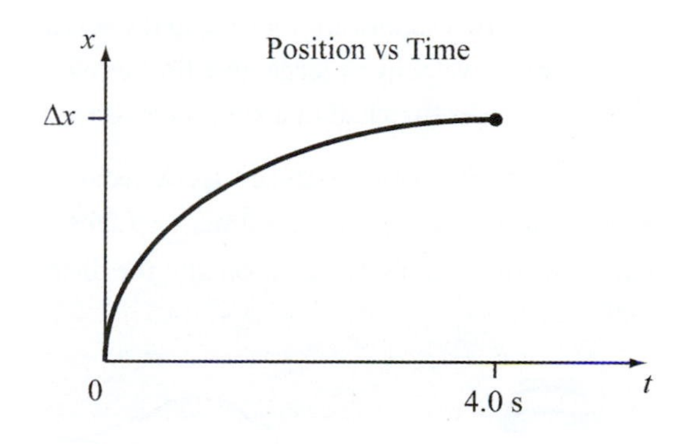

RESEARCH:

(a) To determine the stopping distance, use $\Delta x = t(v_0 + v)/2$.

(b) To determine a, use $v = v_0 + at$.

SIMPLIFY:

(a) With $v = 0$, $\Delta x = v_0 t / 2$.

(b) With $v = 0$, $0 = v_0 + at \Rightarrow a = -v_0 / t$.

CALCULATE:

(a) $\Delta x = \dfrac{(16.67 \text{ m/s})(4.00 \text{ s})}{2} = 33.34 \text{ m}$

(b) $a = -\dfrac{16.67 \text{ m/s}}{4.00 \text{ s}} = -4.167 \text{ m/s}^2$

ROUND:

(a) $\Delta x = 33.3 \text{ m}$

(b) $a = -4.17 \text{ m/s}^2$

DOUBLE-CHECK: The distance traveled while stopping is of an appropriate order of magnitude. A car can reasonably stop from 60 km/h in a distance of about 30 m. The acceleration is negative, indicating that the car is slowing down from its initial velocity.

2.87. **THINK:** The initial velocity is

$$v_0 = 212.809 \text{ mph} \cdot \frac{1 \text{ h}}{3600 \text{ s}} \cdot \frac{1609.3 \text{ m}}{\text{mile}} = 95.1315 \text{ m/s}.$$

The acceleration is $a = -8.0 \text{ m/s}^2$. The final speed is $v = 0$. Determine the stopping distance, Δx.

SKETCH:

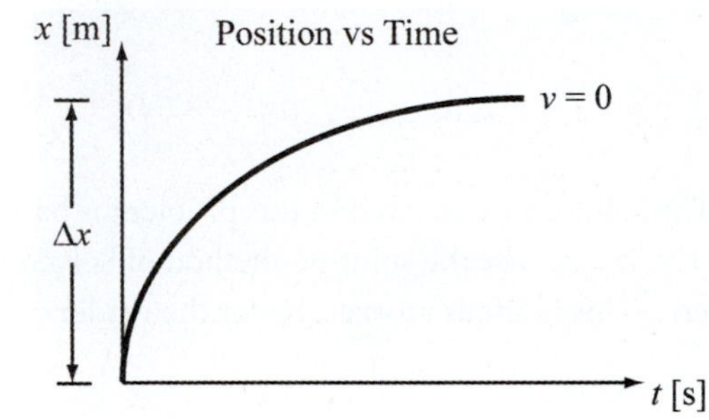

RESEARCH: Use $v^2 = v_0^2 + 2a\Delta x$.

SIMPLIFY: With $v = 0$, $0 = v_0^2 + 2a\Delta x \Rightarrow \Delta x = -v_0^2 / 2a$.

CALCULATE: $\Delta x = -\dfrac{(95.1315 \text{ m/s})^2}{2(-8.0 \text{ m/s}^2)} = 565.6 \text{ m}$

ROUND: The acceleration has two significant figures, so the result should be rounded to $\Delta x = 570 \text{ m}$.

DOUBLE-CHECK: The initial velocity is large and the deceleration has a magnitude close to that of gravity. A stopping distance greater than half of a kilometer is reasonable.

2.91. **THINK:** The police have a double speed trap set up. A sedan passes the first speed trap at a speed of $s_1 = 105.9 \text{ mph}$. The sedan decelerates and after a time, $t = 7.05$ s it passes the second speed trap at a speed of $s_2 = 67.1 \text{ mph}$. Determine the sedan's deceleration and the distance between the police cruisers.

SKETCH:

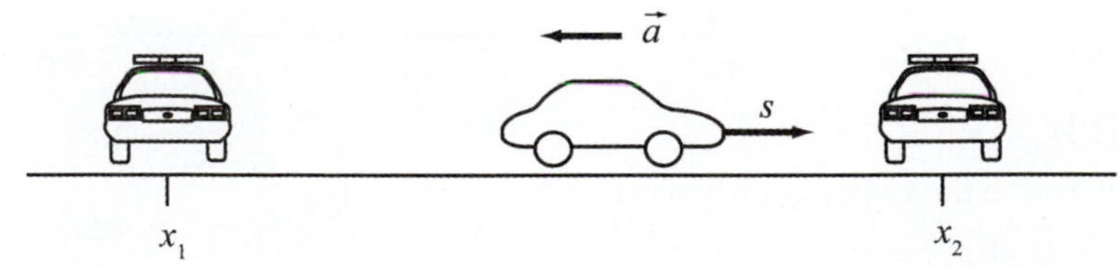

RESEARCH:
(a) Convert the speeds to SI units as follows:

$$s_1 = 105.9 \text{ mph} \cdot \frac{1 \text{ h}}{3600 \text{ s}} \cdot \frac{1609.3 \text{ m}}{\text{mile}} = 47.34 \text{ m/s}$$

$$s_2 = 67.1 \text{ mph} \cdot \frac{1 \text{ h}}{3600 \text{ s}} \cdot \frac{1609.3 \text{ m}}{\text{mile}} = 29.996 \text{ m/s}.$$

The sedan's velocity, v can be written in terms of its initial velocity, v_0 the time t, and its acceleration a: $v = v_0 + at$. Substitute s_1 for v_0 and s_2 for v.
(b) The distance between the cruisers is given by: $\Delta x = x_2 - x_1 = v_0 t + (1/2)at^2$.
SIMPLIFY:
(a) $a = \dfrac{v - v_0}{t} = \dfrac{s_2 - s_1}{t}$
(b) Substitute s_1 for v_0 and the expression from part (a) for a: $\Delta x = s_1 t + (1/2)at^2$
CALCULATE:
(a) $a = \dfrac{29.996 \text{ m/s} - 47.34 \text{ m/s}}{7.05 \text{ s}} = -2.4602 \text{ m/s}^2$
(b) $\Delta x = (47.34 \text{ m/s})(7.05 \text{ s}) + \dfrac{1}{2}(-2.4602 \text{ m/s}^2)(7.05 \text{ s})^2 = 272.6079 \text{ m}$
ROUND: The least number of significant figures provided in the problem are three, so the results should be rounded to $a = -2.46 \text{ m/s}^2$ and $\Delta x = 273 \text{ m}$.
DOUBLE-CHECK: The sedan did not have its brakes applied, so the values calculated are reasonable for the situation. The acceleration would have been larger, and the distance would have been much smaller, if the brakes had been used. The results also have the proper units.

2.95. **THINK:** The distance to the destination is 199 miles or 320 km. To solve the problem it is easiest to draw a velocity versus time graph. The distance is then given by the area under the curve.
SKETCH:

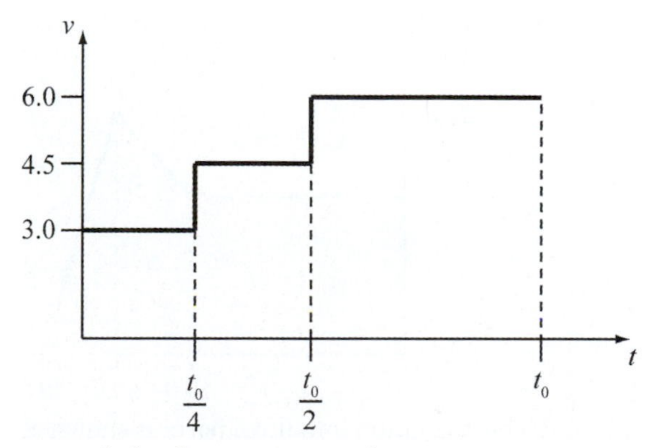

RESEARCH: For a constant speed, the distance is given by $x = vt$.
SIMPLIFY: To simplify, divide the distance into three parts.
Part 1: from $t = 0$ to $t = t_0/4$.
Part 2: from $t = t_0/4$ to $t = t_0/2$.
Part 3: from $t = t_0/2$ to $t = t_0$.
CALCULATE:
(a) The distances are $x_1 = 3.0t_0/4$, $x_2 = 4.5t_0/4$ and $x_3 = 6.0t_0/2$. The total distance is given by

$$x = x_1 + x_2 + x_3 = \frac{(3.0 + 4.5 + 12)t_0}{4} \text{ m} = \frac{19.5t_0}{4} \text{ m} \Rightarrow t_0 = \frac{4x}{19.5} \text{ s}.$$

It follows that

$$t_0 = \frac{4\left(320\cdot 10^3\right)}{19.5}\text{ s} = 65.6410\cdot 10^3\text{ s} = 65641\text{ s} \Rightarrow t_0 = 18.2336\text{ h}$$

(b) The distances are:

$$\left(x_1 = 3.0\left(\frac{65641}{4}\right)\text{ m} = 49.23\text{ km},\quad x_2 = 4.5\left(\frac{65641}{4}\right)\text{ m} = 73.85\text{ km},\quad x_3 = 6.0\left(\frac{65641}{2}\right)\text{ m} = 196.92\text{ km.}\right)$$

ROUND: Since the speeds are given to two significant figures, the results should be rounded to $x_1 = 49\text{ km}$, $x_2 = 74\text{ km}$ and $x_3 = 2.0\cdot 10^2\text{ km}$. $x_1 + x_2 = 123\text{ km} \approx 120\text{ km}$, and then $x = x_1 + x_2 + x_3 = 323\text{ km} \approx 320\text{ km}$.

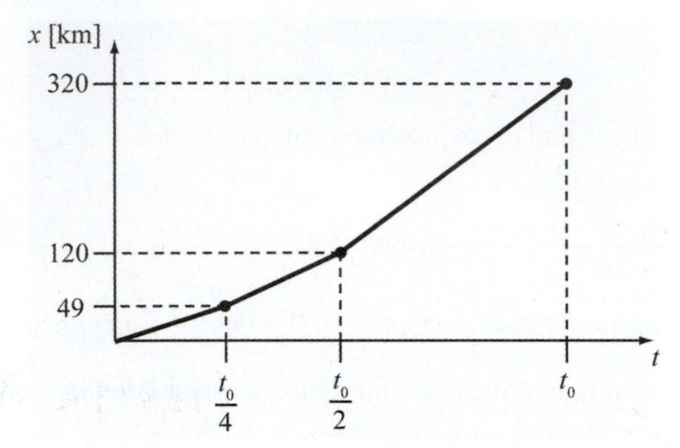

DOUBLE-CHECK: The sum of the distances x_1, x_2 and x_3 must be equal to the total distance of 320 km: $x_1 + x_2 + x_3 = 49.23 + 73.85 + 196.92 = 320\text{ km}$ as expected. Also, note that $x_1 < x_2 < x_3$ since $v_1 < v_2 < v_3$.

2.99. **THINK:** $v_1 = 13.5\text{ m/s}$ for $\Delta t = 30.0\text{ s}$. $v_2 = 22.0\text{ m/s}$ after $\Delta t = 10.0\text{ s}$ (at t = 40.0 s). $v_3 = 0$ after $\Delta t = 10.0\text{ s}$ (at t = 50.0 s). It will be easier to determine the distance from the area under the curve of the velocity versus time graph.

SKETCH:

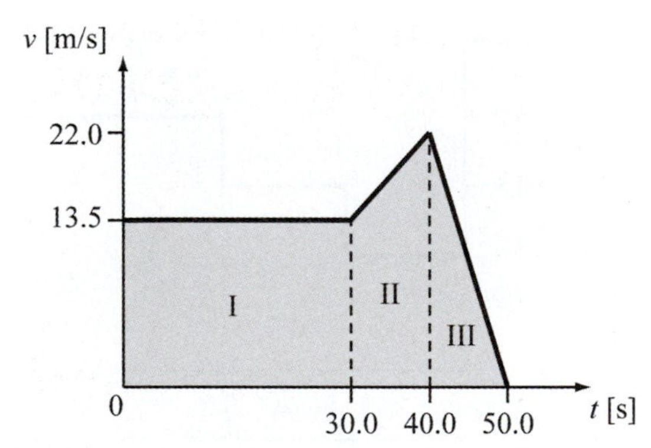

RESEARCH: Divide and label the graph into three parts as shown above.

SIMPLIFY: The total distance, *d* is the sum of the areas under the graph, $d = A_1 + A_2 + A_3$.

CALCULATE: $d = (13.5\text{ m/s})(30.0\text{ s}) + \frac{1}{2}(13.5\text{ m/s} + 22.0\text{ m/s})(10.0\text{ s}) + \frac{1}{2}(22.0\text{ m/s})(10.0\text{ s})$

$$= 405\text{ m} + 177.5\text{ m} + 110\text{ m} = 692.5\text{ m}$$

ROUND: The speeds are given in three significant figures, so the result should be rounded to $d = 693\text{ m}$.

DOUBLE-CHECK: From the velocity versus time plot, the distance can be estimated by assuming the speed is constant for all time, *t*: $d = (13.5\text{ m/s})(50.0\text{ s}) = 675\text{ m}$. This estimate is in agreement with the previous result.

Multi-Version Exercises

2.103. **THINK:** The initial velocity is $v_0 = 28.0$ m/s. The acceleration is $a = -g = -9.81$ m/s^2. The velocity, v is zero at the maximum height. Determine the time, t to achieve the maximum height.

SKETCH:

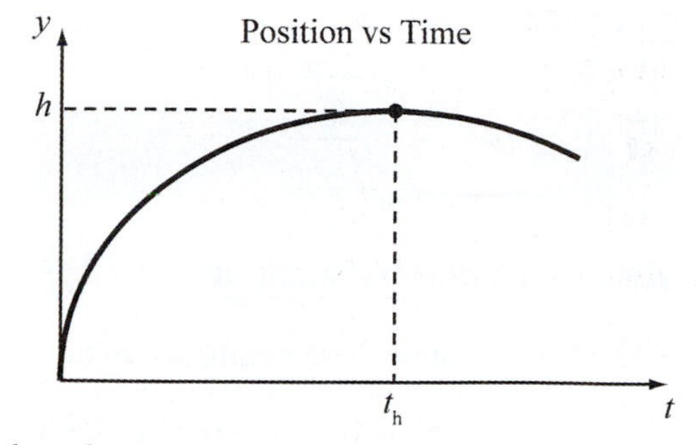

RESEARCH: To determine the velocity use $v = v_0 + at_h$.

SIMPLIFY: $at_h = v - v_0 \Rightarrow t_h = \dfrac{v - v_0}{a} = \dfrac{-v_0}{-g} = \dfrac{v_0}{g}$

CALCULATE: $t_h = \dfrac{28.0 \text{ m/s}}{9.81 \text{ m/s}^2} = 2.8542 \text{ s}$

ROUND: The initial values have three significant figures, so the result should be rounded to $t_h = 2.85$ s.

DOUBLE-CHECK: The initial velocity of the object is about 30 m/s, and gravity will cause the velocity to decrease about 10 m/s per second. It should take roughly three seconds for the object to reach its maximum height.

2.106. **THINK:** Since the rock is dropped from a fixed height and allowed to fall to the surface of Mars, this question involves free fall. It is necessary to impose a coordinate system. Choose $y = 0$ to represent the surface of Mars and $t_0 = 0$ to be the time at which the rock is released.

SKETCH: Sketch the situation at time $t_0 = 0$ and time t, when the rock hits the surface.

RESEARCH: For objects in free fall, equations 2.25 can be used to compute velocity and position. In particular, the equation $y = y_0 + v_{y0}t - \frac{1}{2}gt^2$ can be used. In this case, $y_0 = 1.013$ m. Since the object is not thrown but dropped with no initial velocity, $v_{y0} = 0$ m/s, and $g = 3.699$ m/s^2 on the surface of Mars.

SIMPLIFY: The starting position and velocity ($y_0 = 1.013$ m and $v_{y0} = 0$ m/s), final position ($y = 0$ m) and gravitational acceleration are known. Using the fact that $v_{y0} = 0$ m/s and solving the equation for t gives:

$$0 = y_0 + 0t - \tfrac{1}{2}(g)t^2 =$$
$$\frac{g}{2}t^2 = y_0 \Rightarrow$$
$$t^2 = \frac{2 \cdot y_0}{g}$$
$$t = \sqrt{\frac{2 \cdot y_0}{g}}$$

CALCULATE: On Mars, the gravitational acceleration $g = 3.699$ m/s^2. Since the rock is dropped from a height of 1.013 m, $y_0 = 1.013$ m. Plugging these numbers into our formula gives a time $t = \sqrt{\frac{2.026}{3.699}}$ s.

ROUND: In this case, all measured values are given to four significant figures, so our final answer has four significant digits. Using the calculator to find the square root gives a time $t = 0.7401$ s.

DOUBLE-CHECK: First note that the answer seems reasonable. The rock is not dropped from an extreme height, so it makes sense that it would take less than one second to fall to the Martian surface. To check the answer by working backwards, first note that the velocity of the rock at time t is given by the equation $v_y = v_{y0} - gt = 0 - gt = -gt$ in this problem. Plug this and the value $v_{y0} = 0$ into the equation to find the average velocity $\bar{v}_y = \tfrac{1}{2}(v_y + 0) = \tfrac{1}{2}(-gt)$. Combining this with the expression for position gives:

$$y = y_0 + \bar{v}_y t$$
$$= y_0 + \left(\tfrac{1}{2}(-gt)\right)t$$

Using the fact that the rock was dropped from a height of $y_0 = 1.013$ m and that the gravitational acceleration on Mars is $g = 3.699$ m/s^2, it is possible to confirm that the height of the rock at time $t = 0.7401$ s is $y = 1.013 + \tfrac{1}{2}(-3.699)(0.7401^2) = 0$, which confirms the answer.

2.108. **THINK:** Since the ball is dropped from a fixed height with no initial velocity and allowed to fall freely, this question involves free fall. It is necessary to impose a coordinate system. Choose $y = 0$ to represent the ground. Let $t_0 = 0$ be the time when the ball is released from height $y_0 = 12.37$ m and t_1 be the time the ball reaches height $y_1 = 2.345$ m.

SKETCH: Sketch the ball when it is dropped and when it is at height 2.345 m.

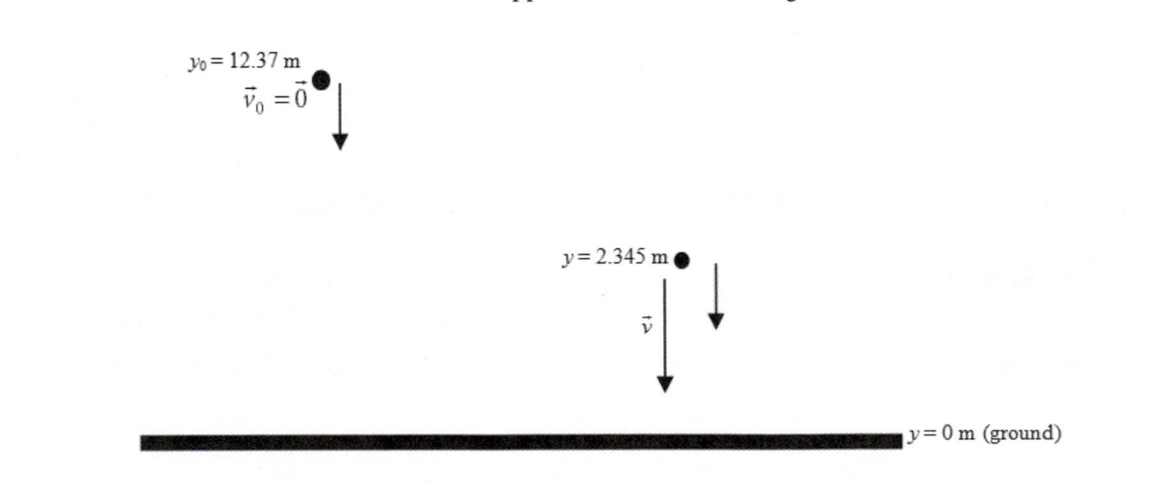

RESEARCH: Equations (2.25) are used for objects in free fall. Since the ball is released with no initial velocity, we know that $v_{0t} = 0$. We also know that on Earth, the gravitational acceleration is 9.81 m/s^2. In this problem, it is necessary to find the time that the ball reaches 2.345 m and find the velocity at that time. This can be done using equations (2.25) part (i) and (iii):

$$\text{(i)} \quad y = y_0 - \tfrac{1}{2}gt^2$$
$$\text{(iii)} \quad v_y = -gt$$

SIMPLIFY: We use algebra to find the time t_1 at which the ball will reach height $v_1 = 2.345$ m in terms of the initial height y_0 and gravitational acceleration g:

$$y_1 = y_0 - \tfrac{1}{2}g(t_1)^2 \Rightarrow$$
$$\tfrac{1}{2}g(t_1)^2 = y_0 - y_1 \Rightarrow$$
$$(t_1)^2 = \tfrac{2}{g}(y_0 - y_1) \Rightarrow$$
$$t_1 = \sqrt{\tfrac{2}{g}(y_0 - y_1)}$$

Combining this with the equation for velocity gives $v_{y1} = -gt_1 = -g\sqrt{\tfrac{2}{g}(y_0 - y_1)}$.

CALCULATE: The ball is dropped from an initial height of 12.37 m above the ground, and we want to know the speed when it reaches 2.345 m above the ground, so the ball is dropped from an initial height of 12.37 m above the ground, and we want to know the speed when it reaches 2.345 m above the ground, so $y_0 = 12.37$ and $y_1 = 2.345$ m. Use this to calculate $v_{y1} = -9.81\sqrt{\tfrac{2}{9.81}(12.37 - 2.345)}$ m/s.

ROUND: The heights above ground (12.37 and 2.345) have four significant figures, so the final answer should be rounded to four significant figures. The speed of the ball at time t_1 is then $-9.81\sqrt{\tfrac{2}{9.81}(12.37 - 2.345)} = -14.02$ m/s. The velocity of the ball when it reaches a height of 2.345 m above the ground is 14.02 m/s towards the ground.

DOUBLE-CHECK: To double check that the ball is going 14.02 m/s towards the ground, we use equation (2.25) (v) to work backwards and find the ball's height when the velocity is 14.02 m/s. We know that:

$$v_y^2 = v_{y0}^2 - 2g(y - y_0) \Rightarrow$$
$$v_y^2 = 0^2 - 2g(y - y_0) = -2g(y - y_0) \Rightarrow$$
$$\frac{v_y^2}{-2g} = y - y_0 \Rightarrow$$
$$\frac{v_y^2}{-2g} + y_0 = y$$

We take the gravitational acceleration $g = 9.81$ m/s^2 and the initial height $y_0 = 12.37$ m, and solve for y when $v_y = -14.02$ m/s. Then $y = \dfrac{v_y^2}{-2g} + y_0 = \dfrac{(-14.02)^2}{-2(9.81)} + 12.37 = 2.352$ m above the ground. Though this doesn't match the question exactly, it is off by less than 4 mm, so we are very close to the given value. In fact, if we keep the full accuracy of our calculation without rounding, we get that the ball reaches a velocity of 14.0246… m/s towards the ground at a height of 2.345 m above the ground.

Chapter 3: Motion in Two and Three Dimensions

Exercises

3.35. **THINK:** To calculate the magnitude of the average velocity between $x = 2.0$ m, $y = -3.0$ m and $x = 5.0$ m, $y = -9.0$ m, the distance between these coordinates must be calculated, then the time interval can be used to determine the average velocity.

SKETCH:

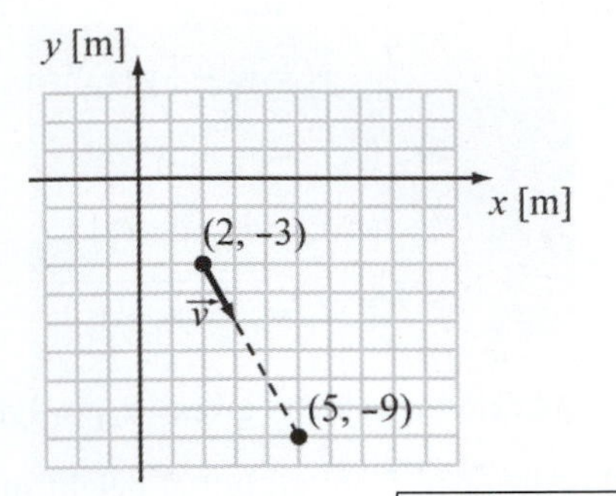

RESEARCH: The equation to find the distance is $d = \sqrt{(x_f - x_i)^2 + (y_f - y_i)^2}$. The average velocity is given by $v = d/t$.

SIMPLIFY: $|\vec{v}| = \dfrac{\sqrt{(x_f - x_i)^2 + (y_f - y_i)^2}}{t}$

CALCULATE: $|\vec{v}| = \dfrac{\sqrt{(5.0\text{ m} - 2.0\text{ m})^2 + ((-9.0\text{ m}) - (-3.0\text{ m}))^2}}{2.4\text{ s}} = 2.7951\text{ m/s}$

ROUND: Rounding to two significant figures, $|\vec{v}| = 2.8$ m/s.

DOUBLE-CHECK: This result is on the same order of magnitude as the distances and the time interval.

3.39. **THINK:** The position components $x(t) = -0.45t^2 - 6.5t + 25$ and $y(t) = 0.35t^2 + 8.3t + 34$ can be used to find the magnitude and direction of the position at $t = 10.0$ s. The velocity and acceleration at $t = 10.0$ s must then be determined.

SKETCH:

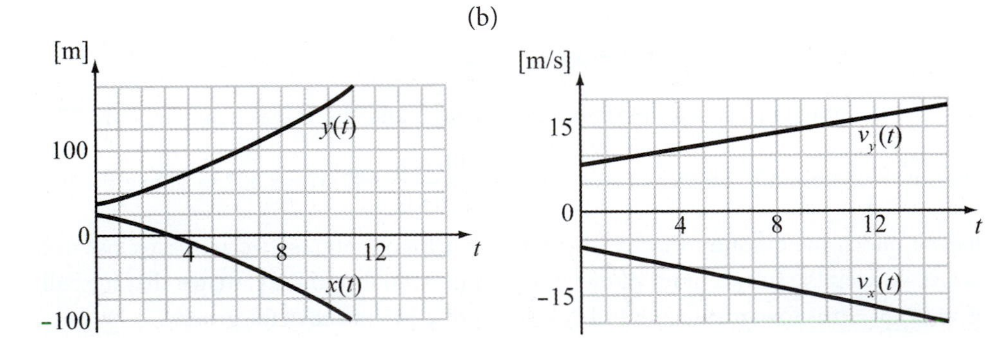

(c)

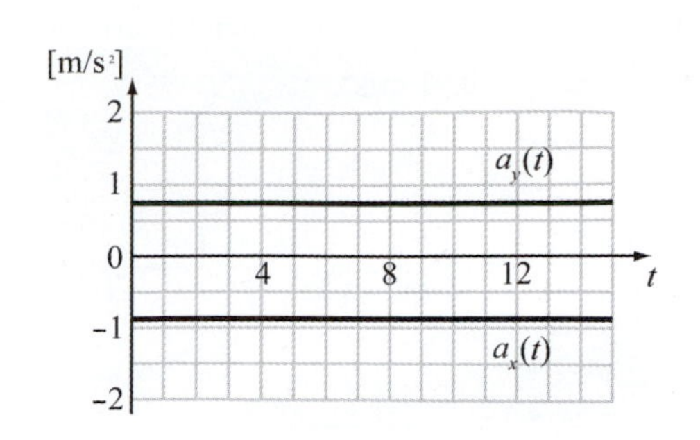

RESEARCH:

(a) Insert $t = 10$ s into the given equations, then use $|\vec{r}| = \sqrt{x^2 + y^2}$ and $\theta = \tan^{-1}(y/x)$.

(b) Differentiate the given components with respect to time to get functions of velocity.

(c) Differentiate the velocity functions with respect to time to get functions of acceleration.

SIMPLIFY:

(a) It is not necessary to simplify.

(b) $v_x(t) = \dfrac{dx(t)}{dt} = \dfrac{d(-0.45t^2 - 6.5t + 25)}{dt} = (-0.90t - 6.5)$ m/s

$v_y(t) = \dfrac{dy(t)}{dt} = \dfrac{d(0.35t^2 + 8.3t + 34)}{dt} = (0.70t + 8.3)$ m/s

(c) $a_x(t) = \dfrac{dv_x(t)}{dt} = \dfrac{d(-0.90t - 6.5)}{dt} = -0.90 \text{ m/s}^2, \; a_y(t) = \dfrac{dv_y(t)}{dt} = \dfrac{d(0.70t + 8.3)}{dt} = 0.70 \text{ m/s}^2$

CALCULATE:

(a) $x(10.0) = -0.45(10.0)^2 - 6.5(10.0) + 25 = -85 \text{ m}, \; y(10.0) = 0.35(10.0)^2 + 8.3(10.0) + 34 = 152 \text{ m}$

Now, insert these values into the magnitude and distance equations:

$$|\vec{r}| = \sqrt{(-85 \text{ m})^2 + (152 \text{ m})^2} = 174 \text{ m}, \; \theta = \tan^{-1}\left(\frac{152}{-85}\right) = -60.786°$$

(b) $v_x(10.0) = -0.90(10.0) - 6.5 = -15.5 \text{ m/s}, \; v_y(10.0) = 0.70(10.0) + 8.3 = 15.3 \text{ m/s}$

$|\vec{v}| = \sqrt{(-15.5)^2 + (15.3)^2} = 21.8 \text{ m/s}, \; \theta = \tan^{-1}\left(\dfrac{15.3}{-15.5}\right) = -44.6°$

(c) Since there is no time dependence, the acceleration is always $\vec{a} = (a_x, a_y) = (-0.90 \text{ m/s}^2, 0.70 \text{ m/s}^2)$.

The $|\vec{a}| = \sqrt{(-0.90 \text{ m/s}^2)^2 + (0.70 \text{ m/s}^2)^2} = 1.140 \text{ m/s}^2, \; \theta = \tan^{-1}\left(\dfrac{0.70}{-0.90}\right) = -37.87°$.

ROUND:

(a) Both distances and the magnitude are accurate to the meter, $|\vec{r}| = 174$ m. Round the angle to three significant figures, 60.8° north of west (note: west is used because x was negative).

(b) The equation's parameters are accurate to a tenth of a meter. The rabbit's velocity is then 21.8 m/s, 44.6° north of west.

(c) It is not necessary to consider the significant figures since the original parameters of the function are used. The rabbit's velocity is 1.14 m/s², 37.9° north of west.

DOUBLE-CHECK:

(a) 174 m in 10 s seems reasonable for a rabbit, considering the world record for the 100 m dash is about 10 s.

(b) The velocity of a rabbit ranges from 12 m/s to 20 m/s. This rabbit would be at the top of that range.

(c) A rabbit may accelerate at this rate but it can not sustain this acceleration for too long.

3.43. **THINK:** Assume the ball starts on the ground so that the initial and final heights are the same. The initial velocity of the ball is $v_{\text{i}} = 27.5 \text{ m/s}$, with $\theta = 56.7°$ and $g = 9.81 \text{ m/s}^2$.

SKETCH:

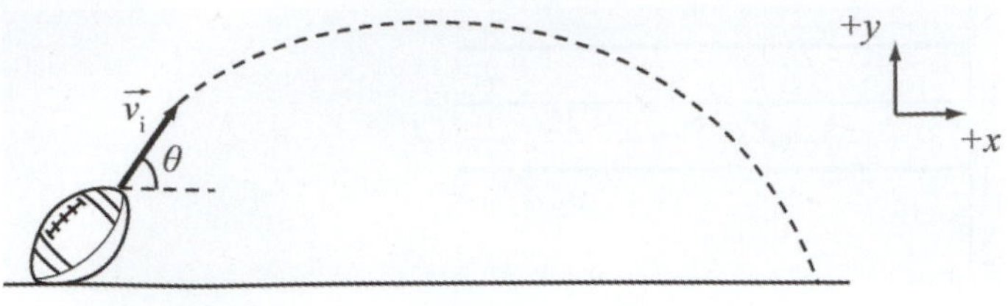

RESEARCH: $y_{\text{f}} - y_{\text{i}} = v_{\text{iy}}t + \frac{1}{2}at^2$ and $v_{\text{iy}} = v_{\text{i}}\sin\theta$.

SIMPLIFY: $0 = v_{\text{i}}(\sin\theta)t - \frac{1}{2}gt^2 \Rightarrow v_{\text{i}}\sin\theta = \frac{1}{2}gt \Rightarrow t = \frac{2v_{\text{i}}\sin\theta}{g}$

CALCULATE: $t = \frac{2(27.5 \text{ m/s})\sin(56.7°)}{9.81 \text{ m/s}^2} = 4.6860 \text{ s}$

ROUND: Rounding to three significant figures, $t = 4.69$ s.

DOUBLE-CHECK: Given the large angle the ball was kicked, about 5 seconds is a reasonable amount of time for it to remain in the air.

3.47. **THINK:** Use the horizontal distance and velocity to determine the time it takes to reach the post. Use the time to determine the height of the ball at that point. Then compare the height of the ball to the height of the goal post. Vertical velocity at this point can be determined as well: $v_{\text{i}} = 22.4$ m/s, $\theta = 49.0°$, $R = 39.0$ m, $h = 3.05$ m, $g = 9.81 \text{ m/s}^2$. Assume the ball is kicked off the ground, $y_{\text{i}} = 0$ m.

SKETCH:

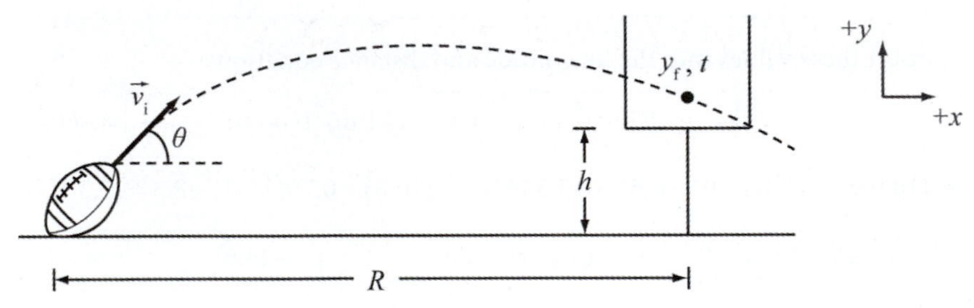

RESEARCH: $R = v_{\text{ix}}t$; $y_{\text{f}} - y_{\text{i}} = v_{\text{iy}}t + \frac{1}{2}at^2$; $v_{\text{fy}} = v_{\text{iy}} + at$; $v_{\text{ix}} = v_{\text{i}}\cos\theta$; and $v_{\text{iy}} = v_{\text{i}}\sin\theta$.

SIMPLIFY:

(a) $t = \frac{R}{v_{\text{i}}\cos\theta}$ and $y_{\text{f}} = (v_{\text{i}}\sin\theta)t - \frac{1}{2}gt^2 = R\tan\theta - \frac{gR^2}{2v_{\text{i}}^2\cos^2\theta}$.

In order to compare the height of the ball to the height of the goal post, subtract h from both sides of the equation,

$$\Delta h = y_{\text{f}} - h = -h + R\tan\theta - \frac{gR^2}{2v_{\text{i}}^2\cos^2\theta}.$$

(b) $v_{\text{fy}} = v_{\text{i}}\sin\theta - \frac{gR}{v_{\text{i}}\cos\theta}$

CALCULATE:

(a) $\Delta h = -3.05\text{ m} + (39.0\text{ m})\tan(49.0°) - \dfrac{(9.81\text{ m/s}^2)(39.0\text{ m})^2}{2(22.4\text{ m/s})^2\cos^2(49.0°)} = 7.2693\text{ m}$

(b) $v_{fy} = (22.4\text{ m/s})\sin(49.0°) - \dfrac{(9.81\text{ m/s}^2)(39.0\text{ m})}{(22.4\text{ m/s})\cos(49.0°)} = -9.1286\text{ m/s}$

ROUND: Round to the appropriate three significant figures:

(a) The ball clears the post by 7.27 m.

(b) The ball is heading downward at 9.13 m/s.

DOUBLE-CHECK: The initial velocity certainly seems high enough to clear the goal post from about 1/3 of the field away. It also makes sense that the vertical velocity at this point is lower than the initial velocity and the ball is heading down.

3.51. **THINK:** Though not explicitly stated, assume the launch angle is not zero. The projectile's height as a function of time is given. This function can be related to a more general function and the specifics of the motion can be determined. $g = 9.81\text{ m/s}^2$ and $v_i = 20.0\text{ m/s}$.

SKETCH:

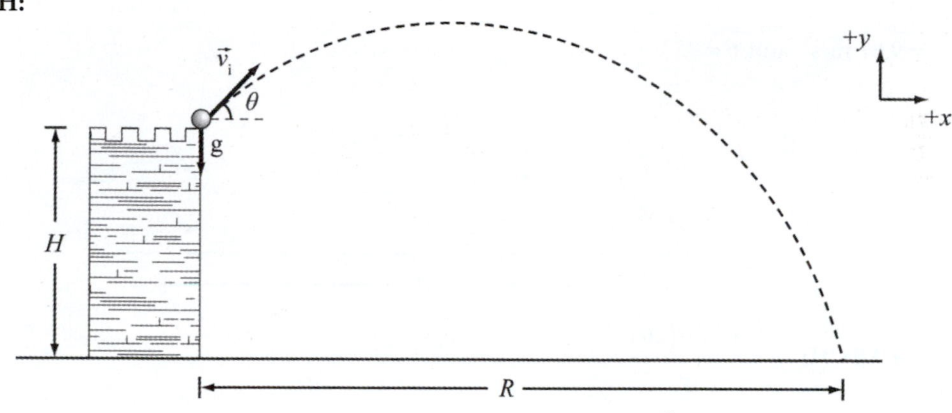

RESEARCH: $y(t) = -4.90t^2 + 19.32t + 60.0$; $R = v_{ix}t$; $y_f - y_i = v_{iy}t + \frac{1}{2}at^2$; $v_{iy} = v_i\sin\theta$; and $v_{ix} = v_i\cos\theta$.

SIMPLIFY:

(a) $y_f = y_i + v_i\sin\theta t - \frac{1}{2}gt^2$, where $(1/2)g = 4.90\text{ m/s}^2$, $v_i\sin\theta = 19.32\text{m/s}$, and $y_i = 60.0\text{ m}$.

(b) $\sin\theta = \dfrac{19.32\text{m/s}}{v_i} \Rightarrow \theta = \sin^{-1}\left(\dfrac{19.32}{v_i}\right)$

(c) $0 = y_i + v_i\sin\theta t - \frac{1}{2}gt^2 = y_i + R\tan\theta - \dfrac{gR^2}{2v_i^2\cos^2\theta}$ $\quad\left(\text{since } t = \dfrac{R}{v_i\cos\theta}\right)$

Therefore, $\left(\dfrac{g}{2v_i^2\cos^2\theta}\right)R^2 - (\tan\theta)R - y_i = 0$. Using the quadratic formula,

$$R = \frac{\tan\theta \pm \sqrt{\tan^2\theta + \dfrac{2y_i g}{v_i^2\cos^2\theta}}}{g/\left(v_i^2\cos^2\theta\right)}.$$

CALCULATE:

(a) $y_i = H = 60.0\text{ m}$

(b) $\theta = \sin^{-1}\left(\dfrac{19.32\text{ m/s}}{20.0\text{ m/s}}\right) = 75.02°$

(c) $$R = \frac{\tan(75.02°) \pm \sqrt{\tan^2(75.02°) + \dfrac{2(60.0\text{ m})(9.81\text{m/s}^2)}{(20.0\text{ m/s})^2\cos^2(75.02°)}}}{\dfrac{9.81\text{m/s}^2}{(20.0\text{ m/s}^2)\cos^2(75.02°)}} = 30.9386\text{ m or } -10.5715\text{ m}$$

ROUND:

(a) The building height is 60.0. m.

(b) The launch angle is $75.0°$.

(c) The object travels 31.0 m (the positive value must be chosen).

DOUBLE-CHECK: Given the large launch angle, it makes sense that the object doesn't travel too far.

3.55. **THINK:** The question does not specify a launch angle. However, for maximum distance, the launch angle is $45°$. Assume the initial and final heights are the same so the range equation can be used. $R = 0.67$ km, $g = 9.81\text{ m/s}^2$ and $\theta = 45°$.

SKETCH:

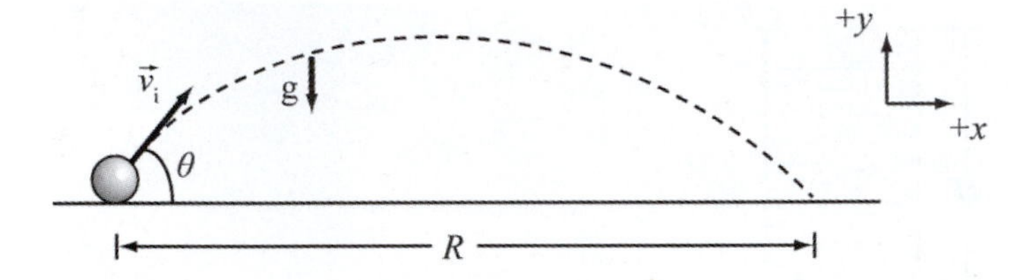

RESEARCH: $R = \dfrac{v_i^2\sin(2\theta)}{g}$

SIMPLIFY: $v_i = \sqrt{\dfrac{gR}{\sin(2\theta)}}$

CALCULATE: $v_i = \sqrt{\dfrac{(9.81\text{ m/s}^2)(670\text{ m})}{\sin(2\cdot 45°)}} = 81.072\text{ m/s}$

ROUND: Rounding to two significant figures: $v_i = 81$ m/s.

DOUBLE-CHECK: This speed is equivalent to about 300 km/h, which seems reasonable for a catapult.

3.59. **THINK:** If the juggler has a ball in her left hand, he can also have one in his right hand, assuming her right hand is throwing the ball up. This means if the juggler has x number of balls, the minimum time between balls is 0.200 s. If the time between any two balls is less than 0.200 s, then the hands can't act fast enough and she'll drop the balls. The given information must be used to determine the time of flight, and thus the total time of one ball going around the loop. Then determine the largest integer, n, which can be multiplied by $t_{\text{pass}} = 0.200$ s which is still less than the total time. H = 90.0 cm, $\theta = 75.0°$ and $g = 9.81\text{ m/s}^2$.

SKETCH:

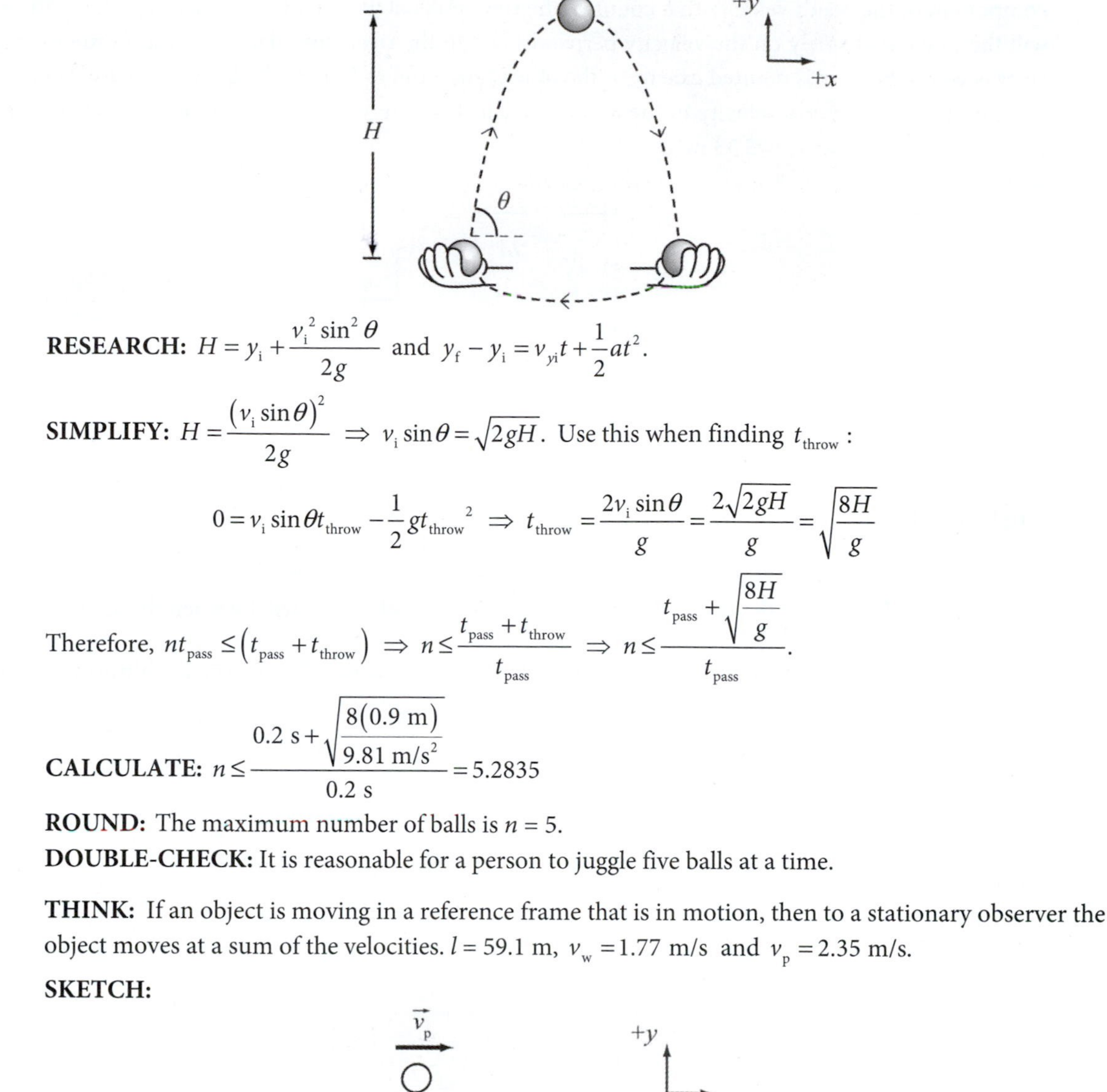

RESEARCH: $H = y_\text{i} + \dfrac{v_\text{i}^2 \sin^2\theta}{2g}$ and $y_\text{f} - y_\text{i} = v_{y\text{i}}t + \dfrac{1}{2}at^2$.

SIMPLIFY: $H = \dfrac{(v_\text{i}\sin\theta)^2}{2g} \Rightarrow v_\text{i}\sin\theta = \sqrt{2gH}$. Use this when finding t_throw:

$$0 = v_\text{i}\sin\theta t_\text{throw} - \frac{1}{2}gt_\text{throw}^{\,2} \Rightarrow t_\text{throw} = \frac{2v_\text{i}\sin\theta}{g} = \frac{2\sqrt{2gH}}{g} = \sqrt{\frac{8H}{g}}$$

Therefore, $nt_\text{pass} \le \left(t_\text{pass} + t_\text{throw}\right) \Rightarrow n \le \dfrac{t_\text{pass} + t_\text{throw}}{t_\text{pass}} \Rightarrow n \le \dfrac{t_\text{pass} + \sqrt{\dfrac{8H}{g}}}{t_\text{pass}}$.

CALCULATE: $n \le \dfrac{0.2 \text{ s} + \sqrt{\dfrac{8(0.9 \text{ m})}{9.81 \text{ m/s}^2}}}{0.2 \text{ s}} = 5.2835$

ROUND: The maximum number of balls is $n = 5$.

DOUBLE-CHECK: It is reasonable for a person to juggle five balls at a time.

3.63. **THINK:** If an object is moving in a reference frame that is in motion, then to a stationary observer the object moves at a sum of the velocities. $l = 59.1$ m, $v_\text{w} = 1.77$ m/s and $v_\text{p} = 2.35$ m/s.

SKETCH:

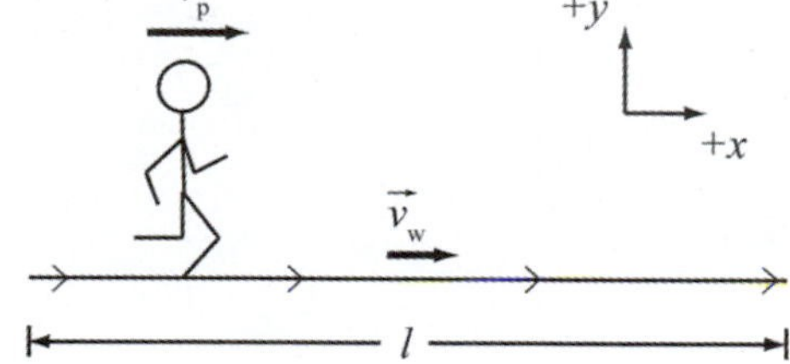

RESEARCH: $x_\text{f} - x_\text{i} = v_x t$

SIMPLIFY: $t = \dfrac{l}{v_\text{w} + v_\text{p}}$

CALCULATE: $t = \dfrac{59.1 \text{ m}}{2.35 \text{ m/s} + 1.77 \text{ m/s}} = 14.345 \text{ s}$

ROUND: Rounding to three significant figures, $t = 14.3$ s.

DOUBLE-CHECK: Given the long length of the walkway and the slow speed, a large time is reasonable.

3.67. **THINK:** If the boaters want to travel directly over to the other side, they must angle the boat so that the component of the boat's velocity that counters the river is equal in magnitude. The time it takes to get over will then be based solely on the velocity perpendicular to the river flow. The minimum time to cross the river is when the boat is pointed exactly at the other side. Also, as long as the boat's velocity component is countering the river, any velocity in the perpendicular direction will get the boat across the river: $l = 127$ m, $v_{\text{B}} = 17.5$ m/s, and $v_{\text{R}} = 5.33$ m/s.

SKETCH:

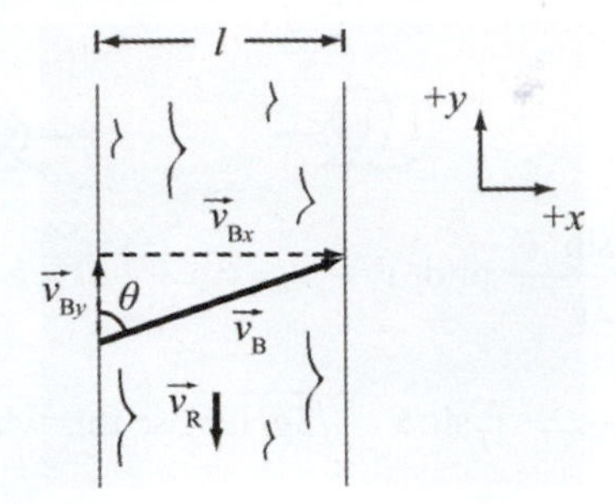

RESEARCH: $\vec{v} = v_x\hat{x} + v_y\hat{y}$; $\left|v_{\text{B}y}\right| = \left|-v_{\text{R}}\right|$; $v_x = v\sin\theta$; $v_y = v\cos\theta$; $x_{\text{f}} - x_{\text{i}} = v_x\Delta t$; $\tan\theta = \left(\frac{v_x}{v_y}\right)$; and $\cos\theta = \left(\frac{v_y}{v}\right)$. In part (e), the minimum speed, technically an infimum, will be when the angle θ is arbitrarily close to 0, and the component of the velocity directly across the stream is arbitrarily close to 0.

SIMPLIFY:

(a) $v_{\text{B}y} = v_{\text{R}} \Rightarrow v_{\text{B}}\cos\theta = v_{\text{B}y} \Rightarrow \theta = \cos^{-1}\left(\frac{v_{\text{R}}}{v_{\text{B}}}\right)$

(b) $l = \left(v_{\text{B}}\sin\theta\right)t \Rightarrow t = \frac{l}{v_{\text{B}}\sin\theta}$

(c) $\theta_{\text{min}} = 90°$

(d) $t_{\text{min}} = \frac{l}{v_{\text{B}}\sin\theta_{\text{min}}}$

(e) $\vec{v} \approx -\vec{v}_{\text{R}}$

CALCULATE:

(a) $\theta = \cos^{-1}\left(\frac{5.33 \text{ m/s}}{17.5 \text{ m/s}}\right) = 72.27°$

(b) $t = \frac{127 \text{ m}}{\left(17.5 \text{ m/s}\right)\sin\left(72.27°\right)} = 7.619 \text{ s}$

(c) $\theta = 90°$

(d) $t_{\text{min}} = \frac{127 \text{ m}}{17.5 \text{ m/s}} = 7.257 \text{ s}$

(e) $v = 5.33 \text{ m/s}$

ROUND:

(a) $\theta = 72.3°$

(b) $t = 7.62$ s

(c) $\theta = 90°$

(d) $t = 7.26$ s

(e) $v = 5.33$ m/s

DOUBLE-CHECK: Given the width of the river and the velocities, these answers are reasonable.

3.71. **THINK:** The horizontal velocity remains unchanged at any point. $v_{\text{i}} = 31.1$ m/s and $\theta = 33.4°$.

SKETCH:

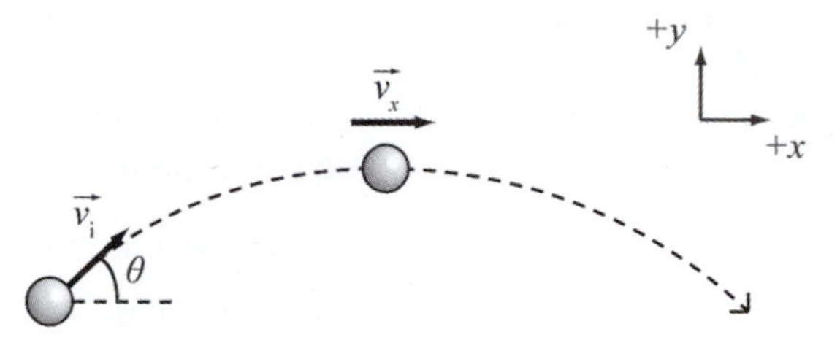

RESEARCH: $v_{\text{ix}} = v_{\text{fx}} = v_{\text{i}} \cos\theta$

SIMPLIFY: $v_x = v_{\text{i}} \cos\theta$

CALCULATE: $v_x = (31.1 \text{ m/s})\cos(33.4°) = 25.964$ m/s

ROUND: Rounding to three significant figures, $v_x = 26.0$ m/s.

DOUBLE-CHECK: The result is smaller than the initial velocity, which makes sense.

3.75. **THINK:** Assuming the box has no parachute, at the time of drop it will have the same horizontal velocity as the velocity of the helicopter, $\vec{v}_{\text{H}}$, and this will remain constant throughout the fall. The initial vertical velocity is zero. The vertical component of the velocity will increase due to the acceleration of gravity. The final speed of the box at impact can be found from the horizontal and final vertical velocity of the box just before impact. $h = 500.$ m, $d = 150.$ m, and $g = 9.81 \text{ m/s}^2$.

SKETCH:

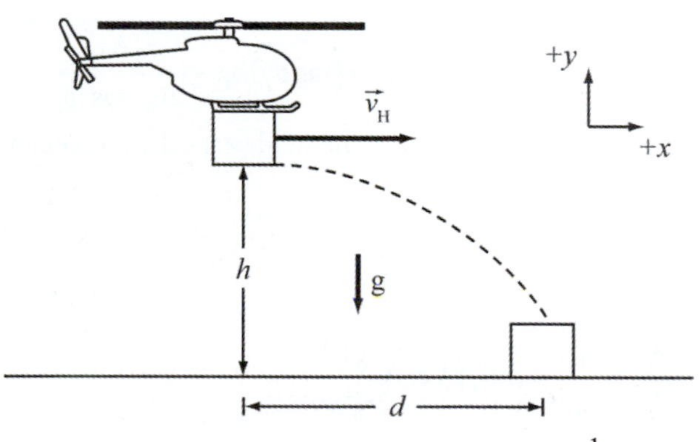

RESEARCH: To find the speed of the helicopter we use: $y_{\text{f}} - y_{\text{i}} = v_{y\text{i}}t + \frac{1}{2}at^2$ and $x_{\text{f}} - x_{\text{i}} = v_x t$. To find the final speed of the box use $v_{\text{fy}} = v_{\text{iy}} + at$ as the vertical component, the helicopter speed as the horizontal component, and $|\vec{v}| = \sqrt{v_y^{\ 2} + v_x^{\ 2}}$ to find the final speed of the box when it hits the ground.

SIMPLIFY: $-h = 0 - \frac{1}{2}gt^2 \Rightarrow t = \sqrt{\frac{2h}{g}}$ and so $v_x = v_{\text{H}} = \frac{d}{t} = d\sqrt{\frac{g}{2h}}$.

$v_y = -gt = -g\sqrt{\frac{2h}{g}} = -\sqrt{2gh}$ and so $|\vec{v}| = \sqrt{\left(-\sqrt{2gh}\right)^2 + \left(d\sqrt{\frac{g}{2h}}\right)^2} = \sqrt{g\left(2h + d^2/[2h]\right)}$

CALCULATE: $v_{\text{H}} = (150.\text{ m})\sqrt{\frac{9.81\text{ m/s}^2}{2(500.\text{ m})}} = 14.857\text{ m/s}$

$|v| = \sqrt{9.81\text{ m/s}^2\left(2(500.\text{ m}) + (150.\text{ m})^2/\left[2(500.\text{ m})\right]\right)} = 100.15\text{ m/s}$

ROUND: $v_{\text{H}} = 14.9\text{ m/s}$ and $|\vec{v}| = 100.\text{ m/s}$.

DOUBLE-CHECK: The helicopter velocity is equivalent to about 50 km/h and the speed that the box has when it hits the ground is equivalent to about 360 km/h. Both are reasonable values.

3.79. **THINK:** Determine which floor of the building the water strikes (each floor is $h_{\text{f}} = 4.00\text{ m}$ high). The horizontal distance between the firefighter and the building is $\Delta x = 60.0\text{ m}$. The initial angle of the water stream is $\theta_0 = 37.0°$. The initial speed is $v_0 = 40.3\text{ m/s}$.

SKETCH:

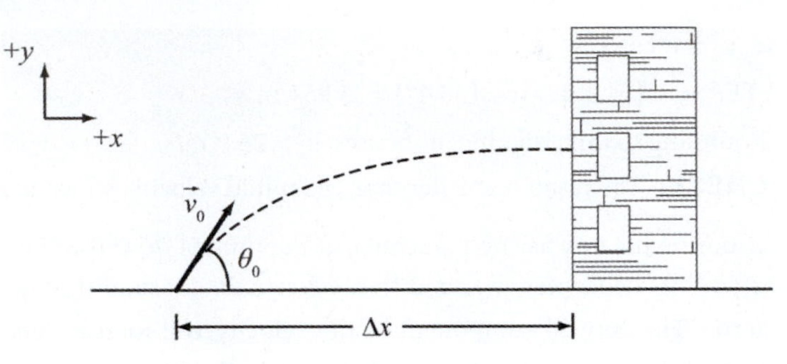

RESEARCH: To determine which floor the water strikes, the vertical displacement of the water with respect to the ground must be determined. The trajectory equation can be used, assuming ideal parabolic motion:

$$\Delta y = (\tan\theta_0)\Delta x - \frac{g(\Delta x)^2}{2v_0^2\cos^2\theta_0}.$$

The floor, at which the water strikes, n, is the total vertical displacement of the water divided by the height of each floor:

$$n = \frac{\Delta y}{h_{\text{f}}}.$$

SIMPLIFY: $n = \frac{\Delta y}{h_{\text{f}}} = \frac{1}{h_{\text{f}}}\left[(\tan\theta_0)\Delta x - \frac{g(\Delta x)^2}{2v_0^2\cos^2\theta_0}\right]$

CALCULATE: $n = \frac{1}{4\text{ m}}\left[\tan(37°)(60\text{ m}) - \frac{(9.81\text{ m/s}^2)(60\text{ m})^2}{2(40.3\text{ m/s})^2\cos^2(37°)}\right] = 7.042$

ROUND: $n = 7.042$ floors above ground level (where $n = 0$) corresponds to ankle height on the 8th floor.

DOUBLE-CHECK: It is reasonable to expect the water from a high-pressure fire hose to reach the 8th floor of a building. Also, the answer is unitless, as it should be.

3.83. **THINK:** Determine g when the range is $R = 2165$ m, $v_0 = 50.0$ m/s and $\theta_0 = 30.0°$.

SKETCH:

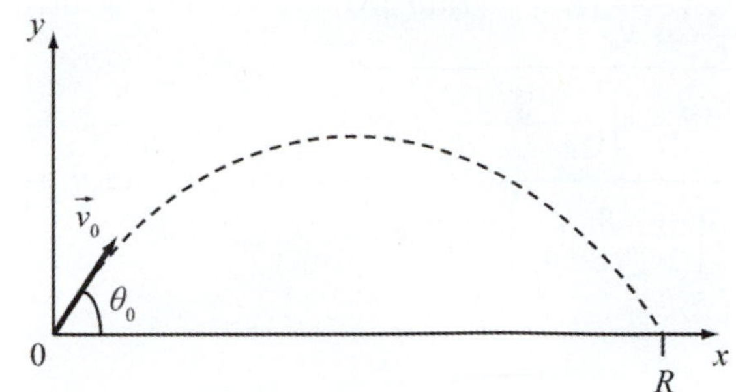

RESEARCH: Since the initial and final heights are equal, the range equation can be used:
$$R = \frac{v_0^2 \sin(2\theta_0)}{g}.$$

SIMPLIFY: $g = \dfrac{v_0^2 \sin(2\theta_0)}{R}$

CALCULATE: $g = \dfrac{(50.0 \text{ m/s})^2 \sin(2(30.0°))}{2165 \text{ m}} = 1.00003 \text{ m/s}^2$

ROUND: $g = 1.00 \text{ m/s}^2$

DOUBLE-CHECK: The units of the calculated g are correct.

3.87. **THINK:** For the shot-put, the initial speed is $v_0 = 13.0$ m/s , the launch angle is $\theta_0 = 43.0°$ and the initial height is $y_0 = 2.00$ m. Determine (a) the horizontal displacement Δx and (b) the flight time t, after the shot hits the ground at $y = 0$.

SKETCH:

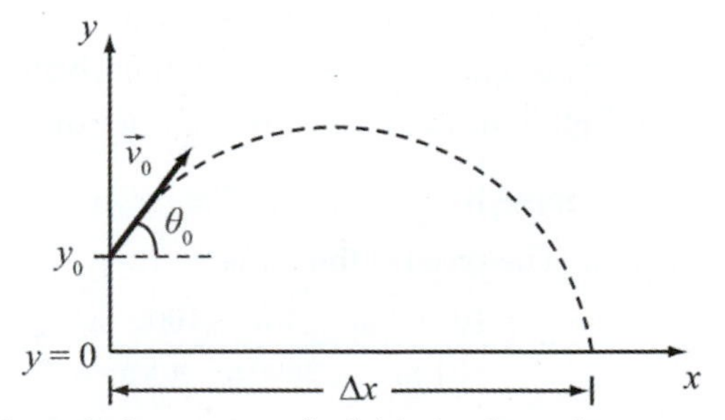

RESEARCH: Assuming ideal parabolic motion, find (a) Δx from the trajectory equation
$$y = y_0 + \tan\theta_0 \Delta x - \frac{g(\Delta x)^2}{2v_0^2 \cos^2\theta_0}$$
and (b) t from the equation $\Delta x = v_0 \cos\theta_0 t$.

SIMPLIFY:

(a) With $y = 0$, $\frac{g}{2v_0^2\cos^2\theta_0}(\Delta x)^2 - (\tan\theta_0)\Delta x - y_0 = 0.$ Solving this quadratic equation yields:

$$\Delta x = \frac{\tan\theta_0 \pm \sqrt{\tan^2\theta_0 - 4\left(\frac{g}{2v_0^2\cos^2\theta_0}\right)(-y_0)}}{2\left(\frac{g}{2v_0^2\cos^2\theta_0}\right)} = \left(\tan\theta_0 \pm \sqrt{\tan^2\theta_0 + \left(\frac{2gy_0}{v_0^2\cos^2\theta_0}\right)}\right)\left(\frac{v_0^2\cos^2\theta_0}{g}\right).$$

(b) $t = \frac{\Delta x}{v_0\cos\theta_0}$

CALCULATE:

(a) $\Delta x = \left(\tan(43°) \pm \sqrt{\tan^2(43°) + \left(\frac{2(9.81\text{ m/s}^2)(2\text{ m})}{(13.0\text{ m/s})^2\cos^2(43°)}\right)}\right)\left(\frac{(13.0\text{ m/s})^2\cos^2 43°}{9.81\text{ m/s}^2}\right)$

$= \left(0.9325 \pm \sqrt{0.8696 + 0.4341}\right)(9.2145\text{ m})$

$= \left(0.9325 \pm \sqrt{1.3037}\right)(9.2145\text{ m})$

$= 19.114\text{ m or } -1.928\text{ m}$

(b) $t = \frac{19.114\text{ m}}{(13.0\text{ m/s})\cos(43°)} = 2.0104\text{ s}$

ROUND:

(a) The sum under the square root is precise to the tenth-place, and so has three significant figures. Then, $\Delta x = 19.1\text{ m}$ (take the positive root).

(b) Since θ_0 and Δx have three significant figures, $t = 2.01$ s.

DOUBLE-CHECK: For near optimal launch angle (optimal being $\theta_0 = 45°$), a horizontal displacement of 19.1 m is reasonable. The flight time of 2.0 s is reasonable for this horizontal displacement.

3.91. **THINK:** The height of the goose is $h_g = 30.0$ m. The height of the windshield is $h_c = 1.00$ m. The speed of the goose is $v_g = 15.0$ m/s. The speed of the car is

$$v_c = \frac{100.0\text{ km}}{1\text{ hr}} \cdot \frac{1\text{ h}}{3600\text{ s}} \cdot \frac{1000\text{ m}}{1\text{ km}} = 27.7778\text{ m/s}.$$

The initial horizontal distance between the goose and the car is $d = 104.0$ m. The goose and the car move toward each other. Determine if (a) the egg hits the windshield and (b) the relative velocity of the egg with respect to the windshield, v_g'. Let θ be the angle the egg makes with the horizontal when it impacts.

SKETCH:

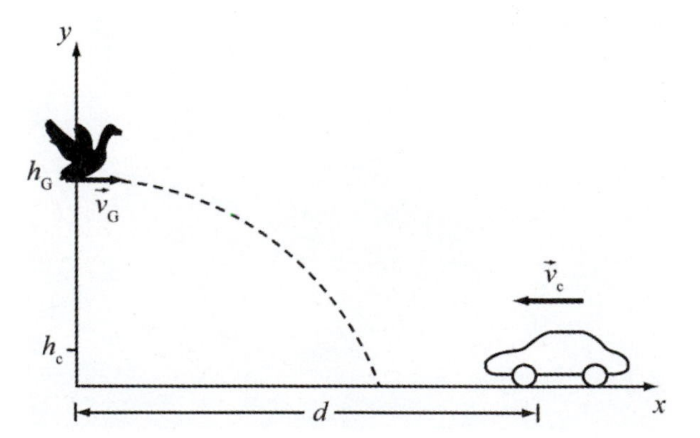

RESEARCH:

(a) The egg must have a vertical displacement $\Delta y_g = h_c - h_g$ to hit the windshield. Use $\Delta y_g = (v_0 \sin\theta_0)t - \frac{1}{2}(gt^2)$ to determine the fall time t. Use $\Delta x_g = (v_0 \cos\theta_0)t$ to determine the horizontal displacement of the egg. In order for the egg to collide with the windshield, the car must travel $\Delta x_c = d - \Delta x_g$ in time t. Use $\Delta x = vt$ to determine the car's travel distance Δx_c. Note that the launch angle of the egg is $\theta_0 = 0°$.

(b) The horizontal component of the egg's speed in the reference frame of the windshield will be $v_{gx}' = v_{gx} + v_c$ because the car is moving toward the egg in the horizontal direction. Because the car has no vertical speed, the vertical speed of the egg in the reference frame of the car is unchanged, $v_{gy}' = v_{gy}$. To determine v_{gy}, use $v_y^2 = v_{y_0}^2 + 2a\Delta y$. Then $v_g' = \sqrt{(v_{gx}')^2 + (v_{gy}')^2}$. The angle of impact is implicitly given by $\tan\theta = \dfrac{v'_{gy}}{v'_{gx}}$.

SIMPLIFY:

(a) Since $\theta_0 = 0°$ and $v_0 = v_g$, $v_{gy_0} = v_g \sin\theta_0 = 0$ and $v_{gx_0} = v_g \cos\theta_0 = v_g$. Then

$$\Delta y_g = -\frac{1}{2}gt^2 \Rightarrow h_c - h_g = -\frac{1}{2}gt^2 \Rightarrow t = \sqrt{\frac{2(h_g - h_c)}{g}},$$

and also $\Delta x_g = v_g t = v_g\sqrt{\dfrac{2(h_g - h_c)}{g}}$. In this time, the car travels $\Delta x_c = v_c t = v_c\sqrt{\dfrac{2(h_g - h_c)}{g}}$.

(b) $v_{gx}' = v_{gx} + v_c = v_g + v_c$ and $v_{gy}' = v_{gy} = \sqrt{2g\Delta y_g} = \sqrt{2g(h_g - h_c)}$. Then, substituting gives: $v_g' = \sqrt{(v_g + v_c)^2 + 2g(h_g - h_c)}$. The angle of impact relative to the car is given by the equation:

$$\theta = \arctan\left(\frac{v'_{gy}}{v'_{gx}}\right) = \arctan\left(\frac{\sqrt{2g(h_g - h_c)}}{v_g + v_c}\right).$$

CALCULATE:

(a) $\Delta x_c = (27.778 \text{ m/s})\sqrt{\dfrac{2(30.0 \text{ m} - 1.00 \text{ m})}{9.81 \text{ m/s}^2}} = 67.542 \text{ m}$

$$d - \Delta x_g = d - v_g\sqrt{\frac{2(h_g - h_c)}{g}} = (104.0 \text{ m}) - (15.0 \text{ m/s})\sqrt{\frac{2(30.0 \text{ m} - 1.00 \text{ m})}{9.81 \text{ m/s}^2}} = 67.527 \text{ m}$$

(b) $v_g' = \sqrt{(15.0 \text{ m/s} + 27.7778 \text{ m/s})^2 + 2(9.81 \text{ m/s}^2)(30.0 \text{ m} - 1.00 \text{ m})} = \sqrt{2398.94 \text{ m}^2/\text{s}^2} = 48.98 \text{ m/s}$

$$\theta = \arctan\left(\frac{\sqrt{2(9.81 \text{ m/s}^2)(30.0 \text{ m} - 1.00 \text{ m})}}{15.0 \text{ m/s} + 27.7778 \text{ m/s}}\right) = 29.144°$$

ROUND: Rounding to three significant figures, Δx_c = 67.5 m and $d - \Delta x_g = 67.5$ m. The egg hits the windshield at a speed of 49.0 m/s relative to the windshield at an angle of 29.1° above the horizontal.

DOUBLE-CHECK: The speed of the egg relative to the windshield is greater than the speed of the car and the goose.

3.95. **THINK:** The bomb's vertical displacement is $\Delta y = -5.00 \cdot 10^3$ m (falling down). The initial speed is $v_0 = 1000.\text{ km/h}(\text{h}/3600\text{ s})(1000\text{ m/km}) = 277.8\text{ m/s}$. The launch angle is $\theta_0 = 0°$. Determine the distance from a target, Δx, and the margin of error of the time Δt if the target is $d = 50.0$ m wide.

SKETCH:

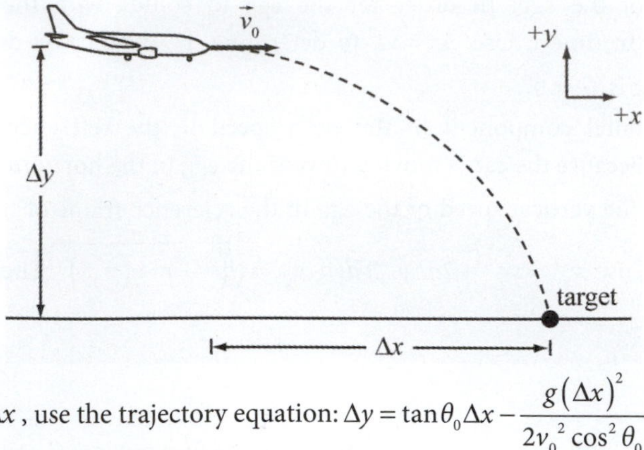

RESEARCH: To find Δx, use the trajectory equation: $\Delta y = \tan\theta_0 \Delta x - \dfrac{g(\Delta x)^2}{2v_0^2\cos^2\theta_0}$.

To find Δt, consider the time it would take the bomb to travel the horizontal distance *d*; this is the margin of error in the time. The margin of error for the release time Δt can be determined from $d = v_0 \cos\theta_0 \Delta t$.

SIMPLIFY: Note, $\tan\theta_0 = 0$ and $\cos\theta_0 = 1$ for $\theta_0 = 0$.

$$\Delta y = -\frac{g(\Delta x)^2}{2v_0^2} \text{ and } \Delta x = \sqrt{\frac{-2v_0^2\Delta y}{g}}$$

For Δt, use $d = v_0\cos\theta_0\Delta t \Rightarrow \Delta t = d / v_0$.

CALCULATE: $\Delta x = \sqrt{\dfrac{-2(277.8\text{ m/s})^2(-5.00\cdot 10^3\text{ m})}{9.81\text{ m/s}^2}} = 8869\text{ m}, \quad \Delta t = \dfrac{50.0\text{ m}}{277.8\text{ m/s}} = 0.1800\text{ s}$

ROUND: $\Delta x = 8.87$ km and $\Delta t = 0.180$ s.

DOUBLE-CHECK: Considering the altitude and the terrific speed of the airplane, these values are reasonable.

Multi-Version Exercises

3.99. **THINK:** The initial height, velocity, and angle of the tennis ball are known. To find the total horizontal distance covered before the tennis ball hits the ground it makes sense to decompose the motion into horizontal and vertical components. Then, find the time at which the tennis ball hits the ground ($y = 0$) and determine the horizontal position at that time. To make the problem as simple as possible, choose $y = 0$ m to be the ground and $x = 0$ m to be the location of the trebuchet where the ball is released.

SKETCH: A sketch helps to see exactly how to decompose the initial velocity into horizontal and vertical components.

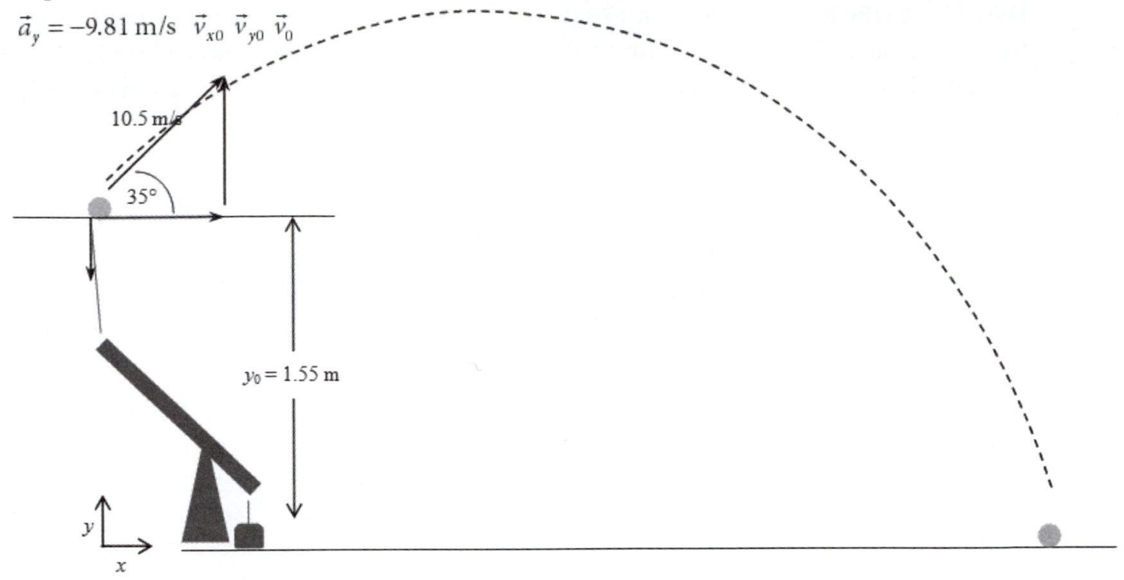

RESEARCH: This problem involves ideal projectile motion. Since there is no horizontal acceleration and the tennis ball starts at $x_0 = 0$, the equation (3.11) for the horizontal position at time t is $x = v_{x0}t$. Equation (3.13) gives the vertical position as $y = y_0 + v_{y0}t - \frac{1}{2}gt^2$. To find a final answer, is necessary to determine the x- and y-components of the initial velocity, given by $v_{y0} = v_0 \sin\theta$ and $v_{x0} = v_0 \cos\theta$.

SIMPLIFY: To find the time when the tennis ball hits the ground, it is necessary to find a non-negative solution to the equation $0 = y_0 + v_{y0}t - \frac{1}{2}gt^2$. The quadratic formula gives a solution of $t = \frac{-v_{y0} \pm \sqrt{(v_{y0})^2 - 4(-\frac{1}{2}g)(y_0)}}{2(-\frac{1}{2}g)} = \frac{v_{y0} \pm \sqrt{(v_{y0})^2 + 2gy_0}}{g}$. It will be necessary to take the positive square root here: the tennis ball cannot land *before* it is released. This time can then be used with the equation for horizontal position to get the position when the tennis ball hits the ground at $x = v_{x0}t = (v_{x0})\frac{v_{y0} + \sqrt{(v_{y0})^2 + 2gy_0}}{g}$. Combining this with the equations for the horizontal and vertical components of the initial velocity ($v_{y0} = v_0 \sin\theta$ and $v_{x0} = v_0 \cos\theta$) gives that the tennis ball lands at $x = (v_0 \cos\theta)\frac{v_0 \sin\theta + \sqrt{(v_0 \sin\theta)^2 + 2gy_0}}{g}$.

CALCULATE: The problem states that the initial height $y_0 = 1.55$ m. The initial velocity $v_0 = 10.5$ m/s at an angle of $\theta = 35°$ above the horizontal. The gravitational acceleration on Earth is –9.81 m/s². Thus the tennis ball lands at

$$
\begin{aligned}
x &= (v_0 \cos\theta)\frac{v_0 \sin\theta + \sqrt{(v_0 \sin\theta)^2 + 2gy_0}}{g} \\
&= \left[10.5\cos(35°)\right]\frac{10.5\sin(35°) + \sqrt{(10.5\sin(35°))^2 + 2 \cdot 9.81 \cdot 1.55}}{9.81} \\
&= 12.43999628
\end{aligned}
$$

ROUND: Since the measured values have 3 significant figures, the answer should also have three significant figures. Thus the tennis ball travels a horizontal distance of 12.4 m before it hits the ground.

DOUBLE-CHECK: Using equation (3.22) for the path of a projectile, it is possible to work backwards from the initial position $(x_0, y_0) = (0, 1.55)$ and the position when the tennis ball lands $(x, y) = (12.4, 0)$, and angle $\theta_0 = 35°$ to find the starting velocity, which should confirm what was given originally.

$$y = y_0 + (\tan\theta_0)x - \frac{g}{2v_0^2\cos^2\theta_0}x^2 \Rightarrow$$

$$0 = 1.55 + (\tan 35°)\cdot 12.4 - \frac{9.81}{2(v_0)^2\cos^2 35°}(12.4)^2 \Rightarrow$$

$$1.55 + (\tan 35°)\cdot 12.4 = \frac{9.81}{2(v_0)^2\cos^2 35°}(12.4)^2 \Rightarrow$$

$$(1.55 + (\tan 35°)\cdot 12.4)\left[\frac{(v_0)^2}{1.55 + (\tan 35°)\cdot 12.4}\right] = \frac{9.81\cdot(12.4)^2}{2(v_0)^2\cos^2 35°}\left[\frac{(v_0)^2}{1.55 + (\tan 35°)\cdot 12.4}\right] \Rightarrow$$

$$(v_0)^2 = \frac{9.81\cdot(12.4)^2}{2\cos^2 35°(1.55 + (\tan 35°)\cdot 12.4)} \Rightarrow$$

$$v_0 = \sqrt{\frac{9.81\cdot(12.4)^2}{2\cos^2 35°(1.55 + (\tan 35°)\cdot 12.4)}} \Rightarrow$$

$$v_0 = 10.4805539$$

Rounded to three significant figures, this becomes $v_0 = 10.5$, which confirms the answer.

3.103. **THINK:** This question involves flying a plane through the air. The speed of the airplane with respect to the wind and the velocity of the wind with respect to the ground are both given, so this problem involves relative motion. The vector sum of the plane's velocity with respect to the air and the velocity of the air with respect to the ground must point in the direction of the pilot's destination. Since the wind is blowing from the west to the east and the pilot wants to go north, the plane should head to the northwest.

SKETCH: The sketch should shows the velocity of the plane with respect to the wind ($\vec{v}_{pa}$), the velocity of the wind (which can be thought of as the velocity of the air with respect to the ground or $\vec{v}_{ag}$), and the velocity of the plane with respect to the ground ($\vec{v}_{pg}$). The distance the airplane is to travel will affect how long it takes the pilot to get to her destination, but will not effect in the $\vec{v}_{pa}$ $\vec{v}_{pg}$ $\vec{v}_{ag}$ direction she flies.

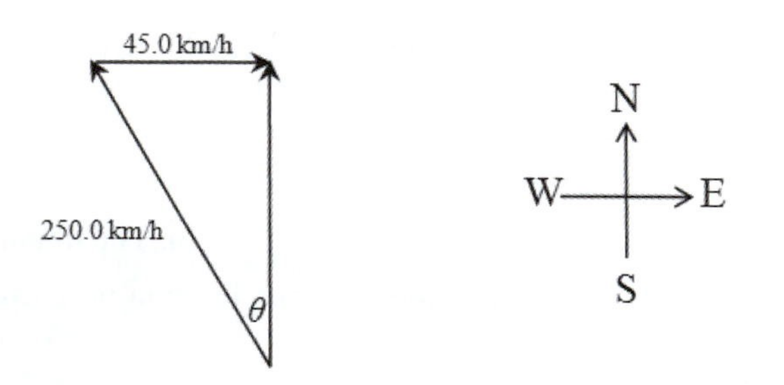

RESEARCH: To solve this problem, it is first necessary to note that the wind, blowing from West to East, is moving in a direction perpendicular to the direction the pilot wants to fly. Since these two vectors form a right angle, it is easy to use trigonometry to find the angle θ with the equation $\sin\theta = \frac{|\vec{v}_{ag}|}{|\vec{v}_{pa}|}$. Since North is 360° and West is 270°, the final answer will be a heading of $(360 - \theta)°$.

SIMPLIFY: To find the final answer, it is necessary to use the inverse sine function. The equation for the angle θ can be found using algebra and trigonometry:

$$\sin^{-1}(\sin\theta) = \sin^{-1}\left(\frac{|\vec{v}_{ag}|}{|\vec{v}_{pa}|}\right)$$

$$\theta = \sin^{-1}\left(\frac{|\vec{v}_{ag}|}{|\vec{v}_{pa}|}\right) \Rightarrow$$

$$360 - \theta = 360 - \sin^{-1}\left(\frac{|\vec{v}_{ag}|}{|\vec{v}_{pa}|}\right)$$

CALCULATE: To find the final answer in degrees, it is necessary first to make sure that the calculator or computer program is in degree mode. Note that the speed of the wind $|\vec{v}_{ag}| = 45.0$ km/h and the speed of the plane with respect to the air $|\vec{v}_{pa}| = 250.0$ km/h are given in the problem. Plugging these in and solving gives a heading of:

$$360 - \sin^{-1}\left(\frac{|\vec{v}_{ag}|}{|\vec{v}_{pa}|}\right) = 360° - \sin^{-1}\left(\frac{45.0 \text{ km/h}}{250.0 \text{ km/h}}\right)$$

$$= 349.6302402°$$

ROUND: Rounding to four significantfigures, the final heading is 349.6°.

DOUBLE-CHECK: Intuitively, this answer seems correct. The pilot wants to fly North and the wind is blowing from West to East, so she should head somewhere towards the Northeast. Since the speed of the airplane with respect to the air is much greater than the speed of the air with respect to the ground (wind speed), the East-West component of the airplane's velocity with respect to the air should be less than the North-South component. Resolving the motion of the plane into horizontal (East-West) and vertical components gives that the horizontal speed of the plane with respect to the ground v_{pax} is $250.0\frac{\text{km}}{\text{h}} \cdot \cos(349.6° - 270°)$ or 45.1 km/h from East to West, which is within rounding the same as the known wind speed.

Chapter 4: Force

Exercises

4.27. **THINK:** The gravitational constant on the Moon is $g_{\text{m}} = g/6$, where g is the Earth's gravitational constant. The weight of an apple on the Earth is w = 1.00 N.

SKETCH:

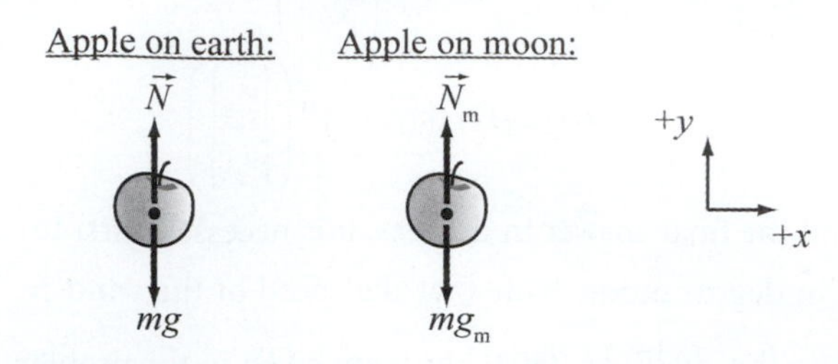

RESEARCH: The gravitational constant on the Earth is $g = 9.81 \text{ m/s}^2$. The weight of the apple, w is given by its mass times the gravitational constant, $w = mg$.

SIMPLIFY:

(a) The weight of the apple on the Moon is $w_{\text{m}} = mg_{\text{m}}$. Simplify this expression by substituting $g_{\text{m}} = g/6$: $w_{\text{m}} = m(g/6)$. Mass is constant so I can write expressions for the mass of the apple on Earth and on the Moon and then equate the expressions to solve for m. On Earth, $m = w/g$. On the Moon, $m = 6w_m/g$. Therefore, $w/g = 6w_{\text{m}}/g \Rightarrow w_{\text{m}} = w/6$.

(b) The expression for the weight of an apple on Earth can be rearranged to solve for m: $m = w/g$.

CALCULATE:

(a) $w_{\text{m}} = \frac{1}{6}(1.00 \text{ N}) = 0.166667 \text{ N}$

(b) $m = \frac{1.00 \text{ N}}{9.81 \text{ m/s}^2} = 0.101931 \text{ kg}$

ROUND: Rounding to three significant figures, (a) $w_{\text{m}} = 0.167 \text{ N}$ and (b) m = 0.102 kg.

DOUBLE-CHECK: It is expected that the weight of the apple on the Moon is much less than the weight of the apple on the Earth. Also, a mass of about 100 g is reasonable for an apple.

4.31. **THINK:** The mass of the elevator cabin is $m_e = 363.7 \text{ kg}$ and the total mass of the people in the elevator is m = 177.0 kg. The elevator is being pulled upward by a cable which has a tension of $T = 7638$ N. The acceleration of the elevator is to be determined.

SKETCH:

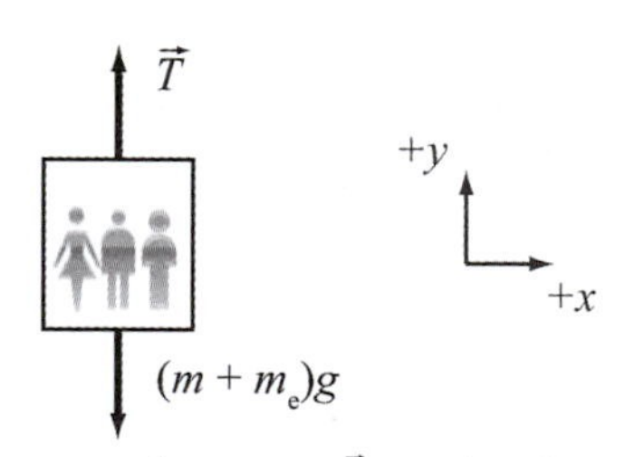

RESEARCH: Force is equal to mass times acceleration, $\vec{F} = m\vec{a}$. The sum of all the forces acting on the elevator will give the net force that acts upon the elevator, $F_{\text{net},y} = \sum F_y$. In this case, $F_{\text{net},y} = ma + m_e a$ $= T - mg - m_e g$. The gravitational acceleration is $g = 9.81 \text{ m/s}^2$.

SIMPLIFY: Group like terms, $a(m+m_e)=T-g(m+m_e)$. Rearrange to solve for a:

$$a=\frac{T-(m+m_e)g}{(m+m_e)}.$$

CALCULATE: $a=\dfrac{7638\text{ N}-(177.0\text{ kg}+363.7\text{ kg})9.81\text{ m/s}^2}{(177.0\text{ kg}+363.7\text{ kg})}=4.3161\text{ m/s}^2$

ROUND: Rounding to three significant figures because that is the precision of g, $a=4.32\text{ m/s}^2$.

DOUBLE-CHECK: The units of the result are correct. Also, the value determined for a is approximately 45% of the acceleration due to gravity, so the answer is reasonable.

SIMPLIFY: The equations can be rearranged to solve for the components of the fourth force: $F_{4,x}=F_2\sin\phi+F_3\cos\alpha-F_1\cos\theta_1$ and $F_{4,y}=F_3\sin\alpha-F_1\sin\theta_1-F_2\cos\phi$.

CALCULATE: $F_{4,x}=(200.)\sin(10.0°)\text{ N}+(100.)\cos(10.0°)\text{ N}-(150.)\cos(60.0°)\text{ N}=58.2104\text{ N}$

$F_{4,y}=(100.)\sin(10.0°)\text{ N}-(150.)\sin(60.0°)\text{ N}-(200.)\cos(10.0°)\text{ N}=-309.500\text{ N}$

$\theta_4=\tan^{-1}\left(\dfrac{-309.500\text{N}}{58.2104\text{ N}}\right)=-79.3483°$ with respect to the positive x-axis.

$|F_4|=\sqrt{(58.2104\text{ N})^2+(-309.500\text{N})^2}=314.9265\text{ N}$

ROUND: The given value for θ_1 has three significant figures, so the answers must be written as $|F_4|=315\text{ N}$ and $\theta_4=79.3°$ below the positive x-axis.

DOUBLE-CHECK: The direction that force F_4 is applied is consistent with the diagram and the magnitude of the force is reasonable.

4.35. **THINK:** The given quantities are the masses of the four weights, $m_1=6.50\text{ kg}$, $m_2=3.80\text{ kg}$, $m_3=10.70\text{ kg}$ and $m_4=4.20\text{ kg}$. Determine the tension in the rope connected m_1 and m_2.

SKETCH: Focus on an arbitrary point between m_1 and m_2.

RESEARCH: The masses are in equilibrium, so the sum of the forces in the vertical direction is equal to zero. Therefore the tension, T in the rope between m_1 and m_2 is equal to the force exerted by gravity due to masses m_2, m_3 and m_4: $T-m_2g-m_3g-m_4g=0$.

SIMPLIFY: $T=(m_2+m_3+m_4)g$

CALCULATE: $T=(3.80\text{ kg}+10.70\text{ kg}+4.20\text{ kg})9.81\text{ m/s}^2=183.447\text{ N}$

ROUND: There are three significant figures provided in the question so the answer should be written $T=183\text{ N}$.

DOUBLE-CHECK: Tension is a force and the result has units of force (Newtons). The value of the tension is also reasonable considering the masses of the objects.

4.39. **THINK:** The given quantities are the hanging masses m_1 and m_2 and the direction of the horizontal forces cause by the hanging masses on the ring. The strings that attach the hanging masses to the ring can be considered massless and the pulleys that the strings are routed through are frictionless. Determine the mass, m_3, and the angle, θ, that will result in the ring being balanced in the middle of the table.

SKETCH: Top-down view:

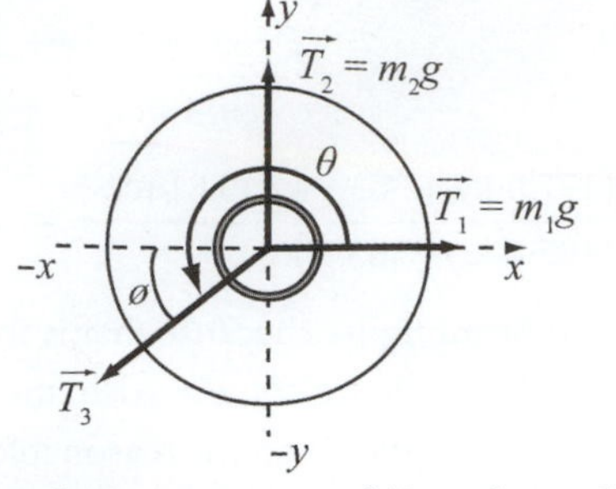

RESEARCH: A sketch of the x and y components of T_3 is shown below.

$+y$, $+x$, $\vec{T}_{3,x}$, $\vec{T}_{3,y}$, ϕ, $\vec{T}_3$

$$T_{3,y} = T_3 \sin\phi = m_3 g \sin\phi$$
$$T_{3,x} = T_3 \cos\phi = m_3 g \cos\phi$$

The angle counterclockwise from the positive x-axis, θ is given by $\theta = 180° + \phi$. For the ring to be balanced, the sum of the forces in the x and y directions must be balanced:

$$\sum F_y = 0 = T_2 - T_3 \sin\phi = m_2 g - m_3 g \sin\phi \qquad (1)$$
$$\sum F_x = 0 = T_1 - T_3 \cos\phi = m_1 g - m_3 g \cos\phi \qquad (2)$$

SIMPLIFY: Solve equation (1) in terms of m_3 and substitute into equation (2) to solve for ϕ. $m_3 = m_2 g / g \sin\phi = m_2 / \sin\phi$ substituted into (2) yields:

$$m_1 g = m_3 g \cos\phi \Rightarrow m_1 = \frac{m_2 \cos\phi}{\sin\phi} \Rightarrow \frac{\sin\phi}{\cos\phi} = \frac{m_2}{m_1} \Rightarrow \tan\phi = \frac{m_2}{m_1} \Rightarrow \phi = \tan^{-1}\left(\frac{m_2}{m_1}\right).$$

CALCULATE: $\phi = \tan^{-1}\left(\frac{0.0300 \text{ kg}}{0.0400 \text{ kg}}\right) = 36.8698°$, $\theta = 180° + 36.8698° = 216.8698°$,

$$m_3 = \frac{0.030 \text{ kg}}{\sin(36.8698°)} = 0.05000 \text{ kg}$$

ROUND: Four significant figures are provided in the question, so the answers should be written $\theta = 216.8°$ and $m_3 = 0.0500$ kg.

DOUBLE-CHECK: By observing the sketch, it can be seen that the value of θ is reasonable to balance the forces. The mass is also a reasonable value.

4.43. **THINK:** Two masses, $M_1 = 100.0$ g and $M_2 = 200.0$ g are placed on an Atwood device. Each mass moves a distance, $\Delta y = 1.00$ m in a time interval of $\Delta t = 1.52$ s. Determine the gravitational acceleration, g_p for the planet and the tension, T in the string. The string is massless and the pulley is frictionless. M_1 and M_2 should be converted to the SI unit of kg.

$$M_1 = (100.0 \text{ g})\left(\frac{1 \text{ kg}}{1000 \text{ g}}\right) = 0.1000 \text{ kg}, \quad M_2 = (200.0 \text{ g})\left(\frac{1 \text{ kg}}{1000 \text{ g}}\right) = 0.2000 \text{ kg}$$

SKETCH:

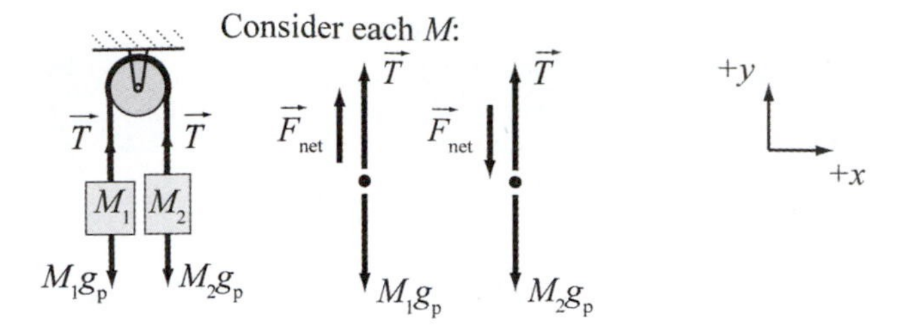

RESEARCH:

(a) The masses are initially at rest, therefore their initial speed $v_0 = 0$. Because the masses are rigidly connected, they accelerate at the same rate, a. The net force for one mass is upward and downward for the other. The value of a can be determined using the kinematic equation $\Delta y = v_0 t + \left(at^2\right)/2$. Because the masses are initially at rest, the equation reduces to $\Delta y = \left(at^2\right)/2$ or $a = 2\Delta y / t^2$. If the forces on mass M_1 are considered, the net force equation is $F_{\text{net}} = M_1 a = T - M_1 g_{\text{p}}$. For mass M_2, the net force equation is $F_{\text{net}} = -M_2 a = T - M_2 g_{\text{p}}$.

(b) Solve for g_{p} and substitute into the force equation to solve for T.

SIMPLIFY:

(a) $M_1 a = T - M_1 g_{\text{p}}$ (1), $-M_2 a = T - M_2 g_{\text{p}}$ (2)

Because the tensions in the ends of the rope are the same, solve equations (1) and (2) in terms of T and equate the expressions.

$$M_2 g_{\text{p}} - M_2 a = M_1 g_{\text{p}} + M_1 a \Rightarrow a\left(M_1 + M_2\right) = g_{\text{p}}\left(M_2 - M_1\right) \Rightarrow g_{\text{p}} = a\left(\frac{M_1 + M_2}{M_2 - M_1}\right)$$

Substitute for a using $\Delta y = \left(at^2\right)/2$ to get $g_{\text{p}} = \dfrac{2\Delta y}{t^2}\left(\dfrac{M_1 + M_2}{M_2 - M_1}\right)$.

(b) $T = M_2\left(g_{\text{p}} - a\right) = M_2\left(g_{\text{p}} - \dfrac{2\Delta y}{t^2}\right)$

CALCULATE:

(a) $g_{\text{p}} = \dfrac{2(1.00 \text{ m})}{(1.52 \text{ s})^2}\left(\dfrac{0.1000 \text{ kg} + 0.2000 \text{ kg}}{0.2000 \text{ kg} - 0.1000 \text{ kg}}\right) = 2.59695 \text{ m/s}^2$

(b) $T = 0.2000 \text{ kg}\left(2.59695 \text{ m/s}^2 - \dfrac{2(1.00 \text{ m})}{(1.52 \text{ s})^2}\right) = 0.346260 \text{ N}$

ROUND: To three significant figures, the answers should be (a) $g_{\text{p}} = 2.60 \text{ m/s}^2$ and (b) $T = 0.346$ N.

DOUBLE-CHECK: The units for the calculated answers were the correct units of acceleration and force. The small tension calculated is reasonable, considering the small masses.

4.47. **THINK:** The problem asks for the force needed to hold the block in place. This means that the net force on the block has to be zero in each case, $F_{\text{net}\,x} = F_{\text{net}\,y} = 0$. The only forces to consider are the force of gravity, which act straight downward, the normal force from the plane, which is perpendicular to the plane, and the third external force we are asked to apply in parts (a) and (b). We do not need to consider friction forces, because the problem stipulates a "frictionless ramp".

SKETCH:

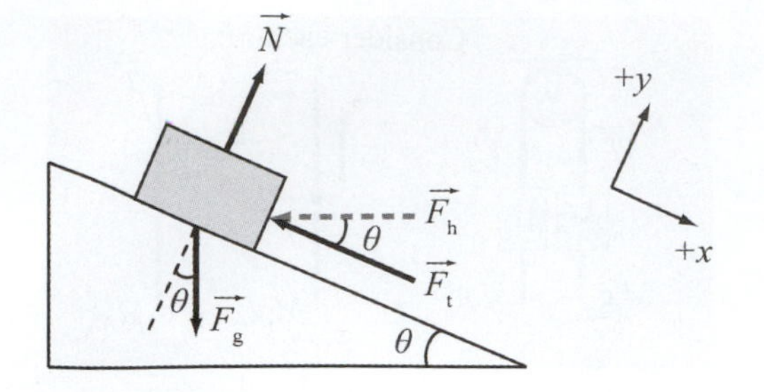

RESEARCH:

(a) To find F_t, the forces acting in the x direction on the block must be balanced.

(b) Note that now F_t and F_h are related by $F_t = F_h \cos\theta$.

SIMPLIFY:

(a) $F_{\text{netx}} = -F_t + F_{gx} = 0,\ \ F_t = F_{gx} = mg\sin\theta$

(b) $F_h = F_t / \cos\theta = mg\sin\theta / \cos\theta = mg\tan\theta$

CALCULATE:

(a) $F_t = (80.0\text{ kg})(9.81\text{ m/s}^2)\sin(36.9°) = 471.2\text{ N}$

(b) $F_h = (80.0\text{ kg})(9.81\text{ m/s}^2)\tan(36.9°) = 589.2\text{ N}$

ROUND: To three significant figures,

(a) $F_t = 471$ N and

(b) $F_h = 589$ N.

DOUBLE-CHECK: With almost all problems involving inclined planes, such as this one, one can obtain great insight and perform easy checks of the algebra by considering the limiting cases of the angle θ of the plane approaching 0 and 90 degrees.

In the case of $\theta \to 0°$ the block will simply sit on a horizontal surface, and no external force should be required to hold it in that position. Our calculations are compatible with this, because $\sin 0° = \tan 0° = 0$. In the case of $\theta \to 90°$ our results for parts (a) and (b) should be very different. In part (a) the force acts long the plane and so will be straight up in this limit, thus balancing the weight of the block all by itself. Therefore, as $\theta \to 90°$, we expect our force to approach $F_t(\theta \to 90°) = mg$. This is satisfied in our solution because $\sin 90° = 1$. In part (b) the external force will act perpendicular to the plane in the limit of $\theta \to 90°$. Thus almost no part of it will be available to balance the weight of the block, and consequently an infinitely big force magnitude should be required. This is also born out by our analytic result for part (b), because $\tan 90° \to \infty$.

4.51. **THINK:** The masses are given as $m_1 = 36.5$ kg, $m_2 = 19.2$ kg and $m_3 = 12.5$ kg. Determine the acceleration of m_1, a_1. As there are no forces in the x-direction, only the y-direction needs to be considered.

SKETCH:

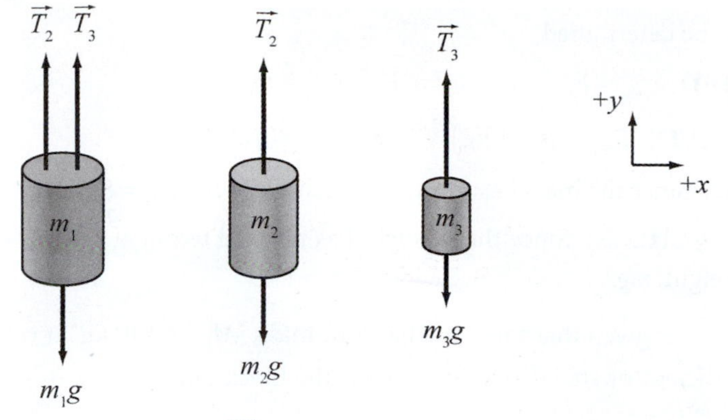

RESEARCH: To determine a_1, use $F_{1\text{net},y} = \sum F_{1y}$. Determine T_2 and T_3 by summing the forces on m_2 and m_3. A key idea is that $a_1 = -a_2 = -a_3$, as all the masses are connected (and ignoring any tipping of m_1). With $m_1 > m_2 + m_3$, it can be seen that m_1 moves downward while m_2 and m_3 move upward.

SIMPLIFY: $m_2: F_{2\text{net}} = T_2 - m_2g \Rightarrow m_2a_2 = T_2 - m_2g \Rightarrow T_2 = m_2(a_2 + g)$.

$m_3: F_3 = T_3 - m_3g \Rightarrow m_3a_3 = T_3 - m_3g \Rightarrow T_3 = m_3(a_3 + g)$

$m_1: F_{1\text{net}} = T_2 + T_3 - m_1g \Rightarrow m_1a_1 = m_2(a_2 + g) + m_3(a_3 + g) - m_1g$

With $a_1 = -a_2 = -a_3$,

$$m_1a_1 = m_2(-a_1 + g) + m_3(-a_1 + g) - m_1g$$
$$m_1a_1 + m_2a_1 + m_3a_1 = m_2g + m_3g - m_1g$$
$$a_1(m_1 + m_2 + m_3) = g(m_2 + m_3 - m_1)$$
$$a_1 = \frac{g(m_2 + m_3 - m_1)}{(m_1 + m_2 + m_3)}$$

CALCULATE: $a_1 = \dfrac{9.81 \text{ m/s}^2 (19.2 \text{ kg} + 12.5 \text{ kg} - 36.5 \text{ kg})}{(36.5 \text{ kg} + 19.2 \text{ kg} + 12.5 \text{ kg})} = -0.69044 \text{ m/s}^2$

ROUND: There are two significant figures in the sum in the numerator, so the answer should be written, $a_1 = 0.69 \text{ m/s}^2$ downward.

DOUBLE-CHECK: a_1 is less than g in magnitude, which it should be for this system.

4.55. **THINK:** The skydiver's total mass is m = 82.3 kg. The drag coefficient is $c_d = 0.533$. The parachute area is $A = 20.11 \text{ m}^2$. The density of air is $\rho = 1.14 \text{ kg/m}^3$. The skydiver has reached terminal velocity $(a_{\text{net}} = 0)$. Determine the drag force of the air, F_{drag}.

SKETCH:

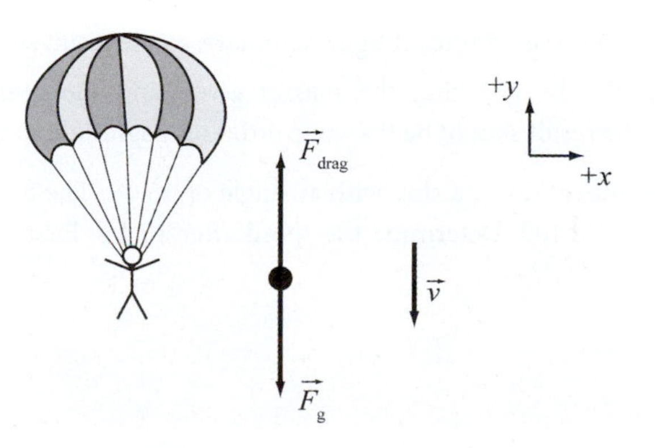

RESEARCH: The skydiver has achieved terminal velocity, that is $F_{\text{net},y}=0$. By balancing the forces in y, F_{drag} can be determined.

SIMPLIFY: $F_{\text{net},y}=F_{\text{drag}}-F_{\text{g}}=0 \Rightarrow F_{\text{drag}}=F_{\text{g}}=mg$

CALCULATE: $F_{\text{drag}}=(82.3\text{ kg})(9.81\text{ m/s}^2)=807.36\text{ N}$

ROUND: Since the mass has three significant figures, $F_{\text{drag}}=807\text{ N}$.

DOUBLE-CHECK: Since the skydiver has reached terminal velocity, the air's drag force should be equal to her weight, mg.

4.59. **THINK:** It is given that there is a block of mass, $M_1=0.640\text{ kg}$ at rest on a cart of mass, $M_2=0.320\text{ kg}$. The coefficient of static friction between the block and the cart is $\mu_{\text{s}}=0.620$. Determine the maximum force on the cart and block such that the block does not slip.

SKETCH:

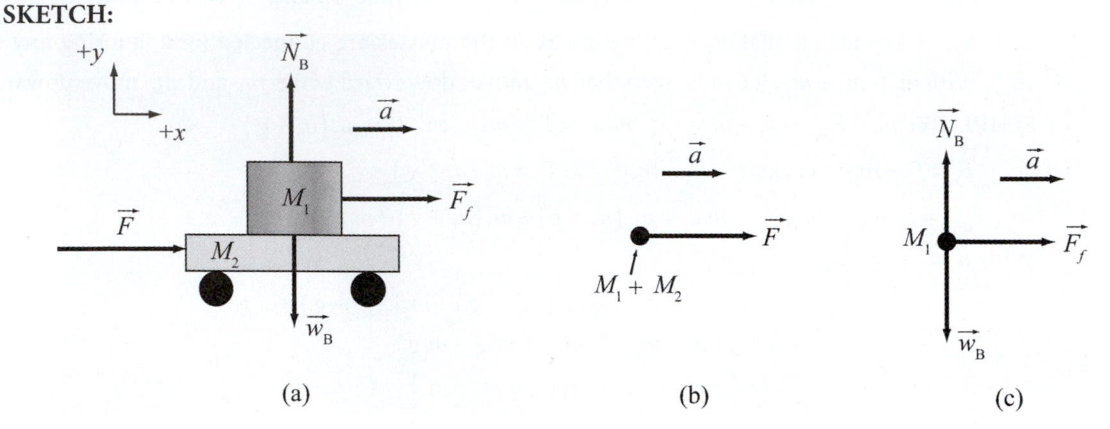

RESEARCH: Use Newton's second law: $\sum F_x=ma_x$, $\sum F_y=ma_y$. The force of friction is given by $F_{\text{f}}=\mu_{\text{s}}N$. First, consider a composite body (block and cart system, free-body diagram (b)). Applying Newton's second law, $\sum F_x=ma \Rightarrow F=(M_1+M_2)a$. Note that both the block and the cart accelerate at the same rate. Second, consider only the block, applying Newton's second law in the horizontal and vertical directions: $\sum F_{\text{B},y}=ma_y \Rightarrow N-w_B=0 \Rightarrow N=w_B=M_1g \ (a_y=0)$, $\sum F_{\text{B},x}=ma_x \Rightarrow F_{\text{f}}=M_1a$.

SIMPLIFY: $F=(M_1+M_2)a$, $F_{\text{f}}=M_1a$, $N=M_1g$. The maximum magnitude of F is when the acceleration is at a maximum. This means also that the force of friction is maximum which is equal to $F_{\text{f}}=\mu_{\text{s}}N=\mu_{\text{s}}M_1g$. Note that when $F_{\text{f}}>\mu_{\text{s}}N$, the block starts to slip. $F_{\text{f}}=\mu_{\text{s}}M_1g=M_1a_{\text{max}} \Rightarrow a_{\text{max}}=\mu_{\text{s}}g$. Therefore, $F_{\text{max}}=(M_1+M_2)a_{\text{max}} \Rightarrow F_{\text{max}}=(M_1+M_2)\mu_{\text{s}}g$.

CALCULATE: $F_{\text{max}}=(0.640\text{ kg}+0.320\text{ kg})(0.620)(9.81\text{ m/s}^2)=5.83891\text{ N}$

ROUND: There are three significant figures initially, so the result should be $F_{\text{max}}=5.84\text{ N}$.

DOUBLE-CHECK: By checking the masses given and the coefficient of static friction, it can be determined that the result should be the same order of magnitude as gravity. This is indeed the case.

4.63. **THINK:** A skier moves down a slop with an angle of 15.0°. The initial speed is 2.00 m/s. The coefficient of kinetic friction is 0.100. Determine the speed after 10.0 s. First, the acceleration of the skier must be determined.

SKETCH:

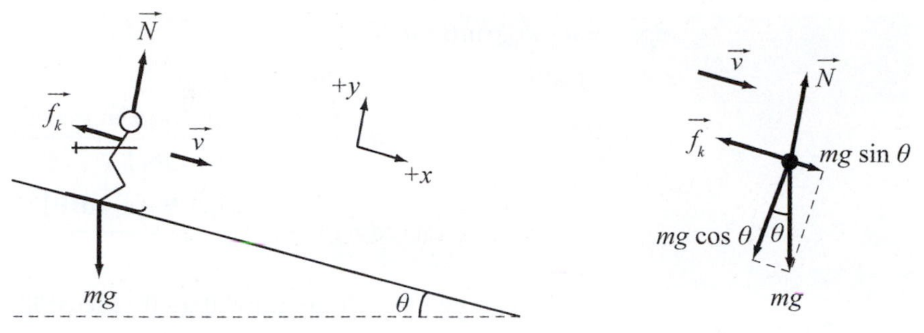

RESEARCH: Assume the direction of motion is the positive direction of the x axis. The force of kinetic friction is given by $f_\text{k} = \mu_\text{k} N$. Use Newton's second law to determine the acceleration of the skier:

$$\sum F_x = ma_x \Rightarrow mg\sin\theta - f_\text{k} = ma_x \Rightarrow ma_x = mg\sin\theta - \mu_\text{k} N$$

$$\sum F_y = ma_y \quad (a_y = 0) \Rightarrow N - mg\cos\theta = 0 \Rightarrow N = mg\cos\theta$$

SIMPLIFY: $ma_x = mg\sin\theta - \mu_\text{k} mg\cos\theta \Rightarrow a_x = g(\sin\theta - \mu_k \cos\theta)$. The speed after the time interval Δt is: $v = v_0 + a_x \Delta t = v_0 + g(\sin\theta - \mu_\text{k}\cos\theta)\Delta t$.

CALCULATE: $v = 2.00 \text{ m/s} + (9.81 \text{ m/s}^2)(\sin 15.0° - 0.100\cos 15.0°)(10.0 \text{ s}) = 17.91 \text{ m/s}$

ROUND: Since v_0 has three significant figures, round the result to v = 17.9 m/s.

DOUBLE-CHECK: 17.9 m/s is equivalent to about 64.4 km/h, which is a reasonable speed.

4.67. **THINK:** The two blocks have masses $m_1 = 0.2500$ kg and $m_2 = 0.5000$ kg. The coefficients of static and kinetic friction are 0.250 and 0.123. The angle of the incline is $\theta = 30.0°$. The blocks are initially at rest.

SKETCH:

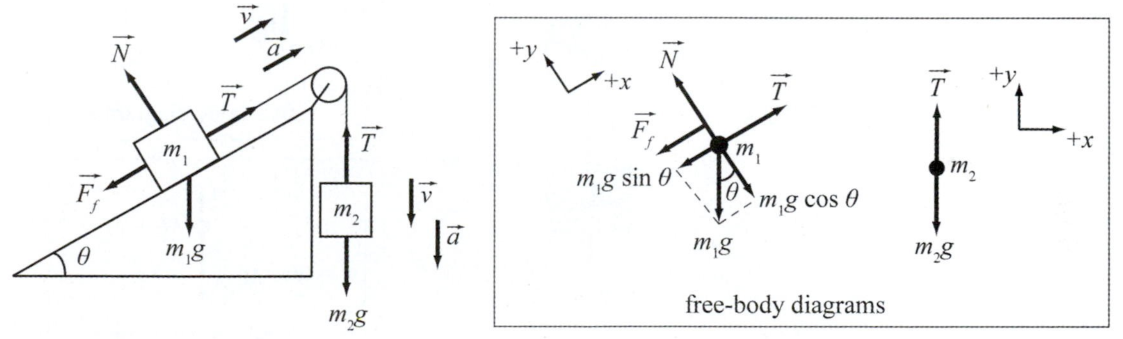

RESEARCH:

(a) If there is no friction, $F_\text{f} = 0$, it is given that $m_2 > m_1$. This would cause block 1 to move up and block 2 to move down. The same motion occurs when there is friction, only the acceleration is less when there is friction.

(b) Use Newton's second law to determine the acceleration:

Body 1: $\sum F_y = 0$ and $a_y = 0$, so $N - m_1 g\cos\theta = 0 \Rightarrow N = m_1 g\cos\theta$.

Also, $\sum F_x = m_1 a$ so $T - m_1 g\sin\theta - F_\text{f} = m_1 a$.

Body 2: $\sum F_y = m_2 a$ so $m_2 g - T = m_2 a \Rightarrow T = m_2 g - m_2 a$.

SIMPLIFY: (b) $T - m_1 g\sin\theta - F_f = m_1 a$ and $F_f = \mu_k N$, so

$$m_2 g - m_2 a - m_1 g\sin\theta - \mu_k N = m_1 a$$

$$m_2 g - m_1 g\sin\theta - \mu_k m_1 g\cos\theta = (m_1 + m_2)a$$

$$a = \frac{m_2 g - m_1 g\sin\theta - \mu_k m_1 g\cos\theta}{(m_1 + m_2)}$$

$$a = g\frac{(m_2 - m_1(\sin\theta + \mu_k\cos\theta))}{(m_1 + m_2)}$$

CALCULATE: (b) $a = (9.81 \text{ m/s}^2)\dfrac{(0.5000 \text{ kg} - 0.2500 \text{ kg}(\sin(30.0°) + 0.123\cos(30.0°)))}{(0.5000 \text{ kg} + 0.2500 \text{ kg})} = 4.5567 \text{ m/s}^2$

ROUND:
(b) Rounding to three significant figures,
$a = 4.56 \text{ m/s}^2$.

DOUBLE-CHECK: The result is reasonable since it is less than the acceleration due to gravity. In addition we find the limit of the acceleration of the Atwood machine in the limit of $\theta = 90°$ (See Example 4.4) and the limit of Example 4.8, Two Blocks Connected by a Rope – with Friction, for $\theta = 0°$ as limiting cases of our answer. This gives us additional confidence in our solution.

4.71. **THINK:** There are two masses, $M_1 = 2.00$ kg and $M_2 = 4.00$ kg. For part (a): a constant velocity means zero acceleration.

SKETCH:

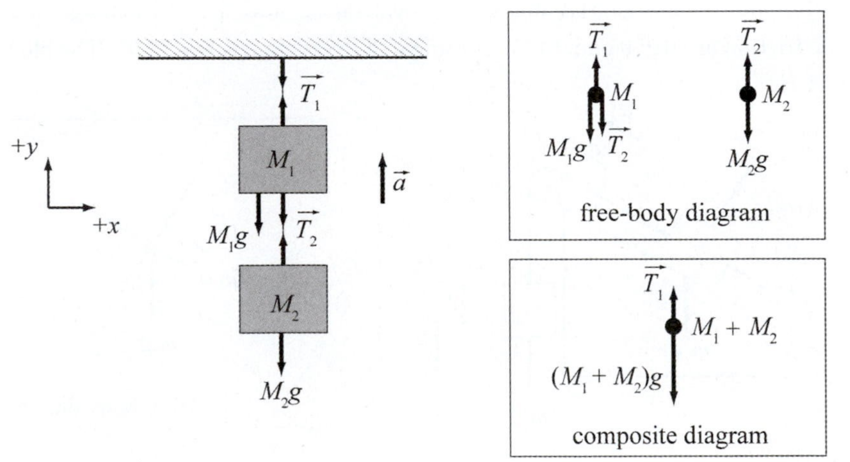

RESEARCH: Using Newton's second law:

Mass 1: $\sum F_y = m_1 a \Rightarrow T_1 - T_2 - m_1 g = m_1 a$

Mass 2: $\sum F_y = m_2 a \Rightarrow T_2 - m_2 g = m_2 a$

SIMPLIFY: $T_2 = m_2(a + g)$. Substitute into the following equation:

$$T_1 = T_2 + m_1 a + m_1 g \Rightarrow T_1 = m_2(a+g) + m_1(a+g) = (m_1 + m_2)(a+g)$$

The composite mass $(m_1 + m_2)$: $\sum F_y = ma \Rightarrow T_1 - (m_1 + m_2)g = (m_1 + m_2)a \Rightarrow T_1 = (m_1 + m_2)(a+g)$.

CALCULATE:

(a) $a = 0$, so $T_1 = (2.00 \text{ kg} + 4.00 \text{ kg})(9.81 \text{ m/s}^2) = 58.86$ N

(b) $a = 3.00 \text{ m/s}^2$, so $T_1 = (2.00 \text{ kg} + 4.00 \text{ kg})(3.00 \text{ m/s}^2 + 9.81 \text{ m/s}^2) = 76.86$ N

ROUND: Since the masses have three significant figures, the results should be rounded to:

(a) $T_1 = 58.9 \text{ N}$

(b) $T_1 = 76.9 \text{ N}$

DOUBLE-CHECK: The tension increases as acceleration increases (assuming the acceleration is upward). As a check, the tension in part (a) is less than the tension in part (b).

4.75. **THINK:** A block of mass $m =$ 20.0 kg is initially at rest and then pulled upward with a constant acceleration, $a = 2.32 \text{ m/s}^2$.

SKETCH:

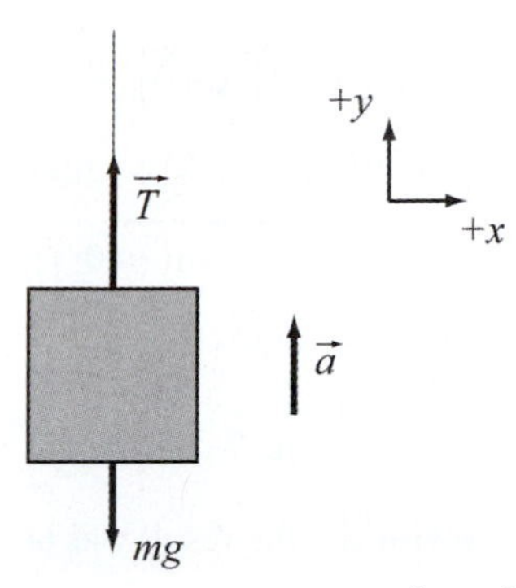

RESEARCH: Using Newton's second law and the equation $v^2 = v_0^{\ 2} - 2ax$: $v_0 = 0 \Rightarrow v^2 = 2ax$.

$\sum F_y = ma_y \Rightarrow T - mg = ma \Rightarrow T = m(a+g)$.

SIMPLIFY: $v^2 = 2ax \Rightarrow v = \sqrt{2ax}$

CALCULATE:

(a) $T = (20.0 \text{ kg})(2.32 \text{ m/s}^2 + 9.81 \text{ m/s}^2) = 242.6 \text{ N}$

(b) $\sum F_y = (20.0 \text{ kg})(2.32 \text{ m/s}^2) = 46.4 \text{ N}$

(c) $v = \sqrt{2(2.32 \text{ m/s}^2)(2.00 \text{ m})} = 3.04631 \text{ m/s}$

ROUND: Rounding to three significant figures, the results are

(a) $T = 243 \text{ N}$,

(b) $\sum F_y = 46.4 \text{ N}$ and

(c) $v = 3.05 \text{ m/s}$.

DOUBLE-CHECK: Since a is about g/4, the net force must be about T/5.

4.79. **THINK:** The sled has a mass, M = 1000. kg. The coefficient of kinetic friction is $\mu_k = 0.600$. The sled is pulled at an angle $\theta = 30.0°$ above the horizontal. Determine magnitude of the tension in the rope when the acceleration is $a = 2.00 \text{ m/s}^2$.

SKETCH:

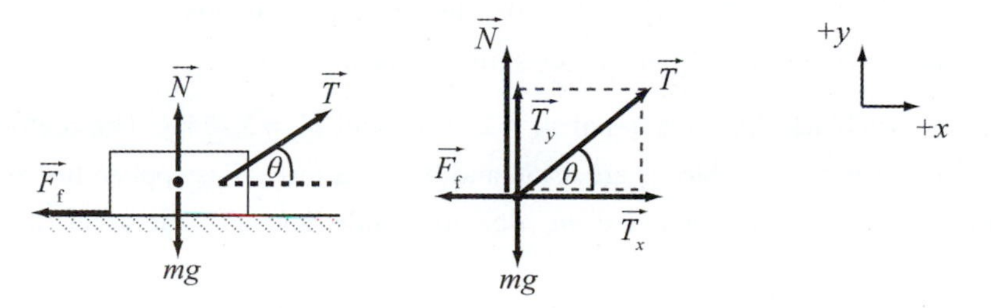

RESEARCH: $T_y = T\sin\theta$, $T_x = T\cos\theta$ and $F_f = \mu_k N$. Using Newton's second law: $\sum F_y = ma_y$ and $a_y = 0$, so $N + T_y - mg = 0 \Rightarrow N = mg - T_y = mg - T\sin\theta.$ Also, $\sum F_x = ma \Rightarrow T_x - F_f = ma$ $\Rightarrow T_x = F_f + ma.$

SIMPLIFY:

$$T\cos\theta = \mu_k N + ma$$
$$T\cos\theta = \mu_k\left(mg - T\sin\theta\right) + ma$$
$$T\cos\theta + \mu_k T\sin\theta = \mu_k mg + ma$$
$$T\left(\cos\theta + \mu_k\sin\theta\right) = \mu_k mg + ma$$
$$T = \frac{m\left(\mu_k g + a\right)}{\left(\cos\theta + \mu_k\sin\theta\right)}$$

CALCULATE: $T = \dfrac{(1000.\text{ kg})\left(0.600\left(9.81\text{ m/s}^2\right) + 2.00\text{ m/s}^2\right)}{\left(\cos(30.0°) + 0.600\sin(30.0°)\right)} = 6763.15\text{ N}$

ROUND: Rounding to three significant figures, $T = 6760$ N.

DOUBLE-CHECK: If there is no friction, $T\cos\theta = ma \Rightarrow T = \dfrac{ma}{\cos\theta} = \dfrac{(1000.\text{ kg})\left(2.00\text{ m/s}^2\right)}{\cos 30.0°} = 2309\text{ N}.$ Since friction was considered previously, the result was larger.

4.83. **THINK:** A book has a mass of $m = 0.500$ kg. The tension on each wire is $T = 15.4$ N. Determine the angle of the wires with the horizontal.

SKETCH:

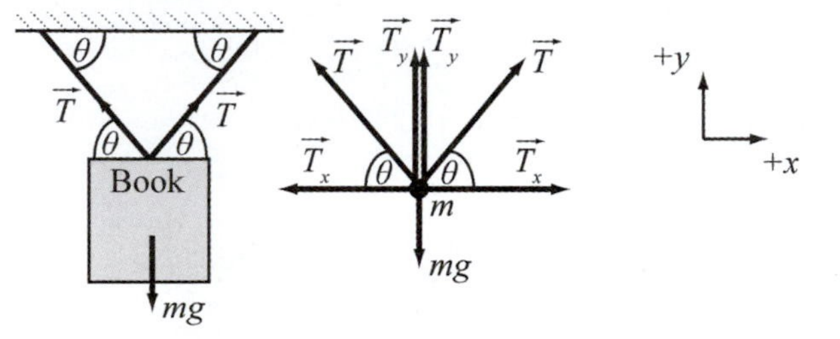

RESEARCH: There is no acceleration in any direction, $a_x = a_y = 0$. Using Newton's second law: $\sum F_y = ma_y = 0 \Rightarrow 2T_y - mg = 0$ and $\sum F_x = ma_x = 0 \Rightarrow T_x - T_x = 0.$

SIMPLIFY: $T_y = T\sin\theta$, so $2T\sin\theta = mg \Rightarrow \sin\theta = \dfrac{mg}{2T} \Rightarrow \theta = \sin^{-1}\left(\dfrac{mg}{2T}\right).$

CALCULATE: $\theta = \sin^{-1}\left(\dfrac{0.500\text{ kg}\left(9.81\text{ m/s}^2\right)}{2(15.4\text{ N})}\right) = 9.1635°$

ROUND: Rounding to three significant figures, $\theta = 9.16°$.

DOUBLE-CHECK: If the angle is $\theta = 90°$, the tension required is $T = mg/2 = 0.500(9.81)/2 = 2.45$ N. It is reasonable that a smaller angle requires more tension.

4.87. **THINK:** Two blocks have masses of $m_1 = 2.50$ kg and $m_2 = 3.75$ kg. The coefficients of static and kinetic friction between the two blocks are 0.456 and 0.380. A force, F is applied horizontally on m_1. Determine the maximum force, F, such that m_1 does not slide, and also the acceleration of m_1 and m_2 when $F = 24.5$ N.

SKETCH:

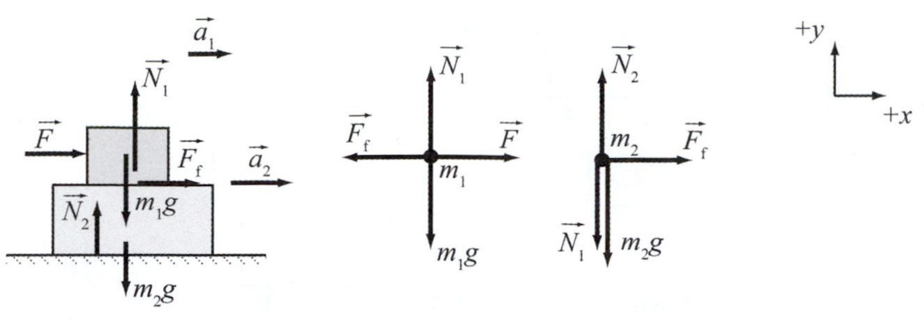

RESEARCH: The force of friction is given by $F_f = \mu_s N_1$. First, consider m_1. Using Newton's second law: $\sum F_x = ma_x \Rightarrow F - F_f = m_1 a_1 \Rightarrow F = F_f + m_1 a_1$ and $\sum F_y = ma_y = 0 \Rightarrow N_1 - m_1 g = 0 \Rightarrow N_1 = m_1 g$. Then, consider m_2: $\sum F_x = ma_x$, $F_f = m_2 a_2$, and $N_2 - N_1 - m_2 g = 0$.

(a) The force is maximum when $F_f = \mu_s N_1$ and $a_1 = a_2 = a$.

(b) If $F = 24.5$ N is larger than F_{max}, then m_1 slides on m_2.

SIMPLIFY:

(a) $F_f = \mu_s N_1 = m_2 a_2 \Rightarrow \mu_s m_1 g = m_2 a_2 \Rightarrow a = \mu_s (m_1 / m_2) g$

$F_{max} = F_f + m_1 a = \mu_s m_1 g + m_1 \mu_s (m_1 / m_2) g = \mu_s m_1 g (1 + m_1 / m_2)$

(b) The force of friction is given by $F_f = \mu_k N_1$. Using the equations, $F = F_f + m_1 a_1$ and $F_f = m_2 a_2$:

$a_2 = \dfrac{F_f}{m_2} = \dfrac{\mu_k m_1 g}{m_2}$ and $a_1 = \dfrac{F - F_f}{m_1} = \dfrac{F - \mu_k m_1 g}{m_1} = \dfrac{F}{m_1} - \mu_k g$.

CALCULATE:

(a) $F_{max} = 0.456(2.50 \text{ kg})(9.81 \text{ m/s}^2)\left(1 + \dfrac{2.50 \text{ kg}}{3.75 \text{ kg}}\right) = 18.639 \text{ N}$

(b) $a_1 = \dfrac{24.5 \text{ N}}{2.50 \text{ kg}} - 0.380(9.81 \text{ m/s}^2) = 6.0722 \text{ m/s}^2$ and $a_2 = \dfrac{(0.380)(2.50 \text{ kg})(9.81 \text{ m/s}^2)}{3.75 \text{ kg}} = 2.4852 \text{ m/s}^2$

ROUND: Rounding to three significant figures,

(a) $F_{max} = 18.6 \text{ N}$,

(b) $a_1 = 6.07 \text{ m/s}^2$ and $a_2 = 2.49 \text{ m/s}^2$.

DOUBLE-CHECK: For part (b), it is known that m_1 slides on m_2. This means that a_1 is larger than a_2.

4.91. **THINK:** Three blocks have masses, $m_1 = 3.50$ kg, $m_2 = 5.00$ kg and $m_3 = 7.60$ kg. The coefficients of static and kinetic friction between m_1 and m_2 are 0.600 and 0.500. Determine the accelerations of m_1 and m_2, and tension of the string. If m_1 does not slip on m_2, then the accelerations of both blocks will be the same. First, make the assumption that the blocks do not slide. Then, it must be determined whether the acting force of friction, F_f, is less than or greater than the maximum force of friction.

SKETCH:

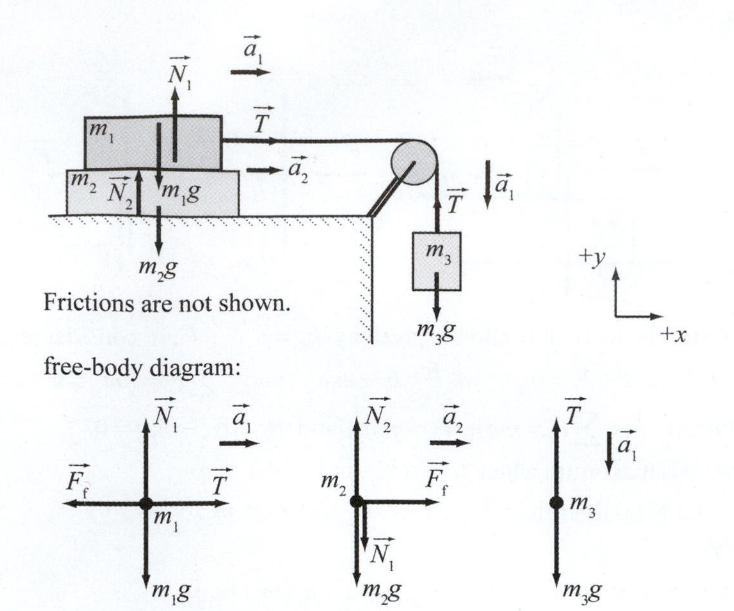

RESEARCH: Simplify the problem by looking at the axis along the string.

$m_1 + m_2$ $\vec{T}$ $\vec{T}$ m_3 m_3g $+y$ $+x$

$m_1 + m_2 + m_3$: m_3g; $m_1 + m_2$: $\vec{T}$; m_3: $\vec{T}$, m_3g

Using Newton's second law: $\sum F = ma \Rightarrow m_3g = (m_1 + m_2 + m_3)a$ and $\sum F = ma \Rightarrow T = (m_1 + m_2)a.$

SIMPLIFY:

(a) $a = \dfrac{m_3g}{(m_1 + m_2 + m_3)}$

(b) $T = \dfrac{(m_1 + m_2)m_3g}{(m_1 + m_2 + m_3)}$

CALCULATE:

(a) $a = \dfrac{7.60 \text{ kg}(9.81 \text{ m/s}^2)}{(3.50 \text{ kg} + 5.00 \text{ kg} + 7.60 \text{ kg})} = 4.631 \text{ m/s}^2$

(b) $T = (3.50 \text{ kg} + 5.00 \text{ kg})4.631 \text{ m/s}^2 = 39.362 \text{ N}$

Now, it must be determined if the force of friction, F_f is less than the maximum force of static friction. From the free-body diagram of m_2: $\sum F_x = m_2a \Rightarrow F_f = m_2a = 5.00(4.631 \text{ m/s}^2) = 23.15 \text{ N}.$ The maximum force of static friction is $f_{s,max} = \mu_s N_1$ where $N_1 = m_1g$. $f_{s,max} = (0.600)(3.50)(9.81) = 20.60 \text{ N}.$ $F_f > f_{s,max}$, so block 1 slips on block 2. Some parts of the question must be reconsidered.

RESEARCH: The force of friction is given by $F_f = \mu_k N_1$. First, consider m_1. Using Newton's second law:

$\sum F_x = ma_x \Rightarrow T - F_f = m_1a_1 \Rightarrow T - \mu_k N_1 = m_1a_1 \Rightarrow T - \mu_k m_1 g = m_1a_1$

$\sum F_y = ma_y = 0 \Rightarrow N_1 - m_1g = 0 \Rightarrow N_1 = m_1g$

Now, consider m_2: $\sum F_x = ma_x \Rightarrow F_f = m_2 a_2 \Rightarrow a_2 = \frac{\mu_k N_1}{m_2} = \mu_k \frac{m_1 g}{m_2}$. Finally, consider m_3: $\sum F_y = ma_y \Rightarrow m_3 g - T = m_3 a_1$.

SIMPLIFY: $T - \mu_k m_1 g = m_1 a_1$ and $m_3 g - T = m_3 a_1$ can be used to eliminate T:

$$m_3 g - m_1 a_1 - \mu_k m_1 g = m_3 a_1 \Rightarrow m_3 g - \mu_k m_1 g = (m_1 + m_3) a_1 \Rightarrow a_1 = g\frac{(m_3 - \mu_k m_1)}{(m_1 + m_3)}$$

Also, $a_2 = \frac{\mu_k m_1}{m_2} g$ and $T = m_1 (a_1 + \mu_k g) = m_3 (g - a_1)$.

CALCULATE:

(a) $a_1 = (9.81 \text{ m/s}^2)\frac{(7.60 \text{ kg} - 0.500(3.50 \text{ kg}))}{(7.60 \text{ kg} + 3.50 \text{ kg})} = 5.1701 \text{ m/s}^2$,

$a_2 = \frac{0.500(3.50 \text{ kg})}{5.00 \text{ kg}}(9.81 \text{ m/s}^2) = 3.4335 \text{ m/s}^2$

(b) $T = 7.60 \text{ kg}(9.81 \text{ m/s}^2 - 5.1701 \text{ m/s}^2) = 35.26 \text{ N}$

ROUND: Rounding to three significant figures,

(a) $a_1 = 5.17 \text{ m/s}^2$, $a_2 = 3.43 \text{ m/s}^2$ and

(b) $T = 35.3$ N.

DOUBLE-CHECK: Because block 1 slips on block 2, a_1 is larger than a_2.

4.95. **THINK:** The three blocks have masses, $M_1 = 0.250$ kg, $M_2 = 0.420$ kg and $M_3 = 1.80$ kg. The coefficient of kinetic friction is $\mu_k = 0.340$.

SKETCH:

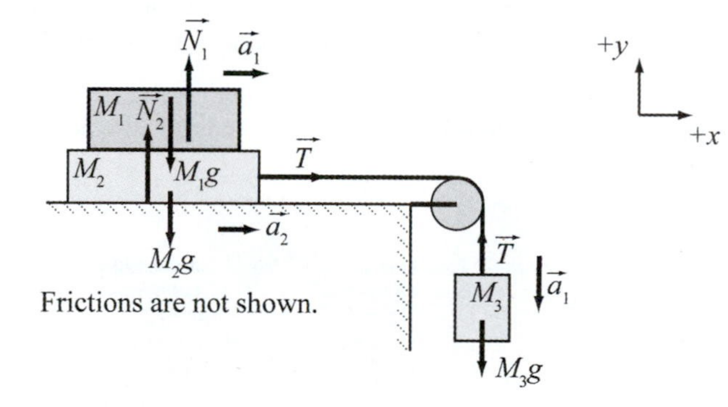

free-body diagram:

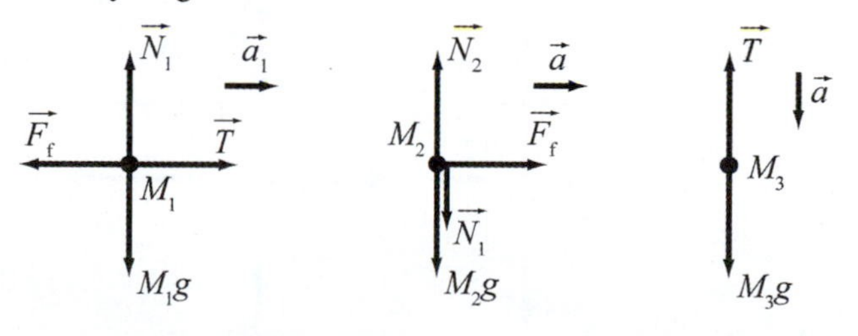

RESEARCH: The force of friction is given by $F_f = \mu_k N$. First, consider M_1. Using Newton's second law:

$$\sum F_x = ma_x \Rightarrow F_{f,1} = M_1 a_1 \text{ and } \sum F_y = ma_y = 0 \Rightarrow N_1 - M_1 g = 0 \Rightarrow N_1 = M_1 g$$

Then, consider M_2: $\sum F_x = ma_x \Rightarrow T - F_{f,1} - F_{f,2} = M_2 a$ and

$$\sum F_y = ma_y = 0 \Rightarrow N_2 - N_1 - M_2 g = 0 \Rightarrow N_2 = (M_1 + M_2) g$$

Finally, consider M_3: $\sum F_y = ma_y \Rightarrow M_3 g - T = M_3 a$.

SIMPLIFY:

(a) $F_{f,1} = \mu_k N_1 = \mu_k M_1 g = M_1 a_1 \Rightarrow a_1 = \mu_k g$, which is the acceleration of the block.

(b) For the slab (M_2) the friction with the table is $F_{f,2} = f_k = \mu_k N_2 = \mu_k (M_1 + M_2) g$; therefore, the equation for the x direction yields $T - \mu_k M_1 g - \mu_k (M_1 + M_2) g = M_2 a$. From the equation for M_3, $T = M_3 (g - a)$. Substituting this for T yields the following.

$$M_3(g-a) - \mu_k M_1 g - \mu_k (M_1 + M_2) g = M_2 a \Rightarrow (M_2 + M_3) a = M_3 g - \mu_k (2M_1 + M_2) g$$

$$\Rightarrow a = \frac{\left[M_3 - \mu_k (2M_1 + M_2)\right] g}{M_2 + M_3}$$

CALCULATE:

(a) $a_1 = (0.340)(9.81 \text{ m/s}^2) = 3.335 \text{ m/s}^2$

(b) $a = \dfrac{\left[1.80 \text{ kg} - (0.340)(2 \cdot 0.250 \text{ kg} + 0.420 \text{ kg})\right](9.81 \text{ m/s}^2)}{0.420 \text{ kg} + 1.80 \text{ kg}} = 6.572 \text{ m/s}^2$

ROUND: Rounding to three significant figures,

(a) $a_1 = 3.34 \text{ m/s}^2$ and

(b) $a = 6.57 \text{ m/s}^2$.

DOUBLE-CHECK: Because M_1 slides on M_2, it is expected that a_1 is less than a. Both a_1 and a must be less than g.

Multi-Version Exercises

4.96. **THINK:** This problem involves two blocks sliding along a frictionless surface. For these types of problems, use Newton's laws. Also note that the tension force from block 1 must be exactly equal and opposite from the force or block 2.

SKETCH: Start with the image from the text to create free body diagrams for each block.

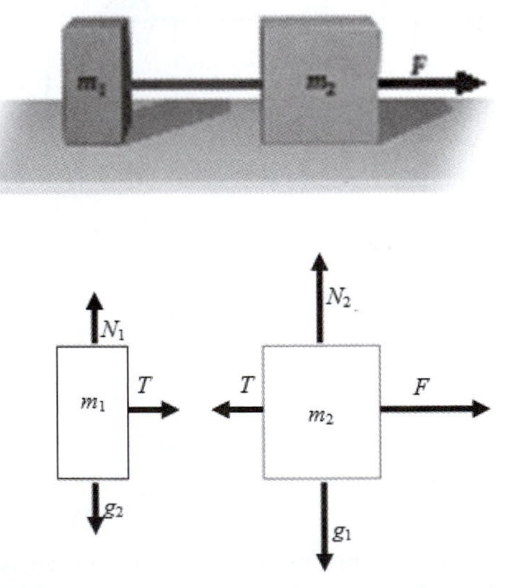

RESEARCH: Since the blocks are lying on a frictionless surface, there is no friction force acting on the blocks. It is necessary to find the tension T on the rope in terms of the force F acting on the second block and the masses m_1 and m_2. Since force F is acting on the blocks and rope as a single system, Newton's Second Law gives that $F = (m_1 + m_2) a$, where a is the acceleration of the blocks. Looking at the horizontal forces on block 1 gives $T = m_1 a$.

SIMPLIFY: It is necessary to find an expression for the tension in terms of the outside force F acting on block 2 and the masses of the two blocks. Rewriting $T = m_1 a$ as $a = \frac{T}{m_1}$ and combining with $F = (m_1 + m_2)a$ gives $F = (m_1 + m_2)\frac{T}{m_1}$. This expression can be re-written to give the tension in terms of known quantities: $T = F\frac{m_1}{m_1 + m_2}$.

CALCULATE: Using the masses and force given in the question statement gives a tension force of:

$$T = 12.61 \text{ N}\left(\frac{1.267 \text{ kg}}{1.267 \text{ kg} + 3.557 \text{ kg}}\right)$$
$$= 3.311954809 \text{ N}$$

ROUND: The masses of the blocks are given to four significant figures, and their sum also has four significant figures. The only other measured quantity is the external force acting on the second block, which also has four significant figures. This means that the final answer should be rounded to four significant figures, giving a total tension of $T = 3.312$ N.

DOUBLE-CHECK: Think of the tension on the rope transmitting the force from block 2 to block 1. Since block 2 is much more massive than block 1, block 1 represents about one fourth of the total mass of the system. So, it makes sense that only about one fourth of the force will be transmitted along the string to the second block. About 12.61/4 or 3.153 N will be transmitted, which is pretty close to our calculated value of 3.312 N.

4.100. **THINK:** This problem involves two hanging masses and two frictionless pulleys. The rope is massless, so it is not necessary to include the gravitational force on the spring. The forces on this system will be the gravitational force and the tension in the string. To solve this problem, it will be necessary to use Newton's laws.

SKETCH: Begin with the sketch from the text. Then, draw free body diagrams for both masses, keeping in mind that the forces exerted by the rope on the masses must be equal.

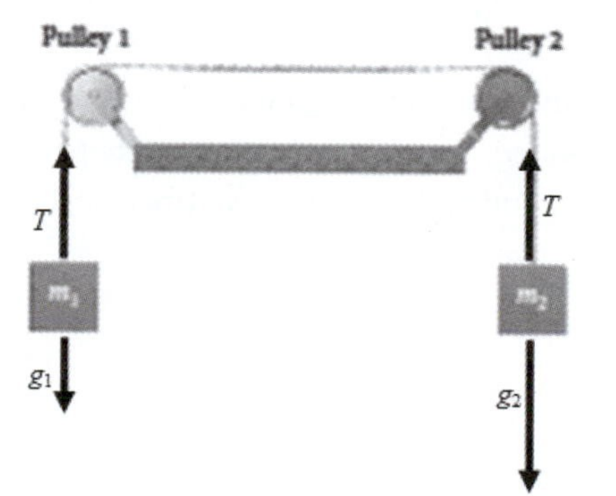

RESEARCH: First note that the two blocks will accelerate at the same rate, but in opposite directions. First note that the gravitational force on each block can be given in terms of the mass and gravitational acceleration: $g_1 = gm_1$ and $g_2 = gm_2$. It is possible to use Newton's Second law on each block individually to get two equations relating the tension T, the masses of the blocks m_1 and m_2, and the acceleration a. For the first block we have that $(T - m_1 g) = m_1 a$. Likewise, $(T - m_2 g) = m_2 \cdot (-a)$ because block 2 is accelerating in the opposite direction from block 1. With these two equations, we should be able to solve for either of the unknown quantities a or T.

SIMPLIFY: Since the problems asks for acceleration a, first find an equation for the tension T in terms of the other quantities. $T - m_1 g = m_1 a$ means that $T = m_1 a + m_1 g$. Substitute this expression for T into the equation $(T - m_2 g) = m_2 \cdot (-a)$ to get:

$$-m_2 a = (m_1 a + m_1 g) - m_2 g \Rightarrow$$
$$-m_1 a - m_2 a = -m_1 a + m_1 a + m_1 g - m_2 g \Rightarrow$$
$$(-m_1 - m_2) a = m_1 g - m_2 g \Rightarrow$$
$$a = \frac{m_1 g - m_2 g}{-m_1 - m_2}$$

CALCULATE: The masses m_1 and m_2 are given in the problem. The gravitational acceleration g is about 9.81 m·s^{-2}. Using these values gives the acceleration a:

$$\begin{aligned} a &= \frac{m_1 g - m_2 g}{-m_1 - m_2} \\ &= \frac{1.183 \text{ kg} \cdot 9.81 \text{ m s}^{-2} - 3.639 \text{ kg} \cdot 9.81 \text{ m s}^{-2}}{-1.183 \text{ kg} - 3.639 \text{ kg}} \\ &= 4.99654915 \text{ m s}^{-2} \end{aligned}$$

ROUND: The masses of the blocks are given to four significant figures, and their sum also has four significant figures. Since they are the only measured quantities in this problem, the final answer should have four significant figures, giving a final answer of 4.997 m/s^2.

DOUBLE-CHECK: In this situation, it is intuitively obvious that the heavier mass will fall towards the ground less quickly than if it were in free fall. So it is reasonable that our calculated value of 4.997 m/s^2 is less than the 9.81 m/s^2, which is the rate at which objects on the Earth's surface generally accelerate towards the ground.

4.103. **THINK:** This problem involves friction, so the only forces acting on the curling stone are gravity, the normal force, and the frictional force. Since gravity and the normal force act in the vertical direction, the only force slowing the horizontal movement of the curling stone is the frictional force. It is necessary to come up with a way to relate the initial velocity to the mass, coefficient of friction, and total distance traveled by the stone. Since the curling stone is slowing to a stop, there is a net external force, so it will be necessary to use Newton's Second law.

SKETCH: It is helpful to draw the free body diagrams for the curling stone at three different times: the moment the curling stone is released, part of the way along its path, and after it has stopped. When the stone is at rest, there is no velocity and no kinetic friction (there is, however, static friction).

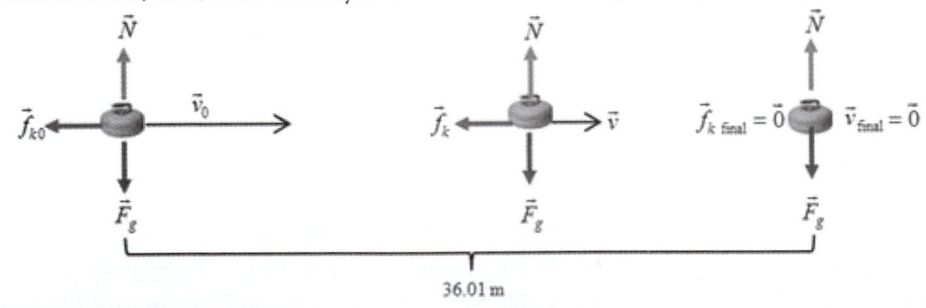

RESEARCH: The only unknown force acting on the curling stone is the kinetic friction force. The magnitude is given by $f_k = \mu_k N$. But the normal force is equal and opposite to the gravitational force ($\vec{N} = -\vec{F}_g = -(mg)$), so the magnitude of the normal force is $N = mg$. The initial kinetic energy of the stone is $K = \frac{1}{2} m v_0^2$ and the kinetic energy of the stone at rest is zero. So, all of the kinetic energy has been dissipated by friction. The energy dissipated by friction is equal to the magnitude of the force times the distance traveled, $f_k d = K$. This will allow us to find the initial velocity in terms of known quantities.

SIMPLIFY: First, it is necessary to combine $f_k d = K$ with the equations for the kinetic energy, normal force, and frictional force. So:

$$f_k d = K \Rightarrow$$
$$(\mu_k N)d = \tfrac{1}{2}mv_0^2 \Rightarrow$$
$$\mu_k(mg)d = \tfrac{1}{2}mv_0^2$$

Use algebra to find an expression for the initial velocity in terms of known quantities:

$$\mu_k(mg)d = \tfrac{1}{2}mv_0^2 \Rightarrow$$
$$\frac{2}{m}\mu_k mgd = \frac{2}{m}\tfrac{1}{2}mv_0^2 \Rightarrow$$
$$2\mu_k gd = v_0^2 \Rightarrow$$
$$\sqrt{2\mu_k gd} = v_0$$

CALCULATE: The mass of the curling stone m = 19.00 kg, the coefficient of kinetic friction between the stone and the ice μ_k = 0.01869 and the total distance traveled d = 36.01 m. The acceleration due to gravity on the surface of the Earth is not given in the problem, but it is g = 9.81 m/s^2. Using these values,

$$\begin{aligned} v_0 &= \sqrt{2\mu_k gd} \\ &= \sqrt{2\cdot 0.01869\cdot 9.81\text{ m/s}^2\cdot 36.01\text{ m}} \\ &= 3.633839261\text{ m/s} \end{aligned}$$

ROUND: The mass of the curling stone, distance traveled, and coefficient of friction between the ice and the stone are all given to four significant figures. These are the only measured values in the problem, so the answer should also be given to four significant figures. The initial velocity was 3.634 m/s.

DOUBLE-CHECK: Since the gravitational and normal forces are perpendicular to the direction of the motion and cancel one another exactly, they will not affect the velocity of the curling stone. Between the time the stone is released and the moment it stops, the frictional force acts in the opposite direction of the velocity and is proportional to the normal force, so this is one-dimensional motion with constant acceleration. It is possible to check this problem by working backward from the initial velocity and force to find an expression for velocity as a function of time, and then use that to find the total distance traveled. Newton's Second law and the equation for the frictional force can be combined to find the acceleration of the curling stone: $f_k = -\mu_k mg = ma_x \Rightarrow a_x = -\mu_k g$. If the spot where the curling stone was released is $x_0 = 0$, then the equations for motion in one dimension with constant acceleration become:

$$\begin{aligned} v &= v_{x0} + at \\ &= 3.634 - 0.01869\cdot 9.81t \end{aligned}$$

and

$$\begin{aligned} d &= x_0 + v_{x0}t + \frac{1}{2}at^2 \\ &= 0 + 3.634t - \frac{1}{2}0.01869\cdot 9.81t^2 \end{aligned}$$

Solving the first equation for $v = 0$ gives $0 = 3.634 - 0.01869\cdot 9.81t \Rightarrow t = \frac{3.634}{0.01869\cdot 9.81} = 19.82$ sec. (The stone was in motion 19.82 seconds, which is reasonable for those who are familiar with the sport of curling.) This value can be used to compute the distance traveled by the stone at the moment it stops as

$$\begin{aligned} d &= 0 + 3.634t - \frac{1}{2}0.01869\cdot 9.81t^2 \\ &= 0 + 3.634(19.82) - \frac{1}{2}0.01869\cdot 9.81(19.82)^2 \\ &= 36.01\text{ m} \end{aligned}$$

This confirms the calculated result.

Chapter 5: Kinetic Energy, Work, and Power

Exercises

5.19. **THINK:** Kinetic energy is proportional to the mass and to the square of the speed. m and v are known for all the objects:

(a) $m = 10.0$ kg, $v = 30.0$ m/s

(b) $m = 100.0$ g, $v = 60.0$ m/s

(c) $m = 20.0$ g, $v = 300.$ m/s

SKETCH:

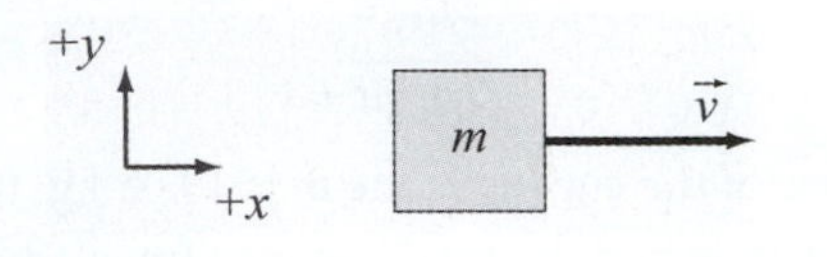

RESEARCH: $K = \frac{1}{2}mv^2$

SIMPLIFY: $K = \frac{1}{2}mv^2$ is already in the right form.

CALCULATE:

(a) $K = \frac{1}{2}(10.0 \text{ kg})(30.0 \text{ m/s})^2 = 4500 \text{ J}$

(b) $K = \frac{1}{2}(100.0 \cdot 10^{-3} \text{ kg})(60.0 \text{ m/s})^2 = 180 \text{ J}$

(c) $K = \frac{1}{2}(20.0 \cdot 10^{-3} \text{ kg})(300. \text{ m/s})^2 = 900 \text{ J}$

ROUND:

(a) 3 significant figures: $K = 4.50 \cdot 10^3$ J

(b) 3 significant figures: $K = 1.80 \cdot 10^2$ J

(c) 3 significant figures: $K = 9.00 \cdot 10^2$ J

DOUBLE-CHECK: The stone is much heavier so it has the greatest kinetic energy even though it is the slowest. The bullet has larger kinetic energy than the baseball since it moves at a much greater speed.

5.23. **THINK:** Given the tiger's mass, $m = 200.$ kg , and energy, $K = 14400$ J , I want to determine its speed. I can rearrange the equation for kinetic energy to obtain the tiger's speed.

SKETCH:

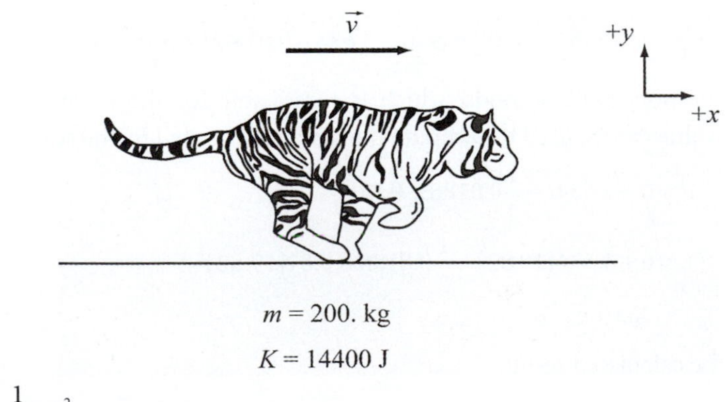

RESEARCH: $K = \frac{1}{2}mv^2$

SIMPLIFY: $v=\sqrt{\dfrac{2K}{m}}$

CALCULATE: $v=\sqrt{\dfrac{2(14400\text{ J})}{200.\text{ kg}}}=12.0\text{ m/s}$

ROUND: Three significant figures: $v=12.0$ m/s.

DOUBLE-CHECK: $(10\text{ m/s})\cdot\dfrac{10^{-3}\text{ km}}{1\text{ m}}\cdot\dfrac{3600\text{ s}}{1\text{ h}}=36\text{ km/h}$. This is a reasonable speed for a tiger.

5.27. **THINK:** The initial speeds are the same for the two balls, so they have the same initial kinetic energy. Since the initial height is also the same for both balls, the gravitational force does the same work on them on their way down to the ground, adding the same amount of kinetic energy in the process. This automatically means that they hit the ground with the same value for their final kinetic energy. Since the balls have the same mass, they consequently have to have the same speed upon ground impact. This means that the difference in speeds that the problem asks for is 0. No further steps are needed in this solution.

SKETCH: Not necessary.

RESEARCH: Not necessary.

SIMPLIFY: Not necessary.

CALCULATE: Not necessary.

ROUND: Not necessary.

DOUBLE-CHECK: Even though our arguments based on kinetic energy show that the impact speed is identical for both balls, you may not find this entirely convincing. After all, most people expect the ball throw directly downward to have a higher impact speed. If you still want to perform a double-check, then you can return to the kinematic equations of chapter 3 and calculate the answer for both cases. Remember that the motion in horizontal direction is one with constant horizontal velocity component, and the motion in vertical direction is free-fall. In both cases we thus have:

$$v_x=v_{0x}$$
$$v_y=\sqrt{v_{0y}^2+2gh}$$

If you now square each equation and add them, you get:

$$v^2=v_x^2+v_y^2=\left(v_{0x}^2\right)+\left(v_{0y}^2+2gh\right)=\left(v_{0x}^2+v_{0y}^2\right)+2gh=v_0^2+2gh$$

Then you see that indeed we have each time for the final speed $v=\sqrt{v_0^2+2gh}$, independent of the direction of the initial velocity vector. What we can learn from this double-check step is two-fold. First, our energy and work considerations yield the exact same results as our kinematic equations from Chapter 3 did. Second, and perhaps more important, the energy and work considerations required much less computational effort to arrive at the same result.

5.31. **THINK:** Only the component of the force parallel to the displacement does work.

SKETCH:

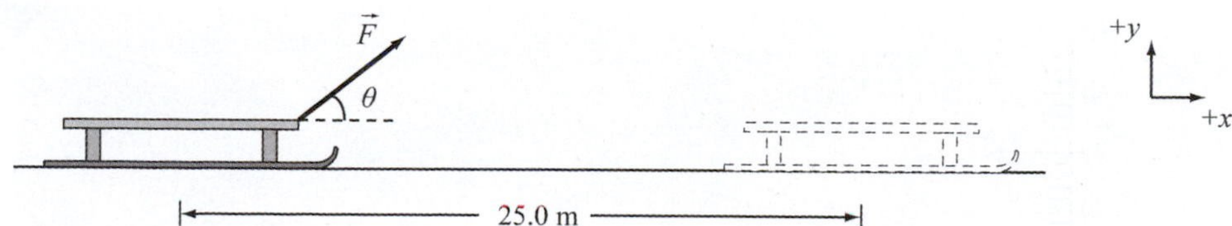

RESEARCH: $W=Fd\cos\theta$

SIMPLIFY: Not applicable.

CALCULATE: $W=Fd\cos\theta=(25.0\text{ N})(25.0\text{ m})\cos30.0°=5.4127\cdot10^2\text{ J}$

ROUND: Three significant figures: $W = 5.41 \cdot 10^2$ J.

DOUBLE-CHECK: The magnitude of the work done by the person is greater than the magnitude of the work done by friction. The units of the work calculations are joules, which are appropriate for work.

5.35. **THINK:** The work done by gravity is mgh. In the absence of friction, the potential energy mgh will be converted to kinetic energy. The actual kinetic energy, when friction is included is less than this. The "missing" energy is the work done by friction. If the work done by friction is known, the frictional force and the coefficient of friction can be determined.

SKETCH:

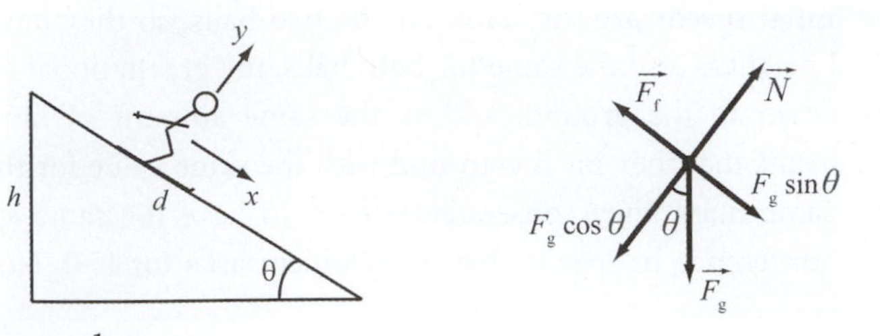

RESEARCH: $W = W_g + W_f = \frac{1}{2}mv^2$, $W_g = mgh$, $W_f = F_f d$

SIMPLIFY: $W_f = W - W_g \Rightarrow F_f d = \frac{1}{2}mv^2 - mgh \Rightarrow -\mu N d = m\left(\frac{1}{2}v^2 - gh\right)$

But $N = F_g \cos\theta = mg\cos\theta \Rightarrow -\mu mg\cos\theta d = m\left(\frac{1}{2}v^2 - gh\right) \Rightarrow \mu = \frac{1}{gd\cos\theta}\left(gh - \frac{1}{2}v^2\right)$

CALCULATE: $g = \left(9.81 \text{ m/s}^2\right)\left(\frac{1 \text{ ft}}{0.3048 \text{ m}}\right) = 32.185 \text{ ft/s}^2$,

$$\mu = \frac{\left(32.185 \text{ ft/s}^2\right)(80.0\sin 30.0° \text{ ft}) - (0.5)(45.0 \text{ ft/s})^2}{\left(32.185 \text{ ft/s}^2\right)(80.0 \text{ ft})\cos 30.0°} = 0.123282$$

ROUND: Three significant figures: $\mu = 0.123$.

DOUBLE-CHECK: This is a reasonable result for the friction coefficient. If I had used SI units, the result would be the same because μ is dimensionless.

5.39. **THINK:** Determine the work necessary to change displacement from 0.730 m to 1.35 m for a force of $F(x) = -kx^4$ with a constant $k = 20.3 \text{ N/m}^4$.

SKETCH:

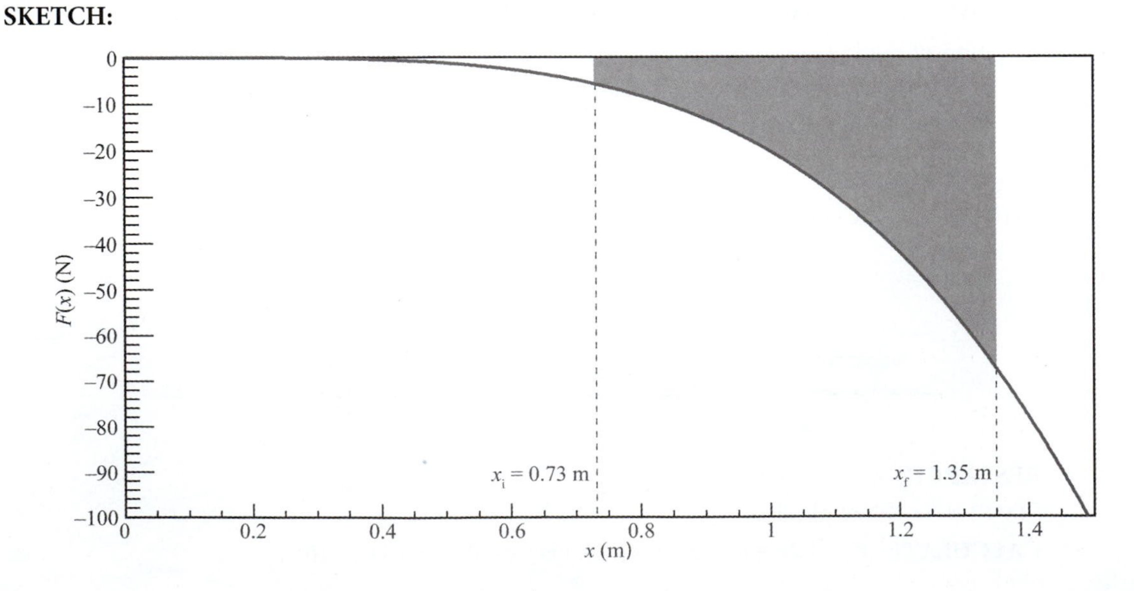

RESEARCH: The work done by the available force is $W = \int_{x_i}^{x_f} F(x)dx$.

SIMPLIFY: $W = \int_{x_i}^{x_f} -kx^4 dx = \left[-\frac{k}{5}x^5 \right]_{x_i}^{x_f} = \frac{k}{5}x_i^5 - \frac{k}{5}x_f^5 = \frac{k}{5}\left(x_i^5 - x_f^5\right)$

CALCULATE: $W = \frac{20.3 \text{ N/m}^4}{5}\left[(0.730 \text{ m})^5 - (1.35 \text{ m})^5\right] = -17.364 \text{ J}$

ROUND: Due to the difference, the answer has three significant figures. The work done *against* the spring force is the negative of the work done *by* the spring force: $W = 17.4$ J.

DOUBLE-CHECK: The negative work in this case is similar to the work done by a spring force.

5.43. **THINK:** The spring constant must be determined given that it requires 30.0 J to stretch the spring 5.00 cm $= 5.00 \cdot 10^{-2}$ m. Recall that the work done by the spring force is always negative for displacements from equilibrium.

SKETCH: Not necessary.

RESEARCH: $W_s = -\frac{1}{2}kx^2$

SIMPLIFY: $k = -\frac{2W_s}{x^2}$

CALCULATE: $k = -\frac{2(-30.0 \text{ J})}{(5.00 \cdot 10^{-2} \text{ m})^2} = 2.40 \cdot 10^4 \text{ N/m}$

ROUND: Variables in the question are given to three significant figures, so the answer remains $k = 2.40 \cdot 10^4$ N/m.

DOUBLE-CHECK: Because the displacement is in the order 10^{-2} m, the spring constant is expected to be in the order of $1/\left(10^{-2}\right)^2 \approx 10^4$.

5.47. **THINK:** Determine the constant speed of a sled drawn by a horse with a power of 1.060 hp. The coefficient of friction is 0.115 and the mass of the sled with a load is 204.7 kg. $P = 1.060 \text{ hp}(746 \text{ W/hp}) = 791 \text{ W}$.

SKETCH:

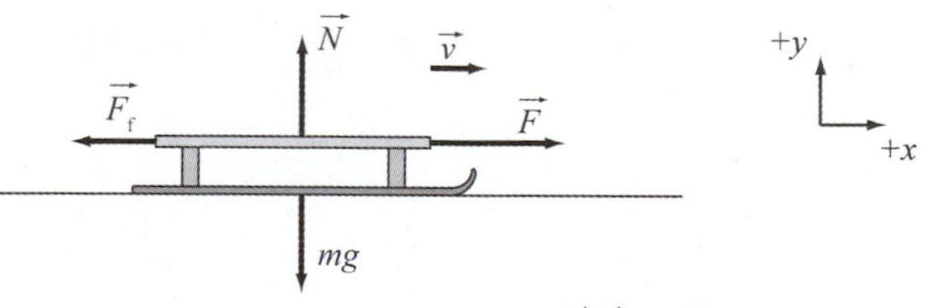

RESEARCH: Use Newton's second law, $F_f = \mu N$ and $P = \vec{F} \cdot \vec{v}$. $\sum F_x = 0$, since $a_x = 0$. So $F - F_f = 0$ $\Rightarrow F = F_f = \mu N$ and $\sum F_y = ma_y = 0 \Rightarrow N - mg = 0 \Rightarrow N = mg$.

SIMPLIFY: $P = Fv \Rightarrow v = \frac{P}{F} = \frac{P}{\mu N} = \frac{P}{\mu mg}$

CALCULATE: $v = \frac{791 \text{ W}}{(0.115)(204.7 \text{ kg})(9.81 \text{ m/s}^2)} = 3.42524 \text{ m/s}$

ROUND: $v = 3.43$ m/s (three significant figures)

DOUBLE-CHECK: $v = 3.43$ m/s $= 12.3$ km/h, which is a reasonable speed.

5.52. **THINK:** If you ride your bicycle on a horizontal surface and stop pedaling, you slow down to a stop. The force that causes this is the combination of friction in the mechanical components of the bicycle, air resistance, and rolling friction between the tires and the ground. In the first part of the problem statement we learn that the bicycle rolls down the hill at a constant speed. This automatically implies that the net force acting on it is zero. (Newton's First Law!) The force along the slope and downward is $mg\sin\theta$ (see sketch). For the net force to be zero this force has to be balanced by the force of friction and air resistance, which acts opposite to the direction of motion, in this case up the slope. So we learn from this first statement that the forces of friction and air resistance have exactly the same magnitude in this case as the component of the gravitational force along the slope. But if you go up the same slope, then gravity and the forces of air resistance and friction point in the same direction. Thus we can calculate the total work done against all forces in this case (and only in this case!) by just calculating the work done against gravity, and then simply multiplying by a factor of 2.

SKETCH: (for just pedaling against gravity)

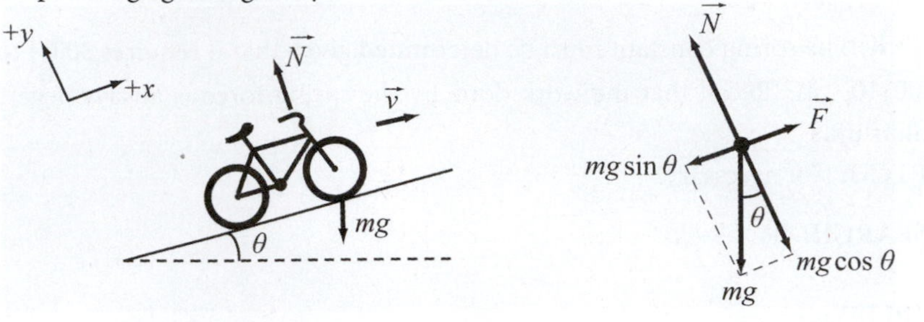

RESEARCH: Again, let's just calculate the work done against gravity, and then in the end multiply by 2. The component of the gravitational force along the slope is $mg\sin\theta$. F is the force exerted by the bicyclist. Power $= Fv$. Using Newton's second law:

$$\sum F_x = ma_x = 0 \Rightarrow F - mg\sin\theta = 0 \Rightarrow F = mg\sin\theta$$

SIMPLIFY: Power $= 2Fv = 2(mg\sin\theta)v$

CALCULATE: $P = 2\cdot 75\text{ kg}\left(9.81\text{ m/s}^2\right)\sin(7.0°)(5.0\text{ m/s}) = 896.654\text{ W}$

ROUND: $P = 0.90\text{ kW}$

DOUBLE-CHECK: $P = 900\text{ W}\cdot\dfrac{1\text{ hp}}{746\text{ W}} = 1.2\text{ hp}$. As this shows, going up a 7 degree slope at 5 m/s requires approximately 1.2 horsepower, which is what a good cyclist can expend for quite some time. (But it's hard!)

5.55. **THINK:** Determine the work done by an athlete that lifted 472.5 kg to a height of 196.7 cm.

SKETCH:

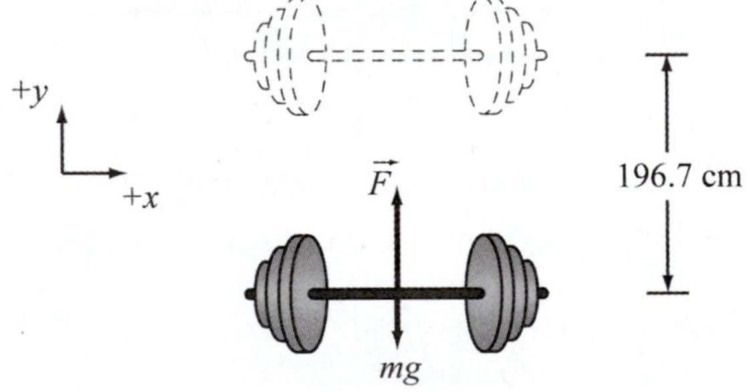

RESEARCH: Use $W = Fd$. F is the combined force needed to the lift the weight, which is $F = mg$.

SIMPLIFY: $W = mgd$

CALCULATE: $W = (472.5\text{ kg})\left(9.81\text{ m/s}^2\right)(1.967\text{ m}) = 9117.49\text{ J}$

ROUND: Rounding to three significant figures, $W = 9.12$ kJ.

DOUBLE-CHECK: A large amount of work is expected for such a large weight.

5.59. **THINK:** A car with mass $m = 1200.$ kg can accelerate from rest to a speed of 25.0 m/s in 8.00 s. Determine the average power produced by the motor for this acceleration.

SKETCH: Not necessary.

RESEARCH: $W = \Delta K, \ P = \dfrac{W}{\Delta t}$

SIMPLIFY: $P = \dfrac{W}{\Delta t} = \dfrac{\Delta K}{\Delta t} = \dfrac{\frac{1}{2}m(v_f^2 - v_i^2)}{\Delta t}$. $v_i = 0$, so $P = \dfrac{\frac{1}{2}mv_f^2}{\Delta t}$.

CALCULATE: $P = \dfrac{\frac{1}{2}(1200.\text{ kg})(25.0\text{ m/s})^2}{8.00\text{ s}} = 46{,}875\text{ W} \cdot \dfrac{1\text{ hp}}{746\text{ W}} = 62.835\text{ hp}$

ROUND: Three significant figures: $P = 62.8$ hp

DOUBLE-CHECK: An average car motor has a power between 100 and 500 hp. This result is reasonable for a small car.

5.63. **THINK:** Determine the initial speed of a sled which is shoved up an incline that makes an angle of 28.0° with the horizontal and comes to a stop at a vertical height of h = 1.35 m.

SKETCH:

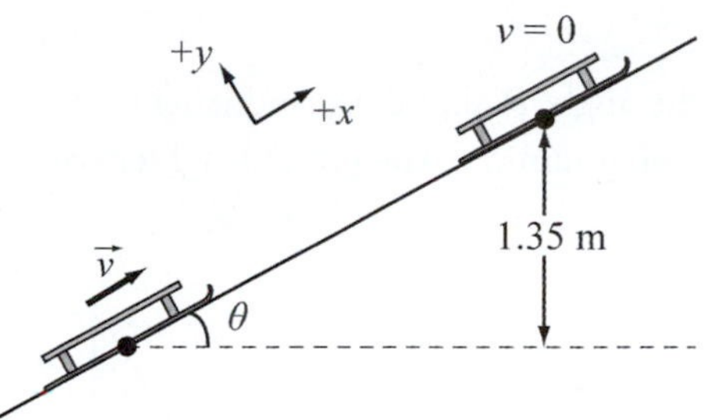

RESEARCH: The work done by gravity must be equal to the change in kinetic energy: $W = \Delta K$.

SIMPLIFY: $W_g = -mgh = K_f - K_i = \frac{1}{2}m(v_f^2 - v_i^2)$. $v_f = 0$, so $-mgh = \frac{1}{2}m(0 - v_i^2) \Rightarrow v_i = \sqrt{2gh}$

CALCULATE: $v_i = \sqrt{2(9.81\text{ m/s}^2)(1.35\text{ m})} = 5.1466\text{ m/s}$

ROUND: The angle was given to three significant figures; so you may think that our result needs to be rounded to three digits. This is not correct, because the angle did not even enter into our calculations. The height was given to three digits, and so we round v_i = 5.15 m/s.

DOUBLE-CHECK: $v_i = 5.15\text{ m/s} = 18.5\text{ km/h}$ is a reasonable value.

5.67. **THINK:** The stack of cement sacks has a combined mass m = 1143.5 kg. The coefficients of static and kinetic friction between the sacks and the bed of the truck are 0.372 and 0.257, respectively. The truck accelerates from rest to $56.6\text{ mph} \cdot \dfrac{0.447\text{ m/s}}{\text{mph}} = 25.3\text{ m/s}$ in $\Delta t = 22.9$ s. Determine if the sacks slide and the work done on the stack by the force of friction.

SKETCH:

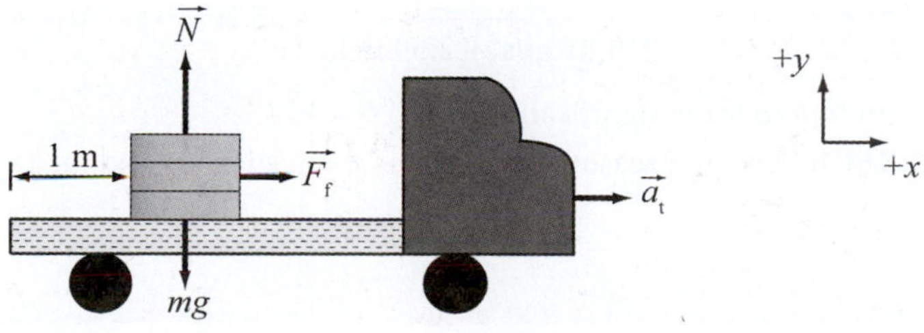

RESEARCH: The acceleration of the truck a_t and the acceleration of the stack a_c must be determined: $a_t = v/\Delta t$. The maximum acceleration that will allow the cement sacks to stay on the truck is calculated by: $F_{f,max} = ma_{c,max} = \mu_s N$.

SIMPLIFY: $F_{f,max} = ma_{c,max} = \mu_s mg \Rightarrow a_{c,max} = \mu_s g$

CALCULATE: $a_t = \dfrac{25.3 \text{ m/s}}{22.9 \text{ s}} = 1.1048 \text{ m/s}^2$, $a_{c,max} = (0.372)(9.81 \text{ m/s}^2) = 3.649 \text{ m/s}^2$

$a_{c,max}$ is larger than a_t. This means that the stack does not slide on the truck bed and $F_f < \mu_s N$. The acceleration of the stack must be the same as the acceleration of the truck $a_c = a_t = 1.10 \text{ m/s}^2$. The work done on the stack by the force of friction is calculated using $W = \Delta K$:

$$W = K_f - K_i = \frac{1}{2}m(v_f^2 - v_i^2). \text{ Since } v_i = 0, \; W = \frac{1}{2}mv_f^2 = \frac{1}{2}(1143.5 \text{ kg})(25.3 \text{ m/s})^2 = 365971 \text{ J}.$$

ROUND: $W = 366$ kJ

DOUBLE-CHECK: The work done by the force of friction can also be calculated by $W = F_f d$; where $F_f = ma_c$ and $d = \frac{1}{2}a_c t^2$:

$$W = ma_c\left(\frac{1}{2}a_c t^2\right) = \frac{1}{2}ma_c^2 t^2 = \frac{1}{2}m(a_c t)^2. \text{ Using } v_f = a_c t, \; W = \frac{1}{2}m(v_f)^2 \text{ as before.}$$

5.71. **THINK:** Determine the angle θ that the granddaughter is released from to reach a speed of 3.00 m/s at the bottom of the swinging motion. The granddaughter has a mass of m = 21.0 kg and the length of the swing is l = 2.50 m.

SKETCH:

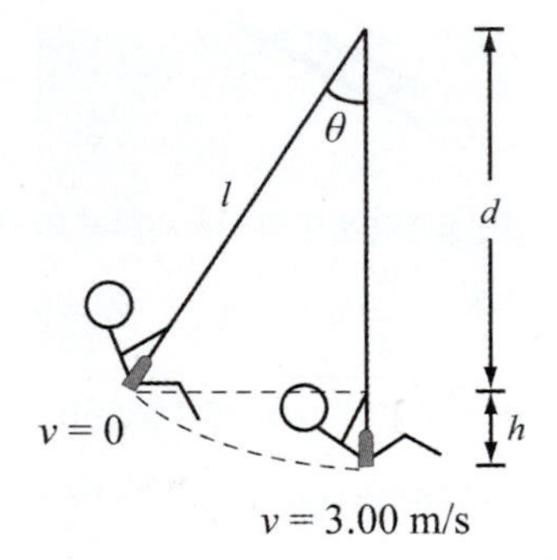

RESEARCH: The energy is given by the change in the height from the top of the swing, mgh. It can be seen from the geometry that $h = l - d = l - l\cos\theta = l(1-\cos\theta)$. At the bottom of the swinging motion, there is only kinetic energy, $K = (1/2)mv^2$.

SIMPLIFY: Equate the energy at the release point to the energy at the bottom of the swinging motion and solve for θ:

$$mgh = \frac{1}{2}mv^2 \Rightarrow gl(1-\cos\theta) = \frac{1}{2}v^2 \Rightarrow \theta = \cos^{-1}\left(1 - \frac{v^2}{2gl}\right)$$

CALCULATE: $\theta = \cos^{-1}\left(1 - \dfrac{(3.00 \text{ m/s})^2}{2(9.81 \text{ m/s}^2)(2.50 \text{ m})}\right) = 35.263°$

ROUND: Rounding to three significant figures, $\theta = 35.3°$.

DOUBLE-CHECK: This is a reasonable angle to attain such a speed on a swing.

5.75. The amount of power required to overcome the force of air resistance is given by $P = F \cdot v$. And the force of air resistance is given by the Ch. 4 formula

$$F_d = \left(\frac{1}{2}c_d A\rho\right)v^2$$

$$\Rightarrow P = \left(\frac{1}{2}c_d A\rho v^2\right)\cdot v = \frac{1}{2}c_d A\rho v^3$$

This evaluates as:

$$P = \frac{1}{2}(0.333)(3.25 \text{ m}^2)(1.15 \text{ kg/m}^3)(26.8 \text{ m/s})^3 = 11{,}978.4 \text{ W} = (11{,}978.4 \text{ W})\left(\frac{1 \text{ hp}}{745.7 \text{ W}}\right) = 16.06 \text{ hp}$$

To three significant figures, the power is 16.1 hp.

Multi-Version Exercises

5.76. **THINK:** This problem involves a variable force. Since we want to find the change in kinetic energy, we can find the work done as the object moves and then use the work-energy theorem to find the total work done.

SKETCH:

RESEARCH: Since the object started at rest, it had zero kinetic energy to start. Use the work-energy theorem $W = \Delta K$ to find the change in kinetic energy. Since the object started with zero kinetic energy, the total kinetic energy will equal the change in kinetic energy: $\Delta K = K$. The work done by a variable force in the x-direction is given by $W = \int_{x_0}^{x} F_x(x')dx'$ and the equation for our force is $F_x(x') = A(x')^6$. Since the object starts at rest at 1.093 m and moves to 4.429 m, we start at x_0 = 1.093 m and end at x = 4.429 m.

SIMPLIFY: First, find the expression for work by substituting the correct expression for the force: $W = \int_{x_0}^{x} A(x')^6\, dx'$. Taking the definite integral gives $W = \frac{A}{7}(x')^7\Big|_{x_0}^{x} = \frac{A}{7}(x^7 - x_0^7)$. Combining this with the work-energy theorem gives $\frac{A}{7}(x^7 - x_0^7) = W = K$.

CALCULATE: The problem states that A = 11.45 N/m^6, that the object starts at x_0 = 1.093 m and that it ends at x = 4.429 m. Plugging these into the equation and calculating gives:

$$K = \frac{A}{7}(x^7 - x_0^7)$$

$$= \frac{11.45 \text{ N/m}^6}{7}\left((4.429 \text{ m})^7 - (1.093 \text{ m})^7\right)$$

$$= 5.467930659 \cdot 10^4 \text{ J}$$

ROUND: The measured values in this problem are the constant A in the equation for the force and the two distances on the x-axis. All three of these are given to four significant figures, so the final answer should have four significant figures: $5.468 \cdot 10^4$ J or 54.68 kJ.

DOUBLE-CHECK: Working backwards, if a variable force in the $+x$-direction changes the kinetic energy from zero to $5.468 \cdot 10^4$ J, then the object will have moved

$$x = \sqrt[7]{\frac{7\left(5.468 \cdot 10^4 \text{ J}\right)}{11.45 \text{ N/m}^6} + 1.093^7} = 4.429008023 \text{ m}.$$

This is, within rounding, the 4.429 m given in the problem, so it seems that the calculations were correct.

5.79. **THINK:** In this problem, the reindeer must pull the sleigh to overcome the friction between the runners of the sleigh and the snow. Express the friction force in terms of the speed and weight of the sleigh, and the coefficient of friction between the sleigh and the ground. It is then possible to find the power from the force and velocity.

SKETCH: Draw a free-body diagram for the sleigh:

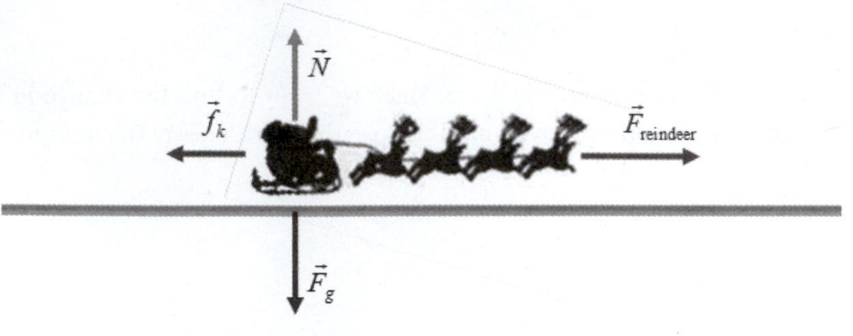

RESEARCH: Since the sleigh is moving with a constant velocity, the net forces on the sleigh are zero. This means that the normal force and the gravitational force are equal and opposite ($\vec{N} = -\vec{F}_g$), as are the friction force and the force from the reindeer ($\vec{F}_{\text{reindeer}} = -\vec{f}_k$). From the data given in the problem, it is possible to calculate the friction force $f_k = \mu_k mg$. The power required to keep the sleigh moving at a constant speed is given by $P = F_{\text{reindeer}} v$. Eventually, it will be necessary to convert from SI units (Watts) to non-standard units (horsepower or hp). This can be cone using the conversion factor 1 hp = 746 W.

SIMPLIFY: To find the power required for the sleigh to move, it is necessary to express the force from the reindeer in terms of known quantities. Since the force of the reindeer is equal in magnitude with the friction force, use the equation for frictional force to find:

$$\left|\vec{F}_{\text{reindeer}}\right| = \left|-\vec{f}_k\right| = f_k = \mu_k mg$$

Use this and the speed of the sleigh to find that $P = F_{\text{reindeer}} v = \mu_k mgv$.

CALCULATE: With the exception of the gravitational acceleration, all of the needed values are given in the question. The coefficient of kinetic friction between the sleigh and the snow is 0.1337, the mass of the system (sleigh, Santa, and presents) is 537. 3 kg, and the speed of the sleigh is 3.333 m/s. Using a gravitational acceleration of 9.81 m/s gives:

$$\begin{aligned} P &= \mu_k mgv \\ &= 0.1337 \cdot 537.3 \text{ kg} \cdot 9.81 \text{ m/s}^2 \cdot 3.333 \text{ m/s} = 2348.83532 \text{ W} \end{aligned}$$

This can be converted to horsepower: $2348.83532 \text{ W} \cdot \dfrac{1 \text{ hp}}{746 \text{ W}} = 3.148572815 \text{ hp}$.

ROUND: The measured quantities in this problem are all given to four significant figures. Though the conversion from watts to horsepower and the gravitational acceleration have three significant figures, they do not count for the final answer. The power required to keep the sleigh moving is 3.149 hp.

DOUBLE-CHECK: Generally, it is thought that Santa has 8 or 9 reindeer (depending on how foggy it is on a given Christmas Eve). This gives an average of between 0.3499 and 0.3936 horsepower per reindeer, which seems reasonable. Work backwards to find that, if the reindeer are pulling the sled with 3.149 hp, then the speed they are moving must be (rounding to four significant figures):

$$v = \frac{3.149 \text{ hp}}{\mu_k mg}$$
$$= \frac{3.149 \text{ hp} \cdot 746 \text{ W/hp}}{0.1337 \cdot 537.3 \text{ kg} \cdot 9.81 \text{ m/s}^2}$$
$$= 3.333452207 \frac{\text{W}}{\text{kg} \cdot \text{m/s}^2}$$
$$= 3.333 \frac{\text{kg} \cdot \text{m}^2/\text{s}^3}{\text{kg} \cdot \text{m/s}^2} = 3.333 \text{ m/s}$$

This matches the constant velocity from the problem, so the calculations were correct.

5.82. **THINK:** In this problem, the energy stored in the spring is converted to kinetic energy as the puck slides across the ice. The spring constant and compression of the spring can be used to calculate the energy stored in the spring. This is all converted to kinetic energy of the puck. The energy is dissipated as the puck slides across the ice. It is necessary to compute how far the puck must slide to dissipate all of the energy that was, originally, stored in the spring.

SKETCH: Sketch the puck when the spring is fully compressed, when it leaves contact with the spring, as it moves across the ice, and at the moment it comes to a stop. Include a free body diagram showing the forces on the puck as it moves across the ice.

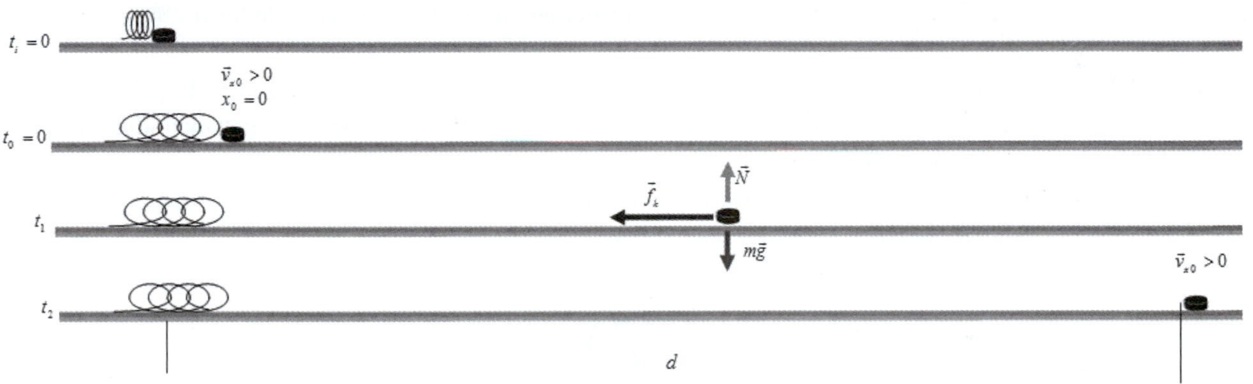

RESEARCH: The potential energy stored in the spring is $U = \frac{1}{2}kx^2$, where x is the compression of the spring. The energy dissipated by the force of friction is $\Delta U = Fd$. The force of friction on the puck is given by $F = \mu_k mg$. It is necessary to find the total distance traveled d.

SIMPLIFY: First, find the energy dissipated by the force of friction in terms of known quantities $\Delta U = \mu_k mgd$. This must equal the energy that was stored in the spring, $U = \frac{1}{2}kx^2$. Setting $\Delta U = U$, solve for the total distance traveled in terms of known quantities:

$$\Delta U = U$$
$$\mu_k mgd = \frac{1}{2}kx^2$$
$$d = \frac{kx^2}{2\mu_k mg}$$

It is important to note that x represents the compression of the spring before the puck was released, and d is the total distance traveled from the time that the puck was released (not from the time the puck left contact with the spring).

CALCULATE: Before plugging the values from the question into the equation above, it is important to make sure that all of the units are the same. In particular, note that it is easier to solve the equation directly if the compression is changed from 23.11 cm to 0.2311 m and the mass used is 0.1700 kg instead of 170.0 g. Then the distance is:

$$\begin{aligned} d &= \frac{kx^2}{2\mu_k mg} \\ &= \frac{15.19\text{ N/m}\cdot(-0.2311\text{ m})^2}{2\cdot 0.02221\cdot 0.1700\text{ kg}\cdot 9.81\text{ m/s}^2} \\ &= 10.95118667\text{ m} \end{aligned}$$

Of this distance, 0.2311 m is the distance the spring was compressed. So the distance traveled by the puck after leaving the spring is 10.95118667 m – 0.2311 m = 10.72008667 m.

ROUND: The measured values are all given to four significant figures, so the final answer is that the hockey puck traveled 10.72 m.

DOUBLE-CHECK: Working backwards, if the hockey puck weighs 0.1700 kg and traveled 10.95 m across the ice (including spring compression) with a coefficient of kinetic friction of 0.02221, then the energy dissipated was $\Delta U = \mu_k mgd = 0.0221\cdot 9.81\text{ m/s}^2\cdot 0.1700\text{ kg}\cdot 10.95\text{ m} = 0.4056\text{ J}$. Since the energy stored in this spring is $U = \frac{1}{2}kx^2 = \frac{15.19\text{ N/m}}{2}x^2$, it is necessary for the spring to have been compressed by $x = \sqrt{\frac{0.4056\text{ J}\cdot 2}{15.19\text{ N/m}}} = 0.231092\text{ m}$, within rounding of the value of 23.11 cm given in the problem.

5.84. **THINK:** Since the bricks travel at a low, constant speed, use the information given in the problem to find the tension force that the crane exerts to raise the bricks. The power can be computed by finding the scalar product of the force vector and the velocity vector.

SKETCH: A free body diagram of the bricks as they are raised to the top of the platform is helpful. The only forces are tension from the crane and gravity.

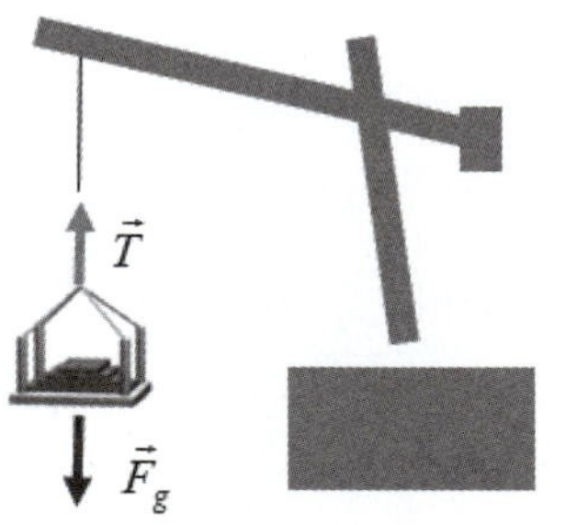

RESEARCH: The average power is the scalar product of the force exerted by the crane on the bricks and the velocity of the bricks: $P = \vec{F}\bullet\vec{v}$, where the force is the tension from the crane. (The speed of the bricks is low, so air resistance is negligible in this case.) The bricks are moving at a constant velocity, so the sum of the forces is zero and $\vec{T} = -\vec{F}_g = -m\vec{g}$. The velocity is constant and can be computed as the distance divided by the time $\vec{v} = \frac{\Delta\vec{d}}{\Delta t}$.

SIMPLIFY: Instead of using vector equations, note that the tension force and the velocity are in the same direction. The equation for the power then becomes $P = \vec{F}\bullet\vec{v} = Fv\cos\alpha$, where α is the angle between the velocity and force. Since $T = mg$ and $v = \frac{d}{t}$, the power is given by the equation $P = \frac{mgd}{t}\cos\alpha$.

CALCULATE: The mass, distance, and time are given in the problem. The velocity of the bricks is in the same direction as the tension force, so $\alpha = 0$.

$$\begin{aligned} P &= \frac{mgd}{t}\cos\alpha \\ &= \frac{75.0 \text{ kg} \cdot 9.81 \text{ m/s}^2 \cdot 45.0 \text{ m}}{52.0 \text{ s}}\cos 0° \\ &= 636.7067308 \text{ W} \end{aligned}$$

ROUND: The mass of the bricks, height to which they are raised, and time are all given to three significant figures, and the answer should have four significant figures. The average power of the crane is 637 W.

DOUBLE-CHECK: To check, note that the average power is the work done divided by the elapsed time: $\bar{P} = \frac{W}{\Delta t}$. Combine this with the equation for the work done by the constant tension force $W = |\vec{F}||\Delta\vec{r}|\cos\alpha$ to find an equation for the average power: $\bar{P} = \frac{|\vec{F}||\Delta\vec{r}|\cos\alpha}{\Delta t}$. Plug in the values for the tension force $\vec{T} = -\vec{F}_g = -mg$ and distance $\Delta\vec{r} = 45.0 \text{ m}$ upward to find:

$$\begin{aligned} P &= \frac{mgd}{t}\cos\alpha \\ &= \frac{75.0 \text{ kg} \cdot 9.81 \text{ m/s}^2 \cdot 45.0 \text{ m}}{52.0 \text{ s}}\cos 0° \\ &= 636.7067308 \text{ W} \end{aligned}$$

When this is rounded to three decimal places, it confirms the calculations.

Chapter 6: Potential Energy and Energy Conservation

Exercises

6.31. **THINK:** The mass of the book is $m = 2.00$ kg and its height above the floor is $h = 1.50$ m. Determine the gravitational potential energy, U_g.

SKETCH:

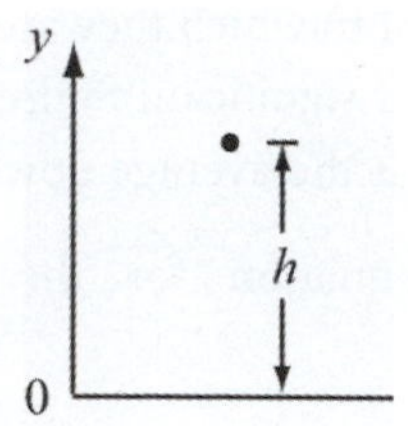

RESEARCH: Taking the floor's height as $U_g = 0$, U_g for the book can be determined from the formula $U_g = mgh.$

SIMPLIFY: It is not necessary to simplify.

CALCULATE: $U_g = (2.00\text{ kg})(9.81\text{ m/s}^2)(1.50\text{ m}) = 29.43\text{ J}$

ROUND: The given initial values have three significant figures, so the result should be rounded to $U_g = 29.4$ J.

DOUBLE-CHECK: This is a reasonable value for a small mass held a small distance above the floor.

6.35. **THINK:** The mass of the car is $m = 1.50 \cdot 10^3$ kg. The distance traveled is $d = 2.50$ km $= 2.50 \cdot 10^3$ m. The angle of inclination is $\theta = 3.00°$. The car travels at a constant velocity. Determine the change in the car's potential energy, ΔU and the net work done on the car, W_{net}.

SKETCH:

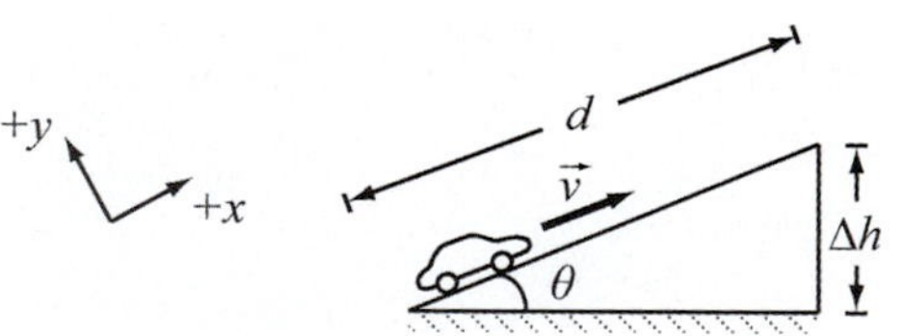

RESEARCH: To determine ΔU the change of height of the car Δh must be known. From trigonometry, the change in height is $\Delta h = d\sin\theta$. Then, $\Delta U = mg\Delta h.$ To determine W_{net} use the work-kinetic energy theorem. Despite the fact that non-conservative forces are at work (friction force on the vehicle), it is true that $W_{\text{net}} = \Delta K$.

SIMPLIFY: $\Delta U = mg\Delta h = mgd\sin\theta$

$$W_{\text{net}} = \Delta K = \frac{1}{2}mv_f^2 - \frac{1}{2}mv_0^2 = \frac{1}{2}m\left(v_f^2 - v_0^2\right)$$

CALCULATE: $\Delta U = \left(1.50 \cdot 10^3\text{ kg}\right)\left(9.81\text{ m/s}^2\right)\left(2.50 \cdot 10^3\text{ m}\right)\sin(3.00°) = 1925309\text{ J}$

$$W_{\text{net}} = \frac{1}{2}m\left(v_f^2 - v_0^2\right) = \frac{1}{2}m(0) = 0$$

ROUND: Since θ has two significant figures, $\Delta U = 1.93 \cdot 10^6$ J, and there is no net work done on the car.

DOUBLE-CHECK: The change in potential energy is large, as the car has a large mass and a large change in height, $\Delta h = \left(2.50 \cdot 10^3\text{ m}\right)\sin(3.00°) = 131$ m. The fact that the net work done is zero while there is a change in potential energy means that non-conservative forces did work on the car (friction, in this case).

6.39. **THINK:** The potential energy functions are (a) $U(y)=ay^3-by^2$ and (b) $U(y)=U_0\sin(cy)$. Determine $F(y)$ from $U(y)$.

SKETCH: A sketch is not necessary.

RESEARCH: $F(y)=-\dfrac{\partial U(y)}{\partial y}$

SIMPLIFY:

(a) $F(y)=-\dfrac{\partial\left(ay^3-by^2\right)}{\partial y}=2by-3ay^2$

(b) $F(y)=-\dfrac{\partial\left(U_0\sin(cy)\right)}{\partial y}=-cU_0\cos(cy)$

CALCULATE: There are no numerical calculations to perform.

ROUND: It is not necessary to round.

DOUBLE-CHECK: The derivative of a cubic polynomial should be a quadratic, so the answer obtained for (a) makes sense. The derivative of a sine function is a cosine function, so it makes sense that the answer obtained for (b) involves a cosine function.

6.43. **THINK:** The initial height of the basketball is $y_0=1.20$ m. The initial speed of the basketball is $v_0=20.0$ m/s. The final height is $y=3.05$ m. Determine the speed of the ball at this point.

SKETCH:

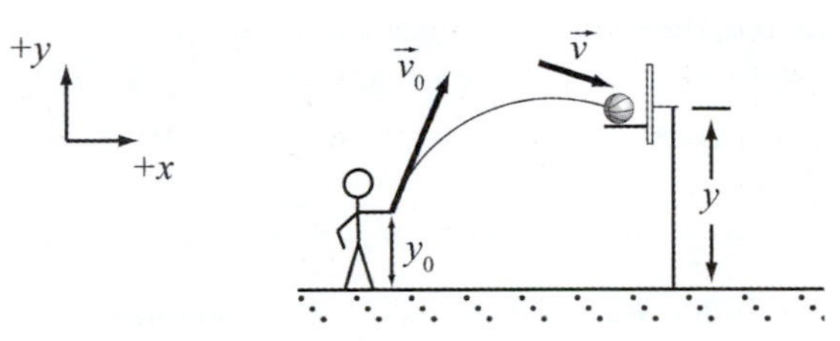

RESEARCH: Neglecting air resistance, there are only conservative forces, so $\Delta K=-\Delta U$. The kinetic energy K can be determined from $K=mv^2/2$ and U from $U=mgh$.

SIMPLIFY: $K_\text{f}-K_\text{i}=U_\text{i}-U_\text{f}$, so $(1/2)mv^2-(1/2)mv_0^2=mgy_0-mgy$. Dividing through by the mass m yields the equation $(1/2)v^2-(1/2)v_0^2=gy_0-gy$. Then solving for v gives

$$v=\sqrt{2\left(g(y_0-y)+\frac{1}{2}v_0^2\right)}.$$

CALCULATE: $v=\sqrt{2\left(\left(9.81\text{ m/s}^2\right)(1.20\text{ m}-3.05\text{ m})+\frac{1}{2}(20.0\text{ m/s})^2\right)}$

$=\sqrt{2\left(-18.1485\text{ m}^2/\text{s}^2+200.0\text{ m}^2/\text{s}^2\right)}$

$=19.071$ m/s

ROUND: The initial height is given with the fewest number of significant figures. Since it has three significant figures the value of v needs to be rounded to three significant figures: $v = 19.1$ m/s.

DOUBLE-CHECK: The final speed should be less than the initial speed since the final height is greater than the initial one.

6.47. **THINK:** The initial height is $h = 40.0$ m. Determine:

(a) the speed v_f at the bottom, neglecting friction,

(b) if the steepness affects the final speed; and

(c) if the steepness affects the final speed when friction is considered.

SKETCH:

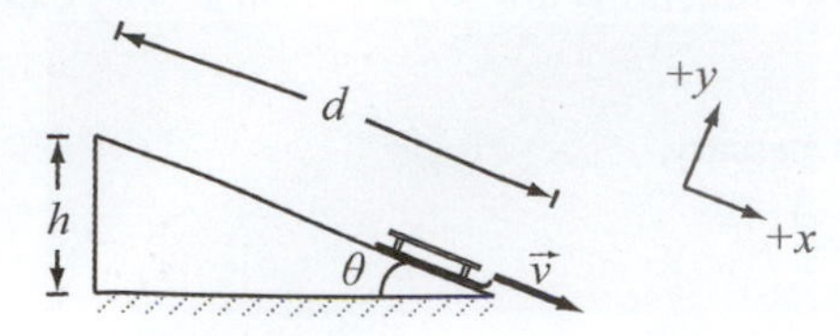

RESEARCH:

(a) With conservative forces, $\Delta K = -\Delta U$. v can be determined from $K = \left(mv_{\text{f}}^{2}\right)/2$ and $U = mgh$.

(b and c) Note that the change in the angle θ affects the distance, d, traveled by the toboggan: as θ gets larger (the incline steeper), d gets smaller.

(c) The change in thermal energy due to friction is proportional to the distance traveled: $\Delta E_{\text{th}} = \mu_{\text{k}} Nd$. The total change in energy of an isolated system is $\Delta E_{\text{tot}} = 0$, where $\Delta E_{\text{tot}} = \Delta K + \Delta U + \Delta E_{\text{th}}$, and ΔE_{th} denotes the non-conservative energy of the toboggan-hill system (in this case, friction).

SIMPLIFY:

(a) With $K_i = 0$ (assuming $v_0 = 0$) and $U_{\text{f}} = 0$ (taking the bottom to be $h = 0$):

$$K_{\text{f}} = U_{\text{i}} \Rightarrow \frac{1}{2}mv_{\text{f}}^{2} = mgh \Rightarrow v_{\text{f}} = \sqrt{2gh}$$

(b) The steepness does not affect the final speed, in a system with only conservative forces, the distance traveled is not used when conservation of mechanical energy is considered.

(c) With friction considered, then for the toboggan-hill system,

$$\Delta E = \Delta K + \Delta U + \Delta E_{\text{th}} = 0 \Rightarrow \Delta K = -\Delta U - \Delta E_{\text{th}} \Rightarrow K_{\text{f}} = U_{\text{i}} - \Delta E_{\text{th}} = mgh - \mu_{\text{k}} Nd$$

The normal force N is given by $N = mg\cos\theta$, while on the hill. With $d = h/\sin\theta$,

$$K_{\text{f}} = mgh - \mu_{\text{k}}\left(mg\cos\theta\right)\left(\frac{h}{\sin\theta}\right) = mgh\left(1 - \mu_{\text{k}}\cot\theta\right).$$

The steepness of the hill does affect K_{f} and therefore v at the bottom of the hill.

CALCULATE:

(a) $v_{\text{f}} = \sqrt{2\left(9.81\text{ m/s}^2\right)\left(40.0\text{ m}\right)} = 28.01\text{ m/s}$

ROUND: Since h has three significant figures, $v = 28.0$ m/s.

DOUBLE-CHECK: This is a very fast, but not unrealistic speed for the toboggan to achieve.

6.51. **THINK:** The spring constant for each spring is $k = 30.0$ N/m. The stone's mass is $m = 1.00$ kg. The equilibrium length of the springs is $l_0 = 0.500$ m. The displacement to the left is $x = 0.700$ m. Determine the system's total mechanical energy, E_{mec} and (b) the stone's speed, v, at $x = 0$.

SKETCH:

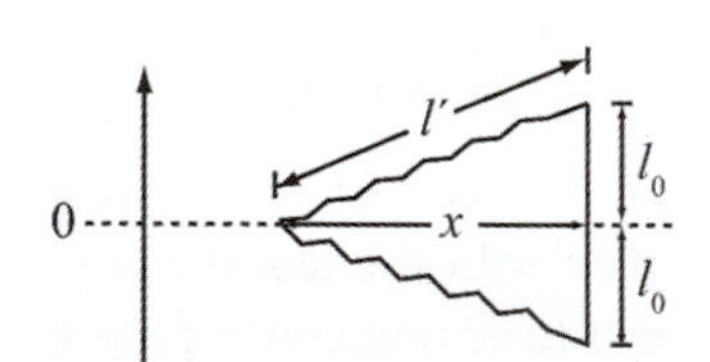

Note: The sketch is a side view. The word "vertical" means that the springs are oriented vertically above the ground. The path the stone takes while in the slingshot is completely horizontal so that gravity is neglected.

RESEARCH:
(a) In order to determine E_{mec}, consider all kinetic and potential energies in the system. Since the system is at rest, the only form of mechanical energy is spring potential energy, $U_s = \left(kx^2\right)/2$.

(b) By energy conservation, ΔE_{mec} (no non-conservative forces). v can be determined by considering $\Delta E_{mec} = 0$.

SIMPLIFY:

(a) $E_{mec} = K + U = U_s = U_{s1} + U_{s2} = \frac{1}{2}k_1\left(l_0 - l'\right)^2 + \frac{1}{2}k_2\left(l_0 - l'\right)^2 = k\left(l_0 - l'\right)^2$. To determine l', use the Pythagorean theorem, $l' = \sqrt{l_0^2 + x^2}$. Then, $E_{mec} = k\left(l_0 - \sqrt{l_0^2 + x^2}\right)^2$.

(b) As the mechanical energy is conserved, $E_{mec\,f} = E_{mec\,i}$ so $K_f + U_{sf} = E_{mec}$ (with $U_f = 0$), and therefore $K_f = E_{mec}$. Solving the equation for kinetic energy, $\frac{1}{2}mv^2 = E_{mec} \Rightarrow v = \sqrt{2E_{mec}/m}$.

CALCULATE:

(a) $E_{mec} = 30.0\text{ N/m}\left(0.500\text{ m} - \sqrt{(0.500\text{ m})^2 + (0.700\text{ m})^2}\right)^2 = 3.893\text{ J}$

(b) $v = \sqrt{2(3.893\text{ J})/1.00\text{ kg}} = 2.790\text{ m/s}$

ROUND: Since all of the given values have three significant figures, the results should be rounded to $E_{mec} = 3.89\text{ J}$ and $v = 2.79\text{ m/s}$.

DOUBLE-CHECK: The values are reasonable considering the small spring constant.

6.55. **THINK:** The skier's mass is m = 55.0 kg. The constant speed is v = 14.4 m/s. The slope length is $l = 123.5\text{ m}$ and the angle of the incline is $\theta = 14.7°$. Determine the mechanical energy lost to friction, ΔE_{th}.

SKETCH:

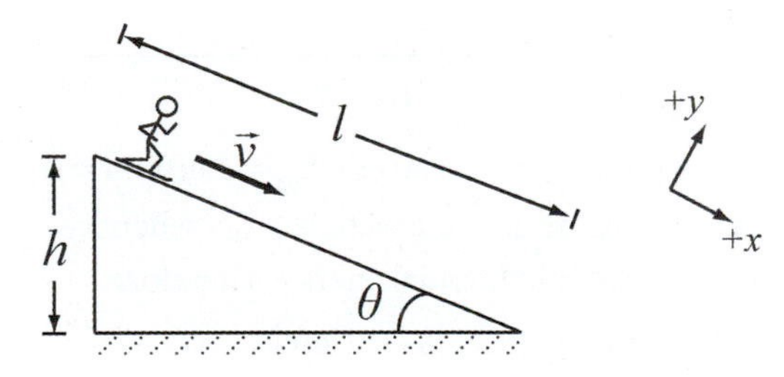

RESEARCH: The skier and the ski slope form an isolated system. This implies that $\Delta E_{tot} = \Delta K + \Delta U + \Delta E_{th} = 0$. Note that $\Delta K = 0$ since v is constant. Use the equation $U = mgh$, where the height of the ski slope can be found using trigonometry: $h = l\sin\theta$.

SIMPLIFY: At the bottom of the slope, $U_f = 0$. Then,

$$\Delta E_{th} = -\Delta U = -\left(U_f - U_i\right) = U_i = mgh = mgl\sin\theta.$$

CALCULATE: $\Delta E_{th} = (55.0\text{ kg})\left(9.81\text{ m/s}^2\right)(123.5\text{ m})\sin 14.7° = 16909\text{ J}$

ROUND: With three significant figures in m, g and θ, the result should be rounded to $\Delta E_{th} = 16.9\text{ kJ}$.

DOUBLE-CHECK: If this energy had been transformed completely to kinetic energy (no friction), and if the skier had started from rest, their final velocity would have been 24.8 m/s at the bottom of the slope. This is a reasonable amount of energy transferred to thermal energy generated by friction.

6.59. **THINK:** The block's mass is m = 1.00 kg. The length of the plank is L = 2.00 m. The incline angle is $\theta = 30.0°$. The coefficient of kinetic friction is $\mu_k = 0.300$. The path taken by the block is $L/2$ upward, $L/4$ downward, then up to the top of the plank. Determine the work , W_b, done by the block against friction.

SKETCH:

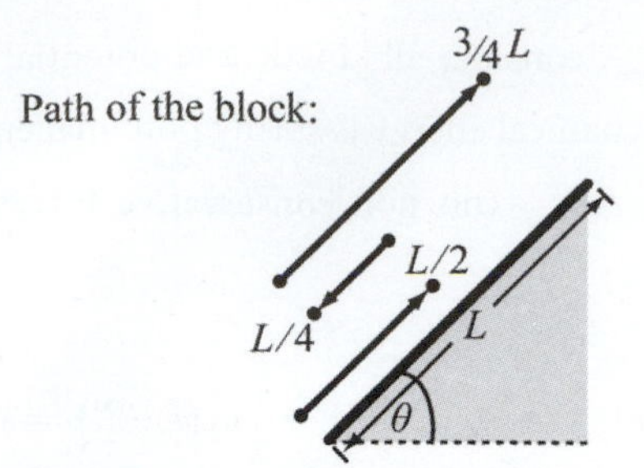

RESEARCH: Friction is a non-conservative force. The work done by friction, W_f, is therefore dependent on the path. It is known that $W_\text{f} = -f_\text{k}d$, and with $W_\text{b} = -W_\text{f}$, the equation is $W_\text{b} = f_\text{k}d.$ The total path of the block is $d = L/2 + L/4 + 3L/4 = 1.5L.$

SIMPLIFY: $W_\text{b} = f_\text{k}d = \mu_\text{k}Nd = \mu_\text{k}mg(\cos\theta)d$ ($N = mg\cos\theta$ on the incline)

CALCULATE: $W_b = (0.300)(1.00 \text{ kg})(9.81 \text{ m/s}^2)\cos(30.0°)(1.50(2.00 \text{ m})) = 7.646 \text{ J}$

ROUND: Each given value has three significant figures, so the result should be rounded to $W_\text{b} = 7.65$ J.

DOUBLE-CHECK: This is a reasonable amount of work done against friction considering the short distance traveled.

6.63. **THINK:** The mass of the cart is 237.5 kg. The initial velocity is $v_0 = 16.5$ m/s. The surface is frictionless. Determine the turning point shown on the graph in the question, sketched below.

SKETCH:

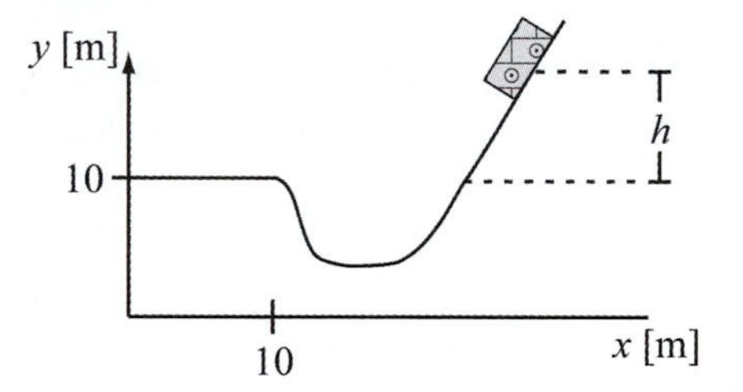

RESEARCH: Since the system is conservative, $E_\text{tot} = \text{constant} = U_\text{max} = K_\text{max}$. Therefore, the kinetic energy at $x = 0$, $y = 10.$ m is the same as the kinetic energy whenever the track is at $y = 10.$ m again. Set $y = 10.$ m as the origin for gravitational potential energy. Therefore,

$$E_\text{tot} = K_\text{max} = \frac{mv_0^2}{2}.$$

This is the available energy to climb the track from $y = 10.$ m. The turning point is when $v = 0$ and

$$U_\text{max} = K_\text{max} \Rightarrow mgh = \frac{mv_0^2}{2}.$$

SIMPLIFY: $h = \frac{v_0^2}{2g}$, $y = 10. \text{ m} + h$

CALCULATE: $h = \frac{(16.5 \text{ m/s})^2}{2(9.81 \text{ m/s}^2)} = 13.9 \text{ m}$, $y = 10. \text{ m} + 13.9 \text{ m} = 23.9 \text{ m}$

ROUND: Reading off the graph is accurate to the nearest integer, so round the value of y to 24 m. Reading off the graph, the value of x at $y = 24$ m is $x = 42$ m.

DOUBLE-CHECK: It is reasonable that the cart will climb about 18 m with an initial velocity of $v_0 = 16.5$ m/s.

6.67. **THINK:** The mass of the car is $m = 987$ kg. The speed is $v = 64.5$ mph. The coefficient of kinetic friction is $\mu_k = 0.301$. Determine the mechanical energy lost.

SKETCH:

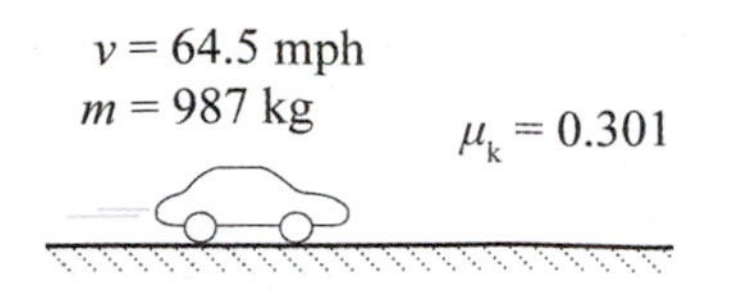

RESEARCH: Since all of the mechanical energy is considered in the form of kinetic energy, the energy lost is equal to the kinetic energy before applying the brakes. Using the conversion 1 mph is equal to 0.447 m/s, the speed can be converted to SI units. Convert the speed: $v = (64.5 \text{ mph})\left(\frac{0.447 \text{ m/s}}{1 \text{ mph}}\right) = 28.8 \text{ m/s}$.

SIMPLIFY: $E_{\text{lost}} = \frac{1}{2}mv^2$

CALCULATE: $E_{\text{lost}} = \frac{1}{2}(987 \text{ kg})(28.8 \text{ m/s})^2 = 4.10 \cdot 10^5 \text{ J}$

ROUND: Rounding to three significant figures, $E_{\text{lost}} = 4.10 \cdot 10^5$ J.

DOUBLE-CHECK: For an object this massive, it is reasonable that it requires such a large amount of energy to stop it.

6.71. **THINK:** The mass of the ball is $m = 1.50$ kg. Its speed is $v = 20.0$ m/s and its height is $h = 15.0$ m. Determine the ball's total energy, E_{tot}.

SKETCH:

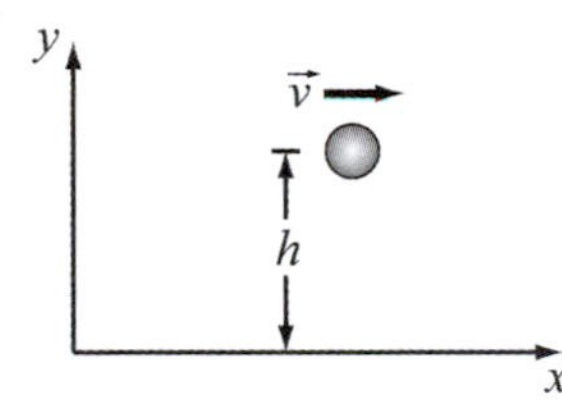

RESEARCH: Total energy is the sum of the mechanical energy and other forms of energy. As there are no non-conservative forces (neglecting air resistance), the total energy is the total mechanical energy. $E_{\text{tot}} = K + U$. Use $K = (mv^2)/2$ and $U = mgh$.

SIMPLIFY: $E_{\text{tot}} = m\left(\frac{1}{2}v^2 + gh\right)$

CALCULATE:

$$E_{\text{tot}} = 1.50 \text{ kg}\left(\frac{1}{2}(20.0 \text{ m/s})^2 + (9.81 \text{ m/s}^2)(15.0 \text{ m})\right)$$
$$= 1.50 \text{ kg}\left(200 \text{ m}^2/\text{s}^2 + 147.15 \text{ m}^2/\text{s}^2\right)$$
$$= 520.725 \text{ J}$$

ROUND: As the speed has three significant figures, the result should be rounded to $E_{\text{tot}} = 521$ J.

DOUBLE-CHECK: The energy is positive and has the correct unit of measurement. It is also on the right order of magnitude for the given values. This is a reasonable energy for a ball.

6.75. **THINK:** The length of the chain is l = 4.00 m and the maximum displacement angle is $\theta = 35°$. Determine the speed of the swing, v, at the bottom of the arc.

SKETCH:

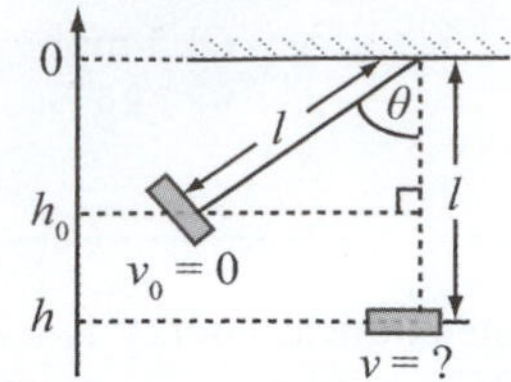

RESEARCH: At the maximum displacement angle, the speed of the swing is zero. Assuming there are no non-conservative forces, to determine the speed, v, the conservation of mechanical energy can be used: $\Delta K = -\Delta U$. Use $K = (mv^2)/2$ and $U = mgh$. The initial height can be determined using trigonometry. Take the top of the swing to be $h = 0$.

SIMPLIFY: $v_0 = 0$ and $K_i = 0$. From the sketch, $h_0 = -l\cos\theta$ and $h = -l$. Then,

$$K_f = U_i - U_f \Rightarrow \frac{1}{2}mv^2 = mg(-l\cos\theta) - mg(-l) \Rightarrow \frac{1}{2}v^2 = g(l - l\cos\theta) \Rightarrow v = \sqrt{2g(l - l\cos\theta)}$$

CALCULATE: $v = \sqrt{2(9.81 \text{ m/s}^2)(4.00 \text{ m} - (4.00 \text{ m})\cos 35.0°)} = 3.767 \text{ m/s}$

ROUND: l and θ have two significant figures, so the result should be rounded to v = 3.77 m/s.

DOUBLE-CHECK: This is a reasonable speed for a swing to achieve when initially displaced from the vertical by 35°.

6.79. **THINK:** A rocket that has a mass of $m = 3.00$ kg reaches a height of $1.00 \cdot 10^2$ m in the presence of air resistance which takes $8.00 \cdot 10^2$ J of energy away from the rocket, so $W_{air} = -8.00 \cdot 10^2$ J. Determine the height the rocket would reach if air resistance could be neglected.

SKETCH:

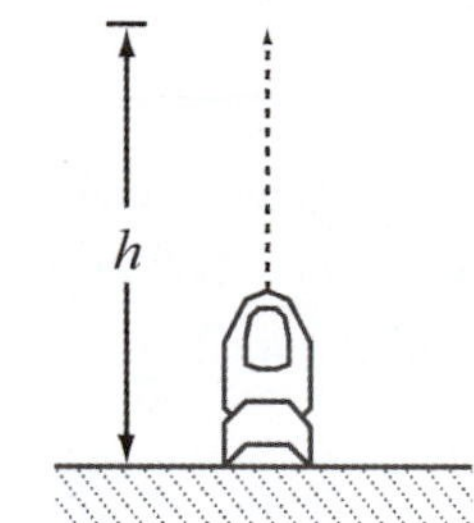

RESEARCH: Air resistance performs $-8.00 \cdot 10^2$ J of work on the rocket. The absence of air resistance would then provide an extra $8.00 \cdot 10^2$ J of energy to the system. If this energy is converted into potential energy, the increase in height of the rocket can be determined.

SIMPLIFY: $U_t = -W_{air} \Rightarrow mgh_t = -W_{air} \Rightarrow h_t = \frac{-W_{air}}{mg}$, where h_t is the added height.

CALCULATE: $h_t = \frac{-(-8.00 \cdot 10^2 \text{ J})}{(3.00 \text{ kg})(9.81 \text{ m/s}^2)} = 27.183 \text{ J/kg} \cdot \text{m/s}^2 = 27.183 \text{ m}$

Therefore, the total height reached by the rocket in the absence of air resistance is

$$h_{tot} = h_0 + h_t = 1.00 \cdot 10^2 \text{ m} + 0.27183 \cdot 10^2 \text{ m} = 1.27183 \cdot 10^2 \text{ m}.$$

ROUND: Since the values are given to three significant figures, the result should be rounded to $h_{tot} = 1.27 \cdot 10^2$ m.

DOUBLE-CHECK: It is reasonable that air resistance will decrease the total height by approximately a fifth.

6.83. **THINK:** A block of mass, $m = 1.00$ kg is against a spring on an inclined plane of angle, $\theta = 30.0°$. The coefficient of kinetic friction is $\mu_k = 0.100$. The spring is initially compressed 10.0 cm and the block moves to 2.00 cm beyond the springs normal length after release (therefore the block moves $d = 12.0$ cm after it is released). Determine (a) the change in the total mechanical energy and (b) the spring constant.

SKETCH:

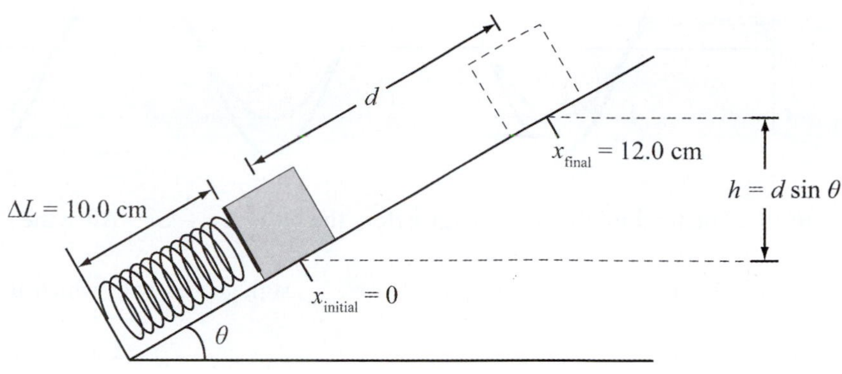

RESEARCH:

(a) Since this is not a conservative system, the change in the total mechanical energy can be related to the energy lost due to friction. This energy can be determined by calculating the work done by the force of friction: $W_{\text{friction}} = F_{\text{friction}} d = \mu_k mg(\cos\theta)d$, and $\Delta E_{\text{tot}} = -W_{\text{friction}} = -\mu_k mg(\cos\theta)d.$

(b) From conservation of energy, the change in total energy, ΔE_{tot} determined in (a), is equal to $\Delta K + \Delta U$. Since $K = 0$ at both the initial and final points it follows that

$$\Delta E_{\text{tot}} = U_{\text{final}} - U_{\text{initial}} = mgd\sin\theta - \frac{1}{2}k\Delta L^2.$$

SIMPLIFY:

(a) $\Delta E_{\text{tot}} = -\mu_k mg(\cos\theta)d$

(b) $k = 2\dfrac{(mgd\sin\theta - \Delta E_{\text{tot}})}{\Delta L^2}$

CALCULATE:

(a) $\Delta E_{\text{tot}} = -(0.100)(1.00 \text{ kg})(9.81 \text{ m/s}^2)\cos(30.0°)(12.0\cdot 10^{-2} \text{ m}) = -0.1019$ J

(b) $k = -2\dfrac{(1.00 \text{ kg})(9.81 \text{ m/s}^2)(0.120 \text{ m})\sin(30.0°) - (-0.1019 \text{ J})}{(0.100 \text{ m})^2} = 138.1$ N/m

ROUND:

(a) Since the lowest number of significant figures is three, the result should be rounded to $\Delta E_{\text{tot}} = -1.02\cdot 10^{-1}$ J (lost to friction).

(b) Since the mass is given to three significant figures, the result should be rounded to $k = 138$ N/m.

DOUBLE-CHECK:

(a) A change of about 0.1 J given away to friction for a distance of 12 cm and with this particular coefficient of friction is reasonable.

(b) The spring constant is in agreement with the expected values.

6.87. **THINK:** The mass hanging vertically from a spring can be treated using a method that is independent of gravitational effects on the mass (see page 185 in the text). The mechanical energy of the mass on a spring is defined in terms of the amplitude of the oscillation and the spring constant. When the mass is pushed, the system gains mechanical energy. This new mechanical energy can be used to calculate the new velocity of the mass at the equilibrium position (b) and the new amplitude (c).

SKETCH: Before the mass is hit, the amplitude of the oscillation is A. After the mass is hit, the amplitude of the oscillation is A_{new}.

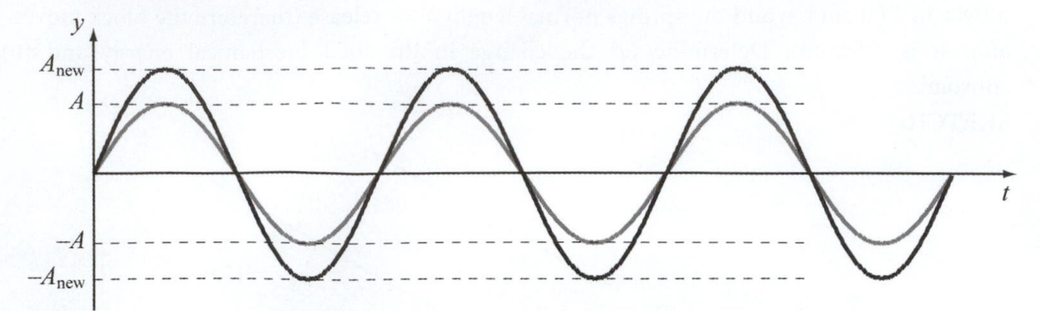

RESEARCH: The total mechanical energy before the hit is $E=\frac{1}{2}kA^2$. After the hit, the total mechanical energy is given by $E_{\text{new}}=\frac{1}{2}kA^2+\frac{1}{2}mv_{\text{push}}^2$ where v_{push} is the speed with which the mass is pushed. The new speed at equilibrium is given by $\frac{1}{2}mv_{\text{new}}^2=E_{\text{new}}$ and the new amplitude of oscillation is given by $\frac{1}{2}kA_{\text{new}}^2=E_{\text{new}}$.

SIMPLIFY:

(a) $E_{\text{new}}=\frac{1}{2}kA^2+\frac{1}{2}mv_{\text{push}}^2$

(b) $v_{\text{new}}=\sqrt{\frac{2E_{\text{new}}}{m}}$

(c) $A_{\text{new}}=\sqrt{\frac{2E_{\text{new}}}{k}}$

CALCULATE:

(a) $E_{\text{new}}=\frac{1}{2}kA^2+\frac{1}{2}mv_{\text{push}}^2=\frac{1}{2}(100.\text{ N/m})(0.200\text{ m})^2+\frac{1}{2}(1.00\text{ kg})(1.00\text{ m/s})^2=2.50\text{ J}$

(b) $v_{\text{new}}=\sqrt{\frac{2E_{\text{new}}}{m}}=\sqrt{\frac{2(2.50\text{ J})}{1.00\text{ kg}}}=2.236\text{ m/s}$

(c) $A_{\text{new}}=\sqrt{\frac{2E_{\text{new}}}{k}}=\sqrt{\frac{2(2.50\text{ J})}{100.\text{ N/m}}}=0.2236\text{ m}$

ROUND: Rounding to three significant figures: $E_{\text{new}}=2.50\text{ J}$, $v_{\text{max,2}}=2.24\text{ m/s}$ and $A_2=22.4\text{ cm}$.

DOUBLE-CHECK: The mechanical energy before the hit was

$$E=(1/2)kA^2=(1/2)(100.\text{ N/m})(0.200\text{ m})^2=2.00\text{ J}.$$

The speed of the mass passing the equilibrium point before the hit was $v=\sqrt{\frac{2E}{m}}=\sqrt{\frac{2(2.00\text{ J})}{1.00\text{ kg}}}=2.00\text{ m/s}$.

It is reasonable that adding 0.5 J to the total energy by means of a hit results in an increase of the speed of the mass at the equilibrium point of 0.24 m/s and an increase of about 2.4 cm to the amplitude.

6.91. **THINK:** The total work can be determined if the path taken and the force applied are known. These are both given as follows: $\vec{F}(x,y)=\left(x^2\hat{x}+y^2\hat{y}\right)$ N and the points are S(10.0 m,10.0 m), P(0 m,10.0 m), Q(10.0 m,0 m) and O(0 m,0 m).

SKETCH:

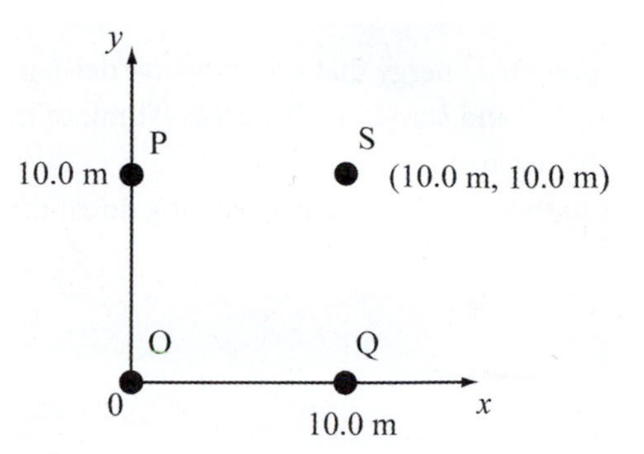

RESEARCH: Work is given by:

$$W=\int_a^b d\vec{l}\bullet\vec{F}=\int_a^b\left(x^2dx+y^2dy\right).$$

The equations of the paths are: along OP, $x = 0$, $dx = 0$; along OQ, $y = 0$, $dy = 0$; along OS, $y = x$, $dy = dx$; along PS, $y = 10$, $dy = 0$; along QS, $x = 10$, $dx = 0$.

SIMPLIFY:

(a) OPS: $W=\int_O^P\left(x^2dx+y^2dy\right)+\int_P^S\left(x^2dx+y^2dy\right)$

$$=\int_0^{10}y^2dy+\int_0^{10}x^2dx=\frac{1}{3}y^3\Big|_0^{10}+\frac{1}{3}x^3\Big|_0^{10}$$

$$=\frac{1}{3}(10)^3+\frac{1}{3}(10)^3=\frac{2}{3}(10)^3$$

$$=W_{OP}+W_{PS}$$

(b) OQS: $W=\int_O^Q d\vec{l}\bullet\vec{F}+\int_Q^S d\vec{l}\bullet\vec{F}$

$$=\int_0^{10}x^2dx+\int_0^{10}y^2dy$$

$$=W_{OQ}+W_{QS}$$

$$=W_{PS}+W_{OP}$$

$$=\left(\frac{2}{3}\right)10^3$$

(c) OS: $W=W_{OS}=\int_O^S\left(x^2dx+y^2dy\right)\Rightarrow\int_O^S\left(x^2dx+x^2dx\right)=\int_0^{10}2x^2dx=2W_{PS}=\frac{2}{3}\left(10^3\right)$

(d) OPSQO: $W=W_{OP}+W_{PS}+W_{SQ}+W_{QO}=\frac{10^3}{3}+\frac{10^3}{3}+\left(-W_{QS}\right)+\left(-W_{OQ}\right)=\frac{2}{3}\left(10^3\right)-\frac{2}{3}\left(10^3\right)=0$

(e) OQSPO: $W=W_{OQ}+W_{QS}+W_{SP}+W_{PO}=\frac{10^3}{3}+\frac{10^3}{3}-\frac{10^3}{3}-\frac{10^3}{3}=0$

CALCULATE: $\frac{2}{3}\left(10.0^3\right)=666.67$. (a), (b) and (c): $W=666.67$. (d) and (e): $W=0$.

ROUND: Rounding to three significant figures, (a), (b) and (c): W = 667 J, and (d), (e): W = 0 J.

DOUBLE-CHECK: The force is conservative and it should not depend on the path. It is expected that $W_{OS}=W_{OP}+W_{PS}=W_{OQ}+W_{QS}$, which is shown to be true in the calculation. It is also expected that the work along a closed path is zero, which is also shown to be true in the calculations.

Multi-Version Exercises

6.93. **THINK:** The gravitational potential energy that the snowboarder has at her highest point is dissipated by friction as she rides down the hill and across the flat area. Think of her motion in two parts: riding down the slope and riding across the flat area.

SKETCH: The sketch needs to show the snowboarder sliding down the hill and on the flat area:

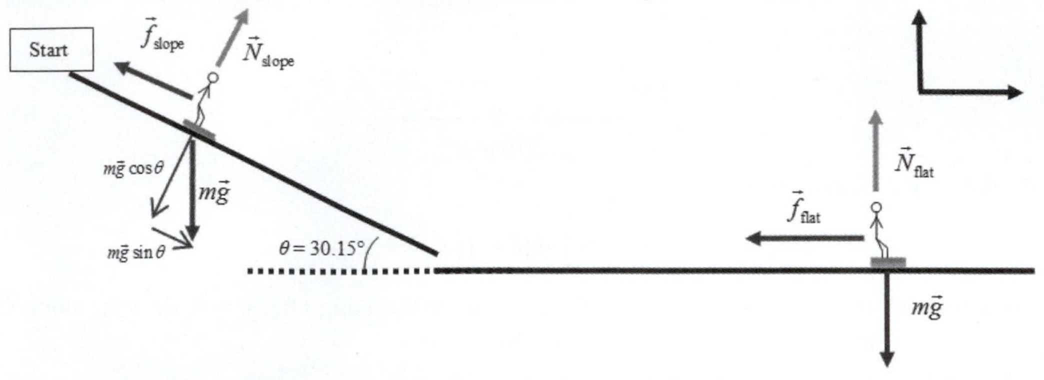

RESEARCH: The energy dissipated by friction must equal the change in gravitational potential energy from her highest point (at the start) to her final position. The work-energy theorem gives $mgh = f_{\text{slope}}d_{\text{slope}} + f_{\text{flat}}d_{\text{flat}}$, where d_{flat} is the distance she travels on the flat snow and d_{slope} is the distance she travels down the slope. Her original starting height is given by $h = d_{\text{slope}}\sin\theta$. The friction force on the slope is given by $f_{\text{slope}} = \mu_k mg\cos\theta$ and the friction force on the flat snow is given by $f_{\text{flat}} = \mu_k mg$.

SIMPLIFY: Since the mass of the snowboarder is not given in the question, it is necessary to find an expression for the distance traveled on the flat snow d_{flat} that does not depend on the mass m of the snowboarder. Substitute the frictional forces $f_{\text{slope}} = \mu_k mg\cos\theta$ and $f_{\text{flat}} = \mu_k mg$ into the work-energy theorem to get

$$mgh = (\mu_k mg\cos\theta)\cdot d_{\text{slope}} + (\mu_k mg)d_{\text{flat}}$$
$$mgh = mg(\mu_k\cos\theta\cdot d_{\text{slope}} + \mu_k d_{\text{flat}})$$
$$h = \mu_k\cos\theta\cdot d_{\text{slope}} + \mu_k d_{\text{flat}}$$

Finally, substitute in $h = d_{\text{slope}}\sin\theta$ for the height h and solve for d_{flat} to get:

$$\mu_k\cos\theta\cdot d_{\text{slope}} + \mu_k d_{\text{flat}} = h$$
$$\mu_k\cos\theta\cdot d_{\text{slope}} + \mu_k d_{\text{flat}} = d_{\text{slope}}\sin\theta$$
$$\mu_k d_{\text{flat}} = d_{\text{slope}}\sin\theta - \mu_k\cos\theta\cdot d_{\text{slope}}$$
$$d_{\text{flat}} = \frac{d_{\text{slope}}\sin\theta - \mu_k\cos\theta\cdot d_{\text{slope}}}{\mu_k}$$

CALCULATE: The question states that the distance the snowboarder travels down the slope is $d_{\text{slope}} = 38.09$ m, the coefficient of friction between her and the snow is 0.02501, and the angle that the hill makes with the horizontal is $\theta = 30.15°$. Plugging these into the equation gives:

$$d_{\text{flat}} = \frac{d_{\text{slope}}\sin\theta - \mu_k\cos\theta\cdot d_{\text{slope}}}{\mu_k}$$
$$= \frac{38.09\text{ m}\cdot\sin 30.15° - 0.02501\cdot\cos 30.15°\cdot 38.09\text{ m}}{0.02501}$$
$$= 732.008853\text{ m}$$

ROUND: The quantities in the problem are all given to four significant figures. Even after performing the addition in the numerator, the calculated values have four significant figures, so the snowboarder travels 732.0 m along the flat snow.

DOUBLE-CHECK: For those who are frequent snowboarders; this seems like a reasonable answer: travel 38.0 m down a slope of more than 30°, and you go quite far: almost three quarters of a kilometer. Working backwards from the answer, the snowboarder traveled 732.0 m along the flat snow and 38.09 m along the slope, so the energy dissipated is

$$f_{\text{slope}} d_{\text{slope}} + f_{\text{flat}} d_{\text{flat}} = 0.02501(mg)\cos(30.15°)38.0 \text{ m} + 0.02501(mg) \cdot 732.0 \text{ m}, \text{ or } 19.13mg.$$

Since this must equal the loss in gravitational potential, we know $mgh = 19.13mg$, so the start was 19.13 m above the flat area. This agrees with the values given in the problem, where the snowboarder traveled 38.09 m at a slope of 30.15°, so she started $38.09\sin 30.15 = 19.13$ meters above the horizontal area.

6.96. **THINK:** At the maximum height, the baseball has no kinetic energy, only gravitational potential energy. We can define zero gravitational potential energy at the point where the catcher gloves the ball. Then the total gravitational potential energy at maximum height equals the total kinetic energy when the ball was caught. The velocity is computed from the kinetic energy.

SKETCH: Sketch the path of the baseball, showing the different heights:

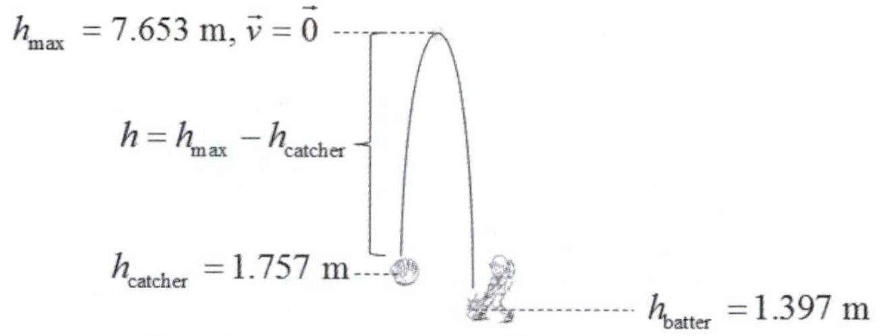

$h_{\text{max}} = 7.653$ m, $\vec{v} = \vec{0}$

$h = h_{\text{max}} - h_{\text{catcher}}$

$h_{\text{catcher}} = 1.757$ m

$h_{\text{batter}} = 1.397$ m

RESEARCH: The gravitational potential energy is given by $K = mgh$ and the total kinetic energy is given by $KE = \frac{1}{2}mv^2$. In this case, the kinetic energy when the baseball lands in the catcher's mitt is equal to the gravitational potential energy difference from the maximum height to the height at which the catcher caught the baseball.

SIMPLIFY: To find the velocity of the baseball when it was caught, it is necessary to note that $K = KE$. This means that $mgh = \frac{1}{2}mv^2$ or $gh = \frac{v^2}{2}$. Since the height h in this problem is really the difference between the maximum height and the height at which the ball was caught ($h = h_{\text{max}} - h_{\text{catcher}}$), the equation can be solved for the velocity when the ball is caught:

$$\frac{v^2}{2} = gh$$

$$v^2 = 2g\left(h_{\text{max}} - h_{catcher}\right)$$

$$v = -\sqrt{2g\left(h_{\text{max}} - h_{catcher}\right)}$$

Since the baseball is moving downward when it was caught, we take the negative square root to indicate that the velocity is in the downward direction.

CALCULATE: The maximum height of the baseball and the height at which it was caught are given in the problem as 7.653 m and 1.757 m, respectively. The velocity is then calculated to be

$$v = -\sqrt{2g\left(h_{\text{max}} - h_{catcher}\right)} = -\sqrt{2 \cdot 9.81 \text{ m/s}^2 \left(7.653 \text{ m} - 1.757 \text{ m}\right)}, \text{ or } -10.75544141 \text{ m/s}$$

ROUND: The measured heights are all given to four significant figures, and the height h calculated by taking their difference also has four significant digits. These are the only measured values used in the problem, so the final answer should also have four significant digits. The velocity of the ball when it was caught was 10.76 m/s towards the ground.

DOUBLE-CHECK: Normally, the speed of pitches and batted balls in baseball are given in terms of miles per hour. It is not uncommon for pitchers to achieve speeds of around 100 mph, but a pop fly rarely travels that quickly. The baseball was going $10.76\frac{\text{m}}{\text{s}}\cdot\frac{1\text{ mile}}{1609.344\text{ m}}\cdot\frac{3600\text{ s}}{\text{hour}}=24.07$ mph when it was caught, which is reasonable in this context.

6.99. **THINK:** This is a projectile motion problem, where it is possible to ignore air resistance. So, the horizontal velocity stays constant. The vertical component of the velocity can be calculated using energy conservation, and then the angel that the ball strikes the ground can be calculated from the horizontal (x-) and vertical (y-) components of the velocity.

SKETCH: Sketch the path of the ball as it is thrown from the building:

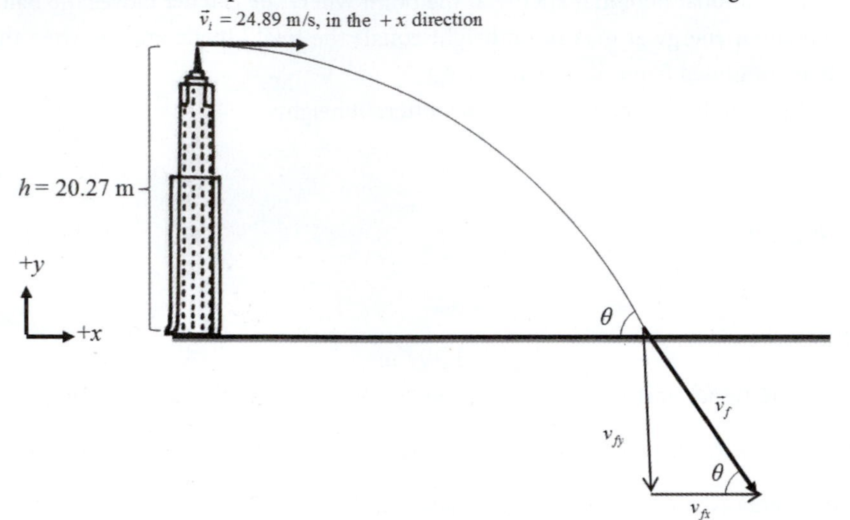

RESEARCH: Since the horizontal velocity is constant, the x–component of the velocity when the ball is released is equal to the x–component of the velocity when the ball lands; $v_{fx}=v_{ix}=v_i$. Since the only change in the velocity is to the y–component, the kinetic energy from the y–component of the velocity must equal the change in gravitational potential energy, $mgh=\frac{1}{2}m\left(v_{fy}^2\right)$. The angle at which the ball strikes the ground can be computed from the x– and y– components of the velocity, plus a little trigonometry: $\theta=\tan^{-1}\left(\frac{v_{fy}}{v_{fx}}\right)$.

SIMPLIFY: To find the final velocity, it is necessary to eliminate the mass term from the equation $mgh=\frac{1}{2}m\left(v_{fy}^2\right)$ and solve for the final velocity, getting $\sqrt{2gh}=v_{fy}$. Since the horizontal velocity does not change, $v_{fx}=v_i$ can also be used. Substitute these into the equation $\theta=\tan^{-1}\left(\frac{v_{fy}}{v_{fx}}\right)$ to get that $\theta=\tan^{-1}\left(\frac{\sqrt{2gh}}{v_i}\right)$.

CALCULATE: The height and initial velocity are given in the problem, and the gravitational acceleration on Earth is about 9.81 m/s² towards the ground. This means that

$$\begin{aligned}\theta &= \tan^{-1}\left(\frac{\sqrt{2gh}}{v_i}\right)\\ &= \tan^{-1}\left(\frac{\sqrt{2\cdot 9.81\text{ m/s}^2\cdot 20.27\text{ m}}}{24.89\text{ m/s}}\right)\\ &= 38.7023859^\circ\end{aligned}$$

ROUND: The measured values in the question were given to four significant figures, and all of the calculations maintain that degree of accuracy. So the final answer should be rounded to four significant figures. The ball lands at an angle of 38.70° from the horizontal.

DOUBLE-CHECK: Working backwards, if the ball lands with a velocity of magnitude $\left|\vec{v}_f\right| = \sqrt{\left|v_{fx}\right|^2 + \left|v_{fy}\right|^2}$, the final velocity has a magnitude $\sqrt{24.89^2 + \sqrt{2gh}^2} = \sqrt{1017.2095}$ m/s. The initial velocity was 24.89 m/s, so the ball gained $\frac{1}{2}m\left(\sqrt{1017.2095}\right)^2 - \frac{1}{2}m(24.89)^2$ J or 198.8487 J in kinetic energy. Since the gravitational potential energy is given by *mgh*, use conservation of energy and algebra to solve for *h*:

$$\begin{aligned}mgh &= \frac{1}{2}m\left(\sqrt{1017.2095}\right)^2 - \frac{1}{2}m(24.89)^2\\ 9.81mh &= m\left(\frac{1}{2}1017.2095 - \frac{1}{2}24.89^2\right)\\ h &= \frac{m}{9.81m}\left(\frac{1}{2}1017.2095 - \frac{1}{2}24.89^2\right)\\ &= \frac{1}{2\cdot 9.81}\left(1017.2095 - 24.89^2\right)\\ &= 20.27\end{aligned}$$

This height (20.27 m) agrees with the value given in the problem, confirming the calculations.

Chapter 7: Momentum and Collisions

Exercises

7.25. **THINK:** Compute the ratios of the momenta and kinetic energies of the car and SUV. $m_{\text{car}} = 1200.$ kg, $m_{\text{SUV}} = 1.5m_{\text{car}} = \frac{3}{2}m_{\text{car}}$, $v_{\text{car}} = 72.0$ mph, and $v_{\text{SUV}} = \frac{2}{3}v_{\text{car}}$.

SKETCH:

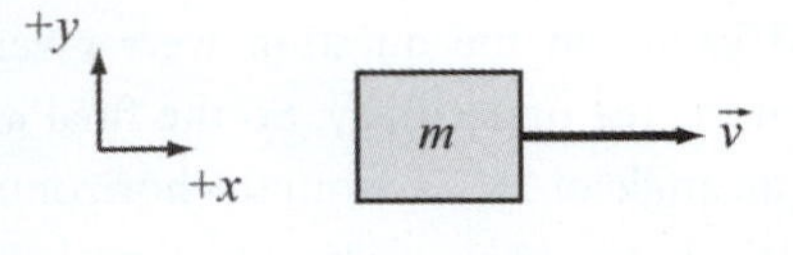

RESEARCH:

(a) $p = mv$

(b) $K = \frac{1}{2}mv^2$

SIMPLIFY:

(a) $\dfrac{p_{\text{SUV}}}{p_{\text{car}}} = \dfrac{m_{\text{SUV}}v_{\text{SUV}}}{m_{\text{car}}v_{\text{car}}} = \dfrac{(3/2)m_{\text{car}}(2/3)v_{\text{car}}}{m_{\text{car}}v_{\text{car}}}$

(b) $\dfrac{K_{\text{SUV}}}{K_{\text{car}}} = \dfrac{(1/2)m_{\text{SUV}}v_{\text{SUV}}^2}{(1/2)m_{\text{car}}v_{\text{car}}^2} = \dfrac{(3/2)m_{\text{car}}\left((2/3)v_{\text{car}}\right)^2}{m_{\text{car}}v_{\text{car}}^2}$

CALCULATE:

(a) $\dfrac{p_{\text{SUV}}}{p_{\text{car}}} = \dfrac{(3/2)(2/3)}{1} = 1$

(b) $\dfrac{K_{\text{SUV}}}{K_{\text{car}}} = \dfrac{(3/2)(2/3)^2}{1} = \dfrac{(3/2)(4/9)}{1} = 2/3 = 0.6667$

ROUND: (a) $\dfrac{p_{\text{SUV}}}{p_{\text{car}}} = 1.0$ (b) $\dfrac{K_{\text{SUV}}}{K_{\text{car}}} = 0.67$

DOUBLE-CHECK: Although the car is lighter, it is moving faster. The changes in mass and speed cancel out for the momentum but not for the kinetic energy because the kinetic energy is proportional to v^2.

7.29. **THINK:** Lois has a mass of 50.0 kg and speed 60.0 m/s. We need to calculate the force on Lois, F_s, when $\Delta t = 0.100$ s. (Subscript *s* means "Superman, mostly, with a small assist from air resistance.") Then we want the value of Δt where acceleration is $a = 6.00g$, which when added to the $1.00g$ required to counteract gravity will mean Lois is subjected to $7.00g$ total. (A person standing motionless on the ground experiences 1 *g*, and any upward acceleration means additional *g*'s.)

SKETCH:

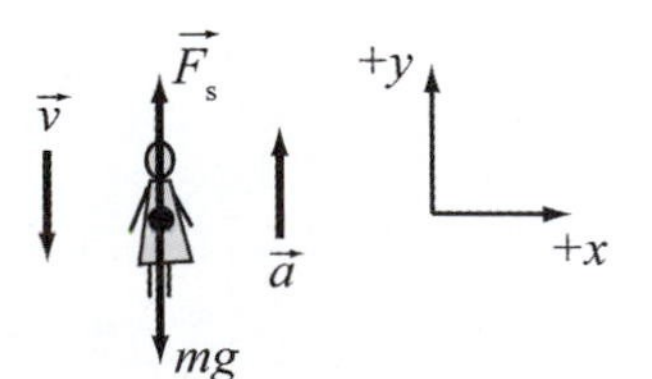

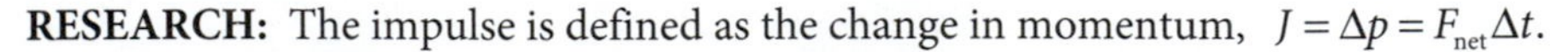

RESEARCH: The impulse is defined as the change in momentum, $J = \Delta p = F_{\text{net}}\Delta t$.

SIMPLIFY: Applying Newton's second law and assuming the force exerted is in the positive y-direction, $\sum F_y = ma_y$.

$$F_{\text{net}} = F_s - mg = ma = \frac{\Delta p}{\Delta t} \Rightarrow F_s = mg + \frac{\Delta p}{\Delta t} = mg + \frac{m(v_f - v_i)}{\Delta t}. \text{ Since } v_f = 0, \ \Delta t = 0.75 \text{ s.}$$

CALCULATE: $v_i = -60.0$ m/s (Note the negative sign as v is in the negative y-direction),

$F_s = (50.0 \text{ kg})(9.81 \text{ m/s}^2) - \dfrac{(50.0 \text{ kg})(-60.0 \text{ m/s})}{0.100 \text{ s}} = 30{,}490.5 \text{ N}, \quad a = 6.00g \Rightarrow F_{\text{net}} = ma = m(6.00g),$

and

$F_{\text{net}}\Delta t = \Delta p = m(v_f - v_i), \ v_f = 0 \Rightarrow \Delta t = \dfrac{-mv_i}{m(6.00g)} = \dfrac{-v_i}{6.00g} = \dfrac{60.0 \text{ m/s}}{6.00(9.81 \text{ m/s}^2)} = 1.0194 \text{ s.}$

ROUND: $F_s = 30{,}500$ N and $\Delta t = 1.02$ s.

DOUBLE-CHECK: The minimal time $\Delta t = 1.02$ s is reasonable.

7.33. **THINK:** The momentum of a photon is given to be $1.30 \cdot 10^{-27}$ kg m/s. The number of photons incident on a surface is $\rho = 3.84 \cdot 10^{21}$ photons per square meter per second. A spaceship has mass $m = 1000.$ kg and a square sail 20.0 m wide.

SKETCH:

RESEARCH: Using impulse, $\vec{J} = F\Delta t = \Delta p = p_f - p_i$. Also, $v = at$.

SIMPLIFY: In $\Delta t = 1$ s, the number of protons incident on the sail is $N = \rho A\Delta t$. The change in momentum in Δt is $\Delta p = N(p_f - p_i) \Rightarrow \Delta p = \rho A\Delta t(p_f - p_i)$. Using $p_f = -p_i$, $F\Delta t = \Delta p = \rho A\Delta t(-p_i - p_i) \Rightarrow F = -2\rho A p_i$.

The actual force on the sail is $F_s = -F = 2\rho A p_i$, so the acceleration is:

$$a = \frac{F_s}{m_s} = \frac{2\rho A p_i}{m_s}.$$

CALCULATE: $t_{\text{hour}} = (1 \text{ hr})\left(\dfrac{3600 \text{ s}}{1 \text{ hr}}\right) = 3600 \text{ s},$

$t_{\text{week}} = (1 \text{ week})\left(\dfrac{24 \text{ hours}}{1 \text{ day}}\right)\left(\dfrac{7 \text{ days}}{1 \text{ week}}\right)\left(\dfrac{3600 \text{ s}}{1 \text{ hr}}\right) = 6.048 \cdot 10^5 \text{ s},$

$t_{\text{month}} = (1 \text{ month})\left(\dfrac{3600 \text{ s}}{1 \text{ hour}}\right)\left(\dfrac{24 \text{ hours}}{1 \text{ day}}\right)\left(\dfrac{365 \text{ days}}{1 \text{ year}}\right)\left(\dfrac{1/12 \text{ year}}{1 \text{ month}}\right) = 2.628 \cdot 10^6 \text{ s},$

$a = \dfrac{2\rho A p_i}{m_s} = \dfrac{2(3.84 \cdot 10^{21} \ /(\text{m}^2 \text{ s}))(20.0 \text{ m} \cdot 20.0 \text{ m})(1.30 \cdot 10^{-27} \text{ kg m/s})}{1000. \text{ kg}} = 3.994 \cdot 10^{-6} \text{ m/s}^2,$

$v_{\text{hour}} = (3.994 \cdot 10^{-6} \text{ m/s}^2)(3600 \text{ s}) = 0.0144 \text{ m/s}, \quad v_{\text{week}} = (3.994 \cdot 10^{-6} \text{ m/s}^2)(6.048 \cdot 10^5 \text{ s}) = 2.416 \text{ m/s},$

$v_{\text{month}} = (3.994 \cdot 10^{-6} \text{ m/s}^2)(2.628 \cdot 10^6 \text{ s}) = 10.496 \text{ m/s},$

$t = \dfrac{8000. \text{ m/s}}{3.994 \cdot 10^{-6} \text{ m/s}^2} = 2.003 \cdot 10^9 \text{ s} = 762.2 \text{ months.}$

ROUND: $v_{\text{hour}} = 0.0144$ m/s, $v_{\text{week}} = 2.42$ m/s, $v_{\text{month}} = 10.5$ m/s, and $t = 762$ months.

DOUBLE-CHECK: The answer for velocities and time are understandable since the acceleration is very small.

7.37. **THINK:** A projectile with mass 7502 kg is fired at an angle of $20.0°$. The total mass of the gun, mount and train car is $1.22 \cdot 10^6$ kg. The speed of the railway gun is initially zero and $v = 4.68$ m/s after finishing. I want to calculate the initial speed of the projectile and the distance it travels.

SKETCH:

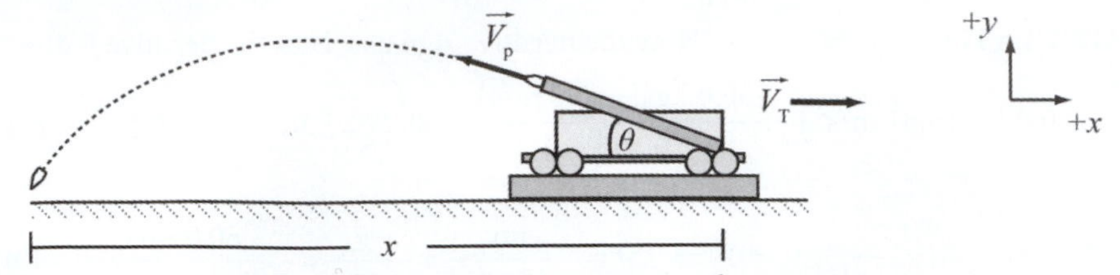

RESEARCH: Use the conservation of momentum. $p_{xi} = p_{xf}$ and $p_{xi} = 0$, so $p_{xf} = 0$.

SIMPLIFY: $m_p v_p \cos\theta + m_T v_T = 0 \Rightarrow v_p = -\dfrac{m_T v_T}{m_p \cos\theta}$

$x = v_{px} t$, where t is twice the time it takes to reach the maximum height. $t_0 = v_{py}/g$, and $t = 2t_0$.

$$x = v_{px}(2t_0) = v_{px}\left(\frac{2v_{py}}{g}\right) = \frac{2v_p^2 \sin\theta\cos\theta}{g} = \frac{v_p^2 \sin 2\theta}{g}$$

CALCULATE: $v_p = -\dfrac{(1.22 \cdot 10^6 \text{ kg})(4.68 \text{ m/s})}{(7502 \text{ kg})\cos 20.0°} = -809.9$ m/s

$$x = \frac{(-809.9 \text{ m/s})^2 \sin(2 \cdot 20.0°)}{9.81 \text{ m/s}^2} = 42979 \text{ m}$$

ROUND: $v_p = -810.$ m/s and $x = 43.0$ km.

DOUBLE-CHECK: The documented muzzle velocity for Gustav was 820 m/s, and its maximum range was approximately 48 km.

7.41. **THINK:** The astronaut's mass is $m_A = 50.0$ kg and the baseball's mass is $m_b = 0.140$ kg. The baseball has an initial speed of 35.0 m/s and a final speed of 45.0 m/s.

SKETCH:

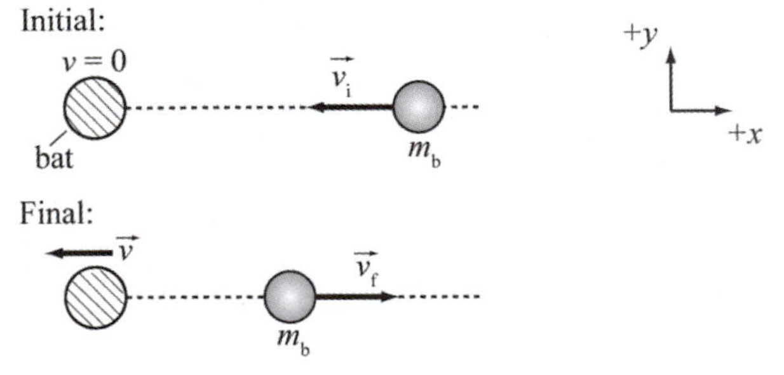

RESEARCH: Use the conservation of momentum. $p_i = p_f$.

SIMPLIFY: $p_i = p_f \Rightarrow m_b v_i + 0 = m_b v_f + m_A v_A \Rightarrow v_A = \dfrac{m_b(v_i - v_f)}{m_A}$

CALCULATE: $v_i = -35.0$ m/s, $v_f = 45.0$ m/s, and $v_A = \dfrac{(0.140 \text{ kg})(-35.0 \text{ m/s} - 45.0 \text{ m/s})}{50.0 \text{ kg}} = -0.224$ m/s.

ROUND: Three significant figures: $v_A = -0.224$ m/s.

DOUBLE-CHECK: The magnitude of v_A is proportional to m_b/m_A, which is about 10^{-3} so it would be expected to find the velocity of the astronaut as relatively small.

7.45. **THINK:** A soccer ball and a basketball have masses $m_1 = 0.400$ kg and $m_2 = 0.600$ kg respectively. The soccer ball has an initial energy of 100. J and the basketball 112 J. After collision, the second ball flew off at an angle of 32.0° with 95.0 J of energy. I need to calculate the speed and angle of the first ball. Let subscript 1 denote the soccer ball, and subscript 2 denote the basketball.

SKETCH:

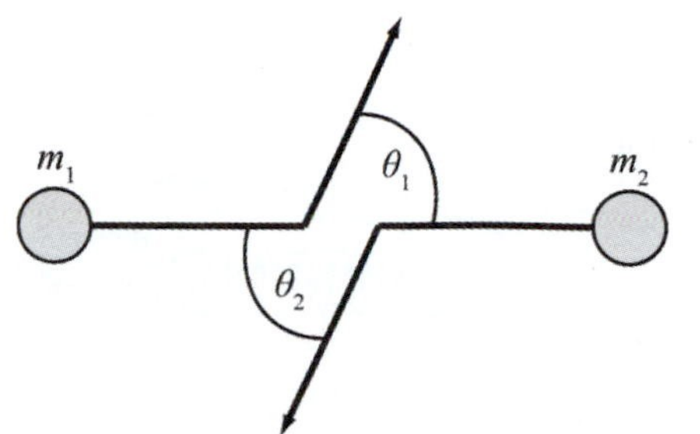

RESEARCH: I need to calculate the speed of the balls using $1/2mv^2 = K$, or $v = \sqrt{2K/m}$, and then apply the conservation of momentum to get $p_{xi} = p_{xf}$ and $p_{yi} = p_{yf}$. I also use $p_{yi} = 0$.

SIMPLIFY: $p_{xi} = p_{xf} \Rightarrow m_1 v_{1i} - m_2 v_{2i} = m_1 v_{1f}\cos\theta_1 - m_2 v_{2f}\cos\theta_2$,

$p_{yi} = p_{yf} = 0 \Rightarrow m_1 v_{1f}\sin\theta_1 - m_2 v_{2f}\sin\theta_2 = 0$,

$v_{1x} = v_{1f}\cos\theta_1$ and $v_{1y} = v_{1f}\sin\theta_1 \Rightarrow v_{1x} = \dfrac{m_1 v_{1i} - m_2 v_{2i} + m_2 v_{2f}\cos\theta_2}{m_1}$ and $v_{1y} = \dfrac{m_2 v_{2f}\sin\theta_2}{m_1}$,

$v_{1i} = \sqrt{\dfrac{2K_{1i}}{m_1}}$, $v_{2i} = \sqrt{\dfrac{2K_{2i}}{m_2}}$, $v_{2f} = \sqrt{\dfrac{2K_{2f}}{m_2}}$, and $v_{1f} = \sqrt{v_{1x}^2 + v_{1y}^2}$.

CALCULATE: $v_{1i} = \sqrt{\dfrac{2(100.\text{ J})}{0.400\text{ kg}}} = 22.36$ m/s, $v_{2i} = \sqrt{\dfrac{2(112\text{ J})}{0.600\text{ kg}}} = 19.32$ m/s, $v_{2f} = \sqrt{\dfrac{2(95.0\text{ J})}{0.600\text{ kg}}} = 17.80$ m/s,

$$v_{1x} = \frac{(0.400\text{ kg})(22.36\text{ m/s}) - (0.600\text{ kg})(19.32\text{ m/s}) + (0.600\text{ kg})(17.80\text{ m/s})\cos 32.0^\circ}{0.400\text{ kg}} = 16.02\text{ m/s},$$

$v_{1y} = \dfrac{(0.600\text{ kg})(17.80\text{ m/s})\sin 32.0^\circ}{0.400\text{ kg}} = 14.15$ m/s, $v_{1f} = \sqrt{(16.02\text{ m/s})^2 + (14.15\text{ m/s})^2} = 21.37$ m/s, and

$\theta_1 = \tan^{-1}\left(\dfrac{14.15\text{ m/s}}{16.02\text{ m/s}}\right) = 41.5^\circ$.

ROUND: $v_{1f} = 21.4$ m/s and $\theta_1 = 41.5^\circ$.

DOUBLE-CHECK: The results for speed and angle are comparable to v_2 and θ_2, which is expected. From energy conservation (assuming elastic collision), the energy is $E_{1f} = E_{1i} + E_{2i} - E_{2f}$ = 100. J + 112 J – 95.0 J = 117 J, which corresponds to a speed of 24.2 m/s for v_{1f}. The result $v_{1f} = 21.4$ m/s is less than this because the energy is not conserved in this case.

7.49. **THINK:** Two blocks with a spring between them sit on an essentially frictionless surface. The spring constant is k = 2500. N/m. The spring is compressed such that $\Delta x = 3.00\text{ cm} = 3.00\cdot 10^{-2}$ m. I need to calculate the speeds of the two blocks. $m_A = 1.00$ kg, and $m_B = 3.00$ kg.

SKETCH:

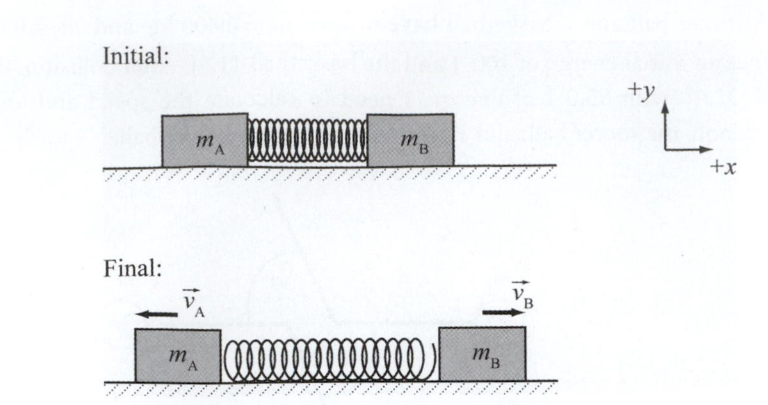

RESEARCH: I use the conservation of momentum and the conservation of energy. Thus $p_i = p_f$, and $E_i = E_f$. I also know that $p_i = 0$ and $E_i = E_s = (1/2)k\Delta x^2$.

SIMPLIFY: $p_i = p_f = 0 \Rightarrow m_A v_A + m_B v_B = 0 \Rightarrow m_A v_A = -m_B v_B$

$$E_i = E_f \Rightarrow \frac{1}{2}k\Delta x^2 = \frac{1}{2}m_A v_A^2 + \frac{1}{2}m_B v_B^2 \Rightarrow k\Delta x^2 = m_A\left(-\frac{m_B}{m_A}v_B\right)^2 + m_B v_B^2 \Rightarrow k\Delta x^2 = \frac{m_B^2}{m_A}v_B^2 + m_B v_B^2$$

Simplifying further gives:

$$\left(\frac{m_B^2}{m_A} + m_B\right)v_B^2 = k\Delta x^2 \Rightarrow v_B = \sqrt{\frac{k\Delta x^2}{\frac{m_B^2}{m_A} + m_B}} = \sqrt{\frac{k\Delta x^2}{m_B\left(1 + \frac{m_B}{m_A}\right)}} \quad \text{and} \quad v_A = -\frac{m_B}{m_A}v_B.$$

CALCULATE: $v_B = \sqrt{\frac{(2500.\text{ N/m})(3.00 \cdot 10^{-2}\text{ m})^2}{(3.00\text{ kg})\left(1 + \frac{3.00\text{ kg}}{1.00\text{ kg}}\right)}} = 0.4330\text{ m/s}$

$v_A = -\frac{3.00\text{ kg}}{1.00\text{ kg}}(0.4330\text{ m/s}) = -1.299\text{ m/s}$

ROUND: $v_B = 0.433\text{ m/s}$ and $v_A = -1.30\text{ m/s}$.

DOUBLE-CHECK: The speed of block A should be larger than the speed of block B since m_A is less than m_B.

7.53. **THINK:** A particle has an initial velocity $v_i = -2.21 \cdot 10^3\text{ m/s}$. I want to calculate the speed after 6 collisions with the left wall (which has a speed of $v_1 = 1.01 \cdot 10^3\text{ m/s}$) and 5 collisions with the right wall (which has a speed of $v_2 = -2.51 \cdot 10^3\text{ m/s}$). The magnetic walls can be treated as walls of mass M.

SKETCH:

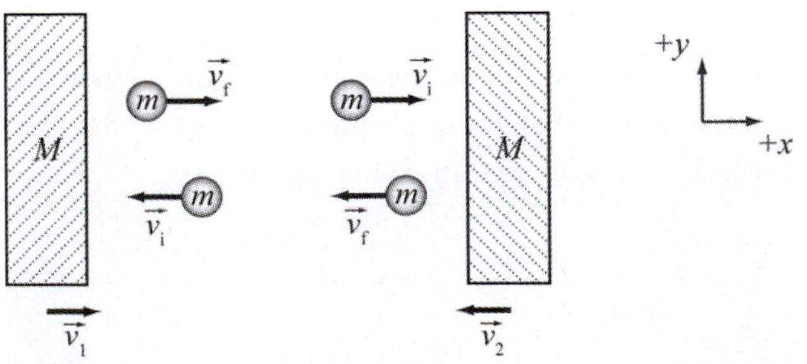

RESEARCH: Consider one wall with speed v_W. Using the conservation of momentum and energy, $p_i = p_f$ and $E_i = E_f$.

SIMPLIFY: $p_i = p_f \Rightarrow mv_i + Mv_{Wi} = mv_f + Mv_{Wf}$

$m(v_i - v_f) = M(v_{Wf} - v_{Wi})$ (1)

$$E_i = E_f$$

$$\frac{1}{2}mv_i^2 + \frac{1}{2}Mv_{Wi}^2 = \frac{1}{2}mv_f^2 + \frac{1}{2}Mv_{Wf}^2$$

$$m(v_i^2 - v_f^2) = M(v_{Wf}^2 - v_{Wi}^2)$$

$$m(v_i - v_f)(v_i + v_f) = M(v_{Wf} - v_{Wi})(v_{Wf} + v_{Wi})$$

$$v_i + v_f = v_{Wf} + v_{Wi}$$

$$v_{Wf} = v_i + v_f - v_{Wi}$$

Substituting back into (1):

$$mv_i + Mv_{Wi} = mv_f + M(v_i + v_f - v_{Wi}) \Rightarrow v_f = \frac{m-M}{m+M}v_i + \frac{2M}{m+M}v_{Wi}$$

If $m \ll M$ then $K_{Sf} = 121$ J.. This means that every collision results in an additional speed of $2v_W$. So after 6 collisions with the left wall and 5 collisions with the right wall, I get $v_f = -v_i + 6(2v_1) - 5(2v_2)$.

CALCULATE: $v_i = -2.21 \cdot 10^3$ m/s, $v_1 = 1.01 \cdot 10^3$ m/s, and $v_2 = -2.51 \cdot 10^3$ m/s.

$v_f = 2.21 \cdot 10^3 \text{ m/s} + 12(1.01 \cdot 10^3 \text{ m/s}) - 10(-2.51 \cdot 10^3 \text{ m/s}) = 3.943 \cdot 10^4 \text{ m/s}$

Since the last collision is with the left wall, the particle is moving to the right and the velocity is positive.

ROUND: $v_f = 3.94 \cdot 10^4$ m/s

DOUBLE-CHECK: Since there have been 11 collisions, it is expected that the resulting speed is about 10 times the original speed.

7.57. **THINK:** I want to find the final velocity of the molecules after they collide elastically. The first molecule has a speed of $v_1 = 672$ m/s at an angle of 30.0° along the positive horizontal. The second has a speed of 246 m/s in the negative horizontal direction. After the collision, the first particle travels vertically. For simplicity, we ignore rotational effects and treat the molecules as simple spherical masses.

SKETCH:

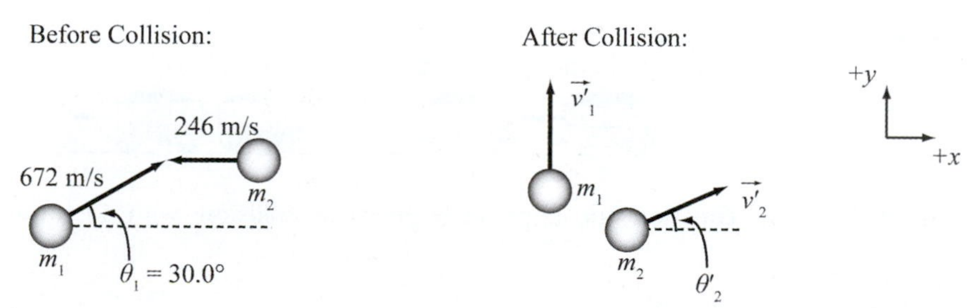

RESEARCH: Since this is an elastic collision, there is conservation of momentum in the x and y components and conservation of energy. $p_{1ix} + p_{2ix} = p_{1fx} + p_{2fx}$, $p_{1iy} + p_{2iy} = p_{1fy} + p_{2fy}$ and $K_i = K_f$.

SIMPLIFY: In the x-direction, the momentum equation gives:

$$mv_{1i}\cos\theta_1 - mv_{2i} = mv_{2f}\cos\theta_2 \Rightarrow v_{1i}\cos\theta_1 - v_{2i} = v_{2fx} \quad (1)$$

The y-component of the momentum gives:

$$mv_{1i}\sin\theta_1 = mv_{1f} + mv_2\sin\theta_2 \Rightarrow v_{1i}\sin\theta_1 = v_{1f} + v_{2fy} \Rightarrow v_{1i}\sin\theta_1 - v_{1f} = v_{2fy} \quad (2)$$

The kinetic energy gives:

$$\frac{1}{2}mv_{1i}^2 + \frac{1}{2}mv_{2i}^2 = \frac{1}{2}mv_{1f}^2 + \frac{1}{2}mv_{2f}^2 \Rightarrow v_{1i}^2 + v_{2i}^2 = v_{1f}^2 + v_{2fx}^2 + v_{2fy}^2 \quad (3)$$

Squaring and adding equations (1) and (2),

$$v^2_{2fx} + v^2_{2fy} = (v_{1i}\cos\theta_1 - v_{2i})^2 + (v_{1i}\sin\theta_1 - v_{1f})^2 \quad (4)$$

Substituting (4) into equation (3),

$$v_{1i}^2 + v_{2i}^2 = v_{1f}^2 + \left(v_{1i}\cos\theta_1 - v_{2i}\right)^2 + \left(v_{1i}\sin\theta_1 - v_{1f}\right)^2$$
$$\Rightarrow v_{1i}^2 + v_{2i}^2 = v_{1f}^2 + \left({v^2}_{1i}\cos^2\theta_1 - 2v_{1i}v_{2i}\cos\theta_1 + {v^2}_{2i}\right) + \left({v^2}_{1i}\sin^2\theta_1 - 2v_{1i}v_{1f}\sin\theta_1 + {v^2}_{1f}\right)$$
$$\Rightarrow v_{1i}^2 + v_{2i}^2 = 2v_{1f}^2 + {v^2}_{1i} + {v^2}_{2i} - 2v_{1i}v_{2i}\cos\theta_1 - 2v_{1i}v_{1f}\sin\theta_1$$
$$\Rightarrow v_{1f}^2 - v_{1i}v_{1f}\sin\theta_1 - v_{1i}v_{2i}\cos\theta_1 = 0$$

There is only one unknown in this equation, so I can solve for v_{1f}.

$$v_{1f} = \frac{v_{1i}\sin\theta_1 \pm \sqrt{\left(-v_{1i}\sin\theta_1\right)^2 + 4v_{1i}v_{2i}\cos\theta_1}}{2}$$

I can solve for v_{2fy} in terms of v_{1f} using $v_{2fy} = v_{1i}\sin\theta_1 - v_{1f}$. The angle $\theta_2' = \arctan\left(\dfrac{v_{2fy}}{v_{2fx}}\right)$.

CALCULATE: $v_{1f} = \dfrac{672 \text{ m/s}\cdot\sin 30.0° \pm \sqrt{\left(-672 \text{ m/s}\cdot\sin 30.0°\right)^2 + 4\cdot 672 \text{ m/s}\cdot 246 \text{ m/s}\cdot\cos 30.0°}}{2}$

$= 581.9908 \text{ m/s}$ or -245.9908 m/s

Since I know that the molecule travels in the positive y direction, $v_{1f} = 581.9908 \text{ m/s}$.

$$v_{2fy} = v_{1i}\sin\theta_1 - v_{1f} = 672 \text{ m/s}\cdot\sin 30° - 581.9908 \text{ m/s} = -245.9908 \text{ m/s}$$
$$v_{2fx} = v_{1i}\cos\theta_1 - v_{2i} = 672 \text{ m/s}\cdot\cos 30° - 246 \text{ m/s} = 335.9691 \text{ m/s}$$

Therefore, $v_{2f} = \sqrt{\left(335.9691 \text{ m/s}\right)^2 + \left(-245.9908 \text{ m/s}\right)^2} = 416.3973 \text{ m/s}$ at an angle of

$\theta_2' = \arctan\left(\dfrac{-245.9908 \text{ m/s}}{335.9691 \text{ m/s}}\right) = -36.211°$.

ROUND: $v_{1f} = 582 \text{ m/s}$ in the positive y-direction and $v_{2f} = 416 \text{ m/s}$ at an angle of $36.2°$ below the positive x-axis.

DOUBLE-CHECK: The results show $v_{1f} < v_1$ and $v_{2f} > v_{2i}$ as expected, so the answers look reasonable.

7.61. **THINK:** I want to know what the speed of the railroad car is after a perfectly inelastic collision occurs. Knowing $m_1 = m_2 = 1439 \text{ kg}$, $v_1 = 12.0 \text{ m/s}$ and $v_2 = 0 \text{ m/s}$.

SKETCH:

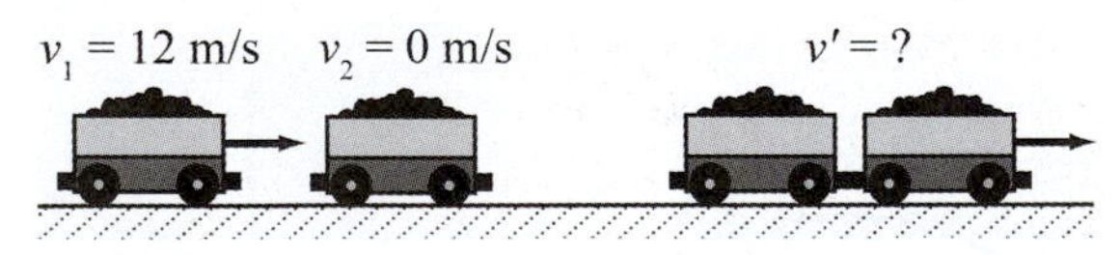

RESEARCH: The equation for a perfectly inelastic collision with identical masses is given by $v_{1i} + v_{2i} = 2v_f$.

SIMPLIFY: $v_f = \left(v_{1i} + v_{2i}\right)/2$

CALCULATE: $v_f = \left(12.0 \text{ m/s} + 0 \text{ m/s}\right)/2 = 6.00 \text{ m/s}$

ROUND: Because the velocity before the collision is given to three significant figures, keep the result to three significant figures. The velocity of the cars after the collision is 6.00 m/s.

DOUBLE-CHECK: This is equivalent to a speed of 22 km/h, which is reasonable for railroad cars.

7.65. **THINK:** The Volkswagen of mass $m_V = 1000.$ kg was going eastward before the collision and the Cadillac had mass $m_C = 2000.$ kg and velocity $v_C = 30.0 \text{ m/s}$ northward, and after the collision both cars stuck together travelling $\theta = 55.0°$ north of east.

SKETCH:

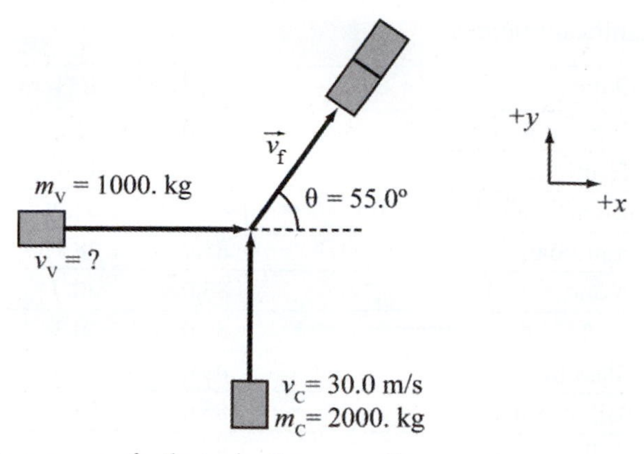

RESEARCH: The collision was perfectly inelastic so use the equation $m_C v_C + m_V v_V = (m_C + m_V) v_f$ for each component of the motion.

SIMPLIFY: In the east-west direction:

$$m_V v_V = (m_C + m_V) v_f \cos\theta \Rightarrow v_f = \frac{m_V v_V}{m_C + m_V} \frac{1}{\cos\theta},$$

and in the north-south direction:

$$m_C v_C = (m_C + m_V) v_f \sin\theta \Rightarrow v_f = \frac{m_C v_C}{m_C + m_V} \frac{1}{\sin\theta}$$

Equating these two expressions for the final velocity gives:

$$\frac{m_V v_V}{m_C + m_V} \frac{1}{\cos\theta} = \frac{m_C v_C}{m_C + m_V} \frac{1}{\sin\theta} \Rightarrow v_V = \frac{m_C}{m_V} v_C \frac{\cos\theta}{\sin\theta} = \frac{m_C}{m_V} v_C \cot\theta.$$

CALCULATE: $v_V = \frac{2000.\text{ kg}}{1000.\text{ kg}}(30.0\text{ m/s})\cot 55.0° = 42.01245\text{ m/s}$

ROUND: The Volkswagen's velocity is 42.0 m/s.

DOUBLE-CHECK: This is a reasonable result. It's in the same order as the Cadillac.

7.69. **THINK:** I want to find the coefficient of restitution for a variety of balls.

SKETCH:

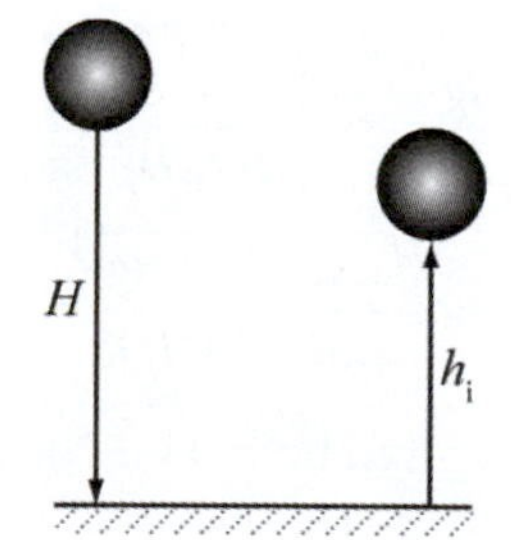

RESEARCH: Using the equation for the coefficient of restitution for heights. $\varepsilon = \sqrt{H/h_1}$.

SIMPLIFY: There is no need to simplify.

CALCULATE: An example calculation: A range golf ball has an initial height $H = 85.0$ cm and a final height $h_1 = 62.6$ cm.

$$\varepsilon = \sqrt{\frac{62.6\text{ cm}}{85.0\text{ cm}}} = 0.85818$$

ROUND: All of the coefficients of restitution will be given to 3 significant figures because all the heights are given to 3 significant figures.

Object	H [cm]	h_1 [cm]	ε
Range golf ball	85.0	62.6	0.858
Tennis ball	85.0	43.1	0.712
Billiard ball	85.0	54.9	0.804
Hand ball	85.0	48.1	0.752
Wooden ball	85.0	30.9	0.603
Steel ball bearing	85.0	30.3	0.597
Glass marble	85.0	36.8	0.658
Ball of rubber bands	85.0	58.3	0.828
Hollow, hard plastic balls	85.0	40.2	0.688

DOUBLE-CHECK: All these values are less than one, which is reasonable.

7.73. **THINK:** I want to know if Jerry will make it over the second fence. Each yard begins and ends with a 2.00 m fence. The range and maximum height of Jerry's initial trajectory are 15.0 m and 5.00 m respectively. Jerry is 7.50 m away from the next fence and he has a coefficient of restitution of 0.80.
SKETCH:

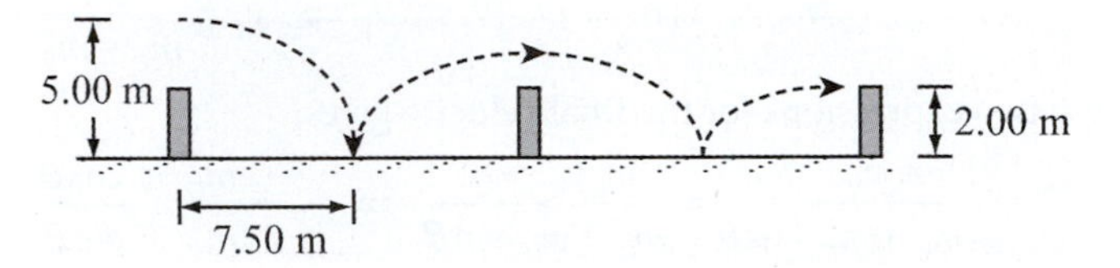

RESEARCH: From the range and maximum height the initial velocity can be found, along with the angle of Jerry's trajectory. $R=\left(v_0^2/g\right)\sin(2\theta)$ and $H=\left(v_0^2/(2g)\right)\sin^2\theta$. With this I can find the x- and y- components of the velocity. Since the coefficient of restitution only acts on the momentum perpendicular to the ground, $v_{yf}=v_{yi}$ and v_x remains constant. With this information the height Jerry attains after travelling another 7.50 m can be found by using $x=v_xt$ and $y=v_yt-\frac{1}{2}gt^2$.

SIMPLIFY: $v_0^2=\dfrac{Rg}{\sin(2\theta)}=\dfrac{2Hg}{\sin^2\theta}\ \Rightarrow\ \dfrac{R}{2\cos\theta}=\dfrac{2H}{\sin\theta}\ \Rightarrow\ \tan\theta=\dfrac{4H}{R}$, $\quad v_0=\sqrt{\dfrac{Rg}{\sin 2\theta}}=\sqrt{\dfrac{2Hg}{\sin^2\theta}}$

$v_{xi}=v_{xf}=v_0\cos\theta=\sqrt{\dfrac{2Hg}{\sin^2\theta}}\cos\theta=\dfrac{\sqrt{2Hg}}{\tan\theta}=\dfrac{\sqrt{2Hg}}{4H/R}=\dfrac{R\sqrt{2Hg}}{4H}$, and $v_{yf}=\varepsilon v_{yi}=\varepsilon v_0\sin\theta=\varepsilon\sqrt{2Hg}$. The time it takes to reach the fence is given by $x=v_xt$, or $t=x/v_x=(4Hx)/\left(R\sqrt{2Hg}\right)$, where x = 7.5 m. The height it attains in this time is:

$$y=v_yt-\frac{1}{2}gt^2=\varepsilon\sqrt{2Hg}\frac{4Hx}{R\sqrt{2Hg}}-\frac{1}{2}g\left(\frac{4Hx}{R\sqrt{2Hg}}\right)^2=\frac{4Hx\varepsilon}{R}-\frac{4Hx^2}{R^2}$$

CALCULATE: $y=\dfrac{4(5\text{ m})(7.50\text{ m})(0.800)}{15.0\text{ m}}-\dfrac{4(5\text{ m})(7.50\text{ m})^2}{(15.0\text{ m})^2}=3\text{ m}$

ROUND: Jerry is at a height of 3 m when he reaches the fence, which means that he does make it over the next fence, with exactly 1 meter to spare.
DOUBLE-CHECK: This is a reasonable answer for the world of cartoon characters.

7.77. **THINK:** I want to find the speed of the tailback and linebacker and if the tailback will score a touchdown. The tailback has mass and velocity $m_t = 85.0$ kg and $v_t = 8.90$ m/s, and the linebacker has mass and velocity $m_l = 110.$ kg and $v_l = -8.00$ m/s.

SKETCH:

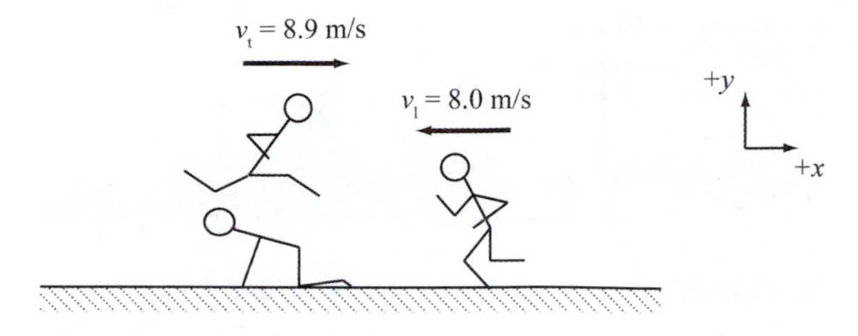

RESEARCH: The conservation of momentum for this perfectly inelastic collision is

$$m_t v_t + m_l v_l = (m_t + m_l)v.$$

SIMPLIFY: Rearranging the equation to solve for the final velocity, $v = \dfrac{m_t v_t + m_l v_l}{m_t + m_l}$.

CALCULATE: $v = \dfrac{(85.0 \text{ kg})(8.90 \text{ m/s}) - (110. \text{ kg})(8.00 \text{ m/s})}{85.0 \text{ kg} + 110. \text{ kg}} = -0.633$ m/s

ROUND:

(a) The values are given to 3 significant figures so the final speed is 0.633 m/s.

(b) Since the velocity is negative, the two go in the direction of the linebacker and the tailback does not score a touchdown.

DOUBLE-CHECK: The speed is quite small, as would be expected of two people opposing each other's motion. Since the initial momentum of the linebacker is greater than the initial momentum of the tailback, the tailback should not be able to score. This is consistent with the calculated result.

7.81. **THINK:** I am looking for the velocity of the nucleus after the decay. The atom starts at rest, i.e. $v_0 = 0$ m/s, and its nucleus has mass $m_0 = 3.68 \cdot 10^{-25}$ kg. The alpha particle has mass $m_a = 6.64 \cdot 10^{-27}$ kg and energy $8.79 \cdot 10^{-13}$ J.

SKETCH:

Before: After:

$\vec{p}_0$ $\vec{p}_a$ +y +x

RESEARCH: I can find the velocity of the alpha particle with the equation $K = \frac{1}{2} m_a v_a^2$. The conservation of momentum gives $p_a = p_0$.

SIMPLIFY: $v_a = \sqrt{\dfrac{2K}{m_a}}$, $p_a = p_0 \Rightarrow m_a v_a = (m_0 - m_a)v_0 \Rightarrow v_0 = \dfrac{m_a}{(m_0 - m_a)} v_a = \dfrac{m_a}{(m_0 - m_a)} \sqrt{\dfrac{2K}{m_a}}$

CALCULATE: $v_0 = \dfrac{6.64 \cdot 10^{-27} \text{ kg}}{(3.68 \cdot 10^{-25} \text{ kg} - 6.64 \cdot 10^{-27} \text{ kg})} \sqrt{\dfrac{2(8.79 \cdot 10^{-13} \text{ J})}{6.64 \cdot 10^{-27} \text{ kg}}} = 298988$ m/s

ROUND: The values are given to three significant figures, so $v_0 = 2.99 \cdot 10^5$ m/s.

DOUBLE-CHECK: Such a high speed is reasonable for such small masses.

7.85. **THINK:** I want to find the average force exerted on the jumper and the number of *g*'s she experiences. She has a mass of $m_{\text{j}} = 55.0 \text{ kg}$ and reaches a speed of $v_{\text{i}} = 13.3 \text{ m/s}$ downwards then goes $v_{\text{f}} = 10.5 \text{ m/s}$ upwards after the cord pulls her back up in $\Delta t = 1.25$ s.

SKETCH:

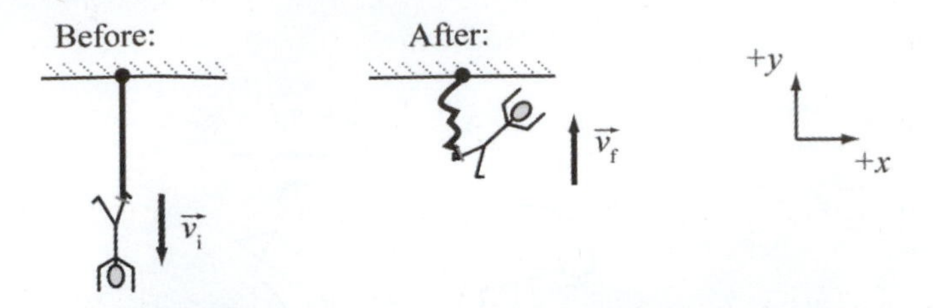

RESEARCH: I use the impulse equation, $F\Delta t = \Delta p$, to find the net force acting on the jumper. I can then use $F = ma$ to find the net force (cord pulling up plus gravity pulling down) and then the number of *g*'s experienced. Number of *g*'s is determined by the action of forces *other* than gravity, so in this case the cord tension. (A person standing motionless on the ground experiences 1 *g* from the upward normal force.)

SIMPLIFY: $F = \dfrac{\Delta p}{\Delta t} = \dfrac{m_{\text{j}}(v_{\text{f}} - v_{\text{i}})}{\Delta t}$ and $a = \dfrac{F}{m_{\text{j}}}$.

CALCULATE: $F_{\text{net}} = \dfrac{(55.0 \text{ kg})(10.5 \text{ m/s} - (-13.3 \text{ m/s}))}{1.25 \text{ s}} = 1047.2 \text{ N}$

$F_{\text{net}} = F_{\text{cord}} - mg \Rightarrow F_{\text{cord}} = F_{\text{net}} + mg = 1047.2 \text{ N} + (55.0 \text{ kg})(9.81 \text{ m/s}^2) = 1586.75 \text{ N}$

Acceleration due to cord: $a = \dfrac{F_{\text{cord}}}{m} = \dfrac{1586.75 \text{ N}}{55.0 \text{ kg}} = 28.85 \text{ m/s}^2$.

Dividing 28.85 by 9.81, the cord subjects the number to 2.9408 *g*'s.

ROUND: The values are given to three significant figures, so the average force is 1590 N and the jumper experiences 2.94 *g*'s.

DOUBLE-CHECK: These numbers are within reasonable levels. A person can experience a few *g*'s without harm and without losing consciousness.

7.89. **THINK:** I have three birds with masses $m_1 = 0.100 \text{ kg}$, $m_2 = 0.123 \text{ kg}$, and $m_3 = 0.112 \text{ kg}$ and speeds $v_1 = 8.00 \text{ m/s}$, $v_2 = 11.0 \text{ m/s}$, and $v_3 = 10.0 \text{ m/s}$. They are flying in directions $\theta_1 = 35.0°$ east of north, $\theta_2 = 2.00°$ east of north, and $\theta_3 = 22.0°$ west of north, respectively. I want to calculate the net momentum.

SKETCH:

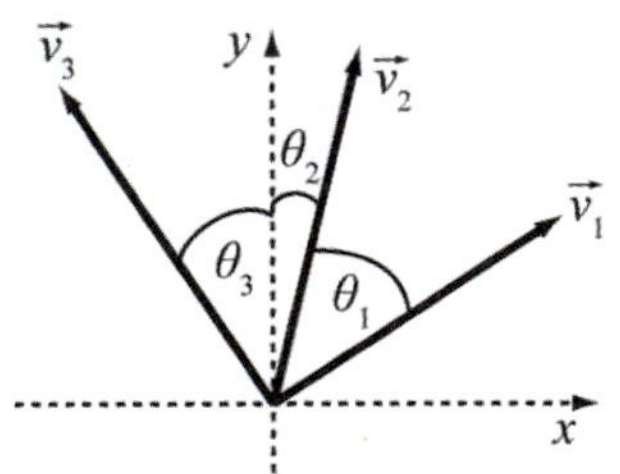

RESEARCH: $\vec{p} = \vec{p}_1 + \vec{p}_2 + \vec{p}_3$, or in component form, $p_x = p_{1x} + p_{2x} + p_{3x}$ and $p_y = p_{1y} + p_{2y} + p_{3y}$.

SIMPLIFY: $p_x = m_1 v_1 \sin\theta_1 + m_2 v_2 \sin\theta_2 - m_3 v_3 \sin\theta_3$, $p_y = m_1 v_1 \cos\theta_1 + m_2 v_2 \cos\theta_2 + m_3 v_3 \cos\theta_3$,

$\vec{p} = p_x \hat{x} + p_y \hat{y}$, and $\vec{v} = \dfrac{\vec{p}}{m}$.

CALCULATE:

$$p_x = (0.100 \text{ kg})(8.00 \text{ m/s})\sin 35.0° + (0.123 \text{ kg})(11.0 \text{ m/s})\sin 2.00° - (0.112 \text{ kg})(10.0 \text{ m/s})\sin 22.0° = 0.0865 \text{ kg m/s}$$

$$p_y = (0.100 \text{ kg})(8.00 \text{ m/s})\cos 35.0° + (0.123 \text{ kg})(11.0 \text{ m/s})\cos 2.00° + (0.112 \text{ kg})(10.0 \text{ m/s})\cos 22.0° = 3.0459 \text{ kg m/s}$$

The speed of a 0.115 kg bird is: $\vec{v} = \dfrac{0.0865 \text{ kg m/s } \hat{x} + 3.0459 \text{ kg m/s } \hat{y}}{0.115 \text{ kg}} = 0.752 \text{ m/s } \hat{x} + 26.486 \text{ m/s } \hat{y}.$

$$\left|\vec{v}\right| = \sqrt{(0.752 \text{ m/s})^2 + (26.486 \text{ m/s})^2} = 26.497 \text{ m/s},$$

$$\tan\theta = \frac{0.752 \text{ m/s}}{26.486 \text{ m/s}} = 0.02839 \Rightarrow \theta = \tan^{-1}(0.02839) = 1.626° \text{ east of north}$$

ROUND: $p_x = 0.0865$ kg m/s , $p_y = 3.05$ kg m/s , $\vec{p} = 0.0865 \text{ kg m/s } \hat{x} + 3.05 \text{ kg m/s } \hat{y}$, $\left|\vec{v}\right| = 26.5$ m/s , $\theta = 1.63°$ east of north

DOUBLE-CHECK: The speed of the fourth bird must be less than the sum of the speeds of the three birds. $v = v_1 + v_2 + v_3 = 8.00 \text{ m/s} + 11.0 \text{ m/s} + 10.0 \text{ m/s} = 29.0 \text{ m/s}$.

7.93. **THINK:** I have a rocket which at the top of the trajectory breaks into two equal pieces. One piece has half the speed of the rocket travelling upward. I want to calculate the speed and angle of the second piece.

SKETCH:

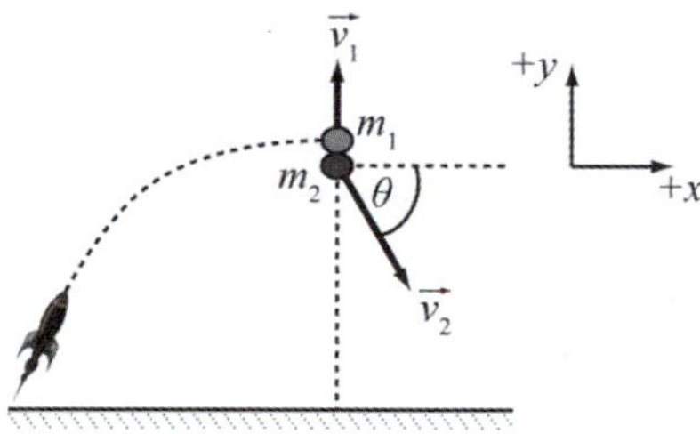

RESEARCH: Use the conservation of momentum. $\vec{p}_{\text{i}} = \vec{p}_{\text{f}}$, or in component form, $p_{\text{xi}} = p_{\text{xf}}$ and $p_{\text{yi}} = p_{\text{yf}}$. I also know that $p_{\text{yi}} = 0$. Let us assume that the speed of the rocket before it breaks is v_0 and mass m_0.

SIMPLIFY: $p_{\text{xi}} = p_{\text{xf}} \Rightarrow m_0 v_0 = m_1 v_{1x} + m_2 v_{2x}$. Since $v_{1x} = 0$ and $m_2 = \frac{1}{2}m_0$, $m_0 v_0 = \frac{1}{2}m_0 v_{2x}$ $\Rightarrow v_{2x} = 2v_0$. $p_{\text{yi}} = p_{\text{yf}} = 0 \Rightarrow 0 = m_1 v_{1y} + m_2 v_{2y}$. Since $m_1 = m_2 = \frac{1}{2}m_0$ and $v_{1y} = \frac{1}{2}v_0$, $v_{2y} = -v_{1y} \Rightarrow v_{2y} = -\frac{1}{2}v_0$. $v_2 = \sqrt{v_{2x}^2 + v_{2y}^2} = \sqrt{2^2 v_0^2 + (-1/2)^2 v_0^2}$; $\theta = \tan^{-1}\left(\dfrac{(-1/2)v_0}{2v_0}\right)$.

Drawing the vector $\vec{v}_2$:

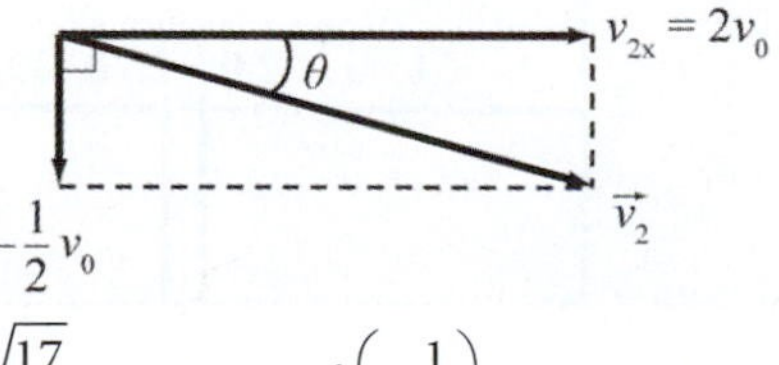

CALCULATE: $v_2 = v_0\sqrt{4 + 1/4} = \dfrac{\sqrt{17}}{2}v_0$, $\theta = \tan^{-1}\left(-\dfrac{1}{4}\right) = -14.04°$

ROUND: Rounding is not needed here.

DOUBLE-CHECK: It makes sense that θ is negative since the first piece is travelling upwards. The y component of v_2 must be in the negative y-direction.

7.97. **THINK:** I have a 170. g hockey puck with initial velocity $v_i = 30.0$ m/s and final velocity $v_f = -25.0$ m/s, changing over a time interval of $\Delta t = 0.200$ s.

SKETCH:

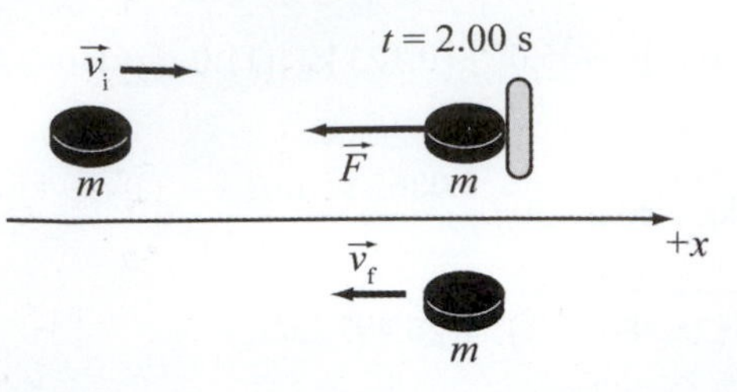

RESEARCH: The initial and final momentums are calculated by $p_i = mv_i$ and $p_f = mv_f$. The force is calculated using $J = F\Delta t = \Delta p = m(v_f - v_i)$.

SIMPLIFY: Simplification is not necessary.

CALCULATE: $p_i = (0.170 \text{ kg})(30.0 \text{ m/s}) = 5.10$ kg m/s, $p_f = (0.170 \text{ kg})(-25.0 \text{ m/s}) = -4.25$ kg m/s, and $F = \frac{p_f - p_i}{\Delta t} = \frac{(-4.25 \text{ kg m/s} - 5.10 \text{ kg m/s})}{0.200 \text{ s}} = -46.75$ N. The position of the puck at t = 2.00 s is: $x_2 = v_i t = (30.0 \text{ m/s})(2.00 \text{ s}) = 60.0$ m. The position of the puck at t = 5.00 s is:

$$x_5 = x_2 + v_f(5.00 \text{ s} - 2.20 \text{ s}) = 60.0 \text{ m} + (-25.0 \text{ m/s})(2.80 \text{ s}) = -10.0 \text{ m}.$$

With all this information I can plot p vs. t, x vs. t and F vs. t.

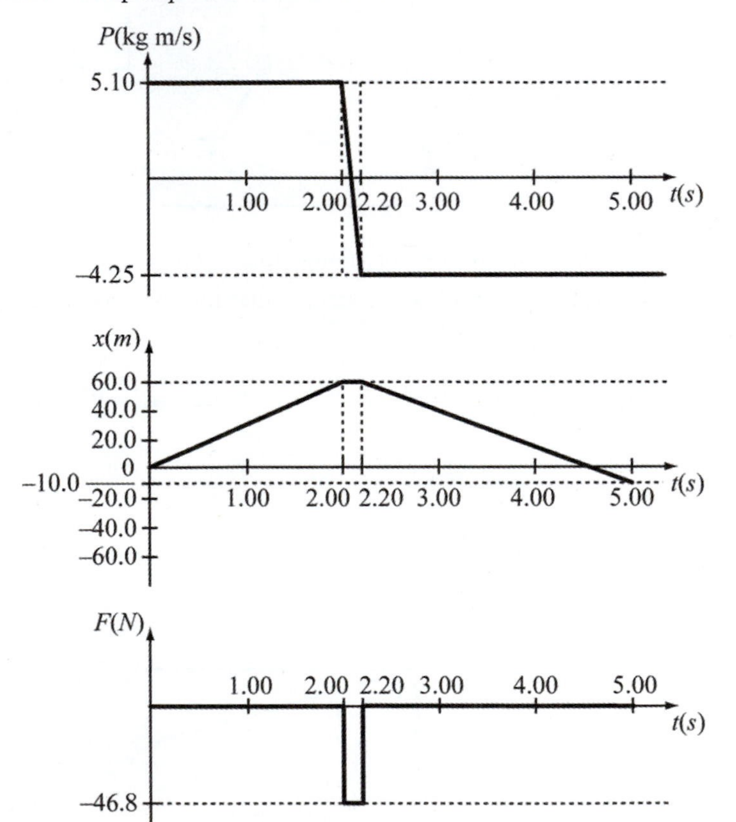

ROUND: $p_i = 5.10$ kg m/s, $p_f = -4.25$ kg m/s, $F = -46.8$ N, $x_2 = 60.0$ m, and $x_5 = -10.0$ m.

DOUBLE-CHECK: The force F is applied only during the interval of 0.200 s. At other times F = 0, or $a = 0$.

7.101. **THINK:** I know the skier's initial speed $v_0 = 22.0$ m/s, the skier's mass $M = 61.0$ kg, the mass of each ski $m = 1.50$ kg and the final velocity of each ski: $\vec{v}_1 = 25.0$ m/s at $\theta_1 = 12.0°$ to the left of the initial direction, and $\vec{v}_2 = 21.0$ m/s at $\theta_2 = 5.00°$ to the right of the initial direction. Calculate the magnitude and direction with respect to the initial direction of the skier's final velocity, $\vec{v}_s$.

SKETCH:

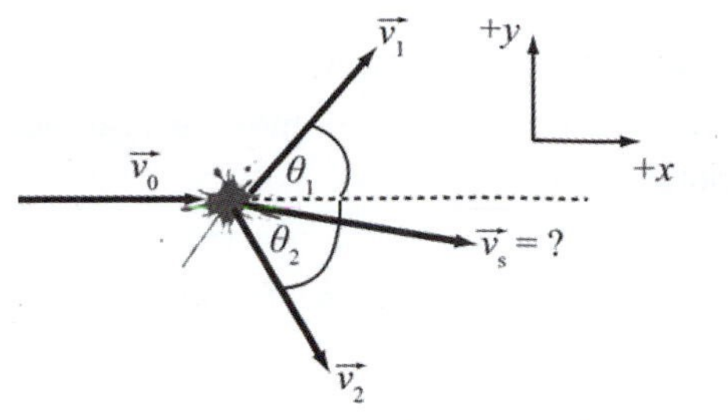

RESEARCH: The conservation of momentum requires $\sum_j (p_{fx})_j = \sum_j (p_{ix})_j$ and $\sum_j (p_{fy})_j = \sum_j (p_{iy})_j$. By conserving momentum in each direction, find $\vec{v}_s$. Take the initial direction to be along the x-axis.

SIMPLIFY: Then, $p_{ix} = m_{\text{total}} v_0 = (M+2m)v_0$, and take $p_{ix} = p_{fx}$ in the equation $p_{fx} = Mv_{sx} + mv_{1x} + mv_{2x} = Mv_{sx} + mv_1\cos\theta_1 + mv_2\cos\theta_2$.

$$(M+2m)v_0 = Mv_{sx} + m(v_1\cos\theta_1 + v_2\cos\theta_2) \Rightarrow v_{sx} = \frac{1}{M}\left((M+2m)v_0 - m(v_1\cos\theta_1 + v_2\cos\theta_2)\right).$$

Similarly,

$$p_{iy} = 0 = p_{fy} = Mv_{sy} + mv_{1y} - mv_{2y} = Mv_{sy} + m(v_1\sin\theta_1 - v_2\sin\theta_2) \Rightarrow v_{sy} = \frac{m}{M}(v_2\sin\theta_2 - v_1\sin\theta_1)$$

With v_{sx} and v_{sy} known, get the direction with respect to the initial direction from $\theta_s = \tan^{-1}(v_{sy}/v_{sx})$. The magnitude of the velocity is $v_s = \sqrt{v_{sx}^2 + v_{sy}^2}$.

CALCULATE:

$$v_{sx} = \frac{1}{61.0\text{ kg}}\left((61.0\text{ kg} + 2(1.50\text{ kg}))(22.0\text{ m/s}) - (1.50\text{ kg})((25.0\text{ m/s})\cos 12.0° + (21.0\text{ m/s})\cos 5.00°)\right)$$
$$= 21.9662\text{ m/s}$$

$$v_{sy} = \frac{1.50\text{ kg}}{61.0\text{ kg}}\left((21.0\text{ m/s})\sin 5.00° - (25.0\text{ m/s})\sin 12.0°\right) = -0.08281\text{ m/s}$$

$$v_s = \sqrt{(21.9662\text{ m/s})^2 + (-0.08281\text{ m/s})^2} = 21.9664\text{ m/s}$$

$\theta_s = \tan^{-1}\left(\dfrac{-0.08281\text{ m/s}}{21.9662\text{ m/s}}\right) = -0.2160°$, where the negative indicates that θ_s lies below the x-axis, or to the right of the initial direction.

ROUND: $\vec{v}_s = 22.0$ m/s at 0.216° to the right of the initial direction.

DOUBLE-CHECK: As the skier's mass is much greater than the mass of the two skis, it is reasonable that the skier carries the majority of the final momentum.

7.105. **THINK:** I want (a) the cannon's velocity $\vec{v}_c$ when the potato has been launched, and (b) the initial and final mechanical energy. There is no friction in the potato-cannon-ice system. Let the cannon's mass be $m_c = 10.0$ kg, the potato's mass be $m_p = 0.850$ kg, the cannon's spring constant be $k_c = 7.06 \cdot 10^3$ N/m, the spring's compression be $\Delta x = 2.00$ m, the cannon and the potato's initial velocities be $\vec{v}_{c,0} = \vec{v}_{p,0} = 0$, and the potato's launch velocity be $\vec{v}_p = 175\text{ m/s}\,\hat{x}$. Take "horizontally to the right" to be the positive $\hat{x}$ direction.

SKETCH:

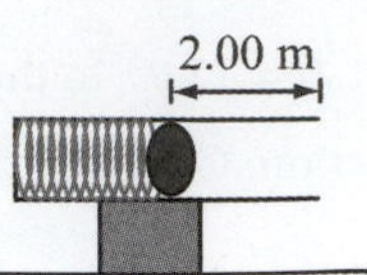

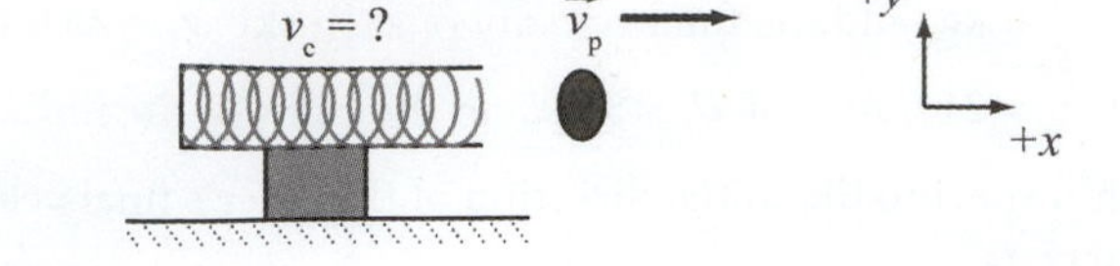

RESEARCH:

(a) Use the conservation of momentum $\Delta\vec{p} = 0$ to determine $\vec{v}_c$ when the potato, cannon and ice are considered as a system. Since the ice does not move, we can neglect the ice in the system and only consider the momenta of the potato and cannon.

(b) The total mechanical energy, E_{mec}, is conserved since the potato-cannon-ice system is isolated. That is, $E_{\text{mec,f}} = E_{\text{mec,i}}$. The value of $E_{\text{mec,i}}$ can be found by considering the spring potential energy of the cannon.

SIMPLIFY:

(a) $\Delta\vec{p} = 0 \Rightarrow \vec{p}_p - \vec{p}_{p,0} + \vec{p}_c - \vec{p}_{c,0} = 0 \Rightarrow \vec{p}_c = -\vec{p}_p \Rightarrow m_c\vec{v}_c = -m_p\vec{v}_p \Rightarrow \vec{v}_c = -\dfrac{m_p}{m_c}\vec{v}_p$

(b) $E_{\text{mec,f}} = E_{\text{mec,i}} = u_{\text{s,i}} = \dfrac{1}{2}k_c(\Delta x)^2$

CALCULATE:

(a) $\vec{v}_c = -\left(\dfrac{0.850 \text{ kg}}{10.0 \text{ kg}}\right)(175 \text{ m/s})\,\hat{x} = -14.875 \text{ m/s}\,\hat{x}$

(b) $E_{\text{mec,f}} = E_{\text{mec,i}} = \dfrac{1}{2}\left(7.06\cdot 10^3 \text{ N/m}\right)(2.00 \text{ m})^2 = 14120 \text{ J}$

ROUND: With three significant figures for all given values,

(a) $\vec{v}_c = -14.9 \text{ m/s}\,\hat{x}$, or $\vec{v}_c = 14.9 \text{ m/s}$ horizontally to the left.

(b) $E_{\text{mec,f}} = E_{\text{mec,i}} = 14.1 \text{ kJ}$.

DOUBLE-CHECK: Note $\vec{v}_c$ and $\vec{v}_p$ are directed opposite of each other, and $v_c < v_p$, as expected. Also, if $E_{\text{mec,f}}$ had been determined by considering the kinetic energies of the potato and the cannon, the same value would have been found.

7.109. **THINK:** Since the collision is elastic, momentum and kinetic energy are conserved. Also, since the alpha particle is backscattered, that means that is reflected 180° back and therefore the collision can be treated as acting in one dimension. The initial and final energies of the alpha particle are given in units of MeV and not J, $K_{\text{i}\alpha} = 2.00 \text{ MeV}$ and $K_{\text{f}\alpha} = 1.59 \text{ MeV}$. I can leave the energy in these units and not convert to Joules. $m_\alpha = 6.65\cdot 10^{-27}$ kg.

SKETCH:

α → X → ← α X →

$K_{i\alpha}$, $V_{i\alpha}$ $K_{ix} = 0\ M_ev$ K_{f2} K_{fx}

$V_{ix} = 0$ m/s V_{f2} V_{fx}

RESEARCH: $v_{\text{f1}} = \left(\dfrac{m_1 - m_2}{m_1 + m_2}\right)v_{\text{1i}}$, $v_{\text{f2}} = \left(\dfrac{2m_1}{m_1 + m_2}\right)v_{\text{1i}}$, and $K_{\text{i}} = K_{\text{f}}$, $E = (1/2)mv^2$.

SIMPLIFY: $K_{\text{i}\alpha} + K_{\text{iX}} = K_{\text{f}\alpha} + K_{\text{fx}} \Rightarrow K_{\text{fX}} = K_{\text{i}\alpha} - K_{\text{f}\alpha}$, $E_{\text{i}\alpha} = \dfrac{1}{2}m_\alpha v_{\text{i}\alpha}^2 \Rightarrow v_{\text{i}\alpha} = \sqrt{\dfrac{2E_{\text{i}\alpha}}{m_\alpha}}$, and

$v_{\text{fX}} = \left(\dfrac{2m_\alpha}{m_\alpha + m_{\text{X}}}\right)v_{\text{i}\alpha} = \left(\dfrac{2m_\alpha}{m_\alpha + m_{\text{X}}}\right)\sqrt{\dfrac{2K_{\text{i}\alpha}}{m_\alpha}}$. Since $K_{\text{fX}} = \dfrac{1}{2}m_{\text{X}}v_{\text{fX}}^2$:

$$K_{\text{i}\alpha}-K_{\text{f}\alpha}=\frac{1}{2}m_{\text{X}}v_{\text{fX}}^2=\frac{1}{2}m_{\text{X}}\left(\left(\frac{2m_\alpha}{m_\alpha+m_{\text{X}}}\right)\sqrt{\frac{2K_{\text{i}\alpha}}{m_\alpha}}\right)^2=\frac{1}{2}m_{\text{X}}\left(\frac{4m_\alpha^2}{\left(m_\alpha+m_{\text{X}}\right)^2}\right)\left(\frac{2K_{\text{i}\alpha}}{m_\alpha}\right)=\frac{4m_{\text{X}}m_\alpha K_{\text{i}\alpha}}{m_\alpha^2+2m_\alpha m_{\text{X}}+m_{\text{X}}^2}.$$

This simplifies to $0=\left(\frac{K_{\text{i}\alpha}-K_{\text{f}\alpha}}{K_{\text{i}\alpha}}\right)m_x^2+\left(2\left(\frac{K_{\text{i}\alpha}-K_{\text{f}\alpha}}{K_{\text{i}\alpha}}\right)m_\alpha-4m_\alpha\right)m_x+\left(\frac{K_{\text{i}\alpha}-K_{\text{f}\alpha}}{K_{\text{i}\alpha}}\right)m_\alpha^2.$

So, $m_{\text{X}}=\dfrac{\left(4-2\left(\frac{K_{\text{i}\alpha}-K_{\text{f}\alpha}}{K_{\text{i}\alpha}}\right)\right)m_\alpha\pm m_\alpha\sqrt{\left(2\left(\frac{K_{\text{i}\alpha}-K_{\text{f}\alpha}}{K_{\text{i}\alpha}}\right)-4\right)^2-4\left(\frac{K_{\text{i}\alpha}-K_{\text{f}\alpha}}{K_{\text{i}\alpha}}\right)^2}}{2\left(\frac{K_{\text{i}\alpha}-K_{\text{f}\alpha}}{K_{\text{i}\alpha}}\right)}$ by the quadratic formula.

CALCULATE:

$$m_{\text{X}}=\frac{\left(4-2\left(\frac{2.00\text{ MeV}-1.59\text{ MeV}}{2.00\text{ MeV}}\right)\right)\left(6.65\cdot10^{-27}\text{ kg}\right)}{2\left(\frac{2.00\text{ MeV}-1.59\text{ MeV}}{2.00\text{ MeV}}\right)}$$

$$\pm\frac{\left(6.65\cdot10^{-27}\text{ kg}\right)\sqrt{\left(2\left(\frac{2.00\text{ MeV}-1.59\text{ MeV}}{2.00\text{ MeV}}\right)-4\right)^2-4\left(\frac{2.00\text{ MeV}-1.59\text{ MeV}}{2.00\text{ MeV}}\right)^2}}{2\left(\frac{2.00\text{ MeV}-1.59\text{ MeV}}{2.00\text{ MeV}}\right)}$$

$$=1.1608\cdot10^{-25}\text{ kg},\ 3.8098\cdot10^{-28}\text{ kg}$$

ROUND: $m_{\text{X}}=1.16\cdot10^{-25}\text{ kg}$. Since the ratio of the masses of atom X and the alpha particle is $m_{\text{X}}/m_\alpha=(1.16\cdot10^{-25}\text{ kg})/(6.65\cdot10^{-27}\text{ kg})=17.5$, and since the alpha particle has 4 nucleons, expect atom X to have 70 nucleons. Consulting the table in Appendix B, germanium is a good guess. Another possibility would be zinc, but zinc-70 is a relatively rare isotope of zinc, whereas germanium-70 is a common isotope of germanium.

DOUBLE-CHECK: Since the alpha particle is reflected back, the mass of atom X must be greater than the mass of the alpha particle, so this value makes sense.

Multi-Version Exercises

7.110. **THINK:** The coefficient of restitution is used to compute the height of the next bounce from the peak of the previous bounce. Since the ball was dropped (not thrown), assume that it started with no velocity, exactly as it would at the peak of a bounce.

SKETCH: The ball hits the floor three times:

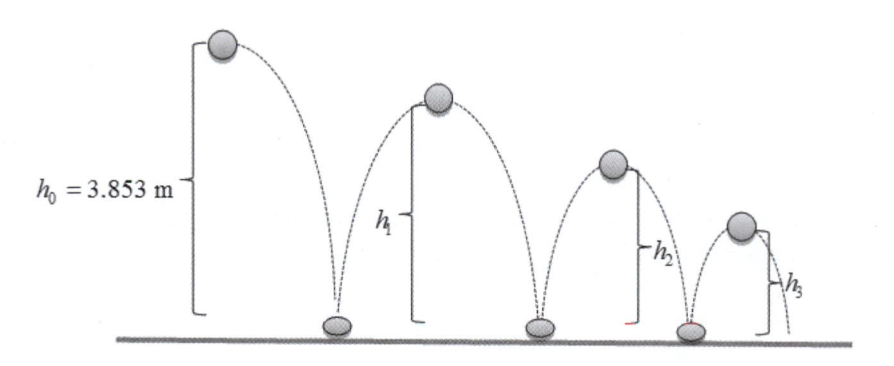

RESEARCH: The coefficient of restitution is defined to be $\varepsilon = \sqrt{\frac{h_f}{h_i}}$. In this case, the ball bounces three times; it is necessary to find expressions relating h_0, h_1, h_2, and h_3.

SIMPLIFY: For the first bounce, $\varepsilon = \sqrt{\frac{h_1}{h_0}}$. For the second bounce, $\varepsilon = \sqrt{\frac{h_2}{h_1}}$, and for the third bounce, $\varepsilon = \sqrt{\frac{h_3}{h_2}}$. Squaring all three equations gives: $\varepsilon^2 = \frac{h_1}{h_0}$, $\varepsilon^2 = \frac{h_2}{h_1}$, and $\varepsilon^2 = \frac{h_3}{h_2}$. Now, solve for h_3 in terms of h_2 and ε: $h_3 = \varepsilon^2 h_2$. Similarly, solve for h_2 in terms of h_1 and ε, then for h_1 in terms of h_0 and ε, to get:

$$h_3 = \varepsilon^2 h_2$$
$$h_2 = \varepsilon^2 h_1$$
$$h_1 = \varepsilon^2 h_0$$

Finally, combine these three equations to get an expression for h_3 in terms of the values given in the problem, h_1 and ε.

$$\begin{aligned} h_3 &= \varepsilon^2 h_2 \\ &= \varepsilon^2\left(\varepsilon^2 h_1\right) \\ &= \varepsilon^2\left(\varepsilon^2\left(\varepsilon^2 h_0\right)\right) \\ &= \varepsilon^6 h_0. \end{aligned}$$

CALCULATE: The coefficient of restitution of the Super Ball is 0.8887 and the ball is dropped from a height of 3.853 m above the floor. So the height of the third bounce is:

$$\begin{aligned} h_3 &= \varepsilon^6 h_0 \\ &= (0.8887)^6\, 3.853 \text{ m} \\ &= 1.89814808 \text{ m} \end{aligned}$$

ROUND: The only numbers used here were the coefficient of restitution and the height. They were multiplied together and were given to four significant figures. Thus the answer should have four figures; the ball reached a maximum height of 1.898 m above the floor.

DOUBLE-CHECK: From experience with Super Balls, this seems reasonable. Double check by working backwards to find the maximum height of each bounce. If the ball bounced 1.898 m on the third bounce, then it reached a height of 1.898 m / 0.8887^2 = 2.403177492 m on the second bounce and 2.403177492 m / 0.8887^2 = 3.042814572 m on the first bounce. From there, the height at which the ball was dropped is computed to be 3.042814572 m / 0.8887^2 = 3.852699416 m. When rounded to four decimal places, this gives an initial height of 3.853 m, which agrees with the number given in the problems and confirms that the calculations were correct.

7.113. **THINK:** This problem uses the properties of conservation of energy. Since the masses and initial speeds of the gliders are given, it is possible to use the fact that the collision is totally elastic and the initial conditions to find the velocity of the glider after the collision.

SKETCH: The sketch shows the gliders before and after the collision. Note that the velocities are all in the x – direction. Define the positive x – direction to be to the right.

BEFORE:

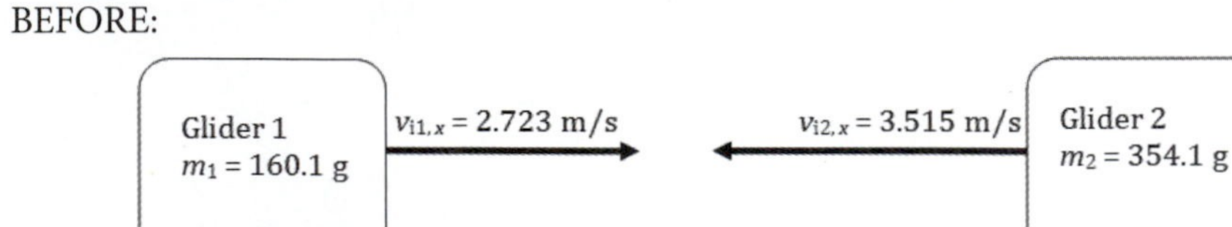

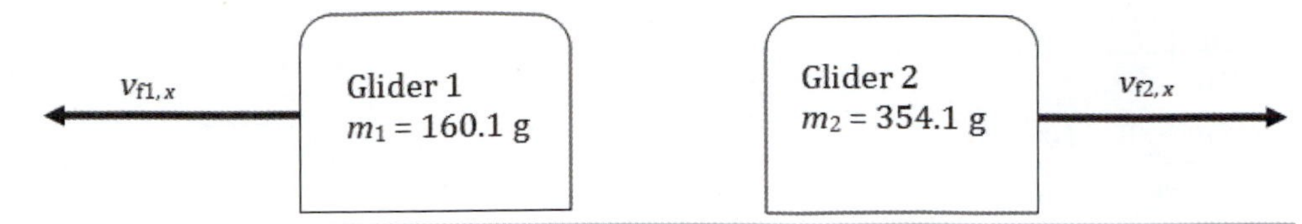

RESEARCH: Since this is a one-dimensional, totally elastic collision, we know that the speed of the first glider after the collision is given by the equation:

$$v_{f1,x} = \left(\frac{m_1 - m_2}{m_1 + m_2}\right) v_{i1\,x} + \left(\frac{2m_2}{m_1 + m_2}\right) v_{i2\,x}$$

SIMPLIFY: Since the masses are given in grams and the velocities in meters per second, there is no need to convert any of the units in this problem. All of the values needed to compute the final velocity of Glider 1.

CALCULATE: The masses and velocities are given in the problem. Substitute them into the equation to get $v_{f1,x} = \left(\frac{160.1\text{ g} - 354.1\text{ g}}{160.1\text{ g} + 354.1\text{ g}}\right)(2.723\text{ m/s}) + \left(\frac{2 \cdot 354.1\text{ g}}{160.1\text{ g} + 354.1\text{ g}}\right) \cdot (-3.515\text{ m/s})$, so the velocity of Glider 1 after the collision is –5.868504473 m/s. The velocity is negative to indicate that the glider is moving to the left.

ROUND: The measured numbers in this problem all have four significant figures, so the final answer should also have four figures. This means that the final velocity of Glider 1 is 5.869 m/s to the left.

DOUBLE-CHECK: Though the speed of Glider 1 is greater after the collision than it was before the collision, which makes sense because Glider 2 was more massive and had a faster speed going into the collision. The problem can also be checked by calculating the speed of Glider 2 after the collision using the equation $v_{f2,x} = \left(\frac{2m_1}{m_1 + m_2}\right) v_{i1\,x} + \left(\frac{m_2 - m_1}{m_1 + m_2}\right) v_{i2\,x}$ and confirming that the energy before the collision is equal to the energy after the collision, $\frac{p_{f1}^2}{2m_1} + \frac{p_{f2}^2}{2m_2} = \frac{p_{i1}^2}{2m_1} + \frac{p_{i2}^2}{2m_2}$.

Before the collision,

$$\begin{aligned}\frac{p_{i1}^2}{2m_1} + \frac{p_{i2}^2}{2m_2} &= \frac{(m_1 v_{i1})^2}{2m_1} + \frac{(m_2 v_{i2})^2}{2m_2} \\ &= \frac{(160.1\text{ g} \cdot 2.723\text{ m/s})^2}{2 \cdot 160.1\text{ g}} + \frac{(354.1\text{ g} \cdot 3.515\text{ m/s})^2}{2 \cdot 354.1\text{ g}} \\ &= 2781.041643\text{ g} \cdot \text{m}^2/\text{s}^2\end{aligned}$$

After the collision,

$$\frac{p_{f1}^2}{2m_1}+\frac{p_{f2}^2}{2m_2}$$

$$=\frac{(m_1v_{f1})^2}{2m_1}+\frac{(m_2v_{f2})^2}{2m_2}$$

$$=\frac{(m_1v_{f1})^2}{2m_1}+\frac{\left(m_2\left[\left(\frac{2m_1}{m_1+m_2}\right)v_{i1x}+\left(\frac{m_2-m_1}{m_1+m_2}\right)v_{i2x}\right]\right)^2}{2m_2}$$

$$=\frac{(160.1\text{ g}\cdot 5.869\text{ m/s})^2}{2\cdot 160.1\text{ g}}$$

$$+\frac{\left(354.1\text{ g}\cdot\left[\left(\frac{2\cdot 160.1\text{ g}}{160.1\text{ g}+354.1\text{ g}}\right)\cdot 2.723\text{ m/s}+\left(\frac{354.1\text{ g}-160.1\text{ g}}{160.1\text{ g}+354.1\text{ g}}\right)\cdot(-3.515\text{ m/s})\right]\right)^2}{2\cdot 354.1\text{ g}}$$

$$=2781.507234\text{ g}\cdot\text{m}^2/\text{s}^2$$

The energies before and after the collision are both close to 2781 g·m^2 / s^2, confirming that the values calculated for the speeds of the gliders were correct.

7.116. **THINK:** For this problem, it will help to think about the components of the momentum that are perpendicular to and parallel to the wall. After the collision, the momentum parallel to the wall is unchanged The perpendicular component is in the opposite direction and is multiplied by the coefficient of restitution after the collision with the wall.

SKETCH: Show the path of the racquetball before and after it hits the wall.

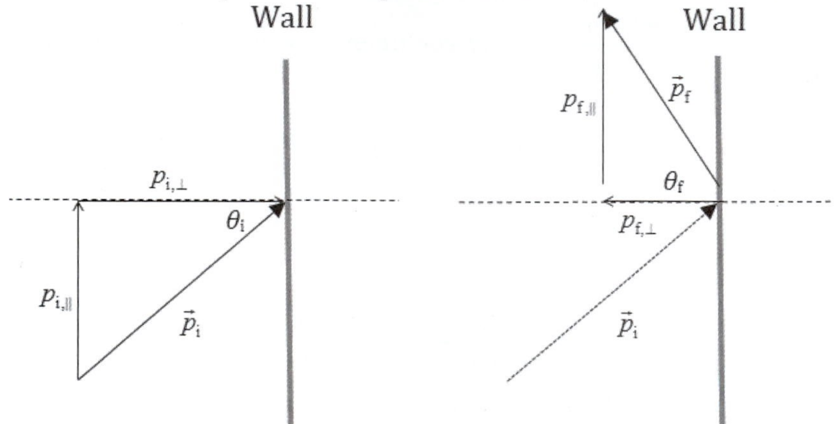

RESEARCH: The mass and initial speed can be used to calculate the initial momentum $\vec{p}_i = m\vec{v}$. The angle at which the racquetball hits the wall is used to calculate the parallel and perpendicular components from the initial momentum: $p_{i,\perp}=p_i\cos\theta_i$ and $p_{i,\parallel}=p_i\sin\theta_i$. The component of the momentum parallel to the wall is unchanged in the collision, so $p_{f,\parallel}=p_{i,\parallel}$. The component of the final momentum perpendicular to the wall has a magnitude equal to the coefficient of restitution times the component of the initial momentum parallel to the wall: $p_{f,\perp}=\varepsilon p_{i,\perp}$ in the opposite direction from $p_{i,\perp}$. With a little trigonometry, the final angle can be calculated from the perpendicular and parallel components of the final momentum: $\tan\theta_f=\frac{p_{f,\parallel}}{p_{f,\perp}}$.

SIMPLIFY: Since $\tan\theta_\text{f} = \dfrac{p_{\text{f},\parallel}}{p_{\text{f},\perp}}$, take the inverse tangent to find $\theta_\text{f} = \tan^{-1}\left(\dfrac{p_{\text{f},\parallel}}{p_{\text{f},\perp}}\right)$. Substitute $p_{\text{f},\parallel} = p_{\text{i},\parallel} = p_\text{i}\sin\theta_\text{i}$ and $p_{\text{f},\perp} = \varepsilon p_\text{i}\cos\theta_\text{i}$ into the equation to get:

$$\begin{aligned}\theta_\text{f} &= \tan^{-1}\left(\frac{p_\text{i}\sin\theta_\text{i}}{\varepsilon p_\text{i}\cos\theta_\text{i}}\right)\\ &= \tan^{-1}\left(\frac{1}{\varepsilon}\cdot\frac{\sin\theta_i}{\cos\theta_i}\right)\\ &= \tan^{-1}\left(\frac{1}{\varepsilon}\cdot\tan\theta_i\right)\end{aligned}$$

CALCULATE: The exercise states that the initial angle is 43.53° and the coefficient of restitution is 0.8199. Using those values, the final angle is:

$$\begin{aligned}\theta_\text{f} &= \tan^{-1}\left(\frac{\tan\theta_i}{\varepsilon}\right)\\ &= \tan^{-1}\left(\frac{\tan 43.53^\circ}{0.8199}\right)\\ &= 49.20289058^\circ\end{aligned}$$

ROUND: The angle and coefficient of restitution are the only measured values used in these calculations, and both are given to four significant figures, so the final answer should also have four significant figures. The racquetball rebounds at an angle of 49.20° from the normal.

DOUBLE-CHECK: This answer is physically realistic. The component of the momentum does not change, but the perpendicular component is reduced by about one fifth, so the angle should increase. To double check the calculations, use the speed and mass of the racquetball to find the initial and final momentum: $p_{\text{i},\parallel} = mv_{\text{i},\parallel} = m\vec{v}_i\sin\theta_\text{i}$ and $p_{\text{i},\perp} = mv_{\text{i},\perp} = m\vec{v}_\text{i}\cos\theta_\text{i}$. Thus $p_{\text{i},\parallel} = 437.9416827 \text{ g}\cdot\text{m/s}$ and $p_{\text{i},\perp} = 461.0105692 \text{ g}\cdot\text{m/s}$. The parallel portion of the momentum is unchanged, and the perpendicular portion is the coefficient of restitution times the initial perpendicular momentum, giving a final parallel component of $p_{\text{f},\parallel} = 437.9416827 \text{ g}\cdot\text{m/s}$ and $p_{\text{f},\perp} = 377.9825657 \text{ g}\cdot\text{m/s}$. The final angle can be computed as $\theta_\text{f} = \tan^{-1}\left(\dfrac{p_{\text{f},\parallel}}{p_{\text{f},\perp}}\right)$ or 49.20°, which confirms the calculations.

7.119. **THINK:** When the boy catches the dodgeball, he holds on to it and does not let go. The boy and the ball stick together and have the same velocity after the collision, so this is a totally inelastic collision. This means that the final velocity of the boy and ball can be calculated from the initial velocities and masses of the boy and dodgeball.

SKETCH: Choose the *x*-axis to run in the same direction as the dodgeball, with the origin at the boy's initial location.

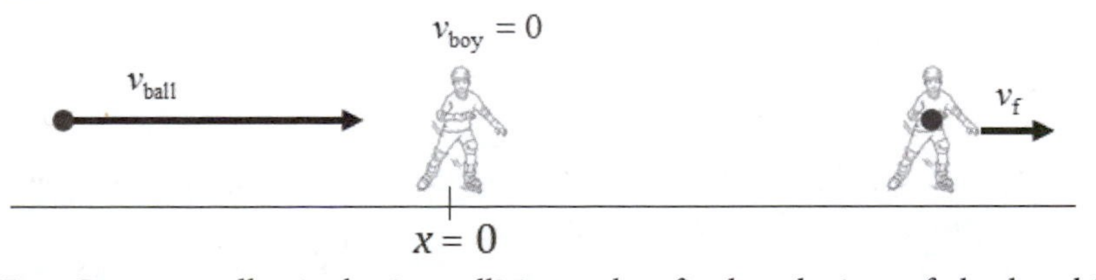

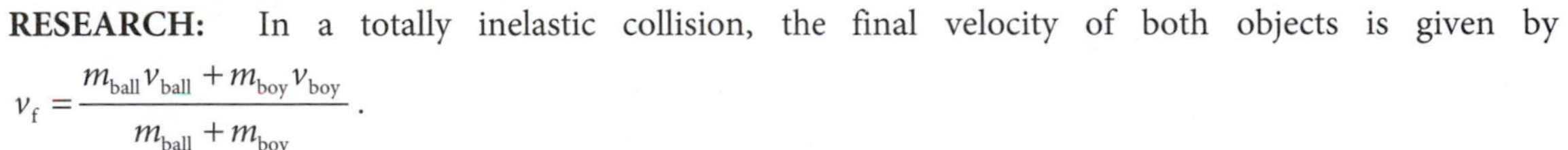

RESEARCH: In a totally inelastic collision, the final velocity of both objects is given by $v_\text{f} = \dfrac{m_\text{ball}v_\text{ball} + m_\text{boy}v_\text{boy}}{m_\text{ball} + m_\text{boy}}$.

SIMPLIFY: Because the initial velocity of the boy $v_{\text{boy}} = 0$, the equation can be simplified to $v_{\text{f}} = \frac{m_{\text{ball}}v_{\text{ball}} + m_{\text{boy}} \cdot 0}{m_{\text{ball}} + m_{\text{boy}}} = \frac{m_{\text{ball}}v_{\text{ball}}}{m_{\text{ball}} + m_{\text{boy}}}$. Since the mass of the ball is given in grams and the mass of the boy is given in kilograms, it is necessary to multiply the mass of the ball by a conversion factor of $\frac{1\text{ kg}}{1000\text{ g}}$.

CALCULATE: The mass of the ball is 511.1 g, or $511.1\text{ g} \cdot \frac{1\text{ kg}}{1000\text{ g}} = 0.5111\text{ kg}$. The mass of the boy is 48.95 kg and the initial velocity of the dodgeball is 23.63 m/s. The final velocity is

$$\begin{aligned} v_{\text{f}} &= \frac{m_{\text{ball}}v_{\text{ball}}}{m_{\text{ball}} + m_{\text{boy}}} \\ &= \frac{0.5111\text{ kg} \cdot 23.63\text{ m/s}}{48.95\text{ kg} + 0.5111\text{ kg}} \\ &= 0.2441776062\text{ m/s}. \end{aligned}$$

ROUND: The measured values in this problem are given to four significant figures, and the sum of the masses also has four significant figures, so the final answer should also have four significant figures. The final velocity of the boy and dodg ball is 0.2442 m/s in the same direction that the dodgeball was traveling initially.

DOUBLE-CHECK: This answer makes sense. The mass of the boy is much greater than the mass of the dodgeball, so a smaller speed of this massive system (boy plus dodgeball) will have the same momentum as the ball traveling much faster. To confirm that the answer is correct, check that the momentum after the collision is equal to the momentum before the collision. Before the collision, the boy is not moving so he has no momentum, and the dodgeball has a momentum of $p_x = mv_x = 0.511\text{ kg} \cdot 23.63\text{ m/s}$ or 12.075 kg · m/s. After the collision, the total momentum is $mv_x = (0.511\text{ kg} + 48.95\text{ kg}) \cdot 0.2442\text{ m/s}$ or 12.078 kg · m/s. These agree within rounding error, so this confirms that the original calculation was correct.

Chapter 8: Systems of Particles and Extended Objects

Exercises

8.29. **THINK:** Determine (a) the distance, d_1, from the center of mass of the Earth-Moon system to the geometric center of the Earth and (b) the distance, d_2, from the center of mass of the Sun-Jupiter system to the geometric center of the Sun. The mass of the Earth is approximately $m_{\text{E}} = 5.9742 \cdot 10^{24}$ kg and the mass of the Moon is approximately $m_{\text{M}} = 7.3477 \cdot 10^{22}$ kg. The distance between the center of the Earth to the center of the Moon is $d_{\text{EM}} = 384,400$ km. Also, the mass of the Sun is approximately $m_{\text{S}} = 1.98892 \cdot 10^{30}$ kg and the mass of Jupiter is approximately $m_{\text{J}} = 1.8986 \cdot 10^{27}$ kg. The distance between the center of the Sun and the center of Jupiter is $d_{\text{SJ}} = 778,300,000$ km.

SKETCH:

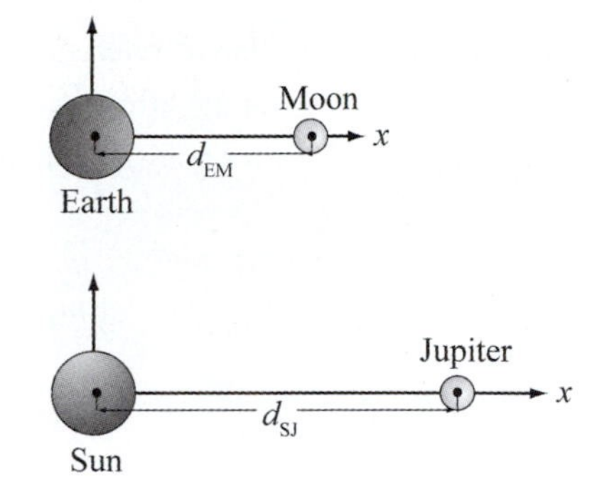

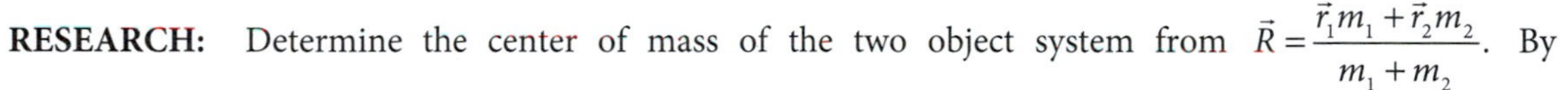

RESEARCH: Determine the center of mass of the two object system from $\vec{R} = \dfrac{\vec{r}_1 m_1 + \vec{r}_2 m_2}{m_1 + m_2}$. By considering the masses on the x-axis (as sketched), the one dimensional equation can be used for x. Assuming a uniform, spherically symmetric distribution of each planet's mass, they can be modeled as point particles. Finally, by placing the Earth (Sun) at the origin of the coordinate system, the center of mass will be determined with respect to the center of the Earth (Sun), i.e. $d_1 (d_2) = x$.

SIMPLIFY:

(a) $d_1 = x = \dfrac{x_1 m_{\text{E}} + x_2 m_{\text{M}}}{m_{\text{E}} + m_{\text{M}}} = \dfrac{d_{\text{EM}} m_{\text{M}}}{m_{\text{E}} + m_{\text{M}}}$

(b) $d_2 = x = \dfrac{x_1 m_{\text{S}} + x_2 m_{\text{J}}}{m_{\text{S}} + m_{\text{J}}} = \dfrac{d_{\text{SJ}} m_{\text{J}}}{m_{\text{S}} + m_{\text{J}}}$

CALCULATE:

(a) $d_1 = \dfrac{(384,400 \text{ km})(7.3477 \cdot 10^{22} \text{ kg})}{(5.9742 \cdot 10^{24} \text{ kg}) + (7.3477 \cdot 10^{22} \text{ kg})} = \dfrac{2.8244559 \cdot 10^{28} \text{ km} \cdot \text{kg}}{6.047677 \cdot 10^{24} \text{ kg}} = 4670.3 \text{ km}$

(b) $d_2 = \dfrac{(7.783 \cdot 10^{8} \text{ km})(1.8986 \cdot 10^{27} \text{ kg})}{(1.98892 \cdot 10^{30} \text{ kg}) + (1.8986 \cdot 10^{27} \text{ kg})} = 742247.6 \text{ km}$

ROUND:

(a) d_{EM} has four significant figures, so $d_1 = 4670.$ km.

(b) d_{SJ} has four significant figures, so $d_2 = 742,200$ km.

DOUBLE-CHECK: In each part, the distance d_1 / d_2 is much less than half the separation distance $d_{\text{EM}} / d_{\text{SJ}}$. This makes sense as the center of mass should be closer to the more massive object in the two body system.

8.33. **THINK:** The mass of the car is $m_c = 2.00$ kg and its initial speed is $v_c = 0$. The mass of the truck is $m_t = 3.50$ kg and its initial speed is $v_t = 4.00$ m/s toward the car. Determine (a) the velocity of the center of mass, $\vec{V}$, and (b) the velocities of the truck, $\vec{v}_t'$ and the car, $\vec{v}_c'$ with respect to the center of mass.

SKETCH:

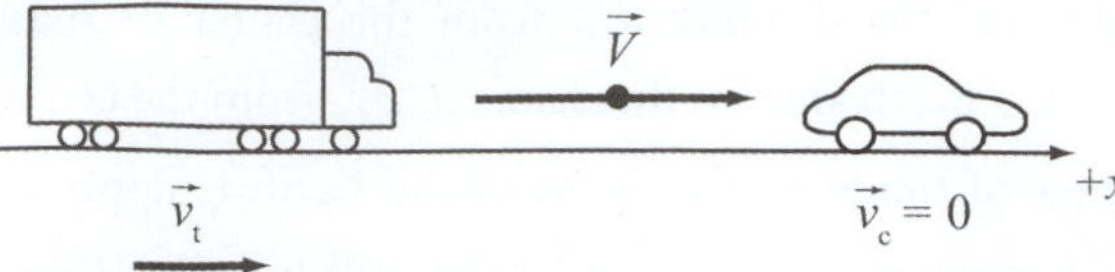

RESEARCH:

(a) The velocity of the center of mass can be determined from $\vec{V} = \frac{1}{M}\sum_{i=1}^{n} m_i \vec{v}_i$.

Take $\vec{v}_t$ to be in the positive x-direction.

(b) Generally, the relative velocity, $\vec{v}'$, of an object with velocity, $\vec{v}$, in the lab frame is given by $\vec{v}' = \vec{v} - \vec{V}$, where $\vec{V}$ is the velocity of the relative reference frame. Note the speeds of the car and the truck relative to the center of mass do not change after their collision, but the relative velocities change direction; that is, $\vec{v}_t'(\text{before collision}) = -\vec{v}_t'(\text{after collision})$ and similarly for the car's relative velocity.

SIMPLIFY:

(a) Substituting $\vec{v}_c = 0$ and $M = m_c + m_t$, $\vec{V} = \frac{1}{M}(m_c\vec{v}_c + m_t\vec{v}_t)$ becomes $\vec{V} = \frac{(m_t\vec{v}_t)}{(m_c + m_t)}$.

(b) $\vec{v}_t'$ and $\vec{v}_c'$ before the collision are $\vec{v}_t' = \vec{v}_t - \vec{V}$ and $\vec{v}_c' = \vec{v}_c - \vec{V} = -\vec{V}$.

CALCULATE:

(a) $\vec{V} = \frac{(3.50 \text{ kg})(4.00\hat{x} \text{ m/s})}{(3.50 \text{ kg} + 2.00 \text{ kg})} = 2.545\hat{x}$ m/s

(b) $\vec{v}_t' = (4.00\hat{x} \text{ m/s}) - (2.545\hat{x} \text{ m/s}) = 1.4545\hat{x}$ m/s, $\vec{v}_c' = -2.545\hat{x}$ m/s

ROUND: There are three significant figures for each given value, so the results should be rounded to the same number of significant figures.

(a) $\vec{V} = 2.55\hat{x}$ m/s

(b) Before the collision, $\vec{v}_t' = 1.45\hat{x}$ m/s and $\vec{v}_c' = -2.55\hat{x}$ m/s. This means that after the collision, the velocities with respect to the center of mass become $\vec{v}_t' = -1.45\hat{x}$ m/s and $\vec{v}_c' = 2.55\hat{x}$ m/s.

DOUBLE-CHECK: $\vec{V}$ is between the initial velocity of the truck and the initial velocity of the car, as it should be.

8.37. **THINK:** The proton's mass is $m_p = 1.6726 \cdot 10^{-27}$ kg and its initial speed is $v_p = 0.700c$ (assumed to be in the lab frame). The mass of the tin nucleus is $m_{sn} = 1.9240 \cdot 10^{-25}$ kg (assumed to be at rest). Determine the speed of the center of mass, v, with respect to the lab frame.

SKETCH: A sketch is not necessary.

RESEARCH: The given speeds are in the lab frame. To determine the speed of the center of mass use $V = \frac{1}{M}\sum_{i=1}^{n} m_i v_i$.

SIMPLIFY: $V = \frac{1}{m_p + m_{sn}}(m_p v_p + m_{sn} v_{sn}) = \frac{m_p v_p}{m_p + m_{sn}}$

CALCULATE: $V = \dfrac{\left(1.6726\cdot 10^{-27}\text{ kg}\right)(0.700c)}{\left(1.6726\cdot 10^{-27}\text{ kg}\right)+\left(1.9240\cdot 10^{-25}\text{ kg}\right)} = 0.0060329c$

ROUND: Since v_{p} has three significant figures, the result should be rounded to $V = 0.00603c$.

DOUBLE-CHECK: Since m_{sn} is at rest and $m_{\text{sn}} >> m_{\text{p}}$, it is expected that $V << v_{\text{p}}$.

8.41. **THINK:** For rocket engines, the specific impulse is $J_{\text{spec}} = \dfrac{J_{\text{tot}}}{W_{\text{expended fuel}}} = \dfrac{1}{W_{\text{expended fuel}}}\int_{t_0}^{t} F_{\text{thrust}}(t')dt'$.

(a) Determine J_{spec} with an exhaust nozzle speed of v.

(b) Evaluate and compare J_{spec} for a toy rocket with $v_{\text{toy}} = 800.\text{ m/s}$ and a chemical rocket with $v_{\text{chem}} = 4.00\text{ km/s}$.

SKETCH: Not applicable.

RESEARCH: It is known that $\vec{F}_{\text{thrust}} = -v_{\text{c}}dm/dt$. Rewrite $W_{\text{expended fuel}}$ as $m_{\text{expended}}g$. With the given definition, J_{spec} can be determined for a general v, and for v_{toy} and v_{chem}.

SIMPLIFY: $J_{\text{spec}} = \dfrac{1}{m_{\text{expended}}g}\int_{m_0}^{m} -v\,dm = -\dfrac{v}{m_{\text{expended}}g}(m-m_0)$. Now, $m - m_0 = -m_{\text{expended}}$, so $J_{\text{spec}} = \dfrac{v}{g}$.

CALCULATE: $J_{\text{spec, toy}} = \dfrac{v_{\text{toy}}}{g} = \dfrac{800.\text{ m/s}}{\left(9.81\text{ m/s}^2\right)} = 81.55\text{ s}$, $J_{\text{spec, chem}} = \dfrac{v_{\text{chem}}}{g} = \dfrac{4.00\cdot 10^3\text{ m/s}}{\left(9.81\text{ m/s}^2\right)} = 407.75\text{ s}$

$$\frac{J_{\text{spec, toy}}}{J_{\text{spec, chem}}} = \frac{v_{\text{toy}}}{v_{\text{chem}}} = \frac{800.\text{ m/s}}{4.00\cdot 10^3\text{ m/s}} = 0.200$$

ROUND:

(a) $J_{\text{spec, toy}} = 81.6\text{ s}$

(b) $J_{\text{spec, chem}} = 408\text{ s}$ and $J_{\text{spec, toy}} = 0.200 J_{\text{spec, chem}}$.

DOUBLE-CHECK: The units of the results are units of specific impulse. Also, as expected $J_{\text{spec, toy}} < J_{\text{spec, chem}}$.

8.45. **THINK:** The thrust force is $F_{\text{thrust}} = 53.2\cdot 10^6\text{ N}$ and the propellant velocity is $v = 4.78\cdot 10^3\text{ m/s}$. Determine (a) dm/dt, (b) the final speed of the spacecraft, v_{s}, given $v_{\text{i}} = 0$, $m_{\text{i}} = 2.12\cdot 10^6\text{ kg}$ and $m_{\text{f}} = 7.04\cdot 10^4\text{ kg}$ and (c) the average acceleration, a_{av} until burnout.

SKETCH:

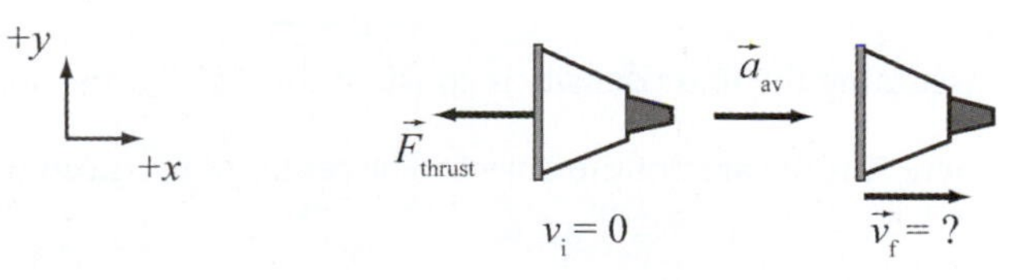

RESEARCH:

(a) To determine dm/dt, use $\vec{F}_{\text{thrust}} = -v_{\text{c}}dm/dt$.

(b) To determine v_{f}, use $v_{\text{f}} - v_{\text{i}} = v_{\text{c}}\ln\left(m_{\text{i}}/m_{\text{f}}\right)$.

(c) Δv is known from part (b). Δt can be determined from the equivalent ratios,

$\dfrac{dm}{dt} = \dfrac{\Delta m}{\Delta t}$, where $\Delta m = m_{\text{i}} - m_{\text{f}}$.

SIMPLIFY:

(a) Since $\vec{F}_{\text{thrust}}$ and $\vec{v}_{\text{c}}$ are in the same direction, the equation can be rewritten as:

$$F_{\text{thrust}} = v_{\text{c}} \frac{dm}{dt} \Rightarrow \frac{dm}{dt} = \frac{F_{\text{thrust}}}{v_{\text{c}}}.$$

(b) $v_{\text{i}} = 0 \Rightarrow v_{\text{f}} = v_{\text{c}} \ln\left(\frac{m_{\text{i}}}{m_{\text{f}}}\right)$

(c) $\frac{dm}{dt} = \frac{\Delta m}{\Delta t} \Rightarrow \Delta t = \frac{\Delta m}{dm/dt}, \; a_{\text{av}} = \frac{\Delta v}{\Delta t} = \frac{v_{\text{f}}}{\Delta m}\left(\frac{dm}{dt}\right) \quad (v_{\text{i}} = 0)$

CALCULATE:

(a) $\frac{dm}{dt} = \frac{(53.2 \cdot 10^6 \text{ N})}{(4.78 \cdot 10^3 \text{ m/s})} = 11129.7 \text{ kg/s}$

(b) $v_{\text{f}} = (4.78 \cdot 10^3 \text{ m/s}) \ln\left(\frac{2.12 \cdot 10^6 \text{ kg}}{7.04 \cdot 10^4 \text{ kg}}\right) = 1.6276 \cdot 10^4 \text{ m/s}$

(c) $a_{\text{av}} = \frac{(1.6276 \cdot 10^4 \text{ m/s})}{(2.12 \cdot 10^6 \text{ kg} - 7.04 \cdot 10^4 \text{ kg})}(11129.7 \text{ kg/s}) = 88.38 \text{ m/s}^2$

ROUND: Each given value has three significant figures, so the results should be rounded to $dm/dt = 11100 \text{ kg/s}$, $v_{\text{f}} = 1.63 \cdot 10^4 \text{ m/s}$ and $a_{\text{av}} = 88.4 \text{ m/s}^2$.

DOUBLE-CHECK: The results all have the correct units. Also, the results are reasonable for a spaceship with such a large thrust force.

8.49. **THINK:** The height is H = 17.3 cm and the base is B = 10.0 cm for a flat triangular plate. Determine the x and y-coordinates of its center of mass. Since it is not stated otherwise, we assume that the mass density of this plate is constant.

SKETCH:

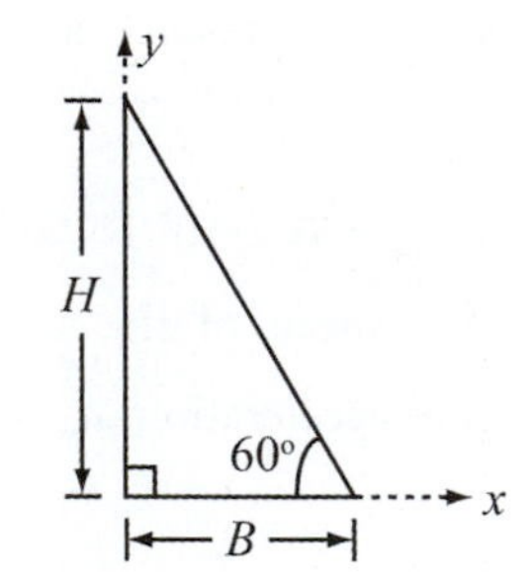

RESEARCH: Assuming the mass density is constant throughout the object, the center of mass is given by $\vec{R} = \frac{1}{A}\int_A \vec{r} dA$, where A is the area of the object. The center of mass can be determined in each dimension.

The x coordinate and the y coordinate of the center of mass are given by $X = \frac{1}{A}\int_A x\,dA$ and $Y = \frac{1}{A}\int_A y\,dA$, respectively. The area of the triangle is $A = HB/2$.

SIMPLIFY: The expression for the area of the triangle can be substituted into the formulae for the center of mass to get

$$X = \frac{2}{HB}\int_A x\,dA \text{ and } Y = \frac{2}{HB}\int_A y\,dA.$$

In the x-direction we have to solve the integral:

$$\int_A x dA = \int_0^B \int_0^{y_m(x)} x dy dx = \int_0^B x dx \int_0^{y_m(x)} dy = \int_0^B x y_m(x) dx = \int_0^B x H(1-x/B) dx = H\int_0^B x - \left(x^2/B\right) dx$$
$$= H\left(\tfrac{1}{2}x^2 - \tfrac{1}{3}x^3/B\right)\Big|_0^B = \tfrac{1}{2}HB^2 - \tfrac{1}{3}HB^2 = \tfrac{1}{6}HB^2$$

Note that in this integration procedure the maximum for the y–integration depends on the value of x: $y_m(x) = H(1-x/B)$. Therefore we arrive at

$$X = \frac{2}{HB}\int_A x\, dA = \frac{2}{HB}\cdot\frac{HB^2}{6} = \tfrac{1}{3}B$$

In the same way we can find that $Y = \tfrac{1}{3}H$.

CALCULATE: $X_{\text{com}} = \tfrac{1}{3}(10.0 \text{ cm}) = 3.33333 \text{ cm}$, $Y_{\text{com}} = \tfrac{1}{3}(17.3 \text{ cm}) = 5.76667 \text{ cm}$

ROUND: Three significant figures were provided in the question, so the results should be written $X = 3.33$ cm and $Y = 5.77$ cm.

DOUBLE-CHECK: Units of length were calculated for both X and Y, which is dimensionally correct. We also find that the center of mass coordinates are inside the triangle, which always has to be true for simple geometrical shape without holes in it. Finally, we can determine the location of the center of mass for a triangle geometrically by connecting the center of each side to the opposite corner with a straight line (see drawing). The point at which these three lines intersect is the location of the center of mass. You can see from the graph that this point has to be very close to our calculated result of $(\tfrac{1}{3}B, \tfrac{1}{3}H)$.

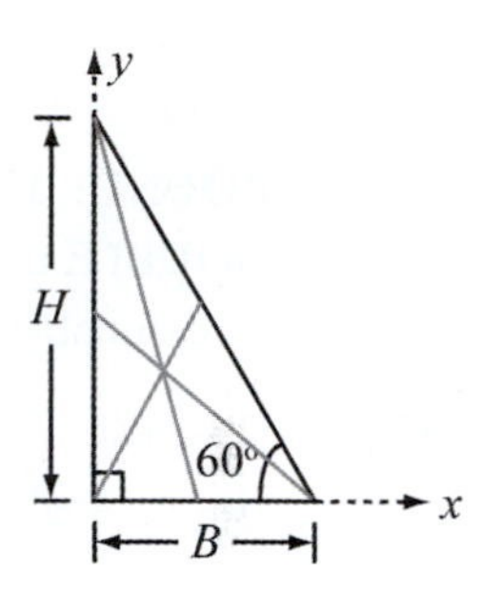

8.53. **THINK:** The linear mass density, $\lambda(x)$, is provided in the graph. Determine the location for the center of mass, X_{com}, of the object. From the graph, it can be seen that

$$\lambda(x) = \begin{cases} \dfrac{\lambda_0}{x_0}x, & 0 \le x < x_0 \\ \lambda_0, & x_0 \le x \le 2x_0 \end{cases}.$$

SKETCH:

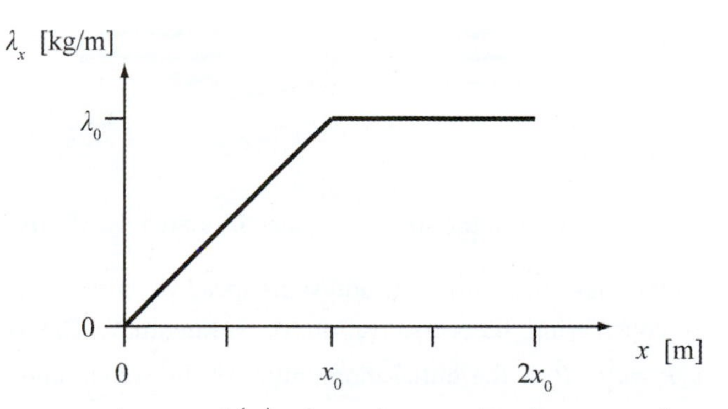

RESEARCH: The linear mass density, $\lambda(x)$, depends on x. To determine the center of mass, use the equation $X_{\text{com}} = \dfrac{1}{M}\int_L x\lambda(x)dx.$ The mass of the system, M, can be determined using the equation $M = \int_L \lambda(x)dx.$ In order to evaluate the center of mass of the system, two separate regions must be considered; the region from $x = 0$ to $x = x_0$ and the region from $x = x_0$ to $x = 2x_0$. The equation for

X_{com} can be expanded to $X_{com} = \frac{1}{M}\int_0^{x_0} x\frac{\lambda_0}{x_0}xdx + \frac{1}{M}\int_{x_0}^{2x_0}\lambda_0 xdx.$ The equation for M is $M = \int_0^{x_0}\frac{\lambda_0}{x_0}xdx + \int_{x_0}^{2x_0}\lambda_0 dx.$

SIMPLIFY: Simplify the expression for M first and then substitute it into the expression for X_{com}.

$$M = \int_0^{x_0}\frac{\lambda_0}{x_0}xdx + \int_{x_0}^{2x_0}\lambda_0 dx = \left[\frac{1}{2}\frac{\lambda_0}{x_0}x^2\right]_0^{x_0} + \left[x\lambda_0\right]_{x_0}^{2x_0} = \frac{1}{2}\lambda_0 x_0 + 2x_0\lambda_0 - x_0\lambda_0 = \frac{3}{2}x_0\lambda_0.$$

Substitute the above expression into the equation for X_{com} to get:

$$X_{com} = \frac{2}{3x_0\lambda_0}\left[\int_0^{x_0} x^2\frac{\lambda_0}{x_0}dx + \int_{x_0}^{2x_0}\lambda_0 xdx\right] = \frac{2}{3x_0\lambda_0}\left[\frac{1}{3}\lambda_0 x_0^2 + 2\lambda_0 x_0^2 - \frac{1}{2}\lambda_0 x_0^2\right] = \frac{2}{3x_0\lambda_0}\left[\lambda_0 x_0^2\left(\frac{2}{6}+\frac{12}{6}-\frac{3}{6}\right)\right]$$

$$= \frac{2}{3x_0\lambda_0}\left(\frac{11}{6}\lambda_0 x_0^2\right) = \frac{11x_0}{9}.$$

CALCULATE: This step does not apply.

ROUND: This step does not apply.

DOUBLE-CHECK: The units for the result are units of length, so the answer is dimensionally correct. It is reasonable that the calculated value is closer to the denser end of the object.

8.57. **THINK:** The mass of the battleship is $m_s = 136{,}634{,}000$ lbs. The ship has twelve 16-inch guns and each gun is capable of firing projectiles of mass, $m_p = 2700.$ lb, at a speed of $v_p = 2300.$ ft/s. Three of the guns fire projectiles in the same direction. Determine the recoil velocity, v_s, of the ship. Assume the ship is initially stationary.

SKETCH:

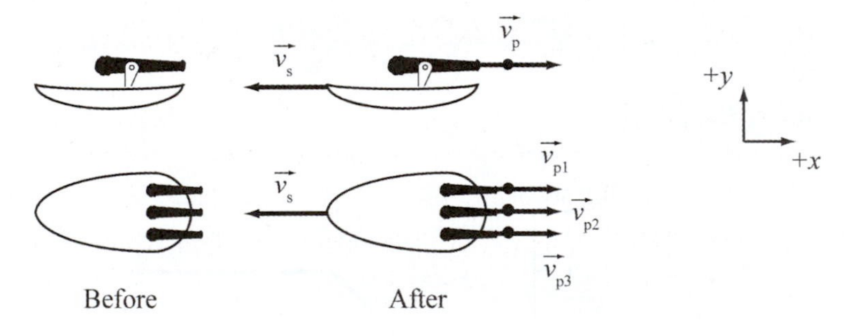

RESEARCH: The total mass of the ship and projectile system is $M = m_s + \sum_{i=1}^{n} m_{pi}$.

All of the projectiles have the same mass and same speed when they are shot from the guns. This problem can be solved considering the conservation of momentum. The equation for the conservation of momentum is $\vec{P}_i = \vec{P}_f$. $\vec{P}_i$ is the initial momentum of the system and $\vec{P}_f$ is the final momentum of the system. Assume the ship carries one projectile per gun. $\vec{P}_i = 0$ because the battleship is initially at rest and $\vec{P}_f = -(m_s + 9m_p)v_s + 3m_p v_p$.

SIMPLIFY: $\vec{P}_i = \vec{P}_f \Rightarrow 0 = -(m_s + 9m_p)v_s + 3m_p v_p \Rightarrow v_s = \frac{3m_p v_p}{(m_s + 9m_p)}$

CALCULATE: $v_s = \frac{3(2700.\text{ lb})(2300.\text{ ft/s})}{(136{,}634{,}000\text{ lb} + 9(2700.\text{ lb}))} = 0.136325\text{ ft/s}$

ROUND: The values for the mass and speed of the projectile that are given in the question have four significant figures, so the result should be rounded to $v_s = 0.1363$ ft/s. The recoil velocity is in opposite direction than the cannons fire.

DOUBLE-CHECK: The mass of the ship is much greater than the masses of the projectiles, so it reasonable that the recoil velocity is small because momentum depends on mass and velocity.

8.61. **THINK:** The student's mass is $m_s = 40.0$ kg, the ball's mass is $m_b = 5.00$ kg and the cart's mass is $m_c = 10.0$ kg. The ball's relative speed is $v_b' = 10.0$ m/s and the student's initial speed is $v_{si} = 0$. Determine the ball's velocity with respect to the ground, $\vec{v}_b$, after it is thrown.

SKETCH:

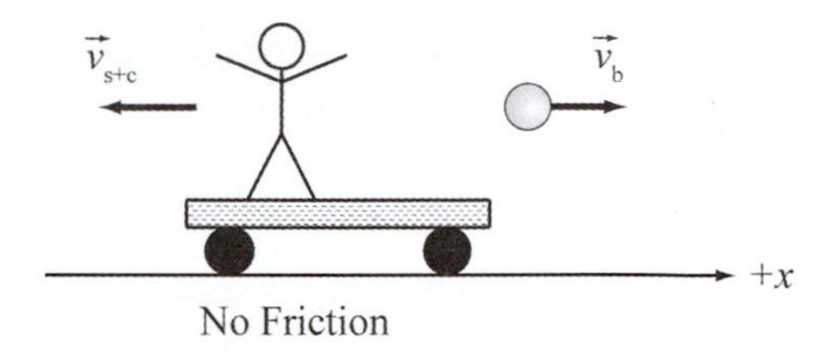

RESEARCH: $\vec{v}_b$ can be determined by considering the conservation of momentum, $\vec{P}_i = \vec{P}_f$, where $p = mv$. Note the ball's relative speed is $\vec{v}_b' = \vec{v}_b - \vec{v}_{s+c}$, where $\vec{v}_b$ and $\vec{v}_{s+c}$ are measured relative to the ground.

SIMPLIFY: $\vec{P}_i = \vec{P}_f \Rightarrow 0 = (m_s + m_c)\vec{v}_{s+c} + m_b\vec{v}_b \Rightarrow 0 = (m_s + m_c)(\vec{v}_b - \vec{v}_b') + m_b\vec{v}_b \Rightarrow \vec{v}_b = \dfrac{\vec{v}_b'(m_s + m_c)}{m_s + m_c + m_b}$

CALCULATE: $\vec{v}_b = \dfrac{(10.0 \text{ m/s})(40.0 \text{ kg} + 10.0 \text{ kg})}{(40.0 \text{ kg} + 10.0 \text{ kg} + 5.00 \text{ kg})} = 9.0909 \text{ m/s}$

ROUND: $\vec{v}_b = 9.09$ m/s in the direction of $\vec{v}_b'$ (horizontal)

DOUBLE-CHECK: It is expected that $v_b < v_b'$ since the student and cart move away from the ball when it is thrown.

8.65. **THINK:** The rocket's initial mass is $M_0 = 2.80 \cdot 10^6$ kg. Its final mass is $M_1 = 8.00 \cdot 10^5$ kg. The time to burn all the fuel is $\Delta t = 160.$ s. The exhaust speed is $v = v_c = 2700.$ m/s. Determine (a) the upward acceleration, a_0, of the rocket as it lifts off, (b) its upward acceleration, a_1, when all the fuel has burned and (c) the net change in speed, Δv in time Δt in the absence of a gravitational force.

SKETCH:

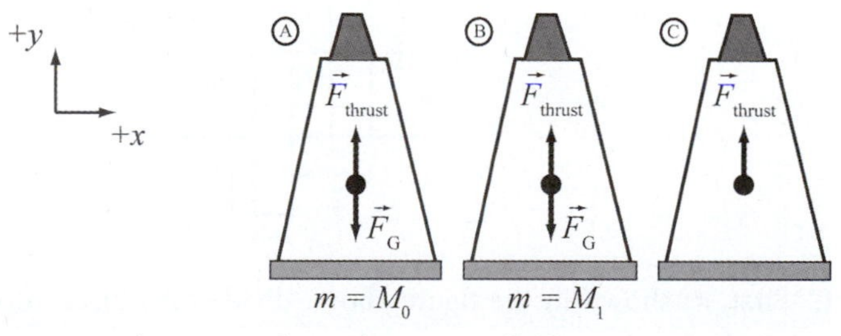

RESEARCH: To determine the upward acceleration, all the vertical forces on the rocket must be balanced. Use the following equations: $\vec{F}_{thrust} = -\vec{v}_c \dfrac{dm}{dt}$, $\vec{F}_g = m\vec{g}$, $\dfrac{dm}{dt} = \dfrac{\Delta m}{\Delta t}$. The mass of the fuel used is $\Delta m = M_0 - M_1$. To determine Δv in the absence of other forces (other than $\vec{F}_{thrust}$), use $v_f - v_i = v_c \ln(m_i / m_f)$.

SIMPLIFY:

(a) $\dfrac{dm}{dt}=\dfrac{M_0-M_1}{\Delta t}$

Balancing the vertical forces on the rocket gives

$$F_{\text{net}}=F_{\text{thrust}}-F_{\text{g}}=ma \Rightarrow M_0a_0=v_{\text{c}}\frac{dm}{dt}-M_0g \Rightarrow a_0=\frac{v_{\text{c}}}{M_0}\left(\frac{M_0-M_1}{\Delta t}\right)-g \Rightarrow a_0=\frac{v_{\text{c}}}{\Delta t}\left(1-\frac{M_1}{M_0}\right)-g.$$

(b) Similarly to part (a):

$$F_{\text{net}}=F_{\text{thrust}}-F_{\text{g}}=ma \Rightarrow M_1a_1=v_{\text{c}}\frac{dm}{dt}-M_1g \Rightarrow a_1=\frac{v_{\text{c}}}{M_1}\left(\frac{M_0-M_1}{\Delta t}\right)-g \Rightarrow a_1=\frac{v_{\text{c}}}{\Delta t}\left(\frac{M_0}{M_1}-1\right)-g.$$

(c) In the absence of gravity, $F_{\text{net}}=F_{\text{thrust}}$. The change in velocity due to this thrust force is $\Delta v=v_{\text{c}}\ln\left(M_0/M_1\right)$.

CALCULATE:

(a) $a_0=\left(\dfrac{2700.\text{ m/s}}{160\text{ s}}\right)\left(1-\dfrac{8.00\cdot 10^5\text{ kg}}{2.80\cdot 10^6\text{ kg}}\right)-9.81\text{ m/s}^2=2.244\text{ m/s}^2$

(b) $a_1=\left(\dfrac{2700.\text{ m/s}}{160.\text{ s}}\right)\left(\dfrac{2.80\cdot 10^6\text{ kg}}{8.00\cdot 10^5\text{ kg}}-1\right)-9.81\text{ m/s}^2=32.38\text{ m/s}^2$

(c) $\Delta v=\left(2700.\text{ m/s}\right)\ln\left(\dfrac{2.80\cdot 10^6\text{ kg}}{8.00\cdot 10^5\text{ kg}}\right)=3382\text{ m/s}$

ROUND:

(a) $a_0=2.24\text{ m/s}^2$

(b) $a_1=32.4\text{ m/s}^2$

(c) $\Delta v=3380\text{ m/s}$

DOUBLE-CHECK: It can be seen that $a_1>a_0$, as it should be since $M_1<M_0$. It is not unusual for Δv to be greater than v_{c}.

8.69. **THINK:** Determine the center of mass of an object which consists of regularly shaped metal of uniform thickness and density. Assume that the density of the object is ρ.

SKETCH:

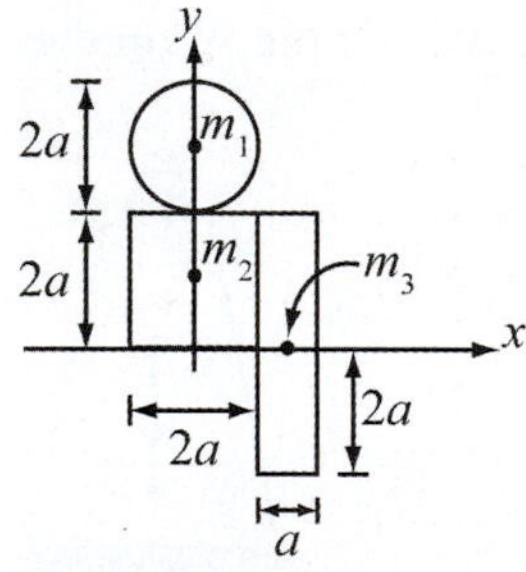

RESEARCH: First, as shown in the figure above, divide the object into three parts, m_1, m_2 and m_3. Determine the center of mass by using $\vec{R}=\dfrac{1}{M}\sum_{i=1}^{3}m_i\vec{r}_i$, or in component form $X=\dfrac{1}{M}\sum_{i=1}^{3}m_ix_i$ and $Y=\dfrac{1}{M}\sum_{i=1}^{3}m_iy_i$. Also, use $m=\rho At$ for the mass, where A is the area and t is the thickness.

SIMPLIFY: The center of mass components are given by:

$$X = \frac{m_1x_1 + m_2x_2 + m_3x_3}{M} \text{ and } Y = \frac{m_1y_1 + m_2y_2 + m_3y_3}{M}$$

The masses of the three parts are $m_1 = \rho\pi a^2 t$, $m_2 = \rho(2a)^2 t$ and $m_3 = \rho 4a^2 t$. The center of mass of the three parts are $x_1 = 0$, $y_1 = 3a$, $x_2 = 0$, $y_2 = a$, $x_3 = 3a/2$ and $y_3 = 0$. The total mass of the object is $M = m_1 + m_2 + m_3 = \rho\pi a^2 t + 4\rho a^2 t + 4\rho a^2 t = \rho a^2 t(8+\pi)$.

CALCULATE: The center of mass of the object is given by the following equations:

$$X = \frac{0 + 0 + 4\rho a^2 t(3a/2)}{\rho a^2 t(8+\pi)} = \left(\frac{6}{8+\pi}\right)a;$$

$$Y = \frac{\rho\pi a^2 t(3a) + 4\rho a^2 t(a) + 0}{\rho a^2 t(8+\pi)} = \left(\frac{4+3\pi}{8+\pi}\right)a.$$

ROUND: Rounding is not required.

DOUBLE-CHECK: The center of mass of the object is located in the area of m_2. By inspection of the figure this is reasonable.

8.73. **THINK:** There are two masses, $m_1 = 2.0$ kg and $m_2 = 3.0$ kg. The velocity of their center of mass and the velocity of mass 1 relative to mass 2 are $\vec{v}_{\text{cm}} = (-1.00\hat{x} + 2.40\hat{y})$ m/s and $\vec{v}_{\text{rel}} = (5.00\hat{x} + 1.00\hat{y})$ m/s. Determine the total momentum of the system and the momenta of mass 1 and mass 2.

SKETCH:

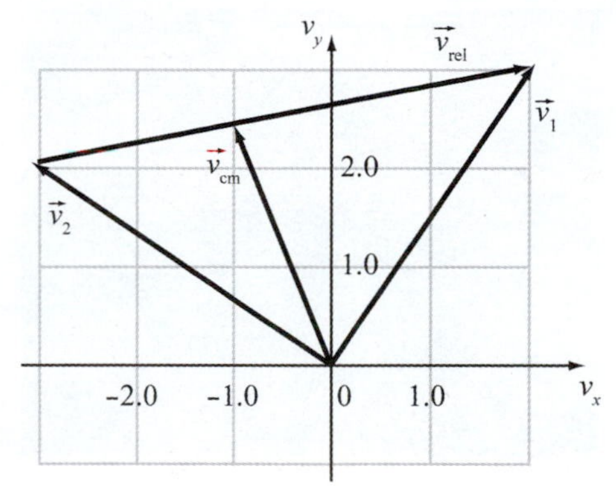

RESEARCH: The total momentum of the system is $\vec{p}_{\text{cm}} = M\vec{v}_{\text{cm}} = m_1\vec{v}_1 + m_2\vec{v}_2$. The velocity of mass 1 relative to mass 2 is $\vec{v}_{\text{rel}} = \vec{v}_1 - \vec{v}_2$.

SIMPLIFY: The total mass M of the system is $M = m_1 + m_2$. The total momentum of the system is given by $\vec{p}_{\text{cm}} = M\vec{v}_{\text{cm}} = (m_1 + m_2)\vec{v}_{\text{cm}} = m_1\vec{v}_1 + m_2\vec{v}_2$. Substitute $\vec{v}_2 = \vec{v}_1 - \vec{v}_{\text{rel}}$ into the equation for the total momentum of the system to get $M\vec{v}_{\text{cm}} = m_1\vec{v}_1 + m_2(\vec{v}_1 - \vec{v}_{\text{rel}}) = (m_1 + m_2)\vec{v}_1 - m_2\vec{v}_{\text{rel}}$. Therefore, $\vec{v}_1 = \vec{v}_{\text{cm}} + \frac{m_2}{M}\vec{v}_{\text{rel}}$. Similarly, substitute $\vec{v}_1 = \vec{v}_2 + \vec{v}_{\text{rel}}$ into the equation for the total momentum of the system to get $M\vec{v}_{\text{cm}} = m_1\vec{v}_{\text{rel}} + (m_1 + m_2)\vec{v}_2$ or $\vec{v}_2 = \vec{v}_{\text{cm}} - \frac{m_1}{M}\vec{v}_{\text{rel}}$. Therefore, the momentums of mass 1 and mass 2 are $\vec{p}_1 = m_1\vec{v}_1 = m_1\vec{v}_{\text{cm}} + \frac{m_1m_2}{M}\vec{v}_{\text{rel}}$ and $\vec{p}_2 = m_2\vec{v}_2 = m_2\vec{v}_{\text{cm}} - \frac{m_1m_2}{M}\vec{v}_{\text{rel}}$.

CALCULATE:

(a)

$\vec{p}_{\text{cm}} = (2.00 \text{ kg} + 3.00 \text{ kg})(-1.00\hat{x} + 2.40\hat{y})\text{ m/s} = (-5.00\hat{x} + 12.0\hat{y}) \text{ kg m/s}$

$\vec{p}_{\text{cm}} = (2.0 \text{ kg} + 3.0 \text{ kg})(-1.0\hat{x} + 2.4\hat{y})\text{ m/s} = (-5.0\hat{x} + 12\hat{y}) \text{ kg m/s}$

(b) $\vec{p}_1 = (2.00 \text{ kg})(-1.00\hat{x} + 2.40\hat{y}) \text{ m/s} + \frac{(2.00 \text{ kg})(3.00 \text{ kg})}{2.00 \text{ kg} + 3.00 \text{ kg}}(5.00\hat{x} + 1.00\hat{y}) \text{ m/s}$

$= (-2.00\hat{x} + 4.80\hat{y}) \text{ kgm/s} + (6.00\hat{x} + 1.20\hat{y}) \text{ kgm/s} = (4.00\hat{x} + 6.00\hat{y}) \text{ kgm/s}$

(c) $\vec{p}_2 = (3.00 \text{ kg})(-1.00\hat{x} + 2.40\hat{y}) \text{ m/s} - \frac{(2.00 \text{ kg})(3.00 \text{ kg})}{2.00 \text{ kg} + 3.00 \text{ kg}}(5.00\hat{x} + 1.00\hat{y}) \text{ m/s}$

$= (-3.00\hat{x} + 7.20\hat{y}) \text{ kgm/s} - (6.00\hat{x} + 1.20\hat{y}) \text{ kgm/s} = (-9.00\hat{x} + 6.00\hat{y}) \text{ kg m/s}$

ROUND: The answers have already been rounded to three significant figures.

DOUBLE-CHECK: It is clear from the results of (a), (b) and (c) that $\vec{p}_{\text{cm}} = \vec{p}_1 + \vec{p}_2$.

Multi-Version Exercises

8.76. **THINK:** This question asks about the fuel consumption of a satellite. This is an example of rocket motion, where the mass of the satellite (including thruster) decreases as the fuel is ejected.

SKETCH: The direction in which the xenon ions are ejected is opposite to the direction of the thrust. The velocity of the xenon with respect to the satellite and the thrust force are shown.

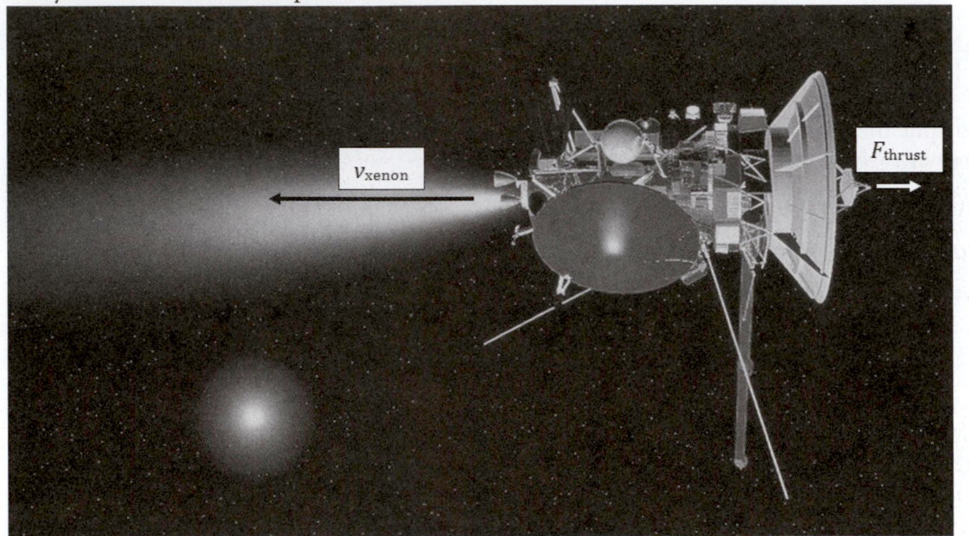

RESEARCH: The equation of motion for a rocket in interstellar space is given by $\vec{F}_{\text{thrust}} = -\vec{v}_c \frac{dm}{dt}$. The velocity of the xenon ions with respect to the shuttle is given in km/s and the force is given in Newtons, or kg · m / s^2. The conversion factor for the velocity is given by $\frac{1000 \text{ m/s}}{1 \text{ km/s}}$.

SIMPLIFY: Since the thrust and velocity act along a single axis, it is possible to use the scalar form of the equation, $F_{\text{thrust}} = -v_c \frac{dm}{dt}$. The rate of fuel consumption equals the change in mass (the loss of mass is due to xenon ejected from the satellite), so solve for $\frac{dm}{dt}$ to get $\frac{dm}{dt} = -\frac{F_{\text{thrust}}}{v_c}$.

CALCULATE: The question states that the speed of the xenon ions with respect to the rocket is $v_c = v_{\text{xenon}} = 21.45$ km/s. The thrust produced is $F_{\text{thrust}} = 1.187 \cdot 10^{-2}$ N. Thus the rate of fuel consumption is:

$$\frac{dm}{dt} = -\frac{F_{\text{thrust}}}{v_c}$$
$$= -\frac{1.187 \cdot 10^{-2} \text{ N}}{21.45 \text{ km/s} \cdot \frac{1000 \text{ m/s}}{1 \text{ km/s}}}$$
$$= -5.533799534 \cdot 10^{-7} \text{ kg/s}$$
$$= -1.992167832 \text{ g/hr}$$

ROUND: The measured values are all given to four significant figures, and the final answer should also have four significant figures. The thruster consumes fuel at a rate of $5.534 \cdot 10^{-7}$ kg/s or 1.992 g/hr.

DOUBLE-CHECK: Because of the cost of sending a satellite into space, the weight of the fuel consumed per hour should be pretty small; a fuel consumption rate of 1.992 g/hr is reasonable for a satellite launched from earth. Working backwards, if the rocket consumes fuel at a rate of $5.534 \cdot 10^{-4}$ g/s, then the thrust is

$$-21.45 \text{ km/s} \cdot \left(-5.534 \cdot 10^{-4} \text{ g/s}\right) = 0.01187 \text{ km} \cdot \text{g/s}^2 = 1.187 \cdot 10^{-2} \text{ N}$$

(the conversion factor is 1 km·g/s^2 = 1 kg·m/s^2). So, this agrees with the given thrust force of $1.187 \cdot 10^{-2}$ N.

8.79. **THINK:** This question asks about the speed of a satellite. This is an example of rocket motion, where the mass of the satellite (including thruster) decreases as the fuel is ejected.

SKETCH: The direction in which the xenon ions are ejected is opposite to the direction of the thrust. The velocity of the xenon with respect to the satellite and the thrust force are shown.

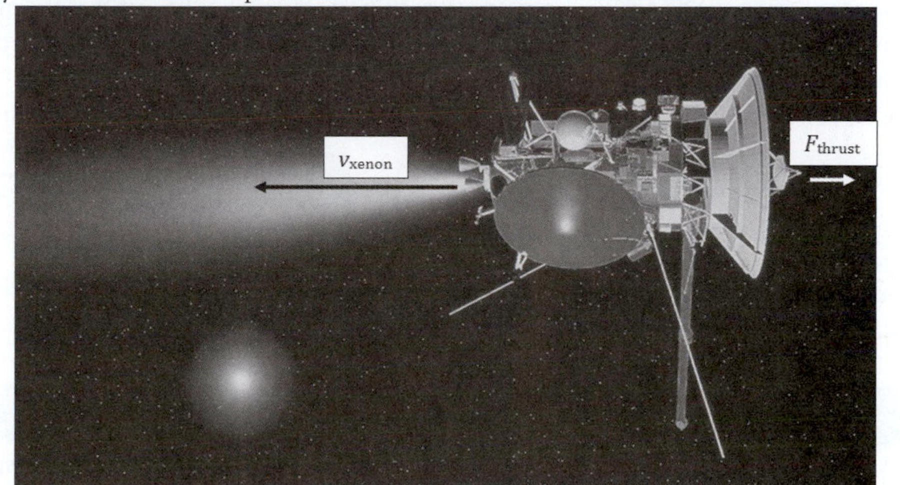

RESEARCH: Initially, the mass of the system is the total mass of the satellite, including the mass of the fuel: $m_{\text{i}} = m_{\text{satellite}}$ After all of the fuel is consumed, the mass of the system is equal to the mass of the satellite minus the mass of the fuel consumed: $m_{\text{f}} = m_{\text{satellite}} - m_{\text{fuel}}$. The change in speed of the satellite is given by the equation $v_{\text{f}} - v_{\text{i}} = v_c \ln(m_{\text{i}} / m_{\text{f}})$, where v_c is the speed of the xenon with relative to the satellite.

SIMPLIFY: To make the problem easier, choose a reference frame where the initial speed of the satellite equals zero. Then $v_{\text{f}} - v_{\text{i}} = v_{\text{f}} - 0 = v_{\text{f}}$, so it is necessary to find $v_{\text{f}} = v_c \ln(m_{\text{i}} / m_{\text{f}})$. Substituting in the masses of the satellite and fuel, this becomes $v_{\text{f}} = v_c \ln\left(m_{\text{satellite}} / \left[m_{\text{satellite}} - m_{\text{fuel}}\right]\right)$.

CALCULATE: The initial mass of the satellite (including fuel) is 2149 kg, and the mass of the fuel consumed is 23.37 kg. The speed of the ions with respect to the satellite is 28.33 km/s, so the final velocity of the satellite is:

$$\begin{aligned} v_{\text{f}} &= v_c \ln\left(m_{\text{satellite}} / \left[m_{\text{satellite}} - m_{\text{fuel}}\right]\right) \\ &= (28.33 \text{ km/s})\ln\left(\frac{2149 \text{ kg}}{2149 \text{ kg} - 23.37 \text{ kg}}\right) \\ &= 3.0977123 \cdot 10^{-1} \text{ km/s} \end{aligned}$$

ROUND: The measured values are all given to four significant figures, and the weight of the satellite minus the weight of the fuel consumed also has four significant figures, so the final answer will have four figures. The change in the speed of the satellite is $3.098 \cdot 10^{-1}$ km/s or 309.8 m/s.

DOUBLE-CHECK: Alhough the satellite is moving quickly after burning all of its fuel, this is not an unreasonable speed for space travel. Working backwards, if the change in speed was $3.098 \cdot 10^{-1}$ km/s, then the velocity of the xenon particles was $v_c = \dfrac{\Delta v_{\text{satellite}}}{\ln(m_{\text{i}} / m_{\text{f}})}$, or

$$v_c = \frac{3.098 \cdot 10^{-1} \text{km /s}}{\ln(2149 \text{ kg} / [2149 \text{ kg} - 23.37 \text{ kg}])} = 28.33 \text{ km /s}.$$

This agrees with the number given in the question, confirming that the calculations are correct.

8.82. **THINK:** The fisherman, boat, and tackle box are at rest at the beginning of this problem, so the total momentum of the fisherman, boat, and tackle box before and after the fisherman throws the tackle box must be zero. Using the principle of conservation of momentum and the fact that the momentum of the tackle box must cancel out the momentum of the fisherman and boat, it is possible to find the speed of the fisherman and boat after the tackle box has been thrown.

SKETCH: The sketch shows the motion of the tackle box, boat, and fisherman after the throw:

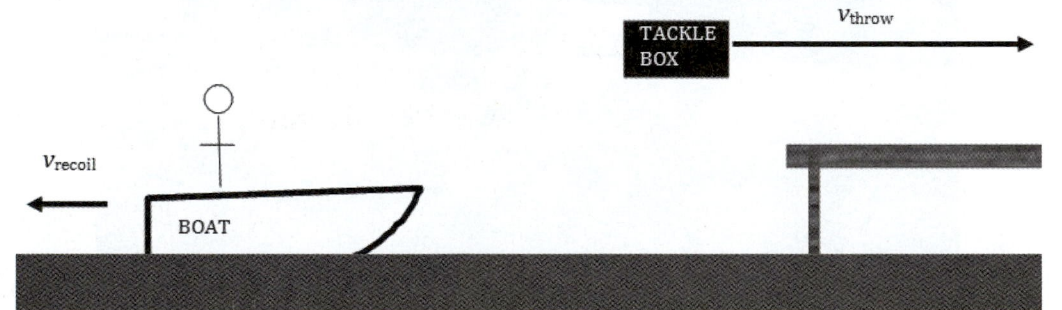

RESEARCH: The total initial momentum is zero, because there is no motion with respect to the dock. After the fisherman throws the tackle box, the momentum of the tackle box is $p_{\text{box}} = m_{\text{box}} v_{\text{box}} = m_{\text{box}} v_{\text{throw}}$ towards the dock. The total momentum after the throw must equal the total momentum before the throw, so the sum of the momentum of the box, the momentum of the boat, and the momentum of the fisherman must be zero: $p_{\text{box}} + p_{\text{fisherman}} + p_{\text{boat}} = 0$. The fisherman and boat both have the same velocity, so $p_{\text{fisherman}} = m_{\text{fisherman}} v_{\text{fisherman}} = m_{\text{fisherman}} v_{\text{recoil}}$ away from the dock and $p_{\text{boat}} = m_{\text{boat}} v_{\text{boat}} = m_{\text{boat}} v_{\text{recoil}}$ away from the dock.

SIMPLIFY: The goal is to find the recoil velocity of the fisherman and boat. Using the equation for momentum after the tackle box has been thrown, $p_{\text{box}} + p_{\text{fisherman}} + p_{\text{boat}} = 0$, substitute in the formula for the momenta of the tackle box, boat, and fisherman: $0 = m_{\text{box}}v_{\text{throw}} + m_{\text{fisherman}}v_{\text{recoil}} + m_{\text{boat}}v_{\text{recoil}}$. Solve for the recoil velocity:

$$m_{\text{box}}v_{\text{throw}} + m_{\text{fisherman}}v_{\text{recoil}} + m_{\text{boat}}v_{\text{recoil}} = 0$$
$$m_{\text{fisherman}}v_{\text{recoil}} + m_{\text{boat}}v_{\text{recoil}} = -m_{\text{box}}v_{\text{throw}}$$
$$v_{\text{recoil}}\left(m_{\text{fisherman}} + m_{\text{boat}}\right) = -m_{\text{box}}v_{\text{throw}}$$
$$v_{\text{recoil}} = -\frac{m_{\text{box}}v_{\text{throw}}}{m_{\text{fisherman}} + m_{\text{boat}}}$$

CALCULATE: The mass of the tackle box, fisherman, and boat, as well as the velocity of the throw (with respect to the dock) are given in the question. Using these values gives:

$$v_{\text{recoil}} = -\frac{m_{\text{box}}v_{\text{throw}}}{m_{\text{fisherman}} + m_{\text{boat}}} = -\frac{13.63 \text{ kg} \cdot 2.911 \text{ m/s}}{75.19 \text{ kg} + 28.09 \text{ kg}} = -0.3841685709 \text{ m/s}$$

ROUND: The masses and velocity given in the question all have four significant figures, and the sum of the mass of the fisherman and the mass of the boat has five significant figures, so the final answer should have four significant figures. The final speed of the fisherman and boat is –0.3842 m/s towards the dock, or 0.3842 m/s away from the dock.

DOUBLE-CHECK: It makes intuitive sense that the much more massive boat and fisherman will have a lower speed than the less massive tackle box. Their momenta should be equal and opposite, so a quick way to check this problem is to see if the magnitude of the tackle box's momentum equals the magnitude of the man and boat. The tackle box has a momentum of magnitude 13.63 kg · 2.911 m/s = 39.68 kg·m/s after it is thrown. The boat and fisherman have a combined mass of 103.28 kg, so their final momentum has a magnitude of 103.28 kg · 0.3842 m/s = 39.68 kg·m/s. This confirms that the calculations were correct.

8.85. **THINK:** The masses and initial speeds of both particles are known, so the momentum of the center of mass can be calculated. The total mass of the system is known, so the momentum can be used to find the speed of the center of mass.

SKETCH: To simplify the problem, choose the location of the particle at rest to be the origin, with the proton moving in the $+x$ direction. All of the motion is along a single axis, with the center of mass (COM) between the proton and the alpha particle.

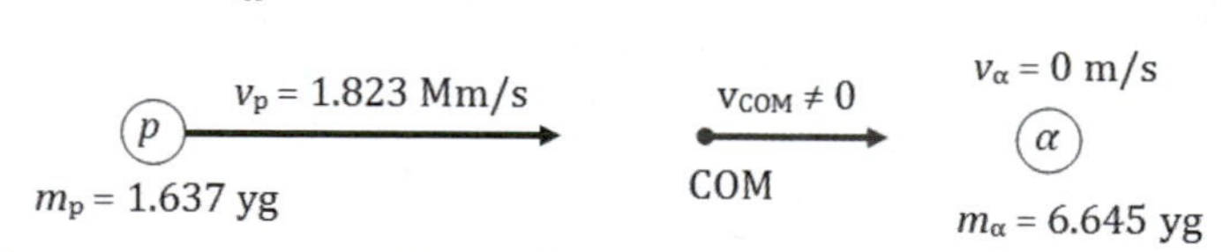

RESEARCH: The masses and velocities of the particles are given, so the momenta of the particles can be calculated as the product of the mass and the speed $p_\alpha = m_\alpha v_\alpha$ and $p_p = m_p v_p$ towards the alpha particle. The center-of-mass momentum can be calculated in two ways, either by taking the sum of the momenta of each particle ($P_{COM} = \sum_{i=0}^{n} p_i$) or as the product of the total mass of the system times the speed of the center of mass ($P_{COM} = M \cdot v_{COM}$).

SIMPLIFY: The masses of both particles are given in the problem, and the total mass of the system M is the sum of the masses of each particle, $M = m_p + m_\alpha$. The total momentum $P_{COM} = \sum_{i=0}^{n} p_i = p_\alpha + p_p$ and $P_{COM} = M \cdot v_{COM}$, so $M \cdot v_{COM} = p_\alpha + p_p$. Substitute for the momenta of the proton and alpha particle (since the alpha particle is not moving, it has zero momentum), substitute for the total mass, and solve for the velocity of the center of mass:

$$M \cdot v_{COM} = p_\alpha + p_p \Rightarrow$$

$$v_{COM} = \frac{p_\alpha + p_p}{M} = \frac{m_\alpha v_\alpha + m_p v_p}{m_\alpha + m_p} = \frac{m_\alpha \cdot 0 + m_p v_p}{m_\alpha + m_p} = \frac{m_p v_p}{m_\alpha + m_p}$$

CALCULATE: The problem states that the proton has a mass of $1.673 \cdot 10^{-27}$ kg and moves at a speed of $1.823 \cdot 10^6$ m/s towards the alpha particle, which is at rest and has a mass of $6.645 \cdot 10^{-27}$ kg. So the center of mass has a speed of

$$v_{COM} = \frac{m_p v_p}{m_p + m_\alpha} = \frac{(1.823 \cdot 10^6 \text{ m/s})(1.673 \cdot 10^{-27} \text{ kg})}{1.673 \cdot 10^{-27} \text{ kg} + 6.645 \cdot 10^{-27} \text{ kg}} = 3.666601346 \cdot 10^5 \text{ m/s}$$

ROUND: The masses of the proton and alpha particle, as well as their sum, have four significant figures. The speed of the proton also has four significant figures. The alpha particle is at rest, so its speed is not a calculated value, and the zero speed does not change the number of figures in the answer. Thus, the speed of the center of mass is $3.667 \cdot 10^5$ m/s, and the center of mass is moving towards the alpha particle.

DOUBLE-CHECK: To double check, find the location of the center of mass as a function of time, and take the time derivative to find the velocity. The distance between the particles is not given in the problem, so call the distance between the particles at an arbitrary starting time $t = 0$ to be d_0. The positions of each particle can be described by their location along the axis of motion, $r_\alpha = 0$ and $r_p = d_0 + v_p t$. Using this, the location of the center of mass is

$$R_{COM} = \frac{1}{m_p + m_\alpha}\left(r_p m_p + r_\alpha m_\alpha\right).$$

Take the time derivative to find the velocity:

$$\begin{aligned}
\frac{d}{dt}R_{\text{COM}} &= \frac{d}{dt}\left[\frac{1}{m_{\text{p}}+m_{\alpha}}\left(r_{\text{p}}m_{\text{p}}+r_{\alpha}m_{\alpha}\right)\right] \\
&= \frac{1}{m_{\text{p}}+m_{\alpha}}\frac{d}{dt}\left[\left(d_0+v_{\text{p}}t\right)m_{\text{p}}+0\cdot m_{\alpha}\right] \\
&= \frac{1}{m_{\text{p}}+m_{\alpha}}\frac{d}{dt}\left(d_0m_{\text{p}}+v_{\text{p}}m_{\text{p}}t+0\right) \\
&= \frac{1}{m_{\text{p}}+m_{\alpha}}\frac{d}{dt}\left(d_0m_{\text{p}}+v_{\text{p}}m_{\text{p}}t\right) \\
&= \frac{1}{m_{\text{p}}+m_{\alpha}}\left(0+v_{\text{p}}m_{\text{p}}\right) \\
&= \frac{v_{\text{p}}m_{\text{p}}}{m_{\text{p}}+m_{\alpha}} \\
&= \frac{\left(1.823\cdot 10^{6}\text{ m/s}\right)\left(1.673\cdot 10^{-27}\text{ kg}\right)}{1.673\cdot 10^{-27}\text{ kg}+6.645\cdot 10^{-27}\text{ kg}} \\
&= 3.666601346\cdot 10^{5}\text{ m/s}
\end{aligned}$$

This agrees with the earlier result.

Chapter 9: Circular Motion

Exercises

9.31. **THINK:** Determine the change in the angular position in radians. Winter lasts roughly a fourth of a year. There are 2π radians in a circle. Consider the orbit of Earth to be circular.

SKETCH:

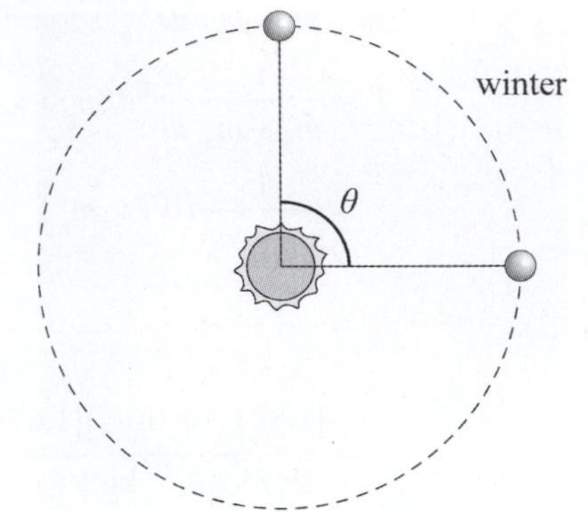

RESEARCH: The angular velocity of the earth is $\omega = 2\pi / \text{yr}$. The angular position is given by $\theta = \theta_0 + \omega_0 t$.

SIMPLIFY: $\Delta\theta = \theta - \theta_0 = \omega_0 t$

CALCULATE: $\Delta\theta = \frac{2\pi \text{ rad}}{\text{yr}}\left(\frac{1}{4}\text{ yr}\right) = \frac{\pi \text{ rad}}{2} = \frac{3.14 \text{ rad}}{2} = 1.57 \text{ rad}$

ROUND: Since π is used to three significant figures, the angle the Earth sweeps over winter is 1.57 rad. It would also be entirely reasonable to leave the answer as $\pi / 2$ radians.

DOUBLE-CHECK: This value makes sense, since there are four seasons of about equal length, so the angle should be a quarter of a circle.

9.35. **THINK:** Determine the average angular acceleration of the record and its angular position after reaching full speed. The initial and final angular speeds are 0 rpm to 33.3 rpm. The time of acceleration is 5.00 s.

SKETCH:

RESEARCH: The equation for angular acceleration is $\alpha = (\omega_{\text{f}} - \omega_{\text{i}}) / \Delta t$. The angular position of an object under constant angular acceleration is given by $\theta = \frac{1}{2}\alpha t^2$.

SIMPLIFY: There is no need to simplify the equation.

CALCULATE: $\alpha = \frac{33.3 \text{ rpm} - 0 \text{ rpm}}{5.00 \text{ s}(60 \text{ s/min})} = 0.111 \text{ rev/s}^2 \cdot \frac{2\pi \text{ rad}}{1 \text{ rev}} = 0.6974 \text{ rad/s}^2$

$$\theta = \frac{1}{2}(0.111 \text{ rev/s})(5.00 \text{ s})^2 = 1.3875 \text{ rev} \cdot \frac{2\pi \text{ rad}}{1 \text{ rev}} = 8.718 \text{ rad}$$

ROUND: To three significant figures, the angular acceleration and position are:

(a) $\alpha = 0.697 \text{ rad/s}^2$

(b) $\theta = 8.72 \text{ rad}$

DOUBLE-CHECK: The calculations yield the correct units of radians and rad/s^2.

9.39. **THINK:** Determine (a) the magnitude and direction of the velocities of the pendulum at position *A* and *B*, (b) the angular speed of the pendulum motion, (c) the period of the rotation and (d) the effects of moving the pendulum to the equator. The latitude of the pendulum is $55.0°$ above the equator. The pendulum swings over a distance of $d = 20.0$ m. The period of the Earth's rotation is $T_\text{E} = 23 \text{ hr} + 56 \text{ min} = 86160 \text{ s}$ and the Earth's radius is $R_\text{E} = 6.37 \cdot 10^6$ m.

SKETCH:

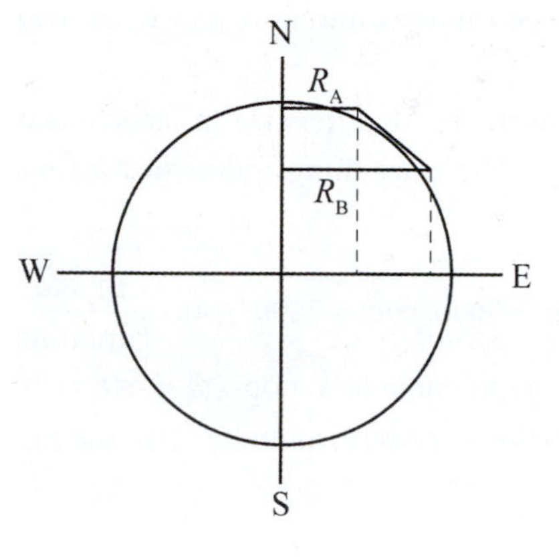

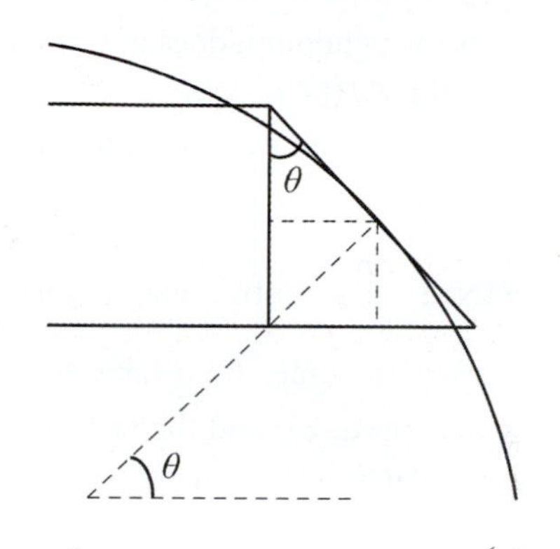

RESEARCH: The following equations can be used: $\omega = \dfrac{2\pi}{T}$, $v = r\omega$, $R_\text{A} = R_\text{E}\cos\theta - \left(\dfrac{d}{2}\sin\theta\right)$ and $R_\text{B} = R_\text{E}\cos\theta + \left(\dfrac{d}{2}\sin\theta\right)$.

SIMPLIFY: The magnitudes of the velocities are:

$$v_\text{A} = R_\text{A}\omega_\text{A} = \frac{2\pi R_\text{A}}{T_\text{E}} = \frac{2\pi\left(R_\text{E}\cos\theta - \left(\frac{d}{2}\sin\theta\right)\right)}{T_\text{E}} \text{ and } v_\text{B} = \frac{2\pi\left(R_\text{E}\cos\theta + \left(\frac{d}{2}\sin\theta\right)\right)}{T_\text{E}}.$$

The angular speed of the rotation is related to the linear speed by $\Delta v = \omega_\text{R} d$. Rearranging gives:

$$\omega_\text{R} = \frac{\Delta v}{d} = \left(\frac{1}{d}\right)\frac{2\pi}{T_\text{E}}\left(\left(R_\text{E}\cos\theta + \left(\frac{d}{2}\sin\theta\right)\right) - \left(R_\text{E}\cos\theta - \left(\frac{d}{2}\sin\theta\right)\right)\right) = \frac{2\pi}{dT_\text{E}}d\sin\theta = \frac{2\pi}{T_\text{E}}\sin\theta.$$

The period is then $T_R = \dfrac{2\pi}{w_R} = \dfrac{2\pi}{\frac{2\pi}{T_\text{E}}\sin\theta} = \dfrac{T_\text{E}}{\sin\theta}$. At the equator, $\theta = 0°$.

CALCULATE:

(a) $$v_\text{A} = 2\pi\left(\frac{\left(6.37 \cdot 10^6 \text{ m}\right)\cos(55.0°) - (10.0 \text{ m})\sin(55.0°)}{86{,}160 \text{ s}}\right) = 266.44277 \text{ m/s}$$

$$v_\text{B} = 2\pi\left(\frac{\left(6.37 \cdot 10^6 \text{ m}\right)\cos(55.0°) + (10.0 \text{ m})\sin(55.0°)}{86{,}160 \text{ s}}\right) = 266.44396 \text{ m/s}$$

$\Delta v = v_\text{B} - v_\text{A} = 266.44396 \text{ m/s} - 266.44277 \text{ m/s} = 0.00119 \text{ m/s}$ or 1.19 mm/s

(b) $$\omega_\text{R} = \frac{2\pi\sin(55.0°)}{86{,}160 \text{ s}} = 5.97 \cdot 10^{-5} \text{ rad/s}$$

(c) $T_{\text{R}} = \dfrac{86{,}160 \text{ s}}{\sin(55.0°)} = 105{,}182 \text{ s}$ or about 29.2 hours

(d) At the equator, $T_R = \lim\limits_{\theta \to 0} \dfrac{T_{\text{E}}}{\sin\theta} = \infty.$

ROUND: The values given in the question have three significant figures, so the answers should also be rounded to three significant figures:

(a) The velocities are $v_A = 266.44277$ m/s and $v_{\text{B}} = 266.44396$ m/s, are in the direction of the Earth's rotation eastward. This means the difference between the velocities is $\Delta v = 1.19$ mm/s.

(b) The angular speed of rotation is $w_{\text{R}} = 1.19 \cdot 10^{-4}$ rad/s.

(c) The period of rotation is about 29.2 hours.

(d) At the equator there is no difference between the velocities at *A* and *B*, so the period is $T_{\text{R}} = \infty$. This means the pendulum does not rotate.

DOUBLE-CHECK: These are reasonable answers. If the difference in velocities was larger, these effects would be seen in everyday life but they are not. These are things pilots deal with when planning a flight path.

9.43. **THINK:** The initial angular speed is $\omega_0 = 3600. \text{ rpm} = 3600. \text{ rpm} \cdot \dfrac{2\pi \text{ rad}}{1 \text{ rotation}} \cdot \dfrac{1 \text{ min}}{60 \text{ s}} = 120\pi \text{ rad/s}.$

Calculate the time, t_1, it takes for the centrifuge to come to a stop $(\omega_1 = \omega(t_1) = 0)$ by using the average angular speed, $\bar{\omega}$, and the fact that that it completes $n = 60.0$ rotations. Use the time taken to stop to find the angular acceleration.

SKETCH:

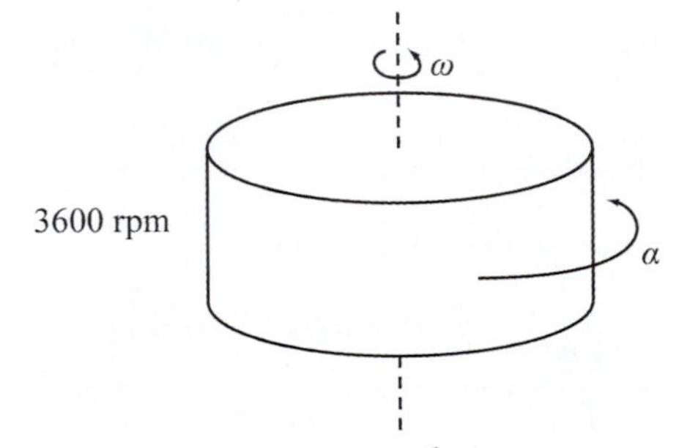

RESEARCH: The average angular speed is given by $\bar{\omega} = \frac{1}{2}(\omega_{\text{f}} + \omega_0)$. Since the centrifuge completes 60.0 turns while decelerating, it turns through an angle of $\Delta\theta = 60.0 \text{ turns} \cdot \dfrac{2\pi \text{ rad}}{\text{turn}} = 120\pi \text{ rad}$. Use the two previous calculated values in the formula $\Delta\theta = \bar{\omega} t_1$ to obtain the time taken to come to a stop. Then, use the equation $\omega(t) = \omega_0 + \alpha t$ to compute the angular acceleration, α.

SIMPLIFY: The time to decelerate is given by, $t_1 = \dfrac{\Delta\theta}{\bar{\omega}} = \dfrac{\Delta\theta}{(\omega_1 + \omega_0)/2}$. Substituting this into the last equation given in the research step gives the equation, $\omega(t_1) = \omega_1 = \omega_0 + \alpha \dfrac{2\Delta\theta}{(\omega_1 + \omega_0)}$. Solving for α yields the equation: $\alpha = \dfrac{(\omega_1 - \omega_0)(\omega_1 + \omega_0)}{2\Delta\theta}$.

CALCULATE: $\alpha = \dfrac{(0 - 120\pi \text{ rad/s})(0 + 120\pi \text{ rad/s})}{2(120\pi \text{ rad})} = -60\pi \text{ rad/s}^2 = -188.496 \text{ rad/s}^2$

ROUND: Since the number of rotations is given to three significant figures, the final result should be also rounded to three significant figures: $\alpha = -188 \text{ rad/s}^2$.

DOUBLE-CHECK: The negative sign of α indicates deceleration, which is appropriate since the centrifuge is coming to a stop. The centrifuge decelerates from 120π rad/s to rest in $t_1 = \frac{2\Delta\theta}{(\omega_1 + \omega_0)} = \frac{2(120\pi \text{ rad})}{(0+120\pi \text{ rad/s})} = 2$ s, and since the angular deceleration is constant, it must be the case that the deceleration is 60π rad/s^2. The answer is therefore reasonable.

9.47. **THINK:** Determine the angular speed of the take-up spool in a tape recorder in the following cases:

(a) When the take-up spool is empty with radius, $r_\text{e} = 0.800$ cm.

(b) When the take-up spool is full with radius, $r_\text{f} = 2.20$ cm.

(c) Determine the average angular acceleration of the take-up spool if the length of the tape is $l = 100.80$ m. The magnetic tape has a constant linear speed of $v = 5.60$ cm/s.

SKETCH:

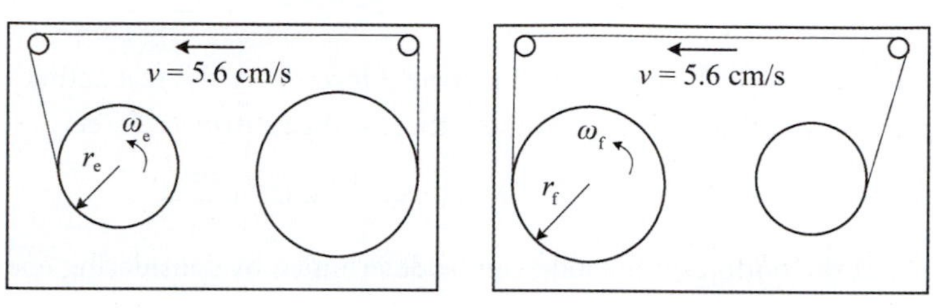

RESEARCH:

(a) & (b) To determine the angular speed, make use of the relationship $v = \omega r \Rightarrow \omega = \frac{v}{r}$.

(c) To determine an average angular acceleration, use the definition, $\alpha = \Delta w / \Delta t$, where the time is determined from $\Delta t = \frac{(\text{distance})}{(\text{speed})} = \frac{l}{v}$.

SIMPLIFY:

(a) $\omega_\text{e} = \frac{v}{r_\text{e}}$

(b) $\omega_\text{f} = \frac{v}{r_\text{f}}$

(c) $\alpha = \frac{\Delta\omega}{\Delta T} = \frac{\omega_\text{f} - \omega_\text{e}}{l / v} = \frac{v(\omega_\text{f} - \omega_\text{e})}{l}$

CALCULATE:

(a) $\omega_e = \frac{5.60 \cdot 10^{-2} \text{ m/s}}{8.00 \cdot 10^{-3} \text{ m}} = 7.00$ rad/s

(b) $\omega_\text{f} = \frac{5.60 \cdot 10^{-2} \text{ m/s}}{2.20 \cdot 10^{-2} \text{ m}} = 2.54$ rad/s

(c) $\alpha = \frac{(5.60 \cdot 10^{-2} \text{ m/s})(2.54 - 7.00)}{100.80 \text{ m}} = -2.48 \cdot 10^{-3} \text{ rad/s}^2$

ROUND: Keep three significant figures:

(a) $\omega_e = 7.00$ rad/s

(b) $\omega_\text{f} = 2.54$ rad/s

(c) $\alpha = -2.48 \cdot 10^{-3}$ rad/s^2

DOUBLE-CHECK: It is reasonable that the angular speed of the spool when it's empty is greater than when it's full. Also, it is expected that the angular acceleration is negative since the angular speed is decreasing as the spool gets full.

9.51. **THINK:** The apparent weight of a rider on a roller coaster at the bottom of the loop is to be determined. From Solved Problem 9.1, the radius is $r = 5.00$ m, and the speed at the top of the loop is 7.00 m/s.

SKETCH:

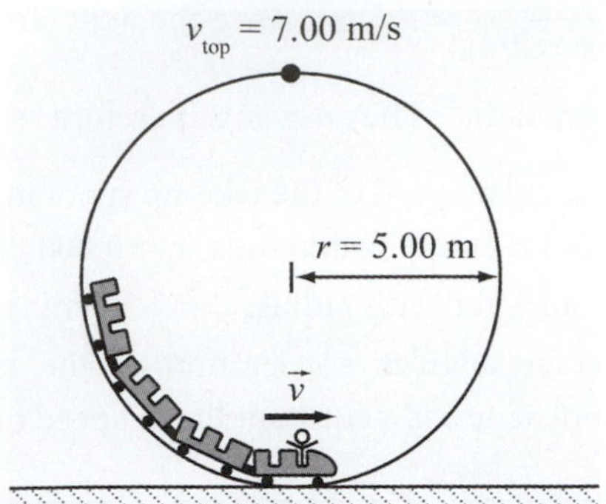

RESEARCH: The apparent weight is the normal force from the seat acting on the rider. At the bottom of the loop the normal force is the force of gravity plus the centripetal force:

$$N = F_g + \frac{mv^2}{r} = mg + \frac{mv^2}{r}.$$

The velocity at the bottom of the loop can be determined by considering energy conservation between the configuration at the top and that at the bottom:

$$\frac{1}{2}mv^2 = mgh + \frac{1}{2}mv_t^2$$

where $h = 2r$. In Solved Problem 9.1 it as determined that the feeling of weightlessness at the top is achieved if $\frac{mv_t^2}{r} = mg$.

SIMPLIFY: Multiply the equation for energy conservation by a factor of $2/r$ and find:

$$\frac{mv^2}{r} = \frac{2mgh}{r} + \frac{mv_t^2}{r}.$$

Since $h = 2r$, this results in:

$$\frac{mv^2}{r} = 4mg + \frac{mv_t^2}{r}.$$

Insert this for the normal force and see

$$N = mg + \frac{mv^2}{r} = mg + 4mg + \frac{mv_t^2}{r} = mg + 4mg + mg = 6mg.$$

CALCULATE: Not needed.

ROUND: Not needed.

DOUBLE-CHECK: Our result means that you experience 6*g* of acceleration at the bottom of the loop, which seems like a large number, if you consider that the maximum acceleration during the launch of a Space Shuttle is kept to 3*g*. However, if you have ever had the opportunity to ride on such a roller coaster, then our result does not seem unreasonable.

9.55. **THINK:** Determine the maximum speed of a car as it goes over the top of a hill such that the car always touches the ground. The radius of curvature of the hill is 9.00 m. As the car travels over the top of the hill it undergoes circular motion in the vertical plane. The only force that can provide the centripetal force for this motion is gravity. Clearly, for small speeds the car remains in contact with the road due to gravity. But the car will lose contact if the centripetal acceleration exceeds gravity.

SKETCH:

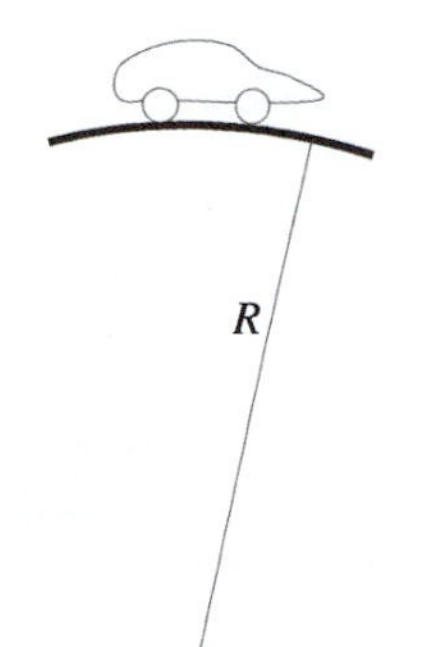

RESEARCH: In the limiting case of the maximum speed we can set the centripetal acceleration equal to g: $g = v_{max}^2 / r$.

SIMPLIFY: Solve for the maximum speed and find $v_{max} = \sqrt{gr}$.

CALCULATE: $v_{max} = \sqrt{gr} = \sqrt{(9.81 \text{ m/s}^2)(9.00 \text{ m})} = 9.40 \text{ m/s}$

ROUND: Since the radius is given to three significant figures, the result is $v_{max} = 9.40$ m/s.

DOUBLE-CHECK: This speed of 9.40 m/s, which is approximately 21.0 mph, seems very small. But on the other hand, this is a very significant curvature at the top of the hill, equivalent to a good-sized speed bump. Going over this type of bump at more than 21 mph makes it likely that your car will lose contact with the road surface.

9.59. **THINK:** A speedway turn has a radius, R, and is banked at an angle of θ above the horizontal. This problem is a special case of Solved Problem 9.4, and the results of that solved problem will be used to obtain a solution to this problem. Determine:

(a) The optimal speed to take the turn when there is little friction present.

(b) The maximum and minimum speeds at which to take the turn if there is now a coefficient of static friction, μ_s.

(c) The value for parts (a) and (b) if $R = 400.$ m, $\theta = 45.0°$, and $\mu_s = 0.700$.

SKETCH:

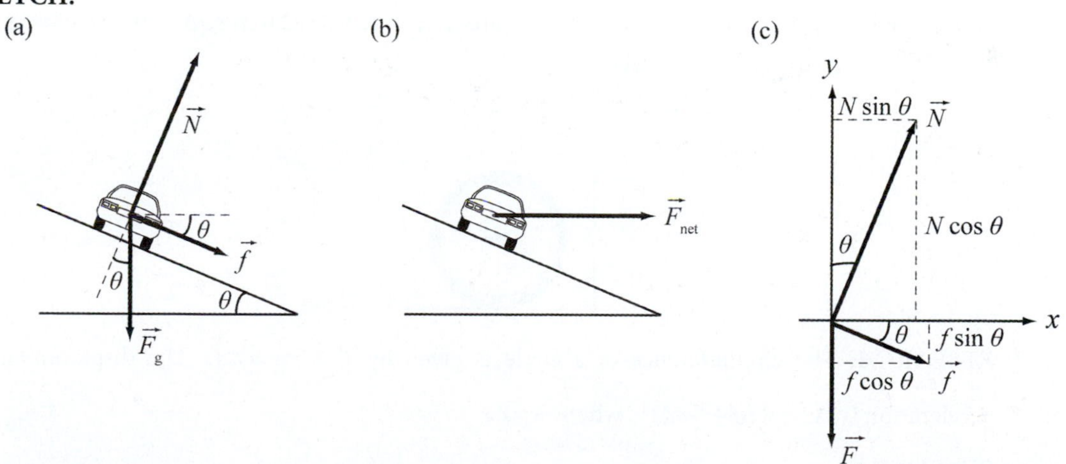

RESEARCH: It was found in Solved Problem 9.4 that the maximum speed a car can go through the banked curve is given by

$$v_{max} = \sqrt{\frac{Rg(\sin\theta + \mu_s \cos\theta)}{\cos\theta - \mu_s \sin\theta}}.$$

SIMPLIFY:

(a) For the case of zero friction the case above approaches the limit of $v_{\text{zero friction}} = \sqrt{\dfrac{Rg\sin\theta}{\cos\theta}} = \sqrt{Rg\tan\theta}$.

(b) For the maximum speed we can use the formula already quoted above. The minimum speed that the car can travel through the curve is given by reversing the direction of the friction force. In this case the friction force points up the bank, because it needs to prevent the car from sliding down. Reversing the sign of the friction force leads to $v_{\min} = \sqrt{\dfrac{Rg(\sin\theta - \mu_s\cos\theta)}{\cos\theta + \mu_s\sin\theta}}$.

CALCULATE:

(c) For the results from part (a):

$$v_{\text{zero friction}} = \sqrt{(400.\text{ m})(9.81\text{ m/s}^2)\tan 45.0^\circ} = 62.64184\text{ m/s}.$$

For the results from part (b), the minimum speed is:

$$v_{\min} = \sqrt{\frac{(400.\text{ m})(9.81\text{ m/s}^2)(\sin 45.0^\circ - 0.700\cos 45.0^\circ)}{\cos 45.0^\circ + 0.700\sin 45.0^\circ}} = 26.31484\text{ m/s}.$$

and the maximum speed is:

$$v_{\max} = \sqrt{\frac{(400.\text{ m})(9.81\text{ m/s}^2)(\sin 45.0^\circ + 0.700\cos 45.0^\circ)}{\cos 45.0^\circ - 0.700\sin 45.0^\circ}} = 149.1174\text{ m/s}.$$

ROUND:

(a) Not applicable.

(b) Not applicable.

(c) $v_{\text{zero friction}} = 62.6$ m/s, $v_{\min} = 26.3$ m/s and $v_{\max} = 149$ m/s.

DOUBLE-CHECK: The results are reasonable considering that the friction-free speed should be within the minimum and maximum speed. The values for the given parameters are consistent with experiment.

9.63. THINK:

(a) If the distance traveled can be determined, then the number of revolutions the tires made can be determined, since the diameter of the tires is known.

(b) The linear speed of the tires and the diameter of the tires are known, so the angular speed can be determined. The known variables are $v_i = 0$, $v_f = 22.0$ m/s, $\Delta t = 9.00$ s, $d = 58.0$ cm. Use $1\dfrac{\text{rev}}{\text{s}} = 2\pi\dfrac{\text{rad}}{\text{s}} \Rightarrow 1\dfrac{\text{rad}}{\text{s}} = \dfrac{1}{2\pi}\dfrac{\text{rev}}{\text{s}}$.

SKETCH:

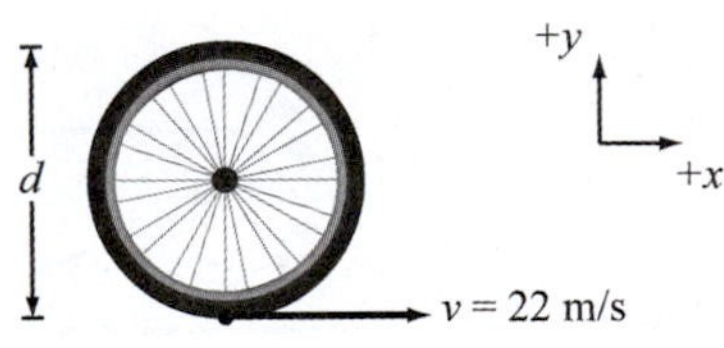

RESEARCH: The circumference of a circle is given by $C = 2\pi r = \pi d$. The displacement at constant acceleration is $\Delta x = v_i\Delta t + \dfrac{1}{2}a\Delta t^2$, where $v = \omega r$.

SIMPLIFY:

(a) $v_i = 0 \Rightarrow \Delta x = \dfrac{1}{2}a\Delta t^2,\ a = \dfrac{\Delta v}{\Delta t} \Rightarrow \Delta x = \dfrac{1}{2}\dfrac{\Delta v}{\Delta t}\Delta t^2 = \dfrac{1}{2}\Delta v\Delta t$

Let N = number of revolutions and the displacement is given by $\Delta x = \left(\dfrac{\text{displacement}}{\text{revolution}}\right)N$. The displacement per revolution is simply the circumference, C, so

$$\Delta x = CN \Rightarrow N = \frac{\Delta x}{C} = \frac{1}{\pi d}\left(\frac{1}{2}\Delta v \Delta t\right) = \frac{\Delta v \Delta t}{2\pi d}.$$

(b) $\omega = \frac{v}{r} = \frac{v}{d/2} = \frac{2v}{d}$

CALCULATE:

(a) $N = \frac{(22.0 \text{ m/s})(9.00 \text{ s})}{2\pi(0.58 \text{ m})} = 54.33$ revolutions

(b) $\omega = \frac{2(22.0 \text{ m/s})}{0.58 \text{ m}} = 75.86 \text{ rad/s} = \frac{75.86}{2\pi} \text{ rev/s} = 12.07 \text{ rev/s}$

ROUND: The results should be rounded to three significant figures.

(a) $N = 54.3$ revolutions

(b) $\omega = 12.1$ rev/s

DOUBLE-CHECK: For the given values, these results are reasonable.

9.67. **THINK:** The acceleration is uniform during the given time interval. The average angular speed during this time interval can be determined and from this, the angular displacement can be determined. **SKETCH:**

RESEARCH: $\omega_{\text{avg}} = \frac{\omega_{\text{f}} + \omega_{\text{i}}}{2}$, $\Delta\theta = \omega_{\text{avg}}\Delta t$

SIMPLIFY: Simplification is not necessary.

CALCULATE: $\omega_{\text{i}} = 33.33 \text{ rpm} = 33.33 \text{ rpm}\left(\frac{2\pi}{60 \text{ s}}\right) = 3.491 \text{ rad/s}$, $\omega_{\text{f}} = 0$

$$\Delta\theta = \left(\frac{3.491}{2} \text{ rad s}^{-1}\right)(15.0 \text{ s}) = 26.18 \text{ rad}$$

number of rotations $= \frac{\Delta\theta}{2\pi} = 4.167$

ROUND: Rounding the result to three significant figures gives the number of rotations to be 4.17 rotations.

DOUBLE-CHECK: These are reasonable results for a turntable.

9.71. **THINK:** Parts (a) and (b) can be solved using the constant angular acceleration equations. For part (c), calculate the angular displacement and, from this, compute the total arc-length, which is equal to the distance traveled. **SKETCH:**

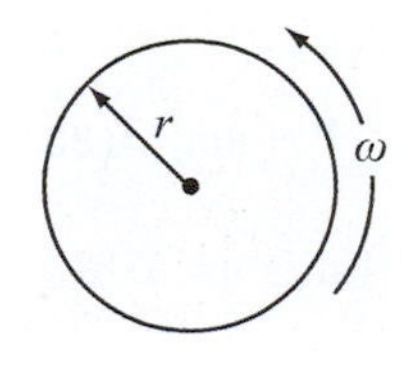

RESEARCH:

(a) $v = \omega r$

(b) $\alpha = \frac{\Delta\omega}{\Delta t}$, $\Delta\theta = \omega_i \Delta t + \frac{1}{2}\alpha \Delta t^2$

(c) $s = r\Delta\theta,\ \Delta\theta = 2\pi(\text{total revs.})$

SIMPLIFY:

(a) $\omega_{\text{i}} = \dfrac{v_{\text{i}}}{r}$

(b) $\Delta\omega = \omega_{\text{f}} - \omega_{\text{i}} = 0 - \dfrac{v_{\text{i}}}{r} = -\omega_{\text{i}},\ \Delta t = \dfrac{\Delta\omega}{\alpha} = -\dfrac{\omega_{\text{i}}}{\alpha}$

$$\Delta\theta = \omega_{\text{i}}\Delta t + \frac{1}{2}\alpha\Delta t^2 = \omega_{\text{i}}\left(\frac{-\omega_{\text{i}}}{\alpha}\right) + \frac{\alpha}{2}\left(\frac{-\omega_{\text{i}}}{\alpha}\right)^2 = \frac{-\omega_{\text{i}}^2}{\alpha} + \frac{\omega_{\text{i}}^2}{2\alpha} = -\frac{\omega_{\text{i}}^2}{2\alpha} \Rightarrow \alpha = \frac{-\omega_{\text{i}}^2}{2\Delta\theta}$$

(c) $s = r\Delta\theta$

CALCULATE:

(a) $\omega_{\text{i}} = \dfrac{35.8 \text{ m/s}}{0.550 \text{ m}} = 65.09 \text{ s}^{-1}$

(b) $\alpha = \dfrac{-(65.09 \text{ s}^{-1})^2}{2(2\pi(40.2))} = -8.387 \text{ s}^{-2}$

(c) $s = (0.550 \text{ m})(2\pi(40.2)) = 138.92 \text{ m}$

ROUND: Rounding the results to three significant figures:

(a) $\omega_{\text{i}} = 65.1 \text{ s}^{-1}$

(b) $\alpha = -8.39 \text{ s}^{-2}$

(c) $s = 139 \text{ m}$

DOUBLE-CHECK: For the parameters given, these are reasonable results.

9.75. **THINK:** From the linear speed and the radius, the centripetal acceleration can be determined. With the pilots' mass, the centripetal force can also be determined. The pilot's apparent weight is the combined effect of gravitational and centripetal accelerations.

SKETCH:

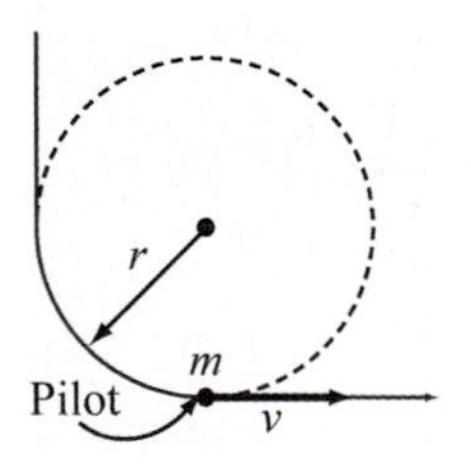

RESEARCH:

(a) $a_{\text{c}} = \dfrac{v^2}{r},\ F_{\text{c}} = \dfrac{mv^2}{r}$

(b) $F_{\text{c}} = \dfrac{mv^2}{r},\ F_{\text{g}} = mg,\ w = \dfrac{mv^2}{r} + mg$

SIMPLIFY: Simplification is not necessary.

CALCULATE:

(a) $a_{\text{c}} = \dfrac{(500. \text{ m/s})^2}{4000. \text{ m}} = 62.50 \text{ m/s}^2,\quad F_{\text{c}} = ma_{\text{c}} = (80.0 \text{ kg})(62.50 \text{ m/s}^2) = 5.00 \cdot 10^3 \text{ N}$

(b) $w = 5.00 \cdot 10^3 \text{ N} + (80.0 \text{ kg})(9.81 \text{ m/s}^2) = 5784.8 \text{ N}$

ROUND: Round the results to three significant figures:

(a) $a_{\text{c}} = 62.5 \text{ m/s}^2$ and $F_{\text{c}} = 5.00 \cdot 10^3 \text{ N}$

(b) $w = 5780 \text{ N}$

DOUBLE-CHECK: These are all reasonable values.

9.79. **THINK:** Use the relationship between angular and centripetal acceleration. The given values are $r_s = 2.75$ m, $r_c = 6.00$ m, $\omega_i = 0$, $\omega_f = 0.600$ rev/s and $\Delta t = 8.00$ s.

SKETCH:

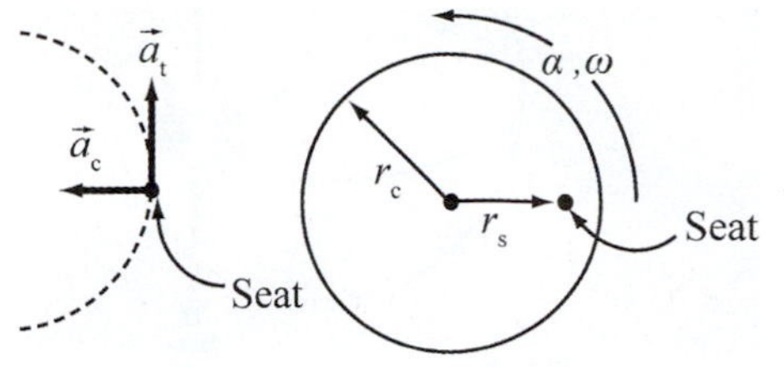

RESEARCH:

(a) $\alpha = \dfrac{\Delta\omega}{\Delta t}$

(b) $a_c = \dfrac{v_s^2}{r_s}, \quad v_s = \omega_s r_s$

(c) $\vec{a} = \vec{a}_c + \vec{a}_t, \quad a_t = \alpha r_s$

SIMPLIFY:

(a) Simplification is not necessary.

(b) $a_c = \omega_s^2 r_s$

(c) $a = \sqrt{a_c^2 + a_t^2}, \quad \tan\theta = \left(\dfrac{a_t}{a_c}\right)$

CALCULATE:

(a) $\alpha = \dfrac{0.600(2\pi)\text{ rad/s}}{8.00\text{ s}} = 0.4712\text{ rad/s}^2$

(b) At 8.00 s, $\omega_s = 0.600$ rev/s, so $a_c = \left(0.600(2\pi)\text{ s}^{-1}\right)^2(2.75\text{ m}) = 39.08\text{ m/s}^2$ and $\alpha = 0.4712\text{ rad/s}^2$.

(c) $a = \sqrt{\left(39.08\text{ m/s}^2\right)^2 + \left(0.4712\text{ s}^{-2}\right)^2(2.75\text{ m})^2} = 39.10\text{ m/s}^2 \quad \theta = \tan^{-1}\dfrac{\left(0.4712\text{ s}^{-2}\right)(2.75\text{ m})}{39.08\text{ m/s}^2} = 1.899°$

If the centripetal acceleration is along the positive x axis, then the direction of the total acceleration is $1.90°$ along the horizontal (rounded to three significant figures).

ROUND: Values are given to three significant figures, so the results should be rounded accordingly.

(a) $\alpha = 0.471\text{ rad/s}^2$

(b) $a_c = 39.1\text{ m/s}^2$ and $\alpha = 0.471\text{ rad/s}^2$.

(c) $a = 39.1\text{ m/s}^2$ at $\theta = 1.90°$.

DOUBLE-CHECK: The total acceleration is quite close to the centripetal acceleration, since the tangential acceleration and the angular acceleration are both quite small.

Multi-Version Exercises

9.82. **THINK:** The only values given in this problem are the radius of the sphere and the coefficient of static friction between the motorcycle and the sphere. The motorcycle will stay on the surface as long as the vertical force exerted by the force of friction is at least as much as the weight of the motorcycle. The friction force is proportional to the normal force exerted by the wall of the dome, which is given by the centripetal force. Combine these to solve for the minimum velocity.

SKETCH:

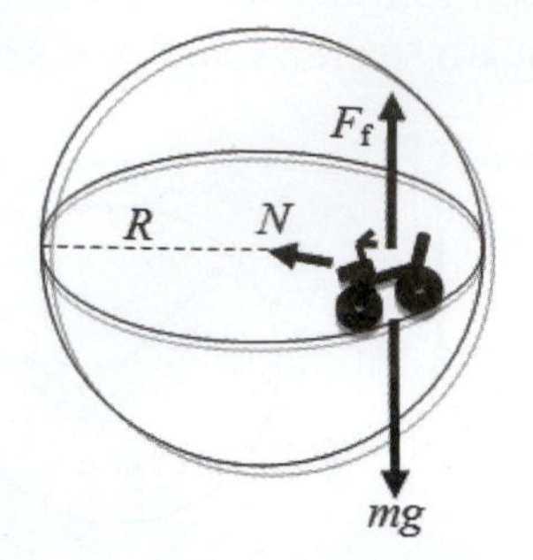

RESEARCH: The centripetal force required to keep the motorcycle moving in a circle is $F_c = \dfrac{mv^2}{R}$. The friction force is $F_f = \mu_s N$, and it must support the weight of the motorcycle, s o $F_f \geq mg$.

SIMPLIFY: Since the normal force equals the centripetal force in this case, substitute F_c for N in the equation $F_f = \mu_s N$ to get $F_f = \mu_s F_c = \mu_s \dfrac{mv^2}{R}$. Combine this with the fact that the frictional force must be enough to support the weight of the motorcycle, so $mg \leq F_f = \mu_s \dfrac{mv^2}{R}$. Finally, solve the inequality for the velocity (keep in mind that the letters represent positive values):

$$\mu_s \frac{mv^2}{R} \geq mg \Rightarrow$$

$$\frac{R}{\mu_s m} \cdot \mu_s \frac{mv^2}{R} \geq \frac{R}{\mu_s m} \cdot mg \Rightarrow$$

$$v^2 \geq \frac{Rg}{\mu_s}$$

$$v \geq \sqrt{\frac{Rg}{\mu_s}}$$

CALCULATE: The radius of the sphere is 12.61 m, and the coefficient of static friction is 0.4601. The gravitational acceleration near the surface of the earth is about 9.81 m/s², so the speed must be:

$$v \geq \sqrt{\frac{Rg}{\mu_s}}$$

$$v \geq \sqrt{\frac{12.61 \text{ m} \cdot 9.81 \text{ m/s}^2}{0.4601}}$$

$$v \geq 16.3970579 \text{ m/s}$$

ROUND: Since the measured values are all given to four significant figures, the final answer will also have four figures. The minimum velocity is 16.40 m/s.

DOUBLE-CHECK: In this case, the motorcycle is traveling at 16.40 m/s, or about 59 kilometers per hour, which is a reasonable speed based on how fast motorcycles *can* go. It needs to travel 12.61(2π) = 79.23 meters to go all the way around the sphere, so it makes one revolution every 4.83 seconds, or between 12 and 13 revolutions per minute. These values all seem reasonable based on past experience with motorcycles.

9.85. **THINK:** The speed of a point on the tip of the propeller can be calculated from the angular speed and the length of the propeller blade. The angular speed of the propeller can be calculated from the frequency. Find the maximum length of the propeller blade such that the angular speed at the tip of the propeller blade is less than the indicated speed of sound.

SKETCH: A view, looking towards the airplane from the front, is shown.

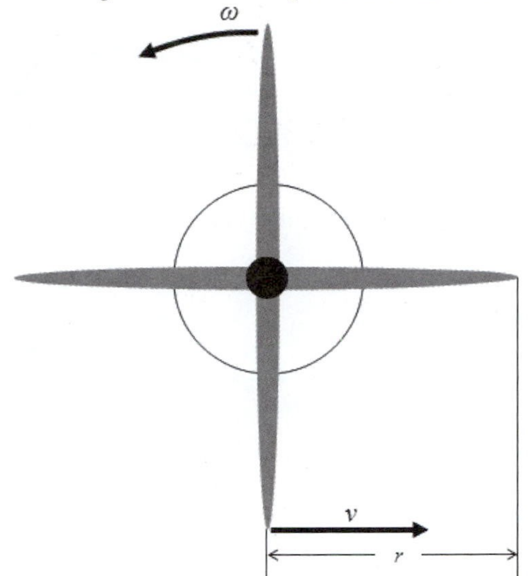

RESEARCH: The linear velocity should be less than the speed of sound $v \le v_{\text{sound}}$. The magnitude of the linear velocity v is equal to the product of the radius of rotation r and the angular speed ω: $v = r\omega$. The angular speed is related to the rotation frequency by $\omega = 2\pi f$. The length of the propeller blade is twice the radius of the propeller ($d = 2r$). Finally, note that the rotation frequency is given in revolutions per minute and the speed of sound is given in meters per second, so a conversion factor of 60 seconds / minute will be needed.

SIMPLIFY: Use the equation for the linear speed ($v = r\omega$) and the equation for the rotation frequency to get $v = 2\pi f \cdot r$. Use this in the inequality $v \le v_{\text{sound}}$ to find that $2\pi f \cdot r \le v_{\text{sound}}$. Solve this for the length of the propeller blade r (note that ω is a positive number of revolutions per minute) to get $r \le \frac{v_{\text{sound}}}{2\pi f}$. The maximum length of the propeller blade is two times the largest possible value of $d = 2r = \frac{v_{\text{sound}}}{\pi f}$.

CALCULATE: The angular frequency f is given in the problem as 2403 rpm and the speed of sound is 343.0 m/s. The maximum length of the propeller blade is thus $d = \frac{343.0 \text{ m/s} \cdot 60 \text{ s/min}}{\pi \cdot 2403 \text{ rev/min}} = 2.726099649 \text{ m}$.

ROUND: The measured values from the problem (the angular frequency and speed of sound) are given to four significant figures, so the final answer should also have four significant figures. The maximum length of a propeller blade is 2.726 m.

DOUBLE-CHECK: For those familiar with propeller-driven aircraft, a total propeller length of about 2.7 m seems reasonable. Working backwards, if the propeller blade is 2.726 m and the linear speed at the tip of the propeller is 343.0 m/s, then the angular speed is $\omega = \frac{v}{r} = \frac{343.0 \text{ m/s}}{1.363 \text{ m}}$. The angular frequency is then $f = \frac{\omega}{2\pi} = \frac{343.0 \text{ m/s}}{2\pi \cdot 1.363 \text{ m}} = 40.05 \text{ rev/sec}$. Since there are 60 seconds in a minute, this agrees with the value of 2403 rev/min given in the problem, and the calculations were correct.

9.87. **THINK:** The linear acceleration can be computed from the change in the speed of the car and the time required to accelerate, both of which are given in the problem. The angular acceleration can be calculated from the linear acceleration and the radius of the tires. Since the car's acceleration is constant and it starts at rest, the motion of the car occurs in only one direction, which can be taken to be the +x direction, and the time that the car starts moving can be taken as time zero.

SKETCH: The car starts at rest, so the constant acceleration and velocity are in the same direction.

RESEARCH: The constant linear acceleration is the change in speed per unit time $a = \frac{\Delta v}{\Delta t}$. The relationship between linear acceleration a and angular acceleration α is given by $a = r\alpha$, where r is the radius of the rotating object.

SIMPLIFY: Since there are two expressions for the linear acceleration, $a = \frac{\Delta v}{\Delta t}$ and $a = r\alpha$, they must be equal to one another: $r\alpha = \frac{\Delta v}{\Delta t}$. Solve for the angular acceleration α to get $\alpha = \frac{\Delta v}{r\Delta t}$. The car starts at rest at time zero, the final velocity is equal to Δv and the total time is equal to Δt, giving $\alpha = \frac{v}{rt}$.

CALCULATE: After 3.945 seconds, the car's final speed is 29.13 m/s. The rear wheels have a radius of 46.65 cm, or $46.65 \cdot 10^{-2}$ m. The angular acceleration is then

$$\alpha = \frac{29.13 \text{ m/s}}{46.65 \cdot 10^{-2} \text{ m} \cdot 3.945 \text{ s}} = 15.82857539 \text{ s}^{-2}$$

ROUND: The time in seconds, radius of the tires, and speed of the car are all given to four significant figures, so the final answer should also have four figures. The angular acceleration of the car is 15.83 s^{-2}.

DOUBLE-CHECK: First note that the units (per second per second) are correct for angular acceleration. Working backwards, if the sports car accelerates with an angular acceleration of 15.83 s^{-2} for 3.945 seconds, it will have a final angular speed of $(15.83 \cdot 3.945) \text{ s}^{-1}$. With a tire radius of 46.65 cm, this means that the car's final speed will be $(46.65 \cdot 15.83 \cdot 3.945)$ cm/s, or 29.13 m/s (when rounded to four significant figures), which agrees with the problem statement. This confirms that the first set of calculations was correct.

9.90. **THINK:** The frequency and radius of the flywheel can be used to calculate the speed at the edge of the flywheel. The centripetal acceleration can be calculated from the linear speed and the radius of the flywheel.

SKETCH:

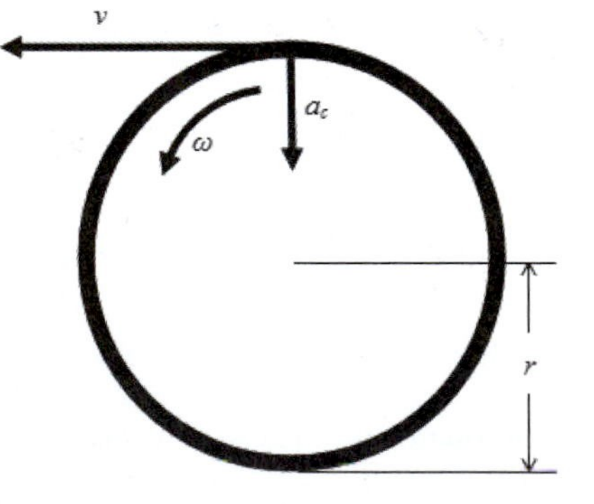

RESEARCH: The centripetal acceleration at the edge of the flywheel is $a_c = \frac{v^2}{r}$, where v is the linear speed at the edge of the flywheel and r is the flywheel's radius. The linear speed v is equal to the angular speed times the radius of the flywheel ($v = r\omega$), and the angular speed ω is related to the frequency f by the equation $\omega = 2\pi f$. The numbers are given in centimeters and revolutions per minute, so conversion factors of $\frac{1\text{ m}}{100\text{ cm}}$ and $\frac{1\text{ min}}{60\text{sec}}$ may be needed.

SIMPLIFY: First, find the equation for the velocity in terms of the angular frequency to get $v = r\omega = r(2\pi f)$. Use this in the equation for centripetal acceleration to find

$$a_c = \frac{v^2}{r} = \frac{(2\pi rf)^2}{r} = 4r(\pi f)^2 .$$

CALCULATE: The radius is 27.01 cm, or 0.2701 m and the frequency of the flywheel is 4949 rpm. So the angular acceleration is $4 \cdot 27.01\text{ cm}\left(\pi \cdot 4949 \frac{\text{rev}}{\text{min}}\right)^2 = 2.611675581 \cdot 10^{10} \frac{\text{cm}}{\text{min}^2}$. Converting to more familiar units, this becomes

$$2.611675581 \cdot 10^{10} \frac{\text{cm}}{\text{min}^2} \cdot \frac{1\text{ m}}{100\text{ cm}} \cdot \left(\frac{1\text{min}}{60\text{sec}}\right)^2 = 7.254654393 \cdot 10^4\text{ m/s}^2 .$$

ROUND: The radius and frequency of the flywheel both have four significant figures, so the final answer should also have four figures. The centripetal acceleration at a point on the edge of the flywheel is $7.255 \cdot 10^4$ m/s^2.

DOUBLE-CHECK: Work backwards to find the frequency from the centripetal acceleration and the radius of the flywheel. The linear velocity is $v = \sqrt{a_c r}$, the angular speed is $\omega = v / r = \frac{\sqrt{a_c r}}{r}$, and the frequency $f = \frac{\omega}{2\pi} = \frac{\sqrt{a_c r}}{2\pi r}$. The radius of the flywheel is 0.2701 m and the centripetal acceleration is $7.255 \cdot 10^4$ m/s^2, so the frequency is

$$\begin{aligned} f &= \frac{\sqrt{a_c r}}{2\pi r} \\ &= \frac{\sqrt{7.255 \cdot 10^4\text{ m/s}^2 \cdot 0.2701\text{ m}}}{2\pi \cdot 0.2701\text{ m}} \\ &= 82.4853\text{ s}^{-1} \cdot \frac{60\text{sec}}{1\text{min}} \\ &= 4949.117882\text{ min}^{-1} \end{aligned}$$

After rounding to four significant figures, this agrees with the frequency given in the problem of 4949 rpm (revolutions per minute).

Chapter 10: Rotation

Exercises

10.39. **THINK:** The children can be treated as point particles on the edge of a circle and placed so they are all the same distance, R, from the center. Using the conversion, 1 kg = 2.205 lbs, the three masses are $m_1 = 27.2$ kg, $m_2 = 20.4$ kg and $m_3 = 36.3$ kg. Using the conversion 1 m = 3.281 ft, the distance is $R =$ 3.657 m.

SKETCH:

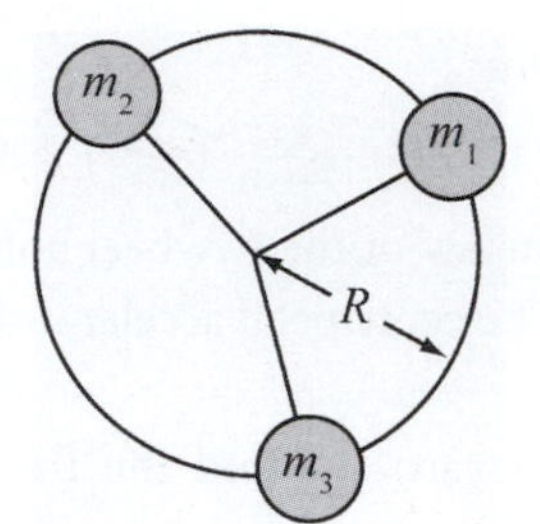

RESEARCH: The moment of inertia for point particles is given by $I = \sum_i m_i r_i^2$.

SIMPLIFY: $I = (m_1 + m_2 + m_3)R^2$

CALCULATE: $I = (27.2 \text{ kg} + 20.4 \text{ kg} + 36.3 \text{ kg})(3.66 \text{ m})^2 = 1123.9 \text{ kg m}^2$

ROUND: $I = 1.12 \cdot 10^3 \text{ kg m}^2$

DOUBLE-CHECK: Since the children are located on the edge of the merry-go-round, a large value for I is expected.

10.43. **THINK:** The change in energy should be solely that of the change in rotational kinetic energy. Assume the pulsar is a uniform solid sphere with $m \approx 2 \cdot 10^{30}$ kg and $R = 12$ km. Initially, the pulsar rotates at $\omega = 60\pi$ rad/s and has a period, T which is increased by 10^{-5} s after 1 y. We calculate the power emitted by the pulsar by taking the time derivative of the rotational kinetic energy. The power output of the Sun is $P_{\text{Sun}} = 4 \cdot 10^{26}$ W.

SKETCH:

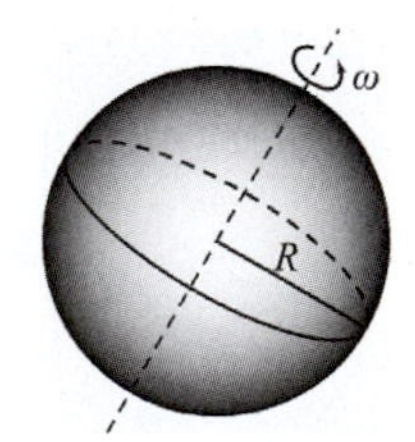

RESEARCH: The kinetic energy is given by $K = I\omega^2 / 2$, so

$$P_{\text{Crab}} = -\frac{dK}{dt} = -I\omega\frac{d\omega}{dt}.$$

The angular velocity is given by $\omega = 2\pi / T$, so

$$\frac{d\omega}{dt} = -\frac{2\pi}{T^2}\frac{dT}{dt} = -\frac{2\pi}{(2\pi/\omega)^2}\frac{dT}{dt} = -\frac{\omega^2}{2\pi}\frac{dT}{dt}.$$

The moment of inertia of a sphere is

$$I = \frac{2}{5}mR^2.$$

SIMPLIFY: Combining our equations gives us

$$P_{\text{Crab}} = \frac{2}{5}mR^2\omega\frac{\omega^2}{2\pi}\frac{dT}{dt} = \frac{mR^2\omega^3}{5\pi}\frac{dT}{dt}.$$

CALCULATE: First we calculate the change in the period over one year,

$$\frac{dT}{dt} = \frac{10^{-5}\text{ s}}{(365\text{ days})(24\text{ hour/day})(3600\text{ s/hour})} = 3.17\cdot 10^{-13}.$$

The power emitted by the pulsar is

$$P_{\text{Crab}} = \frac{dK}{dt} = \frac{(2\cdot 10^{30}\text{ kg})(12\cdot 10^3\text{ m})^2(60\pi\,\text{rad/s})^3}{5\pi}(3.17\cdot 10^{-13}) = 3.89\cdot 10^{31}\text{ W}.$$

So the ratio of the power emitted by the pulsar to the power emitted by the Sun is

$$\frac{P_{\text{Crab}}}{P_{\text{Sun}}} = \frac{3.89\cdot 10^{31}\text{ W}}{4\cdot 10^{26}\text{ W}} = 9.73\cdot 10^4.$$

ROUND: $\frac{P_{\text{Crab}}}{P_{\text{Sun}}} = 1\cdot 10^5.$

DOUBLE-CHECK: Our result for the ratio of the loss in rotational energy of the Crab Pulsar is close to the expected value of 100,000.

10.47. **THINK:** The hanging block with a mass of $M = 70.0$ kg, will cause a tension, T, in the string that will in turn produce a torque, τ, in the wheel with a mass, $m = 30.0$ kg, and a radius, $R = 40.0$ cm. This torque will give the wheel an angular acceleration, α. If there is no slipping, then the angular acceleration of the wheel is directly related to the acceleration of the block.

SKETCH:

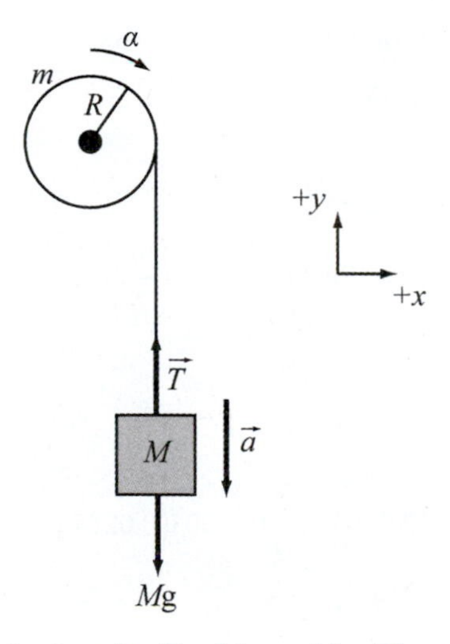

RESEARCH: The balance of forces is given by $T - Mg = -Ma$. The torque produced by the tension, T, is given by $\tau = TR = I\alpha$, where I of the wheel is $mR^2/2$. With no slipping, $R\alpha = a$.

SIMPLIFY: First, determine the tension, $T \Rightarrow T = M(g-a)$. This expression can be substituted into the torque equation to solve for a:

$$M(g-a)R = \frac{1}{2}mR^2\left(\frac{a}{R}\right) \Rightarrow MgR - MaR = \frac{1}{2}mRa \Rightarrow Mg = \left(\frac{1}{2}m + M\right)a \Rightarrow a = \frac{Mg}{\frac{1}{2}m+M}.$$

CALCULATE: $a = \dfrac{70.0\text{ kg}(9.81\text{ m/s}^2)}{\frac{1}{2}(30.0\text{ kg}) + 70.0\text{ kg}} = 8.079\text{ m/s}^2$

ROUND: $a = 8.08 \text{ m/s}^2$

DOUBLE-CHECK: Since there is tension acting opposite gravity, the overall acceleration of the hanging mass should be less than g.

10.51. **THINK:** Each object has its own moment of inertia, I_A and I_B. Disk A with a mass, M = 2.00 kg, and a radius, R = 25.0 cm, rotates about its center of mass while disk B with a mass, m = 0.200 kg and a radius, r = 2.50 cm, rotates a distance, $d = R - r$, away from the axis. This means the parallel axis theorem must be used to determine the overall moment of inertia of disk B, I_B'. The total moment of inertia is the sum of the two. If a torque, $\tau = 0.200$ Nm, is applied then it will cause an angular acceleration, α. If the disk initially rotates at $\omega = -2\pi$ rad/s, then kinematics can be used to determine how long it takes to slow down.

SKETCH:

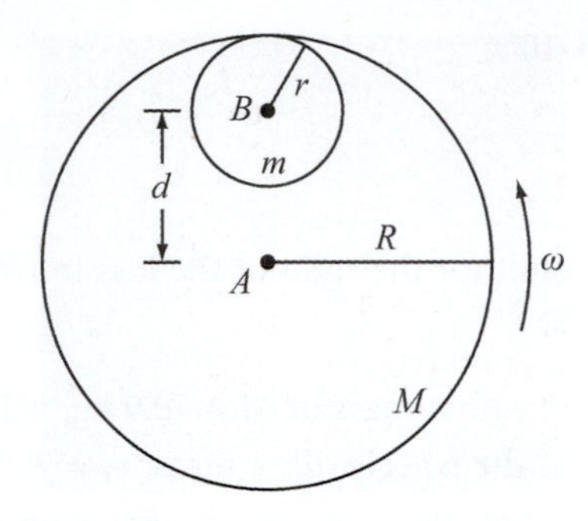

RESEARCH: The moment of inertia of disk A is $I_A = MR^2/2$. The moment of inertia of disk B is $I_B = mr^2/2$. Since disk B is displaced by $d = R - r$ from the axis of rotation, $I_B' = I_B + md^2$, by the parallel axis theorem. Therefore, the total moment of inertia is $I_{tot} = I_A + I_B'$. The torque that is applied produces $\tau = I_{tot}\alpha$, where $\alpha = (\omega_f - \omega_i)/\Delta t$.

SIMPLIFY:

(a) $I_{tot} = I_A + I_B' = I_A + I_B + m(R-r)^2$

$$= \frac{1}{2}MR^2 + \frac{1}{2}mr^2 + mR^2 - 2mRr + mr^2 = \left(\frac{1}{2}M + m\right)R^2 + \frac{3}{2}mr^2 - 2mRr$$

(b) $\tau = I_{tot}\alpha = \dfrac{I_{tot}(0-\omega_i)}{t} \Rightarrow t = -\dfrac{I_{tot}\omega_i}{\tau}$

CALCULATE:

(a) $I_{tot} = \left(\frac{1}{2}(2.00) + 0.200\right)(0.250)^2 + \frac{3}{2}(0.200)(0.0250)^2 - 2(0.200)(0.0250)(0.250) = 0.0726875 \text{ kg m}^2$

(b) $t = -\dfrac{(0.0726875 \text{ kg m}^2)(-2\pi \text{ rad/s})}{0.200 \text{ N m}} = 2.284 \text{ s}$

ROUND:

(a) $I_{tot} = 7.27 \cdot 10^{-2} \text{ kg m}^2$

(b) $t = 2.28$ s

DOUBLE-CHECK: Given the small masses and disk sizes, the moment of inertia should be small. Also, given the small torque and angular velocity, two seconds to come to a stop is reasonable.

10.55. **THINK:** The hanging mass, m = 2.00 kg, will cause a tension, T, in the rope. This tension will then produce a torque, τ, on the wheel with a mass, M = 40.0 kg, a radius, R = 30.0 cm and a c value of 4/9. This torque will then give the wheel an angular acceleration, α. Assuming the rope does not slip, the angular acceleration of the wheel will be directly related to the linear acceleration of the hanging mass.

SKETCH:

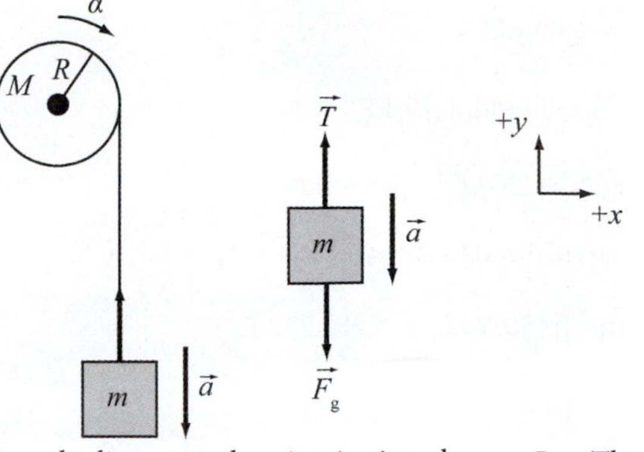

RESEARCH: With no slipping, the linear acceleration is given by $a = R\alpha$. The tension can be determined by $T = m(g-a)$, which in turn produces a torque $\tau = TR = I\alpha$, where the moment of inertia of the wheel is $4MR^2/9$.

SIMPLIFY: To determine the angular acceleration:

$$TR = m(g-a)R = \frac{4}{9}MR^2\alpha \Rightarrow \text{mg}R - maR = \frac{4}{9}MR^2\alpha \Rightarrow \text{mg}R = mR^2\alpha + \frac{4}{9}MR^2\alpha \Rightarrow \alpha = \frac{\text{mg}}{(m + 4M/9)R}.$$

CALCULATE: $\alpha = \dfrac{2.00 \text{ kg}(9.81 \text{ m/s}^2)}{\left(2.00 \text{ kg} + \frac{4}{9}(40.0 \text{ kg})\right)(0.300 \text{ m})} = 3.3067 \text{ rad/s}^2$

ROUND: $\alpha = 3.31 \text{ rad/s}^2$

DOUBLE-CHECK: Given the small hanging mass and the large mass of the wheel, this acceleration is reasonable.

10.59. **THINK:** Assuming a constant angular acceleration, α, and $\Delta t = 25$ s, regular kinematics can be used to determine α and $\Delta\theta$. The total work done by torque, τ, should be converted entirely into rotational energy. $I = 25.0 \text{ kg m}^2$ and $\omega_\text{f} = 150. \text{ rad/s}$.

SKETCH:

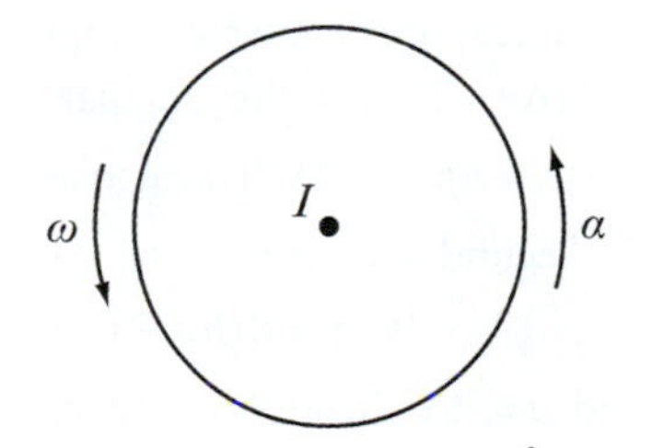

RESEARCH: From kinematics, $\omega_\text{f} - \omega_\text{i} = \alpha\Delta t$ and $\Delta\theta = \alpha(\Delta t)^2/2$. The torque on the wheel is $\tau = I\alpha$. Since the torque is constant, the work done by it is $W = \tau\Delta\theta$. The kinetic energy of the turbine is $K = I\omega^2/2$.

SIMPLIFY:

(a) $\omega_\text{f} = \alpha\Delta t \Rightarrow \alpha = \omega_\text{f}/\Delta t$

(b) $\tau = I\alpha$

(c) $\Delta\theta = \frac{1}{2}\alpha(\Delta t)^2$

(d) $W = \tau\Delta\theta$

(e) $K = \frac{1}{2}I\omega_\text{f}^2$

CALCULATE:

(a) $\alpha = \dfrac{150.\text{ rad/s}}{25.0\text{ s}} = 6.00\text{ rad/s}^2$

(b) $\tau = \left(25.0\text{ kg m}^2\right)\left(6.00\text{ rad/s}^2\right) = 150.\text{ N m}$

(c) $\Delta\theta = \dfrac{1}{2}\left(6.00\text{ rad/s}^2\right)\left(25.0\text{ s}\right)^2 = 1875\text{ rad}$

(d) $W = \left(150.\text{ N m}\right)\left(1875\text{ rad}\right) = 281{,}250\text{ J}$

(e) $K = \dfrac{1}{2}\left(25.0\text{ kg m}^2\right)\left(150.\text{ rad/s}\right)^2 = 281{,}250\text{ J}$

ROUND:

(a) $\alpha = 6.00\text{ rad/s}^2$

(b) $\tau = 150.\text{ N m}$

(c) $\Delta\theta = 1880\text{ rad}$

(d) $W = 281\text{ kJ}$

(e) $K = 281\text{ kJ}$

DOUBLE-CHECK: It is expected that the work and kinetic energy are equal. Since they were each determined independently and they are the same value, the procedure must have been correct.

10.63. **THINK:** The sphere of mass, M, spins clockwise when a horizontal impulse J is exerted at a height h above the tabletop when $R < h < 2R$.

SKETCH:

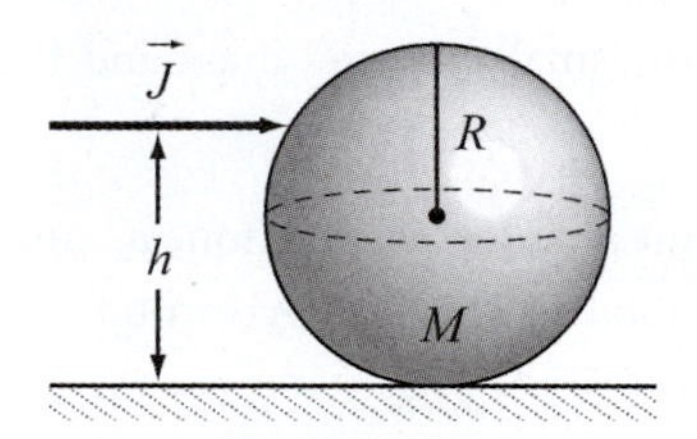

RESEARCH: To calculate the linear speed after the impulse is applied, we use the fact that the impulse J can be written as $J = \Delta p = M\Delta v$. To get the angular velocity, we write the change in the angular momentum of the sphere as $\Delta L = \Delta p(h - R)$. To calculate the height where the impulse must be applied, we have to apply Newton's Second Law for linear motion, $F = Ma$, and Newton's Second Law for rotation, $\tau = I\alpha$. The torque is given by $\tau = F(h - R)$. The object rolls without slipping so from Section 10.3 we know that $v = R\omega$ and $a = R\alpha$. In addition, we can write the impulse as $J = F\Delta t$.

SIMPLIFY: a) Combining these relationships to get the linear velocity gives us

$$J = \Delta p = M\Delta v = Mv \;\Rightarrow\; v = \frac{J}{M}.$$

Combining these relationships to get the angular velocity gives us

$$\Delta L = \Delta p(h - R) = J(h - R)$$

$$\Delta L = I\Delta\omega = I\omega = \frac{2}{5}MR^2\omega$$

$$J(h - R) = \frac{2}{5}MR^2\omega$$

$$\omega = \frac{5J(h - R)}{2MR^2}.$$

b) Combining these relationships to get the height h_0 at which the impulse must be applied for the sphere to roll without slipping we get

$$F = Ma \;\Rightarrow\; \frac{J}{\Delta t} = MR\alpha$$

$$F(h_0 - R) = I\alpha \;\Rightarrow\; \frac{J}{\Delta t}(h_0 - R) = \frac{2}{5}MR^2\alpha.$$

Dividing these two equations gives us

$$h_0 - R = \frac{2}{5}R \;\Rightarrow\; h_0 = \frac{7}{5}R.$$

CALCULATE: Not applicable.

ROUND: Not applicable.

DOUBLE-CHECK: The linear velocity will always be positive. However, the angular velocity can be positive or negative, depending on whether $h > R$ or $h < R$. The fact that $h_0 > R$ is consistent with the ball rolling to the right after the impulse is applied.

10.67. **THINK:** If the disk with radius, R = 40.0 cm, is rotating at 30.0 rev/s, then the angular speed, ω, is 60.0π rad/s. The length of the gyroscope is L = 60.0 cm, so that the disk is located at $r = L/2$ from the pivot.

SKETCH:

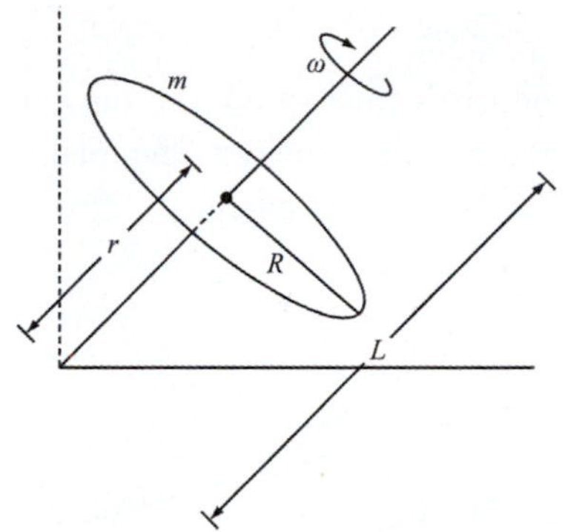

RESEARCH: The precessional angular speed is given by $\omega_p = rmg / I\omega$. The moment of inertia of the disk is $I = mR^2 / 2$.

SIMPLIFY: $\omega_p = \dfrac{\frac{L}{2}mg}{\frac{1}{2}mR^2\omega} = \dfrac{Lg}{R^2\omega}$

CALCULATE: $\omega_p = \dfrac{0.600 \text{ m}\left(9.81 \text{ m/s}^2\right)}{\left(0.400 \text{ m}\right)^2 60.0\pi \text{ rad/s}} = 0.19516 \text{ rad/s}$

ROUND: $\omega_p = 0.195$ rad/s

DOUBLE-CHECK: The precession frequency is supposed to be much less than the frequency of the rotating disk. In this example, the disk frequency is about one thousand times the precession frequency, so it makes sense.

10.71. **THINK:** The oxygen atoms, $m = 2.66 \cdot 10^{-26}$ kg, can be treated as point particles a distance, $d/2$ (where $d = 1.21 \cdot 10^{-10}$ m) from the axis of rotation. The angular speed of the atoms is $\omega = 4.60 \cdot 10^{12}$ rad/s.

SKETCH:

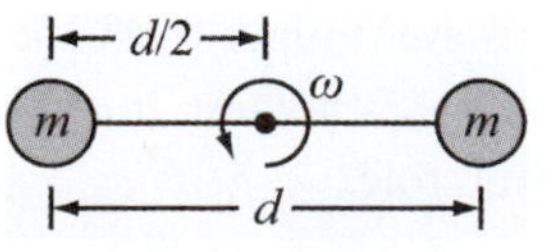

RESEARCH: Since the masses are equal point particles, the moment of inertia of the two is $I = 2m(d/2)^2$. The rotational kinetic energy is $K = I\omega^2/2$.

SIMPLIFY:

(a) $I = 2m\left(\frac{d^2}{4}\right) = \frac{1}{2}md^2$

(b) $K = \frac{1}{2}I\omega^2 = \frac{1}{4}md^2\omega^2$

CALCULATE:

(a) $I = \frac{1}{2}(2.66 \cdot 10^{-26}\text{ kg})(1.21 \cdot 10^{-10}\text{ m})^2 = 1.9473 \cdot 10^{-46}\text{ kg m}^2$

(b) $K = \frac{1}{4}(2.66 \cdot 10^{-26}\text{ kg})(1.21 \cdot 10^{-10}\text{ m})^2(4.60 \cdot 10^{12}\text{ rad/s})^2 = 2.06 \cdot 10^{-21}\text{ J}$

ROUND:

(a) $I = 1.95 \cdot 10^{-46}\text{ kg m}^2$

(b) $K = 2.06 \cdot 10^{-21}\text{ J}$

DOUBLE-CHECK: Since an oxygen molecule is so small, a very small moment of inertia and energy are expected.

10.75. **THINK:** To determine the cart's final speed, use the conservation of energy. The initial gravitational potential energy is converted to kinetic energy. The total kinetic energy at the bottom is the sum of the translational and rotational kinetic energies. Use $m_\text{p} = 8.00$ kg, $m_\text{w} = 2.00$ kg, $L_\text{p} = 1.20$ m, $w_\text{p} = 60.0$ cm, $r = 10.0$ cm, $D = 30.0$ m and $\theta = 15.0°$.

SKETCH:

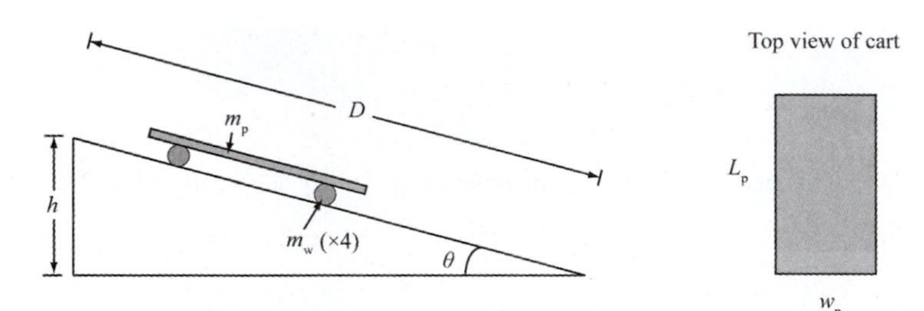

RESEARCH: The initial energy is $E_\text{tot} = U$ (potential energy). The final energy is $E_\text{tot} = K$ (kinetic energy). $U = M_\text{tot}gh$, $h = D\sin\theta$, $M_\text{tot} = m_\text{p} + 4m_\text{w}$, $K = M_\text{tot}v^2/2 + I\omega^2/2$, $\omega = v/r$ and $I = 4(m_\text{w}r^2/2)$.

SIMPLIFY:

$$U = K \Rightarrow M_\text{tot}gh = \frac{1}{2}M_\text{tot}v^2 + \frac{1}{2}I\omega^2 \Rightarrow (m_\text{p} + 4m_\text{w})gD\sin\theta = \frac{1}{2}(m_\text{p} + 4m_\text{w})v^2 + \frac{1}{2}(4)\left(\frac{1}{2}m_\text{w}r^2\right)\left(\frac{v^2}{r^2}\right)$$

$$\Rightarrow (m_\text{p} + 4m_\text{w})gD\sin\theta = \left(\frac{1}{2}m_\text{p} + 2m_\text{w} + m_\text{w}\right)v^2 \Rightarrow v = \sqrt{\frac{(m_\text{p} + 4m_\text{w})gD\sin\theta}{\frac{1}{2}m_\text{p} + 3m_\text{w}}}$$

CALCULATE: $v = \sqrt{\dfrac{(8.00\text{ kg} + 4(2.00\text{ kg}))(9.81\text{ m/s}^2)(30.0\text{ m})\sin 15.0°}{\frac{1}{2}(8.00\text{ kg}) + 3(2.00\text{ kg})}} = 11.04\text{ m/s}$

ROUND: The length of the incline is given to three significant figures, so the result should be rounded to $v = 11.0$ m/s.

DOUBLE-CHECK: This velocity is rather fast. In reality, the friction would slow the cart down. Note also that the radii of the wheels play no role.

10.79. **THINK:** If the angular momentum and the torque are determined, the time can be determined by recalling that torque is the time rate of change of angular momentum. To determine the angular momentum, first determine the angular speed required to produce a centripetal acceleration equal to Earth's gravitational acceleration. From this, the angular momentum, L, of the space station can be determined. Finally, the torque can be determined from the given force and the radius of the space station. $R = 50.0$ m, $M = 2.40\cdot 10^5$ kg and $F = 1.40\cdot 10^2$ N.

SKETCH:

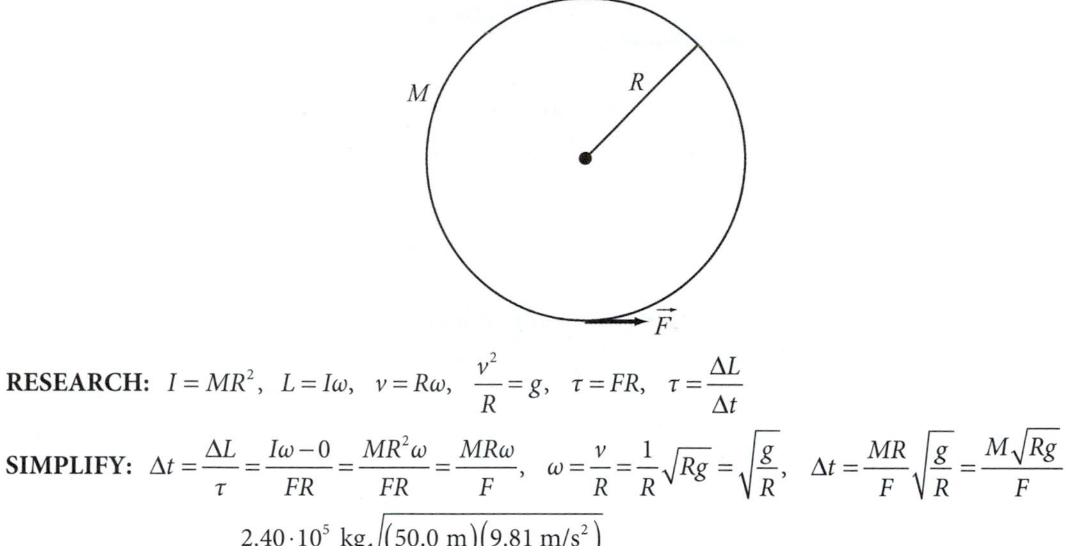

RESEARCH: $I = MR^2$, $L = I\omega$, $v = R\omega$, $\dfrac{v^2}{R} = g$, $\tau = FR$, $\tau = \dfrac{\Delta L}{\Delta t}$

SIMPLIFY: $\Delta t = \dfrac{\Delta L}{\tau} = \dfrac{I\omega - 0}{FR} = \dfrac{MR^2\omega}{FR} = \dfrac{MR\omega}{F}$, $\omega = \dfrac{v}{R} = \dfrac{1}{R}\sqrt{Rg} = \sqrt{\dfrac{g}{R}}$, $\Delta t = \dfrac{MR}{F}\sqrt{\dfrac{g}{R}} = \dfrac{M\sqrt{Rg}}{F}$

CALCULATE: $\Delta t = \dfrac{2.40\cdot 10^5\text{ kg}\sqrt{(50.0\text{ m})(9.81\text{ m/s}^2)}}{1.40\cdot 10^2\text{ N}} = 3.797\cdot 10^4$ s

ROUND: The radius of the space station is given to three significant figures, so the result should be rounded to $\Delta t = 3.80\cdot 10^4$ s.

DOUBLE-CHECK: The result is equal to about 10 hours. For such a relatively small thrust, this result is reasonable. As expected, this time interval increases if either the thrust decreases or the mass increases.

10.83. **THINK:** The quantity of interest can be calculated directly from the given information. $m_r = 5.20$ kg, $m_h = 3.40$ kg, $m_s = 1.10$ kg, $r_1 = 0.900$ m, $r_2 = 0.860$ m, $r_h = 0.120$ m and $l = r_2 - r_h$.

SKETCH:

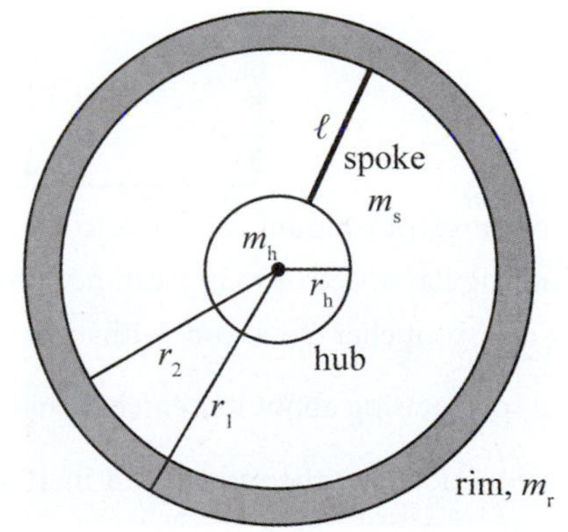

RESEARCH: $I_{\text{rim}} = \dfrac{1}{2}m_r\left(r_1^2 + r_2^2\right)$, $I_{\text{spoke}} = \dfrac{1}{12}m_s l^2 + m_s d^2$, (with $d = \frac{1}{2}l + r_h$), $I_{\text{hub}} = \dfrac{1}{2}m_h r_h^{\,2}$

SIMPLIFY: $I = I_{\text{rim}} + I_{\text{hub}} + 12I_{\text{spoke}}$, $M = m_r + m_h + 12m_s$, $R = r_1$

CALCULATE: $I_{\text{rim}} = \frac{1}{2}5.20 \text{ kg}\left((0.900)^2 + (0.860)^2\right) \text{ m}^2 = 4.029 \text{ kg m}^2$

$I_{\text{hub}} = \frac{1}{2}(3.40 \text{ kg})(0.120 \text{ m})^2 = 2.448 \cdot 10^{-2} \text{ kg m}^2$

$I_{\text{spoke}} = \frac{1}{12}(1.10 \text{ kg})(0.860 \text{ m} - 0.120 \text{ m})^2 + (1.10 \text{ kg})(0.490 \text{ m})^2 = 3.143 \cdot 10^{-1} \text{ kg m}^2$

$I = I_{\text{rim}} + I_{\text{hub}} + 12I_{\text{spoke}} = 7.825 \text{ kg m}^2,\ M = \left[5.20 + 3.40 + 12(1.10)\right] \text{ kg} = 21.8 \text{ kg}$

$c = \frac{I}{MR^2} = \frac{7.825 \text{ kgm}^2}{(21.8 \text{ kg})(0.900 \text{ m})^2} = 0.4431$

ROUND: Rounding to three significant figures, $c = 0.443$.

DOUBLE-CHECK: It is reasonable that the moment of inertia is dominated by the rim and the spokes, and the hub is negligible.

Multi-Version Exercises

10.85. **THINK:** The length and mass of the propeller, as well as the frequency with which it is rotating, are given. To find the kinetic energy of rotation, it is necessary to find the moment of inertia, which can be calculated from the mass and radius by approximating the propeller as a rod with constant mass density.

SKETCH: The propeller is shown as it would be seen looking directly at it from in front of the plane.

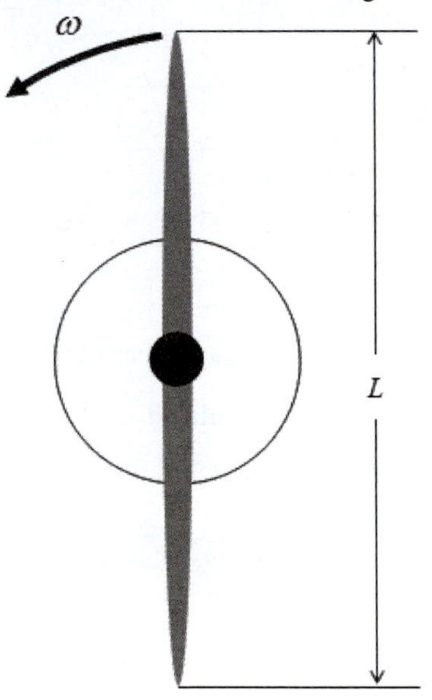

RESEARCH: The kinetic energy of rotation is related to the moment of inertia and angular speed by the equation $K_{\text{rot}} = \frac{1}{2}I\omega^2$. The angular speed $\omega = 2\pi f$ can be computed from the frequency of the propeller's rotation. Approximating the propeller as a rod with constant mass density means that the formula $I = \frac{1}{12}mL^2$ for a long, thin rod rotating about its center of mass can be used.

SIMPLIFY: Combine the equations for the moment of inertia and angular speed to get a single equation for the kinetic energy $K_{\text{rot}} = \frac{1}{2}I\omega^2 = \frac{1}{2}\left(\frac{1}{12}mL^2\right)\cdot(2\pi f)^2$. Using algebra, this can be simplified to $K_{\text{rot}} = \frac{m}{6}(\pi L f)^2$. Since the angular speed is given in revolutions per minute, the conversion 1 minute = 60 seconds will also be needed.

CALCULATE: The propeller weighs $m = 17.36$ kg, it is $L = 2.012$ m long, and it rotates at a frequency of $f = 3280.$ rpm. The rotational kinetic energy is

$$\begin{aligned} K_{\text{rot}} &= \frac{m}{6}(\pi L f)^2 \\ &= \frac{17.36 \text{ kg}}{6}\left(\pi \cdot 2.012 \text{ m} \cdot 3280. \text{ rpm} \cdot \frac{1\text{min}}{60\text{sec}}\right)^2 \\ &= 345{,}461.2621 \text{ J} \end{aligned}$$

ROUND: The values in the problem are all given to four significant figures, so the final answer should have four figures. The propeller has a rotational kinetic energy of $3.455 \cdot 10^5$ J or 345.5 kJ.

DOUBLE-CHECK: Given the large amount of force needed to lift a plane, it seems reasonable that the energy in the propeller would be in the order of hundreds of kilojoules. Working backwards, if a propeller weighing 17.36 kg and having length 2.012 m has rotational kinetic energy 345.5 kJ, then it is turning at $f = \frac{1}{\pi L}\sqrt{\frac{6K_{\text{rot}}}{m}} = \frac{1}{\pi \cdot 2.012 \text{ m}}\sqrt{\frac{6 \cdot 3.455 \cdot 10^5 \text{ J}}{17.36 \text{ kg}}}$. This is 54.667 revolutions per second, which agrees with the given value of $54.667 \cdot 60 = 3280$ rpm. This confirms that the calculations were correct.

10.88. **THINK:** The total kinetic energy of the golf ball is the sum of the rotational kinetic energy and the translational kinetic energy. The translational kinetic energy can be calculated from the mass of the ball and the speed of the center of mass of the golf ball, both of which are given in the question. To find the rotational kinetic energy, it is necessary to find the moment of inertia of the golf ball. Though the golf ball is not a perfect sphere, it is close enough that the moment of inertia can be computed from the mass and diameter of the golf ball using the approximation for a sphere.

SKETCH: The golf ball has both rotational and translational motion.

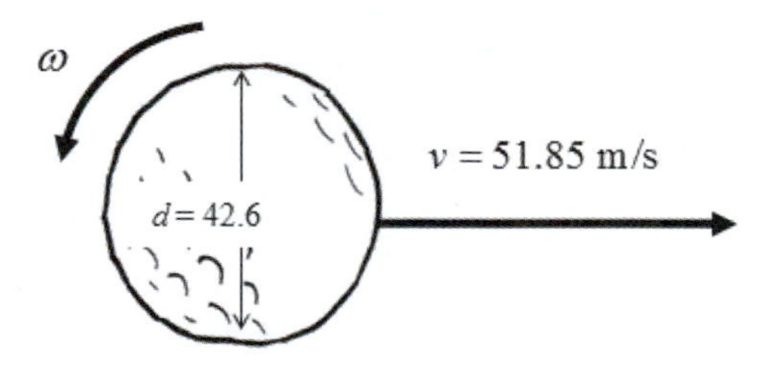

RESEARCH: The total kinetic energy is equal to the translational kinetic energy plus the rotational kinetic energy $K = K_{\text{trans}} + K_{\text{rot}}$. The translational kinetic energy is computed from the speed and the mass of the golf ball using the equation $K_{\text{trans}} = \frac{1}{2}mv^2$. The rotational kinetic energy is computed from the moment of inertia and the angular speed by $K_{\text{rot}} = \frac{1}{2}I\omega^2$. It is necessary to compute the moment of inertia and the angular speed. The angular speed $\omega = 2\pi f$ depends only on the frequency. To find the moment of inertia, first note that golf balls are roughly spherical. The moment of inertia of a sphere is given by $I = \frac{2}{5}mr^2$. The question gives the diameter d which is twice the radius ($d / 2 = r$). Since the frequency is given in revolutions per minute and the speed is given in meters per second, the conversion factor $\frac{1 \text{ min}}{60 \text{ sec}}$ will be necessary.

SIMPLIFY: First, find the moment of inertia of the golf ball in terms of the mass and diameter to get $I = \frac{1}{10}md^2$. Substituted for the angular speed and moment of inertia in the equation for rotational kinetic energy to get $K_{\text{rot}} = \frac{1}{2}I\omega^2 = \frac{1}{2}\left(\frac{1}{10}md^2\right)(2\pi f)^2$. Finally, use the equations $K_{\text{trans}} = \frac{1}{2}mv^2$ and $K_{\text{rot}} = \frac{1}{2}\left(\frac{1}{10}md^2\right)(2\pi f)^2$ to find the total kinetic energy and simplify using algebra:

$$\begin{aligned}K &= K_{\text{trans}} + K_{\text{rot}} \\ &= \tfrac{1}{2}mv^2 + \tfrac{1}{2}\left(\tfrac{1}{10}md^2\right)(2\pi f)^2 \\ &= \tfrac{1}{2}mv^2 + \tfrac{1}{5}m(\pi df)^2\end{aligned}$$

CALCULATE: The mass of the golf ball is 45.90 g = 0.04590 kg, its diameter is 42.60 mm = 0.04260 m, and its speed is 51.85 m/s. The golf ball rotates at a frequency of 2857 revolutions per minute. The total kinetic energy is

$$\begin{aligned}K &= \tfrac{1}{2}mv^2 + \tfrac{1}{5}m(\pi df)^2 \\ &= \tfrac{1}{2}(0.04590\text{ kg})(51.85\text{ m/s})^2 + \tfrac{1}{5}(0.04590\text{ kg})\left(\pi \cdot 0.04260\text{ m} \cdot 2857\text{ rpm} \cdot \frac{1\text{ min}}{60\text{ sec}}\right)^2 \\ &= 62.07209955\text{ J}\end{aligned}$$

ROUND: The mass, speed, frequency, and diameter of the golf ball are all given to four significant figures, so the translational and rotational kinetic energies should both have four significant figures, as should their sum. The total energy of the golf ball is 62.07 J.

DOUBLE-CHECK: The golf ball's translational kinetic energy alone is equal to $\tfrac{1}{2}(0.04590\text{ kg})(51.85\text{ m/s})^2 = 61.7\text{ J},$ and it makes sense that a well-driven golf ball would have much more energy of translation than energy of rotation.

10.91. **THINK:** The gravitational force on the block is transmitted through the rope, causing a torque on the pulley. The torque causes an angular acceleration, and the linear acceleration is calculated from the angular acceleration.

SKETCH: Use the figure from the text:

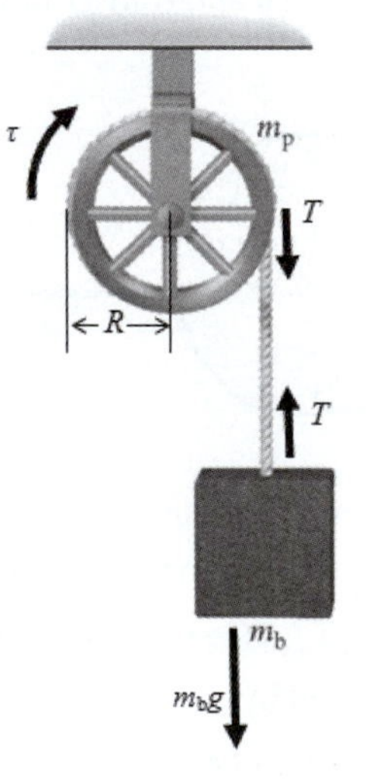

RESEARCH: The torque on the pulley is given by the tension on the rope times the radius of the pulley $\tau = TR$. This torque will cause an angular acceleration $\tau = I\alpha$, where the moment of inertia of the pulley is given by $I = m_{\text{p}}R^2$. The tension on the rope is given by $T - m_{\text{b}}g = -m_{\text{b}}a$ (the minus indicates that the block is accelerating downward). The linear acceleration of the block a is related to the angular acceleration of the pulley α by the equation $a = R\alpha$.

SIMPLIFY: First, substitute for the tension on the pulley $T = m_{\text{b}}g - m_{\text{b}}a$ in the equation for the torque τ to get $\tau = (m_{\text{b}}g - m_{\text{b}}a)R$. Then, substitute for the moment of inertia ($I = m_{\text{p}}R^2$) and angular acceleration ($\alpha = a/R$) in the equation $\tau = I\alpha$ to get $\tau = \left(m_{\text{p}}R^2\right)\left(\dfrac{a}{R}\right) = m_{\text{p}}Ra$. Combine these two expressions for the torque to get $(m_b g - m_b a)R = m_p Ra$. Finally, solve this expression for the linear acceleration a of the block:

$$(m_b g - m_b a)R = m_p aR$$
$$m_b gR - m_b aR + m_b aR = m_p aR + m_b aR$$
$$m_b gR = (m_p R + m_b R)a$$
$$\frac{m_b gR}{m_p R + m_b R} = a$$
$$a = \frac{m_b g}{m_p + m_b}$$

CALCULATE: The mass of the block is m_b = 4.243 kg and the mass of the pulley is m_p = 5.907 kg. The acceleration due to gravity is –9.81 m/s². So, the total (linear) acceleration of the block is

$$a = \frac{m_b g}{m_p + m_b} = \frac{-9.81 \text{ m/s}^2 \cdot 4.243 \text{ kg}}{5.907 \text{ kg} + 4.243 \text{ kg}} = -4.100869951 \text{ m/s}^2.$$

ROUND: The masses of the pulley and block are given to four significant figures, and the sum of their masses has five figures. On the other hand, the gravitational constant g is given only to three significant figures. So, the final answer should have three significant figures. The block accelerates downward at a rate of 4.10 m/s².

DOUBLE-CHECK: A block falling freely would accelerate (due to gravity near the surface of the Earth) at a rate of 9.81 m/s² towards the ground. The block attached to the pulley will still accelerate downward, but the rate of acceleration will be less (the potential energy lost when the block falls 1 meter will equal the kinetic energy of a block in free fall, but it will equal the kinetic energy of the block falling plus the rotational kinetic energy of the pulley in the problem). The mass of the pulley is close to, but a bit larger than, the mass of the block, so the acceleration of the block attached to the pulley should be a bit less than half of the acceleration of the block in free fall. This agrees with the final acceleration of 4.10 m/s², which is a bit less than half of the acceleration due to gravity.

Chapter 11: Static Equilibrium

Exercises

11.27. **THINK:** We know that the weight of the crate is $W = 1000.$ N. The length of the crate is L. Two vertical ropes are pulling the crate upwards. We know that the tension in the left rope is $T_1 = 400.$ N and it is attached a distance $L/4$ from the left edge of the crate. The crate does not move and does not rotate when the platform is lowered. Assume that the crate is of uniform density, so that its center of mass is its geometric center.

SKETCH:

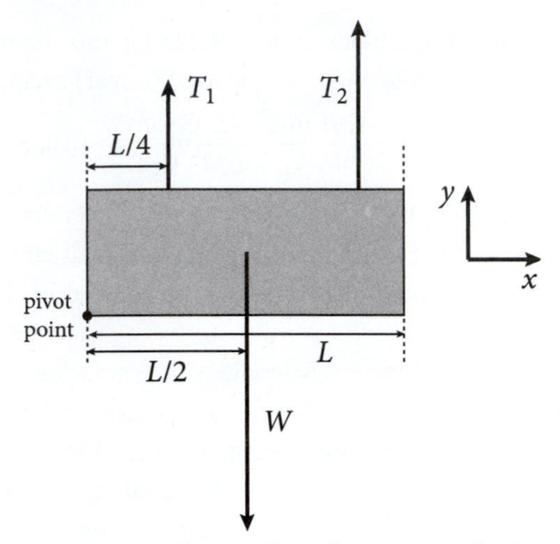

RESEARCH: The crate does not move so the combined tensions, $T_1 + T_2$, must equal the weight of the crate, W = 1000. N. Thus we can write $T_1 + T_2 = W$. The crate does not rotate, so there is no net torque acting on the system. The tension T_1 acts at a distance $x_1 = L/4$ from the assumed pivot point at the lower left corner. The force from the weight of the crate acts at a distance of $L/2$ from the pivot point. The force T_2 acts a distance x_2 from the pivot point. The sum of the torques about the pivot point is given by $\tau_{\text{net}} = -W\frac{L}{2} + T_2x_2 + T_1x_1 = 0$.

SIMPLIFY: Solving for the maximum value of T_2 from the force equations we get

$$T_2 = W - T_1 = 1000.\text{ N} - 400.\text{ N} = 600.\text{ N}.$$

Solving for x_2 from the torque equations gives us

$$0 = -W\frac{L}{2} + T_2x_2 + T_1x_1$$

$$T_2x_2 = W\frac{L}{2} - T_1x_1$$

$$x_2 = \frac{W\frac{L}{2} - T_1x_1}{T_2} = \frac{WL - 2T_1x_1}{2T_2} = \frac{WL - 2T_1(L/4)}{2T_2} = \frac{WL - T_1L/2}{2T_2}.$$

CALCULATE: The tension in the right rope is $T_2 = 600.$ N. The rope on the right is attached at

$$x_2 = \frac{WL - T_1L/2}{2T_2} = \frac{(1000.\text{ N})L - (400.\text{N})L/2}{2(600.\text{N})} = \frac{(800.)L}{1200.} = \frac{2}{3}L.$$

ROUND: Not applicable.

DOUBLE-CHECK: To double-check, we show that the counterclockwise torque has the same magnitude as the clockwise torque when we assume that the pivot point is located at the center of mass of the crate:

$$T_2\left(\frac{2L}{3}-\frac{L}{2}\right)=T_1\left(\frac{L}{2}-\frac{L}{4}\right)$$
$$(600.\text{N})\left(\frac{L}{6}\right)=(400.\text{ N})\left(\frac{L}{4}\right)$$
$$(100.\text{N})L=(100.\text{N})L.$$

11.31. **THINK:** The magnitudes of the four other forces acting on the merry-go-round are $F_1 = 104.9$ N, $F_2 = 89.1$ N, $F_3 = 62.8$ N, and $F_4 = 120.7$ N. I will use the convention that a torque that acts in the clockwise direction is negative and a torque that acts in the counterclockwise direction is positive. Consider the torque about the center as a pivot point.

SKETCH:

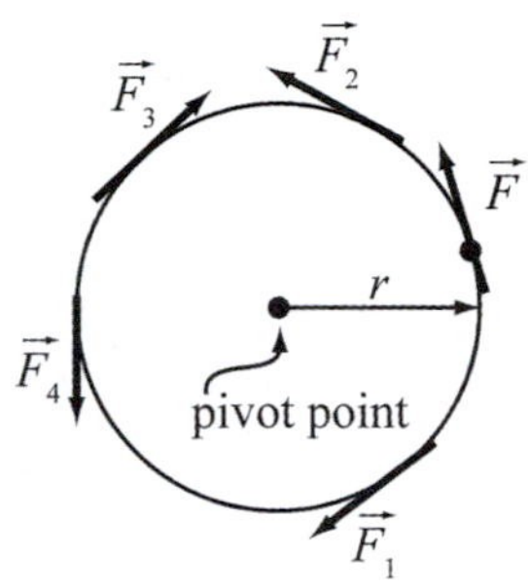

RESEARCH: The merry-go-round has radius, r. The forces are all applied in a tangential direction, so the torque due to each force is given by $\tau_i = F_i r$. In order for the merry-go-round to remain stationary, the sum of the torques about the pivot point must equal zero:

$$\tau_{\text{net}} = \sum_{i=1}^{n}\tau_i = r\sum_{i=1}^{n}F_i.$$

SIMPLIFY: $\tau_{\text{net}} = r(-F_1 + F_2 - F_3 + F_4 + F)$. I want to prevent the merry-go-round from moving, so the net torque is zero. Since the radius is not zero, $F = F_1 - F_2 + F_3 - F_4$.

CALCULATE: $F = 104.9\text{ N} - 89.1\text{ N} + 62.8\text{ N} - 120.7\text{ N} = -42.1\text{ N}$. The negative sign indicates that the force should be applied in the clockwise direction.

ROUND: The least number of significant figures provided in the question is three, so no rounding is necessary. Write the answer as, 42.1 N clockwise.

DOUBLE-CHECK: The calculated force is of the same order of magnitude as the forces which were given. The sum of the magnitudes of the forces in the counter-clockwise direction is more than the sum of the magnitudes of the forces in the clockwise direction. In order to counter-balance this, the fifth child should push in the clockwise direction. This is consistent with the calculated answer.

11.35. **THINK:** I must estimate the distance, *d*, from the shoulder pivot to the point where the deltoid muscle connects to the upper arm. I must also estimate the distance from the shoulder pivot point to the center of mass of the arm, *x*. The deltoid muscle will have to pull up the arm at an angle θ. The mass of the arms is m_{a}. The mass of the weight is $m_{\text{w}} = (10.0\text{ lb})\dfrac{1\text{ kg}}{2.20\text{ lb}} = 4.55\text{ kg}$. I want to calculate the force, F_{D}, the deltoid muscle would have to exert to hold my arm at shoulder lever, and also the force, F_{D}', required to hold the 4.55 kg weight at shoulder level.

SKETCH:

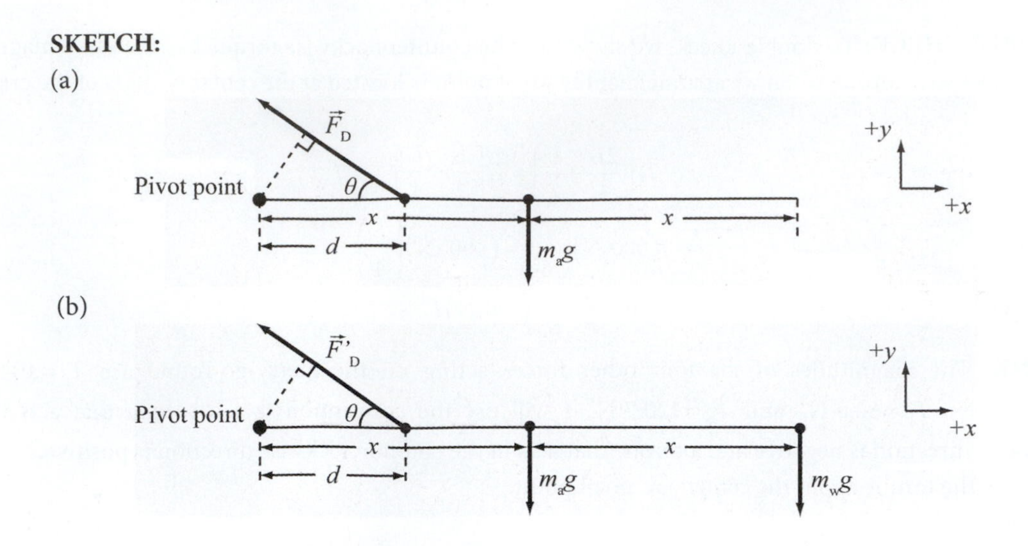

RESEARCH: To hold the arm at shoulder level in either situation, the sum of the torques about the pivot point must equal zero:

$$\sum_{i=1}^{n} \tau_i = 0.$$

In order to calculate values, estimates must be made for d, x, θ, and m_a. I will assume that the center of mass of my arm is in the middle of the arm. I estimate d = 0.120 m, x = 0.300 m, θ = 20.0°, and m_a = 4.00 kg.

SIMPLIFY: For the first situation where the arm is extended at shoulder level, the equation is

$$\sum_{i=1}^{n} \tau_i = 0 = -m_a gx + F_D d\sin\theta \Rightarrow F_D = \frac{m_a gx}{d\sin\theta}.$$

For the second situation, the torque about the pivot due to the force of the weight must be considered:

$$\sum_{i=1}^{n} \tau_i = 0 = -m_a gx - m_W g2x + F'_D d\sin\theta \Rightarrow F'_D = \frac{m_a gx + m_W g(2x)}{d\sin\theta}.$$

CALCULATE: $F_D = \dfrac{(4.00 \text{ kg})(9.81 \text{ m/s}^2)(0.300 \text{ m})}{(0.120 \text{ m})\sin(20.0°)} = 286.83 \text{ N}$

$$F'_D = \frac{(4.00 \text{ kg})(9.81 \text{ m/s}^2)(0.300 \text{ m}) + (4.55 \text{ kg})(9.81 \text{ m/s}^2)(0.600 \text{ m})}{(0.120 \text{ m})\sin(20.0°)} = 939.35 \text{ N}$$

ROUND: $F_D = 287$ N and $F'_D = 939$ N.

DOUBLE-CHECK: Dimensional analysis confirms the answers are given in the correct units of force. It makes sense that the deltoid muscle must exert a much larger force to hold the 10.0 lb weight compared to the force required to hold up just the arm.

11.39. **THINK:** The construction supervisor has mass, $M = 92.1$ kg, and stands a distance, $x_1 = 1.07$ m, away from sawhorse 1. He is standing on a board that is supported by sawhorse 1 and sawhorse 2 on either end. The board has a mass, $m = 27.5$ kg. The two sawhorses are separated by a distance, $l = 3.70$ m. The question asks for the force, F, that the board exerts on sawhorse 1. Use the convention that counter-clockwise torques are positive.

SKETCH:

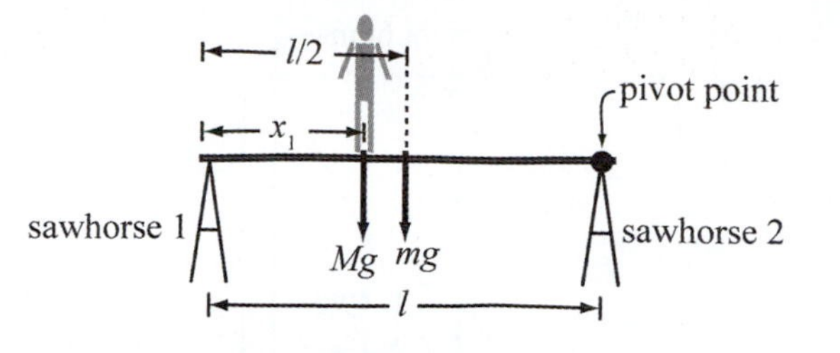

RESEARCH: Assuming the board is of uniform density and thickness, its center of mass should be at $l/2$. The board is in static equilibrium, so the sum of the torques about the pivot point is zero:

$$\tau_{\text{net}} = \sum_{i=1}^{n} \tau_i = 0.$$

The force that the board exerts on sawhorse 1 should be equal in magnitude and opposite in direction to the normal force sawhorse 1 exerts on the board.

SIMPLIFY: $\sum_{i=1}^{n} \tau_i = 0 = -N_1 l + mg\frac{l}{2} + Mg(l - x_1) \Rightarrow N_1 = \dfrac{mg\frac{l}{2} + Mg(l-x_1)}{l}$ and $F = -N_1$, so

$$F = \frac{-mg\frac{l}{2} - Mg(l-x_1)}{l}.$$

CALCULATE:

$$F = \frac{-(27.5\text{ kg})(9.81\text{ m/s}^2)\left(\frac{3.70\text{ m}}{2}\right) - 92.1\text{ kg}(9.81\text{ m/s}^2)(3.70\text{ m} - 1.07\text{ m})}{3.70\text{ m}} = -777.106\text{ N}$$

The negative sign indicates that the force of the board is directed down on sawhorse 1.

ROUND: The question provided three significant figures, so the answer is $F = 777$ N downward.

DOUBLE-CHECK: Dimensional analysis confirms that the answer is in the correct units of force, Newtons. The weight of the board is about 300 N, and the weight of the man is about 900 N. The magnitude and direction of the calculated force seem reasonable for the values given in the question.

11.43. **THINK:** The uniform ladder has length, $l = 10.0$ m, and its mass is $m = (20.0\text{ lb})\dfrac{1\text{ kg}}{2.2046\text{ lb}} = 9.072$ kg. The ladder is leaning against a frictionless wall at an angle $\theta = 60.0°$ with respect to the horizontal. The mass of the boy is

$$m_{\text{b}} = (61.0\text{ lb})\frac{1\text{ kg}}{2.2046\text{ lb}} = 27.67\text{ kg}.$$

The boy climbs a distance $d = 4.00$ m up the ladder. I want to calculate the magnitude of the frictional force exerted on the ladder by the floor. I will use the convention that counterclockwise torque is positive and clockwise torque is negative. Choose the top of the ladder as the pivot point.

SKETCH:

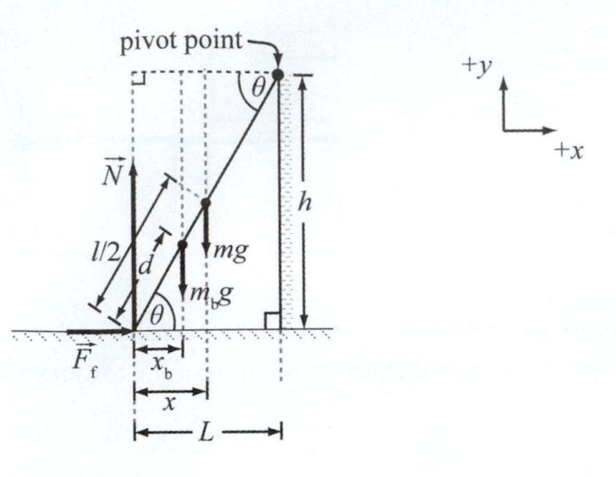

RESEARCH: The system is in static equilibrium, so the sum of the torques about the pivot point is zero:

$$\tau_{\text{net}} = \sum_{i=1}^{n} \tau_i = 0.$$

SIMPLIFY: $\sum_{i=1}^{n} \tau_i = F_f h - NL + m_b g(L - x_b) + mg(L - x) = 0,\ F_f = \dfrac{NL - g\left(m_b(L - x_b) + m(L - x)\right)}{h}$

From the sketch I can derive expressions for L, h, x, and x_b :

$$L = l\cos\theta,\ h = l\sin\theta,\ x_b = d\cos\theta \Rightarrow L - x_b = \cos\theta(l - d),\ x = \frac{l\cos\theta}{2} \Rightarrow L - x = \frac{l\cos\theta}{2}.$$

Substituting these values into the expression for F_f gives:

$$F_f = \frac{Nl\cos\theta - g\left[m_b(l-d)\cos\theta + m\dfrac{l\cos\theta}{2}\right]}{l\sin\theta}.$$

N is unknown, but because the system is in static equilibrium, the sum of the forces in the y direction is zero. $\sum_{i=1}^{n} F_{y,i} = N - m_b g - mg = 0 \Rightarrow N = (m_b + m)g$. Substituting this expression for N into the equation for F_f gives:

$$F_f = \frac{m_b gl\cos\theta + mgl\cos\theta - m_b gl\cos\theta + m_b gd\cos\theta - mg\dfrac{l\cos\theta}{2}}{l\sin\theta} = \frac{mg\dfrac{l\cos\theta}{2} + m_b gd\cos\theta}{l\sin\theta}.$$

CALCULATE:

$$F_f = \frac{9.072\text{ kg}\left(9.81\text{ m/s}^2\right)\dfrac{(10.0\text{ m})\cos 60.0^\circ}{2} + (27.67\text{ kg})\left(9.81\text{ m/s}^2\right)(4.00\text{ m})\cos 60.0^\circ}{(10.0\text{ m})\sin 60.0^\circ} = 88.378\text{ N}$$

ROUND: All of the values given in the question have three significant figures. Therefore the answer should also be rounded to this precision. The final answer is is $F_f = 88.4$ N.

DOUBLE-CHECK: Given that the force of static friction is $f_s = \mu_s N$, the calculated values indicate $\mu_s \approx 88.4\text{ N}/360\text{ N} \approx 0.25$, which is a reasonable value for a coefficient of static friction between the ladder and the floor.

11.47. **THINK:** The beam has length $l = 8.00$ m and mass $m = 100.$ kg. The beam is attached by a bolt to a support a distance $d = 3.00$ m from one end. The beam makes an angle $\theta = 30.0^\circ$ with respect to the horizontal. A mass of $M = 500.$ kg is attached to one end of the beam by a rope. A rope attaches the other end of the beam to the ground. I want to calculate the tension T in the rope, and the force $\vec{F} = F_{bx}\hat{x} + F_{by}\hat{y}$ exerted on the beam by the bolt.

SKETCH:

(a) (b)

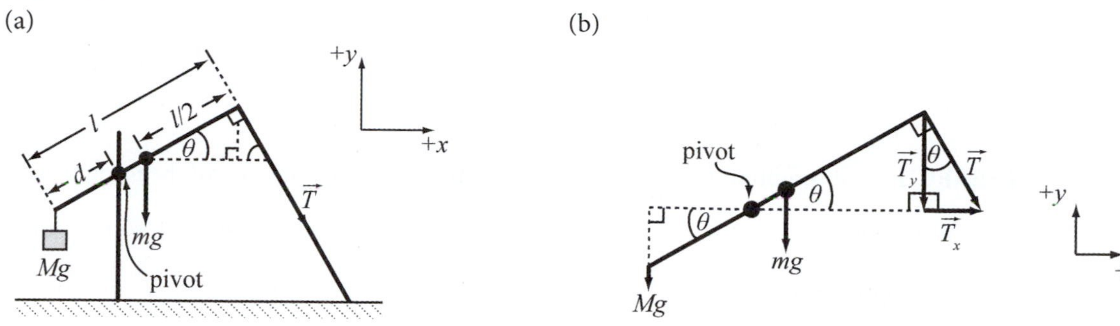

RESEARCH: The sum of the torques about the pivot point is zero because the system is in static equilibrium; $\sum_{i=1}^{n}\tau_i = 0$. The sums of the forces in the x and y directions are also zero:

$$\sum_{i=1}^{n}F_{y,i} = 0;\ \sum_{i=1}^{n}F_{x,i} = 0.$$

Counterclockwise torque is considered positive and clockwise torque is considered negative.

SIMPLIFY: $\sum_{i=1}^{n}\tau_i = 0 = Mgd\cos\theta - mg\left(\frac{l}{2}-d\right)\cos\theta - T(l-d) \Rightarrow T = \dfrac{Mgd\cos\theta - mg\left(\frac{l}{2}-d\right)\cos\theta}{l-d}$

$$\sum_{i=1}^{n}F_{yi} = 0 = -Mg - mg - T_y + F_{by}$$

From sketch (b), $T_y = T\cos\theta$, so, $F_{by} = Mg + mg + T\cos\theta$.

$$\sum_{i=1}^{n}F_{x,i} = 0 = -F_{bx} + T\sin\theta \Rightarrow F_{bx} = T\sin\theta$$

CALCULATE:

$$T = \frac{(500.\text{ kg})(9.81\text{ m/s}^2)(3.00\text{ m})\cos 30.0° - (100.\text{ kg})(9.81\text{ m/s}^2)(4.00\text{ m} - 3.00\text{ m})\cos 30.0°}{8.00\text{ m} - 3.00\text{ m}} = 2378.799\text{ N}$$

$F_{by} = (500.\text{ kg})(9.81\text{ m/s}^2) + (100.\text{ kg})(9.81\text{ m/s}^2) + (2378.799\text{ N})\cos 30.0° = 7946.100\text{ N}$

$F_{bx} = (2378.799\text{ N})\sin 30.0° = 1189.399\text{ N}$

ROUND: To three significant figures: $T = 2380\text{ N}$, $F_{by} = 7950\text{ N}$, $F_{bx} = 1190\text{ N}$.

DOUBLE-CHECK: The calculated values are reasonable given the masses and their configuration in the system.

11.51. **THINK:** The bookcase has height H, mass m, and width $W = H/2$. The bookcase is to be pushed with a constant velocity v across a level floor. The bookcase is pushed with a force F horizontally at its top edge a distance H above the floor. I want to calculate the maximum coefficient of kinetic friction, μ_k, between the bookcase and the floor so that the bookcase does not tip over while being pushed.

SKETCH:

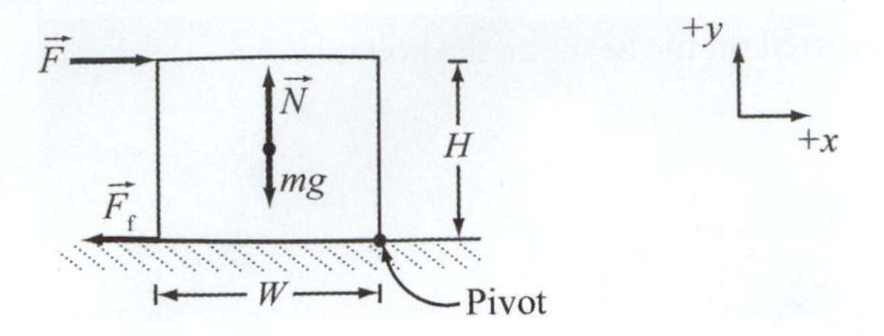

RESEARCH: The condition for the bookcase to not tip over is that the sum of the torques about the pivot point must equal zero, $\sum_{i=1}^{n}\tau_i = 0$. The bookcase stays in contact with the ground, so the sum of forces in the y-direction is zero, $\sum F_y = 0$. The bookcase is being pushed in the x-direction at a constant velocity, which implies its acceleration is zero, therefore the sum of the forces in the x-direction is zero, $\sum F_x = 0$.

SIMPLIFY: $\sum F_y = 0 = -mg + N \Rightarrow N = mg$, $\sum F_x = 0 = F - F_f \Rightarrow F = F_f$. The definition of the frictional force is $F_f = \mu_k N$. Substituting this into the equation gives $F = \mu_k N$.

$$\sum_{i=1}^{n}\tau_i = 0 = mg\frac{W}{2} - FH$$

CALCULATE: Substituting $W = H/2$, $F = \mu_k N$, and $N = mg$ into this equation, and solving for μ_k:

$$mg\frac{H}{4} = \mu_k mgH \Rightarrow \mu_k = 0.25.$$

ROUND: Not applicable.

DOUBLE-CHECK: The coefficient of kinetic friction calculated is reasonable.

11.55. **THINK:** The boy's weight is 60.0 lb, so his mass is $m_b = 27.2$ kg. The plank weighs 30.0 lb, so its mass is $m_p = 13.6$ kg. Its length is $L = 8.00 \text{ ft} = 2.44 \text{ m}$, and lies on two supports, each $d = 2.00 \text{ ft} = 0.6096 \text{ m}$ from each end of the plank. Determine (a) the force exerted by each support, N_L, N_R, when the boy is $a = 3.00 \text{ ft} = 0.9144 \text{ m}$ from the left end, and (b) the distance the boy can go to the right before the plank tips, X_b.

SKETCH:

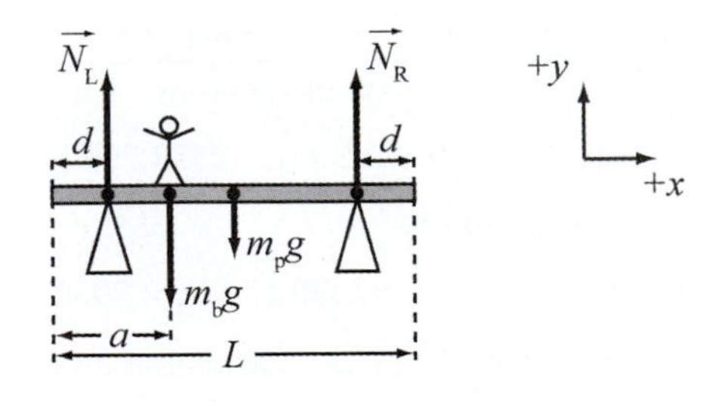

RESEARCH:

(a) Use $\sum F_y = 0$ and $\tau_{net} = \sum \tau_{ccw} - \sum \tau_{cw} = 0$ for the plank in static equilibrium.

(b) The plank will tip when the boy-plank system's center of mass, X is to the right of the right support. Solve for X_b from $X = \frac{1}{M}\sum x_i m_i$. Assume the plank's weight acts at its center, $L/2$.

SIMPLIFY:

(a) Choose the left support as a pivot point:

$$\tau_{net} = -m_b g(a-d) - m_p g\left(\frac{L}{2} - d\right) + N_R(L - 2d) = 0 \Rightarrow N_R = \frac{g\left[m_b(a-d) + m_p\left(\frac{L}{2} - d\right)\right]}{L - 2d}.$$

From $\sum F_y = 0$, $N_R + N_L - m_p g - m_b g = 0 \Rightarrow N_L = g(m_p + m_b) - N_R$.

(b) Choose the left edge of the plank as the origin of the coordinate system. Then

$$X = \frac{1}{M}\left(m_b x_b + m_p \frac{L}{2}\right) \Rightarrow x_b = \frac{XM - m_p \frac{L}{2}}{m_b} = \frac{(L-d)(m_p + m_b) - \frac{1}{2}m_p L}{m_b} = \frac{L\left(\frac{1}{2}m_p + m_b\right) - d(m_p + m_b)}{m_b}.$$

CALCULATE:

(a) $$N_R = \frac{(9.81 \text{ m/s}^2)\left[27.2 \text{ kg}(0.9144 \text{ m} - 0.6096 \text{ m}) + 13.6 \text{ kg}\left(\frac{2.44 \text{ m}}{2} - 0.6096 \text{ m}\right)\right]}{2.44 \text{ m} - 2(0.6096 \text{ m})} = 133.33 \text{ N}$$

$$N_L = (9.81 \text{ m/s}^2)(13.6 \text{ kg} + 27.2 \text{ kg}) - 133.33 \text{ N} = 266.918 \text{ N}$$

(b) $$x_b = \frac{2.44 \text{ m}\left(\frac{13.6 \text{ kg}}{2} + 27.2 \text{ kg}\right) - 0.6096 \text{ m}(13.6 \text{ kg} + 27.2 \text{ kg})}{27.2 \text{ kg}} = 2.1356 \text{ m}$$

ROUND: Since all of the given values have three significant figures,

(a) The right support applies an upward force of $N_R = 133$ N (29.9 lb). The left support applies an upward force of $N_L = 267$ N (60.0 lb).

(b) $x_b = 2.14$ m (7.02 ft) from the left edge of the board.

DOUBLE-CHECK: Since the boy is closer to the left edge, $N_L > N_R$. It is expected that the board will tip when the boy is just past the right support, $x_b > L - d = 2.44 \text{ m} - 0.61 \text{ m} = 1.83 \text{ m}$.

11.59. **THINK:** The ladder's mass is M and its length is $L = 4.00$ m. The coefficient of static friction between the floor and the ladder is $\mu_s = 0.600$. The angle is $\theta = 50.0°$ between the ladder and the floor. The man's mass is $3M$. Determine the distance up the ladder, d, the man can climb before the ladder starts to slip.

SKETCH:

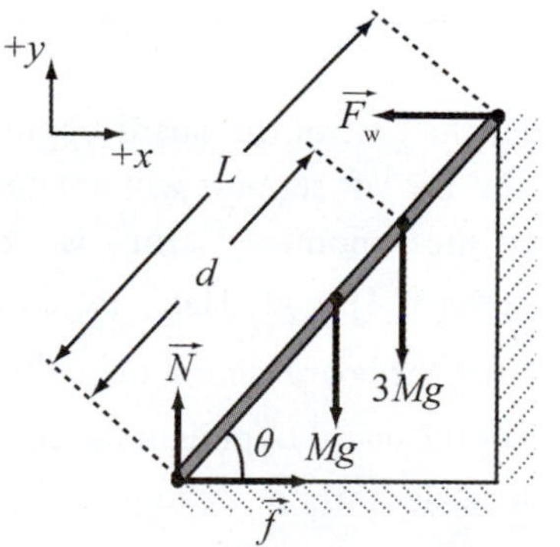

RESEARCH: Use $\tau_{net} = \sum \tau_{ccw} - \sum \tau_{cw} = 0$, $F_{net,x} = \sum F_x = 0$, and $F_{net,y} = \sum F_y = 0$. Assume the ladder's center of mass is at its center, $L/2$. Since the force of the wall on the ladder, F_w, is not known, choose the point where the ladder touches the wall as a pivot point when evaluating τ_{net}.

SIMPLIFY: Determine N first: $F_{net,y} = 0 \Rightarrow N - Mg - 3Mg = 0 \Rightarrow N = 4Mg$.

Consider $\tau_{net} = 0$:

$$Mg\frac{L}{2}\cos\theta + 3Mg(L-d)\cos\theta + fL\sin\theta - NL\cos\theta = 0.$$

$$Mg\frac{L}{2}\cos\theta + 3MgL\cos\theta - 3Mgd\cos\theta + 4\mu_s MgL\sin\theta - 4MgL\cos\theta = 0$$

$$\Rightarrow d = \frac{4\mu_s L\sin\theta - \frac{1}{2}L\cos\theta}{3\cos\theta} = \frac{4}{3}\mu_s L\tan\theta - \frac{1}{6}L.$$

CALCULATE: $d = \frac{4}{3}(0.600)(4.00 \text{ m})\tan(50.0°) - \frac{1}{6}(4.00 \text{ m}) = 3.15 \text{ m}$

ROUND: Since all values have three significant figures, $d = 3.15$ m. The man can go 3.15 m up the ladder before it starts to slip.

DOUBLE-CHECK: It was determined that $d < L$, which it must be.

11.63. **THINK:** Person *B* has a mass twice that of person *A*, that is, $m_B = 2m_A$. The board's mass is $m_b = m_A/2$. Assume the board's weight acts at its center, $L/2$, where L is the length of the board. To determine the distance, *x*, from the edge of the board that person *A* can stand without tipping the board, balance the torques around a pivot point. The two natural choices for the pivot point are the two support points of the board. In this case, we note that if the board tips in a counterclockwise direction, the right support will not contribute to the torque. If the board tips in a clockwise direction, the left support will not contribute to the torque.

SKETCH:

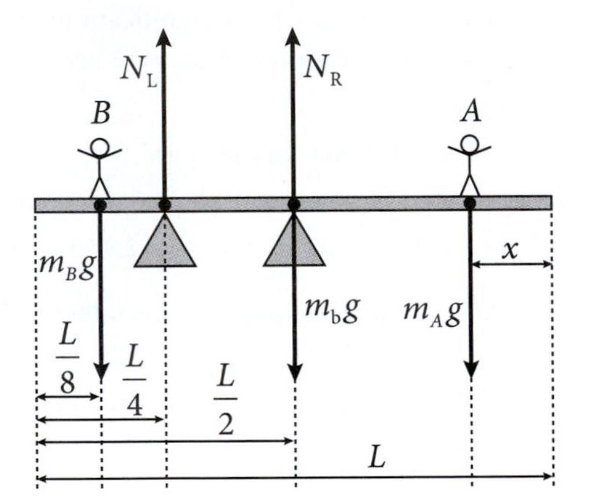

RESEARCH: First, consider the case of the board tipping in the clockwise direction. The pivot point must be the right support and the left support will not contribute. The torque due to the weight of the board is zero because the moment arm is zero. The counterclockwise torque is $m_Bg(3L/8) = (2m_A)g(3L/8) = (3/4)m_AgL.$ The maximum clockwise torque with $x = 0$ is $m_Ag(L/2) = (1/2)m_AgL.$ Thus, the board cannot tip in the clockwise direction.

We then consider the case of the board tipping in the counterclockwise direction. The pivot point is the left support. The counterclockwise torque due to person *B* is $\tau_B = m_Bg(L/8)$. The clockwise torque due to person *A* is $\tau_A = m_Ag((3L/4) - x)$, and the clockwise torque due to the board is $\tau_b = m_bg(L/4)$.

SIMPLIFY: In static equilibrium, we have $\tau_{net} = 0$, so we can write

$$\sum_i \tau_{\text{counterclockwise},i} = \sum_j \tau_{\text{clockwise},j}$$

$$m_Bg(L/8) = m_Ag((3L/4) - x) + m_bg(L/4).$$

Now we express all terms as multiples of $m = m_A$

$$(2m)g(L/8) = mg((3L/4) - x) + (m/2)g(L/4).$$

This gives us

$$(2)(L/8)=((3L/4)-x)+(1/2)(L/4)$$
$$L/4=3L/4-x+L/8$$
$$x=(5/8)L.$$

CALCULATE: Not applicable.

ROUND: Not Applicable

DOUBLE-CHECK: Our result for x is somewhat surprising since this means that person A is standing to the left of the right support. However, it seems reasonable because person B is standing so close to the pivot point at the left support and the center of mass of the board is to the right of the left support. Thus person A must stand closer to the left support to tip the board in a counterclockwise direction.

11.67. **THINK:** The only forces acting on the air freshener are the tension in the string and the force of gravity. The vertical component of the tension must balance the force of gravity while the horizontal component of the tension will cause the air freshener to accelerate in the positive x-direction.

SKETCH:

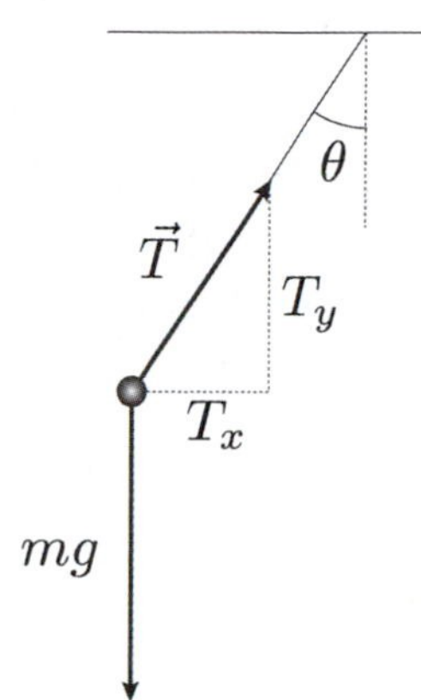

RESEARCH: The sum of the forces in the x-direction is $\sum F_x = T\sin\theta = ma.$ The sum of the forces in the y-direction is $\sum F_y = T\cos\theta - mg = 0 \Rightarrow T\cos\theta = mg.$

SIMPLIFY: Divide the two equations to get: $\dfrac{T\sin\theta}{T\cos\theta}=\dfrac{ma}{mg}\Rightarrow\dfrac{\sin\theta}{\cos\theta}=\dfrac{a}{g}=\tan\theta,$ which implies $\theta=\tan^{-1}\left(\dfrac{a}{g}\right).$

CALCULATE: $\theta=\tan^{-1}\left(\dfrac{5.00\text{ m/s}^2}{9.81\text{ m/s}^2}\right)=27.007°$

ROUND: Since a has three significant figures, $\theta = 27.0°.$

DOUBLE-CHECK: θ is expected to be $0<\theta<90°.$ The size of the angle, $\theta = 27.0°$ is consistent with an angle someone might realistically observe in their own car.

11.71. **THINK:** The given values are $L_1 = 1.00$ m, $M_2 = 0.200$ kg, $L_2 = 0.200$ m, $d = 0.550$ m, $m = 0.500$ kg and $y = 0.707$ m. Determine M_1.

SKETCH:

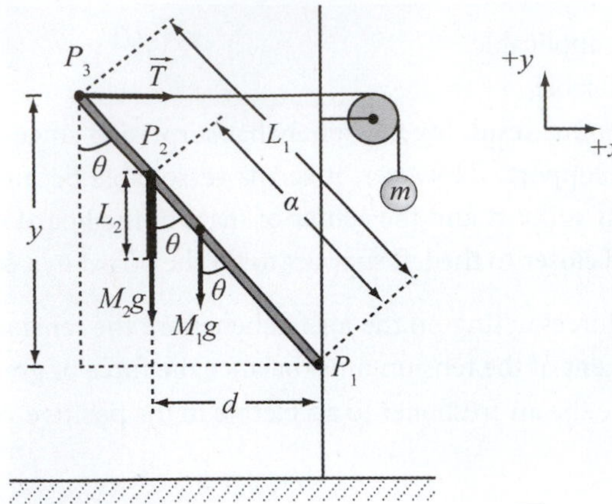

RESEARCH: The beam, B_1, is in static equilibrium, so use $\tau_{\text{net}} = \sum \tau_{\text{ccw}} - \sum \tau_{\text{cw}} = 0$ to determine M_1. Assume M_1 acts at the center of B_1 (a distance $L_1/2$ along the beam).

SIMPLIFY: The force of the support on B_1 at point, P_1, is not known, so choose this as a pivot point. Note $\theta = \cos^{-1}(y/L_1)$ and $T = mg$. Now,

$$\tau_{\text{net}} = M_1 g \frac{L_1}{2} \sin\theta + M_2 g d - T L_1 \cos\theta = 0 \Rightarrow 0 = \frac{1}{2} M_1 g L_1 \sin\theta + M_2 g d - mgy \Rightarrow M_1 = \frac{2(my - M_2 d)}{L_1 \sin\theta}.$$

CALCULATE: $\theta = \cos^{-1}\left(\frac{0.707 \text{ m}}{1.00 \text{ m}}\right) = 45.0°$

$$M_1 = \frac{2\big((0.500 \text{ kg})(0.707 \text{ m}) - (0.200 \text{ kg})(0.550 \text{ m})\big)}{(1.00 \text{ m})\sin(45.0°)} = 2\left(\frac{0.3535 \text{ kg m} - 0.110 \text{ kg m}}{0.707 \text{ m}}\right) = 0.6887 \text{ kg}$$

ROUND: To three significant figures, $M_1 = 0.689$ kg.

DOUBLE-CHECK: Compared to the other masses given in the problem, this result is reasonable.

11.75. **THINK:** Given the mobile sketched in the problem, determine m_1, m_2 and m_3 using the equations of static equilibrium.

SKETCH: Consider the following subsystems of the given mobile.

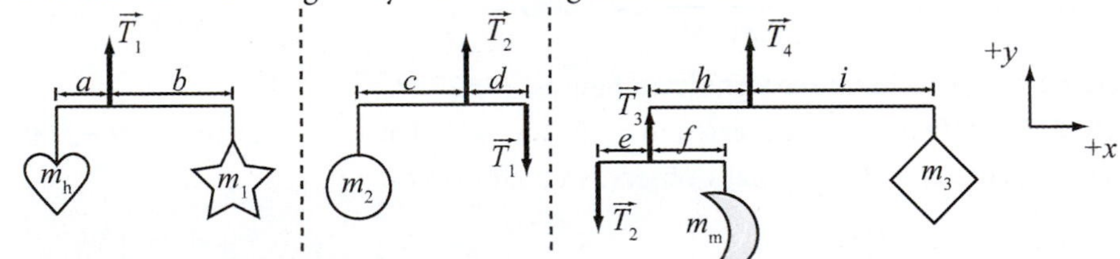

$a = 1.50"$, $b = 3.00"$, $c = 3.00"$, $d = 1.00"$, $e = 1.50"$, $f = 7.50"$, $h = 6.00"$, $i = 9.00"$, $m_{\text{h}} = 0.0600$ kg and $m_{\text{m}} = 0.0240$ kg.

RESEARCH: The mobile is in static equilibrium, so use $\sum F_x = 0$, $\sum F_y = 0$, $\tau_{\text{net}} = \sum \tau_{\text{ccw}} - \sum \tau_{\text{cw}} = 0$. Note all angles are 90°, so $Fd\sin\theta$ becomes Fd.

SIMPLIFY: m_1 : choose T_1 as a pivot point on the a-b bar. Then,

$$\tau_{\text{net}} = -b m_1 g + a m_{\text{h}} g = 0 \Rightarrow m_1 = \frac{a}{b} m_{\text{h}}.$$

m_2 : From $\sum F_y = 0$ on the *a*-*b* bar, it is seen that $T_1 = g(m_1 + m_h)$. Choose T_2 as a pivot point on the *c*-*d* bar. Then, $\tau_{\text{net}} = -T_1 d + m_2 gc = 0 \Rightarrow m_2 = \dfrac{T_1 d}{gc}$.

m_3 : From $\sum F_y = 0$ on the *c*-*d* bar, it can be seen that $T_2 = T_1 + m_2 g$. From $\sum F_y = 0$ on the *e*-*f* bar, it can be seen that $T_3 = T_2 + m_{\text{m}} g = T_1 + g(m_{\text{m}} + m_2)$. Choose T_4 as pivot point on the *h*-*i* bar. Then, $\tau_{\text{net}} = -m_3 gi + T_3 h = 0 \Rightarrow m_3 = \dfrac{T_3 h}{gi}$.

CALCULATE: $m_1 = \dfrac{1.50\text{ "}}{3.00\text{ "}}(0.0600\text{ kg}) = 0.0300\text{ kg}$, $T_1 = (9.81\text{ m/s}^2)(0.0300\text{ kg} + 0.0600\text{ kg}) = 0.8829\text{ N}$

$m_2 = \dfrac{(0.8829\text{ N})(1.00\text{ "})}{(9.81\text{ m/s}^2)(3.00\text{ "})} = 0.0300\text{ kg}$, $T_3 = 0.8829\text{ N} + (9.81\text{ m/s}^2)(0.0240\text{ kg} + 0.0300\text{ kg}) = 1.41264\text{ N}$

$m_3 = \dfrac{(1.41264\text{ N})(6.00\text{ "})}{(9.81\text{ m/s}^2)(9.00\text{ "})} = 0.0960\text{ kg}$

ROUND: Three significant figures: $m_1 = 0.0300\text{ kg}$, $m_2 = 0.0300\text{ kg}$ and $m_3 = 0.0960\text{ kg}$.

DOUBLE-CHECK: Given the mobile arrangement, it is expected that $m_1 < m_{\text{h}}$ and $m_3 > m_1, m_2$.

11.79. **THINK:** Depending on whether the box is pushed or pulled, it will pivot about point *R* or *L*, respectively. This will not affect the solution; so choose *L* as the pivot point. If we pivot about *L*, then the force we apply at the handle should be perpendicular to its moment arm in order to generate maximum torque for a given force, i.e. point in horizontal direction. If we want to find the minimum force needed for tipping, then it would be a good starting point to try this first. However, we need to be able to generate a force of static friction at least as big as this horizontal force; otherwise we cannot prevent slipping. As we will see below, the numbers indeed work out in such a way that the force needed is bigger than the friction force that the weight of the box alone can provide. In order to increase the friction force, we need to increase the normal force between the box and the ground beyond the weight of the box. We can accomplish this by applying some downward force component at the handle. (BTW, this downward force component does not contribute to the torque, because it is parallel to the moment arm.) Let's call the horizontal force component F_x and the vertical component F_y.

SKETCH:

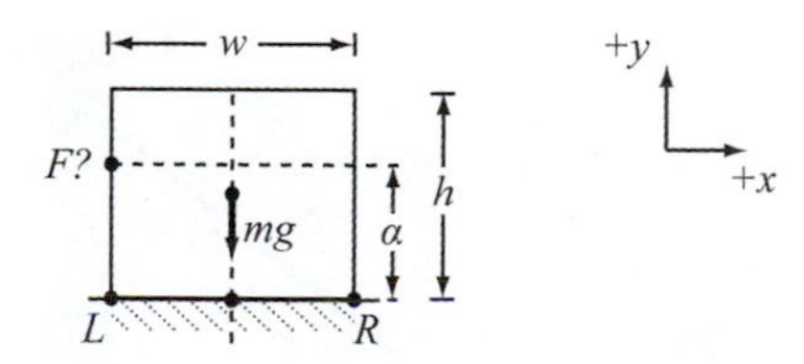

RESEARCH: The equations for equilibrium are for the

- *x*-component of the forces: $-F_x + f = 0$, where f is the friction force between box and ground
- *y*-component of the forces: $F_y + N - mg = 0$
- torques: $\tau_{\text{net}} = F_x \alpha - mg\frac{1}{2}w = 0$

Here we assume that the force is directed to the right and upward. We do not know if the force has a positive or negative *y*-component. We assumed a positive value, but if it turns out to be negative, then our assumption was incorrect, and the *y*-component is negative.

For the case of the maximum friction force without slipping, we have $f = \mu_s N$.

SIMPLIFY: From the equation for zero net torque we obtain

$$\tau_{\text{net}} = F_x \alpha - mg\tfrac{1}{2}w = 0 \Rightarrow F_x = \frac{mgw}{2\alpha}.$$

From $F_y + N - mg = 0$ we can solve for the y-component of the force:

$$F_y = mg - N = mg - f / \mu = mg - F_x / \mu.$$

Then the magnitude of the force is $F = \sqrt{F_x^2 + F_y^2}$.

With F_x and F_y known, the direction of F is given by $\theta = \tan^{-1}\left(F_y / F_x\right)$.

CALCULATE:

(a) $F_x = \dfrac{(20.0 \text{ kg})(9.81 \text{ m/s}^2)(0.300 \text{ m})}{2(0.500 \text{ m})} = 58.86 \text{ N},$

and $F_y = (20.0 \text{ kg})(9.81 \text{ m/s}^2) - \dfrac{58.86 \text{ N}}{0.280} = -14.01 \text{ N}.$

Thus the y-component of the force is in the negative y-direction.

Then, $F = \sqrt{(58.86 \text{ N})^2 + (14.01 \text{ N})^2} = 60.50 \text{ N}.$

(b) $\theta = \tan^{-1}\left(\dfrac{-14.01 \text{ N}}{58.86 \text{ N}}\right) = -13.39°$, so below the horizontal.

ROUND: The least precise value given in the question has two significant figures. The answers should be rounded so they also have two significant figures. Therefore, the minimum force is 60.5 N and is directed at an angle of 13.4° below the horizontal.

DOUBLE-CHECK: Since the y-component of the force turned out to have a negative value, this indeed implies that we had to apply some downward force to prevent the box from slipping. Just to make sure that our solution is consistent, we can calculate the product of the box's weight and the coefficient of friction and make sure that this product is really smaller than our result for the horizontal component of the force, $\mu_s mg = (0.280)(20.0 \text{ kg})(9.81 \text{ m/s}^2) = 54.94 \text{ N}.$ This is indeed smaller than our result for F_x, which shows that some downward force component was indeed needed.

Multi-Version Exercises

11.81. **THINK:** If the hinge where the bar is attached to the wall is the rotation point, then the sum of the clockwise torque and the counterclockwise torque is zero. The counterclockwise torque can be found from the mass and the length of the bar, and the clockwise torque can be used to find the tension in the cable.

SKETCH:

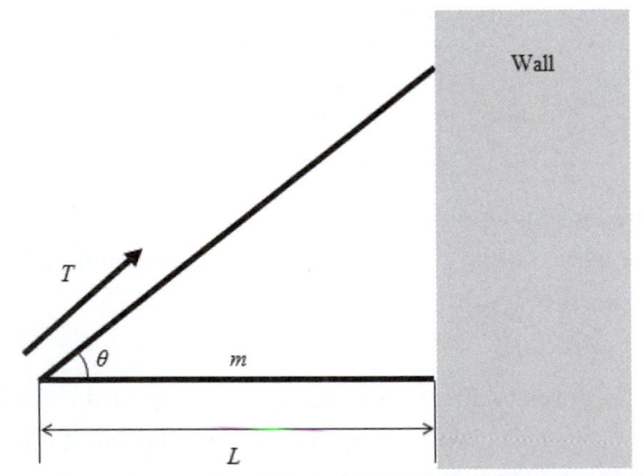

RESEARCH: The net torque is zero, to the magnitude of the counterclockwise torque is equal to the magnitude of the clockwise torque, $\tau_{\text{CCW}} = \tau_{\text{CW}}$. The gravitational force pulling down on the bar ($F_g = mg$) causes a counterclockwise torque that is given by the equation $\tau_{\text{CCW}} = mg(L/2)$. The clockwise torque, due to the tension of the cable, is given by $\tau_{\text{CW}} = TL\sin\theta$.

SIMPLIFY: Since the counterclockwise and clockwise torques are equal, $TL\sin\theta = mg(L/2)$. Solve for the tension on the cable: $T = \dfrac{mg}{2\sin\theta}$.

CALCULATE: The mass of the bar is 81.95 kg and the angle between the bar and the wall is θ = 38.89°. Near the surface of the earth, gravitational acceleration is g = 9.81 m/s². The tension on the cable is thus $T = \dfrac{81.95 \text{ kg} \cdot 9.81 \text{ m/s}^2}{2\sin 38.89°} = 640.2474085$ N.

ROUND: The mass and angle are given to four significant figures, so the final answer should also have four figures. The tension in the cable has a magnitude of 640.2 N.

DOUBLE-CHECK: The bar has a weight of 81.95 kg · 9.81 m/s² = 803.9 N. Since much of the weight of the bar is supported by the wall, it is reasonable to expect the tension on the cable to be less than the weight of the bar, confirming that the calculated value is a reasonable one.

11.84. **THINK:** Since the velocity of the refrigerator is constant, the magnitudes of the friction force and the applied force must be equal (the total force on the refrigerator is zero so $F_{\text{app}} = F_{\text{f}}$). The refrigerator is on the verge of tipping but does not tip over, so the weight will cause a counterclockwise torque that exactly balances the clockwise torque from the force pushing the refrigerator. Find equations for the torques, set them equal to one another, and solve for the maximum coefficient of kinetic friction.

SKETCH: The refrigerator is being pushed from left to right by an applied force, F_{app}.

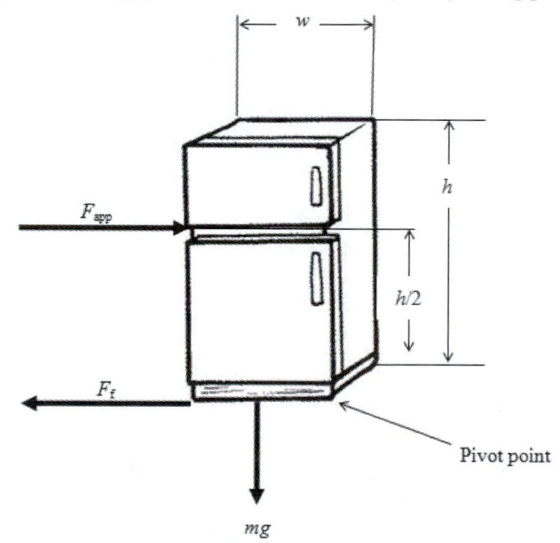

RESEARCH: The counterclockwise torque due to weight has the same magnitude as the clockwise torque due to the force pushing the refrigerator, $\tau_{\text{CCW}} = \tau_{\text{CW}}$. The clockwise torque is due to the applied force on the refrigerator, and is given by $\tau_{\text{CW}} = F_{\text{app}}(h/2)$. The counterclockwise torque is given by $\tau_{\text{CCW}} = mg(w/2)$. The friction force $F_{\text{f}} = \mu_k mg$ is computed from the coefficient of kinetic friction and the normal force (the normal force is equal to the gravitational force on the refrigerator, but opposite in direction).

SIMPLIFY: Since the applied force and frictional force have the same magnitude, the clockwise torque can be expressed in terms of the frictional force as $\tau_{\text{CW}} = F_{\text{f}}(h/2)$. The clockwise and counterclockwise torques must be equal ($\tau_{\text{CCW}} = \tau_{\text{CW}}$), so $mg(w/2) = F_{\text{f}}(h/2)$. Replace the frictional force in this equation with $F_{\text{f}} = \mu_k mg$, and solve for the coefficient of kinetic friction to get:

$$\begin{aligned} mg(w/2) &= (\mu_k mg)(h/2) \\ w &= \mu_k h \\ w/h &= \mu_k \end{aligned}$$

CALCULATE: The width and height of the refrigerator are given in the problem as 1.247 m and 2.177 m, respectively. The maximum coefficient of kinetic friction is then $\mu_k = w/h = 1.247 \text{ m}/2.177 \text{ m} = 0.5728066146$.

ROUND: The dimensions of the refrigerator are given to four significant figures, so the final answer should also have four figures. The maximum coefficient of kinetic friction is 0.5728.

DOUBLE-CHECK: For a refrigerator sliding across the floor, the coefficient of friction must be between 0 and 1. The calculated value is close to the vale for steel sliding on steel, and between the value of rubber sliding or wet and dry concrete. Since the bottom of most refrigerators is made of metal or smooth plastic, and the floor might be linoleum or carpet, a number in this range makes sense. Based on an understanding of how things move in the real world, the calculated value is reasonable.

11.87. **THINK:** In this problem, the maximum angle will occur when the torque due to gravity and the torque due to friction exactly cancel one another. In this case, both the sum of the torques and the sum of the forces will be zero. For this problem, it will be easier to break down the forces into their horizontal and vertical components.

SKETCH: The weight of the ladder is labeled W_l and the weight of the person climbing the ladder is labeled W_p.

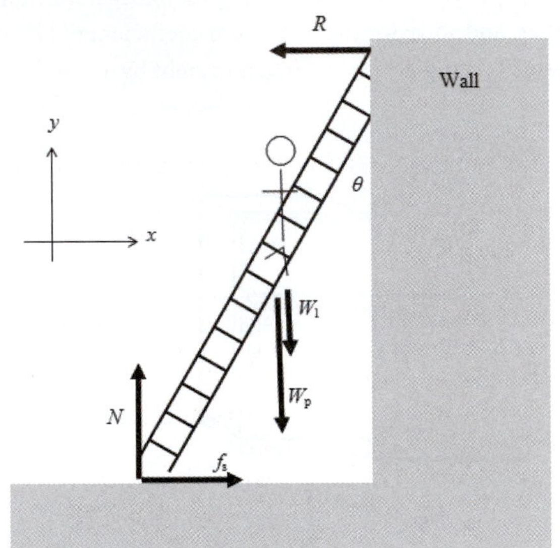

RESEARCH: Both the net force and the net torque are zero in this situation. The only horizontal forces in the x-direction are the force of static friction and the horizontal force of the wall. These must cancel one another, so $f_s - R = 0 \Rightarrow f_s = R$. In the y-direction, the normal force exerted by the floor, the weight of the person, and the weight of the ladder must also cancel out, giving that $N - W_l - W_p = 0 \Rightarrow N = W_l + W_p$. The weight is easily calculated from the mass and the gravitational acceleration: $W_l = m_l g$ and $W_p = m_p g$. The maximum force of static friction is calculated from the normal force using the equation $f_s = \mu_s N$. If l is the length of the ladder, and the pivot point is the place where the ladder touches the floor, then the clockwise torque is given by $\tau_{\text{CW}} = W_l \frac{l}{2}\sin\theta + W_p \frac{l}{2}\sin\theta$. The counterclockwise torque is given by $\tau_{\text{CCW}} = Rl\cos\theta$. Since the net torque is zero, $\tau_{\text{CCW}} - \tau_{\text{CW}} = 0 \Rightarrow \tau_{\text{CCW}} = \tau_{\text{CW}}$.

SIMPLIFY: Since the net torque is zero ($\tau_{\text{CCW}} = \tau_{\text{CW}}$), $Rl\cos\theta = W_l \frac{l}{2}\sin\theta + W_p \frac{l}{2}\sin\theta$. Substitute in for the weights to get $Rl\cos\theta = m_l g \frac{l}{2}\sin\theta + m_p g \frac{l}{2}\sin\theta$. Solve for theta to get:

$$Rl\cos\theta = \left(m_l g + m_p g\right)\frac{l}{2}\sin\theta$$

$$\frac{\cos\theta}{\sin\theta} = \frac{\left(m_l g + m_p g\right)l}{2Rl}$$

$$\cot\theta = \frac{\left(m_l g + m_p g\right)}{2R}$$

$$\theta = \cot^{-1}\left(\frac{m_l g + m_p g}{2R}\right)$$

Use the fact that $R = f_s = \mu_s N$, and the normal force $N = W_l + W_p = m_l g + m_p g$ to get an expression for *R*: $R = \mu_s\left(m_l g + m_p g\right)$. Finally, substitute this into the equation for theta to get a final expression for the angle $\theta = \cot^{-1}\left(\frac{m_l g + m_p g}{2\mu_s\left(m_l g + m_p g\right)}\right) = \cot^{-1}\left(\frac{1}{2\mu_s}\right) = \tan^{-1}\left(2\mu_s\right)$.

CALCULATE: The coefficient of friction is 0.2881, so the maximum angle will be:

$$\theta = \tan^{-1}\left(2\cdot 0.2881\right)$$
$$= \tan^{-1}\left(0.5762\right)$$
$$= 29.9505462^\circ$$

ROUND: The coefficient of friction is given to four significant figures, so the final answer should also have four figures. The maximum angle between the ladder and the wall is 29.95°.

DOUBLE-CHECK: Based on real-world experience, the maximum angle between a ladder of this type and the wall against which it leans is about 30 degrees. Furthermore, it makes intuitive sense that the maximum angle would depend only on the coefficient of friction between the ladder and the floor. If the ladder is at a fixed angle, then it will either slide or not slide regardless of the weight of the person standing at the ladder's midpoint.

Chapter 12: Gravitation

Exercises

12.31. **THINK:** Since the objects are floating in outer space, it can be assumed that the only force on each object is their mutual force of gravity. The space station and the tool exert forces with the same magnitude on each other (Newton's Third Law). In order to find the distance that the objects drift towards each other, we have to calculate the acceleration of the objects, which is the ratio of the gravitational force divided by the mass. Since the space station is MUCH more massive than the tool, it will experience negligible acceleration, and it is sufficient to just calculate the distance the tool moves after an hour needs to be calculated. Even though this force will increase slightly since they will be closer as they accelerate towards each other, it will be assumed that the force of gravity is constant. This assumption can only be made, however, if the distance the tool moves is small compared to the initial separation, and we will have to check this after we are done with our calculation.

SKETCH:

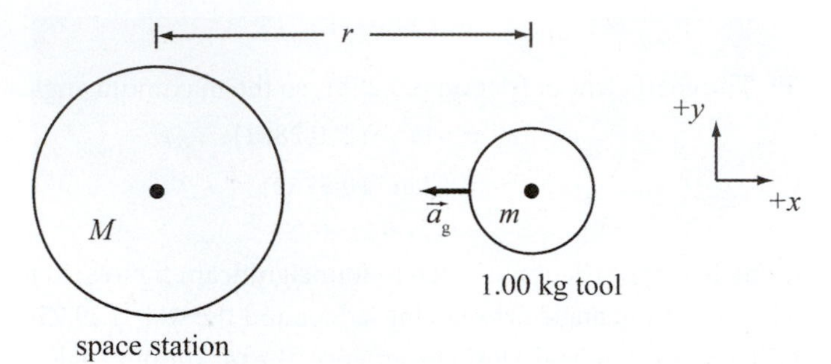

RESEARCH: The two objects attract each other by their mutual gravitational force: $F_g = GMm/r^2$. Using Newton's Second Law, the acceleration of the tool a_{tool} towards the space station due to gravity is $a_{tool} = GM/r^2$. Assuming that the objects were initially at rest and constant acceleration over a time interval t, the distance traveled by the tool is given by: $x = \frac{1}{2}a_{tool}t^2$.

SIMPLIFY: $x = \frac{1}{2}a_{tool}t^2 = \frac{1}{2}\frac{GM}{r^2}t^2$

CALCULATE: Substituting the given values:

$$x = \frac{1}{2}\frac{(6.67\cdot 10^{-11}\text{ N m}^2/\text{kg}^2)(2.00\cdot 10^4\text{ kg})}{(50.0\text{ m})^2}(3600.\text{ s})^2 = 3.45773\cdot 10^{-3}\text{ m}$$

ROUND: $x = 3.46$ mm

DOUBLE-CHECK: Since the displacement x is much smaller than the initial separation $r = 50$ m, the assumption of constant acceleration is justified.

12.35. **THINK:** Consider a uniform rod of mass $M = 333$ kg in the shape of a semi-circle of radius $R = 5.00$ m. To calculate the magnitude of the force on a 77.0 kg point mass m at the centre of this semi-circle it can be assumed that the density of the rod is uniform such that $\rho = M/L$, where L is the length of the rod. If the rod is divided into small elements each of mass $dM = \rho dl$, integration can be used to find the total force on mass m.

SKETCH:

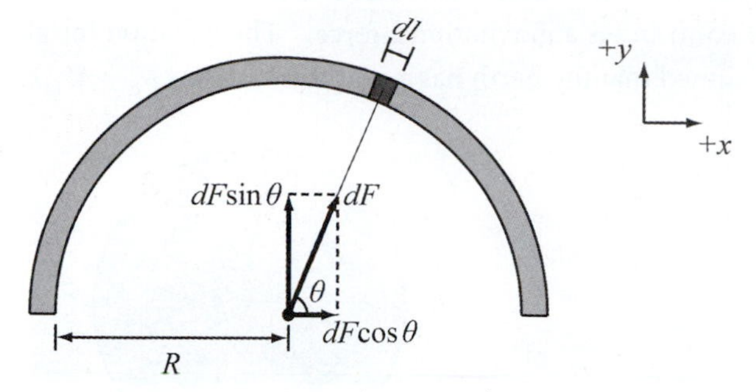

RESEARCH: Use Newton's law of Gravity, $F = G\left(m_1 m_2 / r^2\right)\hat{r}$. The gravitational force on mass m at the center of the semi-circle caused by a small element dM is $d\vec{F} = G\left(\frac{mdM}{R^2}\right)\hat{r}$, or in component form, $dF_x = G\left(mdM / R^2\right)\cos\theta$, $dF_y = G\left(mdM / R^2\right)\sin\theta$.

SIMPLIFY: The components of the force on mass m is then,

$$F_x = \int_0^M \frac{Gm\cos\theta}{R^2} dM \text{ and } F_y = \int_0^M \frac{Gm\sin\theta}{R^2} dM.$$

Using $dM = \rho dl = \rho R d\theta$,

$$F_x = \int_0^\pi \frac{Gm\rho R\cos\theta}{R^2} d\theta \text{ and } F_y = \int_0^\pi \frac{Gm\rho R\sin\theta}{R^2} d\theta.$$

Since G, m, ρ and R are constant, these can be simplified further to

$$F_x = \frac{Gm\rho}{R}\int_0^\pi \cos\theta d\theta \text{ and } F_y = \frac{Gm\rho}{R}\int_0^\pi \sin\theta d\theta.$$

From a mathematical table of integrals,

$$\int_0^\pi \cos\theta d\theta = \left[\sin\theta\right]_0^\pi = \sin\pi - \sin 0 = 0 \text{ and } \int_0^\pi \sin\theta d\theta = \left[-\cos\theta\right]_0^\pi = -\cos\pi + \cos 0 = 2.$$

Therefore, $F_x = 0$ and $F_y = \frac{2Gm\rho}{R}$. Using $\rho = \frac{M}{L}$ and $L = \pi R$ gives $F_y = \frac{2GmM}{\pi R^2}$.

CALCULATE: $F_y = \dfrac{2\left(6.67 \cdot 10^{-11} \text{ N m}^2 / \text{kg}^2\right)(77.0 \text{ kg})(333 \text{ kg})}{\pi(5.00 \text{ m})^2} = 4.355 \cdot 10^{-8} \text{ N}$

ROUND: To three significant figures, $F_y = 4.36 \cdot 10^{-8}$ N. The total net force is in the positive y-direction.

DOUBLE-CHECK: From the symmetry of the shape of the semi-circle, it is clear that $F_x = 0$, since a force due to mass dm on one side will have the same magnitude and opposite direction of a force due to mass dm on the other side.

12.39. **THINK:** The gravitational acceleration decreases towards the center of the Earth since the exterior shell of mass no longer contributes a gravitational force. The equation for gravitational acceleration can be used and it can be assumed that the Earth has a uniform density, $\rho_{\text{E}} = M_{\text{E}} / V_{\text{E}}$.

SKETCH:

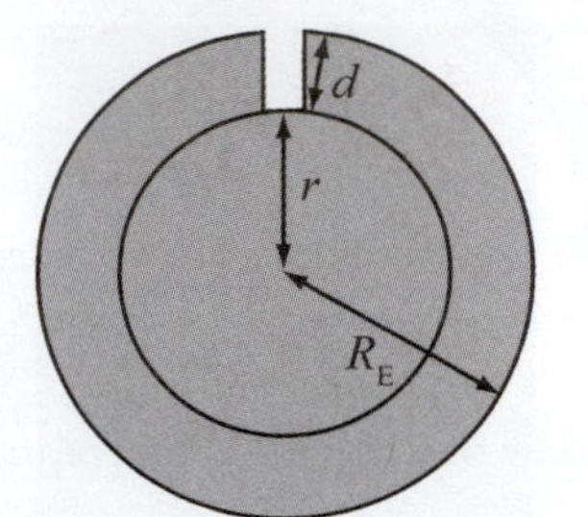

RESEARCH: The gravitational acceleration at the bottom of the mine shaft is given by $a = GM_{\text{int}} / r^2$, where M_{int} is the interior portion of Earth's mass.

SIMPLIFY: $M_{\text{int}} = \frac{4}{3}\pi r^3 \rho_{\text{E}} = \frac{4}{3}\pi r^3 \frac{M_{\text{E}}}{V_{\text{E}}} = \frac{(4/3)\pi r^3 M_{\text{E}}}{(4/3)\pi R_{\text{E}}^3} = M_{\text{E}} \frac{r^3}{R_{\text{E}}^3}$. Therefore, the gravitational acceleration is:

$$a = \frac{GM_{\text{E}}}{r^2}\frac{r^3}{R_{\text{E}}^3} = \frac{GM_{\text{E}}}{R_{\text{E}}^2}\left(\frac{r}{R_{\text{E}}}\right) = g\left(\frac{r}{R_{\text{E}}}\right).$$

CALCULATE:

(a) Substituting $a = g/2$ gives $r = R_{\text{E}}/2 = (6370 \text{ km})/2 = 3185 \text{ km}$. Therefore, the mine depth required for the gravitational acceleration to be reduced by a factor of 2 is $d = R_{\text{E}} - r = 3185 \text{ km}$.

(b) The percentage difference of the gravitational acceleration at the bottom of 3.5 km deep mine relative to that at the Earth's surface is:

$$\frac{a_{\text{surf}} - a_{3.5\text{km}}}{a_{\text{surf}}} = \frac{g - \frac{R_{\text{E}} - d}{R_{\text{E}}} g}{g} = 1 - \left(1 - \frac{d}{R_{\text{E}}}\right) = \frac{d}{R_{\text{E}}} = \frac{3.5 \text{ km}}{6370 \text{ km}} = 5.495 \cdot 10^{-4}$$

ROUND:

(a) To three significant figures, the depth required is 3190 km.

(b) To two significant figures the percentage difference is 0.055%.

DOUBLE-CHECK: Since 3.5 km is a relatively small distance into the surface of the Earth it is expected that the percentage change in the gravitational acceleration is very small.

12.43. **THINK:** The ratio of the escape speed to the satellite speed at the surface of the Moon is needed. The principle of conservation of energy can be used to solve this problem.

SKETCH:

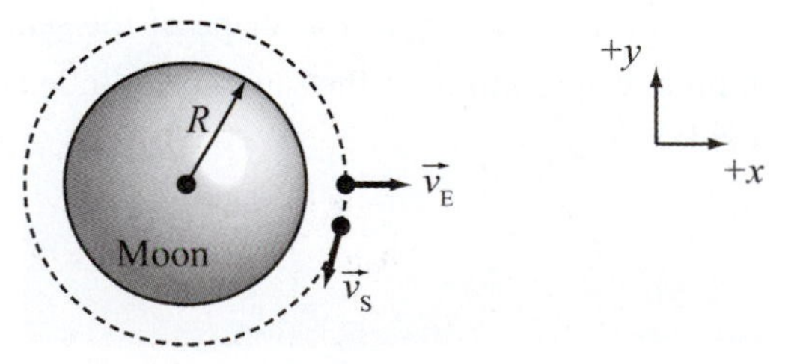

RESEARCH: The escape speed v_{E} is found by using conservation of energy, $\text{E}_{\text{i}} = \text{E}_{\text{f}}$, at the surface of the Moon and at infinity. Because the speed at infinity is zero and the gravitational potential energy is taken to be zero at infinity, the final energy is also zero. Therefore, $(1/2)mv_{\text{E}}^2 - (GMm/R) = 0$. This gives

$v_E = \sqrt{2GM/R}$. The satellite speed v_s is obtained from the condition that the gravitational acceleration is equal to the centripetal acceleration; that is, $GM/R^2 = v_s^2/R$. This yields $v_s = \sqrt{GM/R}$.

SIMPLIFY: Therefore, the ratio of the escape speed to the satellite speed is $\frac{v_E}{v_s} = \frac{\sqrt{2GM/R}}{\sqrt{GM/R}} = \sqrt{2}$.

CALCULATE: No numerical values were given. The solution is algebraic.

ROUND: No rounding is necessary since the answer is algebraic.

DOUBLE-CHECK: In our algebraic simplification, notice that the mass and radius of the Moon dropped out of the equation, and the escape speed was simply a factor of $\sqrt{2}$ larger than the orbital speed. This result confirms the universal result obtained in equation (12.22) in the textbook.

12.47. **THINK:** To solve this problem, a reasonable value of the initial speed of the jump is needed. It is assumed that the density ρ_E of the asteroid is uniform.

SKETCH:

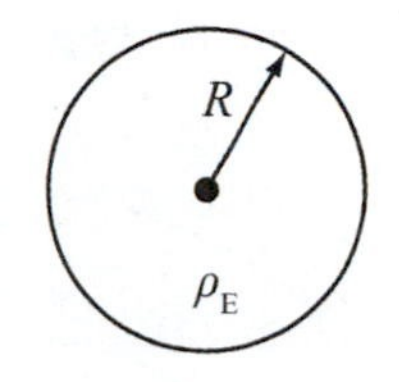

RESEARCH: Take the maximum height of a jump on Earth to be $x = 1.0$ m. Using this height, the initial speed can be found by using $v_f^2 = v^2 - 2gx$ (g can be assumed constant over a distance of 1.0 m). At maximum height on Earth, $v_f = 0$ so: $v = \sqrt{2gx}$. The escape speed from the asteroid is given by $v = \sqrt{2GM/R}$.

SIMPLIFY: Setting the equations equal gives: $2gx = \frac{2GM}{R} \Rightarrow R = \frac{GM}{gx}$. The mass of the asteroid is $M = \frac{4}{3}\pi R^3 \rho_E = M_E\left(\frac{R}{R_E}\right)^3$. Therefore, $R = \frac{R^3}{R_E gx}\left(\frac{GM_E}{R_E^2}\right) = \frac{R^3}{xR_E} \Rightarrow R = \sqrt{xR_E}$.

CALCULATE: $R = \sqrt{(1.0 \text{ m})(6.37 \cdot 10^6 \text{ m})} = 2.524$ km

ROUND: To two significant figures the largest asteroid that one could escape from by jumping is about $R = 2.5$ km.

DOUBLE-CHECK: The initial velocity of a 1.0 m jump on Earth is $v = \sqrt{2(9.81 \text{ m/s}^2)(1.0 \text{ m})} = 4.4$ m/s. Since this is a small velocity, it is reasonable that this would be the escape speed from a very small asteroid.

12.51. **THINK:** The orbit is 111 km above the surface of the Moon. The Moon has a mass and radius of $M_m = 7.35 \cdot 10^{22}$ kg and $R_m = 1.74 \cdot 10^6$ m, respectively.

SKETCH:

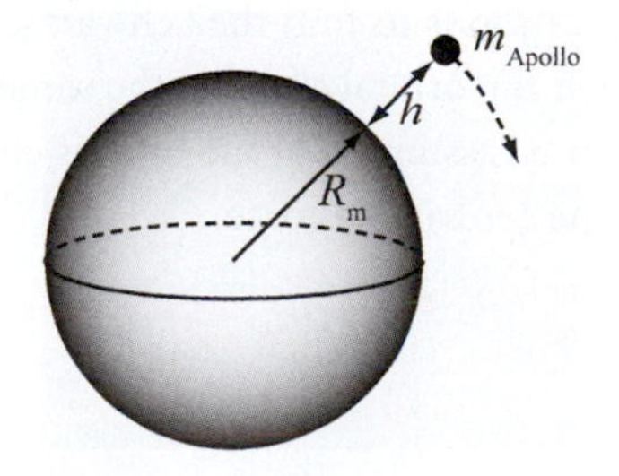

RESEARCH: Using Kepler's third law, the period is $\mathrm{T}=2\pi\sqrt{\frac{r^3}{GM}}$.

SIMPLIFY: $\mathrm{T}=2\pi\sqrt{\frac{(R_\mathrm{m}+h)^3}{GM_\mathrm{m}}}$

CALCULATE: $T=2\pi\sqrt{\frac{(1.74\cdot 10^6\text{ m}+0.111\cdot 10^6\text{ m})^3}{(6.674\cdot 10^{-11}\text{ N m}^2/\text{kg}^2)(7.35\cdot 10^{22}\text{ kg})}}=7144.19\text{ s}$

ROUND: To three significant figures, the period of this orbit is 7140 s.

DOUBLE-CHECK: This trip of about 2 hours around the Moon is fast, but reasonable since the orbit is so close to the surface.

12.55. **THINK:** The goal is to find the orbital speed and period of a satellite $h=700.$ km above the Earth. The speed of a circular orbit is found by setting the force of gravity equal to the centripetal force.

SKETCH:

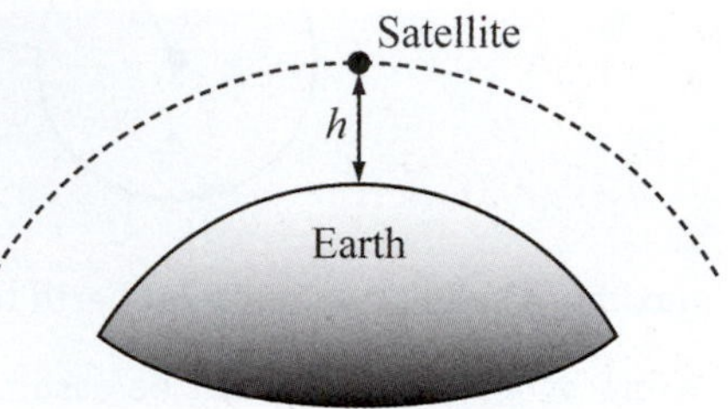

RESEARCH: The orbital speed is equal to $v=\sqrt{GM/r}$, where the radius r is the sum of the radius of the Earth and the height above the surface $r=R_\mathrm{e}+h$. The period is just the distance travelled divided by the speed: $T=\frac{2\pi r}{v}$.

SIMPLIFY: $v=\sqrt{\frac{GM}{r}}=\sqrt{\frac{GM}{R_\mathrm{e}+h}}$ and $T=\frac{2\pi r}{v}$.

CALCULATE: $v=\sqrt{\frac{(6.67\cdot 10^{-11}\text{ N m}^2/\text{kg}^2)(5.97\cdot 10^{24}\text{ kg})}{(6370+700.)\cdot 10^3\text{ m}}}=7504.82\text{ m/s}$ and

$$T=2\pi\frac{(6370+700.)\cdot 10^3\text{ m}}{7504.82\text{ m/s}}=5919.15\text{ s}$$

ROUND: To three significant figures, the orbital speed and period are $v=7.50$ km/s and $T=5920$ s.

DOUBLE-CHECK: A speed of 7.50 km/s is a reasonable value for a satellite in orbit. Notice that the period $T=\frac{2\pi r}{v}=2\pi r\sqrt{\frac{r}{GM}}=2\pi\sqrt{\frac{r^3}{GM}}$, which is Kepler's third law.

12.59. **THINK:** The goal of this question is to find the Schwarzschild radius of a black hole with twice the mass of the Sun, the radius at which the orbital speed is the same as the speed of light, and the radius of a black hole with Earth's mass. It can be assume that the orbit is circular so that the speed can be found by setting the force of gravity equal to the centripetal force.

SKETCH:

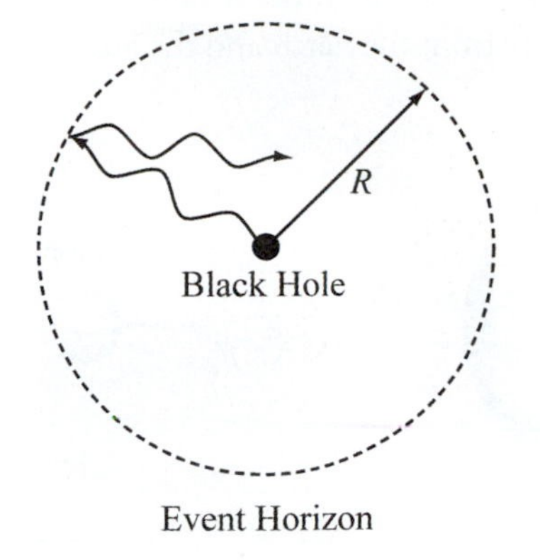

RESEARCH: The escape speed is given by $v = \sqrt{2GM/R}$. The radius of the orbital velocity can be found with $v = \sqrt{GM/R}$. The mass of the Sun is $1.9891 \cdot 10^{30}$ kg. The mass of the Earth is $5.9742 \cdot 10^{24}$ kg.

SIMPLIFY: (a) The Schwarzschild radius can be found by setting the escape speed to the speed of light and solving for *R*: $v = c = \sqrt{2GM/R} \Rightarrow c^2 = 2GM/R \Rightarrow R_s = 2GM/c^2$.

(b) The radius when the orbital speed is the speed of light is given by: $v = c = \sqrt{GM/R} \Rightarrow R = GM/c^2$.

(c) $R_s = 2GM/c^2$

CALCULATE:

(a) $$R_s = \frac{2GM}{c^2} = \frac{2\left(6.674 \cdot 10^{-11} \text{ N m}^2/\text{kg}^2\right)\left((2)1.9891 \cdot 10^{30} \text{ kg}\right)}{\left(2.998 \cdot 10^8 \text{ m/s}\right)^2} = 5907.99 \text{ m}$$

(b) $$R = \frac{GM}{c^2} = \frac{\left(6.674 \cdot 10^{-11} \text{ N m}^2/\text{kg}^2\right)\left((2)1.9891 \cdot 10^{30} \text{ kg}\right)}{\left(2.998 \cdot 10^8 \text{ m/s}\right)^2} = 2953.99 \text{ m}$$

(c) $$R_s = \frac{2GM}{c^2} = \frac{2\left(6.674 \cdot 10^{-11} \text{ N m}^2/\text{kg}^2\right)\left(5.9742 \cdot 10^{24} \text{ kg}\right)}{\left(2.998 \cdot 10^8 \text{ m/s}\right)^2} = 0.0088722 \text{ m}$$

ROUND: To four significant figures:

(a) The radius of the black hole with a mass twice that of the sun is $R_s = 5.908$ km.

(b) The orbital radius when the orbital speed is equal to the speed of light is $R = 2.954$ km.

(c) The radius of a black hole with a mass equal to the mass of the Earth is $R_s = 8.872$ mm.

DOUBLE-CHECK: Such extreme values are typical when dealing with black holes since they are so dense and massive.

12.63. THINK: The force of gravity acting on the Moon due to the Sun and the Earth are calculated separately. The net force due to gravity from the Earth and the Sun can then be found by summing up the two forces.

SKETCH:

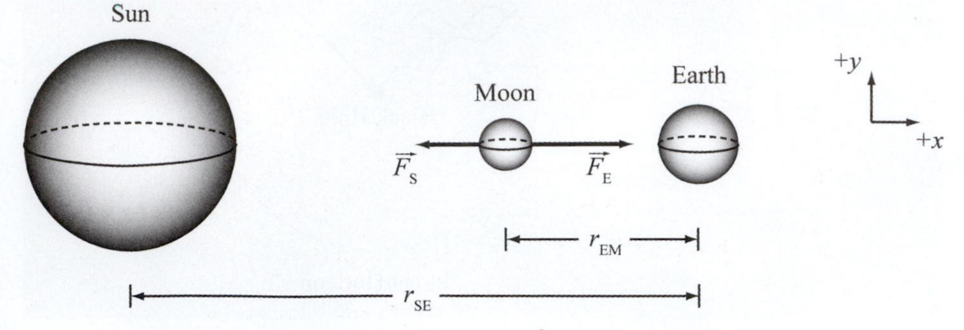

RESEARCH: Use Newton's Law of Gravity, $F = Gm_1m_2/r^2$, to calculate both forces.

SIMPLIFY: $\vec{F}_S = -\frac{Gm_Sm_M}{(r_{SE}-r_{EM})^2}\hat{x}$, $\vec{F}_E = \frac{Gm_Em_M}{(r_{EM})^2}\hat{x}$. The net force acting on the Moon is

$$\vec{F}_M = \sum \vec{F} = \vec{F}_S + \vec{F}_E = Gm_M\left[-\frac{m_S}{(r_{SE}-r_{EM})^2}+\frac{m_E}{(r_{EM})^2}\right]\hat{x}.$$

CALCULATE:

$$\vec{F}_S = -\frac{(6.674\cdot 10^{-11}\text{ N m}^2/\text{kg}^2)(1.99\cdot 10^{30}\text{ kg})(7.344\cdot 10^{22}\text{ kg})}{(1.496\cdot 10^{11}\text{ m}-3.844\cdot 10^{8}\text{ m})^2}\hat{x} = -4.381\cdot 10^{20}\text{ N }\hat{x}$$

$$\vec{F}_E = \frac{(6.674\cdot 10^{-11}\text{ N m}^2/\text{kg}^2)(5.97\cdot 10^{24}\text{ kg})(7.344\cdot 10^{22}\text{ kg})}{(3.844\cdot 10^{8}\text{ m})^2}\hat{x} = 1.980\cdot 10^{20}\text{ N }\hat{x}$$

$$\vec{F}_M = (-4.381\cdot 10^{20}\text{ N}+1.980\cdot 10^{20}\text{ N})\hat{x} = -2.401\cdot 10^{20}\text{ N }\hat{x}$$

ROUND: To three significant figures, the force on the Moon due to the Sun is $F_S = 4.38\cdot 10^{20}$ N towards the Sun. The force on the Moon due to the Earth is $F_E = 1.98\cdot 10^{20}$ N towards the Moon. The total force on the Moon is $F_M = 2.40\cdot 10^{20}$ N towards the Sun. The Moon's orbit never curves away from the Sun toward the Earth.

DOUBLE-CHECK: Use the answer of problem 12.56 as a double-check, the ratio of the magnitude of the two forces was calculated to be approximately 0.5, in accordance with the present result.

12.67. **THINK:** The principle of the conservation of energy can be used to determine the energy supplied by the rockets in order to increase the altitude of the orbit. The energy supplied by the rockets is the difference in orbital energies of the two orbits.

SKETCH:

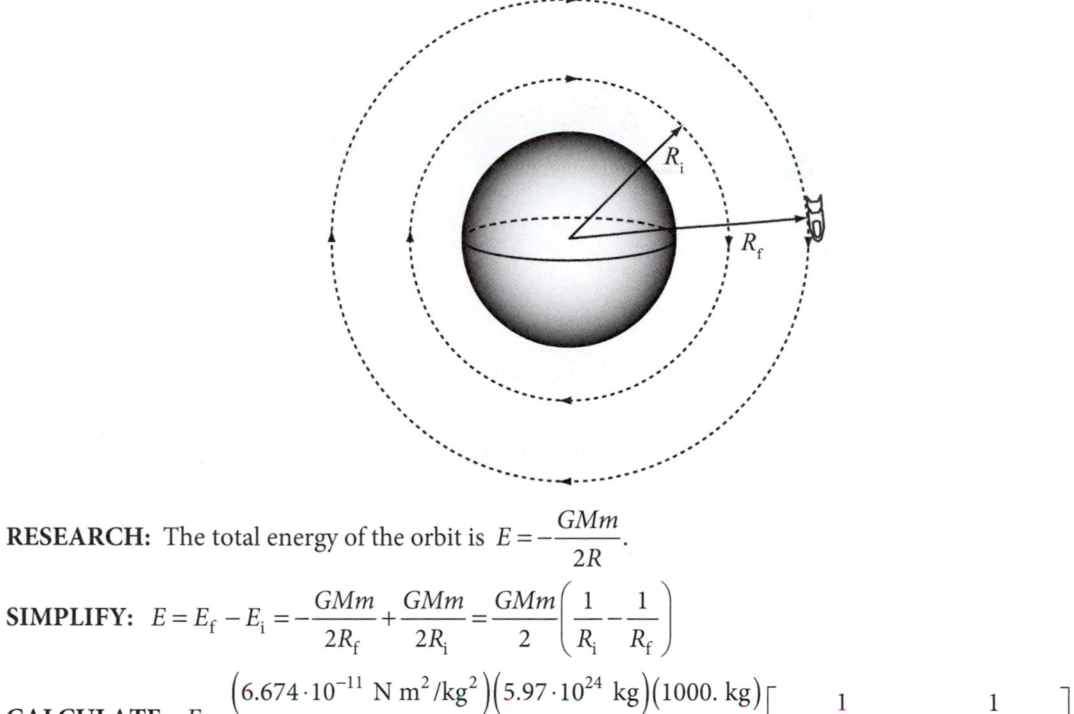

RESEARCH: The total energy of the orbit is $E = -\dfrac{GMm}{2R}$.

SIMPLIFY: $E = E_f - E_i = -\dfrac{GMm}{2R_f} + \dfrac{GMm}{2R_i} = \dfrac{GMm}{2}\left(\dfrac{1}{R_i} - \dfrac{1}{R_f}\right)$

CALCULATE: $E = \dfrac{\left(6.674 \cdot 10^{-11} \text{ N m}^2/\text{kg}^2\right)\left(5.97 \cdot 10^{24} \text{ kg}\right)\left(1000. \text{ kg}\right)}{2}\left[\dfrac{1}{7.00 \cdot 10^6 \text{ m}} - \dfrac{1}{5.00 \cdot 10^7 \text{ m}}\right]$

$= 2.448 \cdot 10^{10} \text{ J}$

ROUND: To three significant figures, the energy supplied by the rockets to move the satellite into a higher orbit is $2.45 \cdot 10^{10}$ J.

DOUBLE-CHECK: This is the same amount of energy required to lift a 2 million kilogram object 1 kilometer into the air (assuming gravitational acceleration of Earth is constant). This large value was expected since the force of gravity between the Earth and the satellite is rather large.

12.71. **THINK:** The effective gravitational force acting on a particle on the Earth's surface at $\lambda = 30.0°$ north of the equator can be found by summing the forces acting on the particle. The only two forces acting on the particle are the force of gravity, pointing toward the center of the Earth, and the force exerted by the surface of the Earth. Because the Earth is rotating and the particle is traveling in a circle, the difference between the x-component of the force of gravity and the x-component of the force exerted by the surface of the Earth must be equal to the centripetal force required to keep the particle traveling in a circle.

SKETCH:

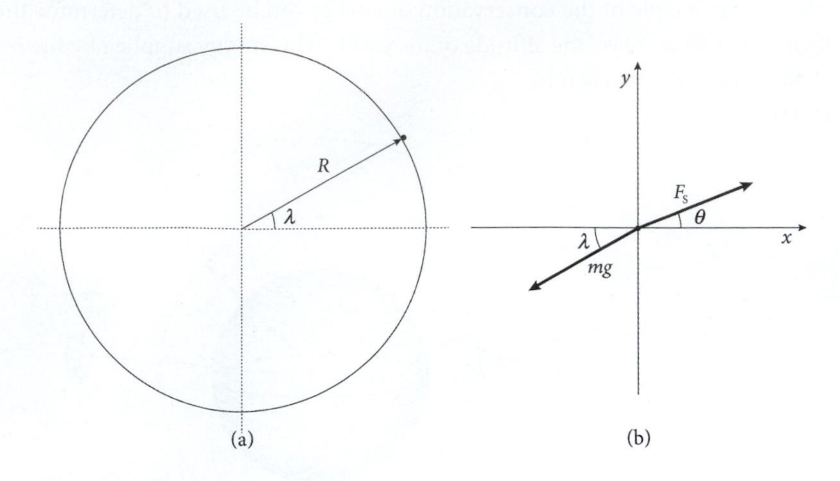

RESEARCH: The y-components of the forces must sum to zero

$F_{S,y} - mg\sin\lambda = 0 \;\Rightarrow\; mg\sin\lambda = F_{S,y}$.

The x-components of the forces must equal the centripetal force required to keep the particle traveling in a circle with radius $r = R\cos\lambda$

$$F_{S,x} - mg\cos\lambda = -mr\omega^2 = -mR\cos\lambda\omega^2$$

$$F_{S,x} = mg\cos\lambda - mR\cos\lambda\omega^2.$$

SIMPLIFY:

We can write the magnitude of the force exerted by the surface as

$$F_S = \sqrt{\left(mg\cos\lambda - mR\cos\lambda\omega^2\right)^2 + \left(mg\sin\lambda\right)^2}$$

$$F_S = \sqrt{\left(mg\cos\lambda - mR\cos\lambda\omega^2\right)^2 + \left(mg\sin\lambda\right)^2} = m\sqrt{\left(g\cos\lambda - R\cos\lambda\omega^2\right)^2 + \left(g\sin\lambda\right)^2}$$

$$F_S = m\sqrt{\left(g - R\omega^2\right)^2\cos^2\lambda + g^2\sin^2\lambda}.$$

The angle of the force exerted by the surface is

$$\theta = \tan^{-1}\left(\frac{mg\sin\lambda}{mg\cos\lambda - mR\cos\lambda\omega^2}\right) = \tan^{-1}\left(\frac{g\sin\lambda}{g\cos\lambda - R\cos\lambda\omega^2}\right) = \tan^{-1}\left(\frac{g\tan\lambda}{g - R\omega^2}\right).$$

The deviation can be expressed as

$$\Delta = \lambda - \tan^{-1}\left(\frac{g\tan\lambda}{g - R\omega^2}\right).$$

CALCULATE:

(a) The magnitude of the force exerted by the surface for $\lambda = 30.0^\circ$ is

$$F_S = m\sqrt{\left(g - R\omega^2\right)^2\cos^2\lambda + g^2\sin^2\lambda}$$

$$F_S = m\sqrt{\left(9.81\text{ m/s}^2 - \left(6.37\cdot 10^6\text{ m}\right)\left(7.27\cdot 10^{-5}\text{ rad/s}\right)^2\right)^2\cos^2 30.0^\circ + \left(9.81\text{ m/s}^2\right)^2\sin^2 30.0^\circ}$$

$$F_S = m\left(9.7848\text{ m/s}^2\right).$$

The angular deviation at $\lambda = 30.0^\circ$ is

$$\Delta = \lambda - \tan^{-1}\left(\frac{g\tan\lambda}{g - R\omega^2}\right) = 30.00° - \tan^{-1}\left(\frac{(9.81 \text{ m/s}^2)\tan 30.0°}{9.81 \text{ m/s}^2 - (6.37 \cdot 10^6 \text{ m})(7.27 \cdot 10^{-5} \text{ rad/s})^2}\right)$$

$\Delta = -0.085365°$.

We can get the maximum by taking the derivative, setting it equal to zero, and solving for λ,

$$\text{Let } a = \frac{g}{g - R\omega^2} = 1.00344$$

$$\frac{d\Delta}{d\lambda} = \frac{d}{d\lambda}\left[\lambda - \tan^{-1}(a\tan\lambda)\right] = 1 - \frac{a\sec^2\lambda}{1 + a^2\tan^2\lambda} = 0$$

$$1 + a^2\tan^2\lambda = a\sec^2\lambda$$

$$\cos^2\lambda + a^2\sin^2\lambda = a$$

Solving numerically (in this case, using Mathematica)

$$\lambda = 2\tan^{-1}\left(\sqrt{1 + 2a - 2\sqrt{a(1+a)}}\right) = 0.784539 \text{ rad} = 44.95°.$$

ROUND:

(a) The effective force of gravitation is $F_S = (9.78 \text{ m/s}^2)m$.

(b) The direction of the effective gravitational force is $29.9°$ above the horizontal.

(c) The angle of λ that maximizes the deviation is $45.0°$ or $\pi/4$ radians.

DOUBLE-CHECK: It is expected that the effective gravitational acceleration is less than *g*. If the Earth were to rotate fast enough, objects would not stay on the surface. The direction points toward the equator from the radial direction. This is the same effect that makes the Earth's oceans bulge at the equator. For $\lambda = 0$, we see that $F_S = m\sqrt{(g - R\omega^2)^2\cos^2 0 + g^2\sin^2 0} = mg - mR\omega^2$, which is what we would expect at the equator.

12.75. **THINK:** Given the distance $r = 1.4960 \cdot 10^{11}$ m and period $T = 3.1557 \cdot 10^7$ s of the Earth's orbit about the Sun, along with the gravitational constant $G = 6.6738 \cdot 10^{-11} \text{ m}^3 \text{ kg}^{-1} \text{ s}^{-2}$, the mass *M* of the Sun can be calculated by using Kepler's third law. The radius and period for the orbit are $1.4960 \cdot 10^{11}$ m and $3.1557 \cdot 10^7$ s, respectively.

SKETCH:

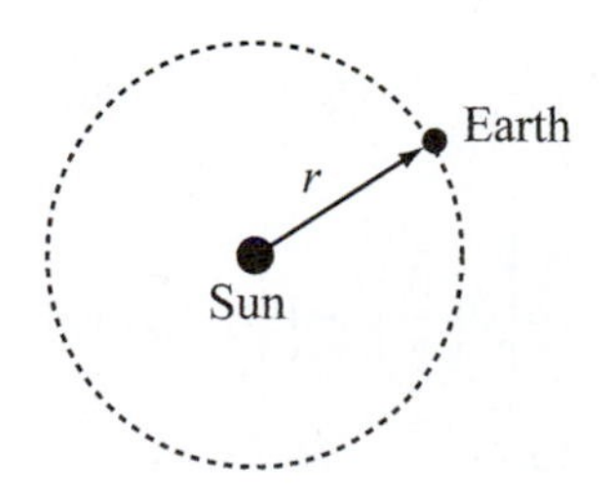

RESEARCH: The mass of the Sun can be found by using Kepler's third law: $\frac{T^2}{r^3} = \frac{4\pi^2}{GM}$.

SIMPLIFY: Solving for *M* gives: $M = \frac{4\pi^2 r^3}{GT^2}$.

CALCULATE: $$M = \frac{4\pi^2(1.4960 \cdot 10^{11} \text{ m})^3}{(6.6738 \cdot 10^{-11} \text{ m}^3\text{kg}^{-1}\text{s}^{-2})(3.1557 \cdot 10^7 \text{ s})^2} = 1.98879 \cdot 10^{30} \text{ kg}$$

ROUND: The values are given to five significant figures, thus, the answer also has this accuracy. The mass of the Sun is $1.9888 \cdot 10^{30}$ kg.

DOUBLE-CHECK: This agrees with the known value of $1.99 \cdot 10^{30}$ kg.

12.79. **THINK:** The space ship starts at 6720 km from the center of the Earth. Instead of slowing the spacecraft, the forward boosters will be applied to put the craft in an elliptical orbit. The current position is the distance at perihelion, r_{p}. The new period should be 1.5 times greater than the present one so that the spaceship meets the station.

SKETCH:

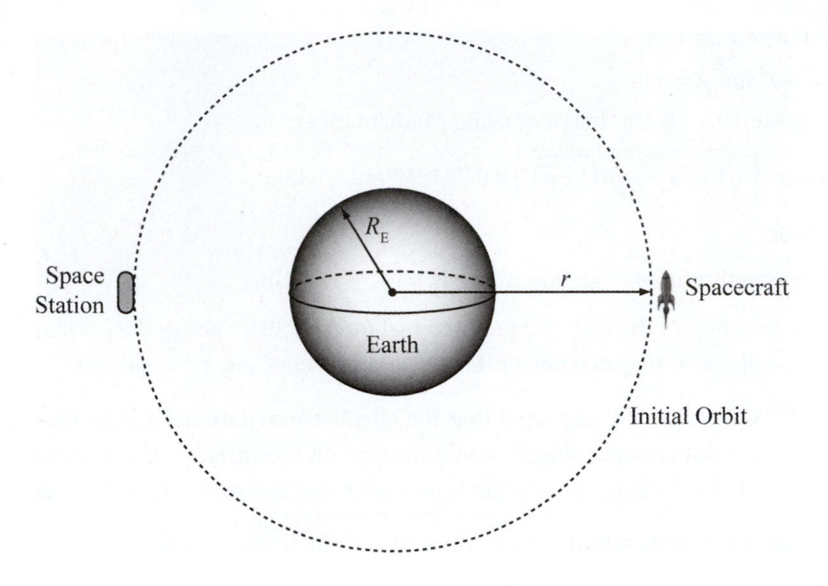

RESEARCH: Use the subscript 1 to refer to the circular orbit, and the subscript 2 to denote the new elliptical orbit. This means that $r = a_1 = r_{\text{p}}$. To find the new semi-major axis we use Kepler's third law, $T_1^2/a_1^3 = T_2^2/a_2^3$. The distance at aphelion, r_{a}, can be found by using the equation $r_{\text{a}} + r_{\text{p}} = 2a$. The period of the new orbit can be found by using $T^2/a^3 = 4\pi^2/GM$.

SIMPLIFY: The distance at perihelion is r, the radius of the original orbit. $a_2 = \left(\dfrac{T_2^2}{T_1^2}\right)^{\frac{1}{3}} a_1 = \left(\dfrac{T_2}{T_1}\right)^{\frac{2}{3}} r = \left(\dfrac{3}{2}\right)^{\frac{2}{3}} r$. Using this, find the distance at aphelion: $r_{\text{a}} + r_{\text{p}} = r_{\text{a}} + r = 2\left(\dfrac{3}{2}\right)^{\frac{2}{3}} r \Rightarrow r_{\text{a}} = \left(2\left(\dfrac{3}{2}\right)^{\frac{2}{3}} - 1\right) r$. The new period is $T_2^2 = \left(\dfrac{4\pi^2}{GM}\right) a_2^3 \Rightarrow T_2 = \sqrt{\left(\dfrac{9\pi^2}{GM}\right) r^3}$.

CALCULATE: $r_{\text{a}} = \left(2\left(\dfrac{3}{2}\right)^{\frac{2}{3}} - 1\right)(6720 \text{ km}) = 10891 \text{ km}$

$$T_2 = \sqrt{\frac{9\pi^2}{(6.674 \cdot 10^{-11} \text{ Nm}^2/\text{kg}^2)(5.97 \cdot 10^{24} \text{ kg})}(6720 \cdot 10^3 \text{ m})^3} = 8225.2 \text{ s}$$

ROUND: The answers should be rounded to three significant figures. For the new orbit, the distance at perihelion is $6.72 \cdot 10^3$ km, the distance at aphelion is $1.09 \cdot 10^4$ km, and the period is $8.23 \cdot 10^3 \text{ s} = 2.28$ hours.

DOUBLE-CHECK: The period of the original orbit was

$$T_1 = \sqrt{\left(\frac{4\pi^2}{GM}\right)r^3} = \sqrt{\frac{4\pi^2}{\left(6.674 \cdot 10^{-11} \text{ Nm}^2/\text{kg}^2\right)\left(5.97 \cdot 10^{24} \text{ kg}\right)}\left(6720 \cdot 10^3 \text{ m}\right)^3} = 5480 \text{ s} = 1.52 \text{ hours}.$$

Since $\frac{T_2}{T_1} = \frac{2.28 \text{ hours}}{1.52 \text{ hours}} = \frac{3}{2}$, the spacecraft will rendezvous with the station, as was required.

Multi-Version Exercises

12.81. **THINK:** When it is launched, the projectile has kinetic energy. At its highest point, it is not moving at all; the energy is now in the form of gravitational potential energy. So, find the kinetic energy of the projectile when it was launched, and that should equal the gravitational potential energy of the projectile at its maximum height.

SKETCH: The projectile when it is launched and at maximum height are shown, with arrows indicating the gravitational force and velocity vectors. Take the surface of the moon to be height $y = 0$ m.

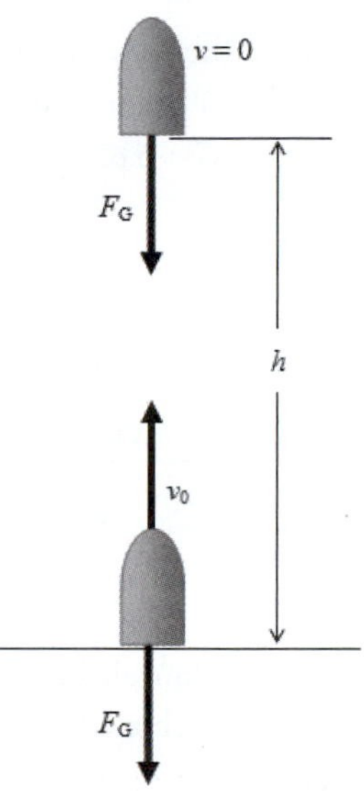

RESEARCH: The acceleration due to gravity on the surface of the moon can be found from Newton's Law of Gravity to be $g_M = \frac{Gm_M}{R_M^2}$. The kinetic energy of the projectile when it is launched is given by $K_0 = \frac{1}{2}mv_0^2$. The change in potential energy of the projectile is given by $\Delta U = -\frac{Gmm_M}{R_M + h} + \frac{Gmm_M}{R_M}$.

Conservation of energy gives us $K_0 = \Delta U$. Since the radius of the moon is given in kilometers and the initial velocity of the projectile is given in meters per second, the conversion factor of 1 km = 1000 m will also be necessary.

SIMPLIFY: Since $K_0 = \Delta U$, substitute in on both sides for the kinetic energy and change in potential energy to get $\frac{1}{2}mv_0^2 = -\frac{Gmm_\text{M}}{R_\text{M}+h} + \frac{Gmm_\text{M}}{R_\text{M}}$. Use algebra to solve for the maximum height h:

$$m\left(\tfrac{1}{2}v_0^2\right) = m\left(\frac{Gm_\text{M}}{R_\text{M}} - \frac{Gm_\text{M}}{R_\text{M}+h}\right)$$

$$\tfrac{1}{2}v_0^2 = \frac{Gm_\text{M}}{R_\text{M}} - \frac{Gm_\text{M}}{R_\text{M}+h}$$

$$\frac{Gm_\text{M}}{R_\text{M}+h} = \frac{Gm_\text{M}}{R_\text{M}} - \frac{v_0^2}{2}$$

$$R_\text{M}+h = \frac{Gm_\text{M}}{\left[\frac{Gm_\text{M}}{R_\text{M}} - \frac{v_0^2}{2}\right]}$$

$$h = \frac{Gm_\text{M}}{\left[\frac{Gm_\text{M}}{R_\text{M}} - \frac{v_0^2}{2}\right]} - R_\text{M}$$

CALCULATE: According to the question, the mass of the moon is $m_\text{M} = 7.348 \cdot 10^{22}$ kg, the radius of the moon is R_M = 1737 km = 1,737,000 m, and the initial speed of the projectile is $v_0 = 114.5$ m/s. The gravitational constant is $G = 6.674 \cdot 10^{-11}\ \text{N} \cdot \text{m}^2 / \text{kg}^2$. Using these values, the maximum height of the projectile is

$$\begin{aligned} h &= \frac{Gm_\text{M}}{\left[\frac{Gm_\text{M}}{R_\text{M}} - \frac{v_0^2}{2}\right]} - R_\text{M} \\ &= \frac{6.674 \cdot 10^{-11}\ \text{N}\cdot\text{m}^2/\text{kg}^2 \cdot 7.348 \cdot 10^{22}\ \text{kg}}{\left[\frac{6.674 \cdot 10^{-11}\ \text{N}\cdot\text{m}^2/\text{kg}^2 \cdot 7.348 \cdot 10^{22}\ \text{kg}}{1.737 \cdot 10^6\ \text{m}} - \frac{(114.5\ \text{m/s})^2}{2}\right]} - 1.737 \cdot 10^6\ \text{m} \\ &= 4042.358098\ \text{m} \end{aligned}$$

ROUND: The numbers used in the calculation all have four significant figures, and the final answer should also have four figures. The projectile reaches a height of 4042 m or 4.042 km.

DOUBLE-CHECK: On Earth, a projectile shot up with an initial velocity $v_0 = 114.5$ m/s would reach a maximum height of about 670 m. The mass of the Moon is about one sixth that of the Earth, so the object will go much higher.

12.83. **THINK:** Kepler's law can be used to compute the semimajor axis from the period. Then, use the geometric fact that the location of the comet at aphelion is equal to twice the semimajor axis minus the location at perihelion to find the aphelion.

SKETCH: The semimajor axis (length $2a$), as well as the location of the comet at perihelion and aphelion are labeled.

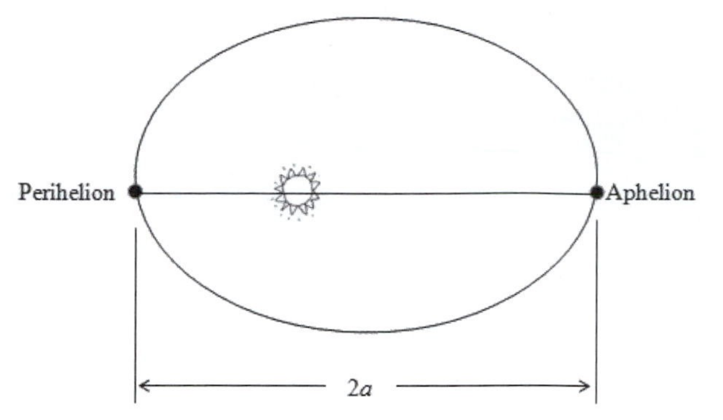

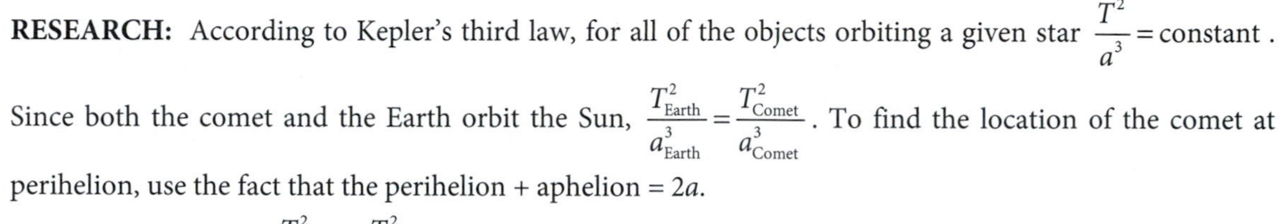

RESEARCH: According to Kepler's third law, for all of the objects orbiting a given star $\frac{T^2}{a^3} = \text{constant}$.

Since both the comet and the Earth orbit the Sun, $\frac{T_{\text{Earth}}^2}{a_{\text{Earth}}^3} = \frac{T_{\text{Comet}}^2}{a_{\text{Comet}}^3}$. To find the location of the comet at perihelion, use the fact that the perihelion + aphelion = $2a$.

SIMPLIFY: Since $\frac{T_{\text{Earth}}^2}{a_{\text{Earth}}^3} = \frac{T_{\text{Comet}}^2}{a_{\text{Comet}}^3}$, the semimajor axis of the comet's orbit is given by $a_{\text{Comet}} = \sqrt[3]{\frac{T_{\text{Comet}}^2 \cdot a_{\text{Earth}}^3}{T_{\text{Earth}}^2}}$. Then the location of the comet at aphelion is given by

$$\begin{aligned}\text{aphelion} &= 2a_{\text{Comet}} - \text{perihelion}\\ &= 2\sqrt[3]{\frac{T_{\text{Comet}}^2 \cdot a_{\text{Earth}}^3}{T_{\text{Earth}}^2}} - \text{perihelion.}\end{aligned}$$

CALCULATE: The comet orbits the sun with a period of 89.17 years, and it is 1.331 AU from the sun at perihelion. The Earth orbits the sun with a period of 1 year and has a semimajor axis of 1 AU. At aphelion, the distance between the comet and the sun is

$$\begin{aligned}\text{aphelion} &= 2\sqrt[3]{\frac{T_{\text{Comet}}^2 \cdot a_{\text{Earth}}^3}{T_{\text{Earth}}^2}} - \text{perihelion}\\ &= 2\sqrt[3]{\frac{(89.17\text{ yr})^2 \cdot (1\text{ AU})^3}{(1\text{ yr})^2}} - 1.331\text{ AU}\\ &= 38.5876495\text{ AU.}\end{aligned}$$

ROUND: Since the numbers used in the calculations all have four figures, the final answer should also have four figures. The distance between the comet and the sun at aphelion is 38.59 AU.

DOUBLE-CHECK: Generally, comets have very eccentric orbits, so it is reasonable that the comet is more than ten times further from the sun at aphelion than it is at perihelion. The comet travels a much greater distance than the Earth, which makes sense as it takes almost 90 times longer to orbit the Sun.

12.86. **THINK:** To escape from the asteroid's gravitational influence, the total kinetic energy of the object must be greater than or equal to the gravitational potential energy.

SKETCH: The escape speed is the minimum speed that a projectile will need to escape the pull of the asteroid's gravity.

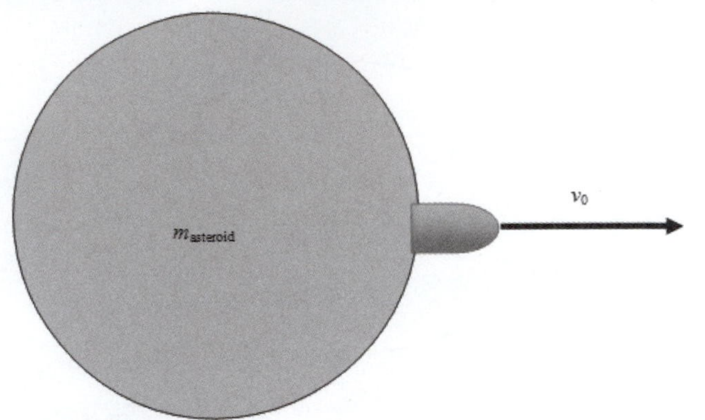

RESEARCH: The kinetic energy of a projectile when it is shot with a speed v from the surface of the asteroid is $K = \frac{1}{2}m_0 v_0^2$. The (absolute value) gravitational potential energy is $U = \frac{Gm_0 m_{\text{asteroid}}}{R_{\text{asteroid}}}$. To escape from the asteroid, the kinetic energy must be greater than or equal to the potential energy, $K \geq U$. The minimum escape speed occurs when they are equal.

SIMPLIFY: First, set the kinetic and potential energies equal to one another $\frac{1}{2}m_0 v_0^2 = K = U = \frac{Gm_0 m_{\text{asteroid}}}{R_{\text{asteroid}}}$.

Solve this for the escape speed:

$$\frac{1}{2}m_0 v_0^2 = \frac{Gm_0 m_{\text{asteroid}}}{R_{\text{asteroid}}}$$

$$v_0^2 = \frac{2Gm_{\text{asteroid}}}{R_{\text{asteroid}}}$$

$$v_0 = \sqrt{\frac{2Gm_{\text{asteroid}}}{R_{\text{asteroid}}}}$$

CALCULATE: The gravitational constant is $G = 6.674 \cdot 10^{-11}$. The mass of the asteroid is $m_{\text{asteroid}} = 1.869 \cdot 10^{20}$ kg, and the asteroid has a radius of $R_{\text{asteroid}} = 358.9$ km $= 358{,}900$ m. The escape speed is

$$\begin{aligned} v_0 &= \sqrt{\frac{2Gm_{\text{asteroid}}}{R_{\text{asteroid}}}} \\ &= \sqrt{\frac{2 \cdot 6.674 \cdot 10^{-11} \text{ N} \cdot \text{m}^2 / \text{kg}^2 \cdot 1.869 \cdot 10^{20} \text{ kg}}{358900 \text{ m}}} \\ &= 263.6489345 \text{ m/s}. \end{aligned}$$

ROUND: Since all of the numbers in this problem have four significant figures, the final answer should also have four significant figures. The escape speed from this asteroid is 263.6 m/s.

DOUBLE-CHECK: 263.6 m/s is faster than most objects on Earth, but it is less than the escape speed from Earth, which makes sense as the asteroid is less massive than Earth. Although the energy that is needed to escape from the asteroid will depend on the mass of the object, the minimum speed that it needs to achieve depends only on the mass and radius of the asteroid (and the gravitational constant). Any object that leaves the asteroid, heading away from the asteroid at the escape speed or greater, will have enough energy to escape the gravitational pull of the asteroid.

Chapter 13: Solids and Fluids

Exercises

13.27. **THINK:** The density at sea-level and the volume can used to determine the total mass. From the molar mass, the number of moles and thus the number of molecules can be determined. V = 0.50 L, $A = 28.95$ g/mol and $\rho = 1.229$ kg/m^3.

SKETCH:

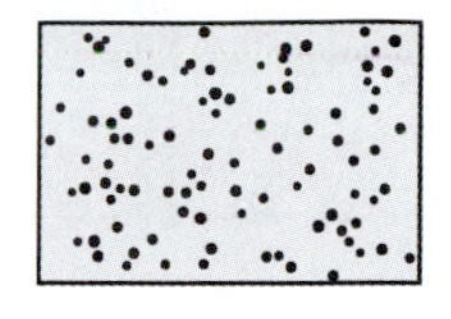

RESEARCH: $m = \rho V$

The number of moles, n, is given by $n = m / A$ (m in grams and A in grams per mole). The number of molecules, N, is given by $N = nN_{\text{A}}$, where $N_{\text{A}} = 6.02 \cdot 10^{23}$ molecules/mol.

SIMPLIFY: $N = nN_{\text{A}} = \dfrac{m}{A}N_{\text{A}} = \dfrac{\rho V N_{\text{A}}}{A}$

CALCULATE: $N = \dfrac{\left(1.229 \text{ kg/m}^3\right)0.50 \text{ L}\left(1 \text{ m}^3 / 10^3 \text{ L}\right)6.02 \cdot 10^{23} \text{ molecules/mol}}{28.95 \cdot 10^{-3} \text{ kg/mol}} = 1.278 \cdot 10^{22}$ molecules

ROUND: Rounding to two significant figures, $N = 1.3 \cdot 10^{22}$ molecules.

DOUBLE-CHECK: N is proportional to both the density and the volume. This is a reasonable result.

13.31. **THINK:** Young's modulus can be used to determine the change of length for the given tension. Use the values: F = 90.0 N, L = 2.00 m, r = 0.300 mm and $Y = 20.0 \cdot 10^{10}$ N/m^2.

SKETCH:

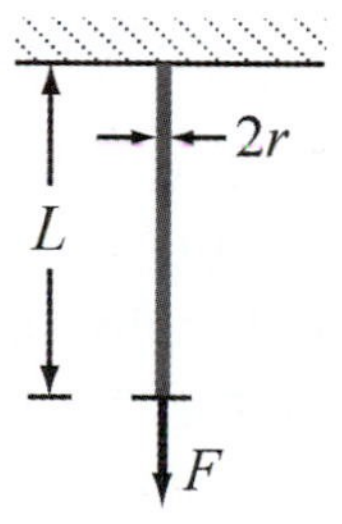

RESEARCH: $F = AY\dfrac{\Delta L}{L}$, and $A = \pi r^2$.

SIMPLIFY: $\Delta L = \dfrac{FL}{YA} = \dfrac{FL}{Y\pi r^2}$

CALCULATE: $\Delta L = \dfrac{(90.0 \text{ N})(2.00 \text{ m})}{\left(20.0 \cdot 10^{10} \text{ N/m}^2\right)\pi\left(0.300 \cdot 10^{-3} \text{ m}\right)^2} = 0.3183$ cm

ROUND: Rounding to three significant figures, $\Delta L = 0.318$ cm.

DOUBLE-CHECK: This is a reasonable result for a wire of the given dimension.

13.35. **THINK:** From the height and the density, the required pressure can be determined. $\rho_{\text{b}} = 1.00$ g/cm^3 (blood) $\rho_{\text{Hg}} = 13.6$ g/cm^3 (mercury).

SKETCH:

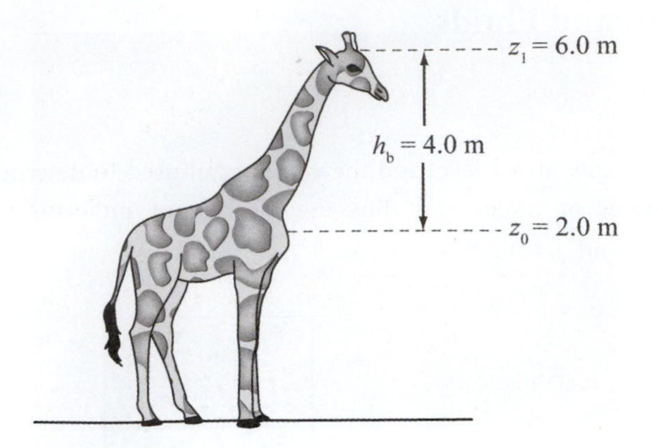

RESEARCH: $p = \rho g h$

SIMPLIFY: $p_b = \rho_b g h_b$, $p_{Hg} = \rho_{Hg} g h_{Hg}$, $p_b = p_{Hg} \Rightarrow h_{Hg} = \dfrac{\rho_b h_b}{\rho_{Hg}}$

CALCULATE: $h_{Hg} = \dfrac{(1.00 \text{ g/cm}^3)(4.0 \cdot 10^2 \text{ cm})}{13.6 \text{ g/cm}^3} = 2.941 \cdot 10^1 \text{ cm} = 294.1 \text{ mm}$

ROUND: Rounding to two significant figures gives $h_{Hg} = 290$ mm.

DOUBLE-CHECK: The systolic blood pressure of a giraffe must be two to three times that of a human. Since the relationship between pressure and height is linear, and a giraffe is about three times the height of a human, this seems like a reasonable result.

13.39. **THINK:** The question asks for the height of Mount McKinley when it is known that the atmospheric pressure on top is 47.7 % of the pressure at sea level and that the air density on Mount Everest is 34.8 % of that at sea level. The height of Mount Everest is given as $h_{\text{Everest}} = 8850$ m.

SKETCH: Not needed.

RESEARCH: We make use of the equations for barometric pressure and for density in a compressible fluid, that is, $P(h) = P_0 e^{-h\rho_0 g/P_0}$, and $\rho(h) = \rho_0 e^{-h\rho_0 g/p_0}$. By solving for the constant $\rho_0 g / p_0$ from the density equation, we need only use the values given by the question without actual knowledge of the constants. We can then use this information in the barometric pressure equation to obtain the height of Mount McKinley.

SIMPLIFY: The air density on Mount Everest is given by $\rho_{\text{Everest}}(h) = 0.348\rho_0 = \rho_0 e^{-h_{\text{Everest}}\rho_0 g/P_0}$, which can be solved for the set of constants as $\dfrac{\rho_0 g}{p_0} = -\dfrac{\ln(0.348)}{h_{\text{Everest}}}$. Now, the pressure on Mount McKinley is given by $P_{\text{McKinley}}(h) = 0.477P_0 = P_0 e^{-h_{\text{McKinley}}\rho_0 g/P_0}$. If we solve for the height of Mount McKinley, h_{McKinley}, we get $h_{\text{McKinley}} = -\dfrac{p_0}{\rho_0 g}\ln(0.477)$. Using the result above, we can write

$$h_{\text{McKinley}} = -\frac{p_0}{\rho_0 g}\ln(0.477) = h_{\text{Everest}}\frac{\ln(0.477)}{\ln(0.348)}.$$

CALCULATE: Inserting the given value for the height of Mount Everest we find

$$h_{\text{McKinley}} = 8850 \text{ m}\frac{\ln(0.477)}{\ln(0.348)} = 6206.33 \text{ m}.$$

ROUND: Since the input values have been given to at least three significant figures, our result for the height of Mount McKinley is $h_{\text{McKinley}} = 6210$ m.

DOUBLE-CHECK: You can look up the height of Mount McKinley and see if what we calculate is reasonably close. (The official height of the mountain is 6194 m, which make it the tallest mountain in North America). However, in an exam situation this check would not be an option. Instead, we can make simple checks of the right unit of our answer, which is meters, and the right order of magnitude. Since we are talking about a mountain, it should be at least a few thousand meters high, which is true for our answer. And of course our answer should come out less than the height of Mt Everest, which is the tallest peak on Earth.

13.43. **THINK:** The question asks for the number of pennies that can be placed in half of a floating racquetball such that the racquetball does not sink. The values given are: the diameter of the racquetball, $d = 5.6$ cm, the mass of the racquetball, $m_{\text{ball}} = 42$ g, the volume of a penny, $v_{\text{penny}} = 0.36 \text{ cm}^3$, and the mass of a penny, $m_{\text{penny}} = 2.5$ g.

SKETCH:

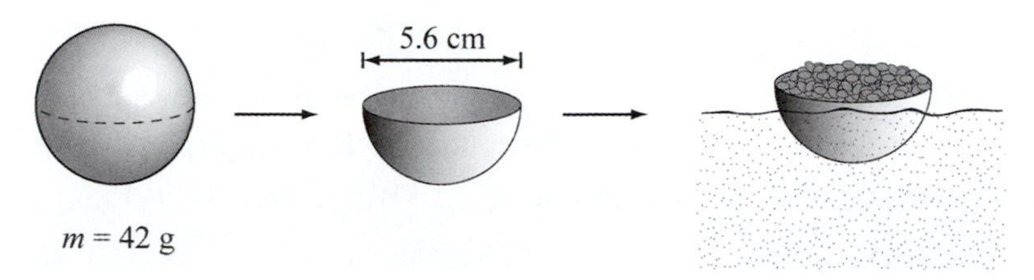

RESEARCH: In static equilibrium, the net force is zero and the buoyant force equals the weight of the racquetball boat and the pennies. The maximum buoyant force occurs when the maximum amount of water is displaced. The maximum amount of displaced water occurs when the top of the boat is level with the surface of the water and $V_{\text{water}} = V_{\text{ball}}/2$. Therefore, $F_{\text{buoyant}} = F_{\text{boat}}$. Recalling that the buoyant force is equal to the weight of the fluid displaced gives $F_{\text{buoyant}} = \rho_{\text{water}} V_{\text{water}} g = \frac{1}{2}\rho_{\text{water}} V_{\text{ball}} g$. Since the weight of the half-racquetball plus the pennies is $F_{\text{boat}} = mg = \left(\frac{m_{\text{ball}}}{2} + m_{\text{pennies}}\right)g$: $\left(\frac{m_{\text{ball}}}{2} + m_{\text{pennies}}\right)g = \frac{1}{2}\rho_{\text{water}} V_{\text{ball}} g$.

SIMPLIFY: Solving for m_{pennies} gives $m_{\text{pennies}} = \frac{1}{2}\rho_{\text{water}} V_{\text{ball}} - \frac{m_{\text{ball}}}{2}$. Now, determine the volume of the racquetball:

$$V_{\text{ball}} = \frac{\pi d^3}{6} = \frac{\pi(5.6 \text{ cm})^3}{6} = 91.95 \text{ cm}^3.$$

The maximum number of pennies the racquetball boat can hold without sinking is then given by:

$$N = \frac{m_{\text{pennies}}}{m_{\text{penny}}}.$$

CALCULATE: First, calculate the volume of the racquetball: $V_{\text{ball}} = \frac{\pi d^3}{6} = \frac{\pi(5.6 \text{ cm})^3}{6} = 91.95 \text{ cm}^3$.

The maximum number of pennies the racquetball boat can hold without sinking is then given by:

$$N = \frac{m_{\text{pennies}}}{m_{\text{penny}}} = \frac{\rho_{\text{water}} V_{\text{ball}} - m_{\text{ball}}}{2m_{\text{penny}}} = \frac{(1.0 \text{ g/cm}^3)(91.95 \text{ cm}^3) - 42 \text{ g}}{2(2.5 \text{ g})} = 9.99.$$

ROUND: Since the number of pennies is an integer, we must round to either 9 or 10. And since rounding up would cause the racquetball boat to sink, the answer must be rounded down, to $N = 9$ pennies.

DOUBLE-CHECK: Based on the given weight of a penny and that of the racquetball (half) compared to the weight of the fluid displaced, the answer is reasonable. It should be noted that the copper and zinc content of an American penny was changed in 1982 and pennies made before 1982 have a different density.

13.47. **THINK:** The question asks for the volume of a piece of iron such that when glued to a piece of wood, this will submerge completely but not sink. The given values are: the length of the piece of wood, l = 20.0 cm, the width of the piece of wood, w = 10.0 cm, the thickness of the piece of wood, t = 2.00 cm, the density of the piece of wood, $\rho_{\text{wood}} = 800.\text{ kg/m}^3$, the density of iron, $\rho_{\text{iron}} = 7860\text{ kg/m}^3$, and the density of water, $\rho_0 = 1000.\text{ kg/m}^3$.

SKETCH:

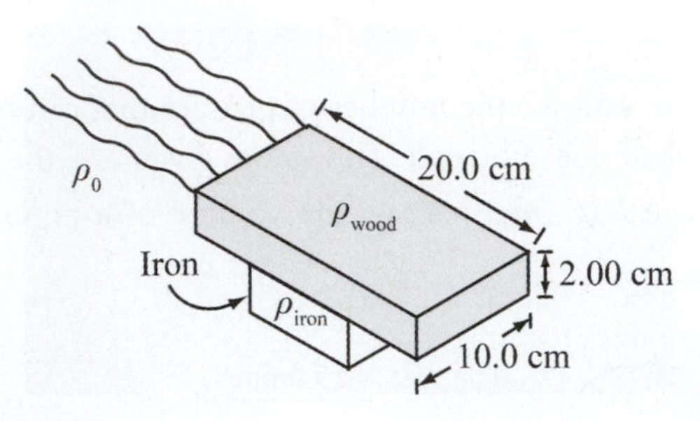

RESEARCH: If the object is to be at equilibrium, the weight of the wood piece plus the weight of the iron piece have to be equal to the buoyant force, that is, $F_{\text{B}} = W_{\text{wood}} + W_{\text{iron}}$. Recall that the buoyant force is equal to the weight of the fluid displaced: $W_{\text{water}} = W_{\text{wood}} + W_{\text{iron}} \Rightarrow \rho_0 V_{\text{displaced}} g = \rho_{\text{wood}} V_{\text{wood}} g + \rho_{\text{iron}} V_{\text{iron}} g$, where $V_{\text{displaced}} = V_{\text{wood}} + V_{\text{iron}}$.

SIMPLIFY: $\rho_0 \left(V_{\text{wood}} + V_{\text{iron}}\right) = \rho_{\text{wood}} V_{\text{wood}} + \rho_{\text{iron}} V_{\text{iron}} \Rightarrow V_{\text{iron}} = V_{\text{wood}} \left(\dfrac{\rho_0 - \rho_{\text{wood}}}{\rho_{\text{iron}} - \rho_0} \right)$

CALCULATE: $V_{\text{iron}} = (0.200\text{ m})(0.100\text{ m})(0.0200\text{ m}) \left(\dfrac{1000.\text{ kg/m}^3 - 800.\text{ kg/m}^3}{7860\text{ kg/m}^3 - 1000.\text{ kg/m}^3} \right) = 1.1662 \cdot 10^{-5}\text{ m}^3$

ROUND: The result should be rounded to three significant figures: $V_{\text{iron}} = 1.17 \cdot 10^{-5}\text{ m}^3$.

DOUBLE-CHECK: This corresponds to a cube of iron with sides about 2.27 cm, and mass approximately 91.7 g. The result is reasonable based on the given values.

13.51. **THINK:** The volume of the Hindenburg zeppelin is $V = 2.000 \cdot 10^5\text{ m}^3$ and the useful lift is $W_{\text{useful}} = 1.099 \cdot 10^6\text{ N}$. The densities of air, hydrogen and helium are $\rho_{\text{air}} = 1.205\text{ kg/m}^3$, $\rho_{\text{H}} = 0.08988\text{ kg/m}^3$ and $\rho_{\text{He}} = 0.1786\text{ kg/m}^3$, respectively. Determine (a) the weight of the structure of the zeppelin and (b) the useful lift capacity of helium compared with that of hydrogen. Since these values are given to four significant figures, the value of the acceleration due to gravity used will be treated as having four significant figures, $g = 9.810\text{ m/s}^2$.

SKETCH:

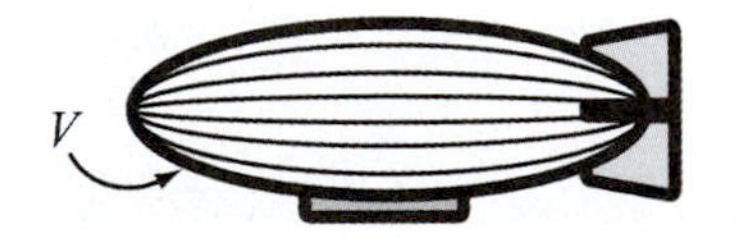

RESEARCH:

(a) Recall that the magnitude of the buoyant force is given by the weight of the fluid displaced (in this case, air), therefore, $F_{\text{B}} = \rho_{\text{air}} Vg$. Then the weight of the zeppelin can be determined from the difference in the total lift capacity and the useful lift capacity. The total lift capacity is given by $W_{\text{tot}} = F_{\text{B}} - W_{\text{H}}$, where W_{H} is the weight of the hydrogen inside the zeppelin. Therefore, the weight of the zeppelin is given by $W_{\text{zep}} = W_{\text{tot}} - W_{\text{useful}}$.

(b) If the zeppelin had been filled with helium instead of hydrogen, the total weight that could be lifted would be $W_1 = F_B - W_{He}$. Therefore, the useful lift would be $W_{He,useful} = W_1 - W_{zep}$. By switching from helium to hydrogen, the useful lift is increased by only

$$\frac{W_{useful} - W_{He,useful}}{W_{He,useful}}.$$

SIMPLIFY:

(a) $W_{zep} = \rho_{air}Vg - \rho_H Vg - W_{useful}$.

(b) $W_{He,useful} = F_B - W_{He} - W_{zep} = \rho_{air}Vg - \rho_{He}Vg - W_{zep}$

CALCULATE:

(a) $W_{zep} = \left(1.205 \text{ kg/m}^3 - 0.08988 \text{ kg/m}^3\right)\left(2.000 \cdot 10^5 \text{ m}^3\right)\left(9.810 \text{ m/s}^2\right) - \left(1.099 \cdot 10^6 \text{ N}\right) = 1.0888 \cdot 10^6 \text{ N}$

(b) $W_{He,useful} = \left(2.000 \cdot 10^5 \text{ m}^3\right)\left(9.810 \text{ m/s}^2\right)\left(1.205 \text{ kg/m}^3 - 0.1786 \text{ kg/m}^3\right) - \left(1.0888 \cdot 10^6 \text{ N}\right)$

$= 9.2493 \cdot 10^5 \text{ N}$

The increase in lift is then $\dfrac{1.099 \cdot 10^6 \text{ N} - 9.2498 \cdot 10^5 \text{ N}}{9.2498 \cdot 10^5 \text{ N}} = 0.188196$ or 18.8196%.

ROUND: Round the results to four significant figures.

(a) $W_{zep} = 1.089 \cdot 10^6 \text{ N}$

(b) $W_{He,useful} = 9.249 \cdot 10^5 \text{ N}$, which is an 18.82% increase from using hydrogen rather than helium.

DOUBLE-CHECK: This result is somewhat counterintuitive, since the density of hydrogen is half the density of helium. The explanation is in the fact that it is the difference between the air density and the filling gas density that matters, which changes by a small fraction when the gas is changed from hydrogen to helium. By this account, filling the Hindenburg with hydrogen rather than helium was a risk not worth taking. The initial plans called for the Hindenburg to be filled with helium. It was the blockade imposed on Germany in the years preceding World War II that determined the use of hydrogen rather than helium.

13.55. **THINK:** The question presents a nozzle with square sides of side length 50.0 cm and 20.0 cm, respectively. While this looks like an awfully big nozzle, it still may be possible to find such a device inside a hydroelectric power plant, for example. Since we are dealing with flowing water, our equations for fluid flow, in particular the continuity equation will come in handy. One word of caution: in part a) we are interested in the fluid speed at the exit end, while the fluid speed at the entrance is given; so it may be tempting to use some kind of kinematic equation to solve for the acceleration in part b). However, this would be wrong, because the condition of constant acceleration is not fulfilled in this case, and so the kinematic equations derived for point particles in Chapters 2 and 3 do not apply.

SKETCH:

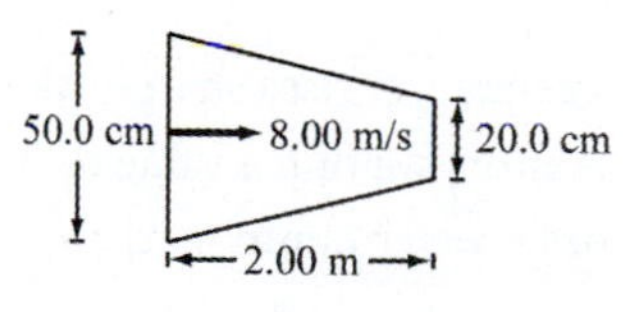

RESEARCH:

(a) From the equation of continuity, $A_1 v_1 = A_2 v_2 = R_V$. Therefore, the flow rate at the exit will be the same as that of the entrance. Since the velocity of the fluid at the entrance is known, the flow rate can be readily obtained.

(b) If we call the coordinate along the direction of the fluid flow x, (measured in units of m) then the area is given as a function of this coordinate as

$$A(x) = (0.500 - 0.150x)^2 \text{ m}^2$$

From the continuity equation we then obtain for the fluid speed as a function of the x-coordinate:

$$v(x) = R_V / A(x) = R_V / (0.500 - 0.150x)^2 \text{ m}^2$$

In order to take the derivative of the fluid speed with respect to time and thus obtain the local acceleration, we have to make use of a change of variables:

$$a \equiv \frac{dv}{dt} = \frac{dv}{dx}\frac{dx}{dt} = \frac{dv}{dx}v$$

For the derivative dv/dx we find

$$dv(x)/dx = 0.300R_{\text{V}}/(0.500-0.150x)^3\,\text{m}^3$$

(c) The increased flow rate increases the velocity, and we can obtain our result by simply inserting a new value of R_{V} at the exit.

SIMPLIFY: Simplification is not necessary for part a). In part b) we obtain:

$$a = \frac{dv}{dx}v = \frac{0.300R_{\text{V}}}{(0.500-0.150x)^3\,\text{m}^3}\frac{R_{\text{V}}}{(0.500-0.150x)^2\,\text{m}^2} = \frac{0.300{R_{\text{V}}}^2}{(0.500-0.150x)^5\,\text{m}^5}$$

CALCULATE:

(a) The flow rate throughout is given by $R_{\text{V}} = (8.00\text{ m/s})(0.500\text{ m})(0.500\text{ m}) = 2.00\text{ m}^3/\text{s}$.

(b) The acceleration at the exit is then given by $a(x{=}2) = \dfrac{0.300(2.00\text{ m}^3/\text{s})^2}{(0.500-0.150{\cdot}2)^5\text{ m}^5} = 3750\text{ m/s}^2$

(c) In part a) we found a flow rate of $2.00\text{ m}^3/\text{s}$, and in part b) we found that the acceleration depends on the square of the flow rate. Increasing the flow rate to $6.00\text{ m}^3/\text{s}$ then increases the acceleration by a factor of 9 to $33{,}750\text{ m/s}^2$.

ROUND:

(a) The flow rate is $2.00\text{ m}^3/\text{s}$.

(b) $a(x=2) = 3{,}750\text{ m/s}^2$

(c) $a(x=2) = 3.38\cdot 10^4\text{ m/s}^2$

DOUBLE-CHECK: Our calculations resulted in extremely high accelerations of hundreds of *g*. This might make us suspicious. However, a nozzle like the one described in the problem is not something we might encounter in our everyday experiences, and so we have no easy reference point to check how reasonable our answer is. It is comforting, though, that the units work out properly. We can also calculate the average acceleration and see how this compares to our answer for part b). The average acceleration is the velocity change between the beginning and end of the nozzle, divided by the average time the water took to cross the nozzle. To do this, we can use our continuity equation and find that the speed at the end of the nozzle is 50 m/s. This means that the velocity change is $\Delta v = 50\text{ m/s} - 8\text{ m/s} = 42\text{ m/s}$. We can obtain the time interval from taking the ratio of the length of the nozzle, divided by the average velocity $\Delta t = \Delta x/\bar{v}$. With $\Delta x = 2\text{ m}$ and $\bar{v} = \frac{1}{2}(v+v_i) = 29\text{ m/s}$ we get $\Delta t = 2/29\text{ s}$ and thus for the average acceleration $\bar{a} = 609\text{ m/s}^2$. This is also a very large number, but not nearly as large as what we found in part b). It indicates that the acceleration rises sharply along the nozzle. For comparison, if we insert $x=0$ into our formula for the acceleration, we find a value of 38.4 m/s^2 at the beginning of the nozzle.

13.59. **THINK:** The question describes a water pump mechanism. By using tap water and drastically reducing the size of the pipe one can create much higher water streaming speed and thus create pressure below atmospheric pressure. (This is similar to putting your thumb on the garden hose, which enables you to spray farther.) The reduced pressure can then be used to suck water out of the sump well and keep the basement dry. In this problem we need to proceed in two steps. First we deal with the horizontal pipe and calculate the speed and pressure in the part of the pipe that has the reduced pipe size. Of course, the Bernoulli equation in the special case of equal height seems tailor-made to do this. Then, in the second step, we take the calculated pressure difference and see to what height we can lift water with it. Again we can use a limit of the Bernoulli equation. We can obtain information on the maximum height we can lift the water to, if we consider the case of very slowly flowing water ($v = 0$), in which we can neglect the v^2 term in the Bernoulli equation).

SKETCH:

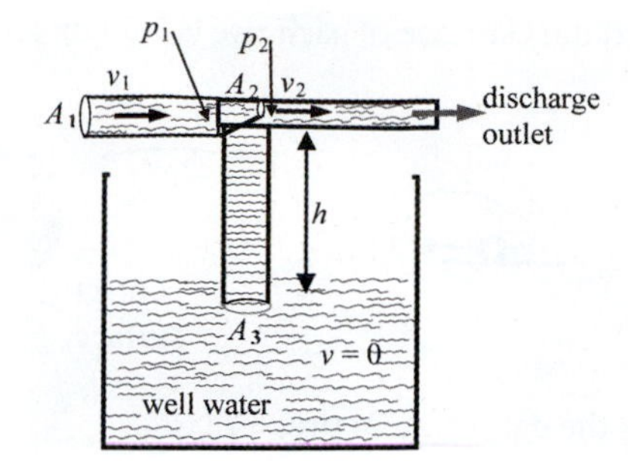

RESEARCH:

(a) To determine the speed at the discharge outlet, recall the continuity equation:

$$A_1 v_1 = A_2 v_2 \Rightarrow v_2 = v_1 \frac{A_1}{A_2}.$$

(b) Recall Bernoulli's equation: $p_1 + \frac{1}{2}\rho v_1^2 + \rho g h_1 = p_2 + \frac{1}{2}\rho v_2^2 + \rho g h_2$. Since $h_1 = h_2$,

$p_1 + \frac{1}{2}\rho v_1^2 = p_2 + \frac{1}{2}\rho v_2^2$.

(c) Making use of Bernoulli's equation:

$$p_3 + \frac{1}{2}\rho v_3^2 + \rho g h_3 = p_2 + \frac{1}{2}\rho v_2^2 + \rho g h_2,$$

where $h_3 = 0$ and $h_2 = h$. For negligible v^2 we then find with $p_3 = p_{\text{atm}}$,

$$p_3 + \rho g h_3 = p_2 + \rho g h_2 \Rightarrow p_3 - p_2 = \rho g (h_2 - h_3) = \rho g h$$

SIMPLIFY:

(a) Since $A_2 = A_1 / 10$, the speed is given by $v_2 = v_1(10)$.

(b) The pressure, p_2, is given by $p_2 = p_1 + \frac{1}{2}\rho\left(v_1^2 - v_2^2\right)$.

(c) $h = \dfrac{p_3 - p_2}{\rho g}$

CALCULATE:

(a) $v_2 = (2.05 \text{ m/s})(10) = 20.5 \text{ m/s}$

(b) $p_2 = 3.03 \cdot 10^5 \text{ Pa} + \frac{1}{2}\left(1000.\text{ kg/m}^3\right)\left((2.05 \text{ m/s})^2 - (20.5 \text{ m/s})^2\right) = 94.98 \text{ kPa}$

(c) $h = \dfrac{(101 \text{ kPa}) - (94.98 \text{ kPa})}{(1000.\text{ kg/m}^3)(9.81 \text{ m/s}^2)} = 0.614 \text{ m}$

ROUND: Round the results to three significant figures.

(a) $v_2 = 20.5 \text{ m/s}$

(b) $p_2 = 95.0 \text{ kPa}$

(c) $h = 0.614 \text{ m}$

DOUBLE-CHECK: The last number is a bit worrisome, because lifting water 2 ft usually does not get it out of the basement. But if you examine the input numbers, you see that the final result depends strongly on the ratio by which the pipe area is reduced and on the initial streaming speed of the water. Adjusting these one can get pumps operating on this principle, which lift water by 10 ft, which is certainly enough for the intended purpose. Water-powered sump pump are a great safety device in case that electricity fails (which is not unusual in some parts of the country during strong thunderstorms …).

13.63. **THINK:** The question asks for the weight of a car if it is known that the tire pressure is 28.0 psi, and the width and length of the contact surface of each tire is 7.50 in, and 8.75 in, respectively.

SKETCH:

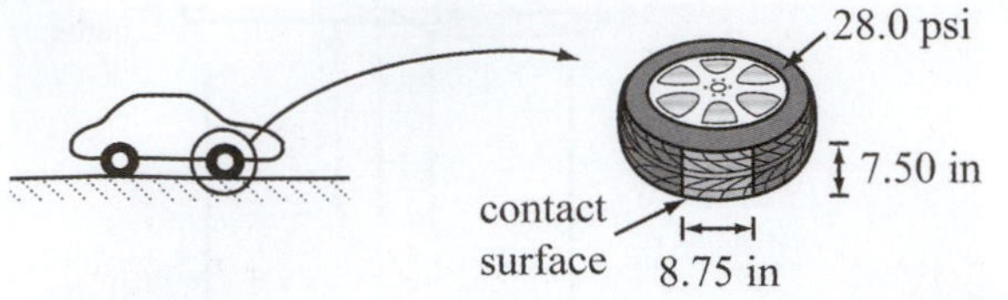

RESEARCH: Recall that the definition of pressure is:

$$P = \frac{\text{Force}}{\text{Area}} = \frac{F}{A}.$$

Therefore, the weight of the car can be taken as the force due to the pressure on all four tires. That is, $F_{\text{car}} = 4PA.$

SIMPLIFY: Simplification is not necessary.

CALCULATE: Recall also that 1 pound per square inch = 6894.75729 Pascals, and that 1 inch = 0.0254 m.

Therefore, $F_{\text{car}} = 4(28.0)(6894.75729 \text{ Pa})(0.0254 \text{ m})^2(7.50)(8.75) = 32694.4 \text{ N}.$

ROUND: To three significant figures, the weight of the car is 32700 N.

DOUBLE-CHECK: Although the result is somewhat high for the weight of an average car, it is reasonable for the given values.

13.67. **THINK:** The question asks about the velocity of water in a pipe when the radius of the pipe is $r_1 = 5.00$ cm and the water is flowing into the pipe at this end with a speed of 2.00 m/s. The pipe narrows down to a radius of $r_2 = 2.00$ cm and we are interested in finding the speed of the water at this end.

SKETCH:

+y
+x
$r_1 = 5.00$ cm
$\vec{v}_1 = 2.00$ m/s
$\vec{v}_2 = ?$
$r_2 = 2.00$ cm

RESEARCH: Using the continuity equation: $A_1v_1 = A_2v_2$. Therefore, $v_1\pi r_1^2 = v_2\pi r_2^2$.

SIMPLIFY: Solving for v_2 : $v_2 = v_1\left(\frac{r_1}{r_2}\right)^2$.

CALCULATE: $v_2 = (2.00 \text{ m/s})\left(\frac{5.00 \text{ cm}}{2.00 \text{ cm}}\right)^2 = 12.5 \text{ m/s}$

ROUND: Since the values are given to three significant figures, the result should remain as $v_2 = 12.5$ m/s.

DOUBLE-CHECK: It is expected that the velocity of the fluid will increase as the cross-sectional area of the pipe decreases.

13.71. **THINK:** On another planet, the pressure under water is determined in the same manner as it is on Earth. The only difference is the new value of gravity, $g' = 0.135g$. The depth is d = 1.00 km and the density of water is $\rho = 1000.$ kg/m^3. Assume the pressure in the atmosphere is zero.

SKETCH:

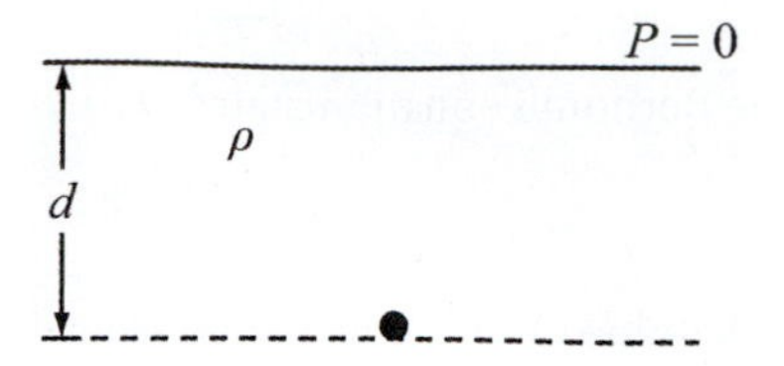

RESEARCH: The pressure at a certain depth is given by $P = P_0 + \rho g d$.

SIMPLIFY: $P = \rho g' d = 0.135 \rho g d$

CALCULATE: $P = 0.135\left(1000.\text{ kg/m}^3\right)\left(9.81\text{ m/s}^2\right)\left(1.00 \cdot 10^3\text{ m}\right) = 1.324 \cdot 10^6\text{ Pa}$

ROUND: $P = 1.32$ MPa

DOUBLE-CHECK: This pressure is still greater than the atmospheric pressure $\left(P_0\right)$ of Earth, which is expected even on another planet.

13.75. **THINK:** Assume the height of the water remains the same. The Bernoulli equation can then be used to determine the velocity at which it boils. The pressure when the water boils is $P' = 2.3388$ kPa. The water is at atmospheric pressure, $P_0 = 101.3$ kPa, when stationary $\left(v_0 \approx 0\text{ m/s}\right)$. The density of water is $\rho_w = 998.2\text{ kg/m}^3$.

SKETCH:

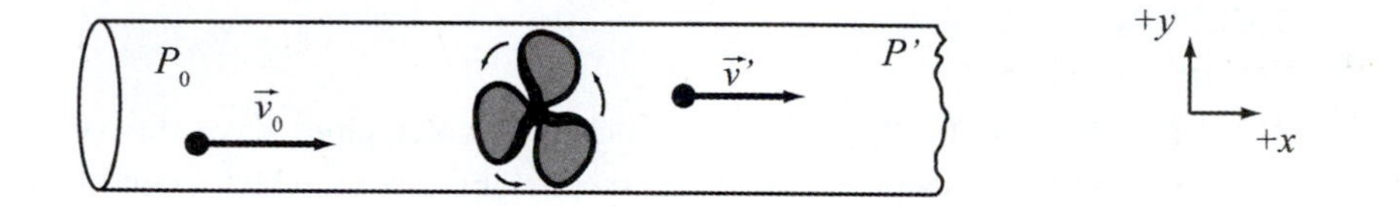

RESEARCH: The Bernoulli equation for constant height is $P_1 + \frac{1}{2}\rho v_1^2 = P_2 + \frac{1}{2}\rho v_2^2$.

SIMPLIFY: $P_0 = P' + \frac{1}{2}\rho_w v'^2 \Rightarrow v' = \sqrt{\frac{2(P - P')}{\rho_w}}$

CALCULATE: $v' = \sqrt{\frac{2\left(101.3\text{ kPa} - 2.3388\text{ kPa}\right)}{998.2\text{ kg/m}^3}} = 14.081\text{ m/s}$

ROUND: $v' = 14.08$ m/s

DOUBLE-CHECK: This velocity is about 32 mph, which is a fairly brisk speed for a boat to travel, but certainly seems in the right order of magnitude.

13.79. **THINK:** The continuity equation states that the volumes per second at the inlet and outlet pipe are the same. The Bernoulli equation can then be used to determine gauge pressure in the outlet pipe. Assume gauge pressure of the inlet pipe is $P_0 = 101.3$ kPa. The diameters of the inlet and the outlet pipes are 2.00 cm and 5.00 cm, respectively. The heights of the pipes are 1.00 m and 6.00 m, respectively. The density of water is $\rho = 1000.\text{ kg/m}^3$. The volume through the pipes is $\Delta V = 0.300\text{ m}^3$ in $\Delta t = 60.0$ s.

SKETCH:

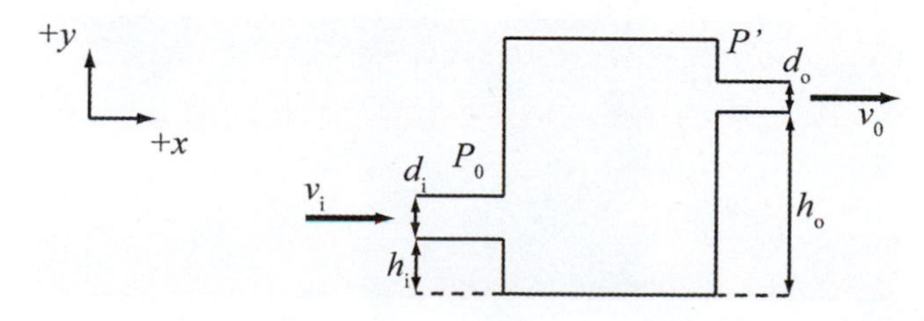

RESEARCH: The area of each pipe is $A_i = \pi(d_i/2)^2$ and $A_o = \pi(d_o/2)^2$. The continuity equation states $A_i v_i = A_o v_o = \Delta V/\Delta t$. The Bernoulli equation states $p_i + \rho g h_i + \frac{1}{2}\rho v_i^2 = p_o + \rho g h_o + \frac{1}{2}\rho v_o^2$.

SIMPLIFY:

(a) The velocity in the outlet pipe can be determined from $A_o v_o = \frac{\Delta V}{\Delta t} \Rightarrow v_o = \frac{\Delta V}{\Delta t}\left(\frac{4}{\pi d_o^2}\right)$.

(b) First, determine the velocity in the inlet pipe: $A_i v_i = \frac{\Delta V}{\Delta t} \Rightarrow v_i = \frac{\Delta V}{\Delta t}\left(\frac{4}{\pi d_i^2}\right)$.

Then, the gauge pressure at the outlet is $p_o = p_i + \rho g(h_i - h_o) + \frac{1}{2}\rho(v_i^2 - v_o^2)$.

CALCULATE:

(a) $v_o = \left(\frac{0.300 \text{ m}^3}{60.0 \text{ s}}\right)\left(\frac{4}{\pi(0.0500 \text{ m})^2}\right) = 2.546 \text{ m/s}$

(b) $v_i = \left(\frac{0.300 \text{ m}^3}{60.0 \text{ s}}\right)\left(\frac{4}{\pi(0.0200 \text{ m})^2}\right) = 15.92 \text{ m/s}$

$$p_o = 101.3 \text{ kPa} + (1000. \text{ kg/m}^3)(9.81 \text{ m/s}^2)(1.00 \text{ m} - 6.00 \text{ m}) + \frac{1}{2}(1000. \text{ kg/m}^3)\left((15.92 \text{ m/s})^2 - (2.546 \text{ m/s})^2\right)$$
$$= 175732.1 \text{ Pa}$$

ROUND: $p_o = 176$ kPa

DOUBLE-CHECK: In the limit that the inlet and outlet pipes have the same diameter, this whole problem only would involve pumping water up a height of 5 m, which would result in a gauge pressure drop of 50 kPa, reducing the outlet pressure to 50 kPa. Since our outlet pipe is larger than the inlet pipe, and since they have to carry the same water flow, the speed in the inlet pipe is greater. This, in turn, means that the pressure in the outlet pipe has to be greater than 50 kPa, which our solution fulfills.

Multi-Version Exercises

13.81. **THINK:** The diving bell is at a fixed depth, so there is no net force moving it up or down (the weight of the diving bell, buoyant force from the water, and the tension from the chain holding the diving bell cancel one another exactly). The net force on the viewing port will depend on the difference in pressure inside and outside of the diving bell. The pressure inside the diving bell is equal to atmospheric pressure, and the pressure outside the diving bell will depend on the depth of the viewing port.

SKETCH: The depth of the diving bell (h) and the diameter of the viewing port (d) are not shown to scale. The forces on the diving bell due to gravity, buoyancy, and the tension on the chain are equal (the diving bell is submerged at a fixed depth and it is not moving), so there is no net upward or downward force.

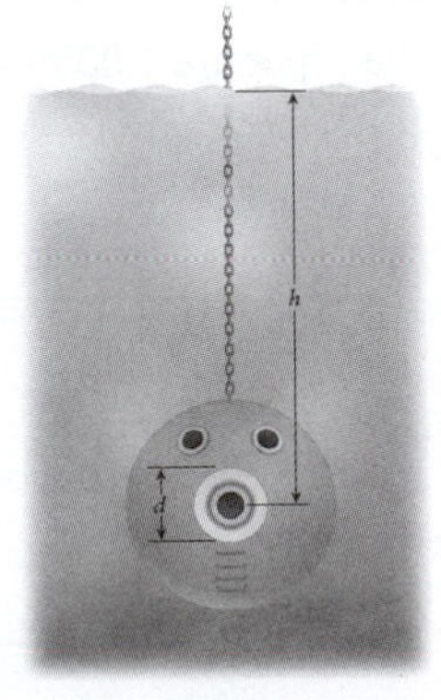

RESEARCH: The pressure at depth h is $P_{ext} = \rho gh + p_{atm}$, where ρ is the density of water and p_{atm} is the atmospheric pressure. The pressure inside the diving bell is $P_{int} = p_{atm}$. The total pressure on the viewing port is $P = P_{ext} - P_{int}$. Since pressure is defined as force per unit area, the total force on the viewing port, surface area of the viewing port, and pressure are related by the equation $F = PA$, where A is the surface area of the viewing port. The area of the round viewing port is given by $A = \pi r^2 = \pi\left(\frac{d}{2}\right)^2$.

SIMPLIFY: The total pressure at depth h is given by

$$\begin{aligned} P &= P_{ext} - P_{int} \\ &= \rho gh + p_{atm} - p_{atm} \\ &= \rho gh. \end{aligned}$$

The net force on the viewing port is then $F = PA = (\rho gh)\left(\frac{\pi}{4}d^2\right)$.

CALCULATE: The depth h = 129.1 m and diameter d = 22.89 cm = 0.2289 m are given in the question. The gravitational acceleration near the surface of the earth is g = 9.81 m/s², and the density of fresh water is ρ = 1000 kg/m³. The net force on the viewing port is:

$$\begin{aligned} F &= (\rho gh)\left(\frac{\pi}{4}d^2\right) \\ &= 1000\text{ kg/m}^3 \cdot 9.81\text{ m/s}^2 \cdot 129.1\text{ m} \cdot \frac{\pi}{4}(0.2289\text{ m})^2 \\ &= 52116.67693\text{ N} \end{aligned}$$

ROUND: All of the numbers in this calculation had four significant figures, so the final answer will also have four figures. The viewing port experiences a force of $5.212 \cdot 10^4$ N = 52.12 kN towards the interior of the diving bell.

DOUBLE-CHECK: As a rule of thumb, divers expect the pressure to increase by 1 atmosphere = $1.103 \cdot 10^5$ Pa for every 10 m of depth, so the expected pressure is about $1.424 \cdot 10^6$ Pa. The window experiences a force of magnitude $5.212 \cdot 10^4$ N, so the pressure is $P = F/A = 1.267 \cdot 10^6$ Pa. This is the correct order of magnitude, so this rough estimate confirms that the calculation is of the correct order of magnitude.

13.84. **THINK:** The balloon experience an upward force from buoyancy and a downward force from gravity. When the balloon is lifting the maximum weight, the upward force and downward force are equal.

SKETCH: The sketch shows the buoyant force and the gravitational force:

RESEARCH: The weight of the air the balloon displaces is $F_B = \rho_{\text{outside}}Vg$. The weight of the hot air filling the balloon is $W_{\text{hot air}} = \rho_{\text{inside}}Vg$. The weight that can be lifted, plus the weight of the hot air filling the balloon, is equal to the weight of the air the balloon displaces: $F_B = W_{\text{hot air}} + W$.

SIMPLIFY: The goal is to find the weight W of the load that the balloon can lift. Using algebra, the weight of the load is $W = F_B - W_{\text{hot air}}$. Substituting in for $F_B = \rho_{\text{outside}}Vg$ and $W_{\text{hot air}} = \rho_{\text{inside}}Vg$ gives $W = \rho_{\text{outside}}Vg - \rho_{\text{inside}}Vg = Vg(\rho_{\text{outside}} - \rho_{\text{inside}})$.

CALCULATE: The acceleration due to gravity is g = 9.81 m/s. The volume of the balloon is V = 2979 m^3. The density of the air outside the balloon is ρ_{outside} = 1.205 kg/m^3 and the density of the air inside the balloon is ρ_{inside} = 0.9441 kg/m^3. The total weight that the balloon can lift is then

$$\begin{aligned} W &= Vg(\rho_{\text{outside}} - \rho_{\text{inside}}) \\ &= 2979 \text{ m}^3 \cdot 9.81 \text{ m/s}^2 \left(1.205 \text{ kg/m}^3 - 0.9441 \text{ kg/m}^3\right) \\ &= 7624.538991 \text{ N} \end{aligned}$$

ROUND: The volume of the balloon and the density of the air are all given to four significant figures. However, the difference in the density outside the balloon and the density inside the balloon $\rho_{\text{outside}} - \rho_{\text{inside}} = 0.2609$, has only three significant figures, so the final answer should also have only three figures. The balloon can lift a maximum of $7.62 \cdot 10^3$ N = 7.62 kN.

DOUBLE-CHECK: To check this, convert the weight from Newton to pounds. The balloon can lift $7.62 \cdot 10^3 \text{ N} \cdot \frac{1 \text{ lb}}{4.448 \text{ N}} = 1710 \text{ lb}$. For a hot air balloon with a wicker basket, nylon balloon, propane or other compressed gas heating mechanism, and a few human passengers, this seems like a realistic weight. (Keep in mind that, if the maximum weight is too low, the balloon will never get off the ground, and if the maximum weight is too high, it will take too long for the balloon to return to earth.)

13.87. **THINK:** The bulk modulus can be used to compute the fractional change in volume from the pressure. The pressure can be computed from the depth of the water, which is given in the question.

SKETCH: The ball is submerged to depth h. The ball is shown before and after it has been submerged.

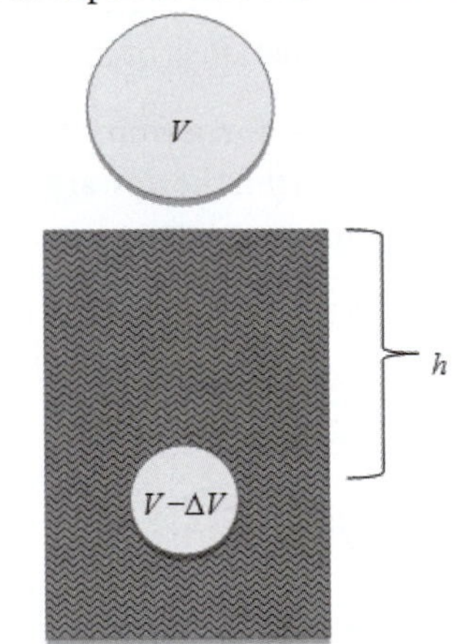

RESEARCH: The pressure at depth h is given by $p = \rho gh$, where ρ is the density of the water in which the ball is submerged. The pressure is defined to be the force per unit area $p = \frac{F}{A}$, and the equation for volume compression is $\frac{F}{A} = B\frac{\Delta V}{V}$.

SIMPLIFY: The goal is to find the fractional change in the volume of the ball, $\frac{\Delta V}{V}$. Replace force per unit area in the equation for volume compression, substitute in for the pressure at depth h, and finally use algebra to solve for the fractional change in volume.

$$B\frac{\Delta V}{V}=\frac{F}{A}\Rightarrow$$
$$B\frac{\Delta V}{V}=p\Rightarrow$$
$$B\frac{\Delta V}{V}=\rho gh\Rightarrow$$
$$\frac{\Delta V}{V}=\frac{\rho gh}{B}$$

CALCULATE: The density of water is 1000 kg/m^3 and the acceleration due to gravity near the surface of the earth is 9.81 m/s^2. The question states that the ball is submerged to a depth of 55.93 m and the bulk modulus is $6.309\cdot 10^7$ N/m^2. The fractional change in volume is then

$$\begin{aligned}\frac{\Delta V}{V}&=\frac{\rho gh}{B}\\&=\frac{1000\text{ kg/m}^2\cdot 9.81\text{ m/s}^2\cdot 55.93\text{ m}}{6.309\cdot 10^7\text{ N/m}^2}\\&=8.696676177\cdot 10^{-3}\end{aligned}$$

ROUND: The height and bulk modulus are both given to four figures, so the final answer should also have four significant figures. The fractional change in volume is $8.697\cdot 10^{-3}$ or 0.8697%.

DOUBLE-CHECK: The pressure is increasing, so it is natural to expect the volume to decrease. From experience, when a ball is submerged at the bottom of a pool or pond, the decrease in radius is minimal. It is reasonable that, even when the ball is submerged to a depth of over 50 meters, the fractional change in volume is less than 1 percent.

13.90. **THINK:** The Betz limit calculated in Example 13.7 applies to seawater. Use the Betz limit, speed of the current, and geometry of the turbine to determine the maximum power that can be extracted.

SKETCH: The diameter of the turbine and velocity of the seawater are shown. The seawater is flowing into the turbine from left to right.

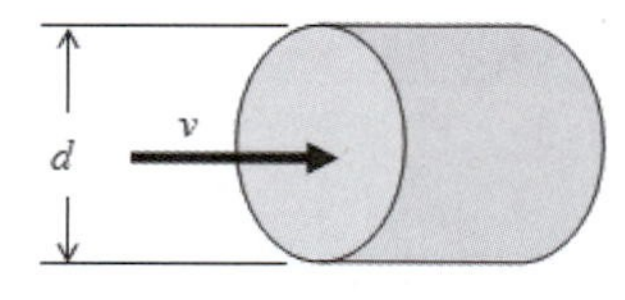

RESEARCH: The Betz limit states that a fraction of 16/27 of the total power can be extracted from the fluid. The total power of a fluid flowing at a velocity v with a density ρ flowing through an area A is $P_{\text{in}}=\frac{1}{2}Av^3\rho$. In this case, the area is the surface area of one end of the turbine and can be expressed in terms of the turbine's diameter as $A=\pi(d/2)^2$. The power extracted by the turbine is 16/27 of the total, so the maximum power extracted is $P=\frac{16}{27}P_{\text{in}}$.

SIMPLIFY: Use the Betz limit to express the maximum power $P=\frac{16}{27}\left(\frac{1}{2}Av^3\rho\right)$. Express the area in terms of the diameter of the turbine ($A=\pi(d/2)^2$) and simplify to get $P=\frac{2}{27}\pi d^2v^3\rho$.

CALCULATE: The density of seawater is given in the problem as ρ = 1024 kg/m³, and it flows through the turbine with a speed of 1.35 m/s. The turbine's rotors have a diameter of 24.5 m. The maximum power that can be extracted is

$$\begin{aligned}P &= \tfrac{2}{27}\pi d^2 v^3 \rho \\ &= \frac{2\pi}{27}(24.5\text{ m})^2(1.35\text{ m/s})^3 1024\text{ kg/m}^3 \\ &= 3.519245266\cdot 10^5\text{ W} \\ &= 351.9245266\text{ kW}.\end{aligned}$$

ROUND: Although the other values have four significant figures, the velocity of the seawater is given to only three figures, so the final answer should have only three significant figures. The maximum power that the turbine can extract under these conditions is 352 kW.

DOUBLE-CHECK: Flowing water, such as a current, river, or waterfall, has a tremendous amount of power. For seawater flowing fairly quickly (a speed greater than 1 m/s), the expected power output is on the order of 10^3 times the square of the diameter. For a turbine with diameter d = 24.5 m, the order of magnitude of the square of the diameter is about 10^2, so it is reasonable to expect that the answer should have an order of magnitude of 10^5 Watts. This agrees with the calculated value (352 kW = $3.52\cdot 10^5$ W).

Chapter 14: Oscillations

Exercises

14.25. The angular frequency is given by $\omega=\sqrt{k/m}$ or $k=m\omega^2$. The spring constant is then $k=m\omega^2=(5.00 \text{ kg})(5.00 \text{ s}^{-1})^2=125 \text{ N/m}.$

14.29. **THINK:** The spring constant can be determined from the mass and the frequency. The mass is $m_1=55.0 \text{ g}$ and bobs with a frequency of $f_1=3.00 \text{ Hz}$. The mass is then changed to $m_2=250. \text{ g}$.

SKETCH:

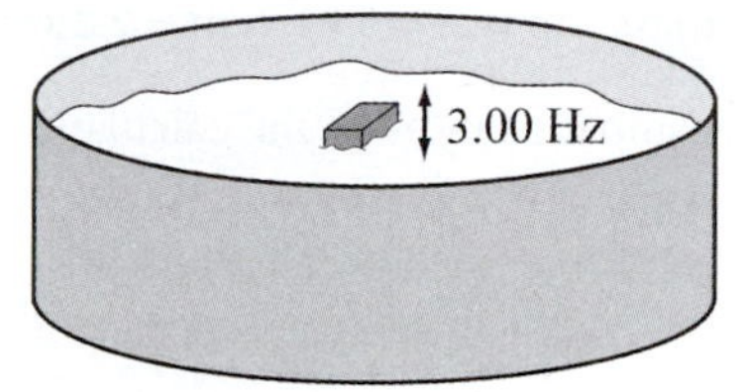

RESEARCH: The angular frequency is given by $\omega=\sqrt{k/m}$ and $f=\omega/2\pi$.

SIMPLIFY: The spring constant is $k=m_1\omega^2=m_1(2\pi f_1)^2=4\pi^2 m_1 f_1^2$. Assume this spring constant does not change with the new mass. The new frequency is:

$$f_2=\frac{\omega}{2\pi}=\frac{1}{2\pi}\sqrt{\frac{k}{m_2}}=\frac{1}{2\pi}\sqrt{\frac{4\pi^2 m_1 f_1^2}{m_2}}=\frac{2\pi f_1}{2\pi}\sqrt{\frac{m_1}{m_2}}=f_1\sqrt{\frac{m_1}{m_2}}.$$

CALCULATE: $k=4\pi^2(0.0550 \text{ kg})(3.00 \text{ Hz})^2=19.54 \text{ kg/s}^2=19.54 \text{ N/m}$. The frequency is:

$$f=3 \text{ Hz}\sqrt{\frac{55.0 \text{ g}}{250. \text{ g}}}=1.407 \text{ Hz}.$$

ROUND: The results should be rounded to three significant figures.

(a) The spring constant is k = 19.5 N/m.

(b) The frequency is 1.41 Hz.

DOUBLE-CHECK: Since the frequency is inversely proportional to the mass, a larger mass should oscillate with a smaller frequency. The results are reasonable.

14.33. **THINK:** Block m_2 will slide off of block m_1 if the maximum acceleration of oscillation produces a force greater than the force of static friction. The spring constant is $k=10.0 \text{ N/m}$ and the masses are $m_1=m_2=20.0 \text{ g}$. The coefficient of static friction is $\mu=0.600$.

SKETCH:

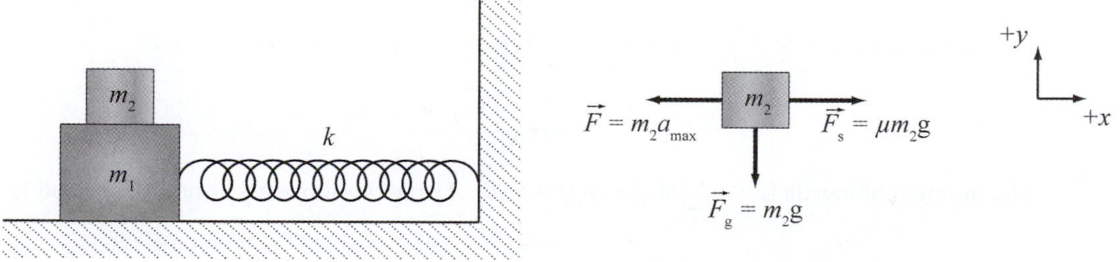

RESEARCH: The maximum acceleration of the oscillation is $a=\omega^2 A$. The angular frequency is $\omega=\sqrt{k/m}$.

SIMPLIFY: The maximum force on the second mass is given by:

$$F_{\text{max}} = m_2 a_{\text{max}} = m_2 \omega^2 A = \frac{m_2 k A}{m_1 + m_2}.$$

The maximum force is equal to the force of static friction:

$$F_s = F_{\text{max}} \Rightarrow \mu m_2 g = \frac{m_2 k A}{m_1 + m_2} \Rightarrow A = \frac{\mu g (m_1 + m_2)}{k}.$$

CALCULATE: $A = \dfrac{(0.600)(9.81 \text{ m/s}^2)(0.0200 \text{ kg} + 0.0200 \text{ kg})}{(10.00 \text{ N/m})} = 0.023544 \text{ m}$

ROUND: The given values have three significant figures, so the maximum amplitude the system can have without having the second mass slip is $A = 0.0235$ m = 2.35 cm.

DOUBLE-CHECK: Dimensional analysis of the calculation shows that the answer is in the correct units of length: $\dfrac{[\text{m/s}^2][\text{kg}]}{[\text{N/m}]} \Rightarrow \dfrac{[\text{m}][\text{kg}][\text{m}][\text{s}^2]}{[\text{s}^2][\text{kg}][\text{m}]} \Rightarrow \text{m}$. This result is reasonable.

14.37. **THINK:** The two balls create a moment of inertia, I, which oscillates about the pivot point, P. Using the sum of the torques acting on the pendulum, the period can be determined. The masses are $m_1 = 1.00$ kg and $m_2 = 2.00$ kg. The two masses are separated by 30.0 cm. The pivot point is 10.0 cm away from the 1.00 kg mass, so $r_1 = 10.0$ cm and $r_2 = 20.0$ cm. A 'slight displacement' implies small values for θ.

SKETCH:

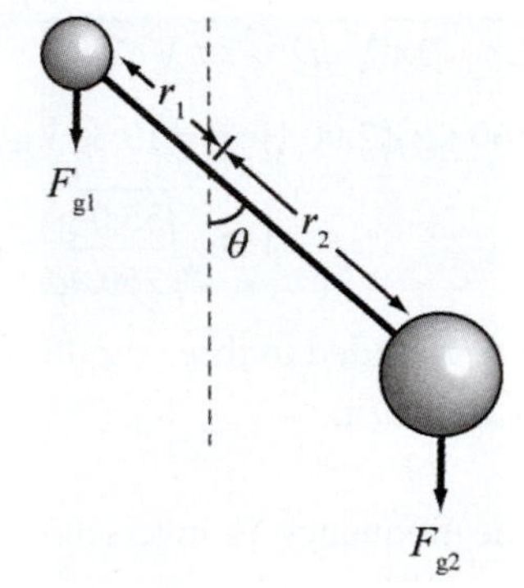

RESEARCH: The torque is given by $\tau = \vec{r} \times \vec{F} = rF\sin\theta$. The sum of the torques is $\tau = I\dfrac{d^2\theta}{dt^2} = \sum_i \tau_i$. The period is given by $T = 2\pi / \omega$.

SIMPLIFY: The torque equation gives $I\dfrac{d^2\theta}{dt^2} = -r_1 F_{g1} \sin\theta + r_2 F_{g2} \sin\theta$. For small angles, $\sin\theta \approx \theta$. The equation thus becomes: $I\dfrac{d^2\theta}{dt^2} = (-r_1 m_1 + r_2 m_2) g\theta$. The angular frequency is then:

$$\omega = \sqrt{\frac{(m_2 r_2 - m_1 r_1) g}{I}}.$$

The moment of inertia is $I = \sum_i m_i r_i^2 = m_1 r_1^2 + m_2 r_2^2$. Using the above equations, the period is:

$$T = \frac{2\pi}{\omega} = 2\pi\sqrt{\frac{I}{(m_2 r_2 - m_1 r_1) g}} = 2\pi\sqrt{\frac{m_1 r_1^2 + m_2 r_2^2}{(m_2 r_2 - m_1 r_1) g}}.$$

CALCULATE: $T = 2\pi\sqrt{\dfrac{1.00 \text{ kg}(0.100 \text{ m})^2 + 2.00 \text{ kg}(0.200 \text{ m})^2}{(2.00 \text{ kg}(0.200 \text{ m}) - 1.00 \text{ kg}(0.100 \text{ m}))(9.81 \text{ m/s}^2)}} = 1.09876 \text{ s}$

ROUND: Rounding to three significant figures, the period of oscillation is $T = 1.10$ s.

DOUBLE-CHECK: A simple pendulum of length 0.3 m will have a period of $2\pi\sqrt{\frac{0.3 \text{ m}}{g}} = 1.09876$ s, the same as that calculated above.

14.41. THINK: To determine the moment of inertia, the expressions for a thin rod and a sphere, along with the parallel axis theorem are needed. Using this result, the period of the pendulum can be determined by direct substitution into what we have derived in Solved Problem 14.2.

SKETCH:

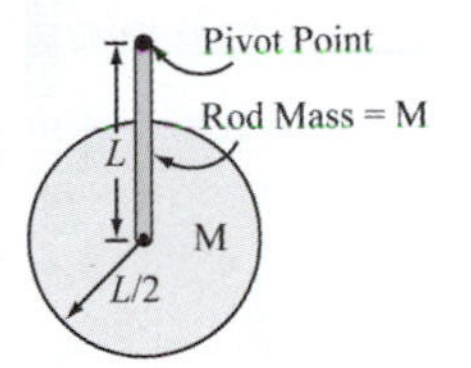

RESEARCH: The moments of inertia are given by: $I_{\text{rod}} = \frac{1}{3}ML^2$ and $I_{\text{sphere}} = \frac{2}{5}Mr^2$. The parallel axis theorem is given by $I = I_{\text{center}} + Mx^2$, where x is the distance of the pivot point to the center of mass. The period is given by:

$$T = 2\pi\sqrt{\frac{I}{M_t gR}},$$

where R is the distance from the pivot point to the center of gravity and M_t is the total mass, which is $2M$ in the present case.

SIMPLIFY:

(a) The total moment of inertia is $I = I_{\text{rod}} + \left(I_{\text{sphere}} + ML^2\right)$. Substituting the moments of inertia gives:

$$I = \frac{1}{3}ML^2 + \frac{2}{5}M\left(\frac{L^2}{4}\right) + ML^2 = \left(\frac{1}{3} + \frac{1}{10} + 1\right)ML^2 = \left(\frac{10+3+30}{30}\right)ML^2 = \frac{43}{30}ML^2.$$

(b) The distance from the pivot point to the center of gravity is $R = 3L/4$, so:

$$T = 2\pi\sqrt{\frac{I}{M_t gR}} \Rightarrow T = 2\pi\sqrt{\frac{43ML^2/30}{2Mg3L/4}} = 2\pi\sqrt{\frac{4(43)L}{6(30)g}} = 2\pi\sqrt{\frac{172L}{180g}} = 2\pi\sqrt{\frac{43L}{45g}}.$$

(c) $T^2 = 4\pi^2\left(\frac{43L}{45g}\right) \Rightarrow L = \frac{45gT^2}{43\left(4\pi^2\right)}$

CALCULATE:

(a) Not necessary.

(b) Not necessary.

(c) $T = 2.0$ s, so, $L = \frac{45\left(9.81 \text{ m/s}^2\right)\left(2.0 \text{ s}\right)^2}{43\left(4\pi^2\right)} = 1.0401915$ m.

ROUND:

(a) Not necessary.

(b) Not necessary.

(c) Since the desired period is given to two significant figures, $L = 1.0$ m.

DOUBLE-CHECK: The results are reasonable, because they work out similar to the case of the pendulum with a point mass at the end, for which we have calculated $T = 2\pi\sqrt{L/g}$. This implies that taking care of the proper distribution of the mass has a noticeable effect relative to what one obtains for a point mass pendulum, but it is not drastically different. [Since the entire exercise involving moments of inertia and center-of-mass coordinates only resulted in a correction factor of $\sqrt{43/45} \approx 0.98$, you may think that it was not worth the effort. However, if a clock is off by this factor, it is slow by approximately half an hour per day and therefore pretty much useless as a time-keeping device.]

14.45. **THINK:** For both parts, use the conservation of energy, along with the expressions for kinetic and potential energy. The mass is m = 2.00 kg, the displacement is $x_0 = 8.00$ cm and the frequency is $f = 4.00$ Hz.

SKETCH:

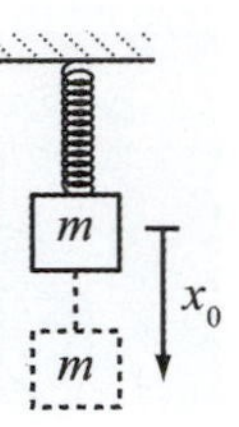

RESEARCH:

(a) $\text{Total energy} = \text{constant} = \frac{1}{2}kx_0^2 = E_{\text{tot}}, \quad k = m\omega^2 = 4\pi^2 mf^2$

(b) $K = \frac{1}{2}mv^2, \quad U = \frac{1}{2}kx^2, \quad K + U = E_{\text{tot}}$

SIMPLIFY:

(a) The energy is constant throughout the oscillation:

$$E_{\text{tot}} = \frac{1}{2}kx_0^2 = \frac{1}{2}m\omega^2 x_0^2 = \frac{1}{2}m(2\pi f)^2 x_0^2 = 2\pi^2 mf^2 x_0^2.$$

(b) $\frac{1}{2}mv^2 + \frac{1}{2}kx^2 = E_{\text{tot}}$

$$\Rightarrow v^2 = \frac{2}{m}\left(E_{\text{tot}} - \frac{1}{2}kx^2\right) = \frac{2}{m}\left(\frac{1}{2}kx_0^2 - \frac{1}{2}kx^2\right) = \frac{k}{m}\left(x_0^2 - x^2\right) = \frac{4\pi^2 mf^2}{m}\left(x_0^2 - x^2\right) \Rightarrow v = 2\pi f\sqrt{x_0^2 - x^2}$$

CALCULATE:

(a) $E_{\text{tot}} = 2\pi^2(2.00 \text{ kg})(4.00 \text{ Hz})^2(.0800 \text{ m})^2 = 4.043 \text{ J}$

(b) $v = 2\pi(4.00 \text{ Hz})\sqrt{(8.00 \text{ cm})^2 - (2.00 \text{ cm})^2} = 194.7 \text{ cm/s} = 1.947 \text{ m/s}$

ROUND: Rounding to three significant figures, $E = 4.04$ J, v = 1.95 m/s

DOUBLE-CHECK: These results are reasonable.

14.49. **THINK:** The equilibrium separation will occur at the minimum of the potential. For small deviations from equilibrium, the potential can be expanded to second order about the minimum. The resulting potential will be formally equivalent to the harmonic potential.

SKETCH:

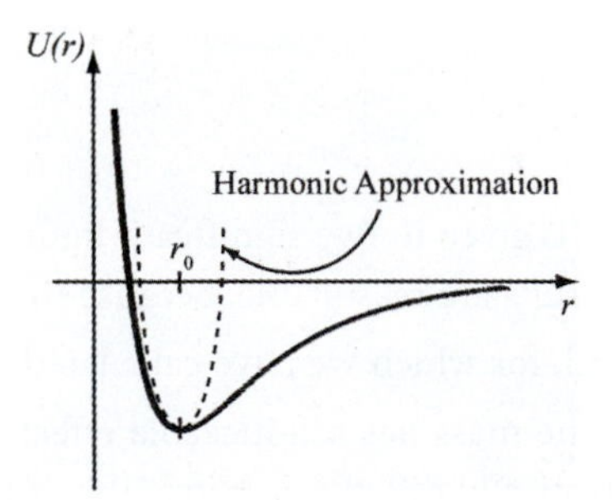

RESEARCH: The minimum is at r_0: $\left[\dfrac{dU(r)}{dr}\right]_{r=r_0}=0.$ The Taylor series about $r=r_0$ is:

$$U(r)\approx U(r_0)+\frac{1}{1!}(r-r_0)^1\left[\frac{dU}{dr}\right]_{r_0}+\frac{1}{2!}(r-r_0)^2\left[\frac{d^2U}{dr^2}\right]_{r_0}+\ldots$$

$$\left[\frac{dU}{dr}\right]_{r_0}=0\Rightarrow U(r)\approx U(r_0)+\frac{1}{2}(r-r_0)^2\left[\frac{d^2U}{dr^2}\right]_{r_0}$$

SIMPLIFY:

(a) $\dfrac{dU}{dr}=\dfrac{d}{dr}\left(\dfrac{A}{r^{12}}-\dfrac{B}{r^6}\right)=-\dfrac{12A}{r^{13}}+\dfrac{6B}{r^7}$

$$\left[\frac{dU}{dr}\right]_{r_0}=0:\ -\frac{12A}{r_0^{13}}+\frac{6B}{r_0^{7}}=0\Rightarrow -\frac{12A}{r_0^{6}}+6B=0\Rightarrow r_0^6=\frac{12A}{6B}\Rightarrow r_0=\left(\frac{2A}{B}\right)^{\frac{1}{6}}$$

(b) $\left[\dfrac{d^2U}{dr^2}\right]_{r_0}=\dfrac{12(13)A}{r_0^{14}}-\dfrac{6(7)B}{r_0^{8}}=\dfrac{156A}{r_0^{14}}-\dfrac{42B}{r_0^{8}}=k$

$$\Rightarrow U(r)\approx U(r_0)+\frac{1}{2}(r-r_0)^2k=U(r_0)+\frac{1}{2}k\left(r^2-2rr_0+r_0^2\right)=\text{constant}+kr_0r+\frac{1}{2}kr^2$$

$$F=-\frac{dU}{dr}\approx -kr_0-kr=\text{constant}-kr.$$

This is Hooke's law with a spring constant, k. The angular frequency is given by $\omega=\sqrt{k/m}$.

$$k=\frac{1}{r_0^{14}}\left[156A-42r_0^{6}B\right]=\left(\frac{B}{2A}\right)^{\frac{14}{6}}\left[156A-42B\left(\frac{2A}{B}\right)\right]=\left(\frac{B}{2A}\right)^{\frac{7}{3}}(156-84)A=72A\left(\frac{B}{2A}\right)^{\frac{7}{3}}=\frac{72}{2^{7/3}}\left(\frac{B^{7/3}}{A^{4/3}}\right)$$

$$\Rightarrow k=9\left(\frac{4B^7}{A^4}\right)^{1/3}$$

$$\omega=\sqrt{\frac{k}{m}}=\frac{3}{\sqrt{m}}\left(\frac{4B^7}{A^4}\right)^{1/6}$$

CALCULATE: This step is not necessary.

ROUND: Rounding is not necessary.

DOUBLE-CHECK: These oscillations yield the vibration spectra of the molecule. Most systems with an equilibrium configuration can exhibit simple harmonic motion for small perturbations from equilibrium.

14.53. **THINK:** From the critical damping condition, determine the value of the damping constant, b. When b is 60.7 % of its full value, use the expression for the underdamped oscillator to determine the period. Model the system as four independent oscillators, each supporting a quarter of the weight of the car. The mass of the car is $m_c=851$ kg and the value of the spring constant is k = 4005 N/m.

SKETCH:

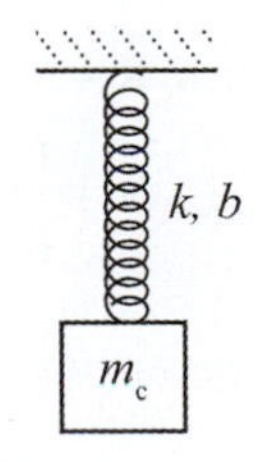

RESEARCH: For critical damping, $b_0=2\sqrt{mk}$. $b=0.607b_0\equiv\alpha b_0\Rightarrow\alpha=0.607,\quad m=\dfrac{m_c}{4}$. The period of underdamped motion is given by $T=2\pi/\omega'$.

$$\omega' = \sqrt{\omega_0^2 - \omega_\gamma^2}, \quad \omega_\gamma = \frac{b}{2m}, \quad \omega_0 = \sqrt{\frac{k}{m}}$$

SIMPLIFY: $\omega_\gamma = \frac{\alpha b_0}{2m} = \frac{\alpha 2\sqrt{mk}}{2m} = \alpha\sqrt{\frac{k}{m}} = \alpha\omega_0, \quad \omega' = \sqrt{\omega_0^2 - \alpha^2\omega_0^2} = \omega_0\sqrt{1-\alpha^2}$

$$T = \frac{2\pi}{\omega_0\sqrt{1-\alpha^2}} = 2\pi\sqrt{\frac{m}{k\left(1-\alpha^2\right)}}$$

CALCULATE: $T = 2\pi\sqrt{\frac{851 \text{ kg}/4}{4005 \text{ N/m}\left(1-(0.607)^2\right)}} = 1.822 \text{ s}$

ROUND: Rounding to three significant figures, $T = 1.82$ s

DOUBLE-CHECK: This is a reasonable result for a car's shock absorbers. The value has seconds as units, which are appropriate for time.

14.57. **THINK:** This is a straightforward exercise in inserting numbers into Eq. 14.33, which states that the amplitude of the damped driven oscillation is

$$A_\gamma = \frac{F_\text{d}}{m\sqrt{\left(\omega_0^2 - \omega_\text{d}^2\right)^2 + 4\omega_\text{d}^2\omega_\gamma^2}}$$

RESEARCH: The problem text specifies that $\omega_0 = 2.40$ rad/s, and that $\omega_\gamma = 0.140$ rad/s.

SIMPLIFY: No simplification necessary.

CALCULATE: Inserting values of (a) $\omega_\text{d} = \frac{1}{2}\omega_0$, (b) $\omega_\text{d} = \omega_0$, and (c) $\omega_\text{d} = 2\omega_0$ then yields:

(a) 1.20 rad/s: $A_\gamma = \frac{2}{3\sqrt{\left((2.40)^2 - (1.20)^2\right)^2 + 4(1.20)^2(0.140)^2}} \text{ m} = 0.1538 \text{ m}$

(b) 2.40 rad/s: $A_\gamma = \frac{2}{3\sqrt{\left((2.40)^2 - (2.40)^2\right)^2 + 4(2.40)^2(0.140)^2}} \text{ m} = 0.9921 \text{ m}$

(c) 4.80 rad/s: $A_\gamma = \frac{2}{3\sqrt{\left((2.40)^2 - (4.80)^2\right)^2 + 4(4.80)^2(0.140)^2}} \text{ m} = 0.03846 \text{ m}$

ROUND: Rounding to three significant figures,

(a) $A_\gamma = 0.154$ m.

(b) $A_\gamma = 0.992$ m.

(c) $A_\gamma = 0.0385$ m.

DOUBLE-CHECK: The values seem reasonable. The largest value is for the case where the driving angular speed is the resonant speed of the system.

14.61. The conservation of energy gives $K_1 + U_1 = K_2 + U_2$. For a mass oscillating on a spring, $K_\text{max} = U_\text{max} = K_1 + U_1$, since $U = 0$, when $K = K_\text{max}$ and $K = 0$, when $U = U_\text{max}$. Therefore, when

$$K = \frac{1}{2}K_\text{max}, \quad U = \frac{1}{2}U_\text{max} = \frac{1}{2}\left(\frac{1}{2}kx^2{}_\text{max}\right) = \frac{1}{2}kx^2,$$

and this is when $x = \frac{x_\text{max}}{\sqrt{2}}$.

14.65. **THINK:** We model the hydrogen atom as two masses connected by a massless spring. Consider two masses m_1 and m_2 located at positions x_1 and x_2 respectively. These two masses are connected by a spring with spring constant k. The equilibrium length of the spring is L. Hooke's Law for a spring tells us that $F=-kx$ and Newton's Second Law tells us that $F=ma$.

RESEARCH: The equations describing the motion of the two masses are

$$m_1a_1=k(x_2-x_1-L)$$
$$m_2a_2=-k(x_2-x_1-L).$$

We can define the quantity $x=x_2-x_1-L$, which is the amount by which the spring is stretched or compressed from its equilibrium length. We can then write

$$m_1a_1=kx$$
$$m_2a_2=-kx.$$

If we add these two equations we get

$$m_1a_1+m_2a_2=kx+(-kx) \Rightarrow m_1a_1+m_2a_2=0.$$

The mass of the system is $M=m_1+m_2$. The x-coordinate of the center of mass of the system is

$$x_{\text{cm}}=\frac{m_1x_1+m_2x_2}{m_1+m_2}.$$

Because the center of mass of the system is at rest, we can write

$$Ma_{\text{cm}}=0.$$

We can re-express our first two equations as

$$a_1=\frac{kx}{m_1}$$
$$a_2=\frac{-kx}{m_2}.$$

SIMPLIFY: We can subtract the second equation from the first to get

$$a_1-a_2=\frac{kx}{m_1}-\frac{-kx}{m_2}=kx\left(\frac{1}{m_1}+\frac{1}{m_2}\right).$$

Remembering that $x=x_2-x_1-L$, we can take the time derivative to get

$$\frac{d^2x}{dt^2}=\frac{d^2}{dt^2}(x_2-x_1-L)=\frac{d^2}{dt^2}(x_2)-\frac{d^2}{dt^2}(x_2)=a_2-a_1.$$

Now combining the last two equations we get

$$\frac{d^2x}{dt^2}=-kx\left(\frac{1}{m_1}+\frac{1}{m_2}\right)=-kx\left(\frac{m_1+m_2}{m_1m_2}\right).$$

We can define the reduced mass μ of our system as

$$\mu=\frac{m_1m_2}{m_1+m_2}.$$

So we then have

$$\frac{d^2x}{dt^2}+\frac{k}{\mu}x=0.$$

We recognize this differential equation as having the same form as simple harmonic motion with an angular speed of

$$\omega = \sqrt{\frac{k}{\mu}}.$$

Each hydrogen atom has $m_1 = m_2 = m_H$ so the reduced mass becomes

$$\mu_{\rm H} = \frac{m_{\rm H} m_{\rm H}}{m_{\rm H} + m_{\rm H}} = \frac{m_{\rm H}}{2}.$$

The angular speed then becomes

$$\omega = \sqrt{\frac{k}{\left(\frac{m_{\rm H}}{2}\right)}} = \sqrt{\frac{2k}{m_{\rm H}}}.$$

Solving for the spring constant gives us

$$k = \frac{m_{\rm H}\omega^2}{2}.$$

The period is related to the angular speed as

$$T = \frac{2\pi}{\omega} \Rightarrow \omega = \frac{2\pi}{T}.$$

We can then write

$$k = \frac{m_{\rm H}\left(\frac{2\pi}{T}\right)^2}{2} = \frac{2\pi^2 m_{\rm H}}{T^2}.$$

CALCULATE: Putting in our numbers we get

$$k = \frac{2\pi^2\left(1.7\cdot 10^{-27}\text{ kg}\right)}{\left(8.0\cdot 10^{-15}\text{ s}\right)^2} = 524.3 \text{ N/m}.$$

ROUND: Rounding to two significant digits gives us 520 N/m.

DOUBLE-CHECK: Our answer has the correct units. This answer is half what we would get if we incorrectly used $\omega = \sqrt{k/m_{\rm H}}$.

14.69. **THINK:** The difference in the respective periods dictates how long they remain out of phase. The smaller this difference (but not zero!), the longer it takes them to come back into phase. l = 1.000 m, $g_{\rm M} = 9.784$ m/s^2 and $g_{\rm O} = 9.819$ m/s^2.

SKETCH:

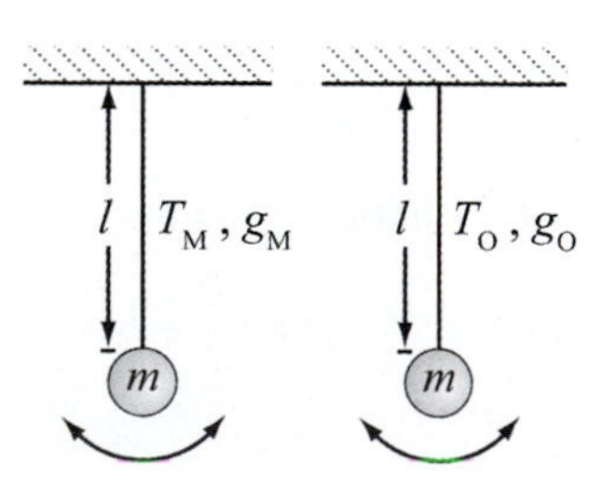

RESEARCH: $T = 2\pi\sqrt{\frac{l}{g}}$

SIMPLIFY: $\Delta T = T_M - T_O = 2\pi\sqrt{l}\left(\frac{1}{\sqrt{g_M}} - \frac{1}{\sqrt{g_O}}\right)$

They will be in phase after n oscillations of the Manila pendulum, such that $nT_M = (n+1)T_O \Rightarrow n(T_M - T_O) = T_O$ and so $n = T_O / \Delta T$. This will take a time of $t = nT_M$ to happen.

CALCULATE: $\Delta T = 2\pi\sqrt{1.000 \text{ m}}\left(\frac{1}{\sqrt{9.784 \text{ m/s}^2}} - \frac{1}{\sqrt{9.819 \text{ m/s}^2}}\right) = 3.58327 \cdot 10^{-3} \text{ s}$

$n = \frac{2\pi\sqrt{1.000 \text{ m}/9.819 \text{ m/s}^2}}{3.58327 \cdot 10^{-3} \text{ s}} = 559.56,\ t = (559.56)2\pi\sqrt{1.000 \text{ m}/(9.784 \text{ m/s}^2)} = 1124 \text{ s}$

ROUND: $n = 559.6$, $t = 1124$ s.

DOUBLE-CHECK: It takes approximately 19 minutes for the pendulums to come back into phase. This result is reasonable.

14.73. **THINK:** A pendulum has a period of 2.00 s and the mass of the bob is 0.250 kg. A weight slowly falls to provide the energy to overcome the frictional damping of the pendulum. The mass of the weight is 1.00 kg and it moves down 0.250 m every day. The Q factor of the pendulum must be determined.

SKETCH:

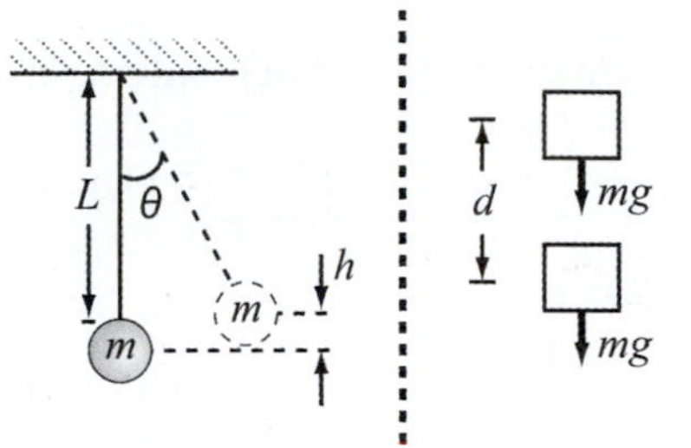

RESEARCH: The Q factor is defined by $Q = 2\pi E / |\Delta E|$, where E is the energy of the pendulum and ΔE is the energy loss. The energy of the pendulum is determined using the maximum height, h, by $E = mgh = mgL(1 - \cos\theta)$. From the period of the pendulum, it is found that:

$$T^2 = 4\pi^2 \frac{L}{g} \text{ or } L = \frac{gT^2}{4\pi^2}.$$

SIMPLIFY: Thus, the energy is given by: $E = mg\left(\frac{gT^2}{4\pi^2}\right)(1 - \cos\theta) = \frac{mg^2T^2}{4\pi^2}(1 - \cos\theta)$. The energy loss in one period of oscillation is: $\Delta E = Mgx = Mg\frac{d}{24(3600 \text{ s})/(2.00 \text{ s})}$. Thus, the Q factor is:

$$Q = 2\pi\frac{mg^2T^2}{4\pi^2}(1 - \cos\theta)\left(\frac{43200}{Mgd}\right) = 43200\left(\frac{m}{M}\right)\frac{gT^2}{2\pi d}(1 - \cos\theta).$$

CALCULATE: $Q = 43200\left(\frac{0.250 \text{ kg}}{1.00 \text{ kg}}\right)\frac{(9.81 \text{ m/s}^2)(2.00 \text{ s})^2}{2\pi(0.250 \text{ m})}(1 - \cos 10.0°) = 4098.78$

ROUND: Keeping three significant figures, $Q = 4.10 \cdot 10^3$.

DOUBLE-CHECK: The Q factor is a dimensionless quantity. $\frac{[\text{kg}]}{[\text{kg}]}\frac{[\text{m/s}^2][\text{s}^2]}{[\text{m}]} \Rightarrow \frac{[\text{kg}][\text{m}][\text{s}]}{[\text{kg}][\text{m}][\text{s}]}$. All the units cancel. The result is reasonable.

14.77. **THINK:** A restoring force of an oscillator is related to its potential energy.

SKETCH:

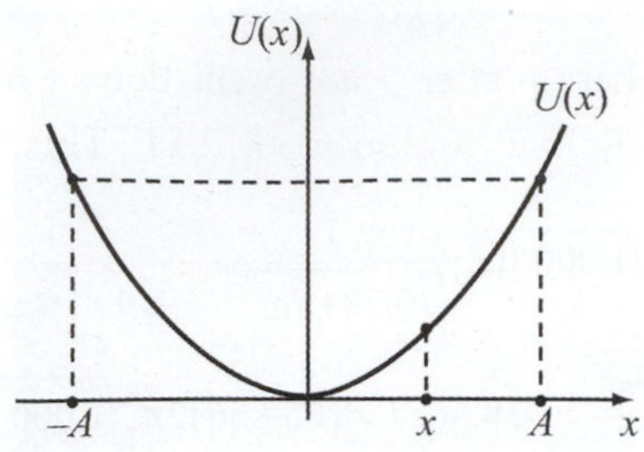

RESEARCH: The restoring force at a position, x, is given by $F(x)=-dU(x)/dx$. Thus, the potential is $U(x)=\int_{x_0}^{x}-F(x)dx$. The velocity of a mass m at the position x is determined using the conservation of energy. The energy at x is equal to the energy at $x = A$, so, $K_f + U_f = K_i + U_i$. Using $K_i = 0$, the equation becomes $(mv^2/2)+U(x)=0+U(A)$.

SIMPLIFY: $v^2 = 2\dfrac{(U(A)-U(x))}{m}$ or $v=\dfrac{dx}{dt}=\sqrt{\dfrac{2}{m}(U(A)-U(x))} \Rightarrow dt=\left[\dfrac{2}{m}(U(A)-U(x))\right]^{-1/2}dx.$

Integrating both sides with intervals $x\in(-A,A)$ and $t\in(0,T/2)$ yields:

$$\int_0^{T/2}dt=\int_{-A}^{A}\left[\frac{2}{m}(U(A)-U(x))\right]^{-1/2}dx.$$

Thus, the period of oscillation is:

$$T=2\int_{-A}^{A}\left[\frac{2}{m}(U(A)-U(x))\right]^{-1/2}dx \qquad (1).$$

(a) Substituting $F(x)=-cx^3$ into $U(x)=\int_{x_0}^{x}-F(x)dx$ gives: $U(x)=\int_{x_0}^{x}cx^3dx=\frac{c}{4}x^4-\frac{c}{4}x_0^4$. For simplicity, it is assumed that $x_0=0$. The potential is therefore given by $U(x)=cx^4/4$. Thus, the expression for the period is:

$$T=2\int_{-A}^{A}\left[\frac{c}{2m}\left(A^4-x^4\right)\right]^{-1/2}dx.$$

(b) Changing the variable, x, with $x = Ay$ yields: $T=2\int_{-1}^{1}\left[\frac{c}{2m}A^4\left(1-y^4\right)\right]^{-1/2}Ady$. Simplifying the previous expression gives: $T=2\sqrt{\frac{2m}{c}}\frac{1}{A}\int_{-1}^{1}\left(1-y^4\right)^{-1/2}dy=\frac{B}{A}$, where B is a constant. Therefore, the period is inversely proportional to A.

(c) Substituting $U(x)=\frac{\gamma}{\alpha}|x|^{\alpha}$ into equation (1) yields:

$$T=2\int_{-A}^{A}\left[\frac{2}{m}\left(\frac{\gamma}{\alpha}|A|^{\alpha}-\frac{\gamma}{\alpha}|x|^{\alpha}\right)\right]^{-1/2}dx.$$

Similarly as above, changing the variable, x, with $x = Ay$ yields:

$$T=2\int_{-1}^{1}\left[\frac{2}{m}\frac{\gamma}{\alpha}|A|^{\alpha}\right]^{-1/2}\left(1-y^{\alpha}\right)^{-1/2}Ady.$$

Thus, $T=2\sqrt{\frac{\alpha m}{2\gamma}}A^{\left(1-\frac{\alpha}{2}\right)}\int_{-1}^{1}\left(1-y^{\alpha}\right)^{-1/2}dy=BA^{\left(1-\frac{\alpha}{2}\right)}$. The period is proportional to $A^{\left(1-\frac{\alpha}{2}\right)}$.

CALCULATE: Not necessary.

ROUND: Not necessary.

DOUBLE-CHECK: If $\alpha = 2$, T is constant and independent of A. If $\alpha = 4$, then $T = B/A$, which is the same as the result in part (b).

Multi-Version Exercises

14.78. **THINK:** This problem involves a block and spring assembly, where the block is sliding back and forth on a frictionless surface. The block in this problem undergoes simple harmonic motion. The mass of the block, spring constant, and displacement are given, so it should be possible to find the displacement as a function of time, using the equations for the motion of a mass on a spring with no damping.

SKETCH: Show the displacement, velocity, and the force on the block due to the spring. Consider only the motion of the block in the x-direction, since there is no net force in the y-direction, and no friction force. Take the start time $t = 0$ to be the moment the block is released, and the equilibrium position of the spring to be the origin where $x = 0$.

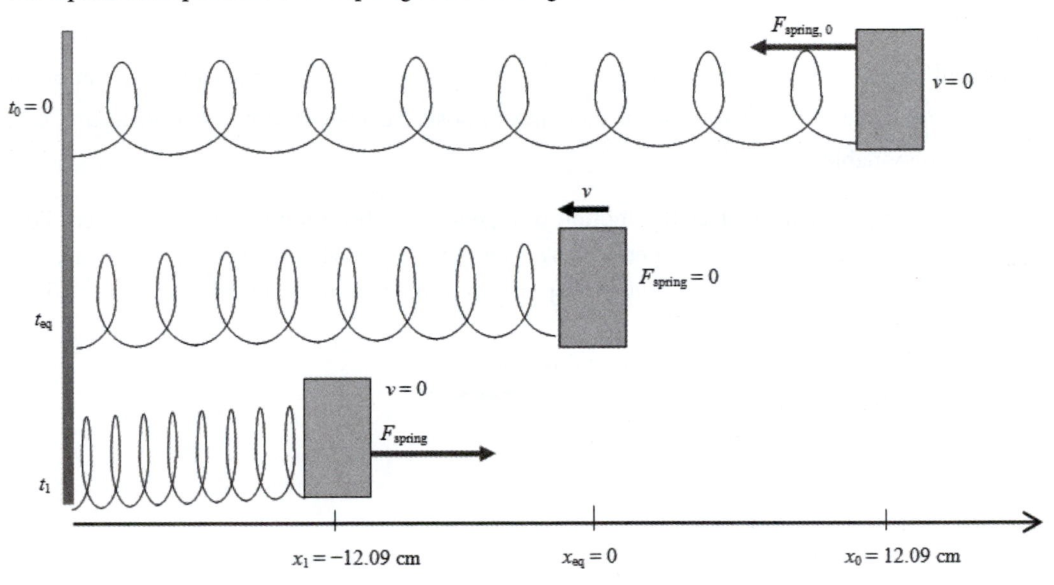

RESEARCH: The equation of motion for a mass on a spring with no damping is $x(t) = A\sin(\omega_0 t + \theta_0)$, where A is the amplitude of the oscillation, ω_0 is the angular speed, and θ_0 is the phase angle. The angular speed can be found from the mass of the block and the spring constant using the equation $\omega_0 = \sqrt{\frac{k}{m}}$, where k is the spring constant and m is the mass. The amplitude of the oscillation is equal to the maximum stretch of the spring, which is the initial position of the block.

SIMPLIFY: The location of the block at time $t = 0$ is $A = 12.09$ cm, so $\theta_0 = \sin^{-1}(1) = \frac{\pi}{2}\text{rad}$. Substituting for the angular speed, the equation for the motion of the block becomes $x(t) = A\sin\left(\sqrt{\frac{k}{m}}t + \theta_0\right)$.

CALCULATE: The amplitude of the oscillation is $A = 12.09$ cm $= 0.1209$ m, the block has mass 1.605 kg, and the spring constant is 14.55 N/m. The phase angle is $\theta_0 = \frac{\pi}{2}\text{rad}$. The location of the block at time $t = 2.834 \cdot 10^{-1}$ s is:

$$x\left(2.834\cdot10^{-1}\text{ s}\right)=A\sin\left(\sqrt{\frac{k}{m}}\left[2.834\cdot10^{-1}\right]+\theta_0\right)$$
$$=0.1209\text{ m}\cdot\sin\left(\sqrt{\frac{14.55\text{ N/m}}{1.605\text{ kg}}}\cdot2.834\cdot10^{-1}\text{ s}+\frac{\pi}{2}\right)$$
$$=7.949320781\cdot10^{-2}\text{ m}$$
$$=7.949320781\text{ cm}$$

ROUND: All of the numbers in this calculation had four significant figures, so the final answer will also have four figures. After $2.834\cdot10^{-1}$ s, the block is 7.949 cm from the equilibrium position, in the same direction that the mass had been pulled at the start of the experiment.

DOUBLE-CHECK: The period of this motion is $\frac{2\pi}{\omega_0}=2.087\text{sec}$. Time t = $2.834\cdot10^{-1}$ s is less than one fourth of the total period, so it is expected that the mass is between the fully stretched position and the equilibrium position. In fact, since the time is closer to one eighth of the period, it is expected that the block will be less than ½ way from the fully stretched position to the equilibrium position, somewhere close to (but more than) 6 cm from the equilibrium position. These estimates confirm that the calculated answer is reasonable.

14.81. **THINK:** This problem involves the motion of a pendulum, but the motion is restricted. For half of its motion, the pendulum will have a period corresponding to the whole length of the string. For the other half of its motion, it will have a period corresponding to the length of the string minus the distance of the peg from the ceiling.

SKETCH: The sketch shows the motion of the pendulum:

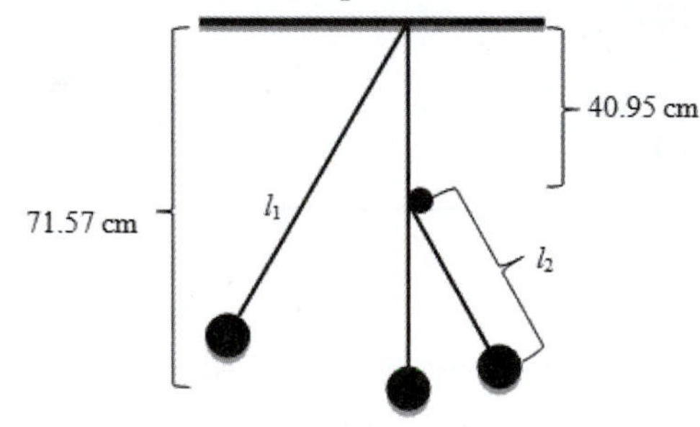

RESEARCH: The period of a pendulum hanging from a string of length l is $T_l=2\pi\sqrt{\frac{l}{g}}$. The pendulum swings for half of the period corresponding the full length of the string l_1, with a half-period of $\frac{1}{2}T_1=\frac{1}{2}2\pi\sqrt{\frac{l_1}{g}}$. Similarly, it swings for half of the period corresponding to the length of the string minus the distance of the peg from the ceiling, $\frac{1}{2}T_2=\frac{1}{2}2\pi\sqrt{\frac{l_2}{g}}$. The total period of the pendulum is the sum of the swing corresponding to length l_1 and the swing corresponding to length l_2, for a total period of $T=\frac{1}{2}T_1+\frac{1}{2}T_2$.

SIMPLIFY: The goal is to find the total period. Substitute the expressions for the periods corresponding to length l_1 and l_2 to find the total period $T=\frac{1}{2}2\pi\sqrt{\frac{l_1}{g}}+\frac{1}{2}2\pi\sqrt{\frac{l_2}{g}}$. This can be simplified to

$T=\frac{\pi\left(\sqrt{l_1}+\sqrt{l_2}\right)}{\sqrt{g}}$, where l_1 is the full length of the string, and l_2 is the length of the string minus the distance from the peg to the ceiling.

CALCULATE: The question states that the length of the string is l_1 = 71.57 cm = 0.7157 m. The distance from the peg to the ceiling is 40.95 cm, so the length corresponding to the second period is l_2 = 71.57 cm – 40.95 cm = 30.62 cm = 0.3062 m. The gravitational acceleration near the surface of the Earth is 9.81 m/s^2. The total period is then

$$T=\frac{\pi\left(\sqrt{l_1}+\sqrt{l_2}\right)}{\sqrt{g}}$$
$$=\frac{\pi\left(\sqrt{0.7157\text{ m}}+\sqrt{0.3062\text{ m}}\right)}{\sqrt{9.81\text{ m/s}^2}}$$
$$=1.403588643\text{ s.}$$

ROUND: The length of the string, the distance from the peg to the ceiling, and their difference all have four significant figures, so the final answer should have four significant figures. The pendulum has a period of 1.404 s.

DOUBLE-CHECK: Imagine that there were two pendulums. A pendulum on a string of length 0.7157 m has a period of 1.697 seconds, while a pendulum on a string of length 0.3062 m has a period of 1.110 seconds. The average of these two periods is 1.404 s. A pendulum swinging for half a cycle on a string of one length and half a cycle on a string of a second length will have a period equal to the average of the periods corresponding to the two lengths (it will NOT have the same period as a string of the average of the two lengths).

14.83. **THINK:** The speed of the object attached to the spring depends only on the distance from the equilibrium position. Since it does not matter if the object is above the equilibrium, below the equilibrium, moving up, or moving down, it is easiest to solve this problem using conservation of energy.

SKETCH: The object is shown at four times: when the spring is stretched down before being released (x_{min}), 1.849 cm below equilibrium ($-x_0$), at equilibrium ($x = 0$), and at maximum height above equilibrium (x_{max}). The spring and gravitational forces are shown. The velocity of the object on the way up and its velocity on the way down are both shown.

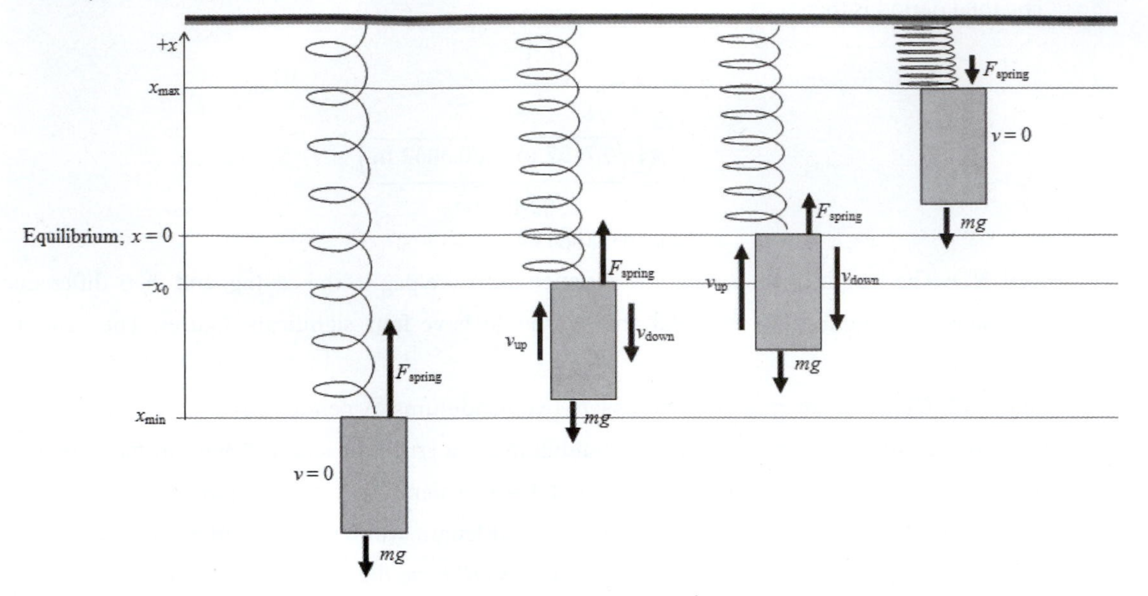

RESEARCH: The potential energy stored in a spring is $U_S = \frac{1}{2}kx^2$, and the kinetic energy of the mass is $K = \frac{1}{2}mv^2$. The total mechanical energy of the mass on the spring is $E = \frac{1}{2}kA^2$. The total mechanical energy should also equal the sum of the kinetic energy and the potential energy ($E = U_S + K$). The spring constant is given in Newtons per meter and the displacements are given in centimeters, so the conversion 100 cm = 1 m will be needed.

SIMPLIFY: Since there are two expressions for the total mechanical energy, set them equal to one another to get $\frac{1}{2}kA^2 = U_S + K$. Then, substitute in the expressions for the potential energy and kinetic energy to get $\frac{1}{2}kA^2 = \frac{1}{2}kx^2 + \frac{1}{2}mv^2$. Finally, solve for the speed of the mass:

$$\frac{1}{2}kA^2 = \frac{1}{2}kx^2 + \frac{1}{2}mv^2$$
$$mv^2 = kA^2 - kx^2$$
$$v^2 = \frac{kA^2 - kx^2}{m}$$
$$v = \sqrt{\frac{kA^2 - kx^2}{m}}$$

CALCULATE: Since the spring was stretched and the object released from rest, the maximum distance from the equilibrium point (the amplitude) is equal to the distance at which it was released. So the amplitude A = 18.51 cm = 0.1851 m. The spring constant k is 23.31 N/m, and the object has a mass of

1.375 kg. The goal is to find the velocity when the mass is a distance of 1.849 cm = 0.01849 meters from the equilibrium point, so $x = \pm 0.01849$ m. Using these values,

$$\begin{aligned} v &= \sqrt{\frac{kA^2 - kx^2}{m}} \\ &= \sqrt{\frac{23.31 \text{ N/m} \cdot (0.1851 \text{ m})^2 - 23.31 \text{ N/m} \cdot (\pm 0.01849 \text{ m})^2}{1.375 \text{ kg}}} \\ &= 0.7583130694 \text{ m/s} \end{aligned}$$

ROUND: The values used in this calculation all have four significant figures, so the final answer should also have four figures. When it is 1.849 cm from the equilibrium point, the mass has a speed of 0.7583 m/s.

DOUBLE-CHECK: The speed of the mass is 0 m/s when it is at the bottom or top of its oscillations. The maximum speed of the mass occurs when the mass passes the equilibrium point ($x = 0$). At this point, the mass achieves a speed of $0.1851 \text{ m}\sqrt{\dfrac{23.31 \text{ N/m}}{1.375 \text{ kg}}} = 0.7621 \text{ m/s}$. This is slightly faster than the speed of the mass when it is 1.849 cm from the equilibrium point, but not by much, confirming that the answer of 0.7583 m/s is reasonable.

Chapter 15: Waves

Exercises

15.23. The time resolution in the air is determined by the time it takes sound to travel 20.0 cm. At a speed of 343 m/s, the resolution time is $t_{\text{max}} = 0.200 \text{ m}/(343 \text{ m/s}) = 5.83 \cdot 10^{-4}$ s. In the water, the speed of sound is $1.50 \cdot 10^3$ m/s, corresponding to a resolution time of $t_{\text{max}} = 0.200 \text{ m}/(1.50 \cdot 10^3 \text{ m/s}) = 1.33 \cdot 10^{-4}$ s. If an individual can only resolve a time difference of $5.83 \cdot 10^{-4}$ s, they will not be able to distinguish a time difference of $1.33 \cdot 10^{-4}$ s. Since our hearing is adapted to land conditions, a sound produced in water seems to reach the listener's ears at the same time regardless of direction. This is why it is impossible for the diver to detect the direction of a motor boat underwater.

15.27. **THINK:**

(a) The question asks for the equation of motion for the masses of an array of n masses each with mass m connected by with springs each with spring constant k. Each mass is located at $x_i = ia$ at equilibrium and we define the displacement of each mass as $x_i = ia + \psi_i$.

(b) The object is to determine the angular frequencies of the normal modes of the array of masses.

SKETCH:

RESEARCH:

(a) Let ψ_i denote the displacement of the ith mass from its equilibrium position. The forces acting on the masses are due to the springs and have the form $F_s = -k\psi$. The angular frequency is given by $\omega_0^2 = k/m$.

(b) For each normal mode, the whole system oscillates with an angular frequency of Ω, so the motion of each particle can be described as $\psi_i = A_i \cos(\Omega t)$. The left hand side is stationary, which implies $\psi_0 = 0$. The right hand side is stationary, which means $\psi_{n+1} = 0$. This suggests an *Ansatz* for the amplitudes $A_i = A\sin(i\phi)$ where A is an arbitrary amplitude, ϕ is a real number that is different for each normal mode, and $i = 1$ to n.

SIMPLIFY:

(a) The net force on the ith mass is

$$F_i = ma_i = m\frac{d^2\psi_i}{dt^2} = -k(\psi_i - \psi_{i-1}) - k(\psi_i - \psi_{i+1}) = k(\psi_{i-1} - 2\psi_i + \psi_{i+1})$$

$$\frac{d^2\psi_i}{dt^2} = \omega_0^2(\psi_{i-1} - 2\psi_i + \psi_{i+1}).$$

All the masses and springs are identical, so this result describes the entire system.

(b) Insert $\psi_i = A_i \cos\Omega t$ into the result of part (a):

$$\frac{d^2\psi_i}{dt^2} = \frac{d^2}{dt^2}(A_i \cos\Omega t) = -A_i\Omega^2 \cos\Omega t$$

$$\omega_0^2(\psi_{i-1} - 2\psi_i + \psi_{i+1}) = \omega_0^2(A_{i-1}\cos\Omega t - 2A_i\cos\Omega t + A_{i+1}\cos\Omega t)$$

$$\Omega^2 A_i + \omega_0^2(A_{i-1} - 2A_i + A_{i+1}) = 0.$$

We have n equations of motion, one for each i from 1 to n. These normal modes look like standing waves. Each mass will oscillate with a sinusoidal form given by $\cos(\Omega t)$ and an amplitude that depends on the normal mode. We take $A_i = A\sin(i\phi)$ as an *Ansatz*. A is an arbitrary amplitude and ϕ is a real number determined by the boundary conditions. For $i = 0$, this form is clearly a solution. For $i = n+1$

this form is a solution if $A_{n+1} = A\sin((n+1)\phi) = 0,$ which is only true for $(n+1)\phi = j\pi,$ where j is an integer, which is true for

$$\phi_j = \frac{j\pi}{n+1}.$$

There are n normal modes so $1 \le j \le n.$ We can write

$$A_i = A\left(\frac{j\pi}{n+1}i\right),\ 1 \le j \le n.$$

Now we put this result into our expression for the normal angular frequencies

$$\Omega^2 A\sin(i\phi) + \omega_0^2\left(A\sin((i-1)\phi) - 2A\sin(i\phi) + A\sin((i+1)\phi)\right) = 0$$

$$\Omega^2 \sin(i\phi) + \omega_0^2\left(\sin(i\phi - \phi) - 2\sin(i\phi) + \sin(i\phi + \phi)\right) = 0.$$

Remembering that

$$\sin(i\phi \pm \phi) = \cos(\phi)\sin(i\phi) \pm \cos(i\phi)\sin(\phi),$$

we can write

$$\Omega^2 \sin(i\phi) + \omega_0^2\left(\cos(\phi)\sin(i\phi) - \cos(i\phi)\sin(\phi) - 2\sin(i\phi) + \cos(\phi)\sin(i\phi) + \cos(i\phi)\sin(\phi)\right) = 0$$

$$\Omega^2 + \omega_0^2\left(\cos(\phi) - \cos(i\phi)\sin(\phi)/\sin(i\phi) - 2 + \cos(\phi) + \cos(i\phi)\sin(\phi)/\sin(i\phi)\right) = 0$$

$$\Omega^2 + \omega_0^2\left(\cos(\phi) - 2 + \cos(\phi)\right) = 0$$

$$\Omega^2 = 2\omega_0^2\left(1 - \cos\phi\right)$$

$$\Omega^2 = 2\omega_0^2\left(2\sin^2\left(\frac{\phi}{2}\right)\right)$$

$$\Omega^2 = 4\omega_0^2\sin^2\left(\frac{\phi}{2}\right)$$

$$\Omega = 2\omega_0\sin\left(\frac{\phi}{2}\right).$$

Putting in out result for ϕ_j we get the angular frequencies for all the normal modes

$$\Omega_j = 2\omega_0\sin\left(\frac{j\pi}{2(n+1)}\right),\ 1 \le j \le n.$$

CALCULATE: This step is not necessary.

ROUND: This step is not necessary.

DOUBLE-CHECK:

Let's double-check our result for $n = 2$ masses. We have two normal modes with angular frequencies given by

$$\Omega_1 = 2\omega_0\sin\left(\frac{\pi}{6}\right) = \omega_0$$

$$\Omega_2 = 2\omega_0\sin\left(\frac{\pi}{3}\right) = \sqrt{3}\omega_0.$$

Ω_1 corresponds to the two masses moving back and forth together while Ω_2 corresponds to the two masses vibrating opposite against each other. This result makes sense because when the two masses are moving together, the middle spring does not contribute and we have two masses with two springs so the angular frequency is just $\Omega_1 = \sqrt{2k/2m} = \omega_0$. For the two masses vibrating against each other, we essentially have three springs for each mass leading to $\Omega_2 = \sqrt{3k/m} = \sqrt{3}\omega_0$. So our result makes sense.

15.31. **THINK:** Determine the equation for a wave traveling in the direction of the positive x-axis with $\lambda = 12.0$ cm, f = 10.0 Hz, A = 10.0 cm and $y(0,0) = 5.00$ cm. Note that during each complete wave oscillation the displacement assumes the same value twice. Since the displacement at (0,0) is given, find two solutions (unless the displacement corresponds to an extremum, which is not the case here). Sketch both cases, 1 and 2, in the next step of the solution.

SKETCH:

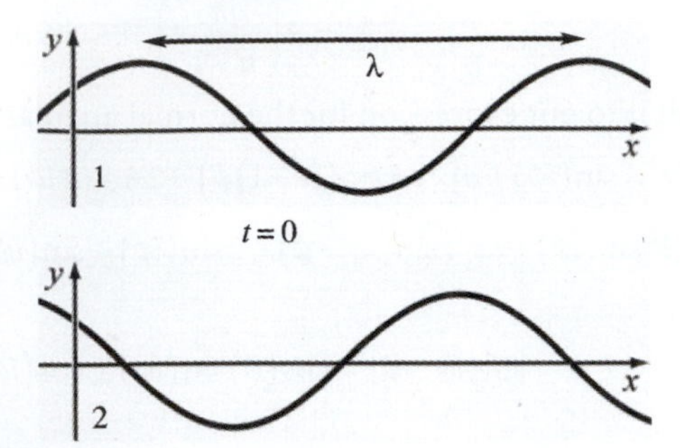

RESEARCH: The wave number and the angular frequency are given by $\kappa = 2\pi / \lambda$ and $\omega = 2\pi f$, respectively. The period is related to the frequency by $T = 1/f$. The speed of the wave is $v = \lambda f$. To determine the phase constant, use the point $y(0,0) = A\sin\phi$. The equation of motion is given by:

$$y(x,t) = A\sin(\kappa x - \omega t + \phi).$$

SIMPLIFY: $\sin\phi = \dfrac{y(0,0)}{A} \Rightarrow \phi = \pm\sin^{-1}\left(\dfrac{y(0,0)}{A}\right)$

CALCULATE:

(a) $\kappa = \dfrac{2\pi}{12.0 \text{ cm}} = \dfrac{2\pi}{0.120 \text{ m}} = 52.36 \text{ m}^{-1}$

(b) $T = \dfrac{1}{10.0 \text{ Hz}} = 0.100 \text{ s}$

(c) $\omega = 2\pi(10.0 \text{ Hz}) = 62.83 \text{ s}^{-1}$

(d) $v = 0.120 \text{ m}(10.0 \text{ Hz}) = 1.20 \text{ m/s}$

(e) $\phi = \pm\sin^{-1}\left(\dfrac{0.0500 \text{ m}}{0.100 \text{ m}}\right) = \pm\dfrac{\pi}{6}$

(f) $y(x,t) = (10.0 \text{ cm})\sin\left(52.36 \text{ m}^{-1}x - 62.83 \text{ s}^{-1}t \pm \dfrac{\pi}{6}\right)$

ROUND:

(a) The wave number is $\kappa = 52.4 \text{ m}^{-1}$.

(b) The period is $T = 0.100$ s.

(c) The angular frequency is $\omega = 62.8 \text{ s}^{-1}$.

(d) The velocity is v = 1.20 m/s.

(e) The phase is $\phi = \pm\pi / 6$.

(f) It is better not to round the coefficients of the equation at this stage, and only round once particular values of x and t are substituted.

DOUBLE-CHECK: Each of the calculated values have the proper SI units.

15.35. **THINK:** Determine the time it takes for sound waves to reach Alice via two different media and compare these values. The first medium is air, in which sound travels at a speed of about 343 m/s. The second medium is a string with a tension of 25.0 N and a linear density of 6.13 g/m or 0.00613 kg/m. The distance between Alice and Bob is 20.0 m.

SKETCH:

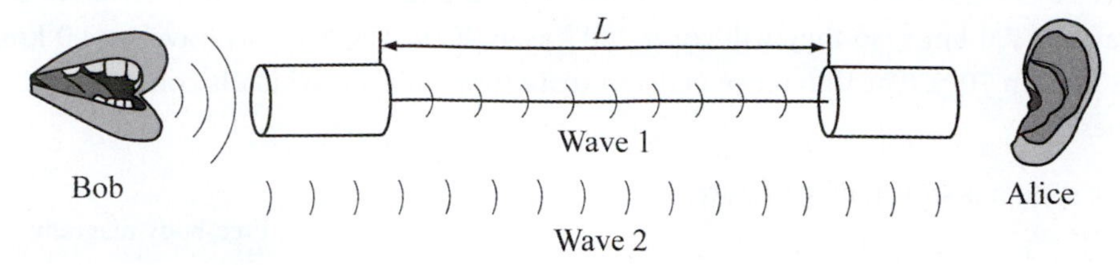

RESEARCH: The time it takes for the sound to travel a distance, *d*, is given by $t = d / v_{\text{a}}$. The velocity of sound in a wire is given by $v_{\text{w}} = \sqrt{T / \mu}$.

SIMPLIFY: The time it takes to travel through air is $t_{\text{a}} = d / v_{\text{a}}$. The time it takes sound to travel through the wire is $t_{\text{w}} = d / v_{\text{w}} = d / \sqrt{T / \mu} = d\sqrt{\mu / T}$. The time difference is given by:

$$\Delta t = t_{\text{a}} - t_{\text{w}} = \frac{d}{v_{\text{a}}} - d\sqrt{\frac{\mu}{T}} = d\left(\frac{1}{v_{\text{a}}} - \sqrt{\frac{\mu}{T}}\right).$$

CALCULATE: $\Delta t = 20.0 \text{ m}\left(\frac{1}{343 \text{ m/s}} - \sqrt{\frac{6.13 \cdot 10^{-3} \text{ kg/m}}{25.0 \text{ N}}}\right) = -0.254868 \text{ s}$, indicating that the sound in air reaches Alice 0.245868 seconds before the sound in the wire.

ROUND: Rounding the result to three significant figures, the sound in the air reaches Alice 0.255 seconds before the sound from the wire does.

DOUBLE-CHECK: $t_{\text{a}} = \frac{20.0 \text{ m}}{343 \text{ m/s}} = 0.0583090 \text{ s}$, and $t_{\text{w}} = 20.0 \text{ m}\sqrt{\frac{6.13 \cdot 10^{-3} \text{ kg/m}}{25.0 \text{ N}}} = 0.313177 \text{ s}$. The difference is –0.255 seconds, confirming the original result.

15.39. **THINK:** The speeds of S waves and P waves of an earthquake are $v_{\text{S}} = 4.0 \text{ km/s}$, and $v_{\text{P}} = 7.0 \text{ km/s}$, respectively. The distance of the dog from the location of the earthquake's epicenter must be determined. The time difference is $\Delta t = 30.0 \text{ s}$.

SKETCH:

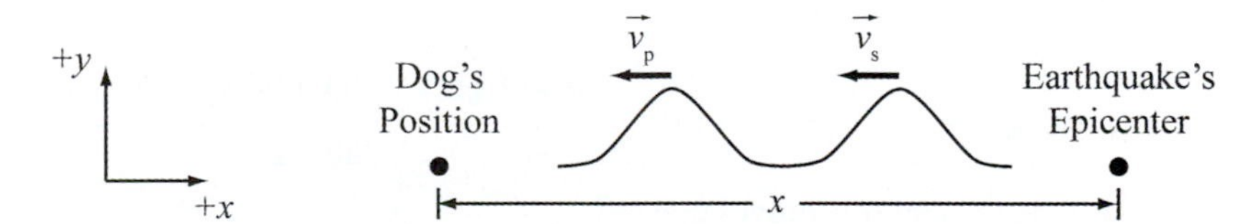

RESEARCH: The time taken for the P waves to reach the dog's position is $t_{\text{P}} = x / v_{\text{P}}$. Similarly, for S waves, the time is $t_{\text{S}} = x / v_{\text{S}}$. Therefore, the interval of time between S waves and P waves at the dog's position is:

$$\Delta t = \frac{x}{v_{\text{S}}} - \frac{x}{v_{\text{P}}}.$$

SIMPLIFY: After manipulation, the distance from the epicenter is $x = \frac{v_{\text{S}} v_{\text{P}} \Delta t}{v_{\text{P}} - v_{\text{S}}}$.

CALCULATE: Substituting the values:

$$x = \frac{(4.0 \text{ km/s})(7.0 \text{ km/s})(30.0 \text{ s})}{7.0 \text{ km/s} - 4.0 \text{ km/s}} = 280. \text{ km}.$$

ROUND: The values in the question were given to two significant figures, so the final answer should be rounded so it also has two significant figures. $x = 280$ km.

DOUBLE-CHECK: Calculate the actual times for the different kinds of waves to arrive. The P waves travel at 7.0 km/s, so they will cover 280 km in 40. s. The S waves travel at 4.0 km/s, so they will cover 280 km in 70. s. The difference between these times is 30. s, which is consistent with what is given in the question.

15.43. The system is shown in the figure.

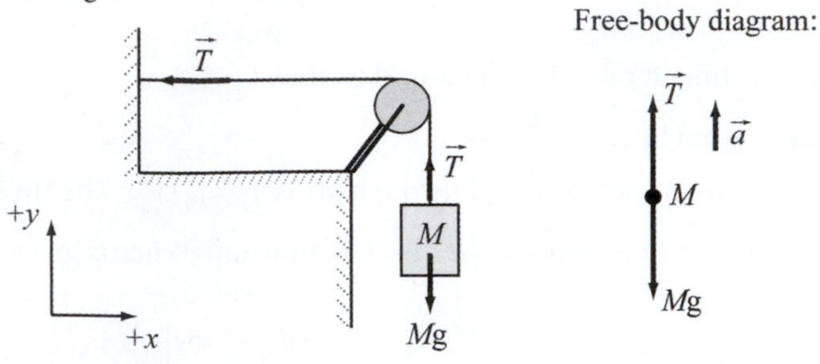

It is given that the mass of a string is $m_S = 5.00\text{ g} = 5.00\cdot 10^{-3}$ kg and its length is L = 70.0 cm = 0.700 m. The mass of the weight is M = 250. kg. Using Newton's second law, the tension is given by $T - Mg = Ma \Rightarrow T = M(a+g)$. The fundamental frequency is given by:

$$f_1 = \frac{v}{2L} = \frac{\sqrt{T/\mu}}{2L}.$$

Substituting $\mu = m_S / L$ and $T = M(a+g)$: $f = \frac{\sqrt{T/(m_s/L)}}{2L} = \frac{1}{2}\sqrt{\frac{T}{m_s L}} = \frac{1}{2}\sqrt{\frac{M(a+g)}{m_s L}}$.

(a) Substituting the values of the variables and a = 0 yields:

$$f = \frac{1}{2}\sqrt{\frac{250.\text{ kg}\left(0+\left(9.81\text{ m/s}^2\right)\right)}{\left(5.00\cdot 10^{-3}\text{ kg}\right)\left(0.700\text{ m}\right)}} = 418.5\text{ Hz} \approx 419\text{ Hz}.$$

(b) It was found previously that $f = \frac{1}{2}\sqrt{\frac{M(a+g)}{m_s L}}$. After rearrangement:

$$a = -g + \frac{4f^2 m_S L}{M} = -9.81\text{ m/s}^2 + \frac{4\left(440.\text{ Hz}\right)^2\left(5.00\cdot 10^{-3}\text{ kg}\right)0.700\text{ m}}{250.\text{ kg}} = 1.03\text{ m/s}^2.$$

Since a is positive, the elevator moves upward.

15.47. **THINK:** It is known that a string has a length of 3.00 m and a mass of 6.00 g. Both ends of the string are fixed. A standing wave on the string has a frequency, f = 300. Hz and three anti-nodes. The tension in the string is to be determined.

SKETCH:

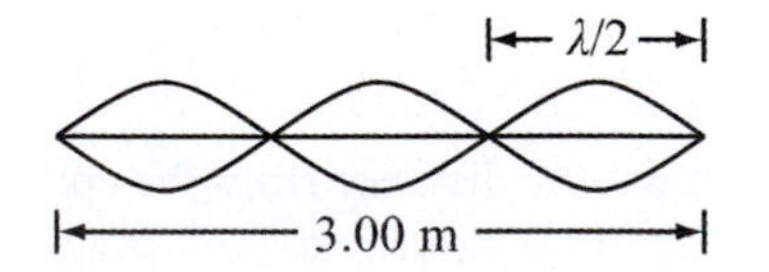

RESEARCH: The speed of a pulse on the string is given by $v = \sqrt{T/\mu}$ or $v = \omega/\kappa$.

SIMPLIFY: Combining these equations gives: $\sqrt{\frac{T}{\mu}} = \frac{\omega}{\kappa} \Rightarrow \frac{T}{\mu} = \left(\frac{\omega}{\kappa}\right)^2$. Using $\mu = m/L$, the tension is:

$T = \left(\frac{m}{L}\right)\left(\frac{\omega}{\kappa}\right)^2$. Substituting $\omega/\kappa = \lambda f$, the tension becomes: $T = \left(\frac{m}{L}\right)(\lambda f)^2$. Because there are three anti-nodes, the wavelength of the standing wave is $\lambda = 2L/3$. Therefore,

$$T=\left(\frac{m}{L}\right)\left(\frac{2}{3}Lf\right)^2=\frac{4}{9}mLf^2.$$

CALCULATE: Substituting $m=6.00\cdot 10^{-3}$ kg, $L=3.00$ m and $f=300.$ Hz yields:

$$T=\left(\frac{4}{9}\right)\left(6.00\cdot 10^{-3}\text{ kg}\right)\left(3.00\text{ m}\right)\left(300.\text{ Hz}\right)^2=720.\text{ N}.$$

ROUND: Rounding to three significant figures, $T=720.$ N

DOUBLE-CHECK: The computed value has correct units for a force, and 720. N is a reasonable tension for a system such as the one given in the question.

15.51. **THINK:** This problem is a superposition of three wave sources with three different frequencies, $\omega_1=2.00$ Hz, $\omega_2=3.00$ Hz and $\omega_3=4.00$ Hz. The speed of each wave is 5.00 m/s. The amplitudes of the waves are the same. The wave sources are located at the edges of a circular pool, as shown below. The displacement of a ball in the center of the pool must be plotted as a function of time.

SKETCH:

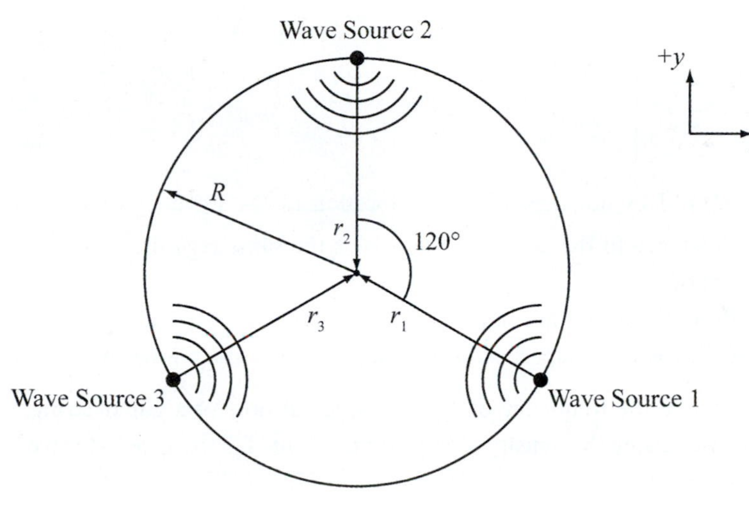

RESEARCH: From the principle of superposition, the displacement of the ball is given by the sum of all the displacements due to the wave sources, that is,

$$z(t)=z_1(t)+z_2(t)+z_3(t)=\frac{C}{\sqrt{r_1}}\sin\left(\kappa_1 r_1-\omega_1 t+\phi_1\right)+\frac{C}{\sqrt{r_2}}\sin\left(\kappa_2 r_2-\omega_2 t+\phi_2\right)+\frac{C}{\sqrt{r_3}}\sin\left(\kappa_3 r_3-\omega_3 t+\phi_3\right).$$

SIMPLIFY: Since $\phi_1=\phi_2=\phi_3=0$ and $r_1=r_2=r_3=R$, the displacement, $z(t)$ is:

$$z(t)=\frac{C}{\sqrt{R}}\left[\sin\left(\kappa_1 R-\omega_1 t\right)+\sin\left(\kappa_2 R-\omega_2 t\right)+\sin\left(\kappa_3 R-\omega_3 t\right)\right].$$

Note that when $t<R/v$, the displacement, $z(t)$, is zero, since the waves have not reached the center of the pool. Using the speeds of the waves and the frequencies, the angular frequencies are $\omega_1=2\pi f_1$, $\omega_2=2\pi f_2$ and $\omega_3=2\pi f_3$, and the wave numbers are $\kappa_1=\omega_1/v$, $\kappa_2=\omega_2/v$ and $\kappa_3=\omega_3/v$. Assuming the amplitudes of the waves are all $C/\sqrt{R}=1$ m, the displacement is given by:

$$z(t)=\left[\sin\left(\frac{2\pi f_1}{v}(R-vt)\right)+\sin\left(\frac{2\pi f_2}{v}(R-vt)\right)+\sin\left(\frac{2\pi f_3}{v}(R-vt)\right)\right],$$

if $R-vt>0$, and $z(t)=0$ if $R-vt<0$. If a new time variable, $t_c=t-R/v$, is used, the above equation simplifies to: $z(t_c)=\sin(2\pi f_1 t_c)+\sin(2\pi f_2 t_c)+\sin(2\pi f_3 t_c)$, if $t_c>0$ and $z(t_c)=0$, if $t_c<0$.

CALCULATE: $z(t_c) = \sin\left(2\pi\left(2.00\text{ s}^{-1}\right)t_c\right) + \sin\left(2\pi\left(3.00\text{ s}^{-1}\right)t_c\right) + \sin\left(2\pi\left(4.00\text{ s}^{-1}\right)t_c\right)$

$$= \sin\left(4\pi\text{ s}^{-1}t_c\right) + \sin\left(6\pi\text{ s}^{-1}t_c\right) + \sin\left(8\pi\text{ s}^{-1}t_c\right)$$

The time for the waves to reach the center of the pool is: $t_0 = \frac{R}{v} = \frac{5.00\text{ m}}{5.00\text{ m/s}} = 1.00\text{ s}$. The plot of the displacement, $z(t)$, is given next.

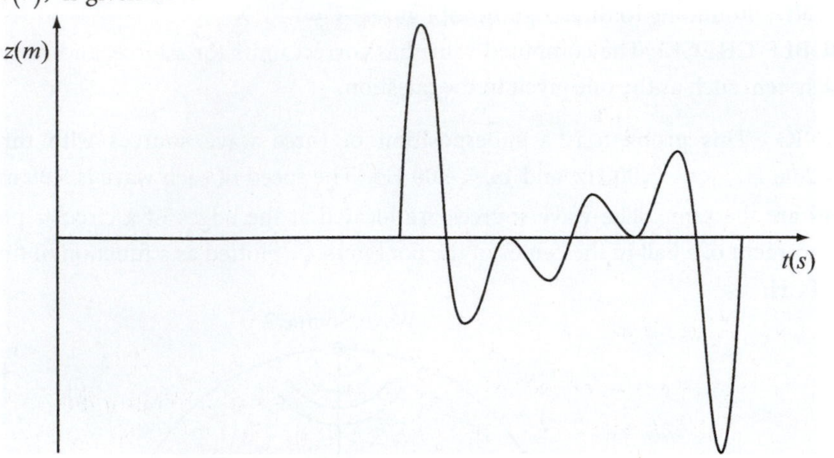

Note that $z(t)$ does not depend on the location of the wave sources at the edges of the pool. This is because the distance to the center of the pool is the same regardless of the location of the sources at the edges of the pool.

ROUND: Rounding is not necessary.

DOUBLE-CHECK: The superposition of three waves is also a wave, which is consistent with the plot.

15.55. (a) To determine the frequency of the fundamental note of a guitar string, the speed of a wave on the string is needed. Using the density of the string, $\mu = m / L_s$, the speed of wave is:

$$v = \sqrt{\frac{T}{\mu}} = \sqrt{\frac{TL_s}{m}}.$$

The fundamental frequency is then: $f = \frac{v}{2L} = \sqrt{\frac{TL_s}{m}}\left(\frac{1}{2L}\right)$. Note that L is the distance between two fixed ends of the guitar string. Substituting m = 10.0 g = 0.0100 kg, L = 0.650 m, $L_s = 1.00$ m and T = 81.0 N into the above equation yields:

$$f = \sqrt{\frac{81.0\text{ N}(1.00\text{ m})}{0.0100\text{ kg}}}\left(\frac{1}{2(0.650\text{ m})}\right) = 69.2\text{ Hz}.$$

(b) Replacing the mass with m = 16.0 g = 0.0160 kg, the frequency becomes:

$$f = \sqrt{\frac{81.0\text{ N}(1.00\text{ m})}{0.0160\text{ kg}}}\left(\frac{1}{2(0.650\text{ m})}\right) = 54.7\text{ Hz}.$$

15.59. The wave speed on a brass wire is given by $v=\sqrt{T/\mu}$. The linear density, μ, is equal to $\mu=\rho A$, where $A=\pi r^2$ is the cross-sectional area of the wire. Therefore, the speed of the wave is:

$$v=\sqrt{\frac{T}{\rho A}}=\sqrt{\frac{T}{\rho\pi r^2}}.$$

Inserting $r=0.500\text{ mm}=0.500\cdot 10^{-3}\text{ m}$, $T=125\text{ N}$ and $\rho=8.60\cdot 10^3\text{ kg/m}^3$ yields:

$$v=\sqrt{\frac{125\text{ N}}{\left(8.60\cdot 10^3\text{ kg/m}^3\right)\pi\left(0.500\cdot 10^{-3}\text{ m}\right)^2}}=136.04\text{ m/s}=136\text{ m/s}.$$

15.63. **THINK:** Two strings are connected and have the same tension, T. The linear mass density of string 2 is $\mu_2=3\mu_1$. If the speed, the frequency and the wavelength of a wave on string 1 are v_1, f_1 and λ_1, respectively. The corresponding variables for string 2, v_2, f_2 and λ_2, can be determined in terms of string 1's variables.

SKETCH:

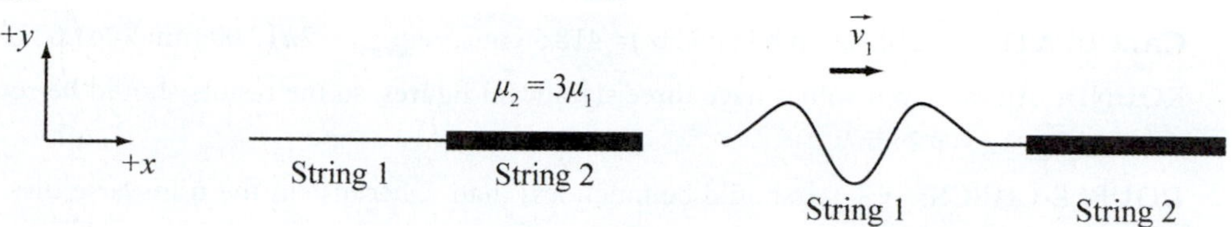

RESEARCH: It is known that when a wave travels to a different material or medium, the frequency of the wave does not change. This means that the frequency of string 2, f_2, is equal to the frequency of string 1, that is, $f_2=f_1$. The speeds of the wave on string 1 and string 2 are $v_1=\sqrt{T/\mu_1}$ and $v_2=\sqrt{T/\mu_2}$.

SIMPLIFY: Since v_1 and μ_1 are known, the tension is given by $T=\mu_1 v_1^2$. Substituting this expression into v_2, the speed of the wave on string 2 becomes:

$$v_2=\sqrt{\frac{\mu_1 v_1^2}{\mu_2}}=v_1\sqrt{\frac{\mu_1}{\mu_2}}.$$

Using $\mu_2=3\mu_1$, the above equation simplifies to $v_2=v_1/\sqrt{3}$. The wavelength of the wave on string 2 is determined using the relation $v_2=\lambda_2 f_2$. Therefore, $\lambda_2=v_2/f_2$. Substituting $f_2=f_1$ and $v_2=v_1/\sqrt{3}$, gives $\lambda_2=v_1/\sqrt{3}f_1$. Since $v_1/f_1=\lambda_1$, this becomes $\lambda_2=\lambda_1/\sqrt{3}$. Therefore, $f_2=f_1$, $v_2=v_1/\sqrt{3}$ and $\lambda_2=\lambda_1/\sqrt{3}$.

CALCULATE: This step is not necessary.

ROUND: This step is not necessary.

DOUBLE-CHECK: It is expected that as the speed of a wave decreases, the wavelength of the wave also decreases.

15.67. **THINK:** The guitar string has length, $l=0.800$ m, and it is oscillating at its fundamental frequency, which means that it has one antinode in the middle, and thus the guitar string length is half of the wavelength (see sketch). The wave speed, v, can be determined from knowing the wavelength and the frequency, which is also given (261.6 Hz). But to find the maximum speed of the midpoint of the string, v_{max}, we have to take the derivative with respect to time.

SKETCH:

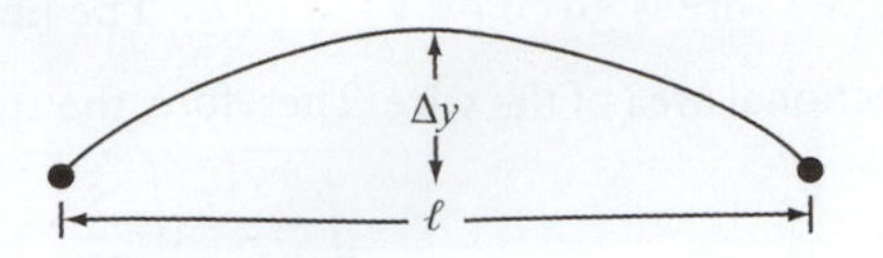

RESEARCH: The wave speed is given by $v = \lambda f$. For $n = 1$, $\lambda_1 = 2l$. To determine v_{max}, consider the standing wave equation $y(x,t) = 2A\sin(\kappa x)\cos(\omega t)$. For the midpoint $\sin(\kappa x) = 1$, and thus the midpoint oscillates in times according to $y_{\text{mid}}(t) = 2A\cos(\omega t)$. Taking the derivative with respect to time, we find $v_{\text{mid}}(t) = -\omega 2A\sin(\omega t)$.

SIMPLIFY: The velocity of the midpoint reaches its maximum value when the sine has a value of -1. We also use $\omega = 2\pi f$. And finally we use the fact that the initial displacement of the midpoint from equilibrium $\Delta y_0 = 2.00 \text{ mm}$ was specified, which means that $2A = \Delta y_0$.

$$v = \lambda_1 f_1 = 2l f_1$$

$$v_{\text{max,mid}} = 2A\omega = \Delta y_0 2\pi f_1$$

CALCULATE: $v = 2(0.800 \text{ m})(261.6 \text{ Hz}) = 418.56 \text{ m/s}$, $v_{\text{max,mid}} = 2\pi(2.00 \text{ mm})(261.6 \text{ Hz}) = 3.287 \text{ m/s}$

ROUND: All the given values have three significant figures, so the results should be rounded to v = 419. m/s and $v_{\text{max,mid}} = 3.29$ m/s.

DOUBLE-CHECK: $v_{\text{max,mid}}$ should be much less than v, because in the transverse direction the string is moving at this speed, but in the wave direction, no part of the string actually moves at this speed.

15.69. The known values for the transverse harmonic wave are $\lambda = 0.200$ m, f = 500. Hz, and A = 0.0300 m. It travels in the $+\hat{z}$ direction and its oscillations occur in the xz-plane. At t = 0 s, $(x_0, z_0) = (A, 0)$.

(a)

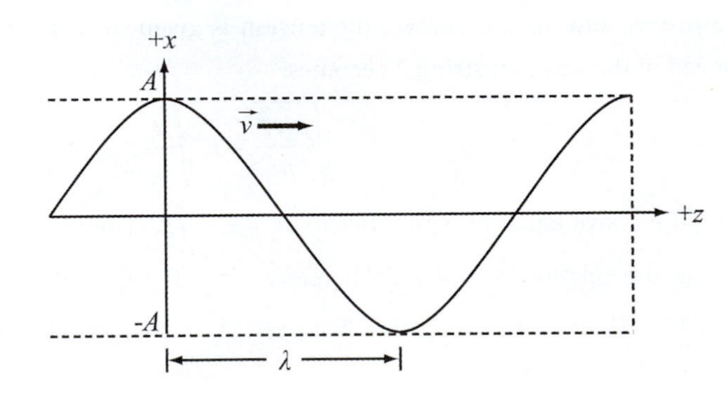

(b) $v = \lambda f = (0.200 \text{ m})(500. \text{ Hz}) = 100. \text{ m/s}$

(c) $\kappa = \dfrac{2\pi}{\lambda} = \dfrac{2\pi}{0.200 \text{ m}} = 10.0\pi \text{ rad/m} \approx 31.4 \text{ rad/m}$

(d) $\mu = 30.0 \text{ g/cm} = 0.0300 \text{ kg/m}$

From $v = \sqrt{T/\mu}$, $T = v^2\mu = (100. \text{ m/s})^2(0.0300 \text{ kg/m}) = 300. \text{ N}$.

(e) For a traveling wave, in general, $y(x,t) = A\sin(\kappa x - \omega t + \phi_0)$. By inspection, $\phi_0 = \pi/2$. Then:

$$x = D(z,t) = (0.0300 \text{ m})\sin\left((10.0\pi \text{ rad/m})x - 2\pi \text{ rad}(500. \text{ Hz})t + \frac{\pi}{2} \text{ rad}\right)$$

$$= (0.0300 \text{ m})\cos((10.0\pi \text{ rad/m})x - (1000.\pi \text{ rad/s})t).$$

Multi-Version Exercises

15.71. **THINK:** The mass and total length of the rubber band can be used to find its mass density. The mass density, tension, and the length of the standing wave can be used to find the lowest-frequency (fundamental frequency) vibration.

SKETCH: Imagine that the standing wave is on the front portion of the rubber band.

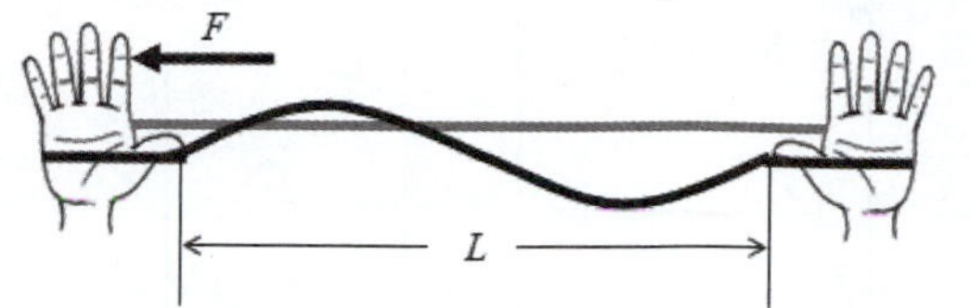

RESEARCH: Since this is a transverse wave on a rubber band, the velocity can be computed from the tension F and linear mass density μ using the equation $v=\sqrt{F/\mu}$. The linear mass density $\mu=m/l$ is computed from the total length and total mass of the rubber band. The resonance frequency of the fundamental frequency is given by $f_1=\frac{v}{2L}$, where L is the length of the vibrating portion of the rubber band.

SIMPLIFY: Since the velocity $v=\sqrt{F/\mu}$, the fundamental frequency is given by $f_1=\frac{v}{2L}=\frac{\sqrt{F/\mu}}{2L}=\frac{\sqrt{F}}{2L\sqrt{\mu}}$. Finally, express the mass density in terms of the total length and total mass of the rubber band to get $f_1=\frac{\sqrt{F}}{2L\sqrt{m/l}}$.

CALCULATE: According to the problem statement, the tension on each side of the rubber band is $F = 1.777$ N, the total length of the rubber band is 20.27 cm = 0.2027 m, the length of the vibrating portion of the rubber band is 8.725 cm = 0.08725 m, and the mass of the rubber band is 0.3491 g = $3.491\cdot10^{-4}$ kg. The fundamental frequency is

$$\begin{aligned} f_1 &= \frac{\sqrt{F}}{2L\sqrt{m/l}} \\ &= \frac{\sqrt{1.777\text{ N}}}{2\cdot0.08725\text{ m}\sqrt{3.491\cdot10^{-4}\text{ kg}/0.2027\text{ m}}} \\ &= 184.0772987\text{ Hz} \end{aligned}$$

ROUND: All of the numbers in this problem have four significant figures and the final answer will also have four figures. The lowest frequency of a vibration on this part of the rubber band is 184.1 Hz.

DOUBLE-CHECK: It is possible to stretch a rubber band with your hands so a low frequency sound is produced when it is plucked. A frequency of 184.1 Hz corresponds approximately to the F-sharp below middle C on a piano. It is possible to reproduce this with a rubber band at home, confirming that the answer is reasonable.

15.74. **THINK:** Assuming that Sun's power is emitted uniformly from every point on the spherical surface, the power per unit area can be computed for a given orbital radius. It is then possible to compute the power intercepted by the solar panel using the area and efficiency of the solar panel.

SKETCH: The sun and satellite are not shown to scale. Only the solar panel portion of the satellite is shown.

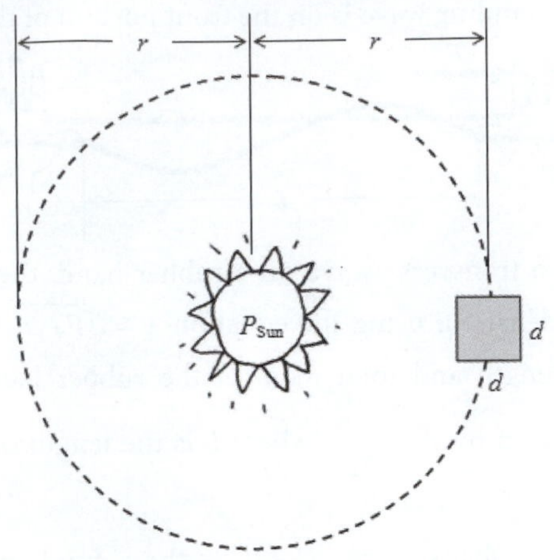

RESEARCH: The area of the solar panel is $A = d^2$. The surface area of a sphere or radius r is $4\pi r^2$. The power of the sun P_{Sun} is distributed evenly, so the power intercepted by the solar panel is $P = \frac{P_{Sun}}{4\pi r^2} \cdot A$. The solar panel is not 100% efficient, so the power delivered is $P_{del} = \varepsilon P$, where ε is the efficiency of the solar panel.

SIMPLIFY: First, use $A = d^2$ to find the power of intercepted by the solar panel, $P = \frac{P_{Sun}}{4\pi r^2} \cdot A = \frac{d^2 P_{Sun}}{4\pi r^2}$.

The power delivered is then $P_{del} = \varepsilon P = \varepsilon \frac{d^2 P_{Sun}}{4\pi r^2}$.

CALCULATE: According to the question, the efficiency of the solar panel is ε = 16.57% = 0.1657, the radius of the satellite's orbit is $r = 4.949 \cdot 10^7$ km = $4.949 \cdot 10^{10}$ m, and the total power output of the Sun is $3.937 \cdot 10^{26}$ W. Since the edges of the square solar panel are each d = 1.459 m long, the power provided by the solar panel is

$$\begin{aligned} P_{del} &= \varepsilon \frac{d^2 P_{Sun}}{4\pi r^2} \\ &= 0.1657 \frac{(1.459 \text{ m})^2 \cdot 3.937 \cdot 10^{26} \text{ W}}{4\pi \left(4.949 \cdot 10^{10} \text{ m}\right)^2} \\ &= 4511.840469 \text{ W} \\ &= 4.511840469 \text{ kW} \end{aligned}$$

ROUND: The numbers in the problem all have four significant figures, so the final answer should also have four figures. The total power provided to the satellite by the solar panel is 4512 W.

DOUBLE-CHECK: The sunlight hitting the solar panel is about $\frac{d^2}{4\pi r^2} = 6.916 \cdot 10^{-23}$ of the total sunlight. The power of the solar panel should have an order of magnitude about 10^{-23} times the power output of the sun. $10^{26} \cdot 10^{-23} = 10^3$, so the final answer ($4.512 \cdot 10^3$) is indeed of an order about 10^{-23} times the power output of the sun ($3.937 \cdot 10^{26}$ W), confirming that the final answer is reasonable.

15.77. **THINK:** The mass density of the string and tension on the string are given in the problem. To find the frequency, it is necessary to determine at which harmonic the string is oscillating, which can be deduced from the image.

SKETCH: Use the image from the text for your sketch, labeling the wavelength and each spot where the amplitude of the wave is maximal:

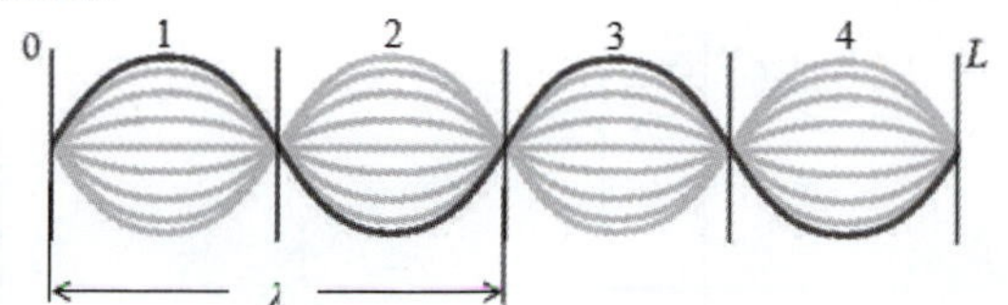

RESEARCH: Looking at the picture, there are four places where the amplitude is maximal, so the string is oscillating in its fourth harmonic. This means that the frequency of the oscillation is given by the equation $f_4 = 4\dfrac{\sqrt{F}}{2L\sqrt{\mu}}$. Since the mass density and length of the string are given in grams and centimeters while the tension is given in Newtons, the conversion factors of 100 cm = 1 m and 1000 g = 1 kg will be needed.

SIMPLIFY: Using algebra, rewrite the equation $f_4 = 4\dfrac{\sqrt{F}}{2L\sqrt{\mu}}$ as $f_4 = \dfrac{2}{L}\sqrt{\dfrac{F}{\mu}}$.

CALCULATE: The problem statement includes the information that the total length of the string L = 116.7 cm = 1.167 m, the string's mass density is 0.2833 g/cm = 0.02833 kg/m, and the tension on the string is 18.25 N. The frequency is thus

$$f_4 = \frac{2}{L}\sqrt{\frac{F}{\mu}}$$
$$= \frac{2}{1.167 \text{ m}}\sqrt{\frac{18.25 \text{ N}}{2.833 \cdot 10^{-2} \text{ kg/m}}}$$
$$= 43.49779947 \text{ Hz}.$$

ROUND: The values in the question all have four significant figures. The harmonic number (an integer) is considered to have infinite precision, so the final answer should also have four figures. The string is vibrating at 43.50 Hz.

DOUBLE-CHECK: The velocity of the wave is given by $v = \sqrt{F/\mu} = 25.38$ m/s. Since $v = f\lambda$, this means that, if the frequency is 43.50 Hz and the velocity is 25.38 m/s, the wavelength is 58.35 cm. This agrees with the observation that the wavelength is half of the total length of the string ($L/2$ = 58.35 cm), confirming that the calculated frequency was correct.

Chapter 16: Sound

Exercises

16.25. Assume $v_{\text{sound}} = 343$ m/s.

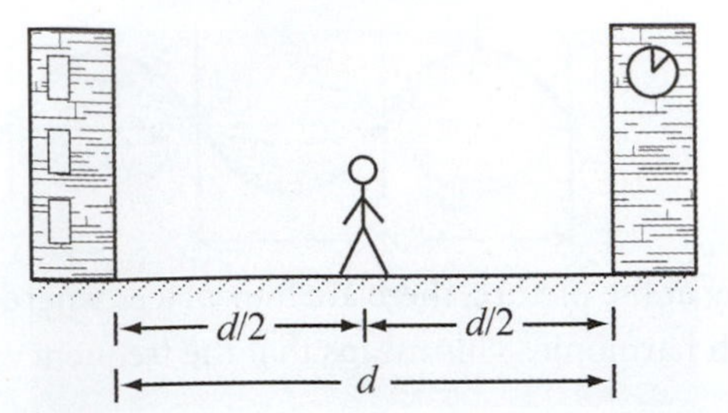

In $t = 0.500$ s, the sound travels a distance, d. It travels from your position, where you first hear it, to the tall building and back to your position. Note the distance between the clock tower and the building is also d. Now, solving gives $d = (343 \text{ m/s})(0.500 \text{ s}) = 172$ m.

16.29. Wave speed is given by $v = \sqrt{B/\rho}$. Solving for the elastic modulus gives:

$$B = v^2\rho = \left(2.0 \cdot 10^8 \text{ m/s}\right)^2 \left(2500 \text{ kg/m}^3\right) = 1.0 \cdot 10^{20} \text{ N/m}^2.$$

This value is some nine orders of magnitude larger than the actual value. Indeed, light waves are electromagnetic oscillations that do not require the motion of glass molecules, or the hypothetical ether for transmission.

16.33. **THINK:** The question asks for the sound intensity, I_2, measured by a detector $r_2 = 4.00$ m away from a source, when the sound intensity at $r_1 = 3.00$ m is $I_1 = 1.10 \cdot 10^{-7}$ W/m^2.

SKETCH:

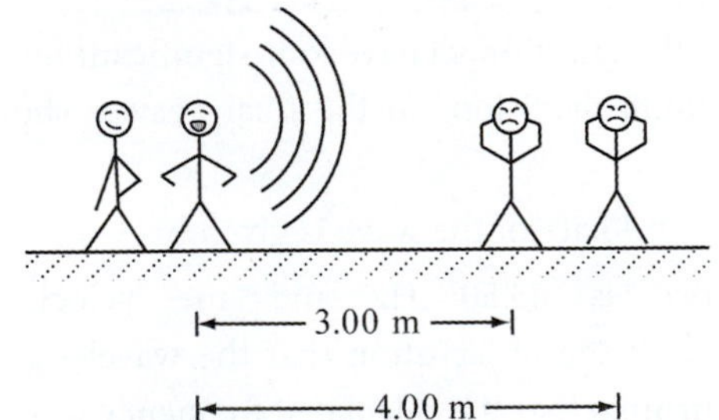

RESEARCH: Recall that the intensity is defined as $I = \text{Power/Area}$. The power emitted by the source can be determined from the intensity at the 3.00 m detector, and then this can be used to determine the intensity at 4.00 m.

SIMPLIFY: $I = \dfrac{\text{Power}}{\text{Area}} = \dfrac{P}{4\pi r^2} \Rightarrow P = I\left(4\pi r^2\right)$. Therefore, $I_2 = \dfrac{I_1 4\pi (r_1)^2}{4\pi (r_2)^2} = \dfrac{I_1 r_1^2}{r_2^2}$.

CALCULATE: $I_{4\text{ m}} = \dfrac{\left(1.10 \cdot 10^{-7} \text{ W/m}^2\right)(3.00 \text{ m})^2}{(4.00 \text{ m})^2} = 6.188 \cdot 10^{-8} \text{ W/m}^2$

ROUND: Since the given values have three significant figures, the result should be rounded to $I_{4\text{ m}} = 6.19 \cdot 10^{-8}$ W/m^2.

DOUBLE-CHECK: It is reasonable that the intensity will decrease as the distance increases.

16.37. The wavelength of the sound is $\lambda = v/f = (343 \text{ m/s})/(490. \text{ Hz}) = 0.700 \text{ m}$. Since the speakers are in phase and are facing each other, their interference will yield a standing wave with an anti-node at the center between them. If she sits a half wavelength away from the center, then she will be at another anti-node. Therefore, the minimum distance away from the center that she can move on the straight line connecting the two speakers and again hear the loudest sound is: $d = \lambda/2 = 0.350 \text{ m}$.

16.41. **THINK:** The question asks for the intensity of the sound wave at the point P_1 in the sketch. The frequency of the sound wave is f = 10,000.0 Hz, and the coordinates at point P_1 are $x_1 = 4.50 \text{ m}$ and $y_1 = 0 \text{ m}$. The distance between the speakers is D = 3.60 m, and the power delivered by the speakers is P = 100.0 W. The question next asks for the sound level due to speaker A at point P_1. Lastly, the question asks for the distance to the first maximum (constructive interference) from the center maximum.

SKETCH:

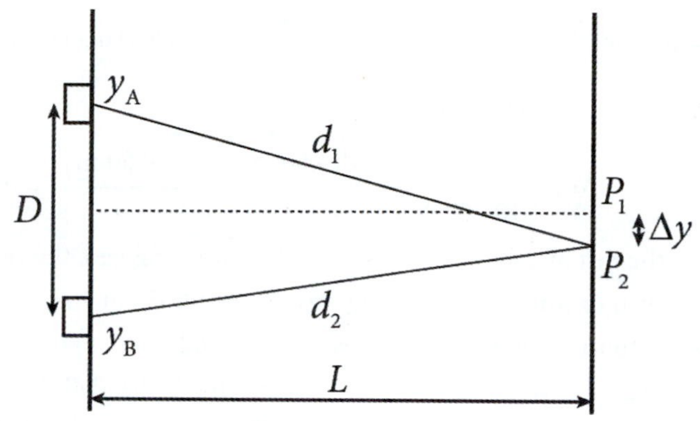

RESEARCH: The definition of intensity is $I = \text{Power}/\text{Area}$. Recall the equation for intensity level is $\beta = 10\log(I/I_0)$. With both speakers on, as one moves toward point P_2, the distances that the sound must travel from speakers A and B change. When the path difference is half a wavelength, the interference is destructive, but when the path difference increases to a full wavelength, the interference is constructive again. The distance from speaker A to P_2 is $d_1 = \sqrt{(D/2 + \Delta y)^2 + L^2}$. Similarly, the distance from speaker B to P_2 is $d_2 = \sqrt{(D/2 - \Delta y)^2 + L^2}$. The first constructive interference peak occurs at P_2 when $d_1 - d_2 = (1)\lambda$. From this, Δy can be determined.

SIMPLIFY: $I_1 = \dfrac{\text{Power}}{\text{Area}} = \dfrac{P}{4\pi r^2}$, $\beta = 10(\log I - \log I_0)$. In order to simplify the calculation of $d_1 - d_2$, let $a = D^2/4 + \Delta y^2 + L^2$ and $b = D\Delta y$. Rewrite the distances as:

$$d_1 = \sqrt{(D/2 + \Delta y)^2 + L^2} = \sqrt{D^2/4 + 2D\Delta y/2 + \Delta y^2 + L^2} = \sqrt{D^2/4 + D\Delta y + \Delta y^2 + L^2}$$
$$d_1 = \sqrt{(D/2 - \Delta y)^2 + L^2} = \sqrt{D^2/4 - 2D\Delta y/2 + \Delta y^2 + L^2} = \sqrt{D^2/4 - D\Delta y + \Delta y^2 + L^2}.$$

Assume that $\Delta y << D$ and $\Delta y << L$, then $a = D^2/4 + \Delta y^2 + L^2 \approx D^2/4 + L^2$. The value of a is much larger than the value of b, so we can apply the approximation given in the statement of the problem

$$\sqrt{a \pm b} \approx \sqrt{a} \pm \frac{b}{2\sqrt{a}}.$$

We can write $d_1 - d_2$ as

$$d_1 - d_2 = \sqrt{a+b} - \sqrt{a-b} = \frac{2b}{2\sqrt{a}} = \frac{D\Delta y}{\sqrt{D^2/4 + L^2}}.$$

Since $d_1 - d_2 = \lambda = v_{\text{sound}}/f$, substituting the equation from the previous line gives:

$$\frac{v_{\text{sound}}}{f} \approx \frac{D\Delta y}{\sqrt{D^2/4+L^2}},$$

which means, solving for Δy,

$$\Delta y \approx \left(\frac{v_{\text{sound}}}{f}\right)\frac{\sqrt{D^2/4+L^2}}{D}.$$

CALCULATE: (a) The intensity at point P_1 is given by:

$$I_1 = \frac{100.0 \text{ W}}{4\pi\left[\left(\frac{3.60 \text{ m}}{2}\right)^2+(4.50 \text{ m})^2\right]} = 0.33877 \text{ W/m}^2.$$

(b) The sound level at point P_1

$$\beta_1 = 10\left(\log\left(0.33877\frac{\text{W}}{\text{m}^2}\right)-\log\left(10^{-12}\frac{\text{W}}{\text{m}^2}\right)\right) = 10\left(\log(0.33877)+12\right) \text{ dB} = 115.299 \text{ dB}.$$

(c) The distance to the first maximum is

$$\Delta y = \left(\frac{343 \text{ m/s}}{10{,}000.0 \text{ Hz}}\right)\frac{\sqrt{(3.60 \text{ m})^2/4+(4.50 \text{ m})^2}}{3.60 \text{ m}} = 0.046178 \text{ m}.$$

ROUND: Since the distances are given to three significant figures, the result is $I_1 = 0.339 \text{ W/m}^2$. Because the intensity is given to three significant figures, the result should be rounded to $\beta_1 = 115 \text{ dB}$. Since there are values given with three significant figures, $\Delta y = 0.0462 \text{ m}$.

DOUBLE-CHECK: The result is obtained directly from the definition of intensity and the units are correct. This sound level is high for a loud speaker but is possible. The value for Δy has appropriate units for a distance, and is consistent with $\Delta y << D$.

16.45. THINK:

(a) The question asks for the speed of a police car as it passes, if the frequency of the siren before it passes is $f_1 = 1300.$ Hz and the frequency after it passes is $f_2 = 1280.$ Hz.

(b) The question next asks for the actual frequency of an ambulance siren if before it passes the stopped car it has a shifted frequency of $f_1 = 1400.$ Hz, and after it passes the car it has a shifted frequency of $f_2 = 1200.$ Hz.

SKETCH:

(a)

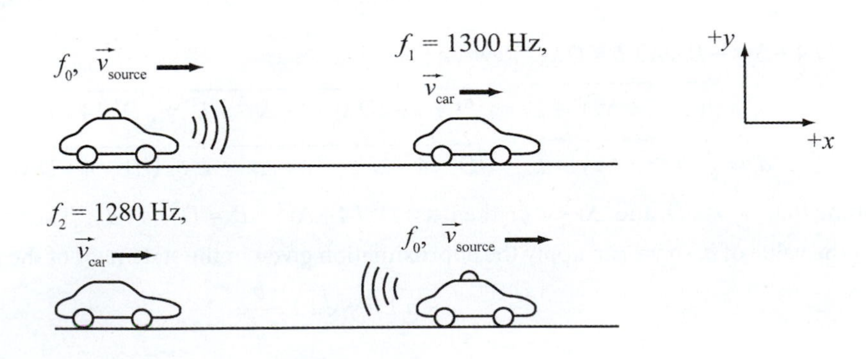

(b)

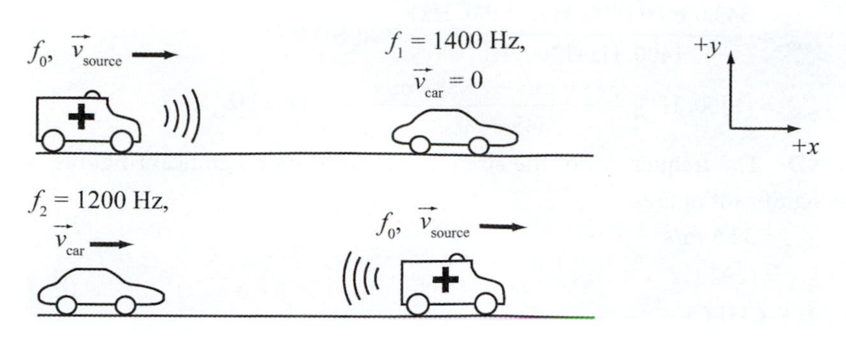

RESEARCH:

Recall that the shift in frequency (Doppler shift) when both source and observer are moving is given by $f_{\text{observer}} = f_{\text{source}} \dfrac{v_{\text{sound}} \pm v_{\text{car}}}{v_{\text{sound}} \pm v_{\text{source}}}$. The speed of the source can be obtained by dividing the corresponding expressions for the before and after frequencies. The speed of the source can be obtained by dividing the corresponding expression for the before and after frequencies. Once the speed is determined, the actual frequency can be calculated.

SIMPLIFY:

(a) From the Doppler shift for the "before" frequency: $f_1 = f_0 \dfrac{v_{\text{sound}} - v_{\text{car}}}{v_{\text{sound}} - v_{\text{source}}}$. From the Doppler shift for the "after" frequency: $f_2 = f_0 \dfrac{v_{\text{sound}} + v_{\text{car}}}{v_{\text{sound}} + v_{\text{source}}}$. Therefore,

$$\frac{f_1}{f_2} = \frac{\left(\dfrac{v_{\text{sound}} - v_{\text{car}}}{v_{\text{sound}} - v_{\text{source}}}\right)}{\left(\dfrac{v_{\text{sound}} + v_{\text{car}}}{v_{\text{sound}} + v_{\text{source}}}\right)} = \left(\frac{v_{\text{sound}} - v_{\text{car}}}{v_{\text{sound}} - v_{\text{source}}}\right)\left(\frac{v_{\text{sound}} + v_{\text{source}}}{v_{\text{sound}} + v_{\text{car}}}\right) = \frac{v_{\text{sound}}^2 - v_{\text{sound}}v_{\text{car}} + v_{\text{source}}\left(v_{\text{sound}} - v_{\text{car}}\right)}{v_{\text{sound}}^2 + v_{\text{sound}}v_{\text{car}} - v_{\text{source}}\left(v_{\text{sound}} + v_{\text{car}}\right)}$$

$$\Rightarrow v_{\text{source}} = \frac{f_1\left(v_{\text{sound}}^2 + v_{\text{sound}}v_{\text{car}}\right) - f_2\left(v_{\text{sound}}^2 - v_{\text{sound}}v_{\text{car}}\right)}{f_1\left(v_{\text{sound}} + v_{\text{car}}\right) + f_2\left(v_{\text{sound}} - v_{\text{car}}\right)}.$$

(b) For the "before" frequency, the Doppler shift gives: $f_1 = f_0 \dfrac{v_{\text{sound}}}{v_{\text{sound}} - v_{\text{source}}}$. For the "after" frequency, the Doppler shift gives: $f_2 = f_0 \dfrac{v_{\text{sound}}}{v_{\text{sound}} + v_{\text{source}}}$. Therefore, $\dfrac{f_1}{f_2} = \dfrac{\left(\dfrac{v_{\text{sound}}}{v_{\text{sound}} - v_{\text{source}}}\right)}{\left(\dfrac{v_{\text{sound}}}{v_{\text{sound}} + v_{\text{source}}}\right)} \Rightarrow v_{\text{source}} = \dfrac{v_{\text{sound}}\left(f_1 - f_2\right)}{f_1 + f_2}$.

The frequency is given by: $f_{\text{source}} = f_1\left(\dfrac{v_{\text{sound}} - v_{\text{source}}}{v_{\text{sound}}}\right)$.

CALCULATE:

(a)

$$v_{\text{source}} = \frac{1300.\text{ Hz}\left((343.0\text{ m/s})^2 + (343.0\text{ m/s})(30.0\text{ m/s})\right) - 1280.\text{ Hz}\left((343.0\text{ m/s})^2 - (343.0\text{ m/s})(30.0\text{ m/s})\right)}{1300.\text{ Hz}(343.0\text{ m/s} + 30.0\text{ m/s}) + 1280.\text{ Hz}(343.0\text{ m/s} - 30.0\text{ m/s})}$$

$$= 32.64\text{ m/s}$$

(b) $v_{\text{source}} = \dfrac{343.0 \text{ m/s}(1400. \text{ Hz} - 1200. \text{ Hz})}{1400. \text{ Hz} + 1200. \text{ Hz}} = 26.385 \text{ m/s}$

$f_{\text{source}} = (1400. \text{ Hz})\left(\dfrac{343.0 \text{ m/s} - 26.385 \text{ m/s}}{343.0 \text{ m/s}}\right) = 1292.3 \text{ Hz}$

ROUND: The frequencies of the siren are given to four significant figures, while the speed is given to three significant figures.

(a) $v_{\text{source}} = 32.6 \text{ m/s}$

(b) $f_{\text{source}} = 1292 \text{ Hz}$

DOUBLE-CHECK:

(a) This is a reasonable speed for the police car as it passes.

(b) This is a reasonable value for the actual frequency, based on the given parameters.

16.49. **(a, b) THINK:** The question presents a stationary car as a train passes by it. A sketch of the time dependent horn frequency heard at the car is shown. Determine (a) the frequency of the horn as it is emitted by the train and (b) the speed of the train.

SKETCH:

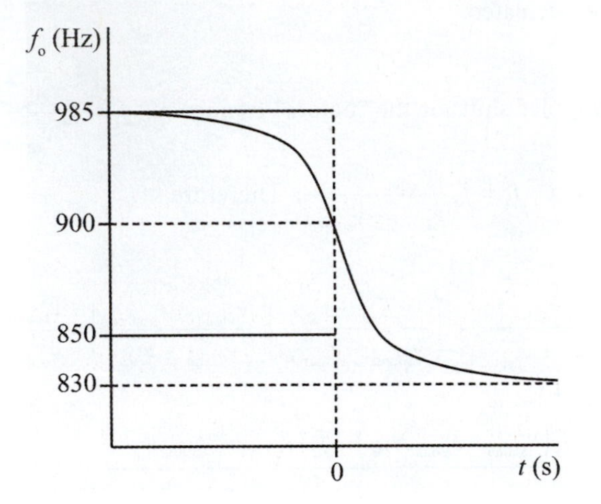

RESEARCH:

(a) From the given sketch, it is clear that the train passes the car at the zero time mark. From the curve, trace to the corresponding frequency at $t = 0$.

(b) From the result of part (a), the speed of the train can be determined by looking at the maximum frequency shift as shown in the sketch and using the Doppler frequency shift equation:

$$f_{\text{detector}} = f_{\text{source}} \frac{v_{\text{sound}} \pm v_{\text{detector}}}{v_{\text{sound}} \pm v_{\text{source}}}.$$

SIMPLIFY:

(a) Simplification is not necessary.

(b) Since the car is not moving, for the case where the train is moving away from the car:

$$f_{\text{car}} = f_{\text{train}}\left(\frac{v_{\text{sound}}}{v_{\text{sound}} + v_{\text{train}}}\right) \Rightarrow v_{\text{train}} = v_{\text{sound}}\left(\frac{f_{\text{train}} - f_{\text{car}}}{f_{\text{car}}}\right).$$

CALCULATE:

(a) From the sketch, the frequency corresponding to $t = 0$ is $f_0 = 900 \text{ Hz}$.

(b) Using the result of part (a), and the maximum frequency shift from the sketch, $f_{\text{car}} = 830 \text{ Hz}$:

$$v_{\text{train}} = (340 \text{ m/s})\left(\frac{900 \text{ Hz} - 830 \text{ Hz}}{830 \text{ Hz}}\right) = 28.7 \text{ m/s}.$$

ROUND:

(a) Since the value was chosen from the sketch, the result remains $f_0 = 900\text{ Hz}$.

(b) Since the values are all obtained by measuring a curve on a rough sketch, round the result to one significant figure, $v_{\text{train}} = 30\text{ m/s}$.

DOUBLE-CHECK: It is reasonable that as the train moves away from the car, the frequency decreases. The results are also reasonable based on the inaccuracy of getting data from a rough sketch.

(c) THINK: In part (a), the value of f_0 is found to be $f_0 = 900\text{ Hz}$. In part (b), the speed of the train is found to be $v_{\text{train}} = 28.7\text{ m/s}$. The speed of sound is about $v_{\text{sound}} = 343\text{m/s}$. From the plot given in the question, the instantaneous slope of f_0 at $t = 0$ can be approximated.

SKETCH: A sketch is not needed.

RESEARCH: The question gives the hint to use the equation $\left.\frac{df_o}{dt}\right|_{t=0} = -\frac{fv^2}{bv_{\text{sound}}}$. Part (c) asks for the value of b. The rest of the quantities in the expression are known or can be found. By inspecting the graph, it appears that the function f_o passes through the points (-0.2, 920) and (0.2, 870). Thus, a reasonable approximation for the instantaneous slope at zero is $\left.\frac{df_o}{dt}\right|_{t=0} \approx \frac{870\text{ Hz} - 920\text{ Hz}}{0.2\text{ s} - (-0.2\text{ s})} = \frac{-50\text{ Hz}}{0.4\text{ s}} = -125\text{ s}^{-2}$.

SIMPLIFY: $b = -\frac{fv^2}{v_{\text{sound}} \left.\frac{df_o}{dt}\right|_{t=0}}$

CALCULATE: $b = -\frac{(900\text{ Hz})(28.7\text{ m/s})^2}{(343\text{ m/s})(-125\text{ s}^{-2})} = 17.2903\text{ m}$

ROUND: Since the frequency can be read off the graph to two significant figures, the distance from the car to the train tracks should be rounded to two significant figures. This means b = 17 m.

DOUBLE-CHECK: Meters are appropriate units for a distance, and a distance of 17 m is a reasonable distance for a car to stop from train tracks while a train passes.

16.53. Recall that the frequencies of standing waves in a pipe with an open end are given by: $f_n = \frac{(2n-1)v}{4L}$. For n = 1 (fundamental node): $f_1 = v/(4L) \Rightarrow 1047\text{ Hz} = 343\text{ m/s}/(4L) \Rightarrow L = 8.19\text{ cm}$, so that the top of the liquid must be 8.19 cm from the top of the bottle.

16.57. The given quantities are the horn frequency, f = 400.0 Hz, the car's speed, $v_c = 20.0\text{ m/s}$ and the speed of sound, $v_s = 343\text{ m/s}$.

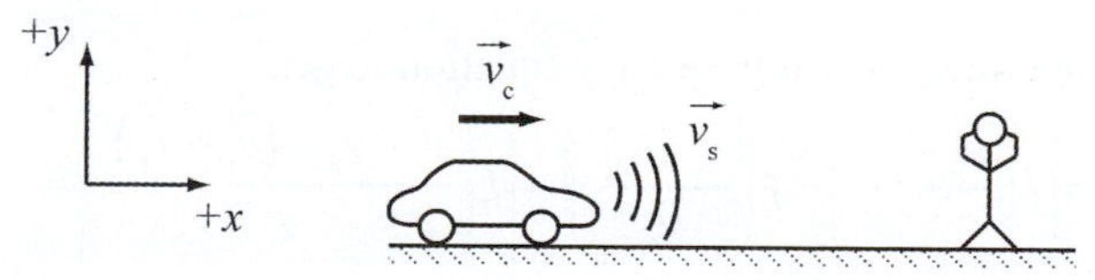

This problem can be solved using the Doppler effect. In this case, the equation is: $f_0 = f\left(\frac{v_s}{v_s - v_c}\right)$. The negative sign indicates that the source is moving towards the observer. Inserting the values:

$$f_0 = (400.0\text{ Hz})\left(\frac{343\text{ m/s}}{343\text{ m/s} - 20.0\text{ m/s}}\right) = 425\text{ Hz}.$$

16.61. The given values are listed in the provided table. Determine the lengths of the pipes required to achieve the listed frequencies. The speed of sound in air is $v_s = 343$ m/s. A wind chime is made from pipes that are open at both ends. The general equation for standing waves in an open pipe is $L = n\lambda / 2$, where λ is the wavelength and n = 1, 2, 3, ... The wavelength can be related to frequency using the equation $\lambda = v / f$. In this case, $\lambda = v_s / f$, and you can substitute this into the equation for L to get:

$$L = \frac{n}{2}\left(\frac{v_s}{f}\right).$$

For the fundamental frequency, n =1. The expression used to fill in the table is $L = v_s / 2f$.

Note	Frequency (Hz)	Length (m)
G4	392	0.438
A4	440	0.390
B4	494	0.347
F5	698	0.246
C6	1046	0.164

16.65. **THINK:** The separation distance is d = 80.0 m. The frequency of the speakers is f = 286 Hz. The beat frequency is $f_B = 10.0$ Hz. Determine the speed, v_R, with which you are running toward one of the speakers. The speed of sound in air is $v_s = 343$ m/s.

SKETCH:

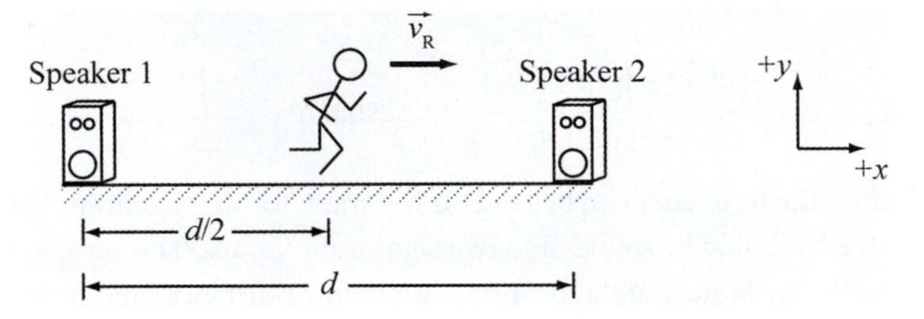

RESEARCH: The equation for the beat frequency is $f_B = |f_1 - f_2|$. The frequencies you detect as you run from the position, $d/2$, towards speaker 2 are f_1 and f_2. These are the observed frequencies due to speaker 1 and speaker 2, respectively. The general equation for the Doppler effect is: $f_o = f\left(\frac{v_s \pm v_{\text{observer}}}{v_s}\right)$, where the minus sign denotes the observer is approaching the source.

SIMPLIFY: You are the observer, so $v_{\text{observer}} = v_R$. $f_1 = f\left(\frac{v_s - v_R}{v_s}\right)$, $f_2 = f\left(\frac{v_s + v_R}{v_s}\right)$. These expressions can be substituted into the beat frequency equation to get:

$$f_B = \left| f\left(\frac{v_s - v_R}{v_s}\right) - f\left(\frac{v_s + v_R}{v_s}\right) \right| = \left| f\left(\frac{v_s - v_R - (v_s + v_R)}{v_s}\right) \right| = f\left| -2\frac{v_R}{v_s} \right| \Rightarrow |v_R| = \frac{f_B v_s}{2f}.$$

CALCULATE: $v_R = \dfrac{10.0 \text{ Hz}(343 \text{ m/s})}{2(286 \text{ Hz})} = 5.997$ m/s

ROUND: The distance and the beat frequency each have three significant figures, so the result should be rounded to $v_R = 6.00$ m/s.

DOUBLE-CHECK: The calculated speed has the proper units. The speed of 6 m/s is reasonable, considering a world class sprinter can run at approximately 10 m/s (e.g. a world class athlete can run the 100 m dash in about 10 s).

16.69. **THINK:** The source travels to the right at a speed of $v_s = 10.00$ m/s and emits a sound wave of frequency, $f_s = 100.0$ Hz. The reflector travels to the left at a speed of $v_R = 5.00$ m/s. Determine the frequency, f, of the reflected sound wave that is detected back at the source. The frequency of the reflected sound wave is f_R. Use v = 343 m/s as the speed of sound, the known value at 20 °C.

SKETCH:

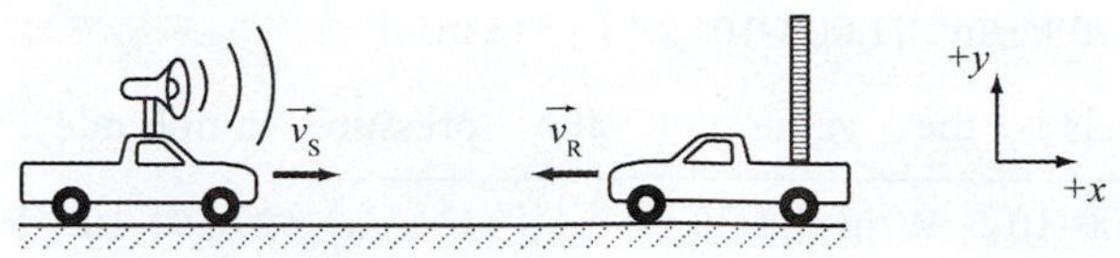

RESEARCH: The general equation describing the Doppler effect when both the source and the observer are moving is:

$$f = f_{\text{source}}\left(\frac{v_{\text{sound}} \pm v_{\text{observer}}}{v_{\text{sound}} \pm v_{\text{source}}}\right).$$

When the observer and source approaching each other, the sign in the numerator is positive and the sign in the denominator is negative. In this problem, consider the frequency that is observed at the reflector. This frequency will be the source frequency for the sound wave that the observer moving to the right with the true source detects.

SIMPLIFY: $f_R = f_s\left(\frac{v+v_R}{v-v_s}\right)$, $f = f_R\left(\frac{v+v_s}{v-v_R}\right)$. Substitute the expression for f_R into the above equation to get:

$$f = f_s\left(\frac{v+v_R}{v-v_s}\right)\left(\frac{v+v_s}{v-v_R}\right).$$

CALCULATE: $f = 100.0 \text{ Hz}\left(\frac{343 \text{ m/s} + 5.00 \text{ m/s}}{343 \text{ m/s} - 10.00 \text{ m/s}}\right)\left(\frac{343 \text{ m/s} + 10.00 \text{ m/s}}{343 \text{ m/s} - 5.00 \text{ m/s}}\right) = 109.1423 \text{ Hz}$

ROUND: The frequency should have three significant figures since the least precise value given in the question has three significant figures. Round the frequency to $f = 109$ Hz.

DOUBLE-CHECK: It is reasonable that the frequency of the reflected wave detected at the source is higher because the source and reflector are approaching each other. Also, the correct value has valid units for frequency.

16.73. **THINK:** Using the results of the previous problems, determine the displacement and pressure amplitudes of a pure tone of frequency $v = 1.000$ kHz in air (density $\rho_0 = 1.20 \text{ kg/m}^3$ and speed of sound $v_s = 343$ m/s) at $\beta = 0.00$ db, and $\beta = 120.$ db. Let $I_0 = 1.00 \cdot 10^{-12} \text{ W/m}^2$ be the standard reference intensity, and $f = \omega/(2\pi)$ be the frequency of the sound wave.

SKETCH: A sketch is not needed to solve this problem.

RESEARCH: From problem 16.72, we learned that $I = \frac{1}{2}\rho_0\omega^2 A^2 v_s$. We will use the formula $\beta = 10\log\frac{I}{I_0}$ to eliminate I. The pressure amplitude is $P = \sqrt{2I\rho_0 v_s}$.

SIMPLIFY: Solve the first formula in RESEARCH to get $A = A_\beta = \sqrt{\frac{2I}{\rho_0\omega^2 v_s}}$, and substitute the expression $I = I_0 10^{\beta/10}$ to eliminate I. The result is $A_\beta = \sqrt{\frac{10^{\beta/10} I_0}{2\pi^2\rho_0 f^2 v_s}}$. The pressure amplitude is $P = P_\beta = \sqrt{2I\rho_0 v_s} = \sqrt{2\left(10^{\beta/10}\right) I_0 \rho_0 v_s}$.

CALCULATE: $A_{0.00} = \sqrt{\dfrac{10^{0.00/10}\cdot\left(1.00\cdot 10^{-12}\text{ W/m}^2\right)}{2\pi^2\left(1.20\text{ kg/m}^3\right)\left(1.000\cdot 10^3\text{ s}^{-1}\right)^2\left(343\text{ m/s}\right)}} = 1.1094\cdot 10^{-11}\text{ m}$ and

$A_{120.} = \sqrt{\dfrac{10^{120./10}\cdot\left(1.00\cdot 10^{-12}\text{ W/m}^2\right)}{2\pi^2\left(1.20\text{ kg/m}^3\right)\left(1.000\cdot 10^3\text{ s}^{-1}\right)^2\left(343\text{ m/s}\right)}} = 1.1094\cdot 10^{-5}\text{ m}.$ At the threshold of hearing ($\beta = 0.00$ decibels) the value of the pressure amplitude is given by the equation $P_{0.00} = \sqrt{2(1)\left(1.00\cdot 10^{-12}\text{ W/m}^2\right)\left(1.20\text{ kg/m}^3\right)\left(343\text{ m/s}\right)} = 2.8691\cdot 10^{-5}\text{ Pa}.$ On the other hand, the value of the pressure amplitude at the threshold of pain is given by the equation $P_{120.} = \sqrt{2\left(10^{120./10}\right)\left(1.00\cdot 10^{-12}\text{ W/m}^2\right)\left(1.20\text{ kg/m}^3\right)\left(343\text{ m/s}\right)} = 28.691\text{ Pa}.$

ROUND: The rounded values are: $A_{0.00} = 1.11\cdot 10^{-11}$ m, $A_{120.} = 1.11\cdot 10^{-5}$ m, $P_{0.00} = 2.87\cdot 10^{-5}$ Pa, and $P_{120.} = 28.7$ Pa.

DOUBLE-CHECK: The calculated displacement of a pure tone frequency of 1.000 kHz at zero decibels is roughly a tenth of an atomic diameter. This is suprisingly small, but it is at least consistent with the fact that such a wave produces no sensation on the skin and only a barely discenible sensation at the eardrum. Increasing from 0 to 120 decibels corresponds to a 10^{12} factor increase in intensity. Since the displacement varies with the square root of the intensity, the calculated value for the displacement at 120 decibels should be larger by a factor of 10^6, which it is. $1.11\cdot 10^{-5}$ m is one hundredth of a millimeter, and it is plausible that such an amplitude would begin to cause pain in the ear. The pressure amplitude at zero decibels is less than three ten-thousandths of an atmosphere, and this is consistent with what would be expected. Again, the pressure amplitude varies with the square root of the intensity, so the calculated value of the displacement at 120 decibels should be larger by a factor of 10^6, which it is. This shows that the calculated answers are reasonable.

Multi-Version Exercises

16.74. **THINK:** The frequency of the sound will be perceived differently by the drivers as a result of the Doppler effect. In this case, the source of the sound is the parked car, which is not moving, while the observer is in the moving car.

SKETCH: The car on the right is parked. The car on the left is moving towards the car on the left at a speed v.

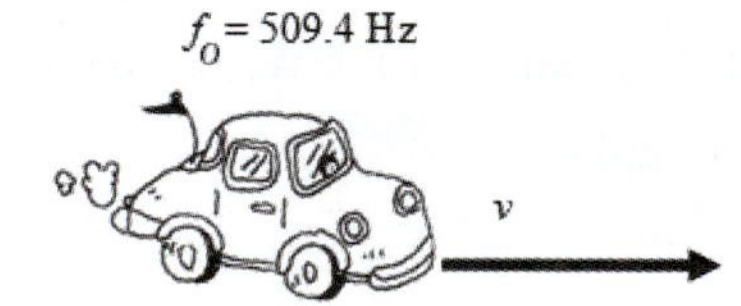

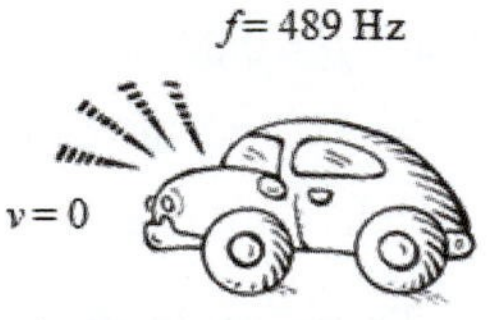

RESEARCH: The source of the sound is stationary and the observer is moving towards the source, so the observed frequency is given by $f_\text{O} = f\left(1+\dfrac{v_\text{observer}}{v_\text{sound}}\right)$.

SIMPLIFY: The goal is to find the speed of the observer, so solve for that variable:

$$f\left(1+\frac{v_{\text{observer}}}{v_{\text{sound}}}\right)=f_{\text{O}}$$

$$1+\frac{v_{\text{observer}}}{v_{\text{sound}}}=\frac{f_{\text{O}}}{f}$$

$$\frac{v_{\text{observer}}}{v_{\text{sound}}}=\frac{f_{\text{O}}}{f}-1$$

$$v_{\text{observer}}=v_{\text{sound}}\left(\frac{f_{\text{O}}}{f}-1\right).$$

CALCULATE: The frequency of the horn is f = 489 Hz, but the driver in the approaching car hears a sound of frequency f_{O} = 509.4 Hz. In this case, the speed of sound is 343 m/s, so the velocity of the observer must be

$$\begin{aligned} v_{\text{observer}} &= v_{\text{sound}}\left(\frac{f_{\text{O}}}{f}-1\right) \\ &= 343 \text{ m/s}\left(\frac{509.4 \text{ Hz}}{489 \text{ Hz}}-1\right) \\ &= 14.30920245 \text{ m/s.} \end{aligned}$$

ROUND: Though the frequency measured by the driver of the moving car has four significant figures, two other measured values, the frequency of the horn and the speed of sound, are given to only three significant figures, so the final answer can have only three significant figures. The car is going 14.3 m/s.

DOUBLE-CHECK: The car is going 14.3 m/s. Since 1 m = $6.214 \cdot 10^{-4}$ miles and 3600 sec = 1 hour, the car is going $14.3 \text{ m/s}\left(6.214\cdot 10^{-4} \tfrac{\text{miles}}{\text{meter}}\right)\left(3600 \tfrac{\text{sec}}{\text{hour}}\right)=32 \text{ mph}$. For a car driving on a paved road, this is a perfectly reasonable speed, so the answer is reasonable.

16.77. **THINK:** The tuba is a brass instrument; it makes music by producing standing waves inside of the coiled tube. The lowest frequency sound that can be produced by the tuba corresponds to half of a wavelength inside the tube.

SKETCH: Think of the tuba, uncoiled, as a pipe that is open at one end.

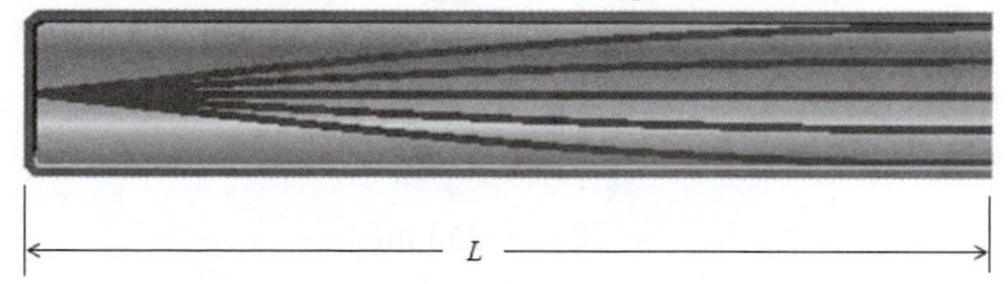

RESEARCH: The lowest frequency sound that can be produced corresponds to half of a wavelength inside the tuba, so the total wavelength is twice the length of the tuba, $\lambda = 2L$. The speed of the sound wave, wavelength, and frequency are related by the equation $v=\lambda f$.

SIMPLIFY: First, rewrite $v=\lambda f$ to find the frequency in terms of the velocity and wavelength $f=v/\lambda$. Then, use the fact that $\lambda=2L$ to find $f=\dfrac{v}{2L}$.

CALCULATE: The length of the tuba is L = 7.373 m and the speed of sound is 343.0 m/s, so the lowest frequency the tuba can produce is

$$\begin{aligned} f &= \frac{v}{2L} \\ &= \frac{343.0 \text{ m/s}}{2\cdot 7.373 \text{ m}} \\ &= 23.26054523 \text{ Hz.} \end{aligned}$$

ROUND: The length of the tuba and speed of sound both have four significant figures, so the final answer should also have four figures. The lowest frequency that the tuba can produce is 23.26 Hz.

DOUBLE-CHECK: A frequency of 23.26 Hz corresponds to a note about 3½ octaves below "middle C" on the piano. The lowest tones that adult humans can hear are at about 20 Hz (children can generally hear a larger range of frequency than adults) so it makes sense that the lowest brass instrument in the orchestra would have a frequency just above 20 Hz. These considerations confirm that 23.26 Hz is a reasonable answer for the minimum frequency produced by an instrument in the orchestra.

16.79. **THINK:** Since the metal bar is solid, the speed of sound will depend on the type and structure of the material. The speed of sound in a solid can be calculated using the density and Young's modulus.

SKETCH: The speed of sound, v, in the metal bar is shown (see Figure 16.2).

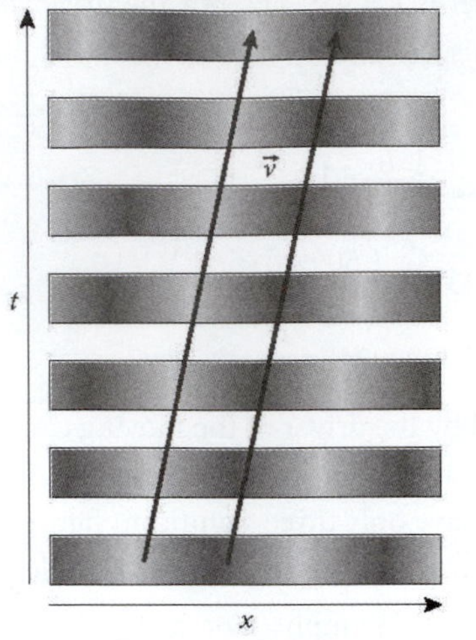

RESEARCH: In general, the speed of sound in a solid is given by $v=\sqrt{Y/\rho}$, where Y is the Young's modulus and ρ is the mass density.

SIMPLIFY: n/a

CALCULATE: The question states that the mass density of the bar is ρ = 3497 kg/m^3 and the Young's modulus is $266.3\cdot 10^9$ N/m^2. The speed of sound in the metal bar is

$$\begin{aligned} v &= \sqrt{Y/\rho} \\ &= \sqrt{\frac{266.3\cdot 10^9 \text{ N/m}^2}{3497 \text{ kg/m}^3}} \\ &= 8726.453263 \text{ m/s}. \end{aligned}$$

ROUND: The values used to calculate the speed of sound all have four significant figures, so the final answer should also have four figures. The speed of sound in the metal bar is 8726 m/s.

DOUBLE-CHECK: The speed of sound in air is 343 m/s, but air is less dense than a solid metal bar. Looking at Table 16.1, the speed of sound in this metal bar is somewhere between the speed of sound in Aluminum and the speed of sound in Diamond, confirming that the answer is physically realistic and of the correct order of magnitude.

Chapter 17: Temperature

Exercises

17.27. The temperature $T_C = -21.8$ °C is three times its equivalent value $T_F = -7.3$ °F. To check this, recall the conversion formula: $T_C = \frac{5}{9}(T_F - 32\text{ °F})$. $T_C = 3T_F \Rightarrow \frac{5}{9}(T_F - 32\text{ °F}) = 3T_F \Rightarrow \frac{5}{9}(-32\text{ °F}) = \left(\frac{27-5}{9}\right)T_F$

$$\Rightarrow T_F = \frac{5(-32\text{ °F})}{22} = -7.2727\text{ °F and } T_C = \frac{5}{9}(-7.3\text{ °F} - 32\text{ °F}) = -21.8\text{ °C}$$

17.31. (a) $T_K = T_C + 273.15\text{ °C} = -79\text{ °C} + 273.15\text{ °C} = 194\text{ K}$. Rounding to two significant figures gives $T_K = 190$ K.

(b) $T_F = \frac{9}{5}T_C + 32\text{ °C} = \frac{9}{5}(-79\text{ °C}) + 32\text{ °C} = -110\text{ °F}$

17.35. We can calculate the bulk expansion coefficient for steel by multiplying the value of the linear expansion coefficient for steel from Table 17.2 by 3, which gives us

$\beta = 3\alpha = 3\left(13.0 \cdot 10^{-6}\text{ °C}^{-1}\right) = 3.90 \cdot 10^{-5}\text{ °C}^{-1}$.

$\rho = \text{density at } 20.0\text{ °C} = 7800.\text{ kg/m}^3$,

$\rho' = \text{density at } 100.0\text{ °C} = \frac{M}{V_0 + \Delta V}$, $V_0 = \text{volume at } 20.0\text{ °C}$

$\Delta V = \beta V_0 \Delta T$;

$$\rho' = \frac{M}{V_0 + \beta V_0 \Delta T} = \frac{M}{V_0(1+\beta\Delta T)} = \frac{\rho}{1+\beta\Delta T} = \frac{7800.\text{ kg/m}^3}{1+(3.90\cdot 10^{-5}\text{ °C}^{-1})(80.0\text{ °C})} = 7776\text{ kg/m}^3.$$

17.39. The cross-sectional area has no relevance-this depends on linear expansion, which is governed by $L_f = L(1+\alpha\Delta T)$, where in this case $L_f = 5.2000$ m, $\Delta T = 60.0$ °C and $\alpha = 13\cdot 10^{-6}$ per degree Celsius from Table 17.2. This gives $L = \frac{L_f}{(1+\alpha\Delta T)} = \frac{5.2000\text{ m}}{1+\left(13\cdot 10^{-6}\text{ °C}^{-1}\right)\left(60.0\text{°C}\right)} = 5.195947\text{ m}$ at -10.0 °C. Thus there will be 4.1 mm between adjacent rails.

17.43. **THINK:** Assume that at 37.0 °C, the pool just begins to overflow. The volume expansion equation can yield the depth of the pool. Let $d = 1.00$ cm = 0.0100 m.
SKETCH:

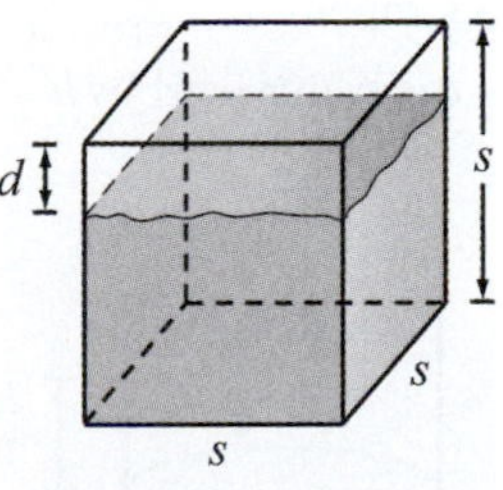

RESEARCH: $V_0 = S^2(S-d)$, $V = S^3$, $\Delta V = \beta V_0 \Delta T$

SIMPLIFY:
$\Delta V = \beta V_0 \Delta T$
$V - V_0 = \beta V_0 \Delta T$
$S^3 - S^2(S-d) = \beta S^2(S-d)\Delta T$
$S - (S-d) = \beta(S-d)\Delta T$
$d = \beta S\Delta T - \beta d\Delta T$
$S = d\dfrac{1+\beta\Delta T}{\beta\Delta T}.$

CALCULATE: $S = (0.0100 \text{ m})\left(\dfrac{1+(207\cdot 10^{-6} \text{ °C}^{-1})(16.0 \text{ °C})}{(207\cdot 10^{-6} \text{ °C}^{-1})(16.0 \text{ °C})}\right) = 3.029 \text{ m}$

ROUND: Three significant figures: S = 3.03 m.

DOUBLE-CHECK: 3.03 m is a realistic depth for a pool.

17.47. **THINK:** Assuming the brass and steel rods, L = 1.00 m each, do not sag, they will increase in length by ΔL_{B} and ΔL_{S}, respectively. The rods will touch when their combined extensions equals the separation, d = 5.00 mm. The linear expansion coefficients of the rods are $\alpha_{\text{B}} = 19\cdot 10^{-6}$ / °C and $\alpha_{\text{S}} = 13\cdot 10^{-6}$ / °C. The initial temperature of the rods is $T_{\text{i}} = 25.0$ °C.

SKETCH:

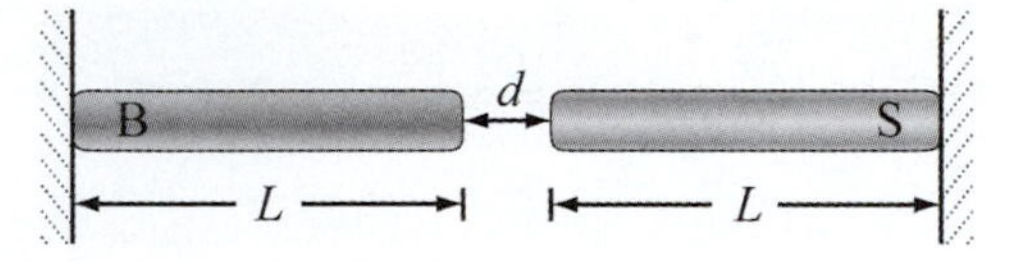

RESEARCH: The brass rod will increase by $\Delta L_{\text{B}} = L\alpha_{\text{B}}\Delta T$. The steel rod will increase by $\Delta L_{\text{S}} = L\alpha_{\text{S}}\Delta T$. The final temperature will be $\Delta T + T_{\text{i}}$. The rods touch when $\Delta L_{\text{B}} + \Delta L_{\text{S}} = d.$

SIMPLIFY: $d = \Delta L_{\text{B}} + \Delta L_{\text{S}} = L(\alpha_{\text{B}} + \alpha_{\text{S}})\Delta T.$ Thus, $\Delta T = \dfrac{d}{L(\alpha_{\text{B}}+\alpha_{\text{S}})} \Rightarrow T_{\text{f}} = \dfrac{d}{L(\alpha_{\text{B}}+\alpha_{\text{S}})} + T_{\text{i}}.$

CALCULATE: $T_{\text{f}} = \dfrac{5.00\cdot 10^{-3} \text{ m}}{(1.00 \text{ m})(19+13)\cdot 10^{-6} \text{ / °C}} + 25.0 \text{ °C} = 181.25 \text{ °C}$

ROUND: Two significant figures: $T_{\text{f}} = 180$ °C.

DOUBLE-CHECK: Given the long length, d, for the total expansion, such a high temperature is not unreasonable.

17.51. **THINK:** When the horseshoe is put in the tank, $r = 10.0$ cm, the water rises by $h = 0.250$ cm. The horseshoe, $T_{\text{i}} = 293.15$ K (room temperature, 20.0 °C) and $T_{\text{f}} = 700.$ K, will increase its volume. When it is put back in water, it will raise the water level by h'. The linear expansion coefficient of the horseshoe is $\alpha = 11.0\cdot 10^{-6}$ K^{-1}.

SKETCH:

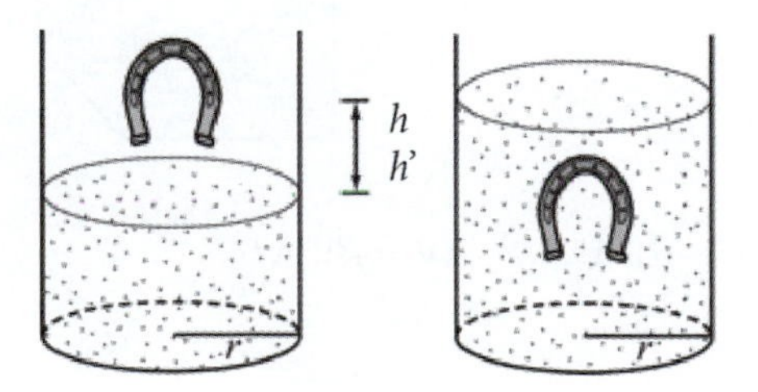

RESEARCH: When water rises by h or h', the volumes displaced are $V=\pi r^2 h$ and $V'=\pi r^2 h'$. The volume of the heated horseshoe is $V'=V(1+3\alpha\Delta T)$. The initial volume of the horseshoe, V_0, is the same as the volume of water it displaced before it was heated, $\pi r^2 h$. The volume of displaced heated water is equal to the volume of the heated horseshoe.

SIMPLIFY: $V'=V(1+3\alpha\Delta T) \Rightarrow \pi r^2 h'=\pi r^2 h(1+3\alpha\Delta T) \Rightarrow h'=h(1+3\alpha\Delta T)$.

CALCULATE: $h'=(0.250\text{ cm})\left(1+3\left(11.0\cdot 10^{-6}\text{ K}^{-1}\right)(293.15\text{ K})\right)=0.252418\text{ cm}$.

ROUND: The least precise value given in the question has three significant figures. Therefore the final answer should be rounded to $h'=0.252$ cm.

DOUBLE-CHECK: The change in the water height is small, which seems reasonable since the change in volume is small and the cross sectional area $(3\cdot 10^4\text{ mm}^2)$ of the tank of water is relatively large.

17.55. **THINK:** For simplicity, define $a=1.00016$, $b=4.52\cdot 10^{-5}$ and $c=5.68\cdot 10^{-6}$. In part (a), a derivative can be used to determine the properties of the water. The volume, V, as a function of temperature, T, is given by $V=a-bT+cT^2$ when the temperature is in the range [0.00 °C, 50.0 °C). In part (b), evaluate β when $T=20.0$ °C.

SKETCH: A sketch is not needed to solve this problem.

RESEARCH: The general function to evaluate the change in volume is $\Delta V=\beta V\Delta T$. The differences can be approximated as differentials, i.e. $\Delta Y/\Delta X\approx dy/dx$.

SIMPLIFY: $\frac{dV}{dT}=\frac{d}{dT}(a-bT+cT^2)=-b+2cT$. Since $\Delta V=\beta V\Delta T$, it follows that:

$$\beta=\frac{1}{V}\left(\frac{\Delta V}{\Delta T}\right)\approx\frac{1}{V}\left(\frac{dV}{dT}\right)=\frac{-b+2cT}{a-bT+cT^2}$$

CALCULATE:

(a) $\beta(T)=\dfrac{-4.52\cdot 10^{-5}+11.36\cdot 10^{-6}T}{1.00016-4.52\cdot 10^{-5}T+5.68\cdot 10^{-6}T^2}$

(b) $\beta(20\text{ °C})=\dfrac{-4.52\cdot 10^{-5}+(11.36\cdot 10^{-6})(20.0\text{°C})}{1.00016-(4.52\cdot 10^{-5})(20.0\text{°C})+(5.68\cdot 10^{-6})(20.0\text{°C})^2}$

$=1.8172\cdot 10^{-4}\text{ /°C}$

ROUND:

(a) Not necessary.

(b) Round to three significant figures: $\beta(T=20.0\text{ °C})=1.82\cdot 10^{-4}\text{ /°C}$

DOUBLE-CHECK: The value for β for water at 20.0 °C from Table 17.3 is $2.07\cdot 10^{-4}$ /°C. Since the calculated value is close, this is a reasonable result.

17.59. When the aluminum container is filled with turpentine, the turpentine will have a volume of $V=5.00$ gal. The volume expansion coefficient of the turpentine, $\beta_{\text{turp}}=900.\cdot 10^{-6}\text{ °C}^{-1}$. The volume expansion coefficient of aluminum is $\beta_{\text{Al}}=66.0\cdot 10^{-6}\text{ °C}^{-1}$. The change in temperature is $\Delta T=12.0$ °C. The change in volume of the turpentine is given by:

$$\Delta V=\beta_{\text{turp}}V\Delta T=(900.\cdot 10^{-6}\text{ °C}^{-1})(5.00\text{ gal})(12.0\text{ °C})\left(\frac{3.785\text{ L}}{1\text{ gal}}\right)=0.2044\text{ L}.$$

The change in volume of the is given by:

$$\Delta V = \beta_{Al} V \Delta T = (66.0\cdot 10^{-6} / {}^\circ\text{C})(5.00 \text{ gal})(12.0 \text{ }^\circ\text{C})\left(\frac{3.785 \text{ L}}{1 \text{ gal}}\right) = 0.01499 \text{ L}.$$

Thus, 0.189 L of turpentine spills out of the container.

17.63. The initial volume of the mercury is $V = 8.00$ mL, the cross-sectional area of the tube is $A = 1.00 \text{ mm}^2$ and the volume expansion coefficient of mercury is $\beta_{Hg} = 181\cdot 10^{-6} / {}^\circ\text{C}$. Consider a change in temperature of $\Delta T = 1.00$ °C. Since the cross-sectional area remains closely the same, $\Delta V = A\Delta L$.

$$\Delta V = \beta_{Hg} V \Delta T = A\Delta L \Rightarrow \Delta L = \frac{\beta_{Hg} V \Delta T}{A} = \frac{(181\cdot 10^{-6} / {}^\circ\text{C})(8.00 \text{ mL})(1.00 \text{ }^\circ\text{C})}{1.00 \text{ mm}^2}\left(\frac{1000. \text{ mm}^3}{\text{mL}}\right) = 1.448 \text{ mm}$$

Thus, the 1.00 °C tick marks should be spaced about 1.45 mm apart.

17.67. The volume expansion coefficient of kerosene is $\beta_k = 990.\cdot 10^{-6} / {}^\circ\text{C}$. If the volume increases by 1.00%, then $\Delta V / V = 0.0100$.

$$\Delta V = \beta_k V \Delta T \Rightarrow \Delta T = \frac{\Delta V}{V}\left(\frac{1}{\beta_k}\right) = \frac{0.0100}{990.\cdot 10^{-6} / {}^\circ\text{C}} = 10.1 \text{ }^\circ\text{C}$$

Thus, the kerosene must be heated up by at least 10.1 °C in order for its volume to increase by 1.00%.

17.71. **THINK:** The steel band has an initial diameter of $d_i = 4.40$ mm, width $w = 3.50$ mm, and thickness $t = 0.450$ mm. As the band cools from $T_i = 70.0$ °C to $T_f = 36.8$ °C its diameter will decrease. Since the circumference of the band is directly proportional to the diameter, both the circumference and the diameter have the same relative change with the decrease in temperature. The tension in the band can be found by considering the Young's modulus of the steel band. Effectively, the band is stretched from its diameter at T_f to the diameter of the tooth.

SKETCH:

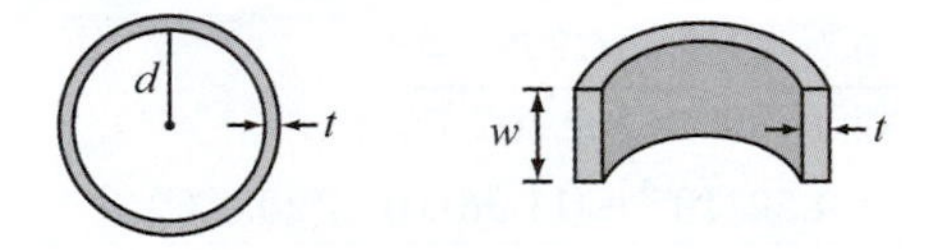

RESEARCH: The change in the area of the band (i.e. area around the tooth) is $\Delta A = 2\alpha_S A\Delta T$ where $A = \pi(d/2)^2$. Young's modulus is the ratio of the stress to the strain where the stress is the force per unit area and the strain is the relative change in length,

$$Y = \frac{F/wt}{\Delta L/L} \Rightarrow \frac{\Delta L}{L} = \frac{F}{wtY}.$$

For steel, $Y = 200.\cdot 10^9 \text{ N/m}^2$. The length of the band is the circumference, so $L = \pi d$. For this problem, use $|\Delta T|$ in place of ΔT. The linear expansion coefficient of steel is

$\alpha_S = 13.0\cdot 10^{-6} \text{ }^\circ\text{C}^{-1}$.

SIMPLIFY: The relative change in area is: $\dfrac{\Delta A}{A} = 2\alpha_S|\Delta T|$. Since the length is proportional to the diameter: $\dfrac{\Delta L}{L} = \dfrac{\Delta d}{d}$. Since $A = \pi d^2/4$, $\Delta A = (\pi d/2)\Delta d$. So we can write

$$\frac{\Delta A}{A} = \frac{(\pi d/2)\Delta d}{\pi d^2/4} = 2\frac{\Delta d}{d}.$$

We can combine these equations to get

$$2\alpha_S|\Delta T| = \frac{2F}{wtY} \Rightarrow F = \alpha_S|\Delta T|wtY.$$

CALCULATE:

$F=(13.0\cdot 10^{-6}\ ^{\circ}\text{C}^{-1})|70.0\ ^{\circ}\text{C}-36.8\ ^{\circ}\text{C}|(3.50\cdot 10^{-3}\text{ m})(0.450\cdot 10^{-3}\text{ m})(200.\cdot 10^{9}\text{ N/m}^2)$

$F=135.954$ N.

ROUND: The answer has to be rounded to three significant figures: $F=136$ N.

DOUBLE-CHECK: Since a tooth is very strong, this large tension that is created will be able to act on the tooth without causing problems. The force must also be large in order to withstand the forces of biting food. Therefore, this is a reasonable result.

17.75. **THINK:** The bugle can be considered a half-closed pipe of length $L=183.0$ cm. The speed of sound in air is dependent on temperature, as is the length of the bugle, so an increase in temperature from $T_i=20.0$ °C to $T_f=41.0$ °C will cause both to change.

SKETCH:

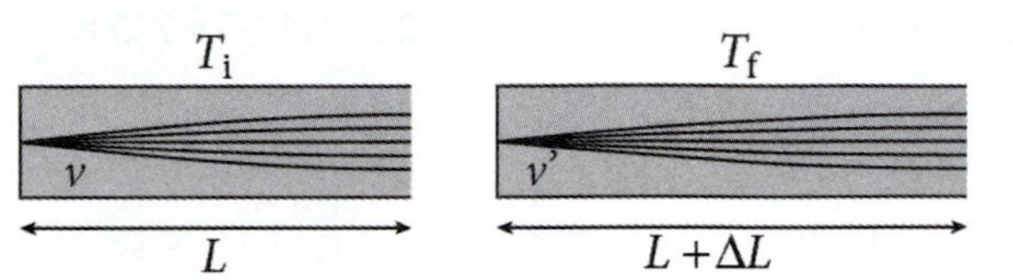

RESEARCH: The fundamental frequency of an open pipe is $f_1=v/(4L)$, where v is the speed of sound. The speed of sound in air as a function of temperature is $v(T)=(331+0.6T)$ m/s, with T in units of °C. The length of the tube increases by $\Delta L=\alpha_B L\Delta T$, with a linear expansion coefficient for brass of $\alpha_B=19.0\cdot 10^{-6}\ ^{\circ}\text{C}^{-1}$.

SIMPLIFY:

(a) If only the change in air temperature is considered, $f_1=\dfrac{v(T_f)}{4L}$.

(b) If only the change in length of the bugle is considered, $f_1=\dfrac{v(T_i)}{4L(1+\alpha_B\Delta T)}$.

(c) If both effects are taken into account, $f_1=\dfrac{v(T_f)}{4L(1+\alpha_B\Delta T)}$.

CALCULATE:

(a) $f_1=\dfrac{(331+(0.6)(41.0))\text{ m/s}}{4(1.830\text{ m})}=48.579$ Hz.

(b) $f_1=\dfrac{(331+(0.6)(20.0))\text{ m/s}}{4(1.830\text{ m})(1+(19.0\cdot 10^{-6}\ ^{\circ}\text{C}^{-1})(41.0\ ^{\circ}\text{C}-20.0\ ^{\circ}\text{C}))}=46.839$ Hz.

(c) $f_1=\dfrac{(331+(0.6)(41.0))\text{ m/s}}{4(1.830\text{ m})(1+(19.0\cdot 10^{-6}/^{\circ}\text{C})(41.0\ ^{\circ}\text{C}-20.0\ ^{\circ}\text{C}))}=48.560$ Hz

ROUND: Three significant figures:

(a) $f_1=48.6$ Hz

(b) $f_1=46.8$ Hz

(c) $f_1=48.6$ Hz

DOUBLE-CHECK: The fundamental frequency of the bugle is fairly insensitive to changes in temperature if the changes in the speed of sound and the expansion of the brass are considered.

Multi-Version Exercises

17.76. The change in length of the steel bar is given by $\Delta L_s = \alpha_s L_s \Delta T$. The change in length of the brass bar is given by $\Delta L_b = \alpha_b L_b \Delta T$. When the two bars have the same length, $L_s + \Delta L_s = L_b + \Delta L_b$. So we can write $L_s + \alpha_s L_s \Delta T = L_b + \alpha_b L_b \Delta T$. Rearranging and solving for the temperature difference gives us

$$\alpha_s L_s \Delta T - \alpha_b L_b \Delta T = L_b - L_s \quad \Rightarrow$$

$$\begin{aligned} \Delta T &= \frac{L_b - L_s}{\alpha_s L_s - \alpha_b L_b} \\ &= \frac{(2.6827 \text{ m}) - (2.6867 \text{ m})}{(13.00 \cdot 10^{-6} \text{ °C}^{-1})(2.6867 \text{ m}) - (19.00 \cdot 10^{-6} \text{ °C}^{-1})(2.6827 \text{ m})} \\ &= 249.311 \text{ °C}. \end{aligned}$$

So the temperature is $T = T_0 + \Delta T = 26.45 \text{ °C} + 249.311 \text{ °C} = 275.8 \text{ °C}$.

17.79. Apply Equation 17.6:

$$\Delta A = 2\alpha A \Delta T = 2(3.749 \cdot 10^{-6} \text{ °C}^{-1})\left(\pi\left(\frac{5.093 \text{ m}}{2}\right)^2\right)(33.37 \text{ °C}) = 5.097 \cdot 10^{-3} \text{ m}^2.$$

17.82. Apply Equation 17.5:

$$\Delta L = L\alpha\Delta T = (501.9 \text{ m})(13.89 \cdot 10^{-6} \text{ °C}^{-1})(-15.91 \text{ °C} - 28.09 \text{ °C}) = -0.3067 \text{ m}.$$

The bar is 0.3067 m shorter.

Chapter 18: Heat and the First Law of Thermodynamics

Exercises

18.27. (a) The work to lift the elephant is $W = mgh$ or $W = (5.0 \cdot 10^3 \text{ kg})(2.0 \text{ m})(9.81 \text{m/s}^2) = 9.8 \cdot 10^4 \text{ J}$.

(b) A food calorie is equal to $4.1868 \cdot 10^3$ J. The task of lifting the elephant consumes $W = 9.8 \cdot 10^4 \text{ J}\left[1 \text{ cal}/\left(4.1868 \cdot 10^3 \text{ J}\right)\right] = 23.4308$ food calories. Assuming that the body converts 100% of food energy into mechanical energy then the number of doughnuts needed is $23.4308 \text{ cal}/(250 \text{ cal}/\text{doughnut}) = 0.093723$. It takes less than one doughnut to power its consumer to lift an elephant. The body usually converts only 30 % of the energy consumed. This corresponds to 0.31 of a doughnut.

18.31. The temperature of the material will be $Q = cm\Delta T = cm(T_\text{f} - T_\text{i})$ or $T_\text{f} = T_\text{i} + Q/cm = T_\text{i} + (Q/cV\rho)$. Since the final temperature is inversely proportional to the specific heat, c, and the density, the material with the largest final temperature will be lead. A large specific heat will give a lower final temperature. The material with the largest specific heat and density, in this case, water, has the smallest final temperature. An example of the calculation for aluminum:

$$T_\text{f}^\text{Al} = 22.0 \text{ °C} + 1.00 \text{ J}/\left[1.00 \text{ cm}^3\left(2.375 \cdot 10^{-3} \text{ kg/cm}^3\right)\left(0.900 \cdot 10^3 \text{ J/(kg K)}\right)\right] = 22.4678 \text{ °C}.$$

Note that we need the density of the material.

Material	Specific Heat (KJ/kg K)	Density (g/cm^3)	Final Temperature °C
Lead	0.129	11.34	22.684
Copper	0.386	8.94	22.290
Steel	0.448	7.85	22.284
Aluminum	0.900	2.375	22.468
Glass	0.840	2.5	22.476
Water	4.19	1.00	22.239

18.35. **THINK:** The problem calls for calculating the change in energy for the copper and the water. The volume of the materials does not change, so the process is isochoric. An isochoric process implies the change in internal energy is equal to the heat transferred. Therefore, to calculate the energies of the materials the final temperatures of the samples are needed. The magnitude of the heat transferred will be equal for both materials. Knowing this and the mass, initial temperatures and the specific heat it is possible to calculate the final temperature. Copper has a mass of $m_\text{c} = 1.00$ kg, an initial temperature of $T_\text{c} = 80.0$ °C, and a specific heat of $c_\text{c} = 386 \text{ J/(kg K)}$. The volume of water is 2.00 L which is equal to a mass of $m_\text{w} = 2.00$ kg, an initial temperature of $T_\text{w} = 10.0$ °C, and a specific heat of $c_\text{w} = 4190 \text{ J/(kg K)}$.

SKETCH:

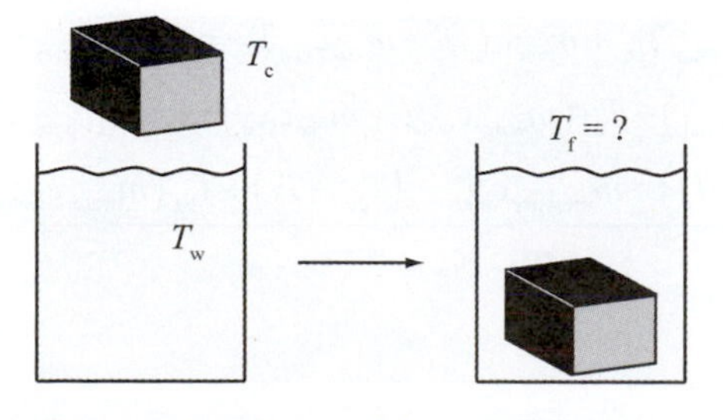

RESEARCH: The heat transferred is equal to $Q = mc\Delta T$.

SIMPLIFY: The temperature change for the water and the copper will be positive and negative, respectively. The heat lost by the copper plus the heat gained by the water equals zero, so $Q_C + Q_W = 0 = m_c c_c (T_f - T_c) + m_w c_w (T_f - T_w)$. Solving this equation for T_f yields $T_f = (m_c c_c T_c + m_w c_w T_w)/(m_w c_w + m_c c_c)$. The magnitude of the change in energy is given by $\Delta E_{int} = Q = mc\Delta T$.

CALCULATE: $T_f = \dfrac{(1.00 \text{ kg})(386 \text{ J/kgK})(80.0 \text{ °C}) + (2.00 \text{ kg})(4190 \text{ J/kgK})(10.0 \text{ °C})}{(1.00 \text{ kg})(386 \text{ J/kgK}) + (2.00 \text{ kg})(4190 \text{ J/kgK})} = 13.0824 \text{ °C}$

$\Delta E_{int,w} = (2.00 \text{ kg})(4190 \text{ J/kgK})(13.0824 \text{ °C} - 10.0 \text{ °C}) = 25830 \text{ J}$

$\Delta E_{int,Cu} = (1.00 \text{ kg})(386 \text{ J/kgK})(13.0824 \text{ °C} - 80.0 \text{ °C}) = -25830 \text{ J}$

ROUND: The energy should be rounded to three significant figures: $\Delta E_{int} = 25800$ J.

DOUBLE-CHECK: The magnitude of the change in energy for the water and copper must be equal since there are no other sources of change in energy. The signs must be opposite so energy is conserved. This is a reasonable amount of heat for a system of this size. Because copper has a much lower specific heat than water, it is expected that the copper will undergo a larger change in temperature.

18.39. THINK: I want to find the original temperature of the water before the thermometer was placed in the water. The vial has a mass of 5.00 g and there is a volume of 6.00 mL of water. The thermometer is composed of 15.0 g of Pyrex and 4.00 g of mercury at a temperature of $T_0 = 20.0$ °C. The thermometer reads $T_{eq} = 29.0$ °C when the thermometer is added to the water. The specific heat of Pyrex is given as 800. J/(kg K), and the specific heat of mercury is given as 140. J/(kg K). The mass of 6.00 mL of water is 6.00 g. The specific heat of water is 4190 J/(kg K).

SKETCH:

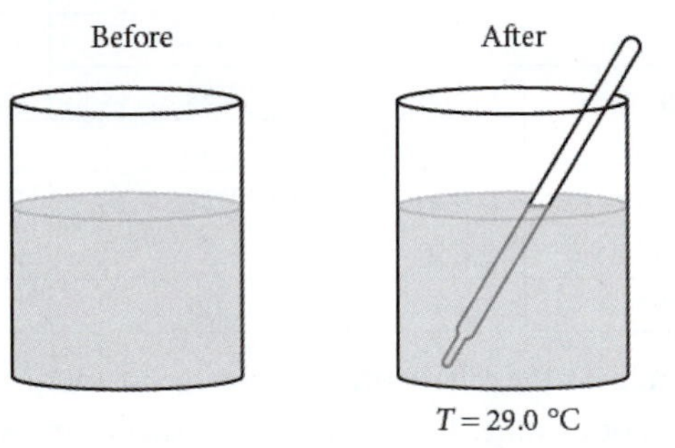

RESEARCH: The heat is given by $Q = mc\Delta T$.

SIMPLIFY: The water will decrease its temperature while the other materials increase their temperature. The heat transfer from the water and the vial must equal the heat transferred to the thermometer

$Q_{water} + Q_{vial} = Q_{pyrex} + Q_{mercury}$.

We can write this as

$$m_{water} c_{water} (T - T_{eq}) + m_{vial} c_{vial} (T - T_{eq}) = m_{pyrex} c_{pyrex} (T - T_0) + m_{mercury} c_{mercury} (T - T_0),$$

where T is the temperature of the water in the vial before the thermometer is inserted. Solving for T:

$$m_{water} c_{water} T - m_{water} c_{water} T_{eq} + m_{vial} c_{vial} T - m_{vial} c_{vial} T_{eq} = m_{pyrex} c_{pyrex} (T_{eq} - T_0) + m_{mercury} c_{mercury} (T_{eq} - T_0)$$

$$T(m_{water} c_{water} + m_{vial} c_{vial}) - T_{eq}(m_{water} c_{water} + m_{vial} c_{vial}) = m_{pyrex} c_{pyrex} (T_{eq} - T_0) + m_{mercury} c_{mercury} (T_{eq} - T_0)$$

$$T = \frac{m_{pyrex} c_{pyrex} (T_{eq} - T_0) + m_{mercury} c_{mercury} (T_{eq} - T_0) + T_{eq}(m_{water} c_{water} + m_{vial} c_{vial})}{m_{water} c_{water} + m_{vial} c_{vial}}.$$

CALCULATE: Putting in our numerical values gives us

$$m_{\text{pyrex}}c_{\text{pyrex}}\left(T_{\text{eq}}-T_0\right)=(0.0150\text{ kg})(800.\text{ J/(kg °C)})(29.0\text{ °C}-20.0\text{ °C})=108\text{ J}$$

$$m_{\text{mercury}}c_{\text{mercury}}\left(T_{\text{eq}}-T_0\right)=(0.00400\text{ kg})(140.\text{ J/(kg °C)})(29.0\text{ °C}-20.0\text{ °C})=5.04\text{ J}$$

$$T_{\text{eq}}\left(m_{\text{water}}c_{\text{water}}+m_{\text{vial}}c_{\text{vial}}\right)=(29.0\text{ °C})\left((0.00600\text{ kg})(4190.\text{ J/(kg °C)})+(0.00500\text{ kg})(800.\text{ J/(kg °C)})\right)=845.06\text{ J}$$

$$m_{\text{water}}c_{\text{water}}+m_{\text{vial}}c_{\text{vial}}=(0.00600\text{ kg})(4190.\text{ J/(kg °C)})+(0.00500\text{ kg})(800.\text{ J/(kg °C)})=29.14\text{ J/°C}$$

$$T=\frac{108\text{ J}+5.04\text{ J}+845.06\text{ J}}{29.14\text{ J/°C}}=32.879\text{ °C}.$$

ROUND: The temperature is reported to three significant figures. The initial temperature of the water is 32.9 °C.

DOUBLE-CHECK: This is a reasonable answer.

18.43. The time needed to vaporize the liquid is given by $t=Q/P=mL_{\text{vap}}/P$. For liquid nitrogen the time of vaporization is calculated to be $t=(1.00\text{ kg})(2.00\cdot10^5\text{ J/kg})/(10.0\text{ W})=2.00\cdot10^4\text{ s}$. The time liquid helium takes to vaporize is given by $t=(1.00\text{ kg})(2.09\cdot10^4\text{ J/kg})/(10.0\text{ W})=2090\text{ s}$. Thus it takes about 10 times longer to vaporize liquid nitrogen than liquid helium.

18.47. **THINK:** The question asks for the amount of water necessary to cool the carbon steel. The steel has a mass of 0.500 kg and must go from a temperature of 1346 °F to 500. °F. These temperatures in Celsius are $T_{\text{h}}=(5/9)(1346\text{ °F}-32\text{ °F})=730.\text{ °C}$ and $T_{\text{c}}=(5/9)(500.\text{ °F}-32\text{ °F})=260.\text{ °C}$. The blade will be surrounded by an unknown quantity of water and 2.000 kg of copper both at room temperature 20.0 °C. The specific heat of copper is 386 J/(kg K). The table associated with the problem gives the specific heat of carbon steel at various temperature ranges.

SKETCH:

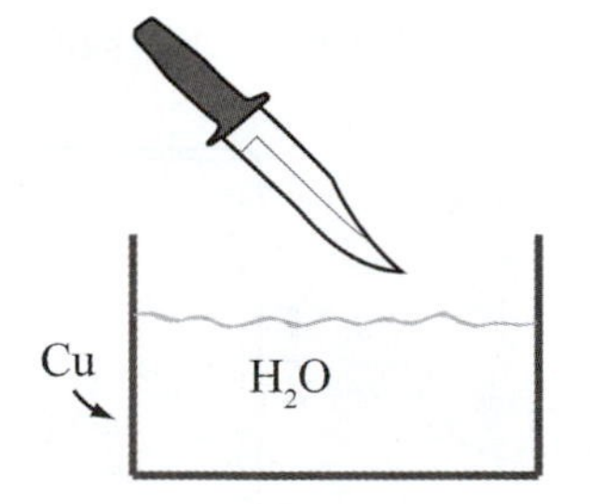

RESEARCH: The heat loss by the carbon steel is equal to the heat gained by the copper and water. The heat is given by $Q=mc\Delta T$.

SIMPLIFY: The heat loss by the carbon steel is:

$$Q_{\text{cs}}=\sum_{\text{i}}m_{\text{cs}}c_{\text{i}}\Delta T_{\text{i}},$$

where the summation goes over the different temperature ranges. The heat transferred to the copper is $Q_{\text{Cu}}=m_{\text{Cu}}c_{\text{Cu}}\Delta T_{\text{b}}$. The water will be brought to its boiling point $Q_{\text{w}}=m_{\text{w}}c_{\text{w}}\Delta T_{\text{b}}$. The water and the copper will reach a temperature of 100. °C. Equating the heat loss of the carbon steel to the heat gain of the water and copper:

$Q_{\text{cs}}=\sum_i m_{\text{cs}}c_i\Delta T_i=Q_{\text{Cu}}+Q_{\text{w}}=m_{\text{Cu}}c_{\text{Cu}}\Delta T_{\text{b}}+m_{\text{w}}c_{\text{w}}\Delta T_{\text{b}}$. Solving for the mass of the water:

$$m_{\text{w}}=\left(\sum_{\text{i}}m_{\text{cs}}c_{\text{i}}\Delta T_{\text{i}}-m_{\text{Cu}}c_{\text{Cu}}\Delta T_{\text{b}}\right)\Big/\left(c_{\text{w}}\Delta T_{\text{b}}\right).$$

CALCULATE:

$$m_{\text{w}} = \frac{0.500\left[(846)(730.-650.)+(754+662+595)100.+553(350.-260.)\right]-2.000(386)(100.-20.0)}{4190(100.-20.0)}$$

$=0.290916,$

with units of

$$\left[m_{\text{w}}\right] = \frac{\text{kg}\left[(\text{J/(kg K)})(^\circ\text{C})+(\text{J/(kg K)})(^\circ\text{C})+(\text{J/(kg K)})(^\circ\text{C})\right]-\text{kg}(\text{J/(kg K)})(^\circ\text{C})}{(\text{J/(kg K)})(^\circ\text{C})} = \text{kg}.$$

Hence, $m_{\text{w}} = 0.290916$ kg.

ROUND: The least precise value is given to three significant figures. If the water does not convert to steam, it takes 291 g of water to cool off the carbon steel.

DOUBLE-CHECK: This is a reasonable answer. It takes relatively little water to accomplish the cooling due to the high specific heat of water. Note that the strength of the carbon steel also depends on the speed at which the steel hardens.

18.51. The question asks for the surface temperature of the Sun. This temperature can be determined by equating the power of a black body to the power reaching the Earth. $P = \sigma\varepsilon A_{\text{s}}T^4 = \int I dA = |I|A_{\text{Earth's orbit}}$. The sun is modeled as a black body, so $\varepsilon = 1$. The area of a sphere is $4\pi r^2$. Using these facts $\sigma 4\pi r_{\text{s}}^2 T^4 = I4\pi r_{\text{ES}}^2$. Solving for the temperature gives $T = \left(Ir_{\text{ES}}^2 / \sigma r_{\text{s}}^2\right)^{1/4}$. Inputting the given values yields:

$$T = \left[(1370.\ \text{W/m}^2)(1.496\cdot 10^8\ \text{km})^2 / (5.67\cdot 10^{-8}\ \text{W/K}^4\text{m}^2)(6.963\cdot 10^5\ \text{km})^2\right]^{1/4} = 5778.99\ \text{K} \text{ or } 5506\ ^\circ\text{C}.$$

The surface temperature of the Sun is 5780 K, or 5510 °C.

18.55. **THINK:** I want to find the rate of heat flow through a window, which is 0.32 cm thick, and has an area of 1.2 m by 1.4 m. The inside and outside temperatures of the window are 8.5 °C and 4.1 °C, respectively.

SKETCH:

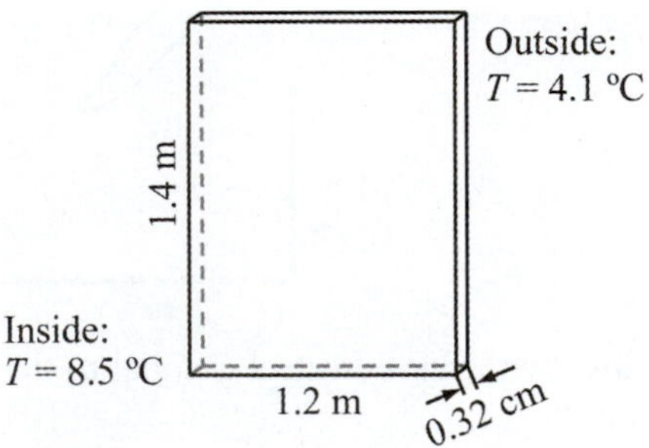

RESEARCH: The power is $P = kA(T_{\text{h}} - T_{\text{c}})/L$.

SIMPLIFY: Not required.

CALCULATE: $P = \dfrac{(0.8\ \text{W/m K})(1.2\ \text{m})(1.4\ \text{m})(281.5\ \text{K} - 277.1\ \text{K})}{0.32\cdot 10^{-2}\ \text{m}} = 1848\ \text{W} = 1.848\ \text{kW}$

ROUND: The heat loss rate is reported to 2 significant figures, so $P = 1.8$ kW.

DOUBLE-CHECK: This is a fairly large amount of power being transmitted through the window, which would lead to significant heating costs. To reduce this power loss, double-pane windows with inert gas between the two panes are used rather than a single-pane window as in this example.

18.59. **THINK:** The Planck spectrum distribution is given by $\varepsilon_{\text{T}}(f) = (2\pi h/c^2)(f^3/(e^{hf/k_{\text{B}}T}-1))$, where h is Planck's constant and c is the speed of light. The frequency of the peak of this distribution is needed. The Boltzmann constant is $k_{\text{B}} = 1.38\cdot 10^{-23}\ \text{m}^2\ \text{kg s}^{-2}\ \text{K}^{-1}$.

SKETCH:

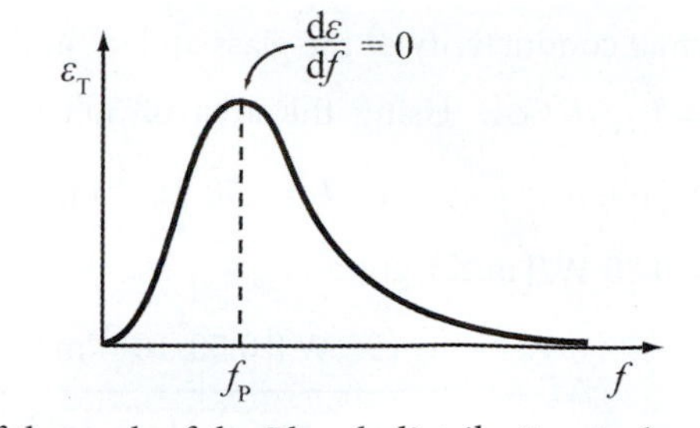

RESEARCH: The frequency of the peak of the Planck distribution is determined from solving $d\varepsilon / df = 0$. The derivative of the Planck distribution is given by:

$$\frac{d\varepsilon}{df} = \left(\frac{2\pi h}{c^2}\right)\left(3f^2\left(e^{hf/k_BT}-1\right)^{-1} - f^3\left(\frac{h}{k_BT}\right)\left(e^{hf/k_BT}-1\right)^{-2}\right).$$

SIMPLIFY: Solving $d\varepsilon / df = 0$, it is found that:

$$3f^2\left(e^{hf/k_BT}-1\right)^{-1} - f^3\left(h/k_BT\right)\left(e^{hf/k_BT}-1\right)^{-2} = 0 \Rightarrow f^2\left(e^{hf/k_BT}-1\right)^{-2}\left[3\left(e^{hf/k_BT}-1\right)-\left(hf/k_BT\right)e^{hf/k_BT}\right]=0.$$

This leads to $3e^x - 3 - xe^x = 0,$ where $x = hf/k_BT.$ Simplifying yields $3 - x = 3e^{-x}$ or $x = 3 - 3e^{-x} = 3\left(1-e^{-x}\right).$ Solving x iteratively with a starting value $x_0 = 3,$ it is found that $x_0 = 3,\ x_1 = 3\left(1-e^{-x_0}\right) = 3\left(1-e^{-3}\right) = 2.8506,\ x_2 = 3\left(1-e^{-x_1}\right) = 3\left(1-e^{-2.8506}\right) = 2.8266$ and after few iterations, the most is $x = 2.8215.$ This means that $hf/k_BT = 2.8215$ or $f = 2.8215\left(k_B/h\right)T.$

CALCULATE:

(a) Substituting $k_B = 1.38\cdot 10^{-23}\ \text{m}^2\ \text{kg s}^{-2}\ \text{K}^{-1}$ and $h = 6.626\cdot 10^{-34}\ \text{J s}$ gives $f = \left(5.8792\cdot 10^{10}\ \text{Hz/K}\right)T$.

(b) At $T = 6.00\cdot 10^3$ K, the frequency of the peak is:

$$f = \left(5.8792\cdot 10^{10}\ \text{Hz/K}\right)\left(6.00\cdot 10^3\ \text{K}\right) = 3.5275\cdot 10^{14}\ \text{Hz}.$$

(c) At $T = 2.735$ K, the frequency of the peak is:

$$f = \left(5.8792\cdot 10^{10}\ \text{Hz/K}\right)\left(2.735\ \text{K}\right) = 1.60796\cdot 10^{11}\ \text{Hz}.$$

(d) At $T = 300.$ K, the frequency of the peak is:

$$f = \left(5.8792\cdot 10^{10}\ \text{Hz/K}\right)\left(300.\ \text{K}\right) = 1.7638\cdot 10^{13}\ \text{Hz} \approx 1.76\cdot 10^{13}\ \text{Hz}.$$

ROUND:

(a) $f = \left(5.88\cdot 10^{10}\ \text{Hz/K}\right)T$.

(b) The temperature of the Sun is given to three significant figures in the question: $f = 3.53\cdot 10^{14}$ Hz.

(c) Boltzmann's constant is given to three significant figures: $f = 1.61\cdot 10^{11}$ Hz. (d) The Earth's temperature is given only to three significant figures in the question: $f = 1.76\cdot 10^{13}$ Hz..

DOUBLE-CHECK: The result in (a), where $f = \text{constant}\left(T\right)$ is known as Wien's displacement law. The rest of the calculated frequencies have Hertz as their units, which is appropriate. As one might expect, the frequencies increase with the temperatures.

18.63. The conduction rate through a spherical glass is given by $P_{\text{cond}} = kA(\Delta T / L)$ where ΔT is the temperature difference, k is the thermal conductivity of the glass and A and L are the area and thickness of the glass. Simplifying gives $\Delta T = P_{\text{cond}} L / kA$. Using the area of sphere, it is found that $\Delta T = P_{\text{cond}} L / \left(k 4\pi r^2\right)$. Substituting $P_{\text{cond}} = 0.95(100.0 \text{ W}) = 95 \text{ W}$, $L = 0.50 \cdot 10^{-3} \text{ m}$, $r = 3.0 \cdot 10^{-2} \text{ m}$ and $k = 0.80 \text{ W/(m K)}$ gives

$$\Delta T = \frac{(95 \text{ W})\left(0.50 \cdot 10^{-3} \text{ m}\right)}{(0.80 \text{ W/mK})(4\pi)\left(3.0 \cdot 10^{-2} \text{ m}\right)^2} = 5.2 \text{ K}.$$

18.67. **THINK:** Solar radiation reaches the Earth's surface at about 1.4 kW/m^2. It is assumed here that the Earth is a black body. This means that there is no reflection due to Earth's atmosphere, and all solar radiation that reaches the Earth is absorbed by the Earth's surface.

SKETCH:

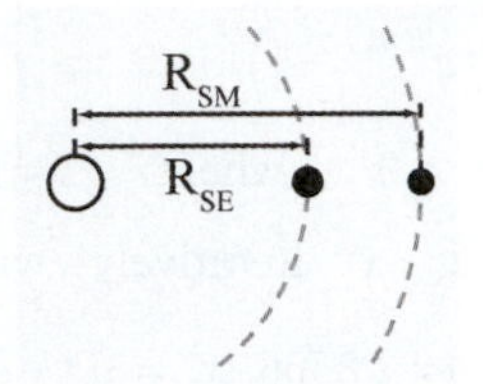

RESEARCH: The Sun emits radiation uniformly in all directions. This means that if the total power of solar radiation is P, then the intensity of radiation at a distance r from the Sun is distributed uniformly over a spherical surface area. Thus, $I = \frac{P}{\text{Spherical Area}} = \frac{P}{4\pi r^2}$. Therefore, the intensity of radiation that reaches the Earth is $I_{\text{E}} = P / 4\pi r_{\text{SE}}^2$. Similarly for Mars, the intensity that reaches Mars is $I_{\text{M}} = P / 4\pi r_{\text{SM}}^2$.

SIMPLIFY: Since I_{E} is known, the power of radiation P can be eliminated from the above equations giving the intensity on Mars as $I_{\text{M}} = \left(4\pi r_{\text{SE}}^2 / 4\pi r_{\text{SM}}^2\right) I_{\text{E}} = \left(r_{\text{SE}}^2 / r_{\text{SM}}^2\right) I_{\text{E}}$.

CALCULATE: Substituting $I_{\text{E}} = 1.4 \text{ kW/m}^2$, $R_{\text{SE}} = 1.496 \cdot 10^{11} \text{ m}$ and $R_{\text{M}} = 2.28 \cdot 10^{11} \text{ m}$ yields:

$$I_{\text{M}} = \left(\frac{1.496 \cdot 10^{11} \text{ m}}{2.28 \cdot 10^{11} \text{ m}}\right)^2 1.4 \cdot 10^3 \text{ W/m}^2 = 6.03 \cdot 10^2 \text{ W/m}^2.$$

ROUND: Keeping only two significant figures gives $I_{\text{M}} = 6.0 \cdot 10^2 \text{ W/m}^2$.

DOUBLE-CHECK: Since R_{SM} is larger than R_{SE} it is expected that I_{M} is less than I_{E}.

18.71. **THINK:** Assume the Gulf Stream is a box shaped object with a length of $8.00 \cdot 10^3$ km, a width of 100. km and a depth of 500. m. Assume also that the temperature inside the box is uniform.

SKETCH:

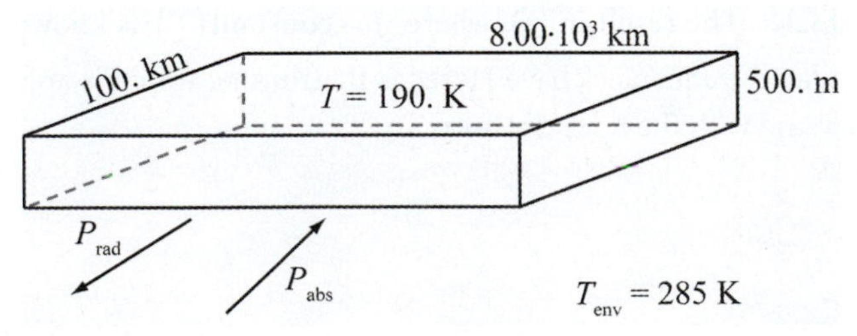

RESEARCH: The power radiated by an object is $P = \sigma\varepsilon A T^4$. The surface area of the Gulf Stream is equal to $A = 2LW + 2Ld + 2Wd$. Here L, W and D are the length, width and depth of the Gulf Stream. The absorbed power is from the Sun, which corresponds to receiving 1400. W/m² for half the day on the surface of the water, $P_{\text{abs}} = (1/2)(1400.\ \text{W/m}^2)(LW) = (700.0\ \text{W/m}^2)(LW)$.

SIMPLIFY: $P_{\text{net}} = P_{\text{rad}} - P_{\text{abs}} = \sigma\varepsilon A T^4 - (700.0\ \text{W/m}^2)(LW)$.

CALCULATE: To make the equation fit on the page, compute P_{rad} first without units:

$$P_{\text{rad}} = (5.6703 \cdot 10^{-8})(0.930)(2)\left[(8.00 \cdot 10^6)(1.00 \cdot 10^5) + (8.00 \cdot 10^6)(500.) + (1.00 \cdot 10^5)(500.)\right](290)^4$$
$$= 5.9978 \cdot 10^{14}.$$

Then, the units of P_{rad} are: $[P_{\text{rad}}] = \left(\text{W}/(\text{K}^4\ \text{m}^2)\right)\left[(\text{m})(\text{m}) + (\text{m})(\text{m}) + (\text{m})(\text{m})\right](\text{K})^4 = \text{W}$, so altogether

$$P_{\text{rad}} = 5.9978 \cdot 10^{14}\ \text{W}.$$
$$P_{\text{abs}} = (700.\ \text{W/m}^2)\left[(8.00 \cdot 10^6\ \text{m})(1.00 \cdot 10^5\ \text{m})\right]$$
$$= 5.60 \cdot 10^{14}\ \text{W}.$$
$$P_{\text{net}} = 5.9978 \cdot 10^{14}\ \text{W} - 5.60 \cdot 10^{14}\ \text{W} = 3.978 \cdot 10^{14}\ \text{W}.$$

ROUND: Rounding to three significant figures gives $P_{\text{net}} = 3.98 \cdot 10^{13}$ W.

DOUBLE-CHECK: It is well known that the Gulf Stream is responsible for significantly warming the waters in the North Atlantic, so one would expect its radiated power to be very large.

18.75. **THINK:** The conduction rate of insulation, P_{cond}, depends on the area and the R value of the material, and the temperature difference on either side of the insulation. Let $T_{\text{H}} = 294$ K, and $T_{\text{C}} = 277$ K.

SKETCH:

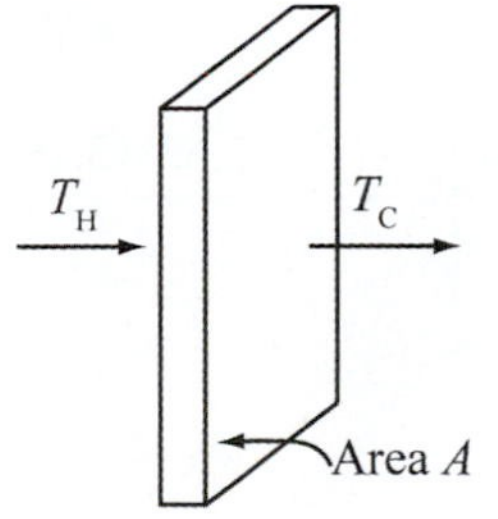

RESEARCH: The conduction rate is given by $P_{\text{cond}} = A(T_{\text{H}} - T_{\text{C}})/R$. The increase in the conduction rate due to the change in the R value is $\Delta P = P_{\text{cond1}} - P_{\text{cond2}} = \left[A(T_{\text{H}} - T_{\text{C}})/R_1\right] - \left[A(T_{\text{H}} - T_{\text{C}})/R_2\right]$.

SIMPLIFY: $\Delta P = A(T_{\text{H}} - T_{\text{C}})\left[(1/R_1) - (1/R_2)\right]$

(a) The change in heat that exits the room in an interval of time is:

$Q = t\Delta P = tA(T_{\text{H}} - T_{\text{C}})\left[(1/R_1) - (1/R_2)\right]$.

(b) In three months, the extra heat that exits the room is $90Q$.

CALCULATE:

(a) Substituting $A = (5.0\ \text{m})(5.0\ \text{m}) = 25\ \text{m}^2$,

$R_1 = 19(0.176\ \text{m}^2\ \text{K/W}) = 3.344\ \text{m}^2\ \text{K/W}$, $R_2 = 30(0.176\ \text{m}^2\ \text{K/W}) = 5.28\ \text{m}^2\ \text{K/W}$ and

$t = 24(3600\ \text{s}) = 86400\ \text{s}$:

$$Q = 86400\ \text{s}(25\ \text{m}^2)(294\ \text{K} - 277\ \text{K})\left[\left(1/(3.344\ \text{m}^2\ \text{K/W})\right) - \left(1/(5.28\ \text{m}^2\ \text{K/W})\right)\right] = 4.03 \cdot 10^6\ \text{J}.$$

(b) $Q=90\left(4.03\cdot 10^6\text{ J}\right)=3.62\cdot 10^8\text{ J}$. Since the electrical energy for heating costs 12 cents/kWh or $3.33\cdot 10^{-6}$ cents/J, the increase in cost of electrical heating is $\text{cost}=3.62\cdot 10^8\text{ J}\left(3.33\cdot 10^{-6}\text{ cents/J}\right)=1206.7\text{ cents}$.

ROUND: Keep only two significant figures.

(a) $Q=4.0\cdot 10^6\text{ J}$

(b) The extra cost is 1200 cents, or 12 dollars.

DOUBLE-CHECK: The units of Q are Joules, and the cost is in cents. These units are expected units for these values. 4 million Joules is a reasonable amount of energy, corresponding to a cost of $12 dollars.

Multi-Version Exercises

18.77. Power = heat /time; solve this for the time:

$$P=Q/t\Rightarrow t=Q/P$$

$$Q=c_{\text{granite}}m_{\text{granite}}\Delta T=c_{\text{granite}}V_{\text{granite}}\rho_{\text{granite}}\Delta T$$

$$t=c_{\text{granite}}V_{\text{granite}}\rho_{\text{granite}}\Delta T/P$$

$$=\frac{(790\text{ J/kg/°C})(0.669\cdot 10^9\text{ m}^3)(2750\text{ kg/m}^3)(64.8\text{°C})}{13.9\cdot 10^6\text{ W}}$$

$$=6.78\cdot 10^9\text{ s}=215\text{ years}$$

18.80.

$$P=A\frac{T_{\text{h}}-T_{\text{c}}}{R_{\text{SI}}}$$

$$A=LW$$

$$R_{\text{SI}}=R/5.678$$

$$P=\frac{LW\left(T_{\text{h}}-T_{\text{c}}\right)}{R/5.678}=\frac{5.678\left(LW\right)\left(T_{\text{h}}-T_{\text{c}}\right)}{R}=\frac{5.678\left(5.183\text{ m}\right)\left(3.269\text{ m}\right)\left(23.37\text{ °C}-1.073\text{ °C}\right)}{29}=74\text{ W}.$$

Chapter 19: Ideal Gases

Exercises

19.29 A tire has an initial gauge pressure of $p_{ig} = 300.\text{ kPa}$, an initial temperature of $T_i = 15.0\text{ °C} = 288\text{ K}$, and a final temperature of $T_f = 45.0\text{ °C} = 318\text{ K}$. The volume change of the tire is negligible, so $V_i = V_f = V$. The number of moles of gas in the tire is constant as well: $n_i = n_f = n$. The final gauge pressure p_{fg} can be found by using Gay-Lussac's Law, $p_i / T_i = p_f / T_f$, with $p_i = p_{ig} + p_{atm}$ and $p_f = p_{fg} + p_{atm}$. Rearranging to solve for p_{fg} gives

$$p_{fg} = \left(p_{ig} + p_{atm}\right)\frac{T_f}{T_i} - p_{atm} = \left(300.\text{ kPa} + 101.3\text{ kPa}\right)\frac{\left(318\text{ K}\right)}{\left(288\text{ K}\right)} - 101.3\text{ kPa} = 342\text{ kPa}.$$

19.33 **THINK:** The pot is filled with steam at a pressure of $p_i = 1.00\text{ atm}$ and a temperature of $T_i = 100.0\text{ °C} = 373.2\text{ K}$. The mass of the pot's lid is $m = 0.500\text{ kg}$. The pot's diameter is $d = 0.150\text{ m}$ and its height is $h = 0.100\text{ m}$. In order for the lid to off of the pot, the force upward from the pressure created by the steam must be greater than the force downward from the weight of the lid. Gay-Lussac's Law can be used to relate the initial and final states of the steam under constant volume.

SKETCH:

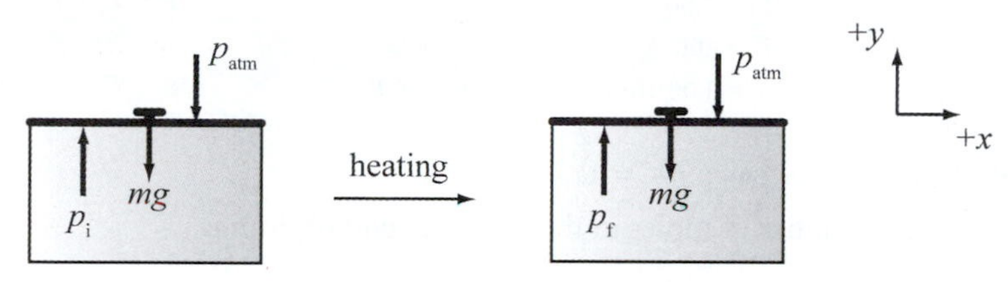

RESEARCH: Pressure is defined as force per unit area: $p = F / A$. The minimum force required to lift the lid off of the pot can be found when the sum of the forces in the y-direction equals zero:

$$\sum_{j=1}^{n} F_{y,j} = 0.$$

Up until the moment when the lid lifts, the volume that the steam occupies is constant. Thus, pressure and temperature can be related using Gay-Lussac's Law: $p_i / T_i = p_f / T_f$. The area of the lid is $A = \pi\left(d/2\right)^2$.

SIMPLIFY: $\sum_{j=1}^{n} F_{y,j} = p_f A - p_{atm} A - mg = 0 \Rightarrow p_f = p_{atm} + \frac{mg}{A}$

$$\frac{p_i}{T_i} = \frac{p_f}{T_f} \Rightarrow T_f = \frac{p_f}{p_i} T_i = \frac{p_{atm} + mg/A}{p_i} T_i$$

CALCULATE:

$$T_f = \frac{\left(101.3\cdot 10^3\text{ Pa}\right) + \left(0.500\text{ kg}\right)\left(9.81\text{ m/s}^2\right) / \left(\pi\left(\left(0.150\text{ m}\right)/2\right)^2\right)}{\left(101.3\cdot 10^3\text{ Pa}\right)}\left(373.2\text{ K}\right) = 374.22\text{ K} = 101.02\text{ °C}$$

ROUND: Rounding to three significant figures, $T_f = 374\text{ K} = 101\text{ °C}$.

DOUBLE-CHECK: It is reasonable that heating the steam slightly above the boiling temperature of water would cause the lid to be lifted off of the pot. You have probably noticed before that a lid on top of a pot rattles as water boils inside.

19.37 The given quantities are the number of moles, $n = 1.00$ mol, the volume, $V = 2.00 \text{ L} = 2.00 \cdot 10^{-3} \text{ m}^3$ and the temperature change, $\Delta T = 100.\ ^\circ\text{C} = 100.$ K. The Ideal Gas Law can be used to find the change in pressure. Since the volume is constant, a change in temperature must cause a change in pressure.

$$\Delta p V = nR\Delta T \Rightarrow \Delta p = \frac{nR\Delta T}{V} = \frac{(1.00 \text{ mol})(8.314 \text{ J/(mol K)})(100. \text{ K})}{(2.00 \cdot 10^{-3} \text{ m}^3)} = 416 \text{ kPa}$$

19.41 **THINK:** Two containers with the volumes, temperatures and pressures shown in the sketch are connected by a tube and allowed to come to equilibrium at pressure p_f. At equilibrium, the final temperature of the system is $T_\text{f} = 300.$ K. The tube that connects the two containers has a negligible volume so the final volume is $V_\text{f} = V_1 + V_2$. The total number of moles in the system remains constant so $n_\text{f} = n_1 + n_2$. The Ideal Gas Law can be used to find the final temperature.

SKETCH:

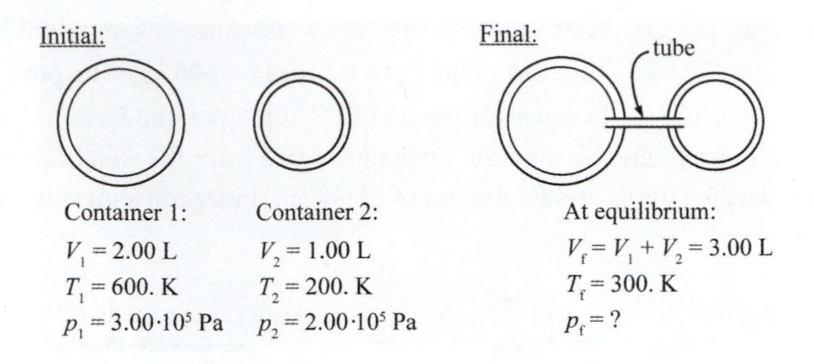

RESEARCH: Ideal Gas Law: $pV = nRT$

SIMPLIFY: The number of moles in the system at equilibrium is:

$$n_1 = \frac{p_1V_1}{RT_1},\ n_2 = \frac{p_2V_2}{RT_2} \Rightarrow n_\text{f} = n_1 + n_2 = \frac{1}{R}\left(\frac{p_1V_1}{T_1} + \frac{p_2V_2}{T_2}\right).$$

Therefore, the final pressure is:

$$p_\text{f} = \frac{n_\text{f}RT_\text{f}}{V_\text{f}} = \frac{1}{R}\left(\frac{p_1V_1}{T_1} + \frac{p_2V_2}{T_2}\right)\left(\frac{RT_\text{f}}{V_1 + V_2}\right) \Rightarrow p_\text{f} = \left(\frac{p_1V_1}{T_1} + \frac{p_2V_2}{T_2}\right)\left(\frac{T_\text{f}}{V_1 + V_2}\right).$$

CALCULATE:

$$p_\text{f} = \left(\frac{(3.00 \cdot 10^5 \text{ Pa})(2.00 \cdot 10^{-3} \text{ m}^3)}{(600. \text{ K})} + \frac{(2.00 \cdot 10^5 \text{ Pa})(1.00 \cdot 10^{-3} \text{ m}^3)}{(200. \text{ K})}\right)\left(\frac{(300. \text{ K})}{(3.00 \cdot 10^{-3} \text{ m}^3)}\right) = 2.00 \cdot 10^5 \text{ Pa}$$

ROUND: $p_\text{f} = 200.$ kPa

DOUBLE-CHECK: Since initially $p_1V_1 / T_1 = p_2V_2 / T_2$, the two containers start out with the same number of gas molecules. Therefore, the average initial kinetic energy per molecule corresponds to an average temperature of $(600 + 200)/2 = 400$ K. Because of higher initial pressure in container 1, gas at first flows from left to right. However, as the temperatures equalize the flow reverses, so that in the end container 1, being twice as large as container 2, holds twice as many molecules. Since the final temperature is less than 400 K, heat has been dissipated into the surrounding environment during the process.

19.45 The number density of atomic hydrogen (H) is

$$\frac{N}{V} = \frac{1.00 \text{ atom}}{\text{cm}^3}\left(\frac{100 \text{ cm}}{1 \text{ m}}\right)^3 = 1 \cdot 10^6 \text{ atoms/m}^3.$$

The temperature is $T = 2.73$ K.

(a) The pressure can be found with the Ideal Gas Law: $pV = Nk_BT$.

$$p = \frac{N}{V}k_BT = \left(1.00\cdot 10^6 \text{ m}^{-3}\right)\left(1.381\cdot 10^{-23} \text{ J/K}\right)\left(2.73 \text{ K}\right) = 3.77\cdot 10^{-17} \text{ Pa}$$

This is a very high vacuum!

(b) The rms speed is $v_{\text{rms}} = \sqrt{3k_BT/m_H}$ and $m_H = 1.008 \text{ u}$, where $\text{u} = 1.661\cdot 10^{-27}$ kg.

$$v_{\text{rms}} = \sqrt{\frac{3\left(1.381\cdot 10^{-23} \text{ J/K}\right)\left(2.73 \text{ K}\right)}{\left(1.008\right)\left(1.661\cdot 10^{-27} \text{ kg}\right)}} = 260. \text{ m/s}.$$

(c) The energy of a monatomic gas is $E_{\text{tot}} = NK_{\text{ave}} = (3/2)Nk_BT$. Given that the number density is $N/V = 10^6$ atoms/m^3, the energy density is:

$$\frac{E_{\text{tot}}}{V} = \frac{3}{2}\frac{N}{V}k_BT \Rightarrow \frac{E_{\text{tot}}}{V} = \frac{3}{2}\left(1.00\cdot 10^6 \text{ m}^{-3}\right)\left(1.381\cdot 10^{-23} \text{ J/K}\right)\left(2.73 \text{ K}\right) = 5.6552\cdot 10^{-17} \text{ J/m}^3.$$

An energy of $E = 1.00$ J would require a cube edge length of:

$$L = \left(\frac{E}{E_{\text{tot}}/V}\right)^{1/3} = \left(\frac{1.00 \text{ J}}{5.6552\cdot 10^{-17} \text{ J/m}^3}\right)^{1/3} = 260{,}525.76 \text{ m} = 261 \text{ km}.$$

19.49 **THINK:** In a period of $\Delta t = 6.00$ s, $N = 9.00\cdot 10^{23}$ nitrogen molecules strike a wall with an area of $A = 2.00\cdot 10^{-4}\text{ m}^2$. The molecules move with a speed of $v = 400.0$ m/s and strike the wall head-on in elastic collisions. Find the pressure p exerted on the wall. Using the principle of conservation of linear momentum for elastic collisions, the pressure on the wall can be calculated.

SKETCH:

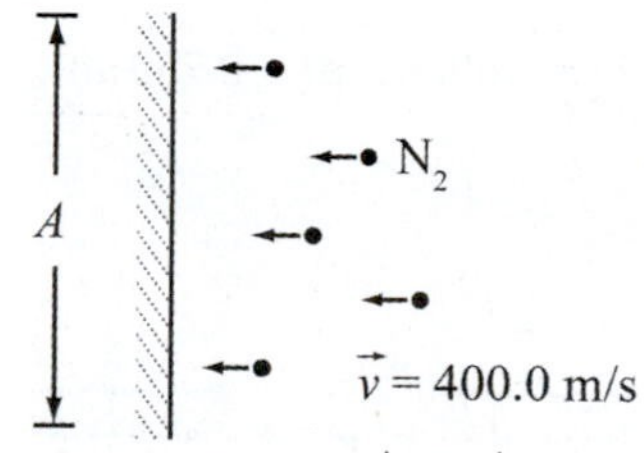

RESEARCH: The pressure is $p = F_{\text{ave}}/A$, where $F_{\text{ave}} = \left|\Delta\vec{p}_{\text{tot}}\right|/\Delta t$ is the average force that acts on the wall, $\left|\Delta\vec{p}_{\text{tot}}\right|$ is the absolute value of the change in the linear momentum of all of the molecules, and A is the area over which the force is acting. In the head-on elastic collisions that the nitrogen molecules undergo, the final velocity has reversed direction with respect to the initial velocity, so

$$\Delta\vec{p} = m\Delta v = m\left(v_f - v_i\right) = m\left(v - (-v)\right) = 2mv.$$

For the total change in linear momentum of N particles, $\Delta\vec{p}_{\text{tot}} = N\Delta p$. The mass of one N_2 molecule is $m = 4.65\cdot 10^{-26}$ kg.

SIMPLIFY: $p = \dfrac{F_{\text{ave}}}{A} = \dfrac{\left|\Delta\vec{p}_{\text{tot}}\right|}{A\Delta t} = \dfrac{2Nmv}{A\Delta t}$

CALCULATE: $p = \dfrac{2\left(9.00\cdot 10^{23}\right)\left(4.65\cdot 10^{-26} \text{ kg}\right)\left(400.0 \text{ m/s}\right)}{\left(2.00\cdot 10^{-4} \text{ m}^2\right)\left(6.00 \text{ s}\right)} = 2.790\cdot 10^4 \text{ Pa}$

ROUND: To three significant figures, the pressure on the wall is $p = 27.9$ kPa.

DOUBLE-CHECK: This pressure is about one quarter of atmospheric pressure so it is reasonable considering the momentum of the N_2 molecules.

19.53 The temperature of $n = 1.00$ mol of a diatomic ideal gas is increased by $\Delta T = 2.00$ K. Assuming a constant volume, the change in internal energy is $\Delta E_{\text{int}} = nC_V\Delta T$ where $C_V = (5/2)R$ for a diatomic gas:

$$\Delta E_{\text{int}} = \frac{5}{2}nR\Delta T = \frac{5}{2}(1.00 \text{ mol})(8.314 \text{ J/(mol K)})(2.00 \text{ K}) = 41.6 \text{ J}.$$

19.57 An initial volume $V_{\text{i}} = 15.0$ L of an ideal monatomic gas at a pressure of $p_{\text{i}} = 1.50 \cdot 10^5$ kPa is expanded adiabatically until the volume **is** doubled, $V_{\text{f}} = 2V_{\text{i}}$. The ratio of the molar specific heats for a monatomic gas is $\gamma = 5/3$.

(a) For an adiabatic gas, pV^{γ} is constant. The pressure of the gas after the adiabatic expansion is:

$$p_{\text{f}}V_{\text{f}}^{\gamma} = p_{\text{i}}V_{\text{i}}^{\gamma} \Rightarrow p_{\text{f}} = p_{\text{i}}\left(\frac{V_{\text{i}}}{V_{\text{f}}}\right)^{\gamma} = (1.50 \cdot 10^5 \text{ kPa})\left(\frac{V_{\text{i}}}{2V_{\text{i}}}\right)^{5/3} = (1.50 \cdot 10^5 \text{ kPa})\left(\frac{1}{2}\right)^{5/3} = 4.72 \cdot 10^4 \text{ kPa}.$$

(b) Suppose the initial temperature of the gas was $T_{\text{i}} = 300.$ K. For an adiabatic expansion, $TV^{\gamma-1}$ is constant. The temperature of the gas after the adiabatic expansion is:

$$T_{\text{f}}V_{\text{f}}^{\gamma-1} = T_{\text{i}}V_{\text{i}}^{\gamma-1} \Rightarrow T_{\text{f}} = T_{\text{i}}\left(\frac{V_{\text{i}}}{V_{\text{f}}}\right)^{\gamma-1} = T_{\text{i}}\left(\frac{V_{\text{i}}}{2V_{\text{i}}}\right)^{\gamma-1} = (300. \text{ K})\left(\frac{1}{2}\right)^{(5/3)-1} = 189 \text{ K}.$$

19.61 **THINK:** In problem 19.60, the initial volume of the monatomic ideal gas was $V_1 = 6.00$ L, the initial temperature was $T_1 = 400.$ K and the initial pressure was $p_1 = 3.00$ atm. The gas underwent an isothermal (constant temperature) expansion to $V_2 = 4V_1$, then an isobaric (constant pressure) compression, and then an adiabatic compression (no heat transfer) back to its original state. The number of moles of gas was $n = 0.5483$ mol and the pressure, volume and temperature of state 2 and state 3 were found to be: $p_2 = 0.750$ atm, $V_2 = 24.0$ L, $T_2 = 400.$ K, $p_3 = 0.750$ atm, $V_3 = 13.78$ L, $T_3 = 229.7$ K (all accurate to three significant figures).

SKETCH:

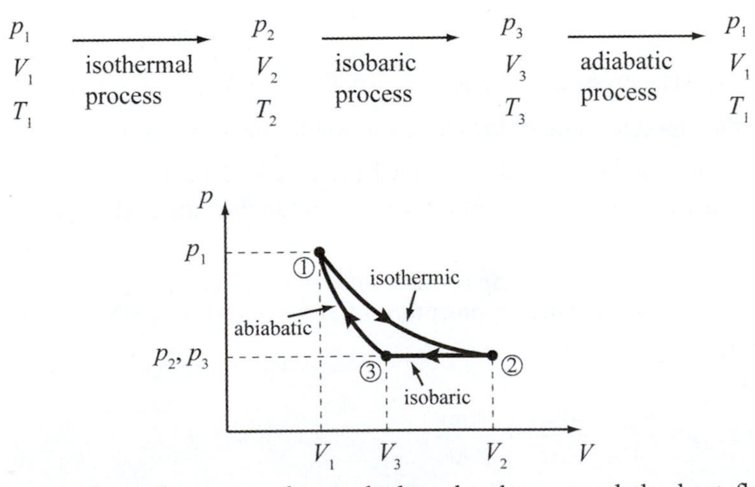

RESEARCH: For an isothermal process, the work done by the gas and the heat flow into the gas is $Q = W = nRT\ln(V_{\text{f}}/V_{\text{i}})$. For an isobaric process, the work done by the gas is $W = p\Delta V$ and the heat flow into the gas is $Q = nC_p\Delta T$. For an adiabatic process, the work done is $W = nR\Delta T/(1-\gamma)$ and the heat flow into the gas is $Q = 0$. For a monatomic ideal gas, $C_p = (5/2)R$ and the ratio of the molar specific heats is $\gamma = 5/3$. To find the number of moles n, use the Ideal Gas Law: $pV = nRT$.

SIMPLIFY: For the isothermal process (state 1 to state 2),

$$Q_{12} = W_{12} = nRT_1 \ln\left(\frac{V_2}{V_1}\right).$$

For the isobaric process (state 2 to state 3),

$$W_{23} = p_2\left(V_3 - V_2\right) \text{ and } Q_{23} = \frac{5}{2}nR\left(T_3 - T_2\right).$$

For the adiabatic process (state 3 to state 1),

$$W_{31} = \frac{nR\left(T_1 - T_3\right)}{1-\gamma} \text{ and } Q_{31} = 0.$$

CALCULATE: For the isothermal process (state 1 to state 2),

$$Q_{12} = W_{12} = \left(0.5483 \text{ mol}\right)\left(8.314 \text{ J/}\left(\text{mol K}\right)\right)\left(400. \text{ K}\right)\ln\left(\frac{\left(24.0 \text{ L}\right)}{\left(6.00 \text{ L}\right)}\right) = 2.5278 \text{ kJ}.$$

For the isobaric process (state 2 to state 3),

$$W_{23} = \left(0.750 \text{ atm}\right)\left(1.013\cdot 10^5 \text{ Pa/atm}\right)\left(13.78 \text{ L} - 24.0 \text{ L}\right)\left(\frac{1 \text{ m}^3}{1000 \text{ L}}\right) = -0.7765 \text{ kJ, and}$$

$$Q_{23} = \frac{5}{2}\left(0.5483 \text{ mol}\right)\left(8.314 \text{ J/}\left(\text{mol K}\right)\right)\left(229.7 \text{ K} - 400. \text{ K}\right) = -1.9408 \text{ kJ}.$$

For the adiabatic process (state 3 to state 1),

$$W_{31} = \frac{\left(0.5483 \text{ mol}\right)\left(8.314 \text{ J/}\left(\text{mol K}\right)\right)\left(400. \text{ K} - 229.7 \text{ K}\right)}{1-\left(5/3\right)} = -1.1645 \text{ kJ, and}$$

$$Q_{31} = 0.$$

ROUND: To three significant figures the answers are: $Q_{12} = W_{12} = 2.53$ kJ, $W_{23} = -0.776$ kJ, $Q_{23} = -1.94$ kJ, $W_{31} = -1.16$ kJ and $Q_{31} = 0$.

DOUBLE-CHECK: The First Law of Thermodynamics is $\Delta E_{\text{int}} = Q - W$. Since the total process is cyclical (starts and ends at state 1), the total change in internal energy must be zero:

$$\begin{aligned}\Delta E_{\text{int},12} + \Delta E_{\text{int},23} + \Delta E_{\text{int},31} &= \left(Q_{12} - W_{12}\right) + \left(Q_{23} - W_{23}\right) + \left(Q_{31} - W_{31}\right) \\ &= \left(2.5278 \text{ kJ} - 2.5278 \text{ kJ}\right) + \left(-1.9408 \text{ kJ} - \left(-0.7765 \text{ kJ}\right)\right) + \left(0 - \left(-1.1645 \text{ kJ}\right)\right) = 0.\end{aligned}$$

Since the total change in internal energy is zero, the answers are reasonable.

19.65 In general, the average speed is $v_{\text{ave}} = \sqrt{\dfrac{8k_{\text{B}}T}{\pi m}}$, and does not depend on the pressure.

(a) The mass of an N_2 molecule is 28.01 amu. At $T = 291$ K, the average speed is:

$$v_{\text{ave},N_2} = \sqrt{\frac{8\left(1.381\cdot 10^{-23} \text{ J/K}\right)\left(291 \text{ K}\right)}{\pi\left(28.01\right)\left(1.661\cdot 10^{-27} \text{ kg}\right)}} = 469 \text{ m/s}.$$

(b) The mass of an H_2 molecule is 2.016 amu. At $T = 291$ K, the average speed is:

$$v_{\text{ave},H_2} = \sqrt{\frac{8\left(1.381\cdot 10^{-23} \text{ J/K}\right)\left(291 \text{ K}\right)}{\pi\left(2.016\right)\left(1.661\cdot 10^{-27} \text{ kg}\right)}} = 1750 \text{ m/s}.$$

19.69 This is an adiabatic process since no heat flow occurs. The temperature of the $n = 1.00$ mol of gas drops from $T_1 = 295$ K to $T_2 = 291$ K. For a diatomic molecule $\gamma = 7/5$, so the work done on the environment is:

$$W = \frac{nR}{1-\gamma}(T_f - T_i) \Rightarrow W = \frac{(1.00 \text{ mol})(8.314 \text{ J/mol K})}{1-(7/5)}(291 \text{ K} - 295 \text{ K}) = 83.1 \text{ J}$$

19.73 The tires of the $3.00 \cdot 10^3$ lb car on a lift have a pressure $p_i = 1 \text{ atm} + 32.0 \text{ lb/in}^2$. The car is then lowered to the ground.

(a) If the volume of the tires does not change appreciably, the weight of the car may deform the tires, but it cannot change the pressure inside the tires. Pressure must vary with either volume or temperature. The pressure in pascals inside the tires after the car has been lowered to the ground is then

$$p_i = 1 \text{ atm} + 32.0 \text{ lb/in}^2 = 1.013 \cdot 10^5 \text{ Pa} + 32.0 \text{ lb/in}^2 \left(\frac{6.894 \cdot 10^3 \text{ Pa}}{\text{lb/in}^2} \right) = 3.219 \cdot 10^5 \text{ Pa}$$

$$p_i = 3.22 \cdot 10^5 \text{ Pa}.$$

(b) The force that supports the weight of the car comes from the tire pressure acting against the ground. The gauge pressure supports the weight of the car. The gauge pressure is

$$p = 32.0 \text{ lb/in}^2 \left(\frac{6.894 \cdot 10^3 \text{ Pa}}{\text{lb/in}^2} \right) = 2.206 \cdot 10^5 \text{ Pa}.$$

The contact area is:

$$A = F/p = \frac{(3.00 \cdot 10^3 \text{ lb})(4.4482 \text{ N/lb})}{(2.206 \cdot 10^5 \text{ Pa})} = 0.06049229 \text{ m}^2$$

$$A = 605 \text{ cm}^2.$$

19.77 At constant temperature of $T = 295$ K, $n = 5.00$ mol of an ideal monatomic gas expands from $V_i = 2.00 \text{ m}^3$ to $V_f = 8.00 \text{ m}^3$.

(a) For an isothermal process (at constant temperature), the work done by the gas is:

$$W = nRT \ln\left(\frac{V_f}{V_i} \right) = (5.00 \text{ mol})(8.314 \text{ J/(mol K)})(295 \text{ K}) \ln\left(\frac{(8.00 \text{ m}^3)}{(2.00 \text{ m}^3)} \right) = 17.0 \text{ kJ}.$$

(b) The final pressure is found from the Ideal Gas Law:

$$p_f V_f = nRT \Rightarrow p_f = \frac{nRT}{V_f} = \frac{(5.00 \text{ mol})(8.314 \text{ J/(mol K)})(295 \text{ K})}{(8.00 \text{ m}^3)} = 1.53 \text{ kPa}.$$

19.79 **THINK:** Find the most probable kinetic energy for a molecule of gas at temperature $T = 300.$ K. The Maxwell kinetic energy distribution for a gas can be used to find the most probable kinetic energy for a molecule at this temperature.

SKETCH:

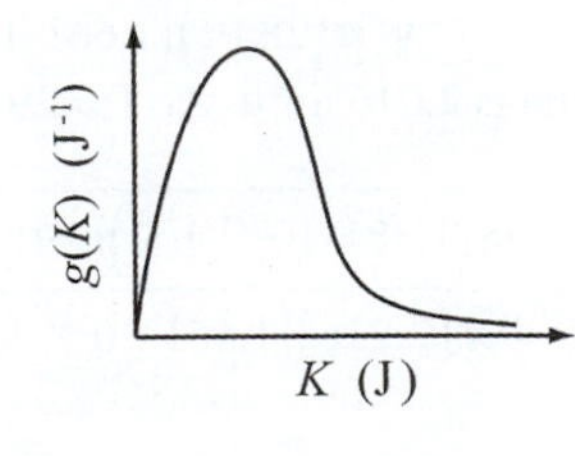

RESEARCH: The Maxwell kinetic energy distribution is

$$g(K)=\frac{2}{\sqrt{\pi}}\left(\frac{1}{k_{\mathrm{B}}T}\right)^{3/2}\sqrt{K}e^{\frac{-K}{k_{\mathrm{B}}T}}$$

The most probable kinetic energy K_{mp} occurs at the maximum value of $g(K)$, and is found by taking the derivative of $g(K)$ with respect to K and setting it to zero.

SIMPLIFY:

$$\frac{d}{dK}g(K)=\frac{2}{\sqrt{\pi}}\left(\frac{1}{k_{\mathrm{B}}T}\right)^{3/2}\left[\frac{1}{2}\frac{1}{\sqrt{K}}e^{-\frac{K}{k_{\mathrm{B}}T}}-\frac{\sqrt{K}}{k_{\mathrm{B}}T}e^{-\frac{K}{k_{\mathrm{B}}T}}\right]_{K=K_{\mathrm{mp}}}=0 \Rightarrow \frac{1}{2\sqrt{K_{\mathrm{mp}}}}=\frac{\sqrt{K_{\mathrm{mp}}}}{k_{\mathrm{B}}T} \Rightarrow K_{\mathrm{mp}}=\frac{k_{\mathrm{B}}T}{2}$$

CALCULATE: $K_{\mathrm{mp}}=\frac{\left(1.381\cdot 10^{-23}\text{ J/K}\right)\left(300.\text{ K}\right)}{2}=2.072\cdot 10^{-21}\text{ J}.$

ROUND: To three significant figures, the most probably kinetic energy is $K_{\mathrm{mp}}=2.07\cdot 10^{-21}$ J.

DOUBLE-CHECK: It is reasonable that the energy depends only on temperature. The answer contains the proper units.

Multi-Version Exercises

19.83 At constant pressure p, the work done is

$$W=p\Delta V=p\left(V_{\mathrm{f}}-V_{\mathrm{i}}\right)=p\left[\left(1-f\right)V_{\mathrm{i}}-V_{\mathrm{i}}\right]=-fpV_{\mathrm{i}}.$$

From the Ideal Gas Law, $V_{\mathrm{i}}=nRT_{\mathrm{i}}/p$. Therefore, the work done by the gas is: $W=-fp\left(\frac{nRT_{\mathrm{i}}}{p}\right)=-fnRT_{\mathrm{i}}$.

$$\begin{aligned}W&=-fnRT_{\mathrm{i}}\\&=-(0.4711)(0.05839\text{ mol})(8.314\text{ J/mol/K})(273.15\text{ K})\\&=-62.47\text{ J}.\end{aligned}$$

19.86 The volume and temperature remain constant. The Ideal Gas Law gives us

$$\begin{aligned}pV&=nRT\\\frac{p}{n}&=\frac{RT}{V}=\text{constant}\\\frac{p_{\mathrm{f}}}{n_{\mathrm{f}}}&=\frac{p_{\mathrm{i}}}{n_{\mathrm{i}}}\\p_{\mathrm{f}}&=p_{\mathrm{i}}\frac{n_{\mathrm{f}}}{n_{\mathrm{i}}}.\end{aligned}$$

Initially there is only air in the bottle, so $n_{\mathrm{i}}=V/22.414\text{ L mol}^{-1}$. After the reaction, there is 1.393 mole of CO_2, so the final number of moles is $n_{\mathrm{f}}=n_{\mathrm{i}}+n_{CO_2}$. So the final pressure is

$p_{\mathrm{f}}=\left(1.013\cdot 10^5\text{ Pa}\right)\frac{n_{\mathrm{i}}+n_{CO_2}}{n_{\mathrm{i}}}$. The initial number of moles is $n_{\mathrm{i}}=\left(2.869\text{ L}\right)/22.414\text{ L mol}^{-1}=0.1280\text{ mol}$.

The final number of moles is

$$p_{\mathrm{f}}=\left(1.013\cdot 10^5\text{ Pa}\right)\frac{0.1280\text{ mol}+1.393\text{ mol}}{0.1280\text{ mol}}=1.204\cdot 10^6\text{ Pa}.$$

(Note that we assumed STP and, to attain four significant figures as suggested by the problem data, used 22.414 L/mol rather than 22.4 L/mol.)

Chapter 20: The Second Law of Thermodynamics

Exercises

20.27. **THINK:** As the water (specific heat $c = 4.19 \text{ kJ/(kg}\cdot\text{K)}$, density $\rho = 1.00 \text{ g/cm}^3 = 1.00 \text{ kg/L}$, and volume $V = 2.00 \text{ L}$) is cooled from $T_{\text{H}} = 25.0 \text{ °C}$ to $T_{\text{L}} = 4.00 \text{ °C}$, a quantity of heat, Q_{L}, is extracted from it by the refrigerator of power $P = 480. \text{ W}$. The coefficient of performance, $K = 3.80$, relates the heat extracted to the work, W, done by the fridge.

SKETCH:

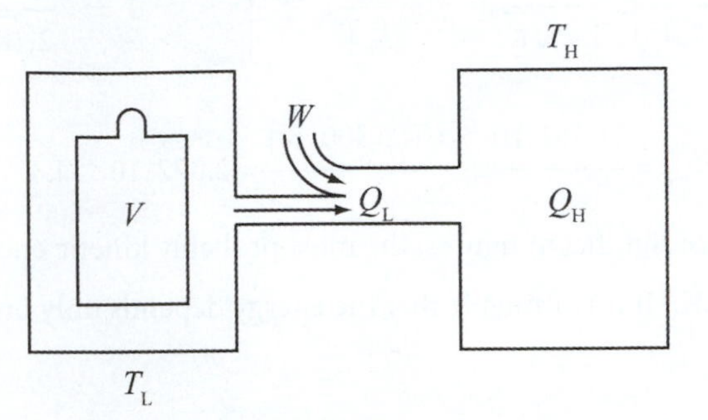

RESEARCH: The mass of the water is $m = \rho V$. The heat then extracted from the water is $Q_{\text{L}} = mc\Delta T$. The coefficient of performance of the fridge is $K = Q_{\text{L}} / W$ and the work done by the fridge to cool the water is $W = P\Delta t$.

SIMPLIFY: The heat extracted from water is $Q_{\text{L}} = \rho V c\left(T_{\text{f}} - T_{\text{i}}\right)$. The work done by the fridge is $W = Q_{\text{L}} / K = P\Delta t$; therefore, $\Delta t = Q_{\text{L}} / (PK) = \rho V c\left(T_{\text{f}} - T_{\text{i}}\right) / (PK)$.

CALCULATE: $\Delta t = \left(1.00 \text{ kg/L}\right)\left(2.00 \text{ L}\right)\left(4.19 \cdot 10^3 \text{ J/kg K}\right)\left(25.0 \text{ °C} - 4.00 \text{ °C}\right) \big/ \left(\left(480. \text{ W}\right)\left(3.80\right)\right) = 96.48 \text{ s}$

ROUND: Three significant figures: $\Delta t = 96.5 \text{ s}$.

DOUBLE-CHECK: A K-value of 3.80 is a bit higher than most conventional fridges, so a relatively short time of cooling is reasonable.

20.31. **THINK:** During the two adiabatic processes (points $4 \rightarrow 3$ and $2 \rightarrow 1$), the heat flow $Q_{43} = Q_{21} = 0 \text{ J}$. Heat flows to and from the system, respectively, during the two isobaric processes (points $1 \rightarrow 4$ and $3 \rightarrow 2$) where $Q_{14} = Q_{\text{L}}$ and $Q_{32} = Q_{\text{H}}$. Since this a closed path process, the total work, W, is equal to the total heat flow, Q. The coefficient of performance can then be calculated solely from the heat flow.

SKETCH:

(a)

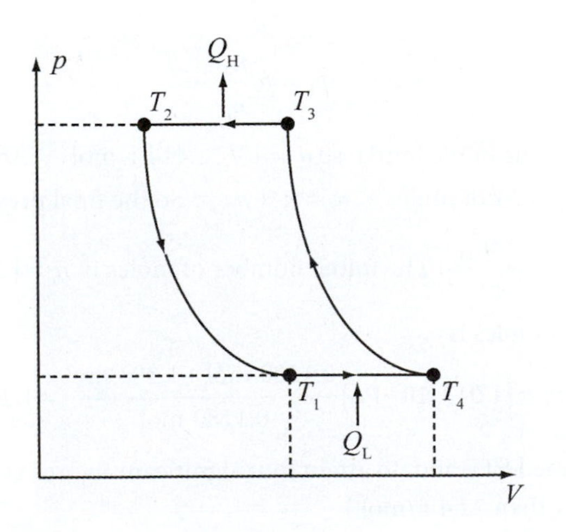

RESEARCH: The heat flowing into the system is $Q_L = Q_{14} = nC_p(T_4 - T_1)$. The heat flowing out of the system is $Q_H = Q_{32} = -nC_p(T_2 - T_3)$. Since it is a closed path $Q = Q_H - Q_L = W$. The coefficient of performance is $K = Q_L / W$.

SIMPLIFY:

(a) Not necessary.

(b) $K = Q_L / W = Q_L / (Q_H - Q_L) = nC_p(T_4 - T_1) / \left[-nC_p(T_2 - T_3) - nC_p(T_4 - T_1)\right]$

Therefore, $K = (T_4 - T_1) / (T_3 - T_2 - T_4 + T_1)$.

CALCULATE: Not required.

ROUND: Not required.

DOUBLE-CHECK: The coefficient of performance will not have any units, since the units of the T values will cancel. The correct coefficient of performance has been derived.

20.35. If such a thermal-energy plant could be designed to operate at maximum efficiency, it would act as an ideal Carnot engine so that the efficiency would be $\varepsilon = (T_H - T_L)/T_H$. A reasonable values for the temperature at sea level would be $T_H = 10.0$ °C (283 K), and the temperature at the bottom of sea would be $T_L = 4.00$ °C (277 K), and since $T_H - T_L = 6.00$ °C (6.00 K), the maximum efficiency of the plant would be $\varepsilon = 6.00 \text{ K} / 283 \text{ K} = 0.0212 = 2.12$ %.

20.39. $K = 5.00$ and $Q_L = 40.0$ cal. In cooling mode:

$$K = \frac{Q_L}{W} = \frac{Q_L}{Q_H - Q_L} = \frac{1}{\frac{Q_H}{Q_L} - 1} \Rightarrow \frac{Q_H}{Q_L} - 1 = \frac{1}{K} \Rightarrow \frac{Q_H}{Q_L} = \frac{1}{K} + 1 \Rightarrow Q_H = Q_L\left(\frac{1}{K} + 1\right).$$

So, $Q_H = 40.0 \text{ cal}\left(\frac{1}{5.00} + 1\right) = 48.0$ cal.

20.43. **THINK:** The thermal efficiency of an engine is the ratio of how much work is output each cycle to how much heat is input.

(a) The pressure and volume can be read off the graph. The temperature can be calculated using the ideal gas law.

(b) The efficiency can be found by computing the total work done and the total heat.

(c) In part (a) the maximum and minimum temperatures were calculated. From this, the maximum efficiency can be computed.

SKETCH:

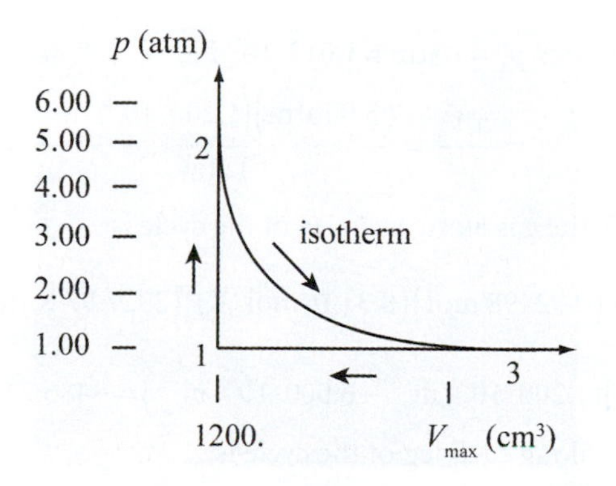

RESEARCH: Ideal gas law: $pV = nRT$. Work done by the gas: $W = 0$ if $\Delta V = 0$; $W = nRT\ln\left(V_f / V_i\right)$ if $\Delta T = 0$; $W = p\Delta V$ if $\Delta p = 0$. Helium is a monatomic gas, so $C_V = (3/2)R$ and $C_p = (5/2)R$. Heat flow: $Q = (3/2)nR\Delta T$ if $\Delta V = 0$; $Q = W$ if $\Delta T = 0$; $Q = (5/2)nR\Delta T$ if $\Delta p = 0$. Thermal efficiency: $\varepsilon = W_{\text{net}} / Q_{\text{H}}$. The maximum thermal efficiency can be calculated as: $\varepsilon_{\text{max}} = 1 - \left(T_{\text{min}} / T_{\text{max}}\right)$.

SIMPLIFY:

(a) At point 1: $V_1 = 1200.\text{ cm}^3$, $p_1 = 1\text{ atm}$, $p_1V_1 = nRT_1 \Rightarrow T_1 = p_1V_1 / nR$. At point 2: $V_2 = 1200.\text{ cm}^3$, $p_2 = 5.00\text{ atm}$, and $T_2 = p_2V_2 / nR$. At point 3: $p_3V_3 = nRT_3$, $p_3 = 1\text{ atm}$. Since $T_3 = T_2$ (isotherm), $p_3V_3 = p_2V_2 \Rightarrow V_3 = p_2V_2 / p_3$.

(b) $\varepsilon = W_{\text{net}} / Q_{\text{H}}$, $W_{\text{net}} = W_{12} + W_{23} + W_{31}$ and $Q_{\text{H}} = Q_{12} + Q_{23} + Q_{31}$ only for $Q > 0$ (only for heat input) since the thermal efficiency is being computed.

$$\left.\begin{aligned} &W_{12} = 0 \\ &W_{23} = nRT_2 \ln\left(\frac{V_3}{V_2}\right) \\ &W_{31} = p_3\left(V_1 - V_3\right) \\ &Q_{12} = \frac{3}{2}nR\left(T_2 - T_1\right) \\ &Q_{23} = W_{23} \\ &Q_{31} = \frac{5}{2}nR\left(T_1 - T_3\right) \end{aligned}\right\} \quad \varepsilon = \frac{W_{12} + W_{23} + W_{31}}{Q_{12} + Q_{23} + Q_{31}}, \text{ only for } Q > 0$$

(c) $\varepsilon_{\text{max}} = 1 - T_{\text{min}} / T_{\text{max}}$

CALCULATE:

(a) First, calculate the number of moles of helium present: $n = \dfrac{0.100\text{ g}}{4.003\text{ g/mol}} = 0.02498\text{ mol}$.

For point 1 the values are: $V_1 = \left(1200.\text{ cm}^3\right)\left(10^{-6}\text{ m}^3 \cdot \text{cm}^{-3}\right) = 1.200 \cdot 10^{-3}\text{ m}^3$, $p_1 = 1\text{ atm} = 1.013 \cdot 10^5\text{ Pa}$,

$$T_1 = \frac{\left(1.013 \cdot 10^5\text{ Pa}\right)\left(1.200 \cdot 10^{-3}\text{ m}^3\right)}{\left(0.02498\text{ mol}\right)\left(8.31\text{ J/(mol} \cdot \text{K)}\right)} = 585.59\text{ K}.$$

For point 2 the values are: $V_2 = 1.200 \cdot 10^{-3}\text{ m}^3$, $p_2 = 5.00\text{ atm} = 5.065 \cdot 10^5\text{ Pa}$,

$$T_2 = \frac{\left(1.200 \cdot 10^{-3}\text{ m}^3\right)\left(5.065 \cdot 10^5\text{ Pa}\right)}{\left(0.02498\text{ mol}\right)\left(8.31\text{ J/(mol} \cdot \text{K)}\right)} = 2927.97\text{ K}.$$

For point 3 the values are: $p_3 = 1\text{ atm} = 1.013 \cdot 10^5\text{ Pa}$, $T_3 = T_2 = 2927.97\text{ K}$,

$$V_3 = \frac{p_2V_2}{p_3} = \frac{\left(5.00\text{ atm}\right)\left(1.200 \cdot 10^{-3}\text{ m}^3\right)}{1\text{ atm}} = 6.0000 \cdot 10^{-3}\text{ m}^3.$$

(b) The work done by the gas along each leg of the cycle is:

$$W_{12} = 0\text{ J},\ W_{23} = \left(0.02498\text{ mol}\right)\left(8.31\text{ J/(mol} \cdot \text{K)}\right)\left(2927.97\text{ K}\right)\ln\left(\frac{6.000 \cdot 10^{-3}\text{ m}^3}{1.200 \cdot 10^{-3}\text{ m}^3}\right) = 978.2150\text{ J, and}$$

$W_{31} = \left(1.013 \cdot 10^5\text{ Pa}\right)\left(1.200 \cdot 10^{-3}\text{ m}^{-3} - 6.000 \cdot 10^{-3}\text{ m}^{-3}\right) = -486.24\text{ J}$. The heat that provided to the engine (absorbed by the gas) along each leg of the cycle is:

$Q_{12} = \frac{3}{2}\left(0.02498\text{ mol}\right)\left(8.31\text{ J/(mol} \cdot \text{K)}\right)\left(2927.97\text{ K} - 585.59\text{ K}\right) = 729.36\text{ J}$, $Q_{23} = W_{23} = 978.2150\text{ J}$, and

$$Q_{31}=\frac{5}{2}(0.02498 \text{ mol})(8.31 \text{ J/(mol}\cdot\text{K)})(585.59 \text{ K}-2927.97 \text{ K})=-1215.60 \text{ J}.$$

Since $Q_{31}<0$, ignore it, because this is heat that is taken from the engine. The efficiency only depends on the heat input to the gas in the engine. Therefore, $\varepsilon=\dfrac{0 \text{ J}+978.2150 \text{ J}-486.24 \text{ J}}{729.36 \text{ J}+978.2149 \text{ J}}=0.288113.$

(c) $\varepsilon_{\text{max}}=1-\dfrac{T_{\text{min}}}{T_{\text{max}}}=1-\dfrac{585.59}{2927.97}=0.8000$

ROUND: Three significant figures:

(a) $p_1=101 \text{ kPa}$, $V_1=1.20\cdot10^{-3} \text{ m}^3$, $T_1=586 \text{ K}$, $p_2=507 \text{ kPa}$, $V_2=1.20\cdot10^{-3} \text{ m}^3$, $T_2=2930 \text{ K}$, $p_3=101 \text{ kPa}$, $V_3=6.00\cdot10^{-3} \text{ m}^3$, $T_3=2930 \text{ K}$

(b) $\varepsilon=0.288$

(c) $\varepsilon_{\text{max}}=0.800$

DOUBLE-CHECK: Of course, it must be that $\varepsilon<\varepsilon_{\text{max}}$ and this is true here. These are reasonable results for the given system

20.47. (a) The theoretical maximum efficiency is provided by a Carnot engine, whose efficiency depends reservoir temperatures. $\varepsilon=1-(T_{\text{L}}/T_{\text{H}})=1-(300. \text{ K}/400. \text{ K})=0.250.$

(b) A Carnot engine consists of two isothermal processes and two isentropic processes. Therefore, after a cycle the total entropy change is zero.

20.51. **THINK:** The entropy is related to the number of states shared by the specified property. This problem can be solved by counting the possibilities.

SKETCH: Possible spin states (5 spins in total, "+"= spin-up, "-" = spin-down).

```
+ + + + +
+ + + + -
+ + + - +     2^5 = 32 possibilites
+ + + - -
    :
- - - - -
```

RESEARCH: $S=k_{\text{B}}\ln w$

SIMPLIFY: 5-up: $w=1$ (there is only 1 way to put 5 up). 3-up: w can be determined by listing the possibilities:

```
+++--
++-+-   ++--+                    10 possibilities.
+-++-   +-+-+   +--++
-+++-   -++-+   -+-++   --+++    ⇒ w = 10
```

Alternate method: Choosing 3 of 5 spin-up: $\Rightarrow w={}_5C_3=10$ (as before). $S_{3\text{ up}}=k_{\text{B}}\ln10.$

CALCULATE: $S_{5\text{ up}}=k_{\text{B}}\ln w=k_{\text{B}}\ln1=0$

$S_{3\text{ up}}=k_{\text{B}}\ln w=k_{\text{B}}\ln10=(1.38\cdot10^{-23} \text{ J/K})\ln10=3.178\cdot10^{-23} \text{ J/K}$

ROUND: $S_{5\text{ up}}=0$ (exactly). For $S_{3\text{ up}}$, round to three significant figures: $S_{3\text{ up}}=3.18\cdot10^{-23} \text{ J/K}.$

DOUBLE-CHECK: By definition the entropy of one microstate is exactly 0. As the number of states increase the entropy of the system increases so it must be that $S_{3\text{ up}}>S_{5\text{ up}}$, as it has been found.

20.55. $\varepsilon=1-\dfrac{T_{\text{L}}}{T_{\text{H}}}=1-\dfrac{288 \text{ K}}{5700 \text{ K}}=0.95.$

20.59. Heat capacity: $C = \Delta E / \Delta T$. Entropy:

$$\Delta S = \int_i^f \frac{dQ}{T} = C\int_{T_i}^{T_f} \frac{dT}{T} = \frac{\Delta E}{\Delta T}\ln\left(\frac{T_f}{T_i}\right) = \frac{0.0700\text{ J}}{0.500\text{ K}}\ln\left(\frac{100.\text{ °C}+273.15\text{ °C}}{10.\text{ °C}+273.15\text{ °C}}\right) = 0.0386\text{ J/K}.$$

20.63. The entropy change for this process is given by $\Delta S = \Delta Q / T$. Dividing both sides by the number of moles, n, yields $(\Delta S / n) = (\Delta Q / n)/T = L_{vap} / T$. So, at a pressure of 100.0 kPa, the boiling temperature is $T = L_{vap} / (\Delta S / n) = (5.568 \cdot 10^3 \text{ J/mol})/(72.1 \text{ J/(mol K)}) = 77.2 \text{ K}$.

20.65. **THINK:** A Carnot engine is the most efficient engine possible that operates between two temperature reservoirs. The efficiency of the Carnot cycle between these two temperature reservoirs depends on the ratio of the temperatures. The question gives the power input and the power output; a ratio of the two will provide the actual efficiency. The unit gal is taken to be American gallons (1 gal = 3.785 L).

SKETCH:

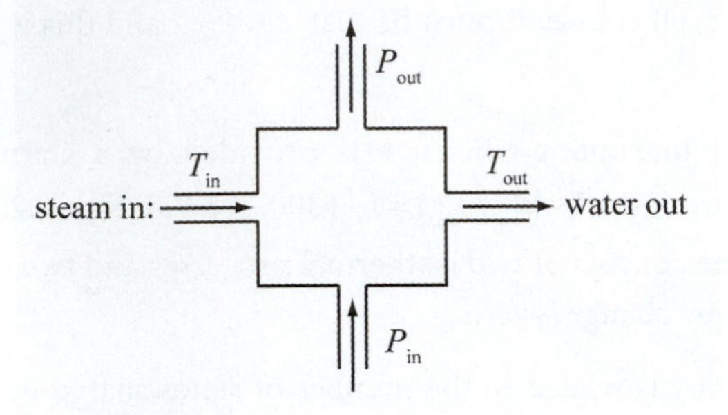

RESEARCH:

(a) The maximum thermal efficiency of an engine (Carnot) is $\varepsilon_{max} = 1 - (T_{out} / T_{in})$.

(b) The actual efficiency is simply $\varepsilon_{act} = P_{out} / P_{in}$.

(c) The heat given to the river water during condensation of the steam is $Q = (P_{in} - P_{out})t = \Delta P t$ over time t and $Q = mc\Delta T$. The volumetric flow rate of the water can be written as $f = (m / \rho)/t$, where m is the mass, ρ is the density and t is time. Therefore, $m = f\rho t$.

SIMPLIFY:

(a) Already in simplified form.

(b) Already in simplified form.

(c) $Q = mc\Delta T = \Delta P t = f\rho t c \Delta T \Rightarrow \Delta T = \dfrac{\Delta P}{f\rho c} \Rightarrow T_f = T_i + \dfrac{\Delta P}{f\rho c}$

CALCULATE:

(a) $\varepsilon_{max} = 1 - \dfrac{30.0\text{ °C} + 273.15\text{ °C}}{300.\text{ °C} + 273.15\text{ °C}} = 0.47108$

(b) $\varepsilon = \dfrac{1000.\text{ MW}}{3000.\text{ MW}} = 0.3333$

(c) $T_f = 20.0\text{ °C} + \dfrac{(3000.\text{ MW} - 1000.\text{ MW})(10^6\text{ W/MW})}{(4.00 \cdot 10^7\text{ gal/h})(1\text{ h/3600 s})(3.785\text{ L/gal})(1\text{ kg/L})(4186\text{ J/(kg °C)})} = 31.361\text{ °C}$

ROUND: Three significant figures:

(a) $\varepsilon_{max} = 0.471$

(b) $\varepsilon = 0.333$

(c) $T_f = 31.4\text{ °C}$

DOUBLE-CHECK: These are reasonable results for the parameters given.

20.69. **THINK:** Since there is no internal energy change during isothermal expansion, the heat absorbed is equal to the work done. Along the isobar, it is only necessary to compute the work done. During the constant volume process, the work is zero, and the heat is equal to the change in internal energy.

SKETCH:

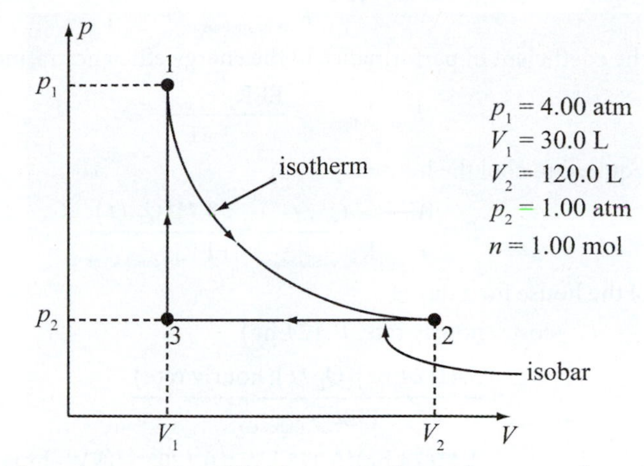

RESEARCH: The ideal gas law: $pV = nRT$. For an isothermal process there is no change in internal energy so the work done by the gas is equal to the heat flow into the gas: $W = Q_H = nRT_2 \ln(V_2/V_1) \equiv W_T$. For an isobaric process: $W = p_2 \Delta V \equiv W_P$. For a constant-volume process: $\Delta E_{int} = nC_V \Delta T \equiv Q_V$. The efficiency of the engine is equal to the ratio of the net work done by the gas to the heat that is input: $\varepsilon = W_{net}/Q_{in}$.

SIMPLIFY: The efficiency of the engine in terms of the work and heat flow over the cycle is: $\varepsilon = \dfrac{W_T + W_P}{Q_H + Q_V} = \dfrac{W_T + W_P}{W_T + Q_V}$.

$$T_2 = \frac{p_2 V_2}{nR} \Rightarrow W_T = nRT_2 \ln\left(\frac{V_2}{V_1}\right) = p_2 V_2 \ln\left(\frac{V_2}{V_1}\right)$$

$$W_P = p_2 (V_1 - V_2)$$

$$Q_V = \frac{3}{2} nR(T_1 - T_3) = \frac{3}{2} nR\left(\frac{p_1 V_1}{nR} - \frac{p_2 V_1}{nR}\right) = \frac{3}{2} V_1 (p_1 - p_2)$$

CALCULATE:

$$W_T = (1.00 \text{ atm})(1.013 \cdot 10^5 \text{ Pa/atm})(120.0 \text{ L})(10^{-3} \text{ m}^3/\text{L}) \ln\left(\frac{120.0 \text{ L}}{30.0 \text{ L}}\right) = 16852 \text{ J}$$

$$W_P = (1.00 \text{ atm})(1.013 \cdot 10^5 \text{ Pa/atm})(30.0 \text{ L} - 120.0 \text{ L})(10^{-3} \text{ m}^3/\text{L}) = -9117 \text{ J}$$

$$Q_V = \frac{3}{2}(30.0 \text{ L})(10^{-3} \text{ m}^3/\text{L})(4.00 \text{ atm} - 1.00 \text{ atm})(1.013 \cdot 10^5 \text{ Pa/atm}) = 13676 \text{ J}$$

Therefore the efficiency is $\varepsilon = \dfrac{16852 \text{ J} - 9117 \text{ J}}{16852 \text{ J} + 13676 \text{ J}} = 0.2534.$

ROUND: Three significant figures: $\varepsilon = 0.253$.

DOUBLE-CHECK: This is a reasonable efficiency for an engine.

Multi-Version Exercises

20.71. The work that the air conditioner is required to do is

$$W = \frac{Q_L}{K_{\text{air conditioner}}}.$$

We can relate the coefficient of performance to the energy efficiency rating by

$$K_{\text{air conditoner}} = \frac{\text{EER}_{\text{air conditioner}}}{3.41}.$$

So the power required to cool the house is

$$P = \frac{W}{t} = \frac{Q_L / t}{K_{\text{air conditioner}}} = \frac{3.41(Q_L / t)}{\text{EER}_{\text{air conditioner}}}.$$

The cost to cool the house for a day is

$$\begin{aligned} \text{cost} &= \text{hourly rate} \cdot P \cdot (24 \text{ hr}) \\ &= \frac{3.41(24 \text{ hr})(Q_L / t)(\text{hourly rate})}{\text{EER}_{\text{air conditioner}}} \\ &= \frac{3.41(24 \text{ hr})(5.375 \text{ kW})(0.1285 \text{ \$/(kW} \cdot \text{hr}))}{10.47} \\ &= \$5.399. \end{aligned}$$

20.74. This system acts like a refrigerator. The maximum coefficient of performance of a refrigerator is

$$K_{\max} = \frac{T_L}{T_H - T_L}.$$

The minimum work that must be done is then

$$\begin{aligned} W_{\min} &= \frac{Q_L}{K_{\max}} \\ &= \frac{Q_L (T_H - T_L)}{T_L} \\ &= \frac{(288.1 \text{ J})(195.3 \text{ °C} - 24.93 \text{ °C})}{24.93 \text{ °C} + 273.15 \text{ °C}} \\ &= 164.7 \text{ J}. \end{aligned}$$

20.77. The efficiency is given by $\varepsilon = W / Q_H$. The first law of thermodynamics tells us that $Q_H = W + Q_L$. We can relate Q_L to the change in temperature of the water $Q_L = mc\Delta T$. The flow rate in terms of volume is

$$f = \frac{V}{t} = \frac{(m / \rho)}{t}.$$

So we can write the mass as $m = f\rho t$. So the efficiency is

$$\varepsilon = \frac{W}{Q_H} = \frac{W}{W + Q_L} = \frac{W}{W + mc\Delta T} = \frac{W}{W + f\rho tc\Delta T}.$$

We can write $W = Pt$, so

$$\begin{aligned} \varepsilon &= \frac{Pt}{Pt + f\rho tc\Delta T} \\ &= \frac{P}{P + f\rho c\Delta T} \\ &= \frac{1833 \text{ W}}{1833 \text{ W} + (132.3 \text{ L/h})(10^{-3} \text{ m}^3/\text{L})\left(\frac{1 \text{ h}}{3600 \text{ s}}\right)(1000 \text{ kg/m}^3)(4186 \text{ J/(kg °C)})(26.69\text{°C} - 11.25\text{°C})} \\ &= 0.4356. \end{aligned}$$

Chapter 21: Electrostatics

Exercises

21.31. The number of atoms or molecules in one mole of a substance is given by Avogadro's number, $N_A = 6.022 \cdot 10^{23}$ mol^{-1}. The faraday unit is $F = N_A e$, where e is the elementary charge of an electron or proton and is equal to $1.602 \cdot 10^{-19}$ C. To calculate the number of coulombs in 1.000 faraday you can multiply N_A by the elementary charge:

$$1.000 \text{ F} = N_A e = (6.022 \cdot 10^{23} \text{ atoms/mol})(1.602 \cdot 10^{-19} \text{ C}) = 96470 \text{ C}.$$

21.35. **THINK:** Protons are incident on the Earth from all directions at a rate of $n = 1245.0 \text{ protons}/\left(\text{m}^2 \text{ s}\right)$. Assuming that the depth of the atmosphere is $d = 120 \text{ km} = 120,000 \text{ m}$ and that the radius of the Earth is $r = 6378 \text{ km} = 6,378,000 \text{ m}$, I want to determine the total charge incident upon the Earth's atmosphere in 5.00 minutes.

SKETCH:

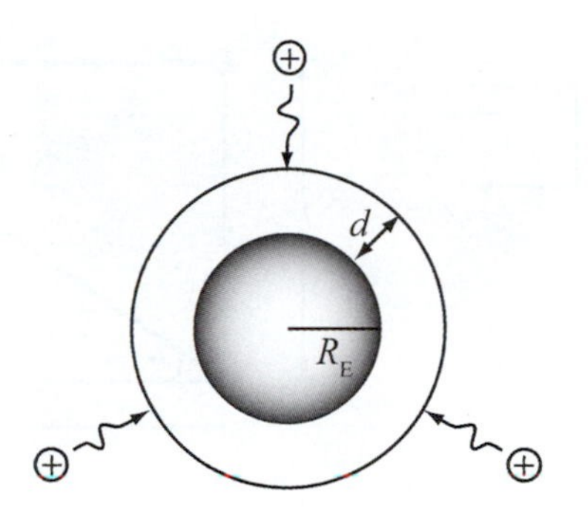

RESEARCH: Modeling the Earth like a sphere, the surface area A can be approximated as $A = 4\pi r^2$. The total number of protons incident on the Earth in the time t can be found by multiplying the rate, n by the surface area of the Earth and the time, t. The total charge Q can be found by multiplying the total number of protons, P by the charge per proton. The elementary charge of a proton is $1.602 \cdot 10^{-19}$ C.

SIMPLIFY: $P = nAT = n4\pi r^2 t, \quad Q = P\left(1.602 \cdot 10^{-19} \text{ C}/P\right)$

CALCULATE:

$P = 1245.0 \text{ protons}/(\text{m}^2\text{s}) \left[4\pi(6,378,000 \text{ m} + 120,000 \text{ m})^2\right](300. \text{ s}) = 1.981800 \cdot 10^{20}$ protons,

$Q = 1.981800 \cdot 10^{20} \text{ protons} \cdot \left(1.602 \cdot 10^{-19} \text{ C/protons}\right) = 31.74844 \text{ C}$

ROUND: To three significant figures 31.7 C

DOUBLE-CHECK: The calculated answer has the correct units of charge. The value seems reasonable considering the values that were provided in the question.

21.39. The charge on each particle is q. When the separation distance is $d = 1.00$ m, the electrostatic force is $F = 1.00$ N. The charge q is found from $F = kq_1q_2/d^2 = kq^2/d^2$. Then,

$$q = \sqrt{\frac{Fd^2}{k}} = \sqrt{\frac{(1.00 \text{ N})(1.00 \text{ m})^2}{8.99 \cdot 10^9 \text{ N m}^2/\text{C}^2}} = 1.05 \cdot 10^{-5} \text{ C}.$$

The sign does not matter, so long as each particle has a charge of the same sign, so that they repel.

21.43. The two up quarks have identical charge $q=(2/3)e=(2/3)\left(1.602\cdot 10^{-19}\text{ C}\right)$. They are $d=0.900\cdot 10^{-15}$ m apart. The magnitude of the electrostatic force between them is

$$F=\frac{kq^2}{d^2}=\frac{\left(8.99\cdot 10^9\text{ N m}^2/\text{C}^2\right)\left[\frac{2}{3}(1.602\cdot 10^{-19}\text{ C})\right]^2}{(0.900\cdot 10^{-15}\text{ m})^2}=127\text{ N}.$$

This is large, however the proton does not 'break apart' because of the strength of the strong nuclear force which binds the quarts together to form the proton. A proton is made of 2 up quarks, each with charge $(2/3)e$, and one down quark with charge $-(1/3)e$. The net charge of the proton is e.

21.47. **THINK:** Identical point charges $Q=32\cdot 10^{-6}C$ are placed at each of the four corners of a rectangle of dimensions $L=2.0$ m by $W=3.0$ m. Find the magnitude of the electrostatic force on any one of the charges. Note that by symmetry the magnitude of the net force on each charge is equal. Choose to compute the net electrostatic force on Q_4.

SKETCH:

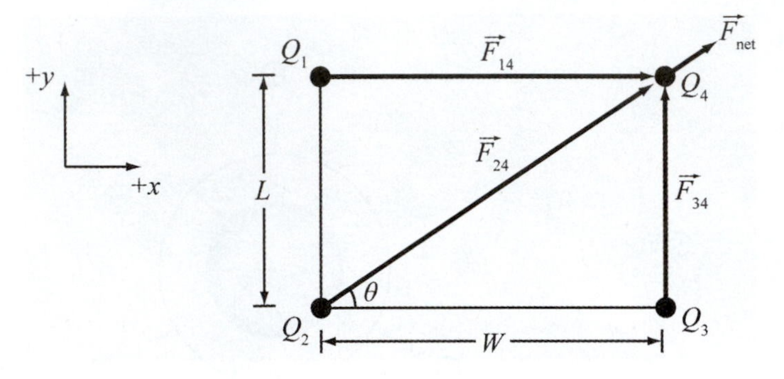

RESEARCH: The magnitude of the force between two charges is $\vec{F}_{12}=\left(kq_1q_2/\left|\vec{r}_{21}\right|^2\right)\hat{r}_{21}$. The total force on a charge is the sum of all the forces acting on that charge. The magnitude of the force is found from $F=\left(F_x^{\,2}+F_y^{\,2}\right)^{1/2}$, where the components F_x and F_y can be considered one at a time.

SIMPLIFY: x-component: $F_x=F_{14,x}+F_{24,x}+F_{34,x}=\dfrac{kQ^2}{W^2}+\dfrac{kQ^2}{W^2+L^2}\cos\theta+0=kQ^2\left(\dfrac{1}{W^2}+\dfrac{W}{\left(W^2+L^2\right)^{3/2}}\right)$

y-component: $F_y=F_{14,y}+F_{24,y}+F_{34,y}=0+\dfrac{kQ^2}{W^2+L^2}\sin\theta+\dfrac{kQ^2}{L^2}=kQ^2\left(\dfrac{W}{\left(W^2+L^2\right)^{3/2}}+\dfrac{1}{L^2}\right)$

$F_{\text{net}}=\sqrt{F_x^{\,2}+F_y^{\,2}}$

CALCULATE: $F_x=\left(8.99\cdot 10^9\text{ N m}^2/C^2\right)(32\cdot 10^{-6}\text{ C})^2\left(\dfrac{1}{\left(3.0\text{ m}\right)^2}+\dfrac{3.0\text{ m}}{\left[\left(3.0\text{ m}\right)^2+\left(2.0\text{ m}\right)^2\right]^{3/2}}\right)=1.612\text{ N}$

$F_y=\left(8.99\cdot 10^9\text{ N m}^2/C^2\right)(32\cdot 10^{-6}\text{ C})^2\left(\dfrac{2.0\text{ m}}{\left[\left(3.0\text{ m}\right)^2+\left(2.0\text{ m}\right)^2\right]^{3/2}}+\dfrac{1}{\left(2.0\text{ m}\right)^2}\right)=2.694\text{ N}$

$F_{\text{net}}=\sqrt{\left(1.612\text{ N}\right)^2+\left(2.694\text{ N}\right)^2}=3.1397\text{ N}$

ROUND: Since each given value has 2 significant figures, $F_{\text{net}} = 3.1$ N

DOUBLE-CHECK: Since L is less than W, the y-component of F_{net} should be greater than the x-component.

21.51. **THINK:** The positions of the three fixed charges are $q_1 = 1.00$ mC at $r_1 = (0,0)$, $q_2 = -2.00$ mC at $r_2 = (17.0 \text{ mm}, -5.00 \text{ mm})$, and $q_3 = +3.00$ mC at $r_3 = (-2.00 \text{ mm}, 11.0 \text{ mm})$. Find the net force on the charge q_2.

SKETCH:

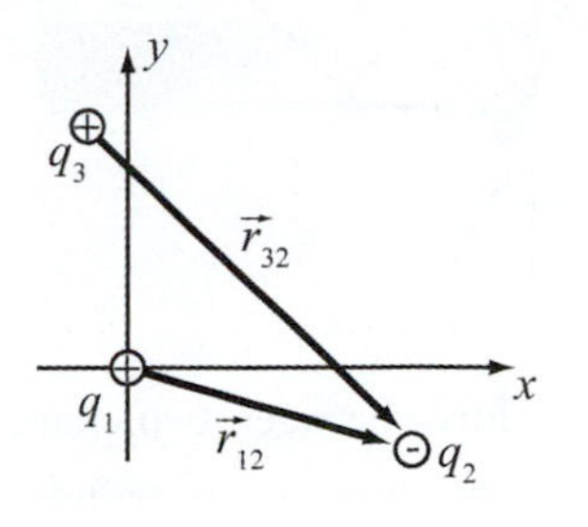

RESEARCH: The magnitude force is $\vec{F}_{12} = kq_1q_2\hat{r}_{12} / \left|\vec{r}_{12}\right|^2 = kq_1q_2\vec{r}_{12} / r_{12}^{\,3}$. The net force on q_2 is the sum of all the forces acting on q_2.

SIMPLIFY: $\vec{F}_{\text{net},2} = \vec{F}_{12} + \vec{F}_{32} = kq_2\left(\dfrac{q_1\left[(x_2 - x_1)\hat{x} + (y_2 - y_1)\hat{y}\right]}{\left[(x_2 - x_1)^2 + (y_2 - y_1)^2\right]^{3/2}} + \dfrac{q_3\left[(x_2 - x_3)\hat{x} + (y_2 - y_3)\hat{y}\right]}{\left[(x_2 - x_3)^2 + (y_2 - y_3)^2\right]^{3/2}}\right)$

CALCULATE: Without units,

$$\begin{aligned}\vec{F}_{\text{net},2} &= \left(8.99\cdot 10^9\right)(-2.00)\left[\frac{(1.00)\left(17.0\hat{x} - 5.00\hat{y}\right)}{\left[(17.0)^2 + (-5.00)^2\right]^{3/2}} + \frac{(3.00)(19.0\hat{x} - 16.0\hat{y})}{\left[(19.0)^2 + (-16.0)^2\right]^{3/2}}\right] \\ &= -1.2181\cdot 10^8\,\hat{x} + 7.2469\cdot 10^7\,\hat{y}.\end{aligned}$$

Then, the units of $\vec{F}_{\text{net},2}$ are:

$$\left[\vec{F}_{\text{net},2}\right] = \left(\text{N m}^2/\text{C}^2\right)(\text{mC})\left[\frac{(\text{mC})(\text{mm} - \text{mm})}{\left[(\text{mm})^2 + (\text{mm})^2\right]^{3/2}} + \frac{(\text{mC})(\text{mm} - \text{mm})}{\left[(\text{mm})^2 + (\text{mm})^2\right]^{3/2}}\right] = \text{N}$$

Altogether , $\vec{F}_{\text{net},2} = \left(-1.2181\cdot 10^8 \text{ N}\right)\hat{x} + \left(7.2469\cdot 10^7 \text{ N}\right)\hat{y}$. The magnitude of the force is

$$F_{\text{net},2} = \sqrt{F_x^{\,2} + F_y^{\,2}} = \sqrt{\left(-1.2181\cdot 10^8 \text{ N}\right)^2 + \left(7.2469\cdot 10^7 \text{ N}\right)^2} = 1.4174\cdot 10^8 \text{ N}$$

ROUND: $\vec{F}_{\text{net},2} = \left(-1.22\cdot 10^8 \text{ N}\right)\hat{x} + \left(7.25\cdot 10^7 \text{ N}\right)\hat{y}$ and $\left|\vec{F}_{\text{net},2}\right| = 1.42\cdot 10^8$ N.

DOUBLE-CHECK: The charges are large and the separation distance are small, so $F_{\text{net},2}$ should be very strong.

21.55. **THINK:** Four point charges, each with charge q, are fixed to the four corners of a square with a sides of length $d = 1.00$ cm. An electron is suspended above a point at which its weight is balanced by the electrostatic force due to the four electrons: $z' = 15.0$ nm above the center of the square. The mass of an electron is $m_e = 9.109\cdot 10^{-31}$ kg, and the charge is $q_e = -e = -1.602\cdot 10^{-19}$ C. Find the value of q of the fixed charges, in Coulombs and as a multiple of the electron charge.

SKETCH:

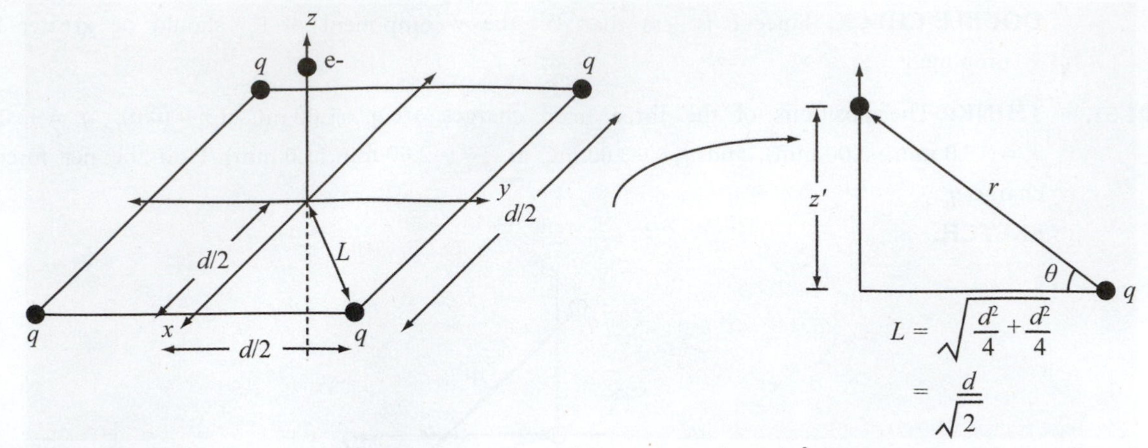

RESEARCH: The electrostatic force between two charges is $F = kq_1q_2 / r^2$. By symmetry, the net force in the horizontal direction is zero, and the problem reduces to a balance of the forces in the vertical direction, with one fixed charge balancing a quarter of the electron's weight. The vertical component of the electrostatic force is $F\sin\theta$. The weight of the electron is $W = m_e g$.

SIMPLIFY: Balancing the forces in the vertical (z) direction yields $F_{\text{coulomb}} = \frac{1}{4}W \Rightarrow \frac{kqq_e}{r^2}\sin\theta = \frac{1}{4}m_e g.$

Solving for q: $q = \frac{1}{4}\frac{m_e g r^2}{kq_e \sin\theta} = \frac{m_e g r^3}{4kq_e z'} = \frac{m_e g(L^2 + z'^2)^{3/2}}{4kq_e z'} = \frac{-m_e g\left(\frac{d^2}{2} + z'^2\right)^{3/2}}{4kez'}.$

CALCULATE: $q = \frac{-\left(9.109\cdot 10^{-31}\text{ kg}\right)(9.81\text{ m/s}^2)\left[\frac{(0.100\text{ m})^2}{2} + (15.0\cdot 10^{-9}\text{ m})^2\right]^{3/2}}{4\left(8.99\cdot 10^9\text{ N m}^2/\text{C}^2\right)(1.602\cdot 10^{-19}\text{ C})(15.0\cdot 10^{-9}\text{ m})}$

$= -3.6561\cdot 10^{-17}\text{ C, or } -228.22e.$

ROUND: With three significant figures in z', $q = -3.66\cdot 10^{-17}\text{ C} = -228e.$

DOUBLE-CHECK: The gravitational force on an electron is small. Each charge q needs to be a few hundred electron charges to balance the gravitational force on the electron.

21.59. The gravitational force between the Earth and Moon is given by $F_g = GM_{\text{Earth}}m_{\text{Moon}} / r_{\text{EM}}^2$. The static electrical force between the Earth and the Moon is $F = kQ^2 / r_{\text{EM}}^2$, where Q is the magnitude of the charge on each the Earth and the Moon. If the static electrical force is 1.00% that of the force of gravity, then the charge Q would be:

$$F = 0.01F_g \Rightarrow \frac{kQ^2}{r_{\text{EM}}^2} = \frac{0.0100GM_{\text{Earth}}m_{\text{Moon}}}{r_{\text{EM}}^2} \Rightarrow Q = \sqrt{\frac{0.0100GM_{\text{Earth}}m_{\text{Moon}}}{k}}.$$

This gives $Q = \sqrt{\frac{0.0100(6.67\cdot 10^{-11}\text{ N m}^2/\text{kg})(5.97\cdot 10^{24}\text{ kg})(7.36\cdot 10^{22}\text{ kg})}{\left(8.99\cdot 10^9\text{ N m}^2/\text{C}^2\right)}} = 5.71\cdot 10^{12}\text{ C.}$

21.63. For the atom described in the previous question, the ratio of the gravitational force between the electron and proton to the electrostatic force is:

$$F_g / F = \frac{\frac{Gm_e m_p}{r^2}}{\frac{k|q_1||q_2|}{r^2}} = \frac{Gm_e m_p}{ke^2}$$

$$= \frac{\left(6.6742 \cdot 10^{-11} \text{m}^3 / (\text{kg s}^2)\right)(9.109 \cdot 10^{-31} \text{ kg})(1.673 \cdot 10^{-27} \text{ kg})}{\left(8.99 \cdot 10^9 \text{ N m}^2 / \text{C}^2\right)(1.602 \cdot 10^{-19} \text{ C})^2}$$

$$= 4.41 \cdot 10^{-40}$$

This value is independent of the radius; if this radius is doubled, the ratio does not change.

21.67. The radius of the nucleus of ^{14}C is $r_0 = 1.505$ fm. The nucleus has charge $q_0 = +6e$.

(a) A proton (charge $q = e$) is placed $d = 3.00$ fm from the surface of the nucleus. Treating the nucleus as a point charge, the distance between the proton and the charge of the nucleus is $r = d + r_0$. The force is repulsive due to the like charges. The magnitude of this force is

$$F = \frac{k|q||q_0|}{r^2} = \frac{k6e^2}{(d+r_0)^2} = \frac{\left(8.99 \cdot 10^9 \text{ N m}^2 / \text{C}^2\right)6(1.602 \cdot 10^{-19} \text{ C})^2}{\left(3.00 \cdot 10^{-15} \text{ m} + 1.505 \cdot 10^{-15} \text{ m}\right)^2} = 68.2097 \text{ N} \approx 68.2 \text{ N}$$

(b) The proton's acceleration is:

$$F = m_p a \Rightarrow a = \frac{F}{m_p} = \frac{68.210 \text{ N}}{1.673 \cdot 10^{-27} \text{ kg}} = 4.077 \cdot 10^{28} \text{ m/s}^2 \approx 4.08 \cdot 10^{28} \text{ m/s}^2$$

21.71. The charge of the Earth is $Q = -6.8 \cdot 10^5$ C. The mass of the object is $m = 1.0$ g. For this object to be levitated near the Earth's surface ($r_E = 6378$ km), the Coulomb force and the force of gravity must be the same. The charge q of the object can be found from balancing these forces:

$$F_g = F_{\text{Coulomb}} \Rightarrow mg = \frac{k|Qq|}{r_E^2} \Rightarrow |q| = \frac{mgr_E^2}{k|Q|}$$

$$|q| = \frac{(0.0010 \text{ kg})(9.81 \text{ m/s}^2)(6.378 \cdot 10^6 \text{ m})^2}{\left(8.99 \cdot 10^9 \text{ N m}^2/\text{C}^2\right)\left|-6.8 \cdot 10^5 \text{ C}\right|} = 6.5278 \cdot 10^{-5} \text{ C} \approx 65 \text{ μC}.$$

Since Q is negative, and the object is levitated by the repulsion of like charges, it must be that $q \approx -65$ μC.

21.75. **THINK:** Three 5.00-g Styrofoam balls of radius 2.00 cm are tied to 1.00 m long threads and suspended freely from a common point. The charge of each ball is q and the balls form an equilateral triangle with sides of 25.0 cm.

SKETCH:

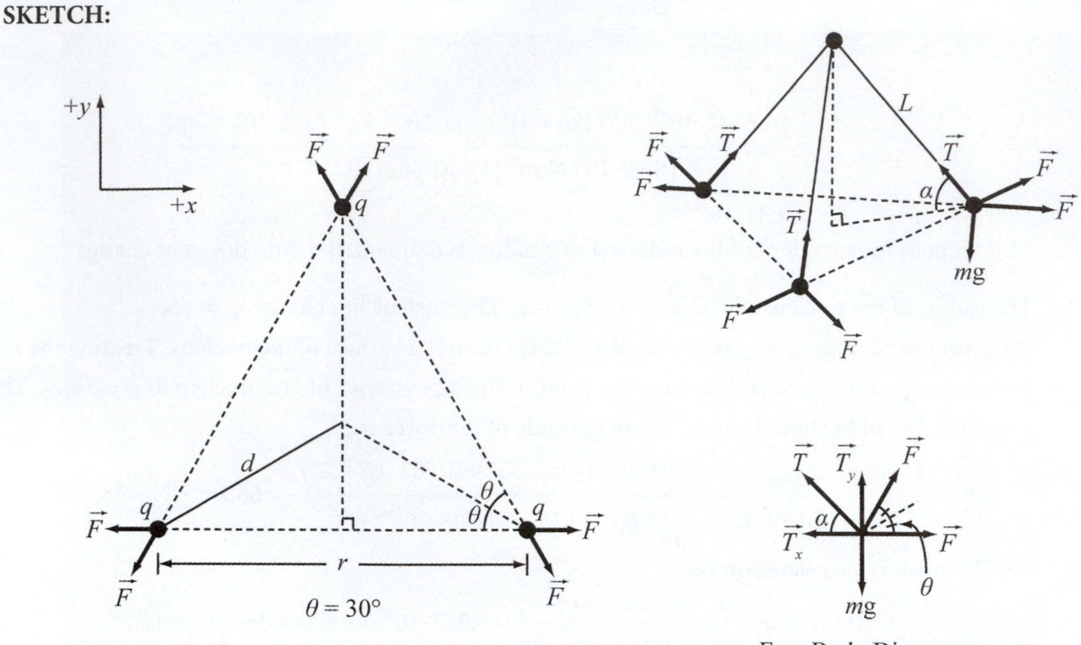

RESEARCH: The magnitude of the force between two charges, q_1 and q_2, is $F_{12} = kq_1q_2 / r^2$. The magnitude of F in the above figure is $F = kq^2 / r^2$. Using Newton's Second Law, it is found that $T_y = T\sin\alpha = mg$ and $T_x = T\cos\alpha = 2F\cos\theta$.

SIMPLIFY: Eliminating T in the above equations yields $\tan\alpha = mg / (2F\cos\theta)$. Rearranging gives, $F = mg / (2\tan\alpha\cos\theta) = kq^2 / r^2$. Therefore, the charge q is

$$q = \sqrt{\frac{mgr^2}{2k\tan\alpha\cos\theta}}.$$

From the sketch, it is clear that the distance of the ball to the center of the triangle is $d = r / (2\cos\theta)$. Therefore $\tan\theta = \sqrt{L^2 - d^2} / d$.

CALCULATE: Substituting the numerical values, $r = 0.250$ m, $m = 5.00 \cdot 10^{-3}$ kg, $g = 9.81$ m/s^2, $L = 1.00$ m and $\theta = 30°$ (exact) gives

$$d = \frac{0.250 \text{ m}}{2\cos(30°)} = 0.1443 \text{ m}$$

$$\tan\alpha = \frac{\sqrt{(1.00 \text{ m})^2 - (0.1443 \text{ m})^2}}{0.1443 \text{ m}} = 6.856$$

$$q = \sqrt{\frac{5.00 \cdot 10^{-3} \text{ kg}(9.81 \text{ m/s}^2)(0.250 \text{ m})^2}{2 \cdot (8.99 \cdot 10^9 \text{ N m}^2/\text{C}^2) 6.856 \cos(30°)}} = 1.69463 \cdot 10^{-7} \text{ C}$$

ROUND: $q = 0.169$ μC

DOUBLE-CHECK: This charge is approximately 11 orders of magnitude larger than the elementary charge e. The charge required to deflect 5.00 g balls by a distance of 25.0 cm would need to be fairly large.

21.79. **THINK:** To solve this problem, the force due to the charges and the tension in the string must balance the gravitational force on the spheres.

SKETCH:

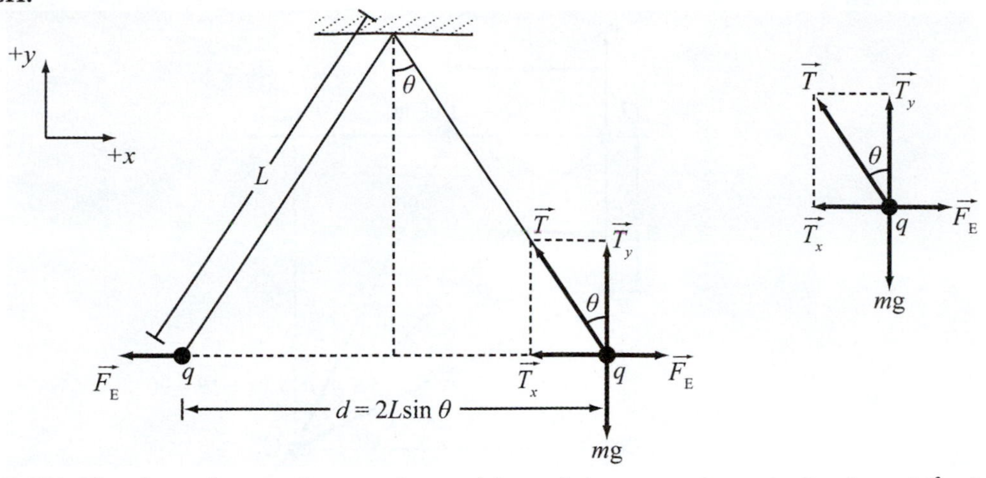

RESEARCH: The force due to electrostatic repulsion of the two spheres is $F_E = kq_1q_2 / d^2 = kq^2 / d^2$. Applying Newton's Second Law yields (I) $T_x = T\sin\theta = F_E$ and (II) $T_y = T\cos\theta = mg$. $L = 0.45$ m, $m = 2.33 \cdot 10^{-3}$ kg, $\theta = 10.0°$.

SIMPLIFY: Dividing (I) by (II) gives $\tan\theta = F_E / (mg) = kq^2 / (d^2 mg)$. After simple manipulation, it is found that the charge on each sphere is $q = \sqrt{d^2 mg\tan\theta / k} = 2L\sin\theta\sqrt{mg\tan\theta / k}$ using $d = 2L\sin\theta$.

CALCULATE: Substituting the numerical values, it is found that

$$q = (2)(0.450 \text{ m})(\sin 10.0°)\sqrt{\frac{2.33 \cdot 10^{-3} \text{ kg}(9.81 \text{ m/s}^2)\tan(10.0°)}{8.99 \cdot 10^9 \text{ N m}^2/\text{C}^2}} = 1.0464 \cdot 10^{-7} \text{ C}.$$

ROUND: Keeping only three significant digits gives $q = 0.105$ μC.

DOUBLE-CHECK: This is reasonable. The relatively small spheres and small distance will mean the charge is small.

21.83. **THINK:** Since this is a two dimensional problem, electrostatic forces are added as vectors. It is assumed that Q_A is a positive charge.

SKETCH:

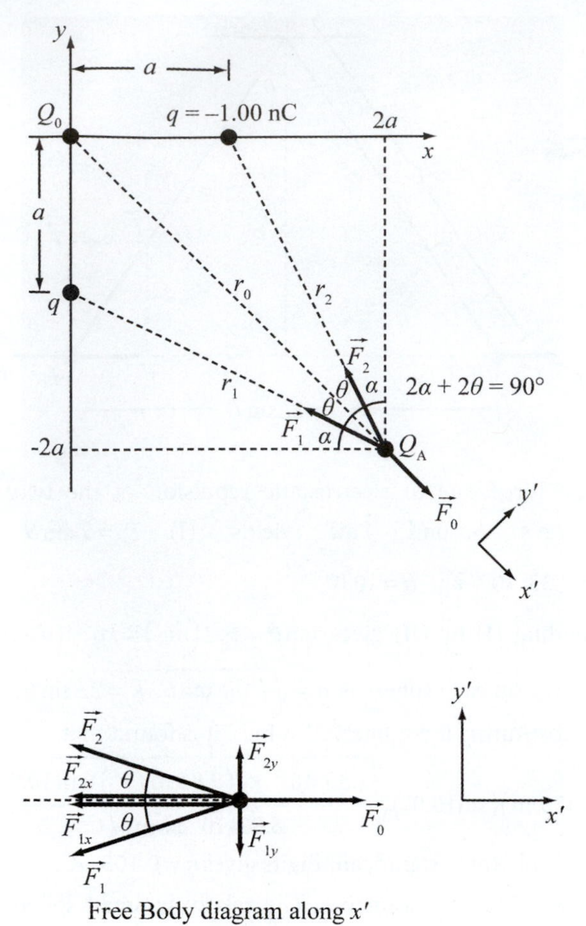

Free Body diagram along x'

RESEARCH: To balance the forces F_1 and F_2, the charge on Q_0 must be positive. The electrostatic forces on Q_A are $F_1 = \dfrac{k|q|Q_A}{r_1^2}$, $F_2 = \dfrac{k|q|Q_A}{r_2^2}$, and $F_0 = \dfrac{kQ_0Q_A}{r_0^2}$. Applying Newton's Second Law, it is found that $F_0 = F_{1x} + F_{2x}$ or $kQ_0Q_A / r_0^2 = F_1\cos\theta + F_2\cos\theta.$ Using $r_1 = r_2$ this becomes $\dfrac{kQ_0Q_A}{r_0^2} = \dfrac{k|q|Q_A}{r_1^2}2\cos\theta.$

SIMPLIFY: Solving the above equation for Q_0 gives the charge Q_0, $Q_0 = (r_0 / r_1)^2 |q| 2\cos\theta.$ From the above figure, it is noted that $r_0 = \sqrt{(2a)^2 + (2a)^2} = 2a\sqrt{2},$ $r_1 = \sqrt{(2a)^2 + a^2} = a\sqrt{5},$ and

$$\cos\theta = \cos(45° - \alpha) = \cos 45° \cos\alpha + \sin 45° \sin\alpha \Rightarrow \cos\theta = \frac{\sqrt{2}}{2}\frac{2a}{a\sqrt{5}} + \frac{\sqrt{2}}{2}\frac{a}{a\sqrt{5}} = \frac{3}{2}\sqrt{\frac{2}{5}} = \frac{3}{10}\sqrt{10}.$$

Therefore the magnitude of charge Q_0 is $|Q_0| = 2|q|\dfrac{8a^2}{5a^2}\dfrac{3}{10}\sqrt{10} = \dfrac{48}{50}\sqrt{10}|q|.$

CALCULATE: Substituting $q = -1.00$ nC yields $|Q_0| = \dfrac{48}{50}\sqrt{10} \cdot |-1.00 \text{ nC}| = 3.036$ nC.

ROUND: Rounding to three significant figures gives $|Q_0| = 3.04$ nC.

DOUBLE-CHECK: Since r_0 is larger than r_1, it is expected that Q_0 is larger than $2|q| = 2$ nC.

Multi-Version Exercises

21.84. The components of the forces in the x-direction give us $T\sin\theta = \dfrac{kq^2}{d^2}$.

The components of the forces in the y-direction give us $T\cos\theta = mg$.

We can divide these two equations to get

$$\tan\theta = \frac{kq^2}{mgd^2} \Rightarrow d = \sqrt{\frac{kq^2}{mg\tan\theta}}.$$

From the figure we can see that

$$\sin\theta = \frac{d/2}{\ell} \Rightarrow \ell = \frac{d}{2\sin\theta}.$$

Combining these two equations gives us

$$\ell = \frac{d}{2\sin\theta} = \frac{\sqrt{\dfrac{kq^2}{mg\tan\theta}}}{2\sin\theta} = \sqrt{\frac{kq^2}{4mg\sin^2\theta\tan\theta}}.$$

Using the known values,

$$\ell = \sqrt{\frac{\left(8.99\cdot 10^9 \text{ N m}^2/\text{C}^2\right)\left(29.59\cdot 10^{-6}\text{ C}\right)^2}{4\left(0.9860 \text{ kg}\right)\left(9.81 \text{m/s}^2\right)\sin^2 29.79°\tan 29.79°}} = 1.211 \text{ m}.$$

21.87. $F_{\text{net}} = F_{13} - F_{23} = \dfrac{kq_1q_3}{\left(x_3 - x_1\right)^2} - \dfrac{kq_2q_3}{\left(x_2 - x_3\right)^2} = 0$

$\left(x_3 - x_1\right)^2 q_2 = \left(x_2 - x_3\right)^2 q_1$

$\left(x_3 - x_1\right)\sqrt{q_2} = \pm\left(x_2 - x_3\right)\sqrt{q_1}.$

We choose the + sign since we know that the force can only balance when $x_1 < x_3 < x_2$.

So we can write

$$x_3 = \frac{\sqrt{q_1}\,x_2 + \sqrt{q_2}\,x_1}{\sqrt{q_1} + \sqrt{q_2}}.$$

Using the known values,

$$x_3 = \frac{\sqrt{3.979\cdot 10^{-6}\text{ C}}\left(14.13 \text{ m}\right) + \sqrt{8.669\cdot 10^{-6}\text{ C}}\left(-5.689 \text{ m}\right)}{\sqrt{3.979\cdot 10^{-6}\text{ C}} + \sqrt{8.669\cdot 10^{-6}\text{ C}}} = 2.315 \text{ m}$$

Chapter 22: Electric Fields and Gauss's Law

Exercises

22.25. The electric field produced by the charge is:

$$E = \frac{kq}{r^2} = \frac{\left(8.99 \cdot 10^9 \text{ N m}^2/\text{C}^2\right)\left(4.00 \cdot 10^{-9} \text{ C}\right)}{\left(0.250 \text{ m}\right)^2} = 575.36 \text{ N/C} \approx 575 \text{ N/C}.$$

22.29. **THINK:** We want to find out where the combined electric field from two point charges can be zero. Since the electric field falls off as the inverse second power of the distance to the charge, and since both charges are located on the x-axis, only points on the same line have any possible chance of canceling the electric field from these two charges, resulting in a net zero electric field. The first charge, $q_1 = 5.0 \text{ C}$, is at the origin. The second charge, $q_2 = -3.0 \text{ C}$, is at $x = 1.0$ m. Let's think about in which region of the x-axis it can be possible to have zero electric field. On the sketch we have marked three regions (I, II, and III). If we place a positive charge anywhere in region 2, then the 5 C will repel it and the -3 C will attract it, resulting in the positive charging moving towards the right. If we place a negative charge into the same reason, it will always move to the left. So we know that the electric field cannot be zero anywhere in region II. Region I is closer to the 5 C charge. Since this is also the charge with the larger magnitude, its electric field will dominate region I, and thus we cannot have any place in region I where the electric field is 0. This leaves only region III, where the two electric fields from the point charges can cancel.

SKETCH:

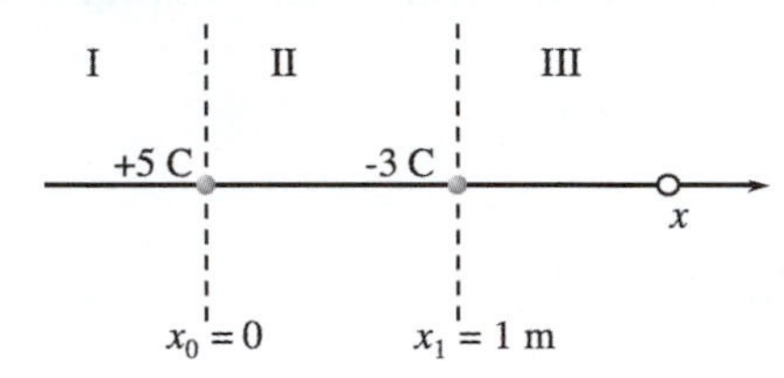

RESEARCH: The electric field due to the charge at the origin is $E_0 = kq_0/x^2$. The other charge produces a field of $E_1 = kq_1/\left(x - x_1\right)^2$.

SIMPLIFY: The combined electric field is $E = kq_0/x^2 + kq_1/\left(x - x_1\right)^2$. Setting the electric field to zero, solve for x:

$$\frac{kq_0}{x^2} + \frac{kq_1}{\left(x - x_1\right)^2} = 0 \Rightarrow \frac{kq_0}{x^2} = -\frac{kq_1}{\left(x - x_1\right)^2} \Rightarrow \left(x - x_1\right)^2 q_0 = -x^2 q_1 \Rightarrow \left(x - x_1\right)^2 \left|q_0\right| = x^2 \left|q_1\right|$$

We could now solve the resulting quadratic equation blindly and would obtain two solutions, each of which we would have to evaluate for validity. Instead, we can make use of the thinking we have done above. In the last step we used the fact that the charge at the origin is positive and the other is negative, replacing them with their absolute values. Now we can take the square root on both sides and choose the positive root, leaving us with

$$\left(x - x_1\right)\sqrt{\left|q_0\right|} = x\sqrt{\left|q_1\right|} \Rightarrow x = \frac{x_1\sqrt{\left|q_0\right|}}{\sqrt{\left|q_0\right|} - \sqrt{\left|q_1\right|}}$$

CALCULATE: $x = \dfrac{(1.00 \text{ m})\sqrt{5.00 \text{ C}}}{\sqrt{5.00 \text{ C}} - \sqrt{3.00 \text{ C}}} = 4.43649 \text{ m}$

ROUND: The positions are reported to three significant figures. The electric field is zero at $x = 4.44$ m.

DOUBLE-CHECK: This is a case where we can simply insert our result and verify that it does what it is supposed to: $E(x{=}4.4 \text{ m}) = k(5 \text{ C})/(4.4 \text{ m})^2 + k(-3 \text{ C})/(4.4 \text{ m} - 1 \text{ m})^2 = 0$.

22.33. **THINK:** As the $m = 4.0\text{ g}$ ball falls the force of gravity acting on it will cause it to accelerate downwards. At the same time, the force due to the electric field acts on the ball causing it to accelerate towards the east. The forces act perpendicular to each other. The problem is solved by finding each component of the velocity. In order to find the velocity due to the electric field, the time required for the ball to travel 30.0 cm downwards is needed.

SKETCH:

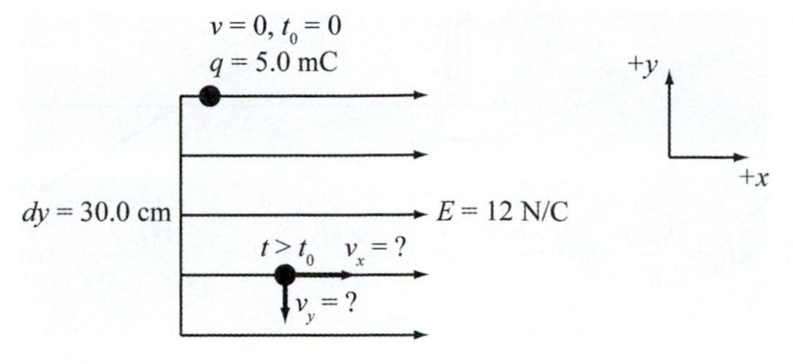

RESEARCH: The velocity in the downward direction is found using $v_y^2 = v_{y0}^2 + 2gdy$. The time it takes to reach this velocity $t = v_y / g$. The acceleration eastward is calculated using $F = ma = qE$. The velocity is then $v_x = a_x t$.

SIMPLIFY: The y-component of the velocity is $v_y = \sqrt{2gdy}$ because the ball starts from rest. The time it takes for the ball to fall 30.0 cm is $t = \sqrt{2gdy} / g$. The acceleration eastward is $a = qE / m$. The velocity eastward is $v_x = a_x t \rightarrow v_x = \left(\frac{qE}{m}\right)\frac{\sqrt{2gdy}}{g} = \frac{qE}{m}\sqrt{\frac{2dy}{g}}$.

CALCULATE: $v_y = \sqrt{2\left(9.81\text{ m/s}^2\right)\left(0.300\text{ m}\right)} = 2.4261\text{ m/s}$ downward

$$v_x = \left(5.0 \cdot 10^{-3}\text{ C}\frac{12\text{ N/C}}{0.0040\text{ kg}}\right)\sqrt{2\left(\frac{0.300\text{ m}}{9.81\text{ m/s}^2}\right)} = 3.7096\text{ m/s eastward}$$

ROUND: The velocity is report to three significant figures. The ball reaches a velocity of $(3.71\text{ m/s})\hat{x} + (2.43\text{ m/s})\hat{y}$.

DOUBLE-CHECK: This is a reasonable answer considering the size of the values given in the question.

22.37. **THINK:** The charge Q is uniformly distributed along the rod of length L. The rod has linear charge density $\lambda = Q / L$. The electric field at a position $x = d$ can be calculated by integrating over the differential electric field due to the differential charge on the rod. The electric field differential $dE = kdq / r^2$, where the differential is along the y-axis, and $R = \sqrt{d^2 + y^2}$. The x- and y-components of the field must be considered individually. The x-component of the field differential is given by $dE_x = dE\cos\theta$, and the y-component is given by $dE_y = dE\sin\theta$.

SKETCH:

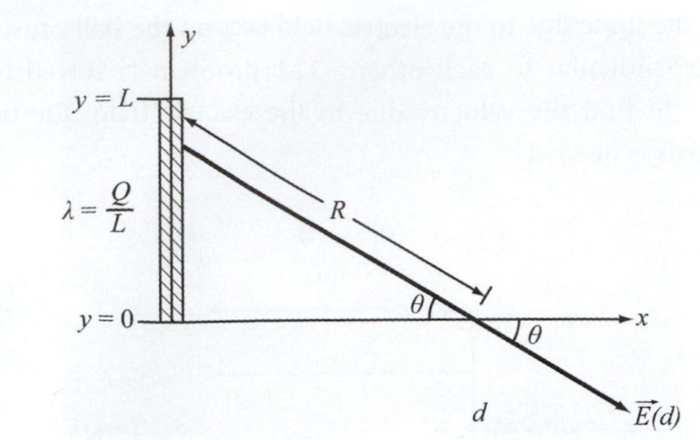

SIMPLIFY: $dE_x = \dfrac{kdQ}{R^2}\cos\theta = \dfrac{k\lambda dy}{R^2}\cos\theta = \dfrac{kQdy}{LR^2}\cos\theta = \dfrac{kQdy}{L\left(d^2+y^2\right)^2}\cos\theta$

$$\cos\theta = \frac{d}{R} = \frac{d}{\sqrt{d^2+y^2}} \Rightarrow dE_x = \left(\frac{kQdy}{L\left(d^2+y^2\right)^2}\right)\left(\frac{d}{\sqrt{d^2+y^2}}\right) = \frac{kdQdy}{L\left(d^2+y^2\right)^{3/2}}$$

$$dE_y = \frac{kQdy}{L\left(d^2+y^2\right)^2}\sin\theta;\ \sin\theta = \frac{y}{R} = \frac{y}{\sqrt{d^2+y^2}} \Rightarrow dE_y = \frac{ykQdy}{L\left(d^2+y^2\right)^{3/2}}$$

Integrate both expressions.

$$E_x = \int_0^L \frac{dkQdy}{L\left(d^2+y^2\right)^{3/2}} = \frac{dkQ}{L}\int_0^L \frac{1}{\left(d^2+y^2\right)^{3/2}}dy = \frac{kQd}{L}\left[\frac{y}{d^2\sqrt{d^2+y^2}}\right]_0^L$$

$$E_y = \int_0^L \frac{ykQdy}{L\left(d^2+y^2\right)^{3/2}} = \frac{kQ}{L}\int_0^L \frac{y}{\left(d^2+y^2\right)^{3/2}}dy = \frac{kQ}{L}\left[\frac{-1}{\sqrt{d^2+y^2}}\right]_0^L$$

$$\vec{E}(d) = E_x\hat{x} - E_y\hat{y}$$

CALCULATE: $E_x = \dfrac{kQd}{L}\left[\dfrac{y}{d^2\sqrt{d^2+y^2}}\right]_0^L = \dfrac{kQd}{L}\left(\dfrac{L}{d^2\sqrt{d^2+L^2}} - 0\right) = \dfrac{kQ}{d\sqrt{d^2+L^2}}$

$$E_y = \frac{kQ}{L}\left[\frac{-1}{\sqrt{d^2+y^2}}\right]_0^L = \frac{kQ}{L}\left(\frac{-1}{\sqrt{d^2+L^2}} - \frac{-1}{d}\right) = \frac{kQ}{dL} - \frac{kQ}{L\sqrt{d^2+L^2}}$$

$$\vec{E}(d) = \left(\frac{kQ}{d\sqrt{d^2+L^2}}\right)\hat{x} - \left(\frac{kQ}{dL} - \frac{kQ}{L\sqrt{d^2+L^2}}\right)\hat{y}$$

ROUND: Not applicable.

DOUBLE CHECK: The magnitude of the electric field decreases as d increases, as expected.

22.41. The torque due to the field is

$$\left|\vec{\tau}\right| = \vec{p}\times\vec{E} = pE\sin\theta = qdE\sin\theta = \left(5.00\cdot 10^{-15}\text{ C}\right)\left(0.400\cdot 10^{-3}\text{ m}\right)\left(2.00\cdot 10^{3}\text{ N/C}\right)\left(\sin 60.0^\circ\right)$$
$$= 3.46\cdot 10^{-15}\text{ N m}.$$

22.45. **THINK:** The net force on falling object in an electric field is the sum of the force due to gravity and the force due to the electric field. If the falling object carries a positive charge, then the force on the object due to the electric field acts in the direction opposite to the force of gravity.

SKETCH:

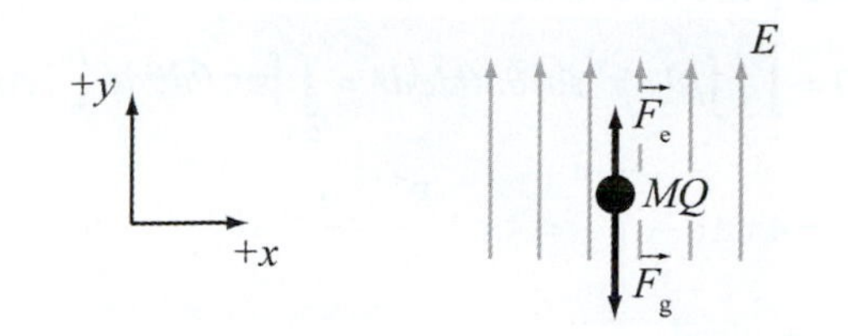

RESEARCH: The net upward force acting on the object is $F = F_e - F_g = QE - Mg = Ma$. This corresponds to a downward acceleration of $a = g - \frac{QE}{M}$. Recall that the speed of an object in free fall is given by $v_f^2 = v_0^2 + 2a\Delta y \Rightarrow v = \sqrt{2ah}$.

SIMPLIFY:

(a) $v = \sqrt{2ah} \Rightarrow a = \frac{v^2}{2h} = g - \frac{QE}{M} \Rightarrow v = \sqrt{2h(g - QE/M)}$

(b) If the value $g - QE/M$ is less than zero, then the argument of the square root is negative. This means the value is non-real and the body does not fall.

CALCULATE: Not applicable.

ROUND: Not applicable.

DOUBLE CHECK: Dimensional analysis confirms that the units of the expression reduce to m/s, the correct units for velocity.

22.49. The sum of the flux through each surface is equal to the charge enclose divided by ε_0. $\sum_i \Phi_i = Q/\varepsilon_0$. The charge is then

$$Q = \varepsilon_0 \sum_i \Phi_i = \left(8.85 \cdot 10^{-12} \text{ C}^2/(\text{N m}^2)\right)(-70.0 - 300.0 - 300.0 + 300.0 - 400.0 - 500.0) \text{ N m}^2$$
$$= -1.124 \cdot 10^{-8} \text{ C} \approx -1.12 \cdot 10^{-8} \text{ C}.$$

22.53. Gauss's Law states that $\oiint E \cdot dA = \frac{q_{enc}}{\varepsilon_0}$. The integral over the sphere gives $\oiint E \cdot dA = EA = E\left[4\pi R^2\right] = \frac{q_{\text{enc}}}{\varepsilon_0} \Rightarrow E = \frac{q_{enc}}{4\pi R^2 \varepsilon_0}$. The electric field outside a uniform distribution of charge is identical to the field created by a point charge of the same magnitude, located at the center of the distribution. Since the radius of the balloon never reaches R, the charge enclosed is constant and the electric field does not change.

22.57. Using Gauss's Law $\oiint E \cdot dA = EA = E\left(4\pi r_{\text{E}}^2\right) = \frac{q_{\text{Earth}}}{\varepsilon_0}$. Solving for the charge gives

$$q_{\text{Earth}} = \varepsilon_0 E\left(4\pi r_{\text{Earth}}^2\right) = \left(8.85 \cdot 10^{-12} \text{ C}^2/\text{N m}^2\right)(-150. \text{ N/C}) 4\pi \left(6371 \cdot 10^3 \text{ m}\right)^2 = -6.7711 \cdot 10^5 \text{ C}$$
$$\approx -6.77 \cdot 10^5 \text{ C}.$$

22.61. **THINK:** A solid sphere of radius R has a non-uniform charge density $\rho = Ar^2$. Integrate the sphere.

SKETCH: Not required.

RESEARCH: The total charge is given by $Q = \int_{\text{Sphere}} \rho dV$.

SIMPLIFY: Integrating in the spherical polar coordinate yields:

$$Q=\int_0^R\int_0^{2\pi}\int_0^{\pi}\rho(r)r^2\sin\theta d\theta d\phi dr=\int_0^{2\pi}\int_0^{\pi}\sin\theta d\theta d\phi\int_0^R\left(Ar^2\right)r^2dr=4\pi A\int_0^R r^4dr$$

$$=4\pi A\left[\frac{r^5}{5}\right]_{r=o}^{r=R}=4\pi A\frac{R^5}{5}=\frac{4}{5}\pi AR^5.$$

CALCULATE: Not required.

ROUND: Not required.

DOUBLE-CHECK: One can check the result by single-variable integration, using spherical shells:

$dV=A_{shell}dr=\left(4\pi r^2\right)dr$

$$Q=\int\rho dV=\int_0^R\left(Ar^2\right)\left(4\pi r^2\right)dr=4\pi A\int_0^R r^4dr=4\pi A\left[\frac{r}{5}\right]_0^R=\frac{4}{5}\pi AR^5$$

Which agrees with the previous answer.

22.65. **THINK:** Use the values from the question: $\sigma_1=3.00\ \mu\text{C/m}^2$, and $\sigma_2=-5.00\ \mu\text{C/m}^2$.

(a) The total field can be determined by superposition of the fields from both plates. The field contributions from the two charged sheets are opposing each other at point *P*, to the left of the first sheet.

(b) The situation is similar to a) except that the fields due to both charged sheets point in the same direction at point P'.

SKETCH:

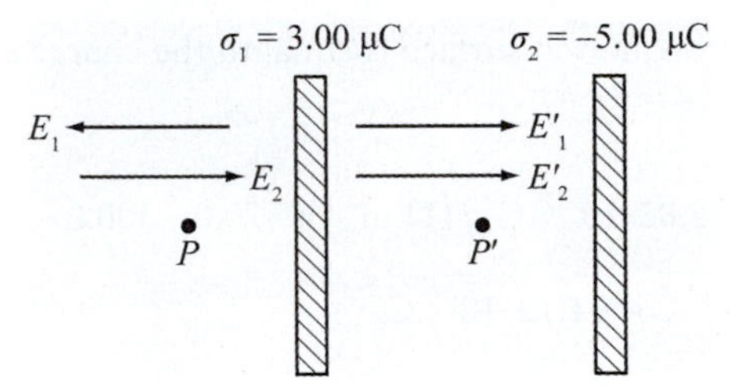

RESEARCH:

(a) At point P, the field due to sheet #1 is given by $E_1=-\left(\sigma_1/2\varepsilon_0\right)\hat{x}$, and the field due to sheet #2 is given by $E_2=-\left(\sigma_2/2\varepsilon_0\right)\hat{x}$. Note that $E_{\text{total}}=E_1+E_2$.

(b) At point P', the field due to sheet #1 is given by $E_1'=\left(\sigma_1/2\varepsilon_0\right)\hat{x}$, and the field due to sheet #2 is given by $E_2'=-\left(\sigma_2/2\varepsilon_0\right)\hat{x}$. Again, $E_{\text{total}}'=E_1'+E_2'$.

SIMPLIFY:

(a) $$E=\left(\frac{-\sigma_1}{2\varepsilon_0}\right)\hat{x}+\left(\frac{-\sigma_2}{2\varepsilon_0}\right)\hat{x}=\frac{-\left(\sigma_1+\sigma_2\right)}{2\varepsilon_0}$$

(b) $$E'=\left(\frac{\sigma_1}{2\varepsilon_0}\right)\hat{x}+\left(\frac{-\sigma_2}{2\varepsilon_0}\right)\hat{x}=\frac{\left(\sigma_1-\sigma_2\right)}{2\varepsilon_0}$$

CALCULATE:

(a) $$E_{\text{total}}=\frac{-(3.00-5.00)\cdot 10^{-6}\ \text{C/m}^2}{2\left(8.85\cdot 10^{-12}\ \text{C}^2/\left(\text{N m}^2\right)\right)}\hat{x}=\left(1.130\cdot 10^5\ \text{N/C}\right)\hat{x}$$

(b) $$E_{\text{total}}=\frac{(3.00-(-5.00))\cdot 10^{-6}\ \text{N/C}}{2\left(8.85\cdot 10^{-12}\ \text{C}^2/\left(\text{N m}^2\right)\right)}\hat{x}=\left(4.520\cdot 10^5\ \text{N/C}\right)\hat{x}$$

ROUND:

(a) $E_{\text{total}} = 1.13 \cdot 10^5$ N/C in the positive x-direction

(b) $E_{\text{total}} = 4.52 \cdot 10^5$ N/C in the positive x-direction

DOUBLE-CHECK: The results are reasonable because the answer in (b) is four times larger than that found in (a) since in (a) the fields are opposing each other and in (b) the fields are in same direction.

22.69. **THINK:** The principle of superposition can be used to find the electric field at the specified point. The electric field at the point $(2.00,1.00)$ is modeled as the sum of a positively charged cylindrical rod with no hole and a negatively charged cylindrical rod whose size and location are identical to those of the cavity. Let's first think about the case of the positively charged cylindrical rod without a hole. Since the point of interest is inside the rod, the entire charge distribution of the rod cannot contribute. Instead we draw our Gaussian surface as a cylinder with our point of interest on its rim (see sketch below, where the dashed circle in the cross-sectional view represents the Gaussian cylinder).

SKETCH:

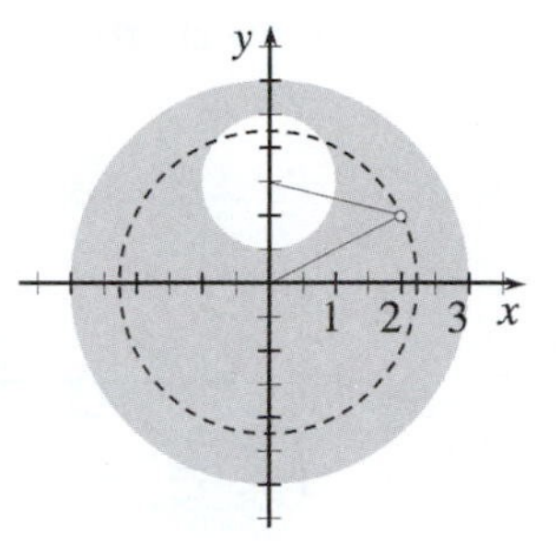

RESEARCH: In section 22.9 of the textbook it was shown that for cylindrical symmetry of the charge distribution the electric field outside the charge distribution can be written as $E = 2k\lambda / r$, where r is the distance to the central axis of the charge distribution and λ is the charge per unit length.

In the problem here the charge was initially uniformly distributed over the entire cross-sectional area, which means that the value of λ for the Gaussian surface and for the hole are proportional to their cross-sectional area: $\lambda_{\text{Gauss}} = \lambda_{\text{rod}}(r/R)^2$, and $\lambda_{\text{hole}} = -\lambda_{\text{rod}}(r_{\text{hole}}/R)^2$.

Now we have the tools to calculate the magnitudes of the individual electric fields of the rod and of the hole. What is left is to add the two, which is a vector addition. So we have to determine the x- and y-components of the fields individually and the combine them.

If E_1 is the field from the dashed cylinder and E_2 is that of the cavity then from considering the geometry the relation are given by: $E_{1x} = E_1 2/\left(2^2+1^2\right)^{1/2}$, $E_{2x} = E_2 2/\left(2^2+1^2\right)^{1/2}$, $E_{1y} = E_1/\left(2^2+1^2\right)^{1/2}$ and $E_{2y} = 0.5E_2/\left(2^2+0.5^2\right)^{1/2}$.

The net electric field is given by the following relations $E_x = E_{1x} + E_{2x}$ and $E_y = E_{1y} + E_{2y}$.

SIMPLIFY:

$E_1 = 2k\lambda_{\text{Gauss}}/r = 2k\lambda_{\text{rod}}(r/R)^2/r = 2k\lambda_{\text{rod}}r/R^2$

$E_2 = 2k\lambda_{\text{hole}}/r_2 = -2k\lambda_{\text{rod}}(r_{\text{hole}}/R)^2/r_2$

where r_2 is the distance between our point of interest and the center of the hole.

$$E_x = E_1\frac{2}{\left(2^2+1^2\right)^{1/2}} + E_2\frac{2}{\left(2^2+0.5^2\right)^{1/2}}, \text{ and } E_y = E_1\frac{1}{\left(2^2+1^2\right)} + E_2\frac{0.5}{\left(2^2+0.5^2\right)^{1/2}}.$$

CALCULATE: $r = \left((0.01\text{ m})^2 + (0.0200\text{ m})^2\right)^{1/2} = 0.02236$ m

$r_2 = \left((0.00500\text{ m})^2 + (0.0200\text{ m})^2\right)^{1/2} = 0.02062$ m

$$E_1 = \frac{2(8.99\cdot 10^9\text{ Nm}^2/\text{C}^2)(6.00\cdot 10^{-7}\text{ C/m})(0.02236\text{ m})}{(0.0300\text{ m})^2} = 2.680\cdot 10^5\text{ N/C}$$

$$E_{21} = -\frac{2(8.99\cdot 10^9\text{ Nm}^2/\text{C}^2)(6.00\cdot 10^{-7}\text{ C/m})}{(0.02062\text{ m})}\left(\frac{0.0100\text{ m}}{0.0300\text{ m}}\right)^2 = -0.581\cdot 10^5\text{ N/C}$$

$E_x = 1.833\cdot 10^5$ N/C, $E_y = 1.339\cdot 10^5$ N/C

ROUND: $E_x = 183$ kN/C, $E_y = 134$ kN/C

DOUBLE-CHECK: We can calculate the magnitude and direction of the combined electric field and find: $E = \sqrt{E_x^2 + E_y^2} = 227$ kN/C, and $\theta = \tan^{-1}(E_y / E_x) = 36.1°$. If the hole would not have been drilled, the magnitude would have been the magnitude we calculated above for E_1, $E_1 = 268$ kN/C, and it would have pointed along the $\vec{r}$ vector with an angle of 26.6°. This means that our result states that the magnitude of the electric field is weakened due to the presence of the hole, and that it does not point radial outward any more, but further away from the *x*-axis. Both of these results are in accordance with expectations and add confidence to our result: the hole modifies the electric field somewhat, but does not do so radically.

22.73.

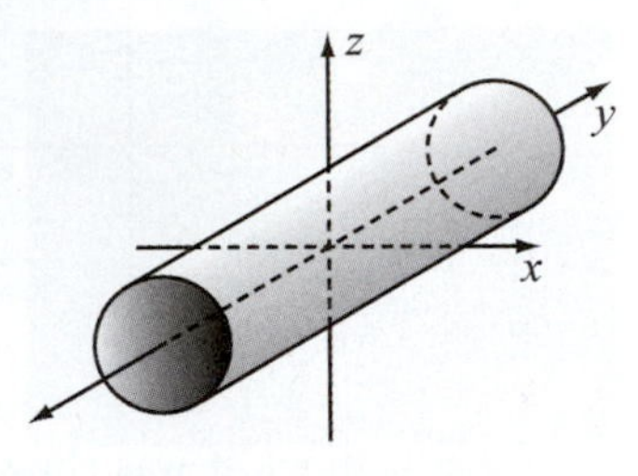

Consider a cylindrical Gaussian surface with a radius of 4.00 cm. By Gauss's Law, $\oiint \vec{E}d\vec{A} = q_{\text{enc}} / \varepsilon_0$. The charge inside the cylinder is $q = \rho\pi r^2 l$, so the field is given by

$$E(2\pi rl) = \frac{\rho\pi r^2 l}{\varepsilon_0} \Rightarrow E = \frac{\rho r}{2\varepsilon_0} = \frac{(6.40\cdot 10^{-8}\text{ C/m}^3)(0.0400\text{ m})}{2(8.854\cdot 10^{-12}\text{ m}^{-3}\text{ kg}^{-1}\text{ s}^4\text{ A}^2)} = 1.45\cdot 10^2\text{ N/C}$$

away from the *y*-axis. The information concerning the radius of the cylinder is irrelevant.

22.77. The sum of the forces on the electron is given by $F_{\text{total}} = F_{\text{gravity}} + F_{\text{coulomb}} = -mg + qE$. $E = -150.$ N/C, $q = -1.602\cdot 10^{-19}$ C,

$m = 9.11\cdot 10^{-31}$ kg. Thus, $F_{net} = qE - mg = ma \Rightarrow a_e = \dfrac{eE}{m_e} - g.$

$$a_e = \frac{(1.602\cdot 10^{-19}\text{ C})(150.\text{ N/C})}{(9.11\cdot 10^{-31}\text{ kg})} - (9.81\text{ m/s}^2) = 2.64\cdot 10^{13}\text{ m/s}^2.$$

22.81. **THINK:** I want to find the charge, q_1 needed to balance out the force of gravity. After finding q, I can determine the number of electrons based on the charge of a single electron.

SKETCH:

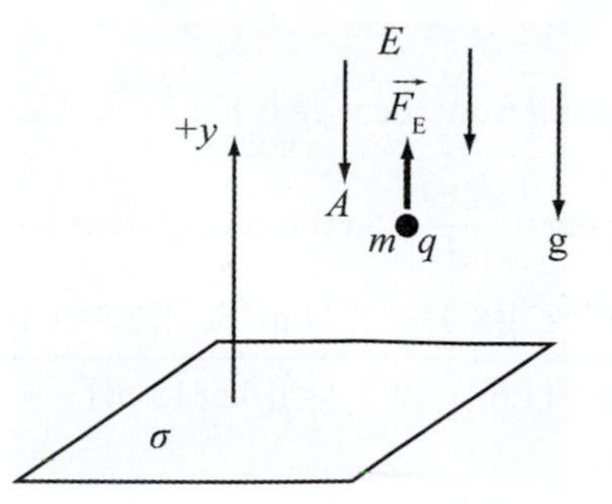

RESEARCH: The net force on the object must equal zero in order for the object to remain motionless. $F_{\text{total}} = F_{\text{gravity}} + F_{\text{coulomb}} = 0,\ F_{\text{gravity}} = -mg,\ F_{\text{coulomb}} = Eq,\ E = \sigma/2\varepsilon_0$ for an infinite plane. The number of electrons is q/q_{electron}.

SIMPLIFY: $F_{\text{total}} = F_{\text{gravity}} + F_{\text{coulomb}} = 0 \Rightarrow F_{\text{total}} = -mg + Eq = 0,\ Eq = mg \Rightarrow \dfrac{\sigma}{2\varepsilon_0}q = mg \Rightarrow q = \dfrac{2mg\varepsilon_0}{\sigma}.$

Number of electrons $= \dfrac{2mg\varepsilon_0}{\sigma q_{\text{electron}}}.\ g = 9.81\text{ m/s}^2,\ \sigma = -3.50\cdot 10^{-5}\text{ C/m}^2,\ m = 1.00\text{ g}.$

CALCULATE:

$$\text{Number of electrons} = \frac{2\left(1.00\cdot 10^{-3}\text{ kg}\right)\left(9.81\text{ m/s}^2\right)\left(8.85\cdot 10^{-12}\text{ m}^{-3}\text{ kg}^{-1}\text{ s}^4\text{ A}^2\right)}{\left(-3.50\cdot 10^{-5}\text{ C/m}^2\right)\left(-1.602\cdot 10^{-19}\text{ C}\right)} = 3.097\cdot 10^{10}\text{ electrons}$$

ROUND: $3.10\cdot 10^{10}$ electrons

DOUBLE-CHECK: This number, though large, is reasonable since the amount of charge on each electron is tiny.

22.83. **THINK:** Using the charge density, Gauss's Law can be used to find the electric field as a function of the radius.

SKETCH: Not required.

RESEARCH: The charge inside a spherical Gaussian surface is given by $q = \rho V_{\text{sphere}}$. $V_{\text{sphere}} = (4/3)\pi r^3$, $\rho = 3.57\cdot 10^{-6}\text{ C/m}^3$ and $r = 0.530$ m. Gauss's Law gives the field $\oiint \vec{E}d\vec{A} = E\left(4\pi r^2\right) = q/\varepsilon_0$.

SIMPLIFY: $E\left(4\pi r^2\right) = \dfrac{q}{\varepsilon_0} \Rightarrow E = \dfrac{1}{4\pi r^2}\left(\dfrac{q}{\varepsilon_0}\right) = \dfrac{1}{4\pi r^2}\left(\dfrac{\rho V}{\varepsilon_0}\right) = \dfrac{1}{4\pi r^2}\left(\dfrac{\rho(4/3)\pi r^3}{\varepsilon_0}\right) = \dfrac{\rho r}{3\varepsilon_0}$

CALCULATE: $E = \dfrac{\left(3.57\cdot 10^{-6}\right)(0.530)}{3\left(8.85\cdot 10^{-12}\right)}\text{ N/C} = 7.127\cdot 10^4\text{ N/C}$

ROUND: $E = 7.13\cdot 10^4$ N/C

DOUBLE-CHECK: The result was independent of the actual radius of the sphere as it should be.

Multi-Version Exercises

22.86. The electric field a distance d from the wire is $E=\frac{2k\lambda}{d}$. The force is then $F=qE=\frac{e2k\lambda}{d}$. From Newton's Second Law we have $F=ma=\frac{e2k\lambda}{d}$. So the acceleration is $a=\frac{e2k\lambda}{md}$.

$$a=\frac{2\left(1.602\cdot 10^{-19}\text{ C}\right)\left(8.99\cdot 10^{9}\text{ N m}^2/\text{C}^2\right)\left(2.849\cdot 10^{-12}\text{ C/m}\right)}{\left(1.673\cdot 10^{-27}\text{ kg}\right)\left(0.6815\text{ m}\right)}=7.198\cdot 10^{6}\text{ m/s}^2$$

22.89. The magnitude of the electric field at the center due to a differential element $d\ell$ is $dE=\frac{k\lambda d\ell}{R^2}$. The x-components add to zero, leaving only a field in the y-direction. The y-component is $dE_y=\frac{k\lambda d\ell}{R^2}\sin\theta$. Taking $d\ell=Rd\theta$ we have $dE_y=\frac{k\lambda R}{R^2}\sin\theta d\theta=\frac{k\lambda}{R}\sin\theta d\theta$. We integrate from 0 to π to get the magnitude of the electric field:

$$\int_0^{\pi}\frac{k\lambda}{R}\sin\theta d\theta=-\frac{k\lambda}{R}\left[\cos\theta\right]_0^{\pi}=2\frac{k\lambda}{R}=\frac{2\pi k\lambda}{L}.$$

So $E=\frac{2\pi k\lambda}{L}=\frac{2\pi\left(8.99\cdot 10^{9}\text{ N m}^2/\text{C}^2\right)\left(5.635\cdot 10^{-8}\text{ C/m}\right)}{\left(0.2213\text{ m}\right)}=1.438\cdot 10^{4}\text{ N/C}.$

Chapter 23: Electric Potential

Exercises

23.25. **THINK:** As the positively charged ball approaches the positively charged plane, its potential energy will increase and its kinetic energy will decrease. At the point when the ball stops, all its initial kinetic energy will have been converted into potential energy. Work must be done on the ball to accomplish this change. The force necessary to do this work is supplied by the electric field created by the charged plane.

SKETCH:

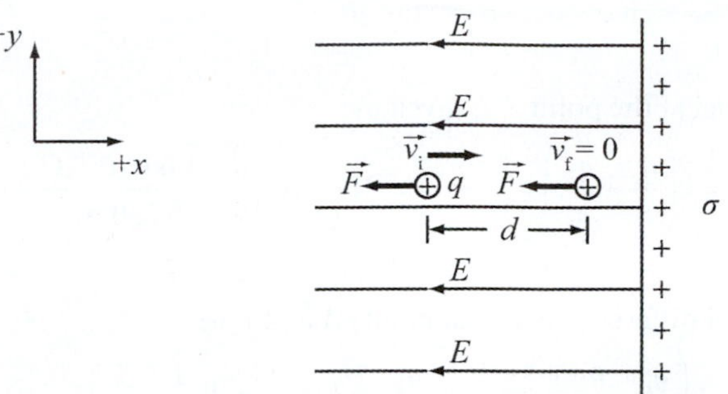

RESEARCH: Since work is force times distance, the stopping distance can be calculated by exploiting the relationship between the work done on the ball, and the force exerted on the ball by the electric field. The electric field due to the charged plane is given by $E = \sigma / \epsilon_0$. The net force acting on the ball is given by $F = qE = q\sigma / \epsilon_0$. The work done on the ball is equal to the change in kinetic energy, $\Delta K = K_f - K_i$. The work-energy relation states $W = -F \cdot d$. $W = \Delta K = K_f - K_i = -F \cdot d = -qE \cdot d = \dfrac{-q\sigma d}{\varepsilon_0}$.

SIMPLIFY: $K_f - K_i = \dfrac{-q\sigma d}{\varepsilon_0} \Rightarrow d = -\dfrac{\varepsilon_0 (K_f - K_i)}{q\sigma}$. Since $K_f = 0$, $d = \dfrac{K_i \epsilon_0}{q\sigma}$.

CALCULATE: $d = \dfrac{(6.00 \cdot 10^8 \text{ J})(8.85 \cdot 10^{-12} \text{ C}^2/\text{N m}^2)}{(5.00 \cdot 10^{-3} \text{ C})(4.00 \text{ C/m}^2)} = 0.2655 \text{ m}$; Therefore, the final distance from the plane is $x = 1 \text{ m} - 0.2655 \text{ m} = 0.7345 \text{ m}$.

ROUND: Keeping three significant figures gives $x = 0.734$ m.

DOUBLE-CHECK: This is a reasonable distance and less than the initial distance of 1 m.

23.29. Using the work-energy relation and $W = -q\Delta V$, it is found that: $W = \Delta K = -q\Delta V \Rightarrow \frac{1}{2}mv_f^2 - \frac{1}{2}mv_i^2 = -q\Delta V$. Since the proton is initially at rest, $v_i = 0$: $\frac{1}{2}mv_f^2 = -q\Delta V \Rightarrow v_f = \sqrt{\dfrac{-2q\Delta V}{m}}$. $q = 1.602 \cdot 10^{-19}$ C, $\Delta V = -500.$ V, and $m = 1.67 \cdot 10^{-27}$ kg, and this means that: $v_f = \sqrt{\dfrac{-2(1.602 \cdot 10^{-19} \text{ C})(-500. \text{ V})}{(1.67 \cdot 10^{-27} \text{ kg})}} = 3.10 \cdot 10^5 \text{ m/s}$.

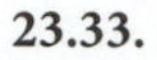

23.33.

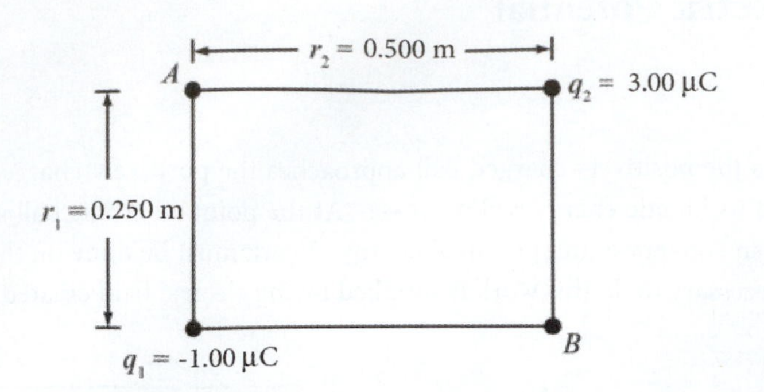

(a) The potential at the point A is given by:

$$V_A = \frac{kq_1}{r_1} + \frac{kq_2}{r_2} = k\left(\frac{q_1}{r_1} + \frac{q_2}{r_2}\right) = \left(8.99 \cdot 10^9 \text{ N m}^2/\text{C}^2\right)\left(\frac{-1.00 \cdot 10^{-6} \text{ C}}{0.250 \text{ m}} + \frac{3.00 \cdot 10^{-6} \text{ C}}{0.500 \text{ m}}\right) = 1.798 \cdot 10^4 \text{ V} \approx 1.80 \text{ kV}.$$

(b) The potential difference between points A and B is:

$$V_{AB} = V_A - V_B = k\left(\frac{q_1}{r_1} + \frac{q_2}{r_2}\right) - k\left(\frac{q_1}{r_2} + \frac{q_2}{r_1}\right) = k(q_1 - q_2)\left(\frac{1}{r_1} - \frac{1}{r_2}\right)$$

$$= \left(8.99 \cdot 10^9 \text{ N m}^2/\text{C}^2\right)\left(-1.00 \cdot 10^{-6} \text{ C} - 3.00 \cdot 10^{-6} \text{ C}\right)\left(\frac{1}{0.250 \text{ m}} - \frac{1}{0.500 \text{ m}}\right) = -7.192 \cdot 10^4 \text{ V} \approx -7.19 \text{ kV}.$$

23.37. All the charge, $Q = 5.60\ \mu\text{C}$, is equidistant from the center, $R = 4.50$ cm, so the electric potential at the center is:

$$V = \frac{kQ}{r} = \frac{\left(8.9875 \cdot 10^9 \text{ N m}^2/\text{C}^2\right)\left(5.60 \cdot 10^{-6} \text{ C}\right)}{(0.0450 \text{ m})} = 1118444 \text{ V} = 1.12 \text{ MV}.$$

23.41. **THINK:** The water droplet can be thought of as a solid insulating sphere of diameter $d = 50.0\ \mu\text{m}$ and a total charge of $q = 20.0$ pC. The potential is then found by integrating the electric field it produces from infinity to the center. The electric fields inside and outside the sphere are different.

SKETCH:

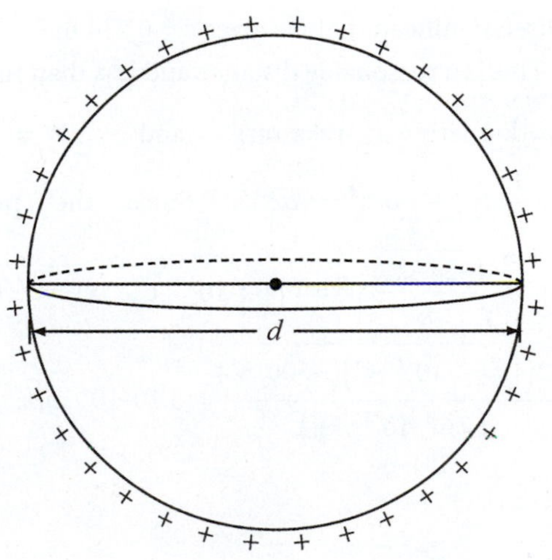

RESEARCH: The electric potential is found by: $V(r)-V(\infty)=-\int_\infty^r E(r)dr$. Since the water droplet is a non-conducting sphere, the electric field outside the sphere is $E_1 = kq/r^2$, while inside the sphere is $E_2 = kqr/R^3$ by Gauss' Law, where $R=d/2$ is the radius of the sphere.

SIMPLIFY:

(a) The potential on its surface, $r=R$, is: $V(R)-0=-\int_\infty^R \frac{kq}{r^2}dr=-kq\int_\infty^R \frac{dr}{r^2}=-kq\left[-\frac{1}{r}\right]_\infty^R=\frac{kq}{R}$.

(b) The potential inside the sphere at center, $r=0$, must be broken into 2 parts.

$$V(0)-0=-\int_\infty^R E_1dr-\int_R^0 E_2dr=-\int_\infty^R \frac{kq}{r^2}dr-\int_R^0 \frac{kqr}{R^3}dr=\frac{kq}{R}-\frac{kq}{R^2}\int_R^0 rdr=\frac{kq}{R}-\frac{kq}{R^3}\left[\frac{1}{2}r^2\right]_R^0$$

$$=\frac{kq}{R}-\frac{kq}{R^3}\left(0-\frac{1}{2}R^2\right)=\frac{kq}{R}+\frac{kq}{2R}\Rightarrow V(0)=\frac{3}{2}\left(\frac{kq}{R}\right)=\frac{3}{2}V(R).$$

CALCULATE:

(a) $V(R)=\frac{\left(8.9875\cdot 10^9 \text{ N m}^2/\text{C}^2\right)(20.0 \text{ pC})}{50.0 \text{ μm}/2}=7190 \text{ V}$

(b) $V(0)=\frac{3}{2}(7190 \text{ V})=10785 \text{ V}$

ROUND:

(a) $V(R)=7.19 \text{ kV}$

(b) $V(0)=10.8 \text{ kV}$

DOUBLE-CHECK: Though these values seem large, the droplet has a charge density of 300 C/m^3, which is quite large for an object. Therefore, the values seem reasonable.

23.45. **THINK:** The electric field, $\vec{E}=E_0xe^{-x}\hat{x}$, has a maximum when its derivative with respect to x is zero. The electric potential is found by integrating the electric field between the two points 0 and x_{max}.

SKETCH:

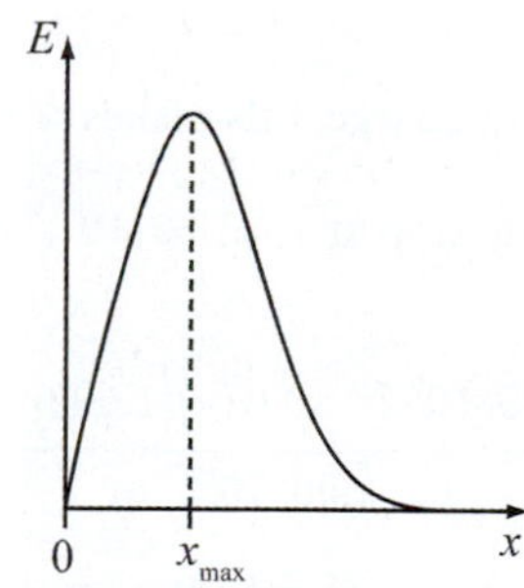

RESEARCH: Electric field is at maximum when $dE/dx=0$. The potential difference between 0 and x_{max} is $V=-\int_0^{x_{\text{max}}}\vec{E}\bullet dx$.

SIMPLIFY:

(a) $\frac{dE}{dx}=\frac{d}{dx}\left(E_0xe^{-x}\right)=E_0\left[\frac{d(x)}{dx}e^{-x}+x\frac{d\left(e^{-x}\right)}{dx}\right]=E_0\left(e^{-x}-xe^{-x}\right)$. If $\frac{dE}{dx}=0$: $e^{-x}=xe^{-x}\Rightarrow x_{\text{max}}=1$

(b) $V=-\int_0^1 E_0xe^{-x}dx=-E_0\int_0^1 xe^{-x}dx=-E_0\left[-(1+x)e^{-x}\right]_0^1=E_0\left[(1+x)e^{-x}\right]_0^1=E_0\left(2e^{-1}-1\right)$

CALCULATE: There is no need to calculate.

ROUND: There is no need to round.

DOUBLE-CHECK: The answer is reasonable.

23.49. The electric field from an electric potential, $V(x)=V_1x^2-V_2x^3$, where $V_1=2.00\text{ V/m}^2$ and $V_2=3.00\text{ V/m}^3$ is found by:

$$E=-\frac{dV}{dx}=-\frac{d}{dx}\left(V_1x^2-V_2x^3\right)=3V_2x^2-2V_1x.$$

This field produces a force on a charge, $q=1.00\ \mu\text{C}$, of $F=qE$. The acceleration of the charge is $a=\frac{F}{m}=\frac{qE}{m}$, where $m=2.50$ mg. Therefore,

$$a=\frac{q\left(3V_2x^2-2V_1x\right)}{m}=\frac{\left(1.00\cdot10^{-6}\text{ C}\right)\left[3\left(3.00\text{ V/m}^3\right)\left(2.00\text{ m}\right)^2-2\left(2.00\text{ V/m}^2\right)\left(2.00\text{ m}\right)\right]}{\left(2.50\cdot10^{-6}\text{ kg}\right)}=11.2\text{ m/s}^2.$$

23.53. **THINK:** The position, r, must be defined for three-dimensional space so that each derivative has a non-zero answer. While the potential is a scalar, each derivative is actually a vector that points in that direction, i.e. E_x points in the x-direction.

SKETCH: Not applicable.

RESEARCH: The position in three-dimensional space is given by $r=\sqrt{x^2+y^2+z^2}$. The electric field in direction $\hat{\alpha}$ is $\vec{E}_\text{i}=-\delta V\hat{\alpha}/\delta\alpha$, where $\alpha=x,y,z$.

SIMPLIFY: $\vec{E}_x=-\frac{\delta V}{\delta x}\hat{x}=-kq\frac{\delta}{\delta x}\left(\sqrt{x^2+y^2+z^2}\right)\hat{x}=\frac{kq}{2}\left(x^2+y^2+z^2\right)^{-3/2}(2x)\hat{x}=\frac{kq}{r^3}x\hat{x}.$ Likewise, $\vec{E}_y=\frac{kq}{r^3}y\hat{y}$ and $\vec{E}_z=\frac{kq}{r^3}z\hat{z}$. Therefore, $\vec{E}(\vec{r})=\frac{kq}{r^3}\left(x\hat{x}+y\hat{y}+z\hat{z}\right)$.

CALCULATE: Not applicable.

ROUND: Not applicable.

DOUBLE-CHECK: In vector notation, $x\hat{x}+y\hat{y}+z\hat{z}$ can be written as $\vec{r}=r\hat{r}$. This is evident if you let $r^2=\vec{r}\bullet\vec{r}$. Therefore, the expression for electric field for a point charge can be written as:

$$\vec{E}(\vec{r})=\frac{kq}{r^3}r\hat{r}=\frac{kq}{r^2}\hat{r}.$$

Since the potential was for a point charge, this makes sense.

23.57. If the proton comes to a complete stop at $r=1.00\cdot10^{-15}$ m, then all of its initial kinetic energy is converted to potential energy:

$$U=\frac{k|e|^2}{r}=\frac{\left(8.9875\cdot10^9\text{ N m}^2/\text{C}^2\right)\left(1.602\cdot10^{-19}\text{ C}\right)^2}{1.00\cdot10^{-15}\text{ m}}=2.31\cdot10^{-13}\text{ J or }1.44\text{ MeV}.$$

23.61. **THINK:** Two balls have masses, $m_1=5.00$ g and $m_2=8.00$ g, and charges, $q_1=5.00$ nC and $q_2=8.00$ nC. Their center separation is $l=8.00$ mm, and although the balls are not point charges, use the center separation to determine the potential energy stored in the two. Conservation of momentum and energy will allow the velocities of each to be determined. Since they are like charges, they repel and so the velocities will be in different directions.

SKETCH:

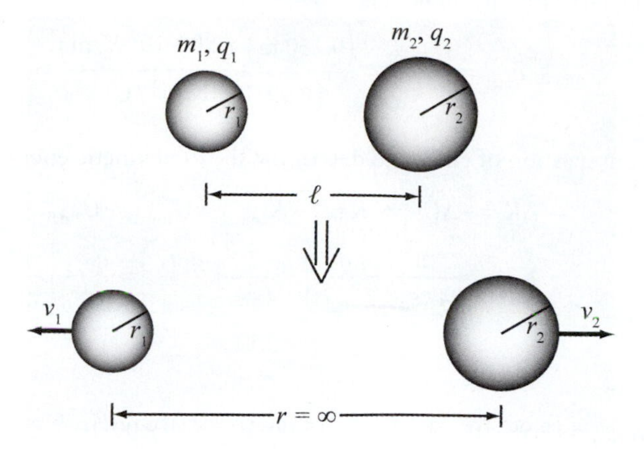

RESEARCH: The balls have no initial momentum, so by the conservation of momentum: $m_1v_1 = m_2v_2$. The initial potential energy of the two balls is given by $U_i = kq_1q_2 / l$. The final kinetic energy of the balls is given by $K_f = (m_1v_1^2/2) + (m_2v_2^2/2)$.

SIMPLIFY: From the conservation of momentum: $v_2 = m_1v_1 / m_2$. Conservation of energy then requires:

$$U_i = K_f \Rightarrow \frac{kq_1q_2}{l} = \frac{1}{2}m_1v_1^2 + \frac{1}{2}m_2v_2^2 = \frac{1}{2}\left[m_1v_1^2 + m_2\left(\frac{m_1v_1}{m_2}\right)^2\right].$$

Therefore, $\frac{2kq_1q_2}{l} = \left[m_1 + \frac{m_1^2}{m_2}\right]v_1^2 \Rightarrow v_1 = \sqrt{\frac{2kq_1q_2}{l}\left(\frac{m_2}{m_1m_2 + m_1^2}\right)}$.

CALCULATE:

$$v_1 = \sqrt{\frac{2(8.9875 \cdot 10^9 \text{ N m}^2/\text{C}^2)(5.00 \text{ nC})(8.00 \text{ nC})}{0.008.00 \text{ m}}\left(\frac{0.00800 \text{ kg}}{0.00500 \text{ kg}(0.00800 \text{ kg}) + (0.00500 \text{ kg})^2}\right)}$$

$$= 0.1052 \text{ m/s}$$

$$v_2 = \frac{5.00 \text{ g}(0.1052 \text{ m/s})}{8.00 \text{ g}} = 0.06575 \text{ m/s}$$

ROUND: $v_1 = 0.105$ m/s and $v_2 = 0.0658$ m/s

DOUBLE-CHECK: The charges are small and the masses relatively large, so the velocities obtained for the masses should be small.

23.65. The infinite plate of surface charge density, $\sigma = 3.5 \cdot 10^{-6}$ C/m^2, produces a constant electric field, $E = \sigma / 2\varepsilon_0$. In going from point A to B, any movement perpendicular to the electric field results in no change in electric potential. Therefore, the only displacement of importance is $\Delta y = -1.0$ m. The change in potential is independent of the charge Q, and since the electric field is constant, it is the product of the electric field times the displacement:

$$\Delta V = -E\Delta y = -\frac{\sigma}{2\varepsilon_0}\Delta y = -\frac{(3.50 \cdot 10^{-6} \text{ C/m}^2)}{2(8.854 \cdot 10^{-12} \text{ C}^2/(\text{N m}^2))}(-1.00 \text{ m}) = 1.98 \cdot 10^5 \text{ V}.$$

23.69. First, determine the relationship between the electric field and the potential. The electric field is given by $E = kq / r^2$. The potential is given by $V = kq / r$. Therefore, the maximum voltage is

$$V_{\text{max}} = E_{\text{max}}r = (2.00 \cdot 10^6 \text{ V/m})(0.250 \text{ m}) = 5.00 \cdot 10^5 \text{ V}.$$

The maximum charge that it can hold is

$$q_{\text{max}} = \frac{r^2 E_{\text{max}}}{k} = \frac{(0.250\text{ m})^2 (2.00 \cdot 10^6\text{ V/m})}{(8.99 \cdot 10^9\text{ N m}^2/\text{C}^2)} = 1.39 \cdot 10^{-5}\text{ C.}$$

23.73. Consider the conservation of energy to determine the final kinetic energy:

$$\Delta K = -\Delta U \Rightarrow K_{\text{final}} - K_{\text{initial}} = U_{\text{initial}} - U_{\text{final}} \Rightarrow$$

$$K_{\text{final}} = \frac{1}{4\pi\varepsilon_0}\left(\frac{q_1 q_2}{r_{\text{initial}}}\right) - \frac{1}{4\pi\varepsilon_0}\left(\frac{q_1 q_2}{r_{\text{final}}}\right) = \frac{q_1 q_2}{4\pi\varepsilon_0}\left[\frac{1}{r_{\text{initial}}} - \frac{1}{r_{\text{final}}}\right].$$

Thus,

$$K_{\text{final}} = (8.99 \cdot 10^9\text{ N m}^2/\text{C}^2)(5.00 \cdot 10^{-6}\text{ C})(9.00 \cdot 10^{-6}\text{ C})\left[\frac{1}{0.100\text{ m}} - \frac{1}{0.200\text{ m}}\right]$$

$$K_{\text{final}} = 2.02275\text{ J} \approx 2.02\text{ J.}$$

23.77. **THINK:**

(a) First determine an expression for the total potential from both charges. After finding the expression, the potential can be determined.

(b) The derivative of the expression determined in part (a) can be used to determine the minimum point.

SKETCH: A sketch is not necessary.

RESEARCH:

(a) Let $q_1 = 0.681\text{ nC}$ and $q_2 = 0.167\text{ nC}$ be the two charges with positions $r_1 = 0$ and $r_2 = 10.9\text{ cm}$, respectively. The total potential is given by:

$$V_{\text{tot}} = \frac{q_1}{4\pi\varepsilon_0}\frac{1}{|r - r_1|} + \frac{q_2}{4\pi\varepsilon_0}\frac{1}{|r - r_2|}.$$

There are three cases, depending on the value of r:

$$V_{\text{tot}} = \frac{1}{4\pi\varepsilon_0}\left(\frac{q_1}{r - r_1} + \frac{q_2}{r - r_2}\right) \text{ for } r > r_1, r_2,\quad V_{\text{tot}} = \frac{1}{4\pi\varepsilon_0}\left(\frac{q_1}{r - r_1} - \frac{q_2}{r - r_2}\right) \text{ for } r_1 < r < r_2 \text{ and}$$

$$V_{\text{tot}} = \frac{1}{4\pi\varepsilon_0}\left(-\frac{q_1}{r - r_1} - \frac{q_2}{r - r_2}\right) \text{ for } r < r_1, r_2.$$

(b) The minima occur each time the derivative is equal to zero: $\partial V_{\text{tot}} / \partial r = 0$.

SIMPLIFY:

(a) There is nothing to simplify.

(b) Take the derivative for all three cases.

$$r > r_1, r_2: \quad \frac{\partial V_{\text{tot}}}{\partial r} = \frac{1}{4\pi\varepsilon_0}\left[-\frac{q_1}{(r - r_1)^2} - \frac{q_2}{(r_1 - r_2)^2}\right].$$ The expression is equal to zero at infinity.

$$r_1 < r < r_2: \quad \frac{\partial V_{\text{tot}}}{\partial r} = \frac{1}{4\pi\varepsilon_0}\left[-\frac{q_1}{(r - r_1)^2} + \frac{q_2}{(r - r_2)^2}\right].$$ The expression is zero when:

$$\frac{q_1}{(r - r_1)^2} = \frac{q_2}{(r - r_2)^2} \Rightarrow \frac{q_1}{r^2} = \frac{q_2}{(r - r_2)^2} \quad (r_1 = 0\text{ cm})$$

$$\Rightarrow \frac{(r - r_2)^2}{r^2} = \frac{q_2}{q_1} \Rightarrow \sqrt{\left(1 - \frac{r_2}{r}\right)^2} = \sqrt{\frac{q_2}{q_1}} \Rightarrow \left|1 - \frac{r_2}{r}\right| = \sqrt{\frac{q_2}{q_1}} \Rightarrow \frac{r_2}{r} - 1 = \sqrt{\frac{q_2}{q_1}} \Rightarrow \frac{r_2}{\sqrt{q_2/q_1} + 1} = r.$$

CALCULATE:

(a) $V_{\text{tot}} = \left(8.99 \cdot 10^9 \text{ N m}^2/\text{C}^2\right)\left(\dfrac{0.681 \text{ nC}}{20.1 \text{ cm} - 0} + \dfrac{0.167 \text{ nC}}{20.1 \text{ cm} - 10.9 \text{ cm}}\right) = 46.78 \text{ V}$

(b) $r = \dfrac{10.9 \text{ cm}}{\sqrt{0.167 \text{ nC}/(0.681 \text{ nC})} + 1} = 7.28997 \text{ cm}$

ROUND:

(a) 46.8 V

(b) 7.29 cm

DOUBLE-CHECK:

(a) The potential is positive and the potential from both charges is the sign that one would expect. This makes sense, since if a test charge was placed at 20.1 cm, it would move away from either one of the charges.

(b) An equilibrium point will exist between the two charges, where the force from one is balanced by the other. Note that $0 < 7.29 \text{ cm} < 10.9 \text{ cm}$.

23.81. (a) Let $q_1 = -3.00 \text{ mC}$ and $q_2 = 5.00 \text{ mC}$ be located at $x_1 = 2.00 \text{ m}$ and $x_2 = -4.00 \text{ m}$, respectively. There are three cases:

$V(x) = \dfrac{1}{4\pi\varepsilon_0}\left(\dfrac{q_1}{x - x_1} + \dfrac{q_2}{x - x_2}\right)$ for $x > x_1, x_2$, $V(x) = \dfrac{1}{4\pi\varepsilon_0}\left(\dfrac{q_1}{x_1 - x} + \dfrac{q_2}{x - x_2}\right)$ for $x_1 < x < x_2$ and

$$V(x) = \frac{1}{4\pi\varepsilon_0}\left(\frac{q_1}{x_1 - x} + \frac{q_2}{x_2 - x}\right) \text{ for } x < x_1, x_2.$$

The three cases stem from $|x - x_1|, |x - x_2| > 0$.

(b) Case $x > x_1, x_2$: $V_{\text{tot}} = 0 \Rightarrow \dfrac{q_1}{x - x_1} = -\dfrac{q_2}{x - x_2} \Rightarrow x = \dfrac{q_1 x_2 + q_2 x_1}{q_1 + q_2}$. Case $x_1 < x < x_2$:

$\dfrac{q_1}{x_1 - x} = -\dfrac{q_2}{x - x_2} \Rightarrow x = \dfrac{q_1 x_2 - q_2 x_1}{q_1 - q_2}$. Case $x < x_1, x_2$: This case yields the same results as the first case.

Zeroes occur at the following points:

$$x = \frac{q_1 x_2 + q_2 x_1}{q_1 + q_2} = \frac{(-3.00 \text{ mC})(-4.00 \text{ m}) + (5.00 \text{ mC})(2.00 \text{ m})}{-3.00 \text{ mC} + 5.00 \text{ mC}} = 11.0 \text{ m},$$

$$x = \frac{q_1 x_2 - q_2 x_1}{q_1 - q_2} = \frac{(-3.00 \text{ mC})(-4.00 \text{ m}) - (5.00 \text{ mC})(2.00 \text{ m})}{-3.00 \text{ mC} - 5.00 \text{ mC}} = -0.250 \text{ m}.$$

(c) $E = -\dfrac{\partial V}{\partial x}$. $E = \dfrac{1}{4\pi\varepsilon_0}\left[\dfrac{q_1}{(x - x_1)^2} + \dfrac{q_2}{(x - x_2)^2}\right]$ for $x > x_1, x_2$, $E = \dfrac{1}{4\pi\varepsilon_0}\left[-\dfrac{q_1}{(x_1 - x)^2} + \dfrac{q_2}{(x - x_2)^2}\right]$ for $x_1 < x < x_2$ and

$$E = \frac{1}{4\pi\varepsilon_0}\left[-\frac{q_1}{(x_1 - x)^2} - \frac{q_2}{(x_2 - x)^2}\right] \text{ for } x < x_1, x_2.$$

Multi-Version Exercises

23.85. When a wire connects the two spheres, they have the same potential at the surface of both spheres:

$$\frac{kQ_1}{R_1}=\frac{kQ_2}{R_2} \Rightarrow \frac{Q_1}{R_1}=\frac{Q_2}{R_2}$$

The charge on the two spheres must sum to the original charge on the first sphere, $Q=Q_1+Q_2$. We can write the charge on the first sphere as $Q_1=Q-Q_2$. Now we can write

$$\frac{Q-Q_2}{R_1}=\frac{Q_2}{R_2}.$$

Solving for Q_2,

$$\frac{Q}{R_1}-\frac{Q_2}{R_1}=\frac{Q_2}{R_2}$$

$$\frac{Q_2}{R_1}+\frac{Q_2}{R_2}=Q_2\left(\frac{R_1+R_2}{R_1R_2}\right)=\frac{Q}{R_1}$$

$$Q_2\left(\frac{R_1+R_2}{R_2}\right)=Q$$

$$Q_2=Q\frac{R_2}{R_1+R_2}=\left(1.953\cdot 10^{-6}\text{ C}\right)\frac{(0.6115\text{ m})}{1.206\text{ m}+0.6115\text{ m}}=6.571\cdot 10^{-7}\text{ C}.$$

23.88. The electric field at the surface of the sphere is given by $E=\frac{kQ}{R^2}$. The potential a distance d from the surface is $V=\frac{kQ}{R+d}$. The charge on the sphere is $Q=\frac{ER^2}{k}$. So we can express the potential a distance d from the surface as

$$V=\frac{kQ}{R+d}=\frac{k\left(ER^2/k\right)}{R+d}=\frac{ER^2}{R+d}$$

$$=\frac{\left(3.165\cdot 10^5\text{ V/m}\right)\left(1.895\text{ m}\right)^2}{1.895\text{ m}+0.2981\text{ m}}=5.182\cdot 10^5\text{ V}.$$

Chapter 24: Capacitors

Exercises

24.29. Assume the supercapacitor is made from parallel plates. The capacitance is $C = \frac{\varepsilon_0 A}{d}$. Rearranging for A yields: $A = \frac{Cd}{\varepsilon_0}$. With $C = 1.00$ F, $d = 1.00 \text{ mm} = 1.00 \cdot 10^{-3}$ m and $\varepsilon_0 = 8.85 \cdot 10^{-12}$ F/m, the area is

$$A = \frac{(1.00 \text{ F})(1.00 \cdot 10^{-3} \text{ m})}{8.85 \cdot 10^{-12} \text{ F/m}} = 1.13 \text{ km}^2.$$

24.33. The capacitance of a spherical conductor is: $C = 4\pi\varepsilon_0 R$. With the radius of the Earth being $R = 6371 \text{ km} = 6.371 \cdot 10^6$ m, the Earth's capacitance is:

$$C = 4\pi(8.85 \cdot 10^{-12} \text{ F/m})(6.371 \cdot 10^6 \text{ m}) = 7.0887 \cdot 10^{-4} \text{ F} \approx 7.089 \cdot 10^{-4} \text{ F}.$$

24.37. The capacitor can be treated like two capacitors in parallel where each has an area equal to half that of the original. One capacitor has the original plate spacing $d_1 = 1.00 \text{ mm} = 0.00100$ m, and the other has a plate spacing $d_2 = 0.500 \text{ mm} = 0.000500$ m. The original area was $A = 1.00 \text{ cm}^2 = 1.00 \cdot 10^{-4} \text{ m}^2$. For capacitors in parallel, the equivalent capacitance is $C_{\text{eq}} = C_1 + C_2$. For this system,

$$C_{\text{eq}} = \frac{\varepsilon_0 A/2}{d_1} + \frac{\varepsilon_0 A/2}{d_2} = \frac{(8.85 \cdot 10^{-12} \text{ F/m})(1.00 \cdot 10^{-4} \text{ m}^2)}{2(0.00100 \text{ m})} + \frac{(8.85 \cdot 10^{-12} \text{ F/m})(1.00 \cdot 10^{-4} \text{ m}^2)}{2(0.000500 \text{ m})}$$
$$= 1.3281 \cdot 10^{-12} \text{ F} \approx 1.33 \text{ pF}.$$

24.41. **THINK:** Six capacitors are arranged as shown in the question.
(a) The capacitance of capacitor 3 is $C_3 = 2.300$ nF. The equivalent capacitance of the combination of capacitors 2 and 3 is $C_{23} = 5.000$ nF. Find the capacitance of capacitor 2, C_2. C_3 and C_2 are in parallel. Use the formula for parallel capacitance.
(b) The equivalent capacitance of the combination of capacitors 1, 2, and 3 is $C_{123} = 1.914$ nF. Find the capacitance of capacitor 1, C_1. C_1 and C_{23} are in series. Use the formula for series capacitance.
(c) The remaining capacitances are $C_4 = 1.300$ nF, $C_5 = 1.700$ nF, and $C_6 = 4.700$ nF. Find the equivalent capacitance of the whole system, C_{eq}.
(d) A battery with a potential difference of $V = 11.70$ V is connected as shown. Find the total charge q deposited on this system of capacitors.
(e) Find the potential drop across capacitor 5 in this case.
SKETCH: Consider the sketch in the question. Sketches of the simplified system are provided in the simplify step.

RESEARCH: For capacitors in series, the equivalent capacitance is $1/C_{\text{eq}} = \sum_{i=1}^{n} 1/C_i$, and the charge on each is the same. For capacitors in parallel, the equivalent capacitance is $C_{eq} = \sum_{i=1}^{n} C_i$, and the potential drop across each is the same. By definition, capacitance is $C = |q/\Delta V|$.

SIMPLIFY:

(a) $C_{23}=C_2+C_3 \Rightarrow C_2=C_{23}-C_3$

(b) $\dfrac{1}{C_{123}}=\dfrac{1}{C_1}+\dfrac{1}{C_{23}} \Rightarrow C_1=\left(\dfrac{1}{C_{123}}-\dfrac{1}{C_{23}}\right)^{-1}$

C_6 C_5 C_4 C_{123}

V

(c) C_4, C_5, and C_6 are in parallel. The equivalent capacitance between these capacitors is $C_{456}=C_4+C_5+C_6$.

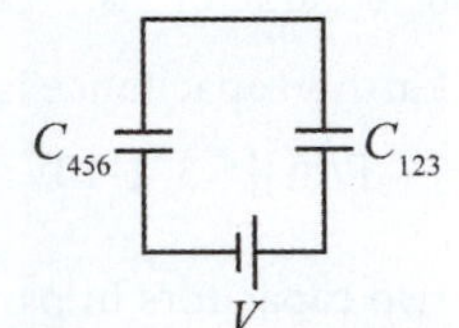

Now, C_{123} and C_{456} are in series. The equivalent capacitance of the entire circuit is

$$C_{eq}=\left(\frac{1}{C_{123}}+\frac{1}{C_{456}}\right)^{-1}.$$

(d) The total charge required to charge the capacitors is $q=C_{eq}\Delta V$.

(e) C_4, C_5, and C_6 are in parallel. The voltage drop across each capacitor is the same: $V_4=V_5=V_6$. The equivalent capacitors C_{123} and C_{456} are in series. The charge on each C_{123} and C_{456} is the same, and they are equal to the total charge in the system: $q=q_{123}=q_{456}$. Since C_4, C_5, and C_6 are in parallel, $q=q_{456}=q_4+q_5+q_6$. Then,

$$q=C_4V_4+C_5V_5+C_6V_6=V_5\left(C_4+C_5+C_6\right),\; V_5=\frac{q}{\left(C_4+C_5+C_6\right)}.$$

CALCULATE:

(a) $C_2=5.000 \text{ nF}-2.300 \text{ nF}=2.700 \text{ nF}$

(b) $C_1=\left(\dfrac{1}{1.914 \text{ nF}}-\dfrac{1}{5.000 \text{ nF}}\right)^{-1}=3.1011 \text{ nF}$

(c) $C_{456}=1.300 \text{ nF}+1.700 \text{ nF}+4.700 \text{ nF}=7.700 \text{ nF},\quad C_{eq}=\left(\dfrac{1}{1.914 \text{ nF}}+\dfrac{1}{7.700 \text{ nF}}\right)^{-1}=1.533 \text{ nF}$

(d) $q=(1.533 \text{ nF})(11.70 \text{ V})=17.94 \text{ nC}$

(e) $V_5=\dfrac{(17.94 \text{ nC})}{(1.300 \text{ nF}+1.700 \text{ nF}+4.700 \text{ nF})}=2.32987 \text{ V}$

ROUND:

(a) To 4 significant figures, $C_2=2.700$ nF.

(b) To 4 significant figures, $C_1=3.101$ nF.

(c) To 4 significant figures, $C_{eq}=1.533$ nF.

(d) To 4 significant figures, $q=17.94$ nC.

(e) To 4 significant figures, $V_5=2.330$ V.

DOUBLE-CHECK: Note for equivalent capacitance C_{23}, where C_2 and C_3 are in parallel, $C_{23} > C_2, C_3$. Similarly, for equivalent capacitance C_{456}, where C_4, C_5, and C_6 are in parallel, $C_{456} > C_4, C_5, C_6$. Finally, the equivalent capacitance of the entire circuit $C_{eq} < C_{123}, C_{456}$, where the equivalent capacitances C_{123} and C_{456} are in series.

24.45. The charge on each plate has a magnitude of $q = 60.0\ \mu\text{C}$ and potential difference of $V = 12.0\ \text{V}$. The capacitance is; therefore, $C = q/V = 60.0\ \mu\text{C}/12.0\ \text{V} = 5.00\cdot 10^{-6}\ \text{F}$. When the potential difference is $V' = 120.\ \text{V}$, the potential energy stored in the capacitor is

$$U = \frac{1}{2}C(V')^2 = \frac{1}{2}\left(5.00\cdot 10^{-6}\ \text{F}\right)\left(120.\ \text{V}\right)^2 = 0.0360\ \text{J}.$$

24.49. **THINK:** Treat the neutron star as a spherical capacitor. The inner radius of the capacitor is the radius of the neutron star, $r_1 = 10.0\ \text{km} = 1.00\cdot 10^4\ \text{m}$. The outer radius is the radius of the neutron star and the $1.00\ \text{cm}$ dipole layer. The charge density is $\sigma = \left(1.00\ \mu\text{C/cm}^2\right)\left(100.\ \text{cm/m}\right)^2 = 0.0100\ \text{C/m}^2$. Find both the capacitance C of the star and electrical energy U stored in the star's dipole layer.

SKETCH:

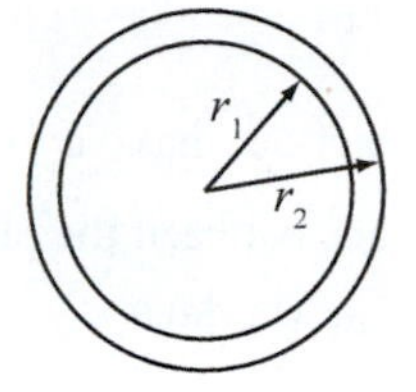

RESEARCH: The capacitance of a spherical capacitor is $C = 4\pi\varepsilon_0 \dfrac{r_1 r_2}{r_2 - r_1}$. The total charge on the dipole layer is $q = \sigma A = \sigma 4\pi r_1^2$. Note since $r_1 \approx r_2$, assume the areas of the inner and outer shells are the same. The potential energy of a capacitor is $U = (1/2)q^2/C$.

SIMPLIFY: $C = 4\pi\varepsilon_0 \dfrac{r_1 r_2}{r_2 - r_1}$, $U = \dfrac{1}{2}\dfrac{q^2}{C} = \dfrac{\left(\sigma 4\pi r_1^2\right)^2}{2C}$.

CALCULATE: $C = \dfrac{4\pi\left(8.854\cdot 10^{-12}\ \text{F/m}\right)\left(1.00\cdot 10^4\ \text{m}\right)\left(10{,}000.01\ \text{m}\right)}{0.0100\ \text{m}} = 1.11263\ \text{F}$

$$U = \frac{\left(\left(0.0100\ \text{C/m}^2\right)4\pi\left(1.00\cdot 10^4\ \text{m}\right)^2\right)^2}{2\left(1.11263\ \text{F}\right)} = 7.096\cdot 10^{13}\ \text{J}.$$

ROUND: To 3 significant figures due to the thickness of the dipole layer, $C = 1.11\ \text{F}$ and $U = 7.10\cdot 10^{13}\ \text{J}$.

DOUBLE-CHECK: There is an enormous amount of charge on the dipole layer, $q \approx 1.30\cdot 10^7\ \text{C}$. Since both C and U are proportional to q, they should also be large (especially U, where $U \propto q^2$).

24.53. In general, the energy stored in a capacitor is $U=(1/2)CV^2$. In terms of its dielectric strength V_{max}/mm, which yields $U=\frac{1}{2}\left(\frac{\kappa\varepsilon_0 A}{d}\right)V^2=\frac{\kappa\varepsilon_0 A}{2d}\left((V_{\text{max}}/\text{mm})d\right)^2=\frac{\kappa\varepsilon_0 Ad}{2}(V_{\text{max}}/\text{mm})^2$. The ratio of U_{Mylar} to U_{air} is $\frac{U_{\text{Mylar}}}{U_{\text{air}}}=\frac{\kappa_{\text{Mylar}}(V_{\text{max}}/\text{mm})^2_{\text{Mylar}}}{\kappa_{\text{air}}(V_{\text{max}}/\text{mm})^2_{\text{air}}}$. From Table 24.1, this becomes

$$\frac{U_{\text{Mylar}}}{U_{\text{air}}}=\frac{(3.1)(280\text{ kV/mm})^2}{(1)(2.5\text{ kV/mm})^2}=3.89\cdot 10^4.$$

24.57. This system is treated as two capacitors in parallel, one with dimensions $L\cdot L/5$ and dielectric κ_1, the other with dimensions $L\cdot 4L/5$ and dielectric κ_2. Then

$$C_{\text{eq}}=C_1+C_2=\frac{\varepsilon_0\kappa_1 A_1}{d}+\frac{\varepsilon_0\kappa_2 A_2}{d}=\frac{\varepsilon_0\kappa_1(L^2/5)}{d}+\frac{\varepsilon_0\kappa_2(4L^2/5)}{d}=\frac{\varepsilon_0 L^2}{d}\left(\frac{\kappa_1}{5}+\frac{4\kappa_2}{5}\right)$$

$$=\frac{(8.85\cdot 10^{-12}\text{ F/m})(0.100\text{ m})^2}{0.0100\text{ m}}\left(\frac{20.0}{5}+\frac{4(5.00)}{5}\right)=7.08310^{-11}\text{ F}\approx 70.8\text{ pF}.$$

24.61. **THINK:** A parallel plate capacitor has a capacitance of $C=120.\text{ pF}$ and plate area of $A=100.\text{ cm}^2=0.0100\text{ m}^2$. The space between the plates is filled with mica of dielectric constant $\kappa=5.40$. The plates of the capacitor are kept at $V=50.0$ V. I want to find:
(a) The strength of the electric field mica, E.
(b) The amount of free charge on the plates, Q.
(c) The amount of induced charge on mica, Q_{ind}.

SKETCH:

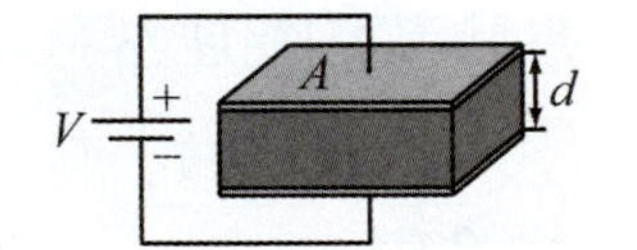

RESEARCH: For a parallel plate capacitor, $C=\varepsilon_0 A\kappa/d$. While connected to a battery, V across the capacitor is equal to that of the battery. The field in the mica is; therefore, just the field between the plates, $E=V/d$. The charge Q is $Q=CV$. The induced charge in the mica is found by considering $E_{\text{net}}=E_0-E_{\text{induced}}$.

SIMPLIFY:

(a) $C=\frac{\varepsilon_0 A\kappa}{d}\Rightarrow d=\frac{\varepsilon_0 A\kappa}{C}$ and $E=\frac{V}{d}=\frac{VC}{\varepsilon_0 A\kappa}$.

(b) $Q=CV$

(c) $E_{\text{net}}=E-E_{\text{ind}}\Rightarrow E_{\text{ind}}=E-E_{\text{net}}$, $\frac{Q_{\text{ind}}}{\varepsilon_0 A}=\frac{Q}{\varepsilon_0 A}-\frac{Q_{\text{net}}}{\varepsilon_0 A}=\frac{Q}{\varepsilon_0 A}-\frac{Q}{\kappa\varepsilon_0 A}$. Then $Q_{\text{ind}}=Q\left(1-\frac{1}{\kappa}\right)$.

CALCULATE:

(a) $E=\frac{(50.0\text{ V})(120.\text{ pF})}{(8.85\cdot 10^{-12}\text{ F/m})(0.0100\text{ m}^2)(5.40)}=12554.9\text{ V/m}$

(b) $Q=(120.\text{ pF})(50.0\text{ V})=6.00\text{ nC}$

(c) $Q_{\text{ind}}=(6.00\text{ nC})\left(1-\frac{1}{5.40}\right)=4.8888\text{ nC}$

ROUND: To three significant figures, $E = 12.6\text{ kV/m}$, $Q = 6.00\text{ nC}$ and $Q_{\text{ind}} = 4.89\text{ nC}$.

DOUBLE-CHECK: The charge induced in the dielectric should be less than the charge on the capacitor plates.

24.65. The largest potential difference that can be sustained without breakdown is about $(2.5\text{ kV/mm})(15\text{ mm}) = 37.5\text{ kV}$. Next, consider the relationship between charge deposited and change in potential:

$$Q = CV = \frac{\varepsilon_0 A}{d}V = \left(8.85\cdot 10^{-12}\text{ C}^2\text{ N}^{-1}\text{ m}^{-2}\right)\left(\frac{0.0025\text{ m}^2}{0.015\text{ m}}\right)\left(3.75\cdot 10^4\text{ V}\right) = 5.5\cdot 10^{-8}\text{ C}.$$

24.69. The capacitance is $C = 1.00\text{ F}$ for a square, parallel-plate capacitor. The separation is $d = 0.100\text{ mm}$ $= 0.000100\text{ m}$, and is filled with paper of $\kappa = 5.00$. For a parallel plate capacitor $C = \frac{\kappa\varepsilon_0 A}{d} = \frac{\kappa\varepsilon_0 L^2}{d}$. Then

$$L = \sqrt{\frac{dC}{\kappa\varepsilon_0}} = \sqrt{\frac{(0.000100\text{ m})(1.00\text{ F})}{(5.00)\left(8.85\cdot 10^{-12}\text{ F/m}\right)}} = 1503\text{ m} \approx 1.50\text{ km}.$$

24.73. The capacitance is given by $C = \frac{\kappa\varepsilon_0 A}{d} = \frac{(9.10)\left(8.85\cdot 10^{-12}\text{ F/m}\right)\left(1.00\cdot 10^{-10}\text{ m}^2\right)}{\left(2.00\cdot 10^{-8}\text{ m}\right)} = 4.03\cdot 10^{-13}\text{ F}.$

24.77. **THINK:** A parallel plate capacitor with an air gap is connected to a 6.00 V battery. The initial energy of the capacitor is $U_{\text{i}} = 72.0\text{ nJ}$. After a dielectric material is inserted, the capacitor has an additional energy of 317 nJ. The final energy stored in the capacitor is $U_{\text{f}} = 72.0\text{ nJ} + 317\text{ nJ} = 389\text{ nJ}$.

SKETCH:

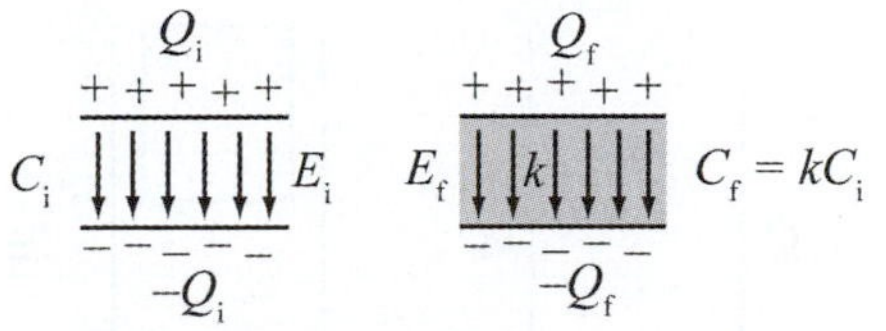

Q_{i}, $+ + + + +$, C_{i}, E_{i}, $- - - - -$, $-Q_{\text{i}}$

Q_{f}, $+ + + + +$, E_{f}, k, $C_{\text{f}} = kC_{\text{i}}$, $- - - - -$, $-Q_{\text{f}}$

RESEARCH: The energy stored in a capacitor is given by $U = (1/2)CV^2$. The initial and final energy are $U_{\text{i}} = (1/2)C_{\text{i}}V^2$ and $U_{\text{f}} = (1/2)C_{\text{f}}V^2$.

SIMPLIFY:

(a) Taking a ratio of U_{f} and U_{i} yields $U_{\text{f}}/U_{\text{i}} = C_{\text{f}}/C_{\text{i}}$. Using $C_{\text{f}} = \kappa C_{\text{i}}$, the dielectric constant is found to be $\kappa = U_{\text{f}}/U_{\text{i}}$.

(b) The charge in the capacitor is given by $Q = CV$. Using $U = (1/2)CV^2$, it is found that the charge is $Q = \left(2U/V^2\right)V = 2U/V$.

(c) The electric field inside a parallel plate capacitor is $E = \frac{Q}{A\kappa\varepsilon_0} = \frac{2U}{\kappa\varepsilon_0 AV}$.

(d) The electric field inside the capacitor after the dielectric material is inserted is $E_{\text{f}} = \frac{2U_{\text{f}}}{\kappa\varepsilon_0 AV} = \frac{2(U_{\text{f}}/\kappa)}{\varepsilon_0 AV}$. Using the result in (a) $U_{\text{i}} = U_{\text{f}}/\kappa$ yields $E_{\text{f}} = 2U_{\text{i}}/(\varepsilon_0 AV) = E_{\text{i}}$. This means that the field does not change.

CALCULATE:

(a) The dielectric constant is $\kappa = \frac{389\text{ nJ}}{72.0\text{ nJ}} = 5.403.$

(b) The charge in the capacitor after the dielectric material has been inserted is

$$Q_f = \frac{2U_f}{V} = \frac{2\left(389 \cdot 10^{-9} \text{ J}\right)}{6.00 \text{ V}} = 0.129 \text{ } \mu\text{C}.$$

(c) The electric field inside the capacitor before the dielectric material is inserted is

$$E_i = \frac{2U_i}{\varepsilon_0 AV} = \frac{2(72.0 \text{ nJ})}{\left(8.85 \cdot 10^{-12} \text{ C}^2/\text{N m}^2\right)\left(5.00 \cdot 10^{-3} \text{ m}^2\right)(6.00 \text{ V})} = 5.424 \cdot 10^5 \text{ N/C}.$$

(d) $E_f = E_i = 5.424 \cdot 10^5$ N/C

ROUND:

(a) $\kappa = 5.40$.

(b) $Q_f = 0.130$ μC.

(c) $E_i = 542$ kV/m.

(d) $E_f = 542$ kV/m.

DOUBLE-CHECK: The numerical results are reasonable.

24.81. **THINK:** A parallel plate capacitor with a squared area of side $L = 10.0$ cm and separation distance $d = 2.50$ mm is charged to a potential difference of $V_0 = 75.0$ V, and then disconnected from the battery. I want to determine the capacitor's capacitance, C_0, and the energy, U_0, stored in it at this point. A dielectric with constant $\kappa = 3.40$ is then inserted into the capacitor such that it fills 2/3 of the volume between the plates. I want to determine the new capacitance, new potential difference between the plates and energy of the capacitor, C', V' and U'. I want to determine how much work, if any, is required to insert the dielectric into the capacitor.

SKETCH:

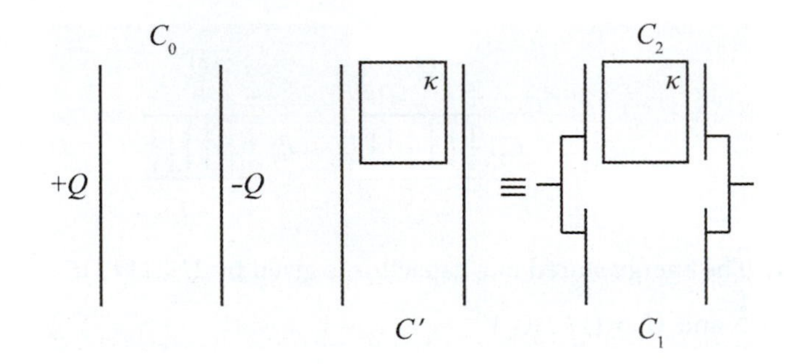

RESEARCH:

(a) The capacitance of a parallel plate capacitor is given by $C_0 = \varepsilon_0 A/d$. The energy stored in the capacitor is $U_0 = (1/2)C_0 V_0^2$.

(b) After a dielectric has been inserted, the capacitor can be treates as two capacitors in parallel, one with a dielectric, and one without. The new capacitance is obtained by adding the contributions of the two parts of the capacitor, i.e., $C' = C_1 + C_2$. Since the charge on the capacitor is unchanged, and the potential is the same across both parts of the new capacitor, $C_0 V_0 = C'V' \Rightarrow V' = \left(C_0/C'\right)V_0$. The new energy stored on the capacitor is $U' = (1/2)C'V'^2$.

SIMPLIFY:

(a) $C_0 = \varepsilon_0 A/d;\;\; U_0 = (1/2)C_0V_0^2$

(b) The new capacitance is $C' = C_1 + C_2 = (1/3)C_0 + (2/3)\kappa C_0 = \left(\frac{1+2\kappa}{3}\right)C_0$. The new potential between the plates is $V' = (C_0/C')V_0$. The new energy stored in the capacitor is

$$U' = (1/2)C'V'^2 = \frac{\left(\frac{1+2\kappa}{3}\right)(C_0)\left(\frac{C_0V_0}{\left(\frac{1+2\kappa}{3}\right)(C_0)}\right)^2}{2} = \frac{1}{2}\left(\frac{3C_0V_0^2}{1+2\kappa}\right) = U_0\left(\frac{3}{1+2\kappa}\right).$$

(c) By using the work energy relation, it is found that the applied work is

$$W = \Delta U = U' - U_0 = \left(\frac{3}{1+2\kappa} - 1\right)U_0 = \frac{2(1-\kappa)}{1+2\kappa}U_0.$$

Since κ is larger than 1, this means that the applied work is negative. Therefore, the external agent does not need to do work to insert the dielectric slab.

CALCULATE: Substituting the numerical values yields,

(a) $C_0 = \frac{\left(8.85\cdot 10^{-12}\text{ C}^2/\text{N m}^2\right)(0.100\text{ m})^2}{0.0025\text{ m}} = 35.42\text{ pF},\;\; U_0 = \frac{1}{2}\left(35.42\cdot 10^{-12}\text{ F}\right)(75.0\text{ V})^2 = 9.961\cdot 10^{-8}\text{ J}$

(b) $C' = \frac{1+2(3.4)}{3}(35.42\text{ pF}) = 92.08\text{ pF},\;\; V' = \frac{35.42\text{ pF}}{92.08\text{ pF}}(75.0\text{ V}) = 28.85\text{ V},$

$U' = \frac{3}{1+2(3.4)}\left(9.961\cdot 10^{-8}\text{ J}\right) = 3.83\cdot 10^{-8}\text{ J}$

(c) Not required.

ROUND: Rounding all results to three significant figures gives:

(a) $C_0 = 35.4$ pF and $U_0 = 0.996\cdot 10^{-8}$ J.

(b) $C' = 92.1$ pF, $V' = 28.9$ V and $U = 3.83\cdot 10^{-8}$ J.

(c) Not required.

DOUBLE-CHECK: The answers are of reasonable magnitudes and their respective units make sense.

24.85. **THINK:** Capacitor $C_1 = 1.00\ \mu\text{F}$ has an electric potential of $\Delta V_1 = 50.0$ V. Capacitor $C_2 = 2.00\ \mu\text{F}$ has an electric potential of $\Delta V_2 = 20.0$ V. The two capacitors are connected positive plate to negative plate. Calculate the final charge, $Q_{1,\text{f}}$ on capacitor C_1 after the two capacitors have come to equilibrium.

SKETCH:

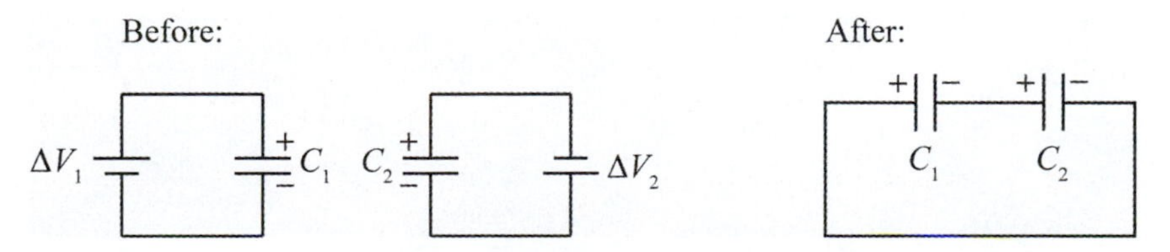

RESEARCH: Because the capacitors are connected in such a way that the positive plate of each is connected to the negative plate of the other, they must be in series. Therefore the final charges on C_1 and C_2 must be equal. Because charge is conserved, the total initial charge must equal the total final charge. The initial charge of the system is $Q_\text{i} = Q_{1,\text{i}} + Q_{2,\text{i}}$ where $Q_{1,\text{i}} = C_1\Delta V_1$ and $Q_{2,\text{i}} = C_2\Delta V_2$.

SIMPLIFY: $Q_\text{i} = Q_{\text{i}1} + Q_{\text{i}2} = C_1V_1 + C_2V_2$, $Q_\text{i} = Q_\text{f}$, $Q_{\text{f}1} = Q_{\text{f}2}$, and $Q_{\text{f}1} = \frac{Q_\text{f}}{2} = \frac{Q_\text{i}}{2} = \frac{C_1V_1 + C_2V_2}{2}$.

CALCULATE: $Q_{f1} = \dfrac{(1.00\cdot 10^{-6}\text{ F})(50.0\text{ V})+(2.00\cdot 10^{-6}\text{ F})(20.0\text{ V})}{2} = 4.50\cdot 10^{-5}\text{ C}$

ROUND: There were three significant figures provided in the question so the answer should be written as $Q_{1,f} = 4.50\cdot 10^{-5}\text{ C}$ or $45.0\ \mu\text{C}$.

DOUBLE-CHECK: It is reasonable that there is less charge stored on capacitor C_1 after it was connected to C_2 because the potential across capacitor C_1 would have to decrease in order for it to come into equilibrium with C_2.

Multi-Version Exercises

24.89. The energy stored in one supercapacitor is $U_1 = \frac{1}{2}C(\Delta V)^2$. The number of required supercapacitors is then

$$n = \frac{U}{U_1} = \frac{U}{\frac{1}{2}C(\Delta V)^2} = \frac{2U}{C(\Delta V)^2}$$
$$= \frac{2(53.63\cdot 10^6\text{ J})}{(3.361\cdot 10^3\text{ F})(2.121\text{ V})^2} = 7094.$$

24.92. The energy stored in the capacitor with dielectric inserted is $U_{\text{in}} = \frac{1}{2}\kappa C(\Delta V)^2$. The energy stored in the capacitor with the dielectric removed is $U_{\text{in}} = \frac{1}{2}C(\Delta V)^2$. The work required is equal to the change in the energy stored in the capacitor:

$$W = \Delta U = \frac{1}{2}\kappa C(\Delta V)^2 - \frac{1}{2}C(\Delta V)^2$$
$$= \frac{1}{2}C(\Delta V)^2(\kappa - 1)$$
$$= \frac{1}{2}(3.547\cdot 10^{-6}\text{ F})(10.03\text{ V})^2(4.617-1) = 6.453\cdot 10^{-4}\text{ J}.$$

Chapter 25: Current and Resistance

Exercises

25.29. The area A is $A = \pi r^2 = 3.14 \cdot 10^{-6}$ m^2, so the current density is

$$J = \frac{i}{A} = \frac{1.00 \cdot 10^{-3} \text{ A}}{3.14 \cdot 10^{-6} \text{ m}^2} = 318.3 \text{ A/m}^2 \approx 318 \text{ A/m}^2.$$

The density of electrons is

$$n = \left(\frac{1 \text{ electron}}{\text{atom}}\right)\left(\frac{6.02 \cdot 10^{23} \text{ atoms}}{26.98 \text{ g}}\right)\left(\frac{2.700 \cdot 10^{6} \text{ g}}{\text{m}^3}\right) = 6.02 \cdot 10^{28} \text{ electrons / m}^3,$$

and the drift speed is

$$v_d = \frac{J}{ne} = \frac{318.3 \text{ A/m}^2}{\left(6.02 \cdot 10^{28} \text{ electrons/m}^3\right)\left(1.602 \cdot 10^{-19} \text{ A s}\right)} = 3.30 \cdot 10^{-8} \text{ m / s}.$$

25.33. The resistances will be the same when their cross-sectional areas are the same.

$$\pi R_B^2 = \pi\left(\frac{d_o}{2}\right)^2 - \pi\left(\frac{d_i}{2}\right)^2 \Rightarrow R_B = \frac{1}{2}\sqrt{d_o^2 - d_i^2}$$

$$\Rightarrow R_B = \frac{1}{2}\sqrt{(3.00 \text{ mm})^2 - (2.00 \text{ mm})^2} = 1.12 \text{ mm}$$

25.37. **THINK:** The copper wire, $L_1 = 1$ m and $r_1 = 0.5$ mm, has an area of A_1. The wire is then stretched to $L_2 = 2$ m. Since the overall volume $(V = AL)$ of the wire remains constant, if the wire doubles in length, the area must be halved.

SKETCH:

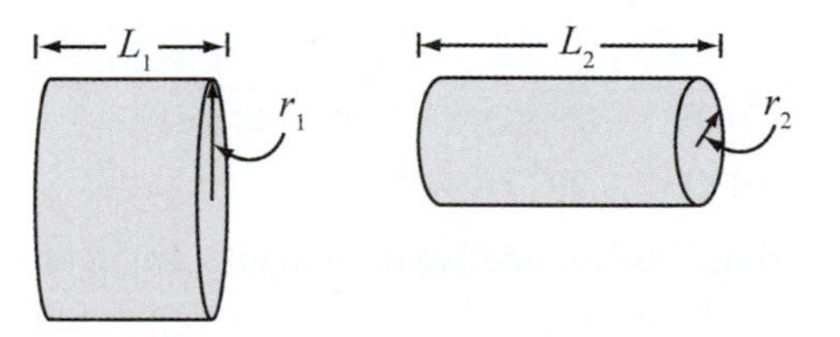

RESEARCH: The resistance of the wire is $R_i = \rho L_i / A_i$. From the conservation of volume, it follows that $V = A_1 L_1 = A_2 L_2$. The fractional change in resistance is $\Delta R / R = (R_2 - R_1)/R_1$.

SIMPLIFY: Since $L_2 = 2L_1$, then $A_2 = (1/2)A_1$. The change in resistance is then $\frac{\Delta R}{R} = \frac{(R_2 - R_1)}{R_1} = \frac{\rho(L_2 / A_2 - L_1 / A_1)}{\rho(L_1 / A_1)} = \frac{2L_1 / (1/2)A_1 - L_1 / A_1}{L_1 / A_1} = \frac{4L_1 / A_1 - L_1 / A_1}{L_1 / A_1} = 3.$ It is the same for aluminum, independent of ρ.

CALCULATE: Not required.

ROUND: Not required.

DOUBLE-CHECK: It would seem to make sense that the fractional change in resistance is the same for all materials, so having the equation independent of ρ makes sense.

25.41. The current is $i = 600.$ A and the potential difference is $\Delta V = 12.0$ V. Therefore, Ohm's Law states $\Delta V = iR \Rightarrow R = \Delta V / i = 12.0 \text{ V} / (600. \text{ A}) = 0.0200\ \Omega.$

25.45. When the external resistor, $R = 17.91\ \Omega$ is connected, the potential drop across it is $\Delta V = 12.68$ V, so the current through the circuit is, by Ohm's Law, $i = \Delta V / R = 12.68\text{ V}/17.91\ \Omega = 0.70798\text{ A} = 0.7080\text{ A}$. This is the same current running through the internal resistor, R_i, which is in series with R, so since the battery has a total emf of $\Delta V_\text{emf} = 14.50$ V, the internal resistance is found using the following calculation:

$$\Delta V_\text{emf} = i(R + R_\text{i}) \Rightarrow R_\text{i} = \frac{\Delta V_\text{emf}}{i} - R = \frac{14.50\text{ V}}{0.7080\text{ A}} - 17.91\ \Omega = 2.5702\ \Omega = 2.570\ \Omega.$$

25.49. **THINK:** The circuit can be redrawn to have the $10.0\ \Omega$ and $20.0\ \Omega$ resistors in series, both of which are parallel to the $30.0\ \Omega$ resistor, and then parallel again with the $40.0\ \Omega$ resistor. These resistors are then put in series with the $50.0\ \Omega$ resistor and the 60.0 V battery.

SKETCH:

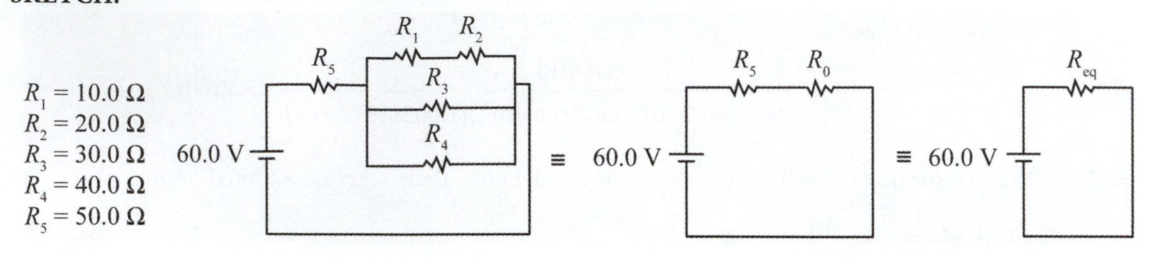

RESEARCH: Resistors in series combine as $R_\text{eq} = \sum_{i=1}^{n} R_i$. Resistors in parallel combine as $\frac{1}{R_\text{eq}} = \sum_{i=1}^{n} \frac{1}{R_i}$.

SIMPLIFY: The combined resistors in parallel become R_p where

$$R_\text{p}^{-1} = (R_1 + R_2)^{-1} + R_3^{-1} + R_4^{-1} \Rightarrow \text{R}_\text{p} = \left(\frac{1}{R_1 + R_2} + \frac{1}{R_3} + \frac{1}{R_4} \right)^{-1}.$$

The equivalent resistance is $R_\text{eq} = R_5 + R_\text{p}$.

CALCULATE: $\text{R}_\text{p} = \left(\frac{1}{10.0\ \Omega + 20.0\ \Omega} + \frac{1}{30.0\ \Omega} + \frac{1}{40.0\ \Omega} \right)^{-1} = 10.909091\ \Omega$

$R_\text{eq} = 50.0\ \Omega + 10.909091\ \Omega = 60.909091\ \Omega$

ROUND: The result should be rounded to three significant figures: $R_\text{eq} = 60.9\ \Omega$.

DOUBLE-CHECK: If you add N equal resistors, R, in parallel, the equivalent resistance is R/N. Since the resistors in parallel are all about $30\ \Omega$ in each branch, the equivalent resistance should be about $10\ \Omega$, which is close to the calculated answer of $11\ \Omega$.. Therefore, the values of R_p and R_eq are reasonable.

25.53. **THINK:** From the circuit, it is clear that resistors $R_1 = 5.00\ \Omega$ and $R_2 = 10.00\ \Omega$ are in series. Resistors $R_3 = R_4 = 5.00\ \Omega$ are in parallel, with equivalent resistance R_{34}. This is also true of resistors $R_5 = R_6 = 2.00\ \Omega$, whose equivalent resistance is R_{56}. This second pair of resistors are in turn connected in series with resistors R_1 and R_2. Ohm's Law can be used to determine the current through the whole circuit which is the same as each resistor in series.

SKETCH:

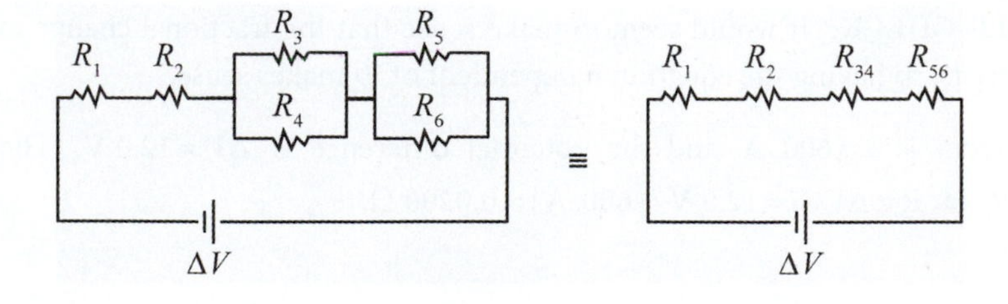

RESEARCH: Equivalent resistances if resistors are in parallel are $R_{34} = \left(1/R_3 - 1/R_4\right)^{-1}$. Total current through 4 resistors in series is $i = \Delta V / \left(R_1 + R_2 + R_{34} + R_{56}\right)$. The potential drop across a resistor is $\Delta V_i = iR_i$.

SIMPLIFY:

(a) The total current is $i = \Delta V / \left(R_1 + R_2 + R_{34} + R_{56}\right)$, $\Delta V_1 = iR_1$, $\Delta V_2 = iR_2$, $\Delta V_3 = \Delta V_4 = iR_{34}$ and $\Delta V_5 = \Delta V_6 = iR_{56}$.

(b) Current through R_1 and R_2 is i. Since $R_3 = R_4$ and $R_5 = R_6$, the current splits evenly among them so $i' = i/2$ through each of them.

CALCULATE:

(a) $R_{34} = \left(1/(5.00\ \Omega) - 1/(5.00\ \Omega)\right)^{-1} = 2.50\ \Omega$ $R_{56} = \left(1/(2.00\ \Omega) - 1/(2.00\ \Omega)\right)^{-1} = 1.00\ \Omega$,

$i = 20.0\ \text{V} / \left(5.00\ \Omega + 10.00\ \Omega + 2.50\ \Omega + 1.00\ \Omega\right) = 1.08108\ \text{A}$, $\Delta V_1 = (1.08108\ \text{A})(5.00\ \Omega) = 5.405\ \text{V}$,

$\Delta V_2 = (1.08108\ \text{A})(10.0\ \Omega) = 10.81\ \text{V}$, $\Delta V_3 = \Delta V_4 = (1.08108\ \text{A})(2.50\ \Omega) = 2.702\ \text{V}$ and

$\Delta V_5 = \Delta V_6 = (1.08108\ \text{A})(1.00\ \Omega) = 1.081\ \text{V}$

(b) $i_1 = i_2 = 1.08108\ \text{A}$, $i_3 = i_4 = i_5 = i_6 = 1.08108/2 = 0.5405\ \text{A}$

ROUND:

(a) $\Delta V_1 = 5.41\ \text{V}$, $\Delta V_2 = 10.8\ \text{V}$, $\Delta V_3 = \Delta V_4 = 2.70\ \text{V}$ and $\Delta V_5 = \Delta V_6 = 1.08\ \text{V}$.

(b) $i_1 = i_2 = 1.08\ \text{A}$ and $i_3 = i_4 = i_5 = i_6 = 0.541\ \text{A}$.

DOUBLE-CHECK: The sum of the four potential drops equals $20\ \text{V}$, so energy is conserved, so the answers make sense.

25.57. (a) If the hair dryer has power $P = 1600.\ \text{W}$ and requires a potential of $V = 110.\ \text{V}$, the current supplied is then $i = P/V = 1600.\ \text{W}/110.\ \text{V} = 14.545\ \text{A} = 14.5\ \text{A}$. i does not exceed $15.0\ \text{A}$, so it will not trip the circuit.

(b) Assuming the hair dryer obeys Ohm's Law, its effective resistance is given by $R = \Delta V / i = (110.\ \text{V})/(14.545\ \text{A}) = 7.56\ \Omega$.

25.61. **THINK:** The overall current through the resistor, R (which takes the values $1.00\ \Omega$, $2.00\ \Omega$ and $3.00\ \Omega$), is found using Ohm's Law for when the load resistance is in series with the internal resistance, $R_i = 2.00\ \Omega$, and the external emf, $V_{\text{emf}} = 12.0\ \text{V}$. I will determine an expression for the power across the load resistor and differentiate with respect to R, and solve this derivative equal to zero, in order to find a maximum in power.

SKETCH:

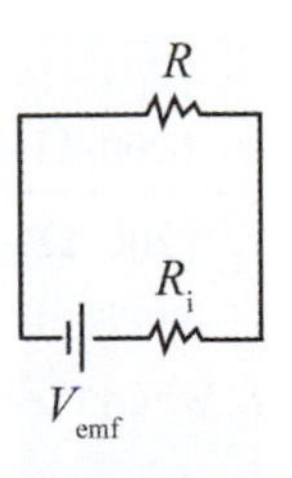

RESEARCH: With R and R_i in series, current through circuit is $i = V_{\text{emf}} / \left(R_i + R\right)$. Power through load resistor is $P = i^2 R$. The power is maximized when $dP/dR = 0$ and $d^2P/dR^2 < 0$.

SIMPLIFY: Power is $P=\left(\frac{V_{\text{emf}}}{R_{\text{i}}+R}\right)^2 R=\frac{V_{\text{emf}}^2}{(R_{\text{i}}+R)^2}R.$ Therefore,

$$\frac{dP}{dR}=V_{\text{emf}}^2\frac{d}{dR}\left[R(R_{\text{i}}+R)^{-2}\right]=V_{\text{emf}}^2\left((R_{\text{i}}+R)^{-2}-2(R)(R_{\text{i}}+R)^{-3}\right)$$

$$=V_{\text{emf}}^2\left(\frac{R_{\text{i}}+R}{(R_{\text{i}}+R)^3}-\frac{2R}{(R_{\text{i}}+R)^3}\right)=V_{\text{emf}}^2\left(\frac{R_{\text{i}}-R}{(R_{\text{i}}+R)^3}\right)=0,$$

and hence $R=R_{\text{i}}$ is a critical point of P. The double-check step will verify that $R=R_{\text{i}}$ leads to a maximum.

CALCULATE: $P_1=\frac{(12.0\text{ V})^2(1.00\ \Omega)}{(1.00\ \Omega+2.00\ \Omega)^2}=16.0\text{ W},$ $P_2=\frac{(12.0\text{ V})^2(2.00\ \Omega)}{(2.00\ \Omega+2.00\ \Omega)^2}=18.0\text{ W}$ and $P_3=\frac{(12.0\text{ V})^2(3.00\ \Omega)}{(3.00\ \Omega+2.00\ \Omega)^2}=17.28\text{ W}.$

ROUND: The values should be rounded to three significant figures each: $P_1=16.0\text{ W},$ $P_2=18.0\text{ W}$ and $P_3=17.3\text{ W}.$

DOUBLE-CHECK: The second derivative of P, $\frac{d^2P}{dR^2}=2\frac{V_{\text{emf}}{}^2(R-2R_{\text{i}})}{(R_i+R)^4}$, is clearly negative when $R=R_{\text{i}}$, which verifies that $R=R_{\text{i}}$ yields a maximum for P.

25.65. It is given $\Delta V=120\text{ V},$ $\Delta t=2.0\text{ min}=120\text{ s}$ and $U_1=48\text{ kJ}.$ The power needed to cook one hot dog is $P_1=U_1/\Delta t=4.8\cdot 10^4\text{ J}/120\text{ s}=4.0\cdot 10^2\text{ W}.$ The current to produces this power is $i_1=P_1/\Delta V=4.0\cdot 10^2\text{ W}/120\text{ V}=3.3\text{ A}.$ The current to cook three hot dogs is $i=3i_1=3(3.3\text{ A})=10.\text{ A}.$

25.69. Consider a circuit with $R_1=200.\ \Omega$ and $R_2=400.\ \Omega.$

(a) What is the power dissipated in R_1 when the two resistors are connected in series?

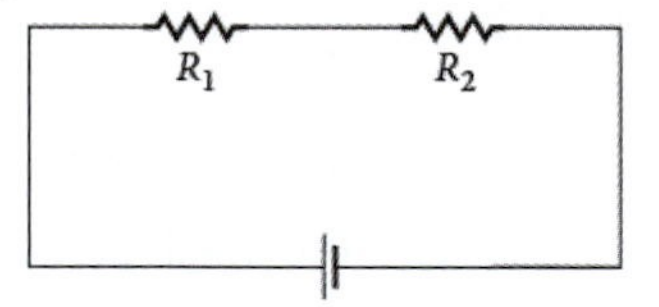

The current in the circuit is given by

$$\Delta V=iR_{\text{eq}}\ \Rightarrow\ i=\frac{\Delta V}{R_{\text{eq}}}=\frac{\Delta V}{R_1+R_2}.$$

The power dissipated in R_1 is then

$$P=i^2R_1=\left(\frac{\Delta V}{R_1+R_2}\right)^2R_1=\frac{R_1(\Delta V)^2}{(R_1+R_2)^2}=\frac{(200.\ \Omega)(9.00\text{ V})^2}{(200.\ \Omega+400.\ \Omega)^2}=0.0450\text{ W}.$$

(b) What is the power dissipated in R_1 when the two resistors are connected in parallel?

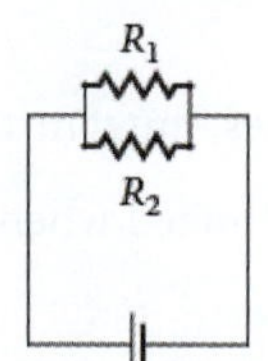

The potential difference across R_1 is 9.00 V so the power dissipated in this case is

$$P = \frac{(\Delta V)^2}{R_1} = \frac{(9.00 \text{ V})^2}{200. \ \Omega} = 0.405 \text{ W}.$$

The ratio of the power delivered to the $200. \ \Omega$ resistor by the 9.00 V battery when the resistors are connected in parallel to the power delivered when connected in series is

$$\frac{P_{\text{parallel}}}{P_{\text{series}}} = \frac{\dfrac{(\Delta V)^2}{R_1}}{\dfrac{R_1(\Delta V)^2}{(R_1+R_2)^2}} = \frac{(R_1+R_2)^2}{R_1^2} = \left(\frac{R_1+R_2}{R_1}\right)^2 = \left(\frac{200. \ \Omega + 400. \ \Omega}{200. \ \Omega}\right)^2 = 9.00.$$

25.73. **THINK:** A battery with emf 12.0 V and internal resistance $R_i = 4.00 \ \Omega$ is attached across an external resistor of resistance R. The maximum power that can be delivered to the resistor R is required.

SKETCH:

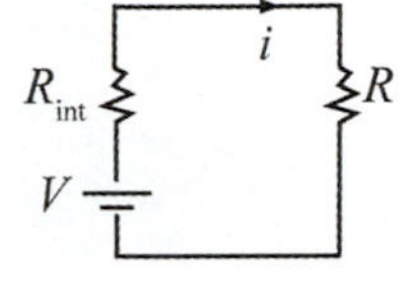

RESEARCH: The power delivered to the resistor R is given by $P = i^2 R$. The current flowing through the circuit is $i = \Delta V / (R + R_i)$. Therefore, the power is $P = \Delta V^2 R / (R + R_i)^2$. The maximum power delivered to the resistor R is given when R satisfies $dP / dR = 0$. That is

$$\frac{dP}{dR} = \frac{\Delta V^2}{(R+R_i)^2} + \frac{\Delta V^2 R(-2)}{(R+R_i)^3} = 0.$$

SIMPLIFY: Solving the above equation for R yields $R + R_i - 2R = 0$ or $R = R_i$. Thus, the maximum power delivered to R is $P = \dfrac{\Delta V^2 R_i}{(2R_i)^2} = \dfrac{\Delta V^2}{4R_i}$.

CALCULATE: Substituting the numerical values gives $P = \dfrac{(12.0 \text{ V})^2}{4(4.00 \ \Omega)} = 9.00 \text{ W}.$

ROUND: $P = 9.00$ W

DOUBLE-CHECK: This is a reasonable amount of power for a 12 V battery to supply.

25.77. **THINK:** For resistors connected in parallel, the potential differences across the resistors are the same.

SKETCH:

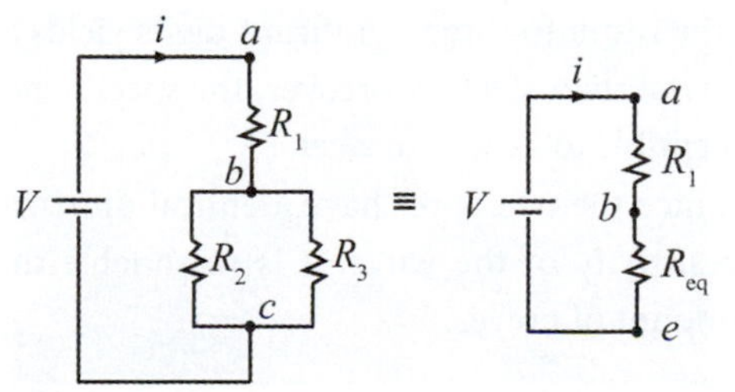

RESEARCH: The equivalent resistance of two resistors in parallel $(R_2 \text{ and } R_3)$ is $R_{\text{eq}} = (R_2)(R_3)/(R_2 + R_3)$.

SIMPLIFY:

(a) The potential difference across R_3 is $V_{\text{bc}} = V_{\text{ac}} - V_{\text{ab}} = V - iR_1 = V\left(\dfrac{R_1}{R_1 + R_{\text{eq}}}\right)$.

(b) Since R_1 and R_{eq} are in series, the current flowing through R_1 and R_{eq} is $i=\frac{V}{R}=\frac{V}{R_1+R_{eq}}$.

(c) The rate thermal energy dissipated from R_2 is $P=\frac{V_{bc}^2}{R_2}$.

CALCULATE: $R_{eq}=\frac{(3.00\ \Omega)(6.00\ \Omega)}{3.00\ \Omega+6.00\ \Omega}=2.00\ \Omega$

(a) The potential difference across R_3 is $V_{bc}=110.\ \text{V}\left(\frac{2.00\ \Omega}{2.00\ \Omega+2.00\ \Omega}\right)=55.0\ \text{V}$.

(b) The current through R_1 is $i=\frac{110.\ \text{V}}{2.00\ \Omega+2.00\ \Omega}=27.5\ \text{A}$.

(c) The thermal energy dissipated from R_2 is $P=\frac{(55.0\ \text{V})^2}{3.00\ \Omega}=1.008\ \text{kW}$.

ROUND: Keeping three significant digits gives:

(a) $V_{bc}=55.0\ \text{V}$

(b) $i=27.5\ \text{A}$

(c) $P=1.01\ \text{kW}$

DOUBLE-CHECK: Each value has appropriate units for what is being measured.

25.81. **THINK:** Two conducting wires have identical length and identical radii of circular cross-sections. I want to calculate the ratio of the power dissipated by the two resistors (copper and steel).

SKETCH: Not required.

RESEARCH: The resistance of a wire is given by $R=\rho L/A$. The power dissipated by the wire is $P=\frac{V^2}{R}=\frac{V^2A}{\rho L}$.

SIMPLIFY: Therefore, the ratio of powers of two wires is

$$\frac{P_{copper}}{P_{steel}}=\left(\frac{V^2A_{copper}}{\rho_{copper}L_{copper}}\right)\div\left(\frac{V^2A_{steel}}{\rho_{steel}L_{steel}}\right).$$

Since $L_{copper}=L_{steel}$ and $A_{copper}=A_{steel}$, the ratio becomes $\frac{P_{copper}}{P_{steel}}=\frac{\rho_{steel}}{\rho_{copper}}$.

CALCULATE: $\frac{P_{copper}}{P_{steel}}=\frac{40.0\cdot10^{-8}\ \Omega\ \text{m}}{1.68\cdot10^{-8}\ \Omega\ \text{m}}=23.8095$

ROUND: Rounding the result to three significant digits yields a ratio of 23.8:1. This is because copper is a better conducting material than steel. Moreover, the specific heat of copper is less than steel. This means that copper is less susceptible to heat than steel.

DOUBLE-CHECK: Since the two wires have identical dimensions, and the power dissipated is inversely proportional to the resistivity of the wires, it is reasonable that the material with the higher resistivity dissipates the larger amount of power.

Multi-Version Exercises

25.84. Following Solved Problem 25.4, we find

$$f=\frac{4P\rho_{\text{Cu}}L}{\pi(\Delta V)^2 d^2}=\frac{4\left(7935\cdot 10^6\text{ W}\right)\left(1.72\cdot 10^{-8}\ \Omega\text{m}\right)\left(643.1\cdot 10^3\text{ m}\right)}{\pi\left(1.177\cdot 10^6\text{ V}\right)^2\left(0.02353\text{ m}\right)^2}=0.1457=14.6\%.$$

25.87. The energy stored in the battery is equal to the power output of the battery multiplied by the time the battery delivers that power. The power delivered by the battery is $P=i\Delta V$. The energy stored in the battery is then $U=Pt=i\Delta Vt$. The time is $t=110.0\text{ min}\frac{60\text{ s}}{\text{min}}=6600\text{ s}$.

$$U=i\Delta Vt=(25.0\text{ A})(10.5\text{ V})(6600\text{ s})=1732500\text{ J}=1.73\text{ MJ}.$$

25.89. The temperature dependence of resistance is $R-R_0=R_0\alpha(T-T_0)$. The resistance at operating temperature is given by $\Delta V=iR \Rightarrow R=\frac{\Delta V}{i}$. Combining these equations gives us

$$\frac{\Delta V}{i}-R_0=R_0\alpha(T-T_0)$$

$$T=T_0+\frac{\frac{\Delta V}{i}-R_0}{R_0\alpha}$$

$$=20.00\text{ °C}+\frac{\frac{3.907\text{ V}}{0.3743\text{ A}}-1.347\ \Omega}{(1.347\ \Omega)\left(4.5\cdot 10^{-3}\text{ °C}^{-1}\right)}$$

$$=1520\text{ °C}.$$

Chapter 26: Direct Current Circuits

Exercises

26.27. The total resistance of the circuit is given by: $R_{\text{total}} = R_1 + R_2$. The current is then $I = \Delta V / R_{\text{total}}$. The potential drop across each resistor is then:

$$V_1 = \left(\frac{R_1}{R_1 + R_2} \right) \Delta V, \quad V_2 = \left(\frac{R_2}{R_1 + R_2} \right) \Delta V.$$

The resistors in series construct a voltage divider. The voltage ΔV is divided between the two resistors with potential drop proportional to their respective resistances.

26.31. **THINK:**

(a) The dead battery is parallel to the starter and the live battery.

(b) Kirchoff's Laws can be used to find the currents. Use the data:

$V_{\text{L}} = 12.00$ V, $V_{\text{D}} = 9.950$ V, $R_{\text{L}} = 0.0100\ \Omega$, $R_D = 1.100\ \Omega$, $R_{\text{S}} = 0.0700\ \Omega$.

SKETCH:

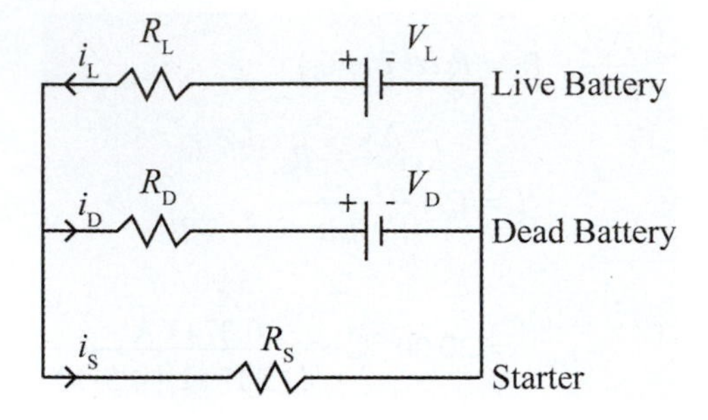

RESEARCH: Kirchoff's Laws give:

$$i_{\text{L}} = i_{\text{D}} + i_{\text{S}} \tag{1}$$

$$i_{\text{D}} = i_{\text{L}} - i_{\text{S}} \tag{1.1}$$

$$V_{\text{L}} - i_{\text{L}} R_{\text{L}} - i_{\text{S}} R_{\text{S}} = 0 \tag{2}$$

$$V_{\text{D}} + i_{\text{D}} R_{\text{D}} - i_{\text{S}} R_{\text{S}} = 0 \tag{3}$$

SIMPLIFY: Substitute (1) into (2) and solve for i_{D}. $V_{\text{L}} - (i_{\text{D}} + i_{\text{S}}) R_{\text{L}} - i_{\text{S}} R_{\text{S}} = 0$ implies $V_{\text{L}} - i_{\text{D}} R_{\text{L}} - i_{\text{S}} (R_{\text{L}} + R_{\text{S}}) = 0$, which in turn implies

$$i_{\text{D}} = \frac{V_{\text{L}} - i_{\text{S}} (R_{\text{L}} + R_{\text{S}})}{R_{\text{L}}} \tag{4}$$

Substitute (4) into (3) and solve for i_{S}. $V_{\text{D}} + \left(\frac{V_{\text{L}} - i_{\text{s}} (R_{\text{L}} + R_{\text{S}})}{R_{\text{L}}} \right) R_{\text{D}} - i_{\text{S}} R_{\text{S}} = 0$ implies

$$i_{\text{S}} = \frac{V_{\text{D}} R_{\text{L}} + V_{\text{L}} R_{\text{D}}}{R_{\text{L}} R_{\text{D}} + R_{\text{S}} R_{\text{D}} + R_{\text{S}} R_{\text{L}}} \tag{5}$$

Substitute (1.1) into (3) and solve for i_{S}. $V_{\text{D}} + (i_{\text{L}} - i_{\text{S}}) R_{\text{D}} - i_{\text{S}} R_{\text{S}} = 0$ implies

$$i_{\text{S}} = \frac{V_{\text{D}} + i_{\text{L}} R_{\text{D}}}{R_{\text{D}} + R_{\text{S}}} \tag{6}$$

Substitute (6) into (2) and solve for i_{L}. $V_{\text{L}} - i_{\text{L}} R_{\text{L}} - \left(\frac{V_{\text{D}} + i_{\text{L}} R_{\text{D}}}{R_{\text{D}} + R_{\text{S}}} \right) R_{\text{S}} = 0$ implies

$$i_{\text{L}} = \frac{V_{\text{L}} R_{\text{D}} + V_{\text{L}} R_{\text{S}} - V_{\text{D}} R_{\text{S}}}{R_{\text{L}} R_{\text{D}} + R_{\text{L}} R_{\text{S}} + R_{\text{D}} R_{\text{S}}} \tag{7}$$

Substitute (5) and (7) into (1) and solve for i_L.

CALCULATE: $i_S = \dfrac{(9.950\text{ V})(0.0100\ \Omega)+(12.00\text{ V})(1.100\ \Omega)}{(0.0100\ \Omega)(1.100\ \Omega)+(0.0700\ \Omega)(1.100\ \Omega)+(0.0700\ \Omega)(0.0100\ \Omega)} = 149.938\text{ A}$

$$i_L = \frac{(12.00\text{ V})(1.100\ \Omega)+(12.00\text{ V})(0.0700\ \Omega)-(9.950\text{ V})(0.0700\ \Omega)}{(0.0100\ \Omega)(1.100\ \Omega)+(0.0100\ \Omega)(0.0700\ \Omega)+(1.100\ \Omega)(0.0700\ \Omega)} = 150.434\text{ A}$$

$150.434\text{ A} = i_D + 149.938\text{ A} \Rightarrow i_D = 150.434\text{ A} - 149.938\text{ A} = 0.496\text{ A}$

ROUND: Three significant figures: $i_S = 150.\text{ A}$, $i_L = 150.\text{ A}$, $i_D = 0.496\text{ A}$.

DOUBLE-CHECK: Inserting the calculated values back into the original Kirchoff's equations;

$$V_L - i_L R_L - i_S R_S = (12\text{ V}) - (150\text{ A})(0.01\ \Omega) - (150\text{ A})(0.07\ \Omega) = 0,$$

and

$$V_D + i_D R_D - i_S R_S = (9.95\text{ V}) + (0.496\text{ A})(1.1\ \Omega) - (150\text{ A})(0.07\ \Omega) = 0,$$

as required.

26.35. **THINK:** Kirchhoff's Laws can be applied to this circuit. We must identify the junctions and the loops. We note that the currents through resistors 1 and 3 are the same and the currents through resistors 6 and 7 are the same. We have five unknowns, i_1, i_2, i_4, i_5, and i_6. We need five equations for the solution.

SKETCH:

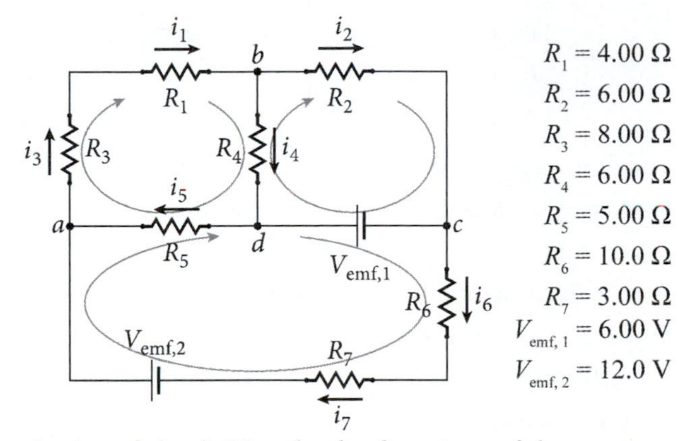

RESEARCH: We have $i_1 = i_3$ and $i_6 = i_7$. We take the directions of the currents as shown in the sketch. There are four junctions giving the following equations

$$a:\ i_5 + i_6 = i_1$$
$$b:\ i_1 = i_2 + i_4.$$

There are three loops that can be analyzed using Kirchoff's loop rule. Analyzing each loop in the clockwise direction:

$$\text{Starting at } a: -i_1R_3 - i_1R_1 - i_4R_4 - i_5R_5 = 0$$
$$\text{Starting at } d:\ -V_{\text{emf},1} - i_6R_6 - i_6R_7 + V_{\text{emf},2} + i_5R_5 = 0$$
$$\text{Starting at } c:\ V_{\text{emf},1} + i_4R_4 - i_2R_2 = 0.$$

The power supplied by each battery is given by $P = Vi.$

SIMPLIFY: Cramer's rule is the most efficient method for solving a system of five equations and five unknowns. Rearranging the equations:

$$-i_1 + i_5 + i_6 = 0$$
$$i_1 - i_2 - i_4 = 0$$
$$-i_1(R_1 + R_3) - i_4R_4 - i_5R_5 = 0$$
$$i_5R_5 - i_6(R_6 + R_7) = V_{\text{emf},1} - V_{\text{emf},2}$$
$$-i_2R_2 + i_4R_4 = -V_{\text{emf},1}.$$

Taking the coefficients of the currents, we can write the matrix equation as:

$$\begin{bmatrix} -1 & 0 & 0 & 1 & 1 \\ 1 & -1 & -1 & 0 & 0 \\ -(R_1+R_3) & 0 & -R_4 & -R_5 & 0 \\ 0 & 0 & 0 & R_5 & -(R_6+R_7) \\ 0 & -R_2 & R_4 & 0 & 0 \end{bmatrix}\begin{bmatrix} i_1 \\ i_2 \\ i_4 \\ i_5 \\ i_6 \end{bmatrix}=\begin{bmatrix} 0 \\ 0 \\ 0 \\ V_{\text{emf},1}-V_{\text{emf},2} \\ -V_{\text{emf},1} \end{bmatrix}$$

CALCULATE: We can use Cramer's rule to solve this system of five equations and five unknowns

$$\begin{bmatrix} -1 & 0 & 0 & 1 & 1 \\ 1 & -1 & -1 & 0 & 0 \\ -12.00 & 0 & -6.00 & -5.00 & 0 \\ 0 & 0 & 0 & 5.00 & -13.00 \\ 0 & -6.00 & 6.00 & 0 & 0 \end{bmatrix}\begin{bmatrix} i_1 \\ i_2 \\ i_4 \\ i_5 \\ i_6 \end{bmatrix}=\begin{bmatrix} 0 \\ 0 \\ 0 \\ -6.00 \\ -6.00 \end{bmatrix}$$

The solution can be calculated by hand using Cramer's rule or using a computer algebra system. The matrix, when evaluated by such a program into reduced row echelon form, gives the numeric solution as:

$$\begin{bmatrix} 1 & 0 & 0 & 0 & 0 \\ 0 & 1 & 0 & 0 & 0 \\ 0 & 0 & 1 & 0 & 0 \\ 0 & 0 & 0 & 1 & 0 \\ 0 & 0 & 0 & 0 & 1 \end{bmatrix}\begin{bmatrix} i_1 \\ i_2 \\ i_4 \\ i_5 \\ i_6 \end{bmatrix}=\begin{bmatrix} 0.250746 \\ 0.625373 \\ -0.374627 \\ -0.152239 \\ 0.402985 \end{bmatrix} \Rightarrow \begin{matrix} i_1 = 0.250746 \text{ A} \\ i_2 = 0.625373 \text{ A} \\ i_4 = -0.374627 \text{ A} \\ i_5 = -0.152239 \text{ A} \\ i_6 = 0.402985 \text{ A} \end{matrix}$$

The current through resistors R_1 and R_3 is $i_1 = i_3 = 0.250746$ A in the assumed direction. The current through resistor R_2 is $i_2 = 0.625373$ A in the assumed direction. The current through resistor R_4 is $i_4 = 0.374627$ A in a direction opposite to the assumed direction. The current through resistor R_5 is $i_5 = 0.152239$ A in a direction opposite to the assumed direction. The current through resistors R_6 and R_7 is $i_6 = i_7 = 0.402985$ A in the assumed direction.

The current flowing through $V_{\text{emf},1}$ is given by $i_2 + i_4 = 0.625373 \text{ A} - 0.374627 \text{ A} = 0.250746 \text{ A}$.

$P(V_{\text{emf},1}) = (6.00 \text{ V})(0.250746 \text{ A}) = 1.504476 \text{ W}$.

The current flowing through $V_{\text{emf},2}$ is given by

$i_1 + i_2 + i_3 + i_6 + i_7 = 0.250746 \text{ A} + 0.625373 \text{ A} + 0.250746 \text{ A} + 0.402985 \text{ A} + 0.402985 \text{ A} = 1.932835 \text{ A}$.

$P(V_{\text{emf},2}) = (12.0 \text{ V})(1.932835 \text{ A}) = 23.19402 \text{ W}$.

ROUND: Rounding to three significant digits and assigning the directions we have:

	Magnitude	Direction
i_1	0.251 A	to the right
i_2	0.625 A	to the right
i_3	0.251 A	upward
i_4	0.375 A	upward
i_5	0.152 A	to the right
i_6	0.403 A	downward
i_7	0.403 A	to the left

$P(V_{\text{emf},1}) = 1.50 \text{ W}$, $P(V_{\text{emf},2}) = 23.2 \text{ W}$.

DOUBLE-CHECK: We can substitute out results for the five currents back into our five equations and show that they are satisfied.

26.39. Let i_A be the maximum current (i.e. full scale value) the ammeter can measure without the shunt. If the shunt is to extend the full scale value by a factor $N = i_{tot} / i_A$, then

$$i_A + i_{shunt} = Ni_A \Rightarrow \frac{i_{shunt}}{i_A} = N - 1.$$

Since the ammeter and shunt have the same voltage across them,

$$R_{i,A} i_A = R_{shunt} i_{shunt} \Rightarrow R_{shunt} = \frac{i_A}{i_{shunt}} R_{i,A} = \frac{R_{i,A}}{N-1}.$$

To allow a current of 100 A, the resistance of the shunt resistor must be

$$R_{shunt} = \frac{(1.00\ \Omega)}{(100.\ \text{A}/1.00\ \text{A})-1} = \frac{1.00\ \Omega}{99.0} = 10.1\ \text{m}\Omega.$$

The fraction of the total current flowing through the ammeter is

$$\frac{i_A}{i_{tot}} = \frac{(1.00\ \text{A})}{(100.\ \text{A})} = 0.0100.$$

The fraction of the total current flowing through the shunt is

$$\frac{i_{shunt}}{i_{tot}} = 1 - \frac{(1.00\ \text{A})}{(100.\ \text{A})} = 0.990.$$

26.43. **THINK:**
(a) The total resistance must first be determined in order to find the current. Since the resistors are in series, the same current flows through both of them.
(b) The current that flows through the circuit is the result of the equivalent resistance including the ammeter. The same current flows through the 1.00 kΩ resistor and the parallel combination of resistor and ammeter. Of the current flowing through this combination, the majority will flow through the lower resistance, i.e., the ammeter. The fraction of the current that goes through the Ammeter can be calculated using the resistances.
SKETCH:

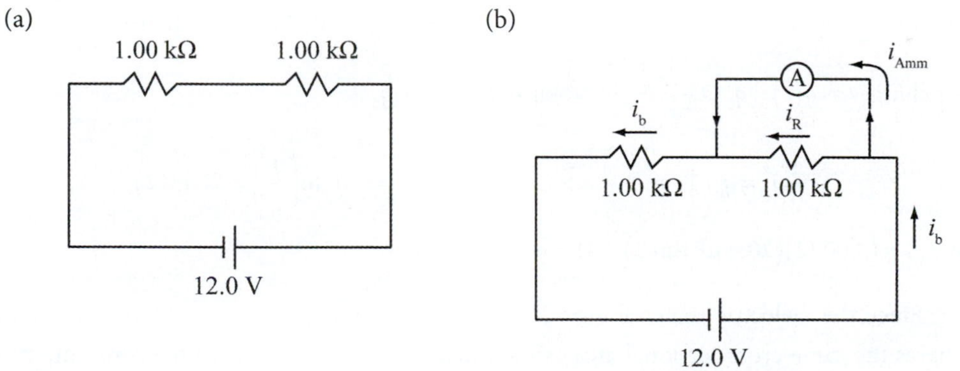

RESEARCH:
(a) $R_{eq} = 2R$, $R = 1.00\ \text{k}\Omega$, $i_a = V / R_{eq}$, $V = 12.0\ \text{V}$

(b) The current that flows through the circuit is $i_b = \dfrac{V}{R_{eq}}$, where $R_{eq} = R + \left(\dfrac{RR_A}{R + R_A}\right)$, $R_A = 1.0\ \Omega$. The current flowing through the resistor/ammeter combination is split into two parts. $i_R = \Delta V_1 / R$, and $i_{Amm} = \Delta V_2 / R_A$.

SIMPLIFY:

(a) $i_a = \dfrac{V}{2R}$

(b) $i_{\text{Amm}} = \dfrac{\Delta V_2}{R_{\text{Amm}}} = \dfrac{i_b RR_{\text{Amm}}}{R_{\text{Amm}}\left(R+R_{\text{Amm}}\right)} = \dfrac{V}{R+\left(\dfrac{RR_{\text{Amm}}}{R+R_{\text{Amm}}}\right)} \cdot \left(\dfrac{RR_{\text{Amm}}}{R_{\text{Amm}}\left(R+R_{\text{Amm}}\right)}\right) = \dfrac{V}{R+2R_{\text{Amm}}}$

CALCULATE:

(a) $i_a = \dfrac{12.0 \text{ V}}{2\left(1.00 \cdot 10^3 \ \Omega\right)} = 6.00 \text{ mA}$

(b) $\dfrac{12.0 \text{ V}}{\left(1.00 \text{ k}\Omega + 2\cdot\left(1.0 \ \Omega\right)\right)} = 0.01198 \text{ A}$

ROUND:

(a) $i_a = 6.00 \text{ mA}$

(b) $i_{\text{Amm}} = 0.012 \text{ A}$

DOUBLE-CHECK: The ammeter measures the current across the other resistor acting like a short across the first resistor, as would be expected.

26.47. Since the position of the resistor with respect to the capacitor is irrelevant, the circuit is simplified to:

$C = 20.0 \ \mu\text{F}$, $R_1 = 1.00 \ \Omega$, $R_2 = 2.00 \ \Omega$, $V = 12.0 \text{ V}$ $\Rightarrow$ $C = 20.0 \ \mu\text{F}$, $R_{eq} = 3.00 \ \Omega$, $V = 12.0 \text{ V}$

$$R_{eq} = R_1 + R_2 = 3.00 \ \Omega$$

The maximum charge of the capacitor is $q_0 = C\Delta V = \left(20.0 \ \mu\text{F}\right)\left(12.0 \text{ V}\right) = 2.40 \cdot 10^{-4} \text{C}$. In general, the capacitor charges as $q(t) = q_0\left(1 - e^{-\frac{t}{RC}}\right)$. When $q(t) = (1/2)q_0$:

$$\frac{1}{2}q_0 = q_0\left(1 - e^{-\frac{t}{RC}}\right) \Rightarrow e^{-\frac{t}{RC}} = \frac{1}{2} \Rightarrow t = -RC\ln\left(\frac{1}{2}\right) = RC\ln(2).$$

Therefore, $t = \left(3.00 \ \Omega\right)\left(20.0 \ \mu\text{F}\right)\ln(2) = 41.6 \ \mu\text{s}$.

26.51. **THINK:** Since the dielectric material ($\kappa = 2.5$, $d = 50.0 \ \mu\text{m}$ and $\rho = 4.0 \cdot 10^{12} \ \Omega \text{ m}$) acts as the resistor and it shares the same cross sectional area as the capacitor, $C = 0.050 \ \mu\text{F}$, a time constant, τ, should be independent of the actual capacitance and resistance, and only depend on the material.

SKETCH:

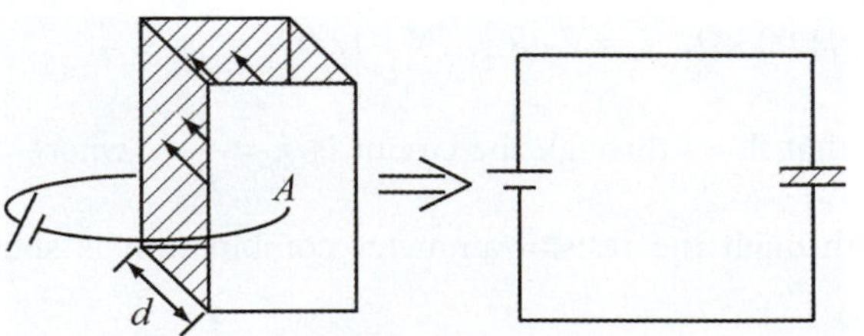

RESEARCH: The capacitance is $C = \kappa\varepsilon_0 A / d$. The resistance is $R = \rho d / A$. The time constant is $\tau = RC$.

SIMPLIFY: The time constant is $\tau = RC = \left(\frac{\rho d}{A}\right)\left(\frac{\kappa\varepsilon_0 A}{d}\right) = \kappa\rho\varepsilon_0$.

CALCULATE: $\tau = 2.5\left(4.0\cdot 10^{12}\ \Omega\ \text{m}\right)\left(8.85\cdot 10^{-12}\ \text{C}^2/\left(\text{N m}^2\right)\right) = 88.5\ \text{s}$

ROUND: $\tau = 89\ \text{s}$

DOUBLE-CHECK: While this value seems relatively high, it is nonetheless perfectly reasonable. The high resistivity and greater than 1 dielectric material, both imply bigger R and C, so a high τ is reasonable.

26.55. **THINK:** The capacitor, $C = 2.00\ \mu\text{F}$, charges via the battery, $\Delta V = 10.0\ \text{V}$, through resistor, $R_1 = 10.0\ \Omega$, so the resistors, $R_2 = 4.00\ \Omega$ and $R_3 = 10.0\ \Omega$, can be simplified to be in parallel. After a long time, the capacitor becomes fully charged and no current goes through it. The potential drop across it is then the same as the drop across R_2 and R_3. The energy of the capacitor is proportional to the square of the potential drop across it. The total energy lost across R_3 is determined by integrating the power across it over time.

SKETCH:

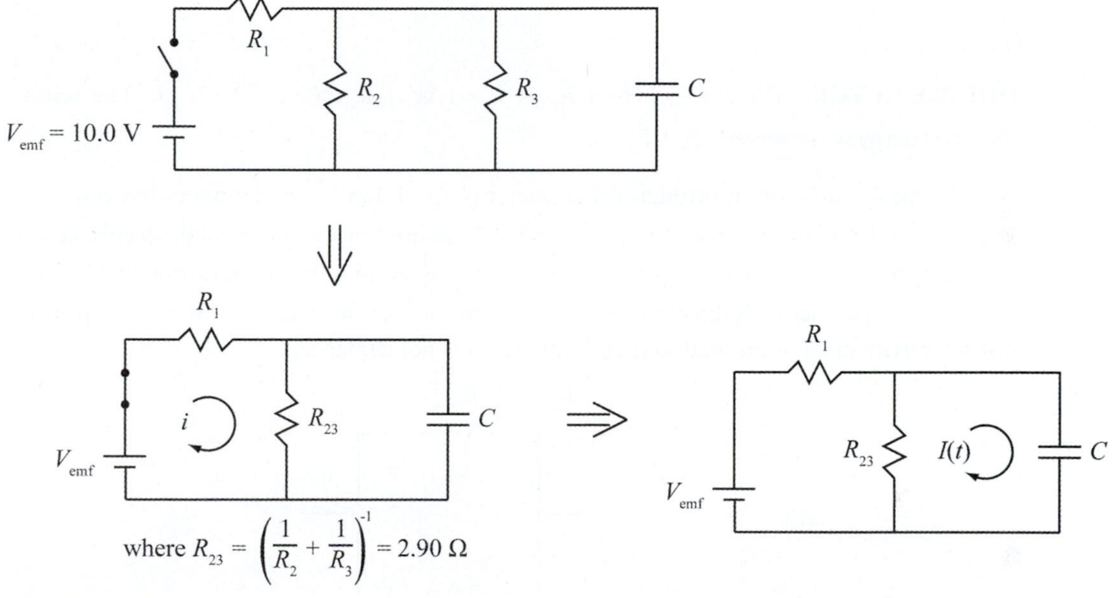

RESEARCH: The current through the circuit after a long time is $i = V_{\text{emf}} / \left(R_1 + R_{23}\right)$. Resistors in parallel add as $R_{23} = \left(R_2^{-1} + R_3^{-1}\right)^{-1}$. The potential drop across the capacitor is $\Delta V_{\text{C}} = iR_{23}$. The energy in the capacitor is given by $E = C\left(\Delta V_{\text{C}}\right)^2 / 2$. When the switch is open, the current through R_3 is $i_3 = \Delta V_{\text{C}} / R_2$. The current across R_3 varies as $i_3(t) = i_3 e^{-t/R_{23}C}$. The power across R_3 is given by $P_3 = i_3^2(t) R_3$. The energy across R_3 is given by $E_3 = \int_0^\infty P_3(t)\,dt$.

SIMPLIFY:

(a) The potential drop across the capacitor is given by: $\Delta V_{\text{C}} = iR_{23} = \frac{V_{\text{emf}} R_{23}}{R_1 + R_{23}}$.

(b) The energy in the capacitor is given by $E = \frac{1}{2}C\left(\Delta V_{\text{C}}\right)^2$.

(c) The energy across R_3 is given by: $E_3 = \int_0^\infty P_3(t)dt = \int_0^\infty R_3 i_3^2 e^{-2t/R_{23}C} dt = \frac{\Delta V_C^2}{R_3}\int_0^\infty e^{-2t/R_{23}C} dt$

$$= \frac{\Delta V_C^2}{R_3}\left[-\frac{R_{23}C}{2}e^{-2t/R_{23}C}\right]_{t=0}^{t=\infty} = \frac{\Delta V_C^2}{R_3}\left[0+\frac{R_{23}C}{2}\right] = \frac{\Delta V_C^2 R_{23}C}{2R_3}.$$

CALCULATE:

(a) $R_{23} = \left((10.0\ \Omega)^{-1} + (4.00\ \Omega)^{-1}\right)^{-1} = 2.86\ \Omega,\ \Delta V_C = \frac{(10.0\ \text{V})(2.86\ \Omega)}{10.0\ \Omega + 2.86\ \Omega} = 2.22\ \text{V}$

(b) $E = \frac{1}{2}(2.00\ \mu\text{F})(2.22\ \text{V})^2 = 4.938\cdot 10^{-6}\ \text{J}$

(c) $E_3 = \frac{(2.22\ \text{V})^2(2.86\ \Omega)(2.00\ \mu\text{F})}{2(10.0\ \Omega)} = 1.411\cdot 10^{-6}\ \text{J}$

ROUND:

(a) $\Delta V_C = 2.22\ \text{V}$

(b) $E = 4.94\ \mu\text{J}$

(c) $E_3 = 1.41\ \mu\text{J}$

DOUBLE-CHECK: The energy across R_2 is $E_2 = (\Delta V_C)^2 R_{23}C/2R_2 = 3.527\ \mu\text{J}$. The result makes sense because energy is conserved: $E_2 + E_3 = E$.

26.59. (a) The maximum current through the ammeter is $i_A = 1.5$ mA. The ammeter has resistance $R_1 = 75\ \Omega$. The current through a resistor is given by $i = V/R$, where V is the potential difference across the resistor. Since current flows through the path of least resistance, when a shunt resistor of small resistance R_{shunt} is connected in parallel with the ammeter, most of the current flows through the shunt resistor. The shunt resistor carries most of the load so that the ammeter is not damaged.

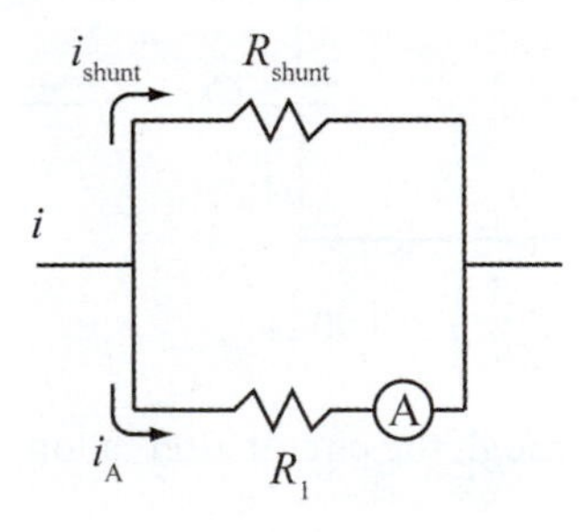

From Kirchoff's rules $i = i_{\text{shunt}} + i_A$ and $i_{\text{shunt}}R_{\text{shunt}} = i_A R_1$. Therefore, $i = \frac{i_A R_1}{R_{\text{shunt}}} + i_A = i_A\left(\frac{R_1}{R_{\text{shunt}}}+1\right)$.

For known current i_A (measured by ammeter) and known resistances R_1 and R_{shunt}, the new maximum current i can be calculated. Note that $i > i_A$. A shunt resistor is added in parallel with an ammeter so the current can be increased without damaging the ammeter.

(b) From Kirchoff's rules shown above,

$$R_{\text{shunt}} = \frac{i_A R_1}{i_{\text{shunt}}} = \frac{i_A R_1}{i - i_A} = \frac{(1.50\ \text{mA})(75.0\ \Omega)}{15.0\ \text{A} - 1.50\ \text{mA}} = 7.501\cdot 10^{-3}\ \Omega = 7.50\ \text{m}\Omega.$$

26.63. **THINK:** When the switch is set to X for a long time, the capacitor, $C = 10.0\ \mu\text{F}$, charges fully so that it has the same potential as the battery, $\Delta V_C = V_{\text{emf}} = 9.00$ V. After placing the switch on Y, the capacitor discharges through resistor $R_2 = 40.0\ \Omega$ and decreases exponentially for both immediately $(t = 0\ \text{s})$ and $t = 1.00$ ms after the switch.

SKETCH:

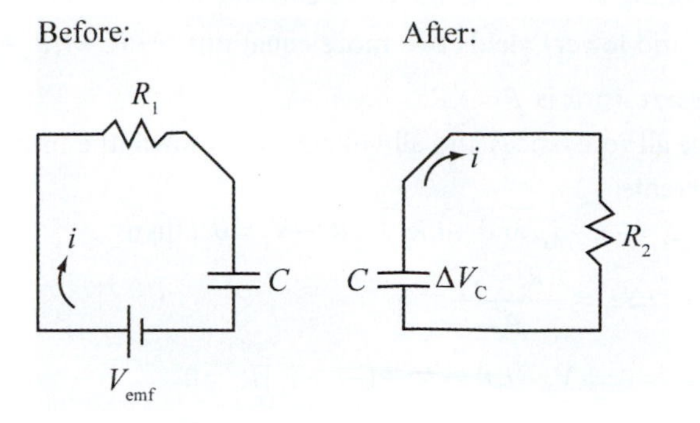

RESEARCH: By Ohm's law, the current initially is $i_0 = \Delta V_C / R_2$, where $\Delta V_C = V_{emf}$. Current decays exponentially as $i(t) = i_0 e^{-t/\tau}$, where $\tau = RC$.

SIMPLIFY:

(a) Initial current is $i_0 = \Delta V_C / R_2 = V_{emf} / R_2$.

(b) After $t = 1$ ms, current is $i(t) = I_0 e^{-t/\tau} = i_0 e^{\frac{-t}{RC}}$.

CALCULATE:

(a) $i_0 = \dfrac{9.00 \text{ V}}{40.0 \ \Omega} = 0.225 \text{ A}$

(b) $i(1.00\text{ms}) = (0.225 \text{ A}) e^{\frac{-1.00 \text{ ms}}{(40.0 \ \Omega)(10.0 \ \mu\text{F})}} = 0.01847 \text{ A}$

ROUND:

(a) $i_0 = 225 \text{ mA}$

(b) $i(1.00 \text{ ms}) = 18.5 \text{ mA}$

DOUBLE-CHECK: After only 1.00 ms, the current decreases by almost 90.0%, which would make this a desirable circuit, so it makes sense.

26.67. **THINK:** From Kirchoff's rules, an equation can be obtained for the sum of the three currents, i_1, i_2 and i_3, and two equations for the two inner loops of the circuit. This will yield 3 equations for 3 unknowns (the currents) and can be solved by simple substitution. Once the currents are known, the power over each resistor is found via Ohm's law. $R_1 = 10.0 \ \Omega$, $R_2 = 20.0 \ \Omega$, $R_3 = 30.0 \ \Omega$, $V_1 = 15.0 \text{ V}$ and $V_2 = 9.00 \text{ V}$.

SKETCH:

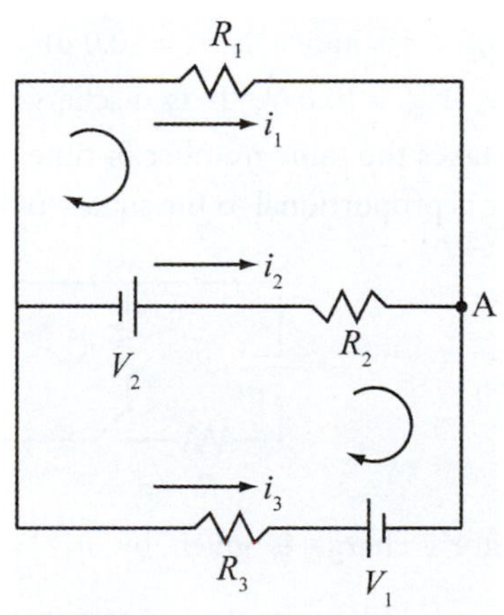

RESEARCH: Looking at point A, the three currents all flow into it, so $i_1+i_2+i_3=0$. Going clockwise in each loop (upper and lower) yields two more equations: $-i_1R_1+i_2R_2-V_2=0$ and $V_2-i_2R_2+V_1+i_3R_3=0$. The power across a resistor is $P=i^2R$.

SIMPLIFY: Since all resistances and all voltages are known, the first three equations can be solved for the three separate currents:

$i_1+i_2+i_3=0 \Rightarrow i_3=-i_1-i_2$, and $-i_1R_1+i_2R_2-V_2=0$, then

$$-i_1R_1+i_2R_2-V_2=0 \Rightarrow i_1=\frac{i_2R_2-V_2}{R_1}.$$

$$V_2-i_2R_2+V_1+i_3R_3=0 \Rightarrow V_2-i_2R_2+V_1+(-i_1-i_2)R_3=0$$

$$\Rightarrow V_2-i_2R_2+V_1+\left(-\left(\frac{i_2R_2-V_2}{R_1}\right)-i_2\right)R_3=0 \Rightarrow i_2=\frac{R_1V_1+V_2(R_1+R_3)}{R_1R_2+R_1R_3+R_2R_3}.$$

$i_3=-i_1-i_2$.

The power across each is then $P_1=i_1^2R_1$, $P_2=i_2^2R_2$ and $P_3=i_3^2R_3$.

CALCULATE:

$$i_2=\frac{R_1V_1+V_2(R_1+R_3)}{R_1R_2+R_1R_3+R_2R_3}=\frac{(10.0\ \Omega)(15.0\ \text{V})+(9.00\ \text{V})(10.0\ \Omega+30.0\ \Omega)}{(10.0\ \Omega)(20.0\ \Omega)+(10.0\ \Omega)(30.0\ \Omega)+(20.0\ \Omega)(30.0\ \Omega)}=0.4636\ \text{A}$$

$$i_1=\frac{i_2R_2-V_2}{R_1}=\frac{(0.4636\ \text{A})(20.0\ \Omega)-(9.00\ \text{V})}{10.0\ \Omega}=0.02727\ \text{A}$$

$i_3=-i_1-i_2=-(0.4636\ \text{A})-(0.02727\ \text{A})=-0.49087\ \text{A}$, or $i_3=0.49087\ \text{A}$ to the left.

$P_1=(0.02727\ \text{A})^2(10.0\ \Omega)=0.00744\ \text{W}$, $P_2=(0.4636\ \text{A})^2(20.0\ \Omega)=4.299\ \text{W}$, and

$P_3=(0.49087\ \text{A})^2(30.0\ \Omega)=7.230\ \text{W}$.

ROUND: $P_1=7.44\ \text{mW}$, $P_2=4.30\ \text{W}$ and $P_3=7.23\ \text{W}$.

DOUBLE-CHECK: Looking back at the values for current, it is found that

$$i_1+i_2+i_3=0.4636\ \text{A}+0.02727\ \text{A}-0.49087\ \text{A}=0,$$

which is what would be expected. Going from left to right on each branch gives

$$-i_1R_1=-\frac{3.00}{11.0}\ \text{V},\ V_2-i_2R_2=-\frac{3.00}{11.0}\ \text{V and } -i_3R_3-V_1=-\frac{3.00}{11.0}\ \text{V}.$$

So the potential drop across each branch in parallel is the same, so the answers make sense.

26.71. **THINK:** The capacitor of capacitance is $C=10.0\ \mu\text{F}$, is charged through a resistor of resistance $R=10.0\ \Omega$, with a battery, $V_{\text{emf}}=10.0\ \text{V}$. It is discharged through a resistor, $R'=1.00\ \Omega$. For either charging or discharging, it takes the same number of time constants to get to half of the maximum value. The energy on the capacitor is proportional to the square of the charge.

SKETCH:

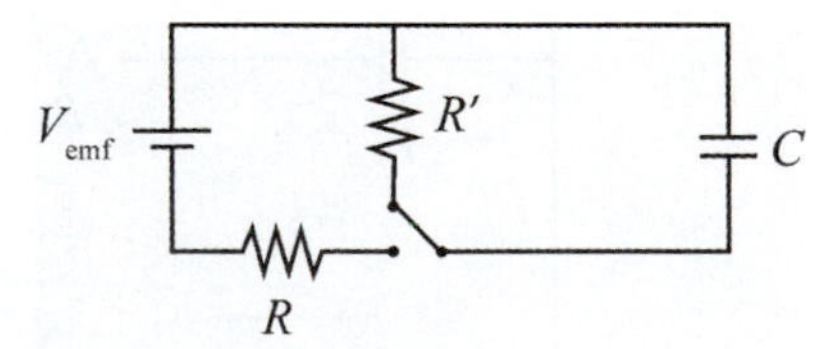

RESEARCH: The capacitor's charge is given by $q(t)=q_0\left(1-e^{-t/\tau}\right)$. In general, the energy on the capacitor is given by $E(t)=q^2(t)/2C$. The time constant is either $\tau=RC$ or $\tau'=R'C$.

SIMPLIFY:

(a) When $q(t)=q_0/2$, then: $q(t)=\frac{1}{2}q_0=q_0\left(1-e^{-t/\tau}\right) \Rightarrow \frac{1}{2}=1-e^{-t/\tau} \Rightarrow t=-\tau\ln\left(\frac{1}{2}\right)=\tau\ln 2.$

(b) If $q(t)=\frac{1}{2}q_0$, the energy is: $E(t)=\frac{q^2(t)}{2C}=\frac{(q_0/2)^2}{2C}=\frac{1}{4}\left(\frac{q_0^2}{2C}\right)=\frac{1}{4}E_{\text{max}}.$

(c) The time constant for discharging is $\tau'=R'C.$

(d) The capacitor discharges to half the original charge in $t=\tau'\ln(2).$

CALCULATE:

(a) $t=\tau\ln 2.00$, or $(0.693)\tau$

(b) 1.00:4.00.

(c) $\tau'=(1.00\ \Omega)(10.0\ \mu\text{F})=10.0\ \mu\text{s}$

(d) $t=(10.0\ \mu\text{s})\ln(2.00)=6.93\ \mu\text{s}$

ROUND:

(a) 0.693τ

(b) 1.00:4.00.

(c) $\tau'=10.0\ \mu\text{s}$

(d) $t=6.93\ \mu\text{s}$

DOUBLE-CHECK: In general, charge decreases exponentially as $q(t)=q_0e^{-t/\tau}$. For $\tau'=10\ \mu\text{s}$ and $t=6.93\ \mu\text{s}$, the charge is $q(6.93\ \mu\text{s})=q_0e^{-6.93/10}=0.497q_0$, which is about half the original charge.

26.75. **THINK:** Consider any square on the grid to have been reduced so that every side has a capacitance, C', which is the equivalent of all the capacitors above, below and along each side. Since the grid is infinite, then no side has more capacitors than any other, so all four are reduced to the same capacitance. Next, consider the same analysis except for only three sides, so that one side is still of capacitance, C, while the others are C'. When those four sides are reduced to one equivalent capacitance, the result should be equal to the original value of C'. This is because the grid is infinite and adding an extra square to the already reduced side should affect nothing, resulting in the same capacitance, giving a recursive relation in C, and thus the total equivalent capacitance, in terms of C, can be determined.

SKETCH:

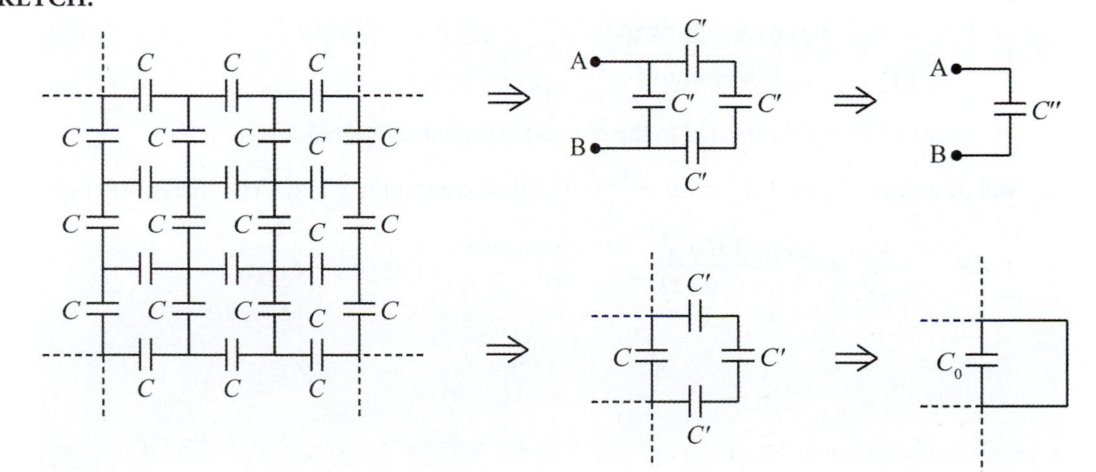

RESEARCH: Capacitors in series add as $C_{\text{eq}}^{-1}=C_1^{-1}+C_2^{-1}$. Capacitors in parallel add as $C_{\text{eq}}=C_1+C_2$.

SIMPLIFY: When all four sides are reduced to C', the equivalent capacitance (across A to B) is:

$$C'' = C' + \left(\frac{1}{C'} + \frac{1}{C'} + \frac{1}{C'}\right)^{-1} = \frac{4C'}{3}.$$

Looking at when one side is reduced using the other three reduced gives C_0 as:

$$C_0 = C + \left(\frac{1}{C'} + \frac{1}{C'} + \frac{1}{C'}\right)^{-1} = C + \frac{C'}{3}.$$

Since $C_0 = C'$: $C' = C + \frac{C'}{3} \Rightarrow C = \frac{2}{3}C'$ and $C' = \frac{3}{2}C.$ Therefore, the total equivalent capacitance is:

$$C'' = \frac{4C'}{3} = \frac{4}{3}\left(\frac{3}{2}C\right) = \frac{4}{2}C = 2C.$$

CALCULATE: Not applicable.

ROUND: Not applicable.

DOUBLE-CHECK: Consider an intersection on the grid. If a voltage was applied to this point, it would see equal capacitance (since it is infinite) in all four directions, meaning it would contribute an equal charge, q, to each direction. If the same voltage with opposite polarity was applied to any adjacent intersection, it would see a $-q$ along each direction. This means the capacitor that joins the two intersections is actually double the charge on one, meaning the potential sees an effective capacitance twice the size of any one capacitor, so an equivalent capacitance of 2C is correct. $C_0 = C + C'/3 \Rightarrow C' = (3/2)C.$ Therefore the total equivalent capacitance is $C'' = (4/3)C' = (4/3)(3/2)C = 2C.$

Multi-Version Exercises

26.76. Using Kirchhoff's Loop Rule we get $-iR_i - V_t + V_e = 0.$ So the required battery charger emf is $V_e = iR_i + V_t = (9.759 \text{ A})(0.1373 \ \Omega) + 11.45 \text{ V} = 12.79 \text{ V}.$

26.79. Kirchhoff's Loop Rule gives us

$$V_{\text{emf},1} - \Delta V_1 - \Delta V_2 - V_{\text{emf},2} = V_{\text{emf},1} - iR_1 - iR_2 - V_{\text{emf},2} = 0.$$

We can rearrange this equation to get

$$V_{\text{emf},1} - i(R_1 + R_2) - V_{\text{emf},2} = 0$$

$$i = \frac{V_{\text{emf},1} - V_{\text{emf},2}}{R_1 + R_2} = \frac{21.01 \text{ V} - 10.75 \text{ V}}{23.37 \ \Omega + 11.61 \ \Omega} = 0.2933 \text{ A}.$$

26.82. When the resistor is connected to the charged capacitor, the initial current i_0 will be given by $V_{\text{emf}} = i_0 R \Rightarrow i_0 = \frac{V_{\text{emf}}}{R}$. The time constant is $\tau = RC$. The current after time t is

$$i = i_0 e^{-t/\tau} = \frac{V_{\text{emf}}}{R} e^{-t/(RC)} = \frac{131.1 \text{ V}}{616.5 \ \Omega} e^{-(3.871 \text{ s})/\left((616.5 \ \Omega)(15.19 \cdot 10^{-3} \text{ F})\right)} = 0.1407 \text{ A}.$$

Chapter 27: Magnetism

Exercises

27.25. $\vec{F}_B = q\vec{v}\times\vec{B},\ F_B = |q|vB\sin\theta,\ \theta = 90°,\ q=-2e \Rightarrow \left|\vec{F}_B\right| = F_B = +2evB \Rightarrow$

$$B = \frac{F_B}{2ev} = \frac{3.00\cdot 10^{-18}\text{ N}}{2\left(1.602\cdot 10^{-19}\text{ C}\right)1.00\cdot 10^{5}\text{ m/s}} = 9.363\cdot 10^{-5}\text{ T} \approx 9.36\cdot 10^{-5}\text{ T}.$$

27.29. $\Delta K = \Delta U$

$$K = eV \Rightarrow \frac{1}{2}mv^2 = eV \Rightarrow v = \sqrt{\frac{2eV}{m}}$$

$$B = \frac{mv}{er} = \frac{m}{er}\sqrt{\frac{2eV}{m}} = \frac{1}{r}\sqrt{\frac{2mV}{e}} = \frac{1}{0.200\text{ m}}\sqrt{\frac{2\left(1.67\cdot 10^{-27}\text{ kg}\right)(400.\text{ V})}{1.602\cdot 10^{-19}\text{ C}}} = 1.44\cdot 10^{-2}\text{ T}.$$

27.33.

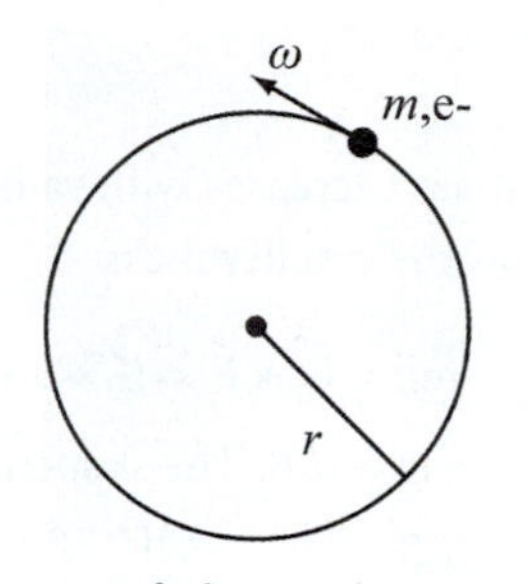

The net force is directed toward the center of the circle. From the right-hand rule, a positive charge requires the magnetic field to be oriented into the plane, in the negative z-direction. Since an electron is negatively charged, it can be concluded that the field points out of the page, along the positive z-direction. The magnitude is given by:

$$\omega = \frac{|q|B}{m} \Rightarrow B = \frac{m\omega}{|q|} = \frac{\left(9.11\cdot 10^{-31}\text{ kg}\right)\left(1.20\cdot 10^{12}\text{ s}^{-1}\right)}{1.602\cdot 10^{-19}\text{ C}} = 6.824\text{ T} \Rightarrow \vec{B} = 6.82\hat{z}\text{ T}.$$

27.37. **THINK:** The particles will move in circular, clockwise paths (in the direction of $\vec{v}\times\vec{B}$) within the magnetic field. The radius of curvature of the path is proportional to the mass of the particle, and inversely proportional to the charge of the particle. Both particles move at the same speed within the same magnetic field. The radii, charges and masses of the particles can then be compared.

SKETCH:

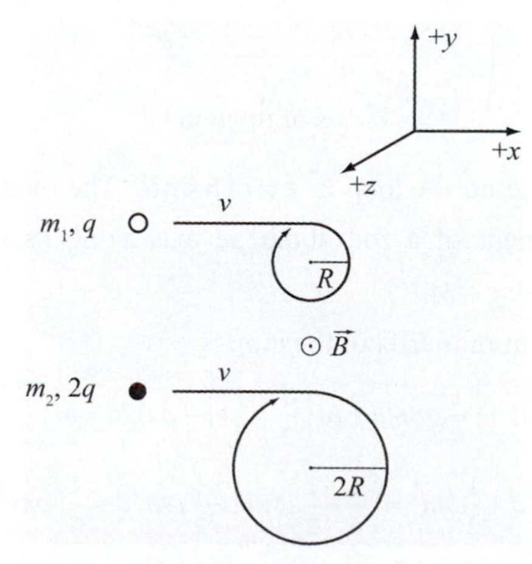

RESEARCH: The radius is related to the mass and charge of the particles by $r = mv / |q| B$. At the instant that the particles enter the magnetic field, the magnetic force acting on them is $\vec{F}_B = q\vec{v} \times \vec{B}$. For the particles to travel in a straight line, the force on the particles due to the electric field must oppose the force due to the magnetic field: $\vec{F}_E = -\vec{F}_B \Rightarrow q\vec{E} = -q\vec{v} \times \vec{B}$.

SIMPLIFY: Since the velocity and the magnetic field is the same for both particles, $\frac{v}{B} = \frac{r_1 q_1}{m_1} = \frac{r_2 q_2}{m_2}$. The ratio of the masses is:

$$\frac{m_1}{m_2} = \frac{r_1 q_1}{r_2 q_2} = \frac{Rq}{2R(2q)} = \frac{1}{4}.$$

For the particles to travel in a straight line,

$$q\vec{E} = -q\vec{v} \times \vec{B} = -qvB(\hat{x} \times \hat{z}) = -qvB(-\hat{y})$$
$$\vec{E} = vB\hat{y}.$$

Therefore, the electric field must have magnitude $E = vB$ and point in the positive y-direction in order for the particles to move in a straight line.

CALCULATE: Not applicable.

ROUND: Not applicable.

DOUBLE CHECK: Since the mass increases with radius and charge, it makes sense that the particle with the smaller charge and radius has the smaller mass.

27.41. The force on the wire is $\vec{F}_{\text{net}} = m\vec{a} = \vec{F}_B + \vec{F}_g = i\vec{L} \times \vec{B} + mg\hat{y} = iLB(-\hat{x} \times -\hat{z}) + mg\hat{y} = -iLB\hat{y} + mg\hat{y}$. For the conductor to stay at rest, $a = 0$ or $mg = iLB$. The suspended mass is then:

$$m = \frac{iLB}{g} = \frac{(20.0\text{ A})(0.200\text{ m})(1.00\text{ T})}{(9.81\text{ m/s}^2)} = 0.408\text{ kg}.$$

27.45. **THINK:** The loop will experience a torque in the presence of a magnetic field as discussed in the chapter. The torque is also equal to the moment of inertia of the loop times its angular acceleration. The loop is square with sides of length, d = 8.00 cm and current, i = 0.150 A. The wire has a diameter of 0.500 mm (which corresponds to a radius of r = 0.250 mm) and a density of $\rho = 8960\text{ kg/m}^3$. The magnetic field is B = 1.00 T and points 35.0° away from the normal of the loop.

SKETCH:

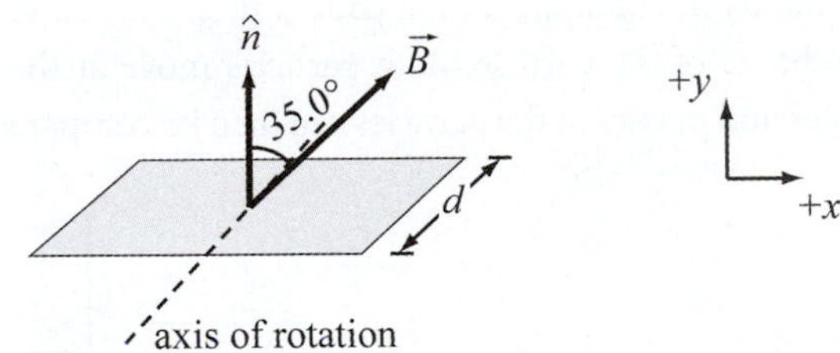

RESEARCH: The torque on the loop is $\tau = iAB\sin\theta$. The moment of inertia for a rod about its center is $I = Md^2/12$. The moment of a rod about an axis along its length is $I = \rho\pi r^4 L/4$. The parallel axis theorem states $I_{\text{off center}} = I_{\text{cm}} + Md^2$.

SIMPLIFY: The moment of inertia of the loop is:

$$I = \frac{1}{12}Md^2 + \frac{1}{12}Md^2 + \left[\frac{1}{4}\rho\pi r^4 d + M\left(\frac{d}{2}\right)^2\right] + \left[\frac{1}{4}\rho\pi r^4 d + M\left(\frac{d}{2}\right)^2\right]$$
$$= \frac{1}{6}Md^2 + \frac{1}{2}\rho\pi r^4 d + \frac{1}{2}Md^2 = \left(\frac{1}{6} + \frac{1}{2}\right)Md^2 + \frac{1}{2}\rho\pi r^4 d = \frac{4}{6}(\rho\pi r^2 d)d^2 + \frac{1}{2}\rho\pi r^4 d = \frac{2}{3}\rho\pi r^2 d^3 + \frac{1}{2}\rho\pi r^4 d.$$

The angular acceleration is $\tau = I\alpha = iAB\sin\theta$ or $\alpha = iAB\sin\theta / I$.

$$\Rightarrow \alpha = \frac{id^2 B\sin\theta}{\rho\pi r^2 d^2\left(2d/3 + r^2/2d\right)} = \frac{iB\sin\theta}{\rho\pi r^2\left(2d/3 + r^2/2d\right)}$$

CALCULATE:

$$\alpha = \frac{(0.150\text{ A})(1.00\text{ T})(\sin 35.0°)}{\left(8960\text{ kg/m}^3\right)\pi(0.000250\text{ m})^2\left\{\left[2(0.0800\text{ m})/3\right] + \left[(0.000250\text{ m})^2/2(0.0800\text{ m})\right]\right\}} = 916.94\text{ rad/s}^2$$

ROUND: The values are given to two significant figures, thus the loop experiences an initial angular acceleration of $\alpha = 917\text{ rad/s}^2$.

DOUBLE-CHECK: One Tesla represents a magnetic field of large magnitude, resulting in a correspondingly large acceleration. This result is reasonable.

27.49. The torque on the loop due to the magnetic field is $\tau = \text{N}iAB\sin\theta = NiAB$. This is equal to the applied torque, $\tau = rF$. Equating the torques gives the magnetic field:

$$B = \frac{rF}{NiA} = \frac{rF}{Ni\pi r^2} = \frac{F}{Ni\pi r} = \frac{1.2\text{ N}}{120.(0.490\text{ A})\pi(0.0480\text{ m})} = 0.1353358\text{ T} = 0.135\text{ T}.$$

27.53. Assume the electron orbits the hydrogen with speed, v. The current of the electron going around its orbit is:

$$i = \frac{q}{t} = -\frac{e}{d/v} = -\frac{ev}{d} = -\frac{ev}{2\pi r}.$$

The magnetic moment of the orbit is: $\mu = iA = i\pi r^2 = -\frac{ev}{2\pi r}\left(\pi r^2\right) = -\frac{1}{2}evr$. Angular momentum is given by $L = rp = rmv$. Using the angular momentum, the moment is:

$$\vec{\mu} = -\frac{1}{2}er\vec{v} = -\frac{1}{2}e\left(\frac{m}{m}\right)r\vec{v} = -\frac{erm\vec{v}/2}{m} = -\frac{e\vec{L}}{2m}.$$

27.57. **THINK:** The question asks for the carrier density of the thin film, and the nature of the carriers. The film has a thickness of $h = 1.50\text{ μm}$. The current is i = 12.3 mA and the voltage reads V = -20.1 mV. The magnetic field is B = 0.900 T.

SKETCH:

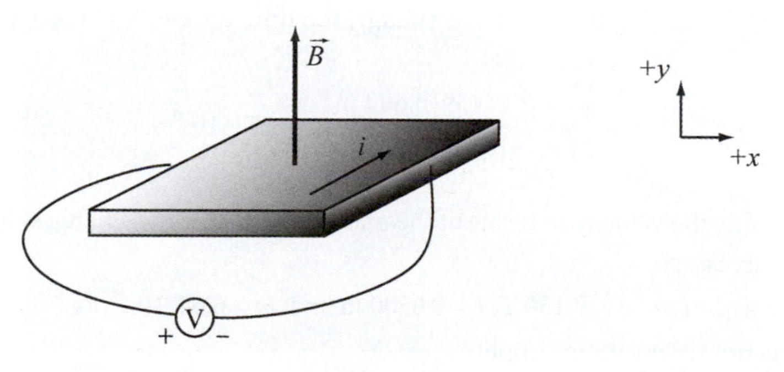

RESEARCH:

(a) Due to the magnetic force, the charge carriers are accumulated on the visible edge of the sample. Since the polarity of the Hall potential is negative, the charge carriers are holes.

(b) The Hall voltage magnitude is given by $\Delta V_{\text{H}} = \frac{iB}{neh}$.

SIMPLIFY:

(b) The charge carrier density is $n = \frac{iB}{he\Delta V_{\text{H}}}$.

CALCULATE:

(b) $$n=\frac{12.3\cdot 10^{-3}(0.900\text{ T})}{(1.50\cdot 10^{-6}\text{ m})(1.602\cdot 10^{-19}\text{ C})(20.1\cdot 10^{-3}\text{ V})}=2.2919\cdot 10^{24}\text{ holes/m}^3$$

ROUND:

(b) The values are given to three significant figures, so the carrier density of the film is $n=2.29\cdot 10^{24}$ holes/m^3.

DOUBLE-CHECK: This is a reasonable value for a carrier density.

27.61. The force on a current carrying wire in a magnetic field is $F=ilB\sin\theta$. To determine the minimum current, set $\theta=90°$:

$$i=\frac{F}{lB}=\frac{1.00\text{ N}}{0.100\text{ m}(0.430\cdot 10^{-4}\text{ T})}=232{,}558\text{ A}\approx 2.33\cdot 10^5\text{ A}.$$

The minimum current required for the wire to experience a force of 1.0 N is $i=2.33\cdot 10^5$ A.

27.63. (a) The electron must travel in a circular path with a radius of 60.0 cm, as shown in the figure below.

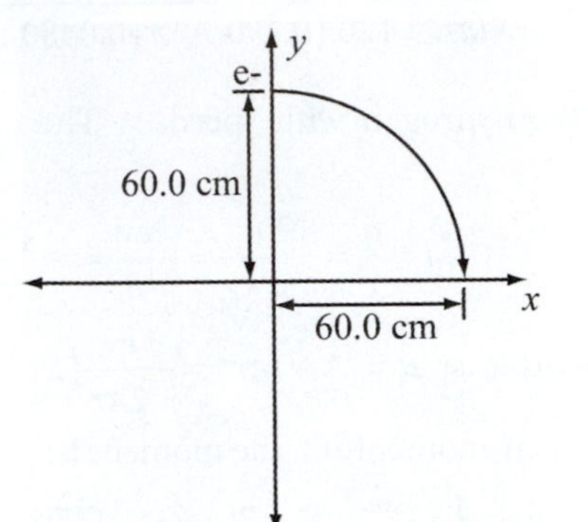

By the right-hand rule, B must be in the negative z-direction.

$$B=\frac{mv}{q_e r}=\frac{(9.11\cdot 10^{-31}\text{ kg})(2.00\cdot 10^5\text{ m/s})}{(1.60\cdot 10^{-19}\text{ C})(0.600\text{ m})}=1.89\cdot 10^{-6}\text{ T}$$

(b) The magnetic force is perpendicular to the motion and does no work.

(c) Since the speed of an electron does not change, the time the electron takes to travel a quarter-circle is given by:

$$t=\frac{\pi r}{2v}=\frac{3.14159(0.600\text{ m})}{2(2.00\cdot 10^5\text{ m/s})}=4.71\cdot 10^{-6}\text{ s}.$$

27.67. **THINK:** Determine the velocity in terms of the mass and see how this changes the answer.

SKETCH: Not necessary.

RESEARCH: $v=qBr/m$, $B=0.150$ T, $r=0.0500$ m and $m=6.64\cdot 10^{-27}$ kg.

SIMPLIFY: It is not necessary to simplify.

CALCULATE: $$v=\frac{(1.602\cdot 10^{-19}\text{ C})(0.150\text{ T})(0.0500\text{ m})}{6.64\cdot 10^{-27}\text{ kg}}=1.809\cdot 10^5\text{ m/s}$$

Note that for $m'=\frac{3}{4}m$, $v'=\frac{qBr}{3m/4}=\frac{4}{3}v$; the velocity increases by a factor of $4/3$.

ROUND: $v=1.81\cdot 10^5$ m/s

DOUBLE-CHECK: Since there is an inverse relationship between v and m, it makes sense that decreasing m by a factor of 3/4 increases v by factor of $(3/4)^{-1}=4/3$.

27.71. **THINK:** Equilibrium occurs when the net torque on the coil is zero. Use the values: $A=d^2$, $d=0.200$ m, $i=5.00$ A, $m=0.250$ kg, $B=0.00500$ T, and $N=30$.

SKETCH:

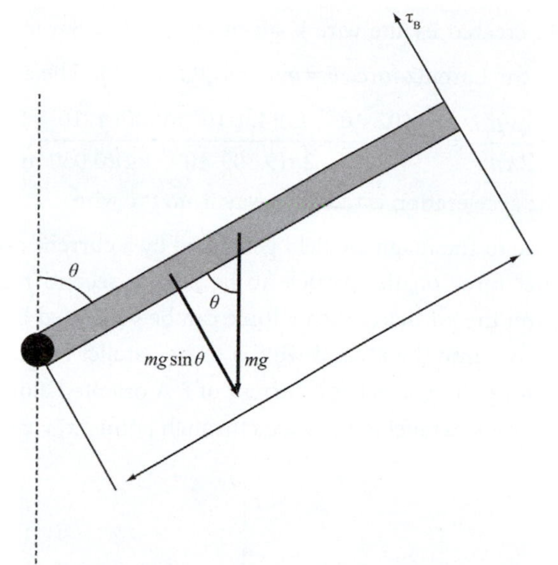

RESEARCH: $\tau_g = mg\sin\theta(d/2)$, $\tau_B = NiAB\cos\theta$. It is required that $\tau_g = \tau_B$.

SIMPLIFY: $\tau_g = \tau_B \Rightarrow \frac{1}{2}mgd\sin\theta = Nid^2B\cos\theta$

$$\frac{\sin\theta}{\cos\theta} = \frac{2NdiB}{mg} \Rightarrow \tan\theta = \frac{2NdiB}{mg} \Rightarrow \theta = \tan^{-1}\left(\frac{2NdiB}{mg}\right)$$

CALCULATE: $\theta = \tan^{-1}\left[\frac{2(30)(0.200\text{ m})(5.00\text{ A})(0.00500\text{ T})}{(0.250\text{ kg})(9.81\text{ m/s}^2)}\right] = 6.9740°$

ROUND: The number of turns is precise, so it does not limit the precision of the answer. The rest of the values are given to three significant figures of precision, so it is appropriate to round the final answer to: $\theta = 6.97°$.

DOUBLE-CHECK: It makes sense that θ is inversely proportional to m, since the less the coil weighs, the more vertical it must be.

Multi-Version Exercises

27.74. For the ball to travel in a circle with radius r, we have $r = \frac{mv}{|q|B}$.

Then $B = \frac{mv}{|q|r} = \frac{(5.063\cdot 10^{-3}\text{ kg})(3079\text{ m/s})}{(11.03\text{ C})(2.137\text{ m})} = 0.6614\text{ T}.$

27.77. The electric force is given by $F_E = qE$. The magnetic force is given by $F_B = vBq$. Setting these forces equal to each other gives us

$$qE = vBq$$

$$v = \frac{E}{B} = \frac{1.749\cdot 10^4\text{ V/m}}{46.23\cdot 10^{-3}\text{ T}} = 3.783\cdot 10^5\text{ m/s}.$$

Chapter 28: Magnetic Fields of Moving Charges

Exercises

28.31. The magnetic field created by the wire is given by the Biot-Savart Law $B = \mu_0 i / (2\pi r)$. The force on the electron is given by the Lorentz force $F = qvB = qv\mu_0 i / (2\pi r)$. The acceleration of the electron is

$$a = \frac{F}{m} = \frac{qv\mu_0 i}{2\pi m r} = \frac{(1.602 \cdot 10^{-19}\text{ C})(4.0 \cdot 10^5\text{ m/s})(4\pi \cdot 10^{-7}\text{ T m/A})(15\text{ A})}{2\pi(9.109 \cdot 10^{-31}\text{ kg})(0.050\text{ m})} = 4.2 \cdot 10^{12}\text{ m/s}^2$$

The direction of the acceleration is radially away from the wire.

28.35. **THINK:** A force due to the magnetic field generated by a current carrying wire acts on a moving particle. In order for the net force on the particle to be zero, a second force of equal magnitude and opposite direction must act on the particle. Such a force can be generated by another current carrying wire placed near the first wire. Assume the second wire is to be parallel to the first and has the same magnitude of current. The wire along the x-axis has a current of 2 A oriented along the x-axis. The particle has a charge of $q = -3\ \mu\text{C}$ and travels parallel to the y-axis through point $(x, y, z) = (0,2,0)$.

SKETCH:

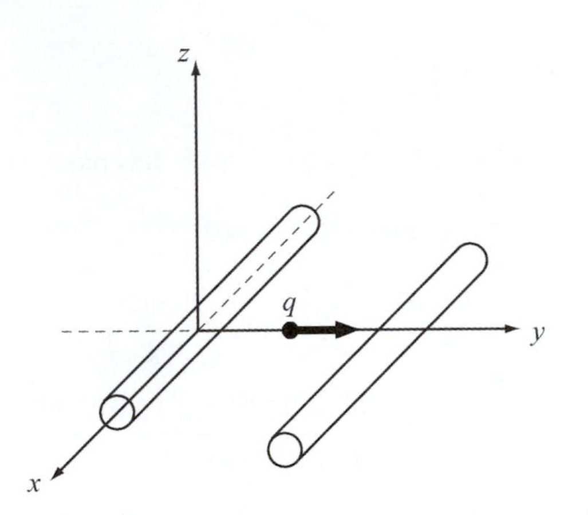

RESEARCH: The magnetic field produced by the current is given by the $B = \mu_0 i / (2\pi r)$. The force on the particle is given by the Lorentz force, $F = qv_0 B$.

SIMPLIFY: If the wires carry the same current then the new wire must be equidistant from the point that the particle passes through the xy-plane. Only then will the magnetic force on the particle due to each wire be equal. By the right hand rule, the currents will be in the same direction. This means that $r_1 = r_2$.

CALCULATE: The requirement $r_1 = r_2$ means that the second wire should be placed parallel to the first wire (parallel to the x-axis) so that it passes through the point $(x, y, z) = (0,4,0)$.

ROUND: Not necessary.

DOUBLE-CHECK: It is reasonable that two wires carrying the same current need to be equidistant from a point in order for the magnitude of the force to be the same.

28.39. **THINK:** The current carrying wires along the x- and y-axes will each generate a magnetic field. The superposition of these fields generates a net field. The magnitude and direction of this net field at a point on the z-axis is to be determined.

SKETCH:

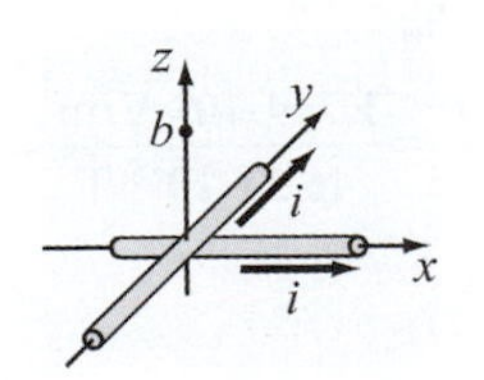

RESEARCH: Both currents produce a magnetic field with magnitude $B=\mu_0 i/(2\pi r)$. The magnetic field produced by the wire along the x-axis gives $\vec{B}_1=\mu_0 i(-\hat{y})/(2\pi b)$. The wire along the y-axis creates a magnetic field of $\vec{B}_2=\mu_0 i\hat{x}/(2\pi b)$.

SIMPLIFY: The total magnetic field is then $\vec{B}_{\text{net}}=\vec{B}_1+\vec{B}_2=\dfrac{\mu_0 i}{2\pi b}\hat{x}-\dfrac{\mu_0 i}{2\pi b}\hat{y}$. The magnitude of the field is $B=\dfrac{\mu_0 i}{2\pi b}\sqrt{1^2+(-1)^2}=\dfrac{\sqrt{2}\mu_0 i}{2\pi b}=\dfrac{\mu_0 i}{\sqrt{2}\pi b}$. The direction of the field is $\theta=\tan^{-1}\left(\dfrac{-\mu_0 i}{\sqrt{2}\pi b}\Big/\dfrac{\mu_0 i}{\sqrt{2}\pi b}\right)$ in the x-y plane at a height of b.

CALCULATE: $\tan^{-1}(-1)=-45°$ in the x-y plane at point b.

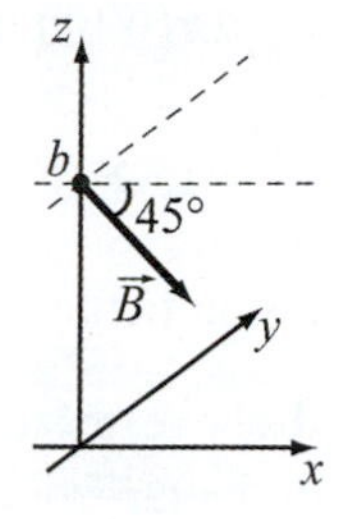

ROUND: Not applicable.

DOUBLE CHECK: Both the right hand rule and the symmetry of the problem indicates that the net field should be in the fourth quadrant.

28.43. **THINK:** The wire creates a magnetic field that produces a Lorentz force on the moving charged particle. The question asked for the force if the particle travels in various directions. The velocity is 3000 m/s in various directions.

SKETCH:

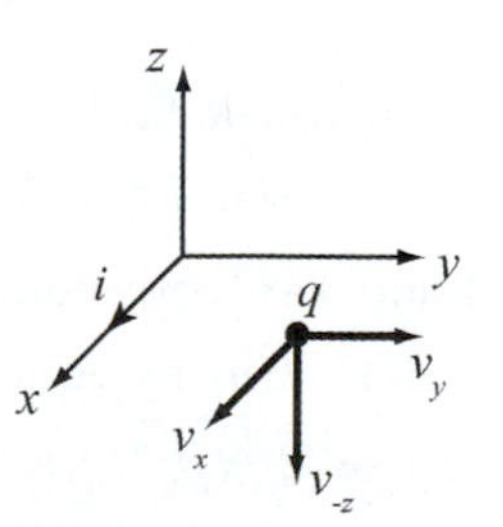

RESEARCH: The magnetic field produced by an infinite wire is $B=\mu_0 i/(2\pi d)$. By the right hand rule the field points in the positive z-direction. The force produced by the magnetic field is $\vec{F}=q\vec{v}\times\vec{B}$.

SIMPLIFY: The force is given by $\vec{F}=q\vec{v}\times\vec{B}=\dfrac{q\mu_0 i}{2\pi d}\vec{v}\times\hat{z}=\dfrac{q\mu_0 i}{2\pi d}\left(|\vec{v}|\cdot\hat{n}\times\hat{z}\right)$ where $\hat{n}$ is the direction of the particle.

CALCULATE: $\vec{F}=\dfrac{(9.00\text{ C})(4\pi\cdot 10^{-7}\text{ T m/A})(7.00\text{ A})}{2\pi(2.00\text{ m})}(3000.\text{ m/s}\cdot\hat{n}\times\hat{z})=1.89\cdot 10^{-2}\text{ N}(\hat{n}\times\hat{z})$

Note that $\hat{x}\times\hat{z}=-\hat{y}$, $\hat{y}\times\hat{z}=\hat{x}$, and $-\hat{z}\times\hat{z}=0$.

ROUND: The force should be reported to 3 significant figures.

(a) The force is $\vec{F}=-1.89\cdot 10^{-2}\text{ N}\,\hat{y}$ if the particle travels in the positive x-direction.

(b) The force is $\vec{F}=1.89\cdot 10^{-2}\text{ N}\,\hat{x}$ if the particle travels in the positive y-direction.

(c) The force is $F=0$ if the particle travels in the negative z-direction.

DOUBLE-CHECK: The right hand rule confirms the directions of the forces for each direction of motion of the particle.

28.47. Using Ampere's Law, the magnetic field at various points can be determined. $\oint \vec{B} \bullet d\vec{s} = \mu_0 i_{\text{enclosed}}$. For the cylinder, assuming the current is distributed evenly, $B2\pi r = \mu_0 i_{\text{enc}}$ or $B = \mu_0 i_{\text{enc}} / (2\pi r)$. The field at $r = r_{\text{a}} = 0$ is zero since is does not enclose any current $B_{\text{a}} = 0$. The field at $r = r_{\text{b}} < R$ is $B_{\text{b}} = \frac{\mu_0 i_{\text{enc}}}{2\pi r_{\text{b}}} = \frac{\mu_0}{2\pi r_{\text{b}}}\left(i_{\text{tot}} \frac{\pi r_{\text{b}}^2}{\pi R^2} \right) = \frac{\mu_0 i r_{\text{b}}}{2\pi R^2} = \frac{\left(4\pi \cdot 10^{-7} \text{ T m/A}\right)(1.35 \text{ A})(0.0400 \text{ m})}{2\pi (0.100 \text{ m})^2} = 1.08 \cdot 10^{-6}$ T. Note that i is equal to the fraction of total area of the conductor's cross section and the total current. The field at $r_{\text{c}} = R$ is $B_{\text{c}} = \frac{\mu_0 i_{\text{enc}}}{2\pi r_{\text{c}}} = \frac{\mu_0 i_{\text{tot}}}{2\pi R} = \frac{\left(4\pi \cdot 10^{-7} \text{ T m/A}\right)(1.35 \text{ A})}{2\pi (0.100 \text{ m})} = 2.70 \cdot 10^{-6}$ T. The field at $r_{\text{d}} > R$ is $B_{\text{d}} = \frac{\mu_0 i_{\text{enc}}}{2\pi r_{\text{d}}} = \frac{\left(4\pi \cdot 10^{-7} \text{ T m/A}\right)(1.35 \text{ A})}{2\pi (0.160 \text{ m})} = 1.69 \cdot 10^{-6}$ T. By inspection it can be seen that the magnetic field at r_b, r_c and r_d the magnetic field will point to the right.

28.49. **THINK:** To find the magnetic field above the center of the surface of a current carrying sheet, use Ampere's Law. The path taken should be far from the edges and should be rectangular as shown in the diagram. The current density of the sheet is $J = 1.5$ A/cm.

SKETCH:

+z
+x
$\vec{B}_2$
4
3
1
2
$\vec{B}_1$

RESEARCH: The direction of the magnetic field is found using the right hand rule to be $+x$ above the surface of the conductor. Ampere's Law states $\oint \vec{B} \bullet d\vec{s} = B2\pi r = \mu_0 i_{\text{enclosed}}$.

SIMPLIFY: Note that sections 1 and 3 are perpendicular the field. $B \cdot ds = 0$ for these two sections. If the path of 4 and 2 has a length of L, then by Ampere's Law, $\oint B \cdot ds = B_1 L + B_2 L = \mu_0 i_{\text{enclosed}} = \mu_0 JL$. By symmetry $B_1 = B_2$. Thus, $2B_1 = \mu_0 J$ or $B_1 = \mu_0 J / 2$.

CALCULATE: $B_1 = \frac{\left(4\pi \cdot 10^{-7} \text{ T m/A}\right)(1.5 \text{ A/cm})(100 \text{ cm/m})}{2} = 9.42478 \cdot 10^{-5}$ T

ROUND: The magnetic field is accurate to two significant figures. The magnetic field near the surface of the conductor is $B_1 = 9.4 \cdot 10^{-5}$ T.

DOUBLE-CHECK: The form for the magnetic field is similar to that of a solenoid. It is divided by a factor of 2, which makes sense when considering the setup of a solenoid. The form of the equation is similar to that of question 28.12. This makes sense because the magnetic field inside a solenoid is generated by a current carrying wire on both sides of the Amperian loop, whereas the field generated by the flat conducting surface originates on one side of the Amperian loop only. In effect, the flat conductor can be seen as similar to half a solenoid, flattened out. See figure 28.21 in the text for a visual.

28.51. The magnetic field in a solenoid is given by $B = \mu_0 in$. Let the magnetic field of solenoid B be $B_B = \mu_0 in$. The magnetic field of solenoid A is $B_A = \mu_0 i(4 \text{ N})/(3 \text{ L}) = (4/3)\mu_0 in = (4/3)B_B$. The ratio of solenoid A magnetic field to that of solenoid B is 4:3.

28.55. **THINK:** If the perpendicular momentum of a particle is not large enough, its radius of motion will not be large enough to enter the detector. The minimum momentum perpendicular to the axis of the solenoid is determined by a condition such that the centripetal force is equal to the force due to the magnetic field.

SKETCH:

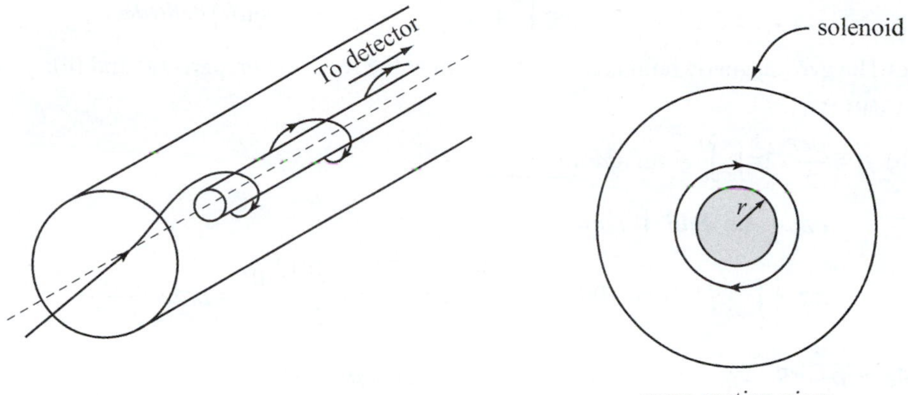

RESEARCH: Since the particle originates from the axis of the detector, the minimum radius of the circular motion of the particle must be equal to the radius of the detector as shown above. The magnetic force on the particle is $F = qvB$. Centripetal acceleration is $a_C = v^2 / r$. The magnetic field due to the solenoid is $B = \mu_0 in$.

SIMPLIFY: Using Newton's Second Law, the momentum is $qvB = mv^2 / r \Rightarrow mv = p = qrB$. Therefore, the minimum momentum is $p = \mu_0 qrin$.

CALCULATE: Substituting the numerical values yields.

$$p = \left(4\pi \cdot 10^{-7} \text{ T m/A}\right)\left(1.602 \cdot 10^{-19} \text{C}\right)\left(0.80 \text{ m}\right)\left(22 \text{ A}\right)\left(550 \cdot 10^2 \text{ m}^{-1}\right) = 1.949 \cdot 10^{-19} \text{ kg m/s}$$

ROUND: Rounding the result to two significant figures gives $p = 1.9 \cdot 10^{-19}$ kg m/s.

DOUBLE-CHECK: This is a reasonable value.

28.59. The magnitude of the magnetic field inside a solenoid is given by $B = \mu in = \kappa_m \mu_0 i(N / L)$. Thus the relative magnetic permeability κ_m is given by the equation:

$$\kappa_m = \frac{BL}{\mu_0 iN} = \frac{(2.96 \text{ T}) \cdot \left(3.50 \cdot 10^{-2} \text{ m}\right)}{\left(4\pi \cdot 10^{-7} \text{ T m/A}\right) \cdot \left(3.00 \text{ A}\right) \cdot (500.)} = 54.96 \approx 55.0.$$

28.63. **THINK:** The classical angular momentum of rotating object is related to its moment of inertia. To get the magnetic dipole of a uniformly changed sphere, the spherical volume is divided into small elements. Each element produces a current and a magnetic dipole moment. The dipole moment of all elements is then added to get the net dipole moment.

SKETCH:

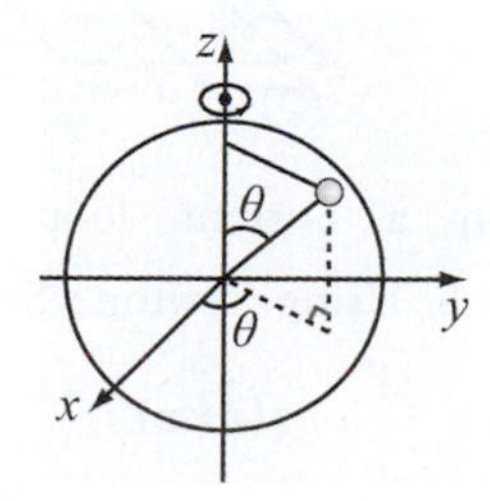

RESEARCH:

(a) The classical angular momentum of the sphere is given by $L = I\omega = \left(2/5\right)mR^2\omega$.

(b) The current produced by a small volume element dV is $i = \rho dV\omega / (2\pi)$. Thus the magnetic dipole moment of this element is $d\mu = \frac{\rho\omega dV}{2\pi}\pi(r\sin\theta)^2$. Integrating all the elements gives

$$\mu = \int_0^{2\pi}\int_0^{\pi}\int_0^{R}\frac{\rho\omega r^2}{2}\left(\sin^2\theta\right)\left(r^2\sin\theta\right)dr\,d\theta\,d\phi.$$

(c) The gyromagnetic ratio is simply the ratio of the results from parts (a) and (b): $\gamma_e = \mu / L$.

SIMPLIFY:

(b) $\mu = \frac{\rho\omega}{2}\cdot 2\pi\int_0^{\pi}\int_0^{R} r^4\sin^3\theta drd\theta$

$$= \rho\pi\omega\int_0^{\pi}\sin^3\theta d\theta\cdot\int_0^{R} r^4 dr$$

$$= \rho\pi\omega\left[\int_{\cos 0}^{\cos\pi} -(1-\cos^2\theta)d\cos\theta\right]\frac{R^5}{5} = \rho\pi\omega\left[-x+\frac{x^3}{3}\right]_1^{-1}\frac{R^5}{5} = \rho\pi\omega\left(\frac{4}{3}\right)\frac{R^5}{5}$$

Since $\rho\frac{4}{3}\pi R^3 = q$, the magnetic moment becomes $\mu = q\omega R^2 / 5$.

(c) Taking the ratio of the magnetic dipole moment and the angular momentum yields: $\gamma_e = \frac{\mu}{L} = \frac{\frac{q\omega R^2}{5}}{\frac{2}{5}mR^2\omega} = \frac{q}{2m}$. Substituting $q = -e$ gives: $\gamma_e = -e/(2m)$.

CALCULATE: Not required

ROUND: Not required

DOUBLE-CHECK: The magnetic dipole and the angular momentum should both be quadratic in R, so it is logical that the ratio of these two quantities is independent of R.

28.67. The magnetic dipole moment is defined as $\mu = iA = i\pi R^2$. This means the current that produces this magnetic dipole moment is $i = \mu/\left(\pi R^2\right)$. Substituting the numerical values gives the current of

$$i = \frac{8.0\cdot 10^{22}\text{ A m}^2}{\pi(2.5\cdot 10^6\text{ m})^2} = 4.07\cdot 10^9\text{ A} \approx 4.1\cdot 10^9\text{ A}.$$

28.71. **THINK:** The torque due to the current in a loop of wire in a magnetic field must balance the torque due to weight.

SKETCH:

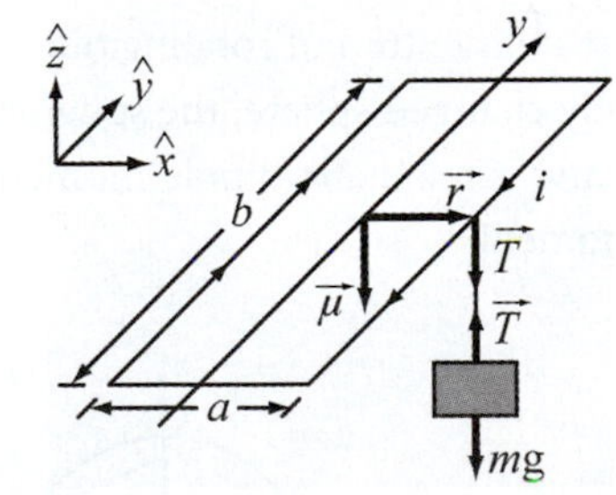

RESEARCH: The torque on a current loop in a uniform magnetic field is given by $\tau_B = \vec{\mu}\times\vec{B} = iN\vec{A}\times\vec{B} = iNA(-\hat{z})\times\vec{B}$. Using Newton's Second Law, the torque due to the weight is found to be $\tau_W = \vec{r}\times\vec{T} = \left(\frac{1}{2}a\hat{x}\right)\times mg\left(-\hat{z}\right) = -\frac{1}{2}amg(\hat{x}\times\hat{z})$.

SIMPLIFY: Since the system is in equilibrium, the net torque must be zero: $\sum \tau = \tau_{\text{B}} + \tau_{\text{w}} = 0.$ Thus,

$$\tau_{\text{B}} = -\tau_{\text{w}}$$
$$-iNA\hat{z} \times \vec{B} = \frac{1}{2}amg(\hat{x} \times \hat{z}) = -\frac{1}{2}amg(\hat{z} \times \hat{x}).$$

This means that the magnetic field vector is in positive $\hat{x}$. Substituting $\vec{B} = B\hat{x}$ gives $iNAB = \frac{1}{2}amg$. After simplifying and using $A = ab$, $\vec{B} = \frac{1}{2}\frac{amg}{iNA}\hat{x} = \frac{1}{2}\frac{amg}{iN(ab)}\hat{x} = \frac{mg}{2iNb}\hat{x}.$

CALCULATE: Substituting the numerical values produces $\vec{B} = \frac{(0.0500 \text{ kg})(9.81 \text{ m/s}^2)}{2(1.00 \text{ A}) \cdot 50 \cdot (0.200 \text{ m})}\hat{x} = 0.02453 \text{ T}.$

ROUND: Three significant figures yields, $\vec{B} = 24.5$ mT.

DOUBLE-CHECK: The magnetic force must be in the positive *z*-direction to balance gravity. By the right hand rule, it can be seen that the magnetic field must point in the positive *x*-direction for this to occur. This is consistent with the result calculated above. The result is reasonable.

28.75. **THINK:** Assuming the inner loop is sufficiently small such that the magnetic field due to the larger loop is same across the surface of the smaller loop, the torque on the small loop can be determined by its magnetic moment.

SKETCH:

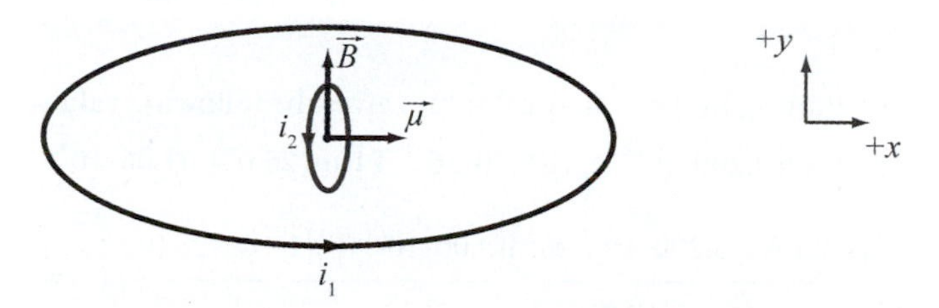

RESEARCH: The torque experienced by the small loop is given by $\vec{\tau} = \vec{\mu} \times \vec{B}$. The magnetic field in the center of the loop is given by $\vec{B} = \frac{\mu_0 i_1}{2R}\hat{y}$. The magnetic dipole moment of the small loop is $\vec{\mu} = i_2\vec{A}_2 = i_2\pi r^2\hat{x}$.

SIMPLIFY: Combining all the above expressions yields the torque.

$$\tau = |\vec{\tau}| = \left|(i_2\pi r^2\hat{x}) \times \left(\frac{\mu_0 i_1}{2R}\hat{y}\right)\right| = \frac{\pi\mu_0 i_1 i_2 r^2}{2R}|\hat{x} \times \hat{y}| = \frac{\pi\mu_0 i_1 i_2 r^2}{2R}$$

CALCULATE: Putting in all the numerical values gives

$$\tau = \frac{\pi(4\pi \cdot 10^{-7} \text{ T m/A})(14.0 \text{ A})(14.0 \text{ A})(0.00900 \text{ m})^2}{2(0.250 \text{ m})} = 1.254 \cdot 10^{-7} \text{ N m}.$$

ROUND: Rounding to 3 significant figures gives, $\tau = 1.25 \cdot 10^{-7}$ N m.

DOUBLE-CHECK: The units are correct: $\tau = \frac{[\text{T m/A}][\text{A}][\text{A}][\text{m}^2]}{[\text{m}]} = \frac{[\text{N}][\text{A}][\text{m}^2]}{[\text{A}][\text{m}]} = [\text{N m}].$

28.79. **THINK:** To do this problem, the inertia of a long thin rod is required. The torque on a wire is also needed. The measure of the angle θ is 25.0°, and the current is $i = 2.00$ A. Let $A = 0.200 \cdot 10^{-4}$ m^2 and $B = 9.00 \cdot 10^{-2}$ T.

SKETCH:

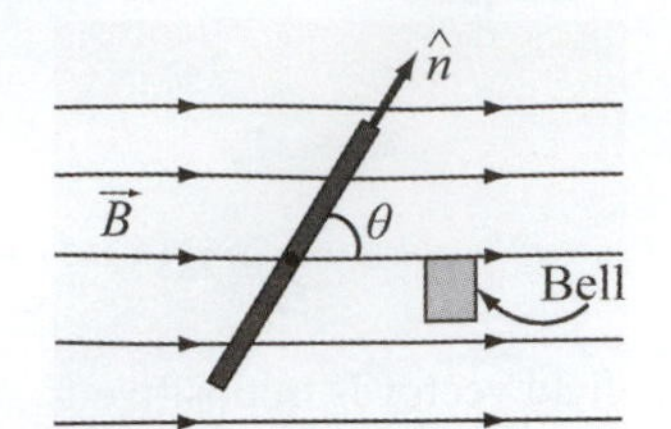

RESEARCH: The magnetic dipole moment of the wire is given by $\vec{\mu} = NiA\hat{n}$.

(a) The torque on the wire is $\vec{\tau} = \vec{\mu} \times \vec{B}$. The magnitude of this torque is $\tau = \mu B \sin\theta = NiAB\sin\theta$.

(b) The angular velocity of the rod when it strikes the bell is determined by using conservation of energy, that is, $E_i = E_f$ or $U_i + K_i = U_f + K_f$.

SIMPLIFY:

(a) $\tau = \mu B \sin\theta = NiAB\sin\theta$.

(b) Since $K_i = 0$, the final kinetic energy is

$$K_f = U_i - U_f$$

$$\frac{1}{2}I\omega^2 = -\mu B\cos\theta + \mu B\cos(0°) = -\mu B\cos\theta + \mu B = \mu B(1-\cos\theta)$$

Thus the angular velocity is $\omega = \sqrt{\frac{2\mu B(1-\cos\theta)}{I}} = \sqrt{\frac{2NiAB(1-\cos\theta)}{(1/12)mL^2}}$, using $I = \frac{1}{12}mL^2$, the inertia of a thin rod.

CALCULATE: Putting in the numerical values gives the following values.

(a) $\tau = (70)(2.00 \text{ A})(0.200\cdot 10^{-4} \text{ m}^2)(9.00\cdot 10^{-2} \text{ T})\sin(25.0°) = 1.06\cdot 10^{-4} \text{ N m}$

(b) $$\omega = \left[\frac{2(70)(2.00 \text{ A})(0.200\cdot 10^{-4} \text{ m}^2)(9.00\cdot 10^{-2} \text{ T})(1-\cos 25.0°)}{(1/12)(0.0300 \text{ kg})(0.0800 \text{ m})^2}\right]^{1/2} = 1.72 \text{ rad/s}$$

ROUND: Rounding to 3 significant figures yields $\tau = 1.06\cdot 10^{-4}$ N m, $\omega = 1.72$ rad/s.

DOUBLE-CHECK: The torque should have units of Newton-meters, while the angular velocity should have units of radians per second.

28.83. **THINK:** To solve this problem, the current enclosed by an Amperian loop must be determined.

SKETCH:

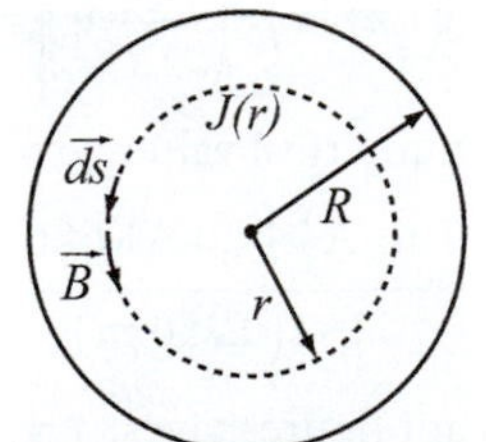

RESEARCH: Applying Ampere's Law on a loop as shown above gives $\oint \vec{B}\cdot d\vec{s} = B(2\pi r) = \mu_0 i_{\text{enc}}$. i_{enc} is the current enclosed by the Amperian loop, that is $i_{\text{enc}} = \iint J(r)dA = \int_0^{2\pi}\int_0^r J(r')r'dr'd\theta$.

SIMPLIFY: Since $J(r)$ is a function of r only, the above integral becomes $i_{\text{enc}} = 2\pi\int_0^r J(r')r'dr'$. Substituting $J(r) = J_0(1 - r/R)$ yields

$$i_{\text{enc}} = 2\pi J_0 \int_0^r \left[r' - \frac{r'^2}{R}\right]dr' = 2\pi J_0\left[\frac{r'^2}{2} - \frac{r'^3}{3R}\right]_0^r = 2\pi J_0\left[\frac{r^2}{2} - \frac{r^3}{3R}\right].$$

Thus, the magnetic field is $B = \frac{\mu_0 2\pi J_0}{2\pi r}\left[\frac{r^2}{2} - \frac{r^3}{3R}\right] = \mu_0 J_0 \left[\frac{r}{2} - \frac{r^2}{3R}\right].$

CALCULATE: Not required.

ROUND: Not required.

DOUBLE-CHECK: The form of the answer is reasonable.

Multi-Version Exercises

28.85. The magnetic field at the center of a portion of a circle with radius R subtended by an angle Φ is

$$B_\Phi = \int dB = \int_0^\Phi \frac{\mu_0}{4\pi}\frac{iRd\phi}{R^2} = \frac{\mu_0 i \Phi}{4\pi R}.$$

In this loop we have three sections:

$1:\ R = r,\ \Phi = \pi/2$

$2:\ R = 2r,\ \Phi = \pi/2$

$3:\ R = 3r,\ \Phi = \pi.$

The segments of wire running directly toward/away from point P have no effect. So the magnetic field at the center of the loop is

$$B = B_1 + B_2 + B_3 = \frac{\mu_0 i\left(\frac{\pi}{2}\right)}{4\pi(r)} + \frac{\mu_0 i\left(\frac{\pi}{2}\right)}{4\pi(2r)} + \frac{\mu_0 i(\pi)}{4\pi(3r)} = \frac{\mu_0 i}{8r} + \frac{\mu_0 i}{16r} + \frac{\mu_0 i}{12r}$$

$$B = \frac{6\mu_0 i}{48r} + \frac{3\mu_0 i}{48r} + \frac{4\mu_0 i}{48r} = \frac{13\mu_0 i}{48r} = \frac{13(4\pi \cdot 10^{-7}\text{ T m/A})(3.857\text{ A})}{48(1.411\text{ m})} = 9.303 \cdot 10^{-7}\text{ T}.$$

28.88. The magnetic field inside a toroidal magnet is given by $B = \frac{\mu_0 N i}{2\pi r}.$

$$N = \frac{2\pi r B}{\mu_0 i} = \frac{2\pi(1.985\text{ m})(66.78 \cdot 10^{-3}\text{ T})}{(4\pi \cdot 10^{-7}\text{ T m/A})(33.45\text{ A})} = 19{,}814$$

To four significant figures, the toroid has 19,810 turns.

Chapter 29: Electromagnetic Induction

Exercises

29.29. The potential difference around the loop is:

$$V_{\text{emf}}=-\frac{d\Phi}{dt}\approx-\frac{\Delta\Phi}{\Delta t}=-\frac{\Delta(AB)}{\Delta t}=-A\frac{\Delta B}{\Delta t}=-\pi r^2\frac{\Delta B}{\Delta t}=-\pi(0.0100\text{ m})^2\left(\frac{0\text{ T}-1.20\text{ T}}{20.0\text{ s}}\right)=1.89\cdot10^{-5}\text{ V}.$$

Note that the area of the ring is perpendicular to the field. Thus, the normal of the area is parallel to the field and $\cos\theta=1$.

29.33. **THINK:** The current in the outer loop generates a magnetic field. Because the magnitude of the current in the outer loop changes with time, the magnetic field it generates also changes. The changing magnetic field, in turn, induces a potential difference and thus a current in the inner loop. Let I be the current in the outer loop and i be the induced current in the inner loop.

SKETCH:

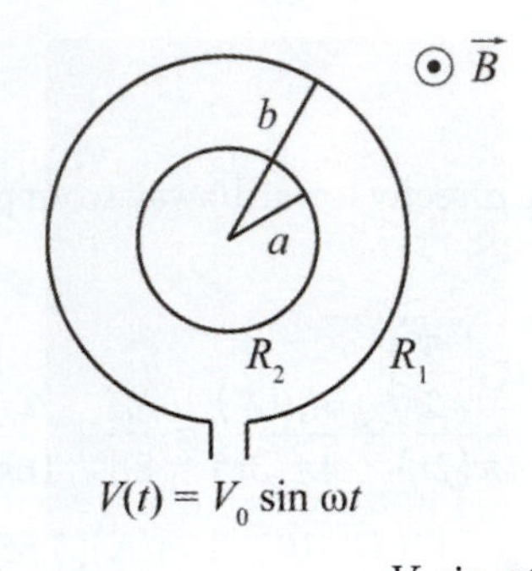

RESEARCH: The current through the large loop is $I=\dfrac{V_0\sin\omega t}{R_1}$. This creates a magnetic field at the center of the loop of:

$$B_1=\frac{\mu_0 I}{2b}$$

which is derived from the Biot-Savart Law. Since the radius of the inner loop is much smaller than the radius of the outer loop, the magnetic field through the inner loop is $B_1=\mu_0V_0\sin\omega t/2bR_1$. This magnetic field creates a flux of:

$$\Phi_B=B_1A=B_1\pi a^2=\frac{\mu_0\pi a^2V_0}{2bR_1}\sin\omega t.$$

The induced potential across the inner loop is then:

$$\Delta V_{\text{ind}}=-\frac{d\Phi_B}{dt}=-\frac{d}{dt}\left(\frac{\mu_0\pi a^2V_0}{2bR_1}\sin\omega t\right).$$

This voltage corresponds to a current in the inner loop of:

$$i=\frac{\Delta V_{\text{ind}}}{R_2}=-\frac{1}{R_2}\frac{d}{dt}\left(\frac{\mu_0\pi a^2V_0}{2bR_1}\sin\omega t\right).$$

SIMPLIFY: The potential difference induced in the inner loop is:

$$\Delta V_{\text{ind}}=-\frac{d}{dt}\left(\frac{\mu_0\pi a^2V_0}{2bR_1}\sin\omega t\right)=-\frac{\mu_0\pi a^2V_0}{2bR_1}\frac{d}{dt}(\sin\omega t)=-\frac{\mu_0\pi a^2V_0\omega}{2bR_1}\cos\omega t,$$

and the induced current in the inner loop is:

$$i=\frac{\Delta V_{\text{ind}}}{R_2}=-\frac{1}{R_2}\frac{d}{dt}\left(\frac{\mu_0\pi a^2V_0}{2bR_1}\sin\omega t\right)=-\frac{\mu_0\pi a^2V_0}{2bR_1R_2}\frac{d}{dt}(\sin\omega t)=-\frac{\mu_0\pi a^2V_0\omega}{2bR_1R_2}\cos\omega t.$$

CALCULATE: Not applicable.

ROUND: Not applicable.

DOUBLE CHECK: The time dependence on the current for the outer loop and inner loop is shown in the plot below. For example, for $\omega t < \pi / 2$ (taking positive values to be the counterclockwise direction) if the current in the outer loop is moving counterclockwise and increasing then the current in the inner loop is increasing in the clockwise direction. This is consistent with Lenz's Law.

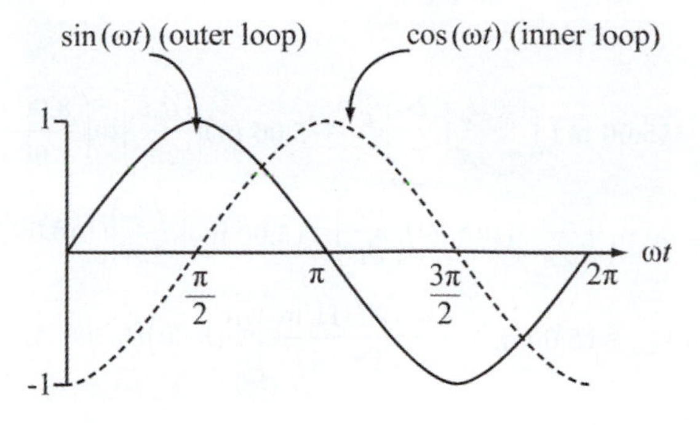

29.37. **THINK:** As a conductor travels through a magnetic field, perpendicular to the ground, of intensity $B = 0.426$ G, it creates a voltage difference between its ends. The length of metal of interest is L = 5.00 m and rotates at $1.00 \cdot 10^4$ rpm.

SKETCH:

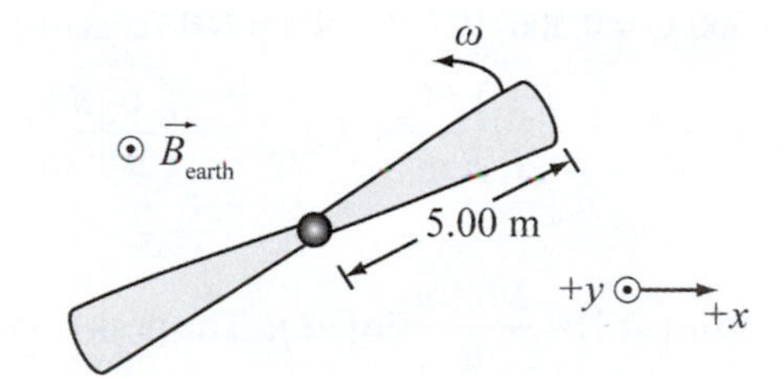

RESEARCH: The potential difference across a wire moving in a magnetic field is $\Delta V_{\text{ind}} = vLB$. Each element of the blade travels at a different speed, $v = r\omega$. To calculate the potential difference, the length must be divided into pieces of length, dl, which travel at $v = l\omega$. The value should be integrated over the total length, from 0 to L.

SIMPLIFY: $\int \Delta V = \int_0^L vBdl = \int_0^L l\omega Bdl = \frac{1}{2}\omega BL^2$

In terms of the blade's rpm, the potential difference is $V = \frac{1}{2}\left(\frac{2\pi(\text{rpm})}{60 \text{ s}}\right)BL^2$.

CALCULATE: $V = \left(\frac{\pi\left(1.00 \cdot 10^4 \text{ rpm}\right)}{60.0 \text{ s/rpm}}\right)\left(0.426 \cdot 10^{-4} \text{ T}\right)\left(5.00 \text{ m}\right)^2 = 0.557633 \text{ V} \approx 0.558 \text{ V}$

ROUND: The potential difference from the hub of the helicopter's blade to its far end is $\Delta V_{\text{ind}} = 0.558$ V.

DOUBLE-CHECK: We can double-check this result by assuming that the blade moves with a constant speed equal to the speed of the middle of the blade, $v = \left(\frac{L}{2}\right)\omega = \left(\frac{L}{2}\right)2\pi f = 2\pi Lf$. The induced potential difference would be $\Delta V = vB\frac{L}{2} = \left(2\pi Lf\right)B\frac{L}{2} = \pi fBL^2$, which is the same answer we got by integrating over the length of the blade.

29.41. **THINK:** The current in the wire will cause a magnetic field. The changing current will cause a changing flux through the loop, inducing a potential.

SKETCH: Provided with question.

RESEARCH: For a wire: $B=\frac{\mu_0}{4\pi}\left(\frac{2i}{r}\right)$. $\Delta V_{\text{ind}}=\frac{d\Phi_{\text{B}}}{dt}$, $i=2.00\text{ A}+(0.300\text{ A/s})t$, A = 7.00 m by 5.00 m, $\Phi_{\text{B}}=\oiint \vec{B}\bullet d\vec{A}$.

SIMPLIFY: $\Phi_{\text{B}}=(5.00\text{ m})\int_{1\text{ m}}^{8\text{ m}}\frac{\mu_0}{4\pi}\left(\frac{2i}{r}\right)dr=(5.00\text{ m})\left(\frac{\mu_0 i}{2\pi}\right)\ln\left(\frac{8.00\text{ m}}{1.00\text{ m}}\right)=(5.00\text{ m})\left(\frac{\mu_0 i}{2\pi}\right)\ln 8.00$

$$\Delta V_{\text{ind}}=\frac{d\Phi_{\text{B}}}{dt}=(5.00\text{ m})\left(\frac{\mu_0}{2\pi}\right)(\ln 8.00)\left(\frac{di}{dt}\right)=(5.00\text{ m})\left(\frac{\mu_0}{2\pi}\right)(\ln 8.00)(0.300\text{ A/s})$$

CALCULATE: $\Delta V_{\text{ind}}=(5.00\text{ m})\left(\frac{4\pi\cdot 10^{-7}\text{ H/m}}{2\pi}\right)(\ln 8.00)(0.300\text{ A/s})=6.238\cdot 10^{-7}\text{ V}$

ROUND: $\Delta V_{\text{ind}}=6.24\cdot 10^{-7}\text{ V}$

DOUBLE-CHECK: It makes sense that the larger the rate of change of the current, the larger the induced voltage.

29.45. **THINK:** First determine an expression for the magnetic flux, and then use Faraday's law to determine the induced voltage.

SKETCH: A sketch is not necessary.

RESEARCH: $B_{\text{Earth}}=0.300\text{ G}=0.300\cdot 10^{-4}\text{ T}$, $\Phi_{\text{B}}=NBA\cos(\omega t)$, $A=\pi r^2$, $r=0.250\text{ m}$, $N=1.00\cdot 10^5$, $\omega=2\pi(150.\text{ Hz})$, $i_{\text{ind}}=\frac{\Delta V_{\text{ind}}}{R}=-\left(\frac{1}{R}\right)\frac{d\Phi_{\text{B}}}{dt}$, $i_{\text{ind,peak}}=-\left(\frac{1}{R}\right)\frac{d\Phi_{\text{B}}}{dt}_{\text{peak}}$, $R=1500.\ \Omega$

SIMPLIFY:

(a) $i_{\text{ind}}=-\left(\frac{1}{R}\right)\left(-NBA\omega\sin(\omega t)\right)=\frac{NBA\omega}{R}\sin(\omega t)$; The peak occurs at $|\sin(\omega t)|=1$: $i_{\text{ind,peak}}=\frac{NBA\omega}{R}$.

(b) $i_{\text{avg}}=0.7071\left(i_{\text{ind,peak}}\right)$, $P_{\text{avg}}=i_{\text{avg}}^2R$

CALCULATE:

(a) $i_{\text{ind,peak}}=\frac{(1.00\cdot 10^5)(0.300\cdot 10^{-4}\text{ T})(0.250\text{ m})^2 2\pi^2(150.\text{ Hz})}{(1500.\ \Omega)}=0.3701\text{ A}$

(b) $i_{\text{avg}}=0.7071(0.3701\text{ A})=0.2617\text{ A}$, $P_{\text{avg}}=(0.2617\text{ A})^2(1500.\ \Omega)=102.7\text{ W}$

ROUND:

(a) $i_{\text{ind,peak}}=0.370\text{ A}$

(b) $i_{\text{avg}}=0.262\text{ A}$, $P_{\text{avg}}=103\text{ W}$

DOUBLE-CHECK: The answer seems reasonable since there are a very large number of turns for the generator turning at a very fast rate.

29.49. (a) $\tau_{\text{L}}=\frac{L}{R}=\frac{1.00\text{ H}}{1.00\text{ M}\Omega}=1.00\ \mu\text{s}$

(b) $i(t)=\frac{V_{\text{emf}}}{R}\left(1-e^{-t/\tau_{\text{L}}}\right)$. At t = 0, $i(t)=0$. At $t=2.00\ \mu\text{s}$, $i(t)=\frac{10.0\text{ V}}{1.00\text{ M}\Omega}\left(1-e^{-(2.00\ \mu\text{s})/(1.00\ \mu\text{s})}\right)=8.65\ \mu\text{A}$.

At steady state. $t\to\infty$: $i(\infty)=\frac{V_{\text{emf}}}{R}=10.0\ \mu A$.

29.53. **THINK:** After a long time, the inductor acts like a short-circuit. The circuit is in steady state, so the current is no longer changing. $V_{\text{emf}} = 18$ V, $R_1 = R_2 = 6.0\ \Omega$, $L = 5.0$ H.

SKETCH: An equivalent sketch when the circuit is in steady-state is as follows.

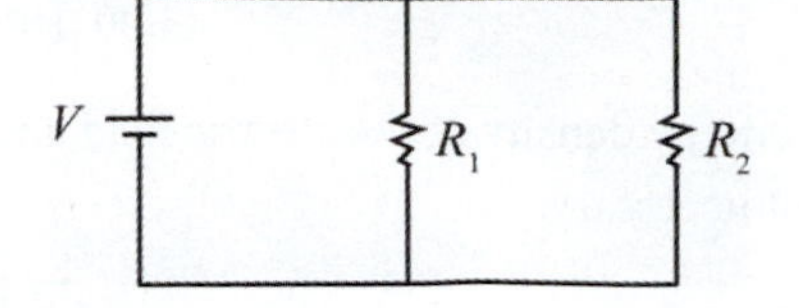

RESEARCH: The current from the battery is given by $i_{\text{tot}} = \dfrac{V_{\text{emf}}}{R_{\text{net}}}$, where $R_{\text{net}} = \left(\dfrac{1}{R_2} + \dfrac{1}{R_1}\right)^{-1}$. The current through each resistor is given by Ohm's Law, $i = V / R$. The sum of the potentials around any loop must be zero: $V_{\text{emf}} + V_{R_1} = 0$, $V_{\text{emf}} + V_{R_2} + V_L = 0$.

SIMPLIFY:

(a) $i_{\text{tot}} = \dfrac{V_{\text{emf}}}{R_1 R_2}(R_1 + R_2)$

(b) $i_{R_1} = \dfrac{V_{R_1}}{R_1}$

(c) $i_{R_2} = \dfrac{V_{R_2}}{R_2}$

(d) $V_{\text{emf}} + V_{R_1} = 0 \Rightarrow V_{R_1} = -V_{\text{emf}}$

(e) $V_{\text{emf}} + V_{R_2} + V_L = 0 \Rightarrow V_{R_2} = -V_{\text{emf}} - V_L = -V_{\text{emf}} - L\dfrac{di}{dt}$

(f) $V_L = L\dfrac{di}{dt}$

(g) $\dfrac{di_{R_1}}{dt} = \dfrac{di_L}{dt} = \dfrac{V_L}{L}$

CALCULATE:

(a) $i_{\text{tot}} = \dfrac{18\text{ V}}{(6.0\ \Omega)(6.0\ \Omega)}(6.0\ \Omega + 6.0\ \Omega) = 6.0$ A

(b) $i_{R_1} = \dfrac{18\text{ V}}{6.0\ \Omega} = 3.0$ A

(c) $i_{R_2} = \dfrac{18\text{ V}}{6.0\ \Omega} = 3.0$ A

(d) $V_{R_1} = -18$ V

(e) $V_{R_2} = -18\text{ V} - (5.0\text{ H})(0) = -18$ V

(f) $V_L = 5.0\text{ H}(0) = 0$

(g) $\dfrac{di_{R_1}}{dt} = \dfrac{0}{5.0\text{ H}} = 0$

ROUND: Not necessary.

DOUBLE CHECK: Evaluating the loop containing the inductor using equation 29.29 shows that after a long time, $i_2(t) = \dfrac{V_{\text{emf}}}{R_2}\left(1 - e^{-t/(L/R)}\right) = \dfrac{V_{\text{emf}}}{R_2}$, as found above. Kirchoff's rules can be used to show that $i_{R_1} = i_{R_2}$, also as found above.

29.57. (a) $u_{\text{B}} = \dfrac{1}{2\mu_0}B^2 = \dfrac{1}{2\left(4\pi \cdot 10^{-7}\text{ H}\cdot\text{m}^{-1}\right)}\left(4.00\cdot 10^{10}\text{ T}\right)^2 = 6.366\cdot 10^{26}\text{ J/m}^3$

(b) The associated mass density is then: $\dfrac{u_{\text{B}}}{c^2} = \rho_{\text{rest}} = \dfrac{6.366\cdot 10^{26}\text{ J/m}^3}{(3.00\cdot 10^8\text{ m/s}^2)^2} = 7.07\cdot 10^9\text{ kg/m}^3$

29.61. **THINK:** Determine the energy density of the electric field and the magnetic field separately.
SKETCH: A sketch is not necessary.

RESEARCH: $u_{\text{B}} = \dfrac{1}{2\mu_0}\left|B^2\right|,\quad u_{\text{E}} = \dfrac{1}{2}\varepsilon_0\left|E^2\right|,\quad \left|\vec{B}_0\right| = \left|\dfrac{k\times\vec{E}_0}{\omega}\right|,\quad \omega = \dfrac{\left|\vec{k}\right|}{\sqrt{\mu_0\varepsilon_0}},\ \vec{E}(\vec{x},t) = \vec{E}_0\cos\left(\vec{k}\bullet\vec{x}-\omega t\right),$
$\vec{B}(\vec{x},t) = \vec{B}_0\cos\left(\vec{k}\bullet\vec{x}-\omega t\right).$

SIMPLIFY:

$$\frac{u_{\text{B}}}{u_{\text{E}}} = \frac{\left|B^2\right|}{2\mu_0}\left(\frac{1}{2}\varepsilon_0\left|E^2\right|\right)^{-1} = \frac{1}{\mu_0\varepsilon_0}\frac{\left|B^2\right|}{\left|E^2\right|} = \frac{1}{\mu_0\varepsilon_0}\left[\frac{\left|\vec{B}_0\right|^2\cos^2\left(\vec{k}\bullet\vec{x}-\omega t\right)}{\left|\vec{E}_0\right|^2\cos^2\left(\vec{k}\bullet\vec{x}-\omega t\right)}\right] = \frac{1}{\mu_0\varepsilon_0}\left(\frac{1}{\omega^2}\right)\frac{\left|k\times\vec{E}_0\right|^2}{\left|\vec{E}_0\right|^2} = \frac{\left|k\times\vec{E}_0\right|^2}{\left|\vec{k}\right|^2\left|\vec{E}_0\right|^2}$$

Note that $\vec{k}$ is perpendicular to $\vec{E}_0$ so $\left|\vec{k}\times\vec{E}_0\right|^2 = \left|\vec{k}\right|^2\left|\vec{E}_0\right|^2$, so the above expression becomes $\dfrac{u_{\text{B}}}{u_{\text{E}}} = 1.$

CALCULATE: No calculations are necessary.
ROUND: Rounding is not necessary.
DOUBLE-CHECK: This result shows that the energy in this type of wave is partitioned equally between the electric and magnetic fields.

29.65. The energy stored in a solenoid is given by $U_{\text{B}} = Li^2/2.$ The energy is dependent only on the magnitude, not the direction of the current. Therefore the energy stored in the magnetic field does not change.

29.67. The current of an RL circuit is given by: $i(t) = \dfrac{V_{\text{emf}}}{R}\left(1-e^{-t/\tau}\right),$ where $\tau = L/R.$ For $t = 20.0\ \mu\text{s}$:

$$\frac{1}{2}\left(\frac{V_{\text{emf}}}{R}\right) = \frac{V_{\text{emf}}}{R}\left(1-e^{-t/\tau}\right) \Rightarrow 1-\frac{1}{2} = e^{-t/\tau} \Rightarrow \tau\ln\left(\frac{1}{2}\right) = -t \Rightarrow \frac{L}{R}\ln\left(\frac{1}{2}\right) = -t \Rightarrow L = -\frac{Rt}{\ln(1/2)}$$

$$\Rightarrow L = -\frac{\left(3.00\cdot 10^3\ \Omega\right)\left(20.0\cdot 10^{-6}\text{ s}\right)}{\ln(1/2)} = 0.0866\text{ H.}$$

29.71. **THINK:** A solenoid of length, l = 3.0 m, and n = 290 turns/m has a current of i = 3.0 A, and stores an energy of $U_{\text{B}} = 2.8$ J. Find the cross-sectional area, A, of the solenoid.
SKETCH:

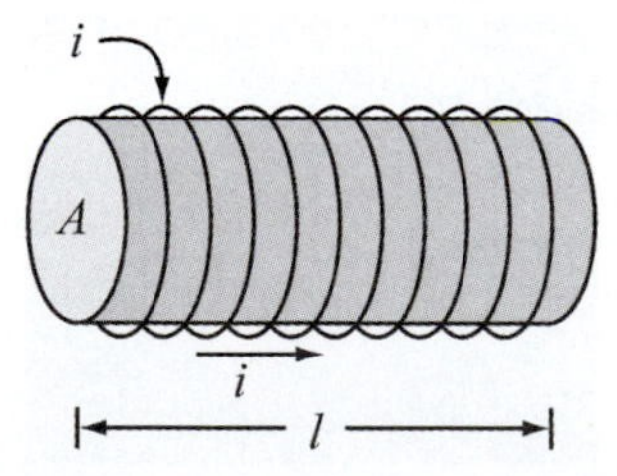

RESEARCH: The energy stored in the magnetic field of an ideal solenoid is $U_{\text{B}} = \mu_0 n^2 lAi^2/2.$

SIMPLIFY: Solving for A yields: $A = \dfrac{2U_{\text{B}}}{\mu_0 n^2 li^2}.$

CALCULATE:

$$A=\frac{2(2.80\text{ J})}{\left(4\pi\cdot 10^{-7}\text{ T m/A}\right)\left(290\text{ m}^{-1}\right)^2(3.00\text{ m})(3.00\text{ A})^2}=1.9625\text{ J/T A}=1.9625\,\frac{\text{N m}}{\left(\text{V s/m}^2\right)(\text{J/V s})}=1.9625\text{ m}^2$$

ROUND: Rounding to three significant figures, $A=1.96\text{ m}^2$.

DOUBLE-CHECK: Considering the length, l, of the solenoid, this is a reasonable cross-sectional area. The units of the result are also correct.

29.75. **THINK:** A conducting rod of length, L = 0.500 m, slides over a frame of two metal bars placed in a magnetic field of strength, B = 1000. gauss = 0.1000 T. The ends of the rods are connected by two resistors, $R_1=100.\ \Omega$ and $R_2=200.\ \Omega$. The conducting rod moves with a constant velocity of v = 8.00 m/s. Determine (a) the current flowing through the two resistors, i_1 and i_2, (b) the power, P, delivered to the resistors, and (c) the force, F, needed for the motion of the rod with constant velocity.

SKETCH:

RESEARCH:

(a) The induced potential difference across the resistors is $V_{\text{ind}}=-d\Phi_{\text{B}}/dt$. Since B is constant while A varies in time at a velocity of v, this expression becomes $V_{\text{ind}}=-B(dA/dt)=-BLv$. The current in each resistor can be determined from $V_{\text{ind}}=i_{\text{ind}}R$.

(b) The power delivered to the resistors is $P=i_1^2R_1+i_2^2R_2$.

(c) The force needed to move the rod with a constant velocity is obtained by calculating the total force acting on the rod. The magnetic force on the rod, F_{mag}, is given by $F_{\text{mag}}=BiL=B\left(V/R_{\text{eq}}\right)L$, where R_{eq} is the equivalent resistance. Note for n resistors in parallel, the equivalent resistance is:

$$\frac{1}{R_{\text{eq}}}=\frac{1}{R_1}+\frac{1}{R_2}+\ldots+\frac{1}{R_n}.$$

SIMPLIFY:

(a) $V_{\text{ind}}=-BLv,\ i_1=\dfrac{\left|V_{\text{ind}}\right|}{R_1},\ i_2=\dfrac{\left|V_{\text{ind}}\right|}{R_2}$

(b) $P=i_1^2R_1+i_2^2R_2$

(c) $F_{\text{mag}}=BV_{\text{ind}}L\left(\dfrac{1}{R_1}+\dfrac{1}{R_2}\right)=B^2L^2v\left(\dfrac{1}{R_1}+\dfrac{1}{R_2}\right)$

CALCULATE:

(a) $V_{\text{ind}}=-(0.100\text{ T})(0.500\text{ m})(8.00\text{ m/s})=-0.400\text{ V},\ i_1=\dfrac{\left|-0.400\text{ V}\right|}{100.\ \Omega}=0.00400\text{ A},$

$i_2=\dfrac{\left|-0.400\text{ V}\right|}{200.\ \Omega}=0.00200\text{ A}$

(b) $P=(0.00400\text{ A})^2(100.\ \Omega)+(0.00200\text{ A})^2(200.\ \Omega)=0.00240\text{ W}$

(c) $F_{mag} = (0.100\text{ T})^2 (0.500\text{ m})^2 (8.00\text{ m/s})\left(\frac{1}{100.\ \Omega} + \frac{1}{200.\ \Omega}\right) = 0.000300\text{ N}$

ROUND:

(a) $i_1 = 4.00\text{ mA}$, $i_2 = 2.00\text{ mA}$

(b) $P = 2.40\text{ mW}$

(c) $F_{mag} = 0.300\text{ mN}$

DOUBLE-CHECK: The calculated values are consistent with the given values. Dimensional analysis confirms all the units are correct.

29.79. **THINK:** The rectangular circuit loop has length, L = 0.600 m, and width, w = 0.150 m, with resistance, $R = 35.0\ \Omega$. It is held parallel to the xy-plane with one end inside a uniform magnetic field as shown in the figure. The magnetic field is $\vec{B}_R = 2.00\hat{z}$ T along the positive z-axis to the right of the dotted line; $\vec{B}_L = 0$ T to the left of the dotted line. Determine the magnitude of the force, F_{app}, required to move the loop to the left at a constant speed of v = 0.100 m/s, while the right end of the loop is still in the magnetic field. Determine the power, P, used by an agent to pull the loop out of the magnetic field at this speed, and the power, P_R, dissipated by the resistor.

SKETCH:

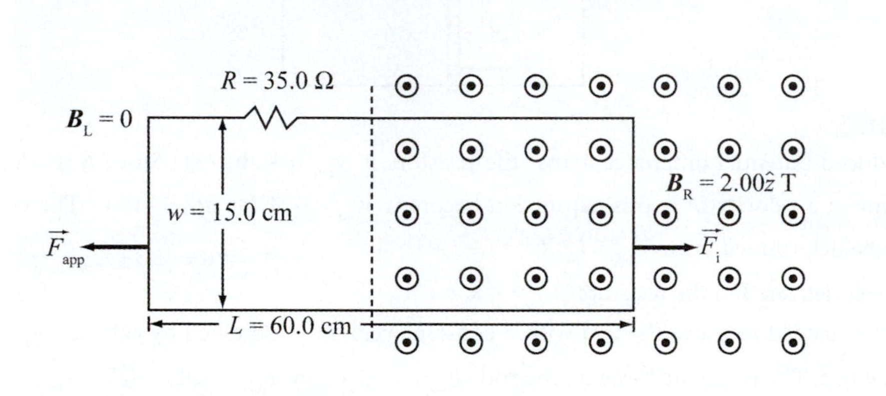

RESEARCH: The magnitude of the force required to move the loop will be equal to the magnitude of the force, F_i, on the current induced in the segment of the loop that lies along the y-axis in the magnetic field. That is, $F_{app} = F_2 = iwB\sin\theta$. Since the angle, θ, between the loop segment of length, w, and the magnetic field is 90°: $\sin\theta = 1$. The induced current, i, is $i = V_{ind}/R$, where $V_{ind} = vwB$ (see equation 29.15). The power, P, used by an agent to pull the loop out of the magnetic field is given by $P = F_{app}v$. The power dissipated by the resistor is given by $P_R = i^2R$.

SIMPLIFY: The current is $i = \frac{V_{ind}}{R} = \frac{vwB}{R}$.

(a) $F_{app} = iwB$

(b) $P = F_{app}v$

(c) $P_R = i^2R$

CALCULATE:

(a) $i = \dfrac{(0.100\text{ m/s})(0.150\text{ m})(2.00\text{ T})}{35.0\ \Omega} = 0.85714\text{ mA}$

$F_{\text{app}} = (0.85714\text{ mA})(0.150\text{ m})(2.00\text{ T}) = 0.25714\text{ mN}$

(b) $P = (0.25714\text{ m N})(0.100\text{ m/s}) = 25.7714\ \mu\text{W}$

(c) $P_{\text{R}} = (0.85714\text{ mA})^2(35.0\ \Omega) = 25.714\ \mu\text{W}$

ROUND:

(a) $F_{\text{app}} = 0.257\text{ mN}$

(b) $P = 25.8\ \mu\text{W}$

(c) $P_{\text{R}} = 25.7\ \mu\text{W}$

DOUBLE-CHECK: All the power used to move the loop while in the magnetic field is dissipated in the resistor: $P = P_{\text{R}}$.

Multi-Version Exercises

29.80. $i(t) = i_{\max}\left(1 - e^{-t/\tau}\right),\ \tau = L/R$

$$\frac{3}{4}i_{\max} = i_{\max}\left(1 - e^{-t/\tau}\right) \Rightarrow -\frac{t}{\tau} = -\frac{tR}{L} = \ln\left(\frac{1}{4}\right)$$

$$\Rightarrow R = -\frac{L}{t}\ln\left(\frac{1}{4}\right) = \frac{L}{t}\ln(4) = \frac{33.03\cdot 10^{-3}\text{ H}}{3.35\cdot 10^{-3}\text{ s}}\ln(4) = 13.7\ \Omega$$

29.83. $\Phi_{\text{B}} = NAB\cos(2\pi ft)$, which means that $V_{\text{ind}} = -d\Phi_{\text{B}}/dt = NAB(2\pi f)\sin(2\pi ft)$.

The maximum occurs when $\sin(2\pi ft) = 1$. $N = 1$, so

$$V_{\max} = AB(2\pi f) = (0.25\pi d^2)B(2\pi f) = 0.5\pi^2(0.0195\text{ m})^2\left(4.77\cdot 10^{-5}\text{ T}\right)\left(13.3\text{ s}^{-1}\right) = 1.19\cdot 10^{-6}\text{ V}.$$

Chapter 30: Alternating Current Circuits

Exercises

30.27. From the inductance and capacitance, $L = 32.0$ mH and $C = 45.0$ μF, the frequency of oscillation is $\omega_0 = (LC)^{-1/2}$. The total energy is constant at $U = q_0^2/2C$ where $q_0 = 10.0$ μC, and the charge varies as $q = q_0\cos(\omega_0 t)$. Since energy remains constant, when the energy in both is the same, it is $(1/2)U$.

$$U_E = \frac{1}{2}U \Rightarrow \frac{q_0^2\cos^2(\omega_0 t)}{2C} = \frac{1}{2}\left(\frac{q_0^2}{2C}\right) \Rightarrow \cos^2(\omega_0 t) = \frac{1}{2} \Rightarrow t = \frac{1}{\omega_0}\cos^{-1}\left(\frac{1}{\sqrt{2}}\right)$$

$$\Rightarrow t = \sqrt{LC}\cos^{-1}\left(\frac{1}{\sqrt{2}}\right) = \sqrt{(32.0 \text{ mH})(45.0 \text{ μF})}\cos^{-1}\left(\frac{1}{\sqrt{2}}\right) = 9.42\cdot 10^{-4} \text{ s}$$

Note that the result does not depend on the original amount of charge, q_0. Dimensional analysis shows that the result has the correct units: $\sqrt{\left[\frac{\text{s}^2}{\text{F}}\right]\cdot[\text{F}]} = [\text{s}]$.

30.31. **THINK:** The charge on the capacitor will oscillate with time as a cosine function with a period determined by the inductance, $L = 0.200$ H, and capacitance, $C = 10.0$ μF. The potential, $V_{emf} = 12.0$ V, will give the initial charge on the capacitor. Ignoring the sign of the charge, the charge on the capacitor will equal $Q = 80.0$ μC periodically.

SKETCH:

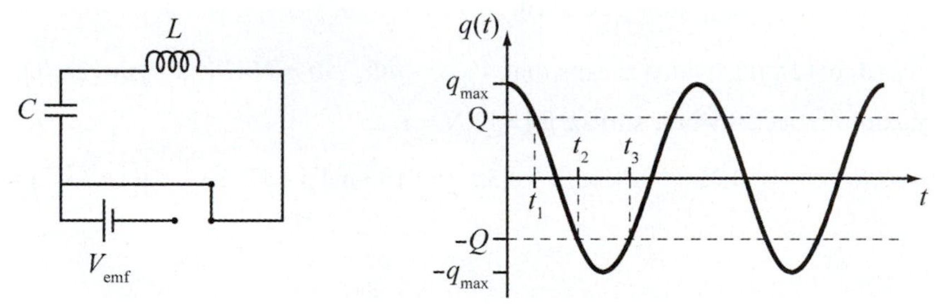

RESEARCH: The initial (and maximum) charge on the capacitor is $q_{max} = CV$. The charge will oscillate as $q = q_{max}\cos(\omega_0 t)$, where $\omega_0 = 1/\sqrt{LC}$.

SIMPLIFY: The first time t_1 when the charge on the capacitor is equal to Q is

$$q = Q = q_{max}\cos(\omega_0 t_1) \Rightarrow t_1 = \left(\frac{1}{\omega_0}\right)\cos^{-1}\left(\frac{Q}{q_{max}}\right);$$

By symmetry, the second time t_2 and the third time t_3 are given by

$$\omega_0 t_2 = \pi - \omega_0 t_1 = \omega_0\left(\frac{\pi}{\omega_0} - t_1\right) \Rightarrow t_2 = \frac{\pi}{\omega_0} - t_1 \text{ and}$$

$$\omega_0 t_3 = \pi + \omega_0 t_1 = \omega_0\left(\frac{\pi}{\omega_0} + t_1\right) \Rightarrow t_3 = \frac{\pi}{\omega_0} + t_1.$$

CALCULATE: $\omega_0 = \dfrac{1}{\sqrt{(0.200 \text{ H})(10.0 \text{ μF})}} = 707.107$ rad/s

$q_{max} = (10.0 \text{ μF})(12.0 \text{ V}) = 120. \text{ μC}$

$$t_1 = \left(\frac{1}{(707.107 \text{ rad/s})}\right)\cos^{-1}\left(\frac{80.0 \text{ μC}}{120. \text{ μC}}\right) = 0.0011895 \text{ s}$$

$$t_2 = \frac{\pi}{(707.107 \text{ rad/s})} - 0.0011895 \text{ s} = 0.0032534 \text{ s}$$

$$t_3 = \frac{\pi}{(707.107 \text{ rad/s})} + 0.0011895 \text{ s} = 0.0056323 \text{ s}$$

ROUND: Rounding the times to three significant figures gives: $t_1 = 1.19$ ms, $t_2 = 3.25$ ms, and $t_3 = 5.63$ ms.

DOUBLE-CHECK: Given the high frequency, small times are expected, so the answers are reasonable.

30.35. **THINK:** The frequency of the damped oscillation is independent of the initial charge, and hence potential. It then only depends on the inductance, $L = 0.200$ H, the resistance, $R = 50.0\ \Omega$, capacitance $C = 2.00\ \mu$F.

SKETCH:

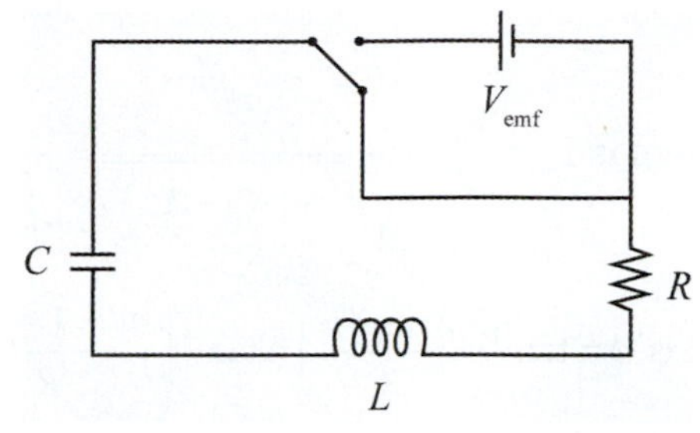

RESEARCH: In general the charge on the capacitor is

$$q = q_{max} e^{-Rt/2L} \cos(\omega t),$$

where $\omega = \sqrt{\omega_0^2 - (R/2L)^2}$ and where $\omega_0 = 1/\sqrt{LC}$. The frequency of oscillation is $f = \omega/(2\pi)$.

SIMPLIFY: $f = \dfrac{\omega}{2\pi} = \dfrac{\sqrt{(\omega_0^2 - R/2L)^2}}{2\pi} = \dfrac{\sqrt{1/(LC) - R^2/(4L^2)}}{2\pi}$

CALCULATE: $f = \dfrac{\sqrt{1/((0.200 \text{ H})(2.00 \cdot 10^{-6} \text{ F})) - (50.0\ \Omega)^2/(4(0.200 \text{ H})^2)}}{2\pi} = 250.858$ Hz

ROUND: Rounding to three significant figures, $f = 251$ Hz.

DOUBLE-CHECK: This frequency is typical of RLC circuits. Additionally, dimensional analysis yields:

$$\left[\left[\frac{\text{F}}{\text{s}^2}\right]\frac{1}{[\text{F}]} - \frac{[\Omega]^2}{[\Omega \cdot \text{s}]^2}\right]^{\frac{1}{2}} = \left[\frac{1}{[\text{s}^2]}\right]^{\frac{1}{2}} = \frac{1}{[\text{s}]}.$$

30.41. Given the frequency, $f = 1.00$ kHz, the angular frequency is $\omega = 2\pi f$. The phase constant for an RLC circuit is given by

$$\tan\phi = \frac{X_L - X_C}{R} = \frac{\omega L - 1/(\omega C)}{R} \Rightarrow \phi = \tan^{-1}\left(\frac{2\pi f \cdot L - 1/(2\pi f \cdot C)}{R}\right)$$

For the values $R = 100.\ \Omega$, $L = 10.0$ mH and $C = 100.\ \mu$F, the phase constant is

$$\phi = \tan^{-1}\frac{\left((2\pi)(1.00 \cdot 10^3 \text{ Hz})(10.0 \cdot 10^{-3} \text{ H}) - ((2\pi)(1.00 \cdot 10^3 \text{ Hz})(100. \cdot 10^{-6} \text{ F}))^{-1}\right)}{(100.\ \Omega)} = 0.549 \text{ rad.}$$

The impedance for this circuit is

$$Z = \sqrt{R^2 + (\omega L - 1/(\omega C))^2} = \sqrt{R^2 + (2\pi f \cdot L - 1/(2\pi f \cdot C))^2}$$

$$= \sqrt{(100.\ \Omega)^2 + \left((2\pi)(1.00 \cdot 10^3 \text{ Hz})(10.0 \cdot 10^{-3} \text{ H}) - ((2\pi)(1.00 \cdot 10^3 \text{ Hz})(100. \cdot 10^{-6} \text{ F}))^{-1}\right)^2} = 117\ \Omega.$$

30.45. **THINK:** The maximum current occurs for when the AC voltage is at its peak, $V_{\text{m}} = 110$ V. The angular frequency of the oscillation is $\omega = 377$ rad/s. The voltage acts across the total impedance of the circuit where $R = 2.20\ \Omega$, $L = 9.30$ mH and $C = 2.27$ mF. The maximum current I_{m}' occurs for a capacitance C' that puts the RLC circuit in resonance with the supplied voltage. At resonance, the phase constant is zero and only the resistor influences the current.

SKETCH: Not required.

RESEARCH: The impedance of inductor and capacitor are $X_L = \omega L$ and $X_C = 1/\omega C$. The maximum current is $I_{\text{m}} = V_{\text{m}}/Z$, where $Z = \sqrt{R^2 + (X_L - X_C)^2}$. The phase angle for the circuit is $\phi = \tan^{-1}\left((X_L - X_C)/R\right)$. When at resonance, $\omega = \omega_0 = 1/\sqrt{LC}$. The maximum current at resonance is $I_{\text{m}} = V_{\text{m}}/R$.

SIMPLIFY:

(a) The maximum current is $I_{\text{m}} = \dfrac{V_{\text{m}}}{Z} = \dfrac{V_{\text{m}}}{\sqrt{R^2 + (X_L - X_C)^2}} = \dfrac{V_{\text{m}}}{\sqrt{R^2 + \left(\omega L - (\omega C)^{-1}\right)^2}}$.

(b) The phase constant is $\phi = \tan^{-1}\left(\dfrac{X_L - X_C}{R}\right) = \tan^{-1}\left(\dfrac{\omega L - (\omega C)^{-1}}{R}\right)$.

(c) The required capacitance for resonance is

$$\omega = \frac{1}{\sqrt{LC'}} \Rightarrow C' = \frac{1}{\omega^2 L},$$

and maximum current at resonance is $I_{\text{m}} = \dfrac{V_{\text{m}}}{R}$.

CALCULATE:

(a) $$I_{\text{m}} = \frac{(110\ \text{V})}{\sqrt{(2.20\ \Omega)^2 + \left((377\ \text{rad/s})(9.30 \cdot 10^{-3}\ \text{H}) - \left((377\ \text{rad/s})(2.27 \cdot 10^{-3}\ \text{F})\right)^{-1}\right)^2}} = 34.2675\ \text{A}$$

(b) $$\phi = \tan^{-1}\frac{(377\ \text{rad/s})(9.30 \cdot 10^{-3}\ \text{H}) - \left((377\ \text{rad/s})(2.27 \cdot 10^{-3}\ \text{F})\right)^{-1}}{(2.20\ \Omega)} = 0.8157\ \text{rad}$$

(c) $$C' = \frac{1}{(377\ \text{rad/s})^2 (9.30 \cdot 10^{-3}\ \text{H})} = 7.565 \cdot 10^{-4}\ \text{F} \text{ and } I_{\text{m}}' = \frac{(110\ \text{V})}{(2.20\ \Omega)} = 50.0\ \text{A}$$

ROUND: To three significant figures:

(a) $I_{\text{m}} = 34.3$ A

(b) $\phi = 0.816$ rad

(c) $C' = 757\ \mu\text{F}$, $I_{\text{m}}' = 50.0$ A, $\phi' = 0$ rad

DOUBLE-CHECK: The current is at a maximum when at resonance, so having $I_{\text{m}}' > I_{\text{m}}$ is reasonable.

30.49. The quality factor for an RLC circuit is defined as $Q = \omega_0(\text{Energy stored/Power lost})$. For the RLC circuit, the resonant frequency is $\omega_0 = 1/\sqrt{LC}$. In general, the energy stored in the circuit is $U = U_0 e^{-Rt/L}$. The power lost is defined as $P = -\left|dU/dt\right|$. Therefore,

$$Q = \frac{\omega_0 \left(U_0 e^{-Rt/L}\right)}{-\dfrac{d}{dt}\left(U_0 e^{-Rt/L}\right)} = \frac{\left(U_0 e^{-Rt/L}\right)}{\sqrt{LC}(R/L)\left(U_0 e^{-Rt/L}\right)} = \frac{1}{R}\sqrt{\frac{L}{C}}.$$

30.53. **THINK:** In order to receive the best signal, the radio should be tuned at resonance with the incoming frequency, $f_0 = 88.7$ MHz. The inductance of the radio receiver is $L = 8.22$ μH. Signal strength is $V_m = 12.9$ μV. The similar signal with frequency $f = 88.5$ MHz, is not at resonance, so its total impedance is influenced by the resistor, capacitor and inductor such that its current is half that of the current for the frequency at resonance.

SKETCH: Not required.

RESEARCH: At resonance, $\omega = \omega_0 = 1/\sqrt{LC}$, where the angular frequency is $\omega_0 = 2\pi f_0$. The impedances of the inductor and capacitor are $X_L = \omega L$ and $X_C = 1/(\omega c)$, respectively. The impedance of the RLC circuit is $Z = \sqrt{R^2 + (X_L - X_C)^2}$. At resonance, the current amplitude is $I_m = V_m / R$, and when not at resonance, the current amplitude is $I'_m = V_m / Z$.

SIMPLIFY:

(a) At resonance, $\omega_0 = \dfrac{1}{\sqrt{LC_0}} = 2\pi f_0 \Rightarrow C_0 = \dfrac{1}{4\pi^2 f_0^2 L}$.

(b) When $I'_m = \dfrac{1}{2} I_m$, then

$$\frac{V_m}{Z} = \frac{1}{2}\frac{V_m}{R_0} \Rightarrow 2R_0 = Z = \sqrt{R_0^2 + (X_L - X_C)^2}$$

$$4R_0^2 = R_0^2 + (X_L - X_C)^2 \Rightarrow R_0 = \left|\frac{(X_L - X_C)}{\sqrt{3}}\right| = \left|\frac{2\pi fL - (2\pi fc)^{-1}}{\sqrt{3}}\right|.$$

CALCULATE:

(a) $$C = \frac{1}{4\pi^2 (88.7 \cdot 10^6 \text{ Hz})^2 (8.22 \cdot 10^{-6} \text{ H})} = 3.9167 \cdot 10^{-13} \text{ F}$$

(b) $$R_0 = \left|\frac{2\pi(88.5 \cdot 10^6 \text{ Hz})(8.22 \cdot 10^{-6} \text{ H}) - \left(2\pi(88.5 \cdot 10^6 \text{ Hz})(3.9167 \cdot 10^{-13} \text{ F})\right)^{-1}}{\sqrt{3}}\right| = 11.941 \ \Omega$$

ROUND: To three significant figures,

(a) $C = 0.392$ pF

(b) $R_0 = 11.9 \ \Omega$

DOUBLE-CHECK: This is an RLC circuit, so the current across the circuit decays exponentially with a time constant $\tau = 2L / R = 1.38$ μs. Assuming that the time constant represents the delay from when the radio picks up the signal to when it transmits it as sound, it is reasonable that the value is small.

30.57. The primary coil has $N_P = 200$ turns and the secondary coil has $N_S = 120$ turns. The secondary coil drives a current I through a resistance of $R = 1.00$ kΩ. The input voltage applied across the primary coil is $V_{rms} = 75.0$ V. The voltage across the secondary coil is $V_S = V_{rms} N_S / N_P$. The power dissipated in the resistor is $P = \dfrac{(V_{rms} N_S / N_P)^2}{R} = \dfrac{((75.0 \text{ V})(120)/(200))^2}{(1.00 \cdot 10^3 \ \Omega)} = 2.03$ W.

30.61. This question deals with an LC circuit. The given quantities are the frequency, $f = 1000.$ kHz and the inductance, $L = 10.0$ mH. What is the capacitance, C, of the capacitor when the station is properly tuned? Equating the expressions $\omega = 2\pi f$ and $\omega = 1/\sqrt{LC}$ and solving for C: $2\pi f = 1/\sqrt{LC}$,

$$2\pi f = \frac{1}{\sqrt{LC}} \Rightarrow C = \frac{1}{(2\pi f)^2 L}$$

$$C = \frac{1}{(2\pi f)^2 L} = \frac{1}{\left((2\pi)(1.000 \cdot 10^6 \text{ Hz})\right)^2 (1.00 \cdot 10^{-2} \text{ H})} = 2.53 \cdot 10^{-12} \text{ F}.$$

30.65. In an RLC circuit, the inductance is $L = 65.0$ mH and the capacitance is $C = 1.00$ μF. The circuit loses electromagnetic energy at a rate of $\Delta U = -3.50\%$ per cycle. The energy stored in the electric field of the capacitor is expressed by $\Delta U_E = q_{max}^2 e^{-Rt/L} \cos(\omega_0 t)$. The rate of energy loss is $\frac{\Delta U_E}{U_E} = \frac{U_{E, final} - U_{E, initial}}{U_{E, initial}}$, where time $t_{initial}$ is zero and t_{final} is the time to complete one cycle, $t_{final} = 2\pi / \omega_0$. The rate of energy loss per cycle can now be written as

$$\frac{\Delta U_E}{U_E} = -0.035 = \frac{\frac{q_{max}^2}{2C} e^{-2\pi R/\omega_0 L} \cos^2(2\pi) - \frac{q_{max}^2}{2C} e^{-0} \cos^2(0)}{\frac{q_{max}^2}{2C} e^{-0} \cos^2(0)} = e^{-2\pi R/\omega L} - 1.$$

Since $\omega_0 = 1/\sqrt{LC}$,

$$1 - 0.0350 = e^{-\left(2\pi R\sqrt{LC}\right)/L} \Rightarrow \ln(0.9650) = -\frac{2\pi R\sqrt{LC}}{L}$$

$$R = -\frac{\ln(0.9650)}{2\pi}\sqrt{\frac{L}{C}} = -\frac{\ln(0.9650)}{2\pi}\sqrt{\frac{(65.0 \cdot 10^{-3} \text{ H})}{(1.00 \cdot 10^{-6} \text{ F})}} = 1.45 \ \Omega.$$

30.69. **THINK:** The unknown wire-wound resistor R is initially connected to a DC power supply. When there is a voltage of $V_{emf} = 10.0$ V across the resistor, the current is $I = 1.00$ A. Next the resistor is connected to an AC power source with $V_{rms} = 10.0$ V. When the AC power source is operated at frequency $f = 20.0$ kHz, a current of $I_{rms} = 0.800$ A is measured. Find:

(a) the resistance, R;

(b) the inductive reactance, X_L, of the resistor;

(c) the inductance, L, of the resistor; and

(d) the frequency, f', of the AC power source at which $X_L = R$.

SKETCH: Not required.

RESEARCH:

(a) The resistance of the resistor when used with the DC source can be found using Ohm's law, $R = V / I$.

(b) When connected to the AC power source, the resistor can be treated as an RL series circuit. The impedance of the RL circuit is $Z = \sqrt{R^2 + X_L^2} = V_{rms} / I_{rms}$.

(c) The inductance can be found with $X_L = \omega L$ and $\omega = 2\pi f$.

(d) $\omega L = R$, $\omega = 2\pi f'$

SIMPLIFY:

(a) $R=\frac{V}{I}$

(b) $Z=\sqrt{R^2+X_L^2} \Rightarrow X_L^2=Z^2-R^2 \Rightarrow$ Substituting $Z=\frac{V_{\text{rms}}}{I_{\text{rms}}}$ gives, $X_L=\sqrt{\left(\frac{V_{\text{rms}}}{I_{\text{rms}}}\right)^2-R^2}$.

(c) $L=\frac{X_L}{\omega}=\frac{X_L}{2\pi f}$

(d) $f'=\frac{R}{2\pi L}$

CALCULATE:

(a) $R=\frac{(10.0\text{ V})}{(1.00\text{ A})}=10.0\ \Omega$

(b) $X_L=\sqrt{\left(\frac{10.0\text{ V}}{0.800\text{ A}}\right)^2-(10.0\ \Omega)^2}=7.50\ \Omega$

(c) $L=\frac{(7.50\ \Omega)}{2\pi(20.0\cdot 10^3\text{ Hz})}=5.968\cdot 10^{-5}\text{ H}$

(d) $f'=\frac{10.0\ \Omega}{2\pi(5.968\cdot 10^{-5}\text{ H})}=26667\text{ Hz}$

ROUND: The answers should be reported to three significant figures.

(a) $R=10.0\ \Omega$

(b) $X_L=7.50\ \Omega$

(c) $L=5.97\cdot 10^{-5}\text{ H}$

(d) $f'=26.7\text{ kHz}$

DOUBLE-CHECK: Since the current decreased when the power supply was changed from DC to AC, the resistance must have increased. This additional resistance is explained by the inductive reactance of the resistor. The units for all calculated values are correct.

30.73. **THINK:** A resistor R is connected across an AC source which oscillates at angular frequency ω. Show that the power dissipated in R oscillates with frequency 2ω.

SKETCH:

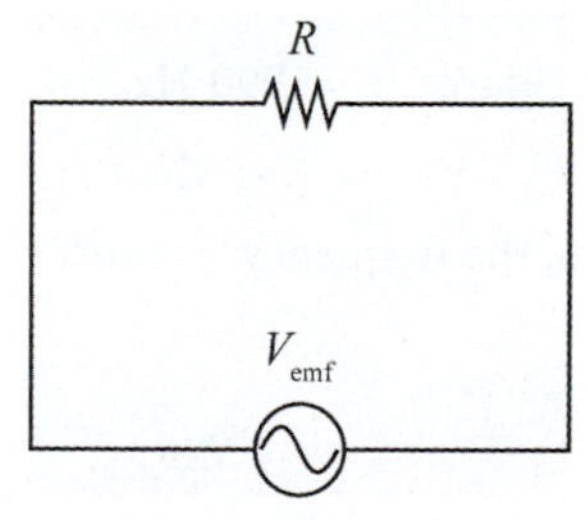

RESEARCH: The power is $P=i_R^2R$. For an AC power supply, $i_R=I_R\sin(\omega t)$. Also useful is the trigonometric identity $\cos(2\theta)=1-2\sin^2\theta$.

SIMPLIFY: $P=i_R^2R=I_R^2\sin^2(\omega t)R=I_R^2\frac{1-\cos(2(\omega t))}{2}R=\frac{1}{2}I_R^2R(1-\cos(2\omega t))$. It can be seen from the above equation that the power oscillates with a frequency twice that of the voltage.

CALCULATE: Not required.

ROUND: Not required.

DOUBLE-CHECK: Power, P, is proportional to i^2. Since i varies proportionately with V, it must be the case that i^2 varies proportionately with V^2. Since V varies proportionately with ω, it must be the case that V^2 varies proportionately with ω^2. Therefore by transitivity, P is proportional to ω^2. Therefore, there exists a constant, c, such that $P = c\omega^2$. So the change in P with respect to time, dP/dt, will be proportional to $d\omega^2/dt$, or 2ω.

30.77. **THINK:** The RC low-pass filter has a breakpoint frequency of $f_\text{B} = 200.$ Hz. Find the frequency at which the output voltage divided by the input voltage is $V_\text{out}/V_\text{in} = 0.100$.

SKETCH:

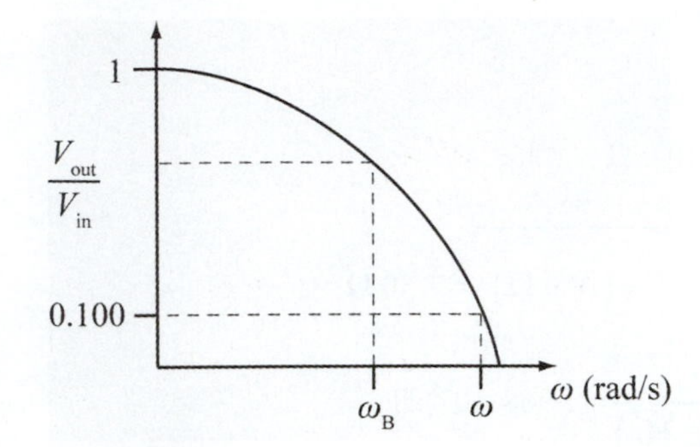

RESEARCH: For a RC low-pass filter, the breakpoint frequency is: $\omega_\text{B} = 1/(RC)$, where $\omega_\text{B} = 2\pi f_\text{B}$. The ratio of the input voltage to output voltage is

$$\frac{V_\text{out}}{V_\text{in}} = \frac{1}{\sqrt{1+\omega^2 R^2 C^2}}.$$

SIMPLIFY: $\omega_\text{B} = 2\pi f_\text{B} = \dfrac{1}{RC} \Rightarrow RC = \dfrac{1}{2\pi f_\text{B}}$

$$\frac{V_\text{in}}{V_\text{out}} = \sqrt{1+\omega^2(RC)^2} = \sqrt{1+\omega^2\left(\frac{1}{2\pi f_\text{B}}\right)^2} \Rightarrow \left(\frac{V_\text{in}}{V_\text{out}}\right)^2 = 1+\left(\frac{\omega}{2\pi f_\text{B}}\right)^2 \Rightarrow \left(\frac{V_\text{in}}{V_\text{out}}\right)^2 - 1 = \frac{\omega^2}{(2\pi f_\text{B})^2}$$

$$\omega = 2\pi f_\text{B}\sqrt{\left(\frac{V_\text{in}}{V_\text{out}}\right)^2 - 1} \Rightarrow f = f_\text{B}\sqrt{\left(\frac{V_\text{in}}{V_\text{out}}\right)^2 - 1}$$

CALCULATE: $f = (200.\text{ Hz})\sqrt{\left(\dfrac{1}{0.100}\right)^2 - 1} = 1989.97\text{ Hz}$

ROUND: To three significant figures, $f = 1990$ Hz.

DOUBLE-CHECK: Since V_out/V_in is less than $1/\sqrt{2}$ (the value associated with the breakpoint frequency), by the above sketch, the frequency f must be greater than the breakpoint frequency f_B.

Multi-Version Exercises

30.78. $X_L = 2\pi fL = 2\pi(605\text{ Hz})(42.1\text{ mH}) = 160.\ \Omega$

Chapter 31: Electromagnetic Waves

Exercises

31.23. **THINK:** A magnetic field can be produced by a current and by induction due to a change in an electric flux. To solve this problem, use the Maxwell-Ampere law. There is no current between the plates, but there is a change in the electric flux. The wire carries a current $i = 20.0$ A. The parallel plate capacitor has radius $R = 4.00$ cm, and separation $s = 2.00$ mm. The radius of interest is $r = 1.00$ cm from the center of the parallel plates.

SKETCH:

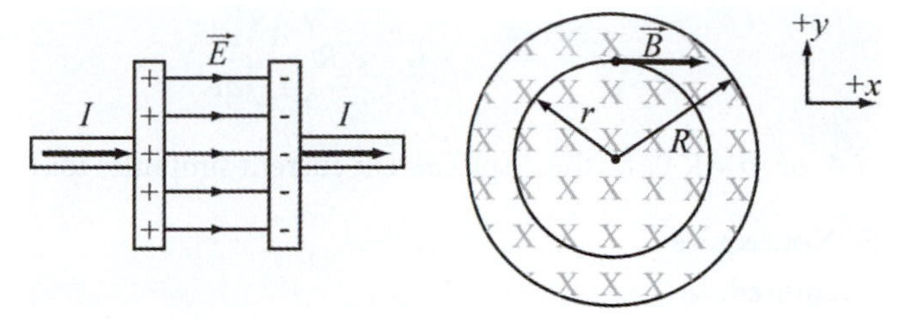

RESEARCH: Since there is no current between the capacitor plates, the Maxwell-Ampere law becomes:

$$\oint \vec{B} \cdot d\vec{S} = \mu_0 \varepsilon_0 \frac{d\Phi_E}{dt}.$$

SIMPLIFY: Applying this law along a circular Amperian loop with a radius, $r \le R$, as shown above. Since $\vec{B}$ is parallel to $d\vec{S}$, the left-hand side of the above equation is $\oint \vec{B} \cdot d\vec{S} = B \oint dS = B2\pi r$. Assuming the electric field, $\vec{E}$, is uniform between the capacitor plates and directed perpendicular to the plates, the electric flux through the loop is $\Phi_E = EA_r = E\pi r^2$. Thus, the Ampere-Maxwell law becomes:

$$B(2\pi r) = \mu_0 \varepsilon_0 \pi r^2 \frac{dE}{dt}.$$

Therefore, the magnetic field is: $B = \left(\frac{\mu_0 \varepsilon_0 r}{2}\right)\frac{dE}{dt}$. Since the electric field of the capacitor is $E = \sigma / \varepsilon_0$, the rate of change of the electric field is given by:

$$\frac{dE}{dt} = \frac{d}{dt}(\sigma / \varepsilon_0) = \left(\frac{1}{\varepsilon_0}\right)\frac{d\sigma}{dt} = \left(\frac{1}{\varepsilon_0}\right)\frac{d}{dt}(q / A_R) = \left(\frac{1}{\varepsilon_0 A_R}\right)\frac{dq}{dt}.$$

Since $i = dq/dt$, $\frac{dE}{dt} = \frac{i}{\varepsilon_0 \pi R^2}$. Using this result, the magnetic field is: $B = \left(\frac{\mu_0 \varepsilon_0 r}{2}\right)\frac{i}{\varepsilon_0 \pi R^2} = \left(\frac{\mu_0 i}{2\pi R^2}\right) r.$

CALCULATE: $B = \frac{(4\pi \cdot 10^{-7} \text{ H/m})(20.0 \text{ A})}{2\pi (0.0400 \text{ m})^2}(0.0100 \text{ m}) = 2.50 \cdot 10^{-5} \text{ T}$

ROUND: Three significant figures are required: $B = 2.50 \cdot 10^{-5}$ T.

DOUBLE-CHECK: This is the same as calculating a magnetic field inside a wire with radius, R. Applying Ampere's law gives:

$$B = \frac{\mu_0}{2\pi r}\left(\frac{\pi r^2}{\pi R^2}\right) i = \left(\frac{\mu_0 i}{2\pi R^2}\right) r.$$ This is the same result as above.

31.27. **THINK:** To determine the displacement current, the electric field inside the conductor is needed.

SKETCH:

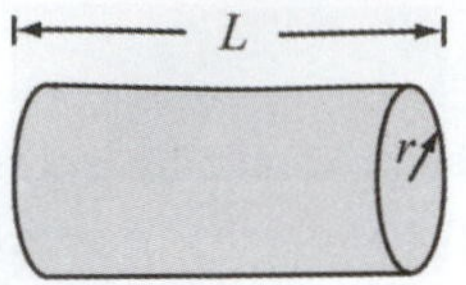

RESEARCH: The displacement current is defined as: $i_d = \varepsilon_0 d\Phi_E / dt$. The electric flux inside the conductor is: $\Phi_E = EA = (V/L)A$.

SIMPLIFY: Since $V = iR$, the electric flux becomes $\Phi = iRA/L$. Therefore, the displacement current is:

$$i_d = \varepsilon_0 R\left(\frac{A}{L}\right)\frac{di}{dt}.$$

Using $R = \rho L/A$ or $\rho = RA/L$, the displacement current simplifies to: $i_d = \varepsilon_0 \rho \dfrac{di}{dt}$.

CALCULATE: Not required.

ROUND: Not required.

DOUBLE-CHECK: Since the current depends in part on the resistance of the current carrying conductor, and the resistance depends on the geometry and resistivity of the material, it makes sense that the current is some function of the resistivity.

31.31. (a) The time delay from New York to Baghdad by cable is $\Delta t = \dfrac{d}{c} = \dfrac{1\cdot 10^7 \text{ m}}{3.00\cdot 10^8 \text{ m/s}} = 0.03$ s.

(b) The time delay via satellite is given by $\Delta t = d/c$. The distance, d, is given by twice the distance from New York to the satellite, that is, $d = 2\sqrt{(36000 \text{ km})^2 + (5000 \text{ km}/2)^2} = 2\cdot 36345 \text{ km} = 72691 \text{ km}$. The time delay is:

$$\Delta t = \frac{7.269\cdot 10^7 \text{ m}}{3.00\cdot 10^8 \text{ m/s}} = 0.24 \text{ s}.$$

When the signal travels by the cable, the time delay is very short, so it is not noticeable. However, the time delay for the signal traveling via satellite is about a quarter of a second. This means in a conversation, Alice will find that she receives a response from her fiancé after 0.5 s, which is quite noticeable.

31.35. **THINK:** To solve this problem, the frequency of oscillation of an RLC circuit must be determined. The circuit has a capacitor $C = 2.0\cdot 10^{-12}$ F, and must have a resonance frequency such that it will generate a radio wave with wavelength $\lambda = 150$ m.

SKETCH: A sketch is not required.

RESEARCH: The angular frequency of the *RLC* circuit in resonance is $\omega_0 = 1/\sqrt{LC}$.

SIMPLIFY: Using $\omega_0 = 2\pi f$ and $f = c/\lambda$, the above equation becomes: $\dfrac{2\pi c}{\lambda} = \dfrac{1}{\sqrt{LC}}$. The inductance required in the *RLC* circuit is: $L = \dfrac{\lambda^2}{(2\pi c)^2 C}$.

CALCULATE: $$L = \frac{(150 \text{ m})^2}{\left[2\pi\left(3.00\cdot 10^8 \text{ m/s}\right)\right]^2\left(2.0\cdot 10^{-12} \text{ F}\right)} = 0.00317 \text{ H}$$

ROUND: Rounding to two significant figures yields $L = 3.2$ mH.

DOUBLE-CHECK: A wavelength of 150 m corresponds to a frequency of $2\cdot 10^6$ Hz. Such a large frequency necessarily requires a fairly small inductance.

31.39. The intensity of the laser beam is $I = P/A$. This intensity is related to the amplitude of the electric field by $I = E^2/(2\mu_0 c)$. Therefore, the amplitude of the electric field in the beam is:

$$E = \sqrt{\frac{2\mu_0 cP}{A}} = \sqrt{\frac{2(4\pi\cdot 10^{-7}\text{ H/m})(3.00\cdot 10^{8}\text{ m/s})(3.00\cdot 10^{3}\text{ W})}{\pi(0.500\cdot 10^{-3}\text{ m})^2}} = 1.697\cdot 10^{6}\text{ V/m} \approx 1.70\cdot 10^{6}\text{ V/m}.$$

31.43. **THINK:** A laser beam has a power of 10.0 W and a beam diameter of 1.00 mm. Assume the intensity of the beam is the same throughout the cross section of the beam.

SKETCH: A sketch is not required.

RESEARCH:

(a) The intensity of the laser beam is given by $I = P/A$. Area $A = \pi r^2$.

(b) The intensity is related to the rms electric field by $I = E_{\text{rms}}^2/(\mu_0 c) \Rightarrow E_{\text{rms}} = \sqrt{\mu_0 cI}$.

(c) The time-averaged Poynting vector is equal the intensity of the beam, $S_{\text{ave}} = I$.

(d) $S(x,t) = \dfrac{[E(x,t)]^2}{\mu_0 c}$ and $E(x,t) = E_{\text{m}}\sin(kx - \omega t + \phi)$.

(e) The rms magnetic field is $B_{\text{rms}} = E_{\text{rms}}/c$.

SIMPLIFY:

(d) Substituting the expression for $E(x,t)$ gives: $S(x,t) = \dfrac{1}{\mu_0 c}E_{\text{m}}^2\sin^2(kx - \omega t + \phi)$. Because $S(0,0) = 0$, take $\phi = 0$. Therefore, $S(x,t) = 2I\sin^2(kx - \omega t)$. Note that $\omega = 2\pi f = 2\pi c/\lambda$ and $k = 2\pi/\lambda$.

CALCULATE:

(a) $I = \dfrac{10.0\text{ W}}{\pi(0.500\cdot 10^{-3}\text{ m})^2} = 1.2732\cdot 10^{7}\text{ W/m}^2$

(b) $E_{\text{rms}} = \sqrt{(4\pi\cdot 10^{-7}\text{ H/m})(3.00\cdot 10^{8}\text{ m/s})(1.2732\cdot 10^{7}\text{ W/m}^2)} = 6.932809\cdot 10^{4}\text{ V/m}$

(c) $S_{\text{ave}} = 1.2732\cdot 10^{7}\text{ W/m}^2$

(d)
$$S(x,t) = 2(1.2732\cdot 10^{7}\text{ W/m}^2)\sin^2\left[\left(\frac{2\pi}{514.5\cdot 10^{-9}\text{ m}}\right)x - \left(\frac{2\pi(3.00\cdot 10^{8}\text{ m/s})}{514.5\cdot 10^{-9}\text{ m}}\right)t\right]$$
$$= 2.5464\cdot 10^{7}\text{ W/m}^2\sin^2\left[(1.22122\cdot 10^{7}\text{ m}^{-1})x - (3.66366\cdot 10^{15}\text{ Hz})t\right]$$

(e) $B_{\text{rms}} = \dfrac{6.932809\cdot 10^{4}\text{ V/m}}{3.00\cdot 10^{8}\text{ m/s}} = 2.30936\cdot 10^{-4}\text{ T}$

ROUND:

(a) $I = 1.27\cdot 10^{7}\text{ W/m}^2$. This intensity is much larger than the intensity of sunlight on Earth (1400 W/m^2).

(b) $E_{\text{rms}} = 6.93\cdot 10^{4}\text{ V/m}$

(c) $S_{\text{ave}} = 1.27\cdot 10^{7}\text{ W/m}^2$

(d) $S(x,t) = 2.5464\cdot 10^{7}\text{ W/m}^2\sin^2\left((1.22122\cdot 10^{7}\text{m}^{-1})x - (3.66366\cdot 10^{15}\text{Hz})t\right)$.

Rounding the coefficients to three significant figures,

$S(x,t) = 2.55\cdot 10^{7}\text{ W/m}^2\sin^2\left((1.22\cdot 10^{7}\text{m}^{-1})x - (3.66\cdot 10^{15}\text{Hz})t\right)$.

Note that for given values for x and t, it would be better to keep the unrounded coefficients and then round the calculted value of S.

(e) $B_{\text{rms}} = 2.31\cdot 10^{-4}$ T

DOUBLE-CHECK: The laser has a very high power output in a very narrow beam. This is a desirable property in a laser. The results make sense.

31.47. The net force on the sail is $F = A\Delta P$. The area, A, is $A = \pi R^2 = \pi\left(10.0\cdot 10^3 \text{ m}\right)^2 = 3.142\cdot 10^8 \text{ m}^2$. The differential pressure, ΔP, is:

$$\Delta P = \frac{2I}{c} - \frac{I}{c} = \frac{I}{c}.$$

The intensity, I, is given by the Stefan-Boltzman law as:

$$I = \sigma T^4 = \left(5.67\cdot 10^{-8} \text{ W/m}^2 \text{ K}^4\right)\left(2.725 \text{ K}\right)^4 = 3.126\cdot 10^{-6} \text{ W/m}^2.$$

$$\Rightarrow \Delta P = \frac{3.126\cdot 10^{-6} \text{ W/m}^2}{3.00\cdot 10^8 \text{ m/s}} = 1.042\cdot 10^{-14} \text{ Pa}$$

$$\Rightarrow F = \left(3.142\cdot 10^8 \text{ m}^2\right)\left(1.042\cdot 10^{-14} \text{ Pa}\right) = 3.27\cdot 10^{-6} \text{ N}$$

31.51. **THINK:** Given the density and volume, the mass can be determined. Given the power and spot size of the laser, the intensity and the radiation pressure of the laser can be determined. To determine how many lasers are needed, calculate the total force required and divide this by the force per laser applied. $\rho = 1.00 \text{ mg/cm}^3$, $D = 2.00$ mm, $t = 0.100$ mm, $P = 5.00$ mW, $d = 2.00$ mm.

SKETCH:

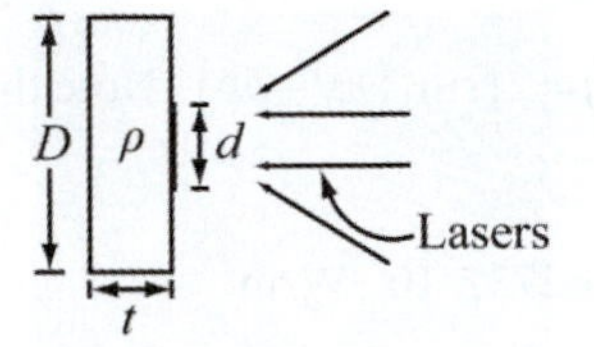

RESEARCH:

(a) The weight is given by $w = mg$, where $m = \pi\left(D/2\right)^2 t\rho$. Note that $1 \text{ mg/cm}^3 = 1 \text{ kg/m}^3$.

(b) $P_{\text{r}} = \dfrac{I}{c}$ (absorbing material), $I = \dfrac{P}{A}$, $A = \pi\left(\dfrac{d}{2}\right)^2$.

(c) $F_{\text{las}} = P_{\text{r}}A$. The number of lasers needed is given by $N = w / F_{\text{las}}$.

SIMPLIFY:

(a) $w = \dfrac{\pi D^2 t\rho g}{4}$

(b) $I = P\dfrac{4}{\pi d^2} = \dfrac{4P}{\pi d^2}$, $P_{\text{r}} = \dfrac{I}{c} = \dfrac{4P}{\pi d^2 c}$

(c) $F_{\text{las}} = P_{\text{r}}A = \dfrac{IA}{c} = \dfrac{P}{cA}A = \dfrac{P}{c}$, $N = \dfrac{w}{F_{\text{las}}} = \dfrac{wc}{P}$

CALCULATE:

(a) $w = \dfrac{\pi\left(2.00\cdot 10^{-3} \text{ m}\right)^2\left(0.100\cdot 10^{-3} \text{ m}\right)\left(1.00 \text{ kg/m}^3\right)\left(9.81 \text{ m/s}^2\right)}{4} = 3.082\cdot 10^{-9}$ N

(b) $I = \dfrac{4\left(5.00\cdot 10^{-3} \text{ W}\right)}{\pi\left(2.00\cdot 10^{-3} \text{ m}\right)^2} = 1.592\cdot 10^3 \text{ W/m}^2$, $P_{\text{r}} = \dfrac{I}{c} = \dfrac{1.592\cdot 10^3 \text{ W/m}^2}{3.00\cdot 10^8 \text{ m/s}} = 5.30510^{-6} \text{ N/m}^2$

(c) $N = \dfrac{\left(3.082\cdot 10^{-9} \text{ N}\right)\left(3.00\cdot 10^8 \text{ m/s}\right)}{\left(5.00\cdot 10^{-3} \text{ W}\right)} = 184.9$

ROUND:

(a) $w = 3.08 \cdot 10^{-9}$ N $= 3.08$ nN

(b) $I = 1.59$ kW/m^2, $P_r = 5.31$ μN/m^2

(c) $N = 185$ lasers

DOUBLE-CHECK: Even though the object is very light, it would still require a large power output to produce enough radiation pressure to overcome the force of gravity.

31.55. **THINK:** First calculate the intensity of the light after it first passes through the two polarizers. Once the intensity is calculated, the magnitude of the electric and magnetic fields can be determined. The angles of the first and second polarizers are $\theta_1 = 35°$ and $\theta_2 = 55°$, respectively. The laser spot size diameter is $d = 1.00$ mm and the laser power is $P = 15.0$ mW.

SKETCH:

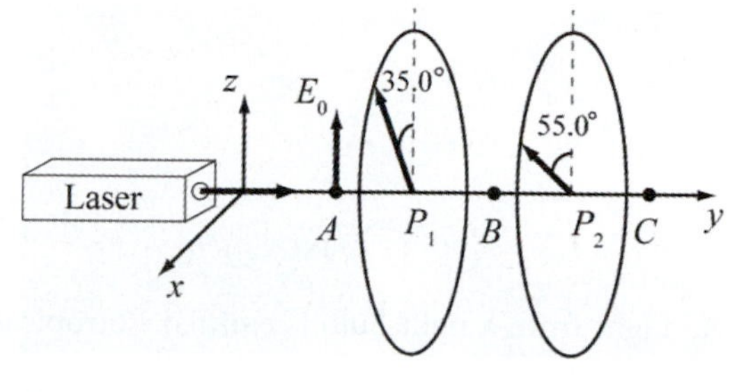

RESEARCH: $I_1 = I_0 \cos^2\theta_1$, $I_2 = I_1 \cos^2(\theta_2 - \theta_1)$, $I_0 = \dfrac{P}{A}$, $A = \pi\left(\dfrac{d}{2}\right)^2$, $I = \dfrac{1}{2}\left(\dfrac{E^2}{c\mu_0}\right)$, $\dfrac{E}{B} = c$

SIMPLIFY: $I_2 = I_1\cos^2(\theta_2-\theta_1) = I_0\cos^2\theta_1\cos^2(\theta_2-\theta_1)$, $I_0 = \dfrac{P}{A} = \dfrac{4P}{\pi d^2} \Rightarrow I_2 = \dfrac{4P}{\pi d^2}\cos^2\theta_1\cos^2(\theta_2-\theta_1)$

$E = \sqrt{2Ic\mu_0}$, $B = E/c$

CALCULATE: $I_2 = \dfrac{4(15.0\cdot 10^{-3}\text{ W})}{\pi(1.00\cdot 10^{-3}\text{ m})^2}\cos^2(35°)\cos^2(55°-35°) = 1.132\cdot 10^4\text{ W/m}^2$

$E = \sqrt{2(1.132\cdot 10^4\text{ W/m}^2)(3.00\cdot 10^8\text{ m/s})(4\pi\cdot 10^{-7}\text{ T m A}^{-1})} = 2.921\cdot 10^3\text{ V/m}$

$B = \dfrac{2.921\cdot 10^3\text{ V/m}}{3.00\cdot 10^8\text{ m/s}} = 9.737\cdot 10^{-6}\text{ T}$

ROUND: $I_2 = 1.13\cdot 10^4\text{ W/m}^2$, $E = 2.92\cdot 10^3\text{ V/m}$, $B = 9.74\cdot 10^{-6}\text{ T}$

DOUBLE-CHECK: The initial intensity of the laser light is about $1.9\cdot 10^4\text{ W/m}^2$. The initial electric and magnetic fields are also significantly larger. It is expected that some of the intensity of the laser beam would be blocked by the polarizers.

31.59. $S = \left(\dfrac{\text{power}}{\text{area}}\right) = \left(\dfrac{200.\text{ W}}{1.00\cdot 10^6\text{ m}^2}\right) = 2.00\cdot 10^8\text{ W/m}^2 = \dfrac{E^2}{2c\mu_0} \Rightarrow E = \sqrt{2c\mu_0 S} = 3.88\cdot 10^5\text{ V/m}$

Note the 2 is in the denominator from: $E_{rms} = \dfrac{E}{\sqrt{2}}$. Therefore, $E_{rms}^2 = \dfrac{E^2}{2}$ and $S = \dfrac{E_{rms}^2}{c\mu_0} = \dfrac{E^2}{2c\mu_0}$. The wavelength has nothing to do with the solution.

31.63. $E_{\text{rms}} = \sqrt{c\mu_0 I} = \sqrt{c\mu_0\left(\dfrac{P}{A}\right)} = \sqrt{\dfrac{c\mu_0 P}{\pi(d/2)^2}} = \sqrt{\dfrac{4c\mu_0 P}{\pi d^2}}$

$$= \sqrt{\frac{4\left(3.00\cdot 10^8 \text{ m/s}\right)\left(4\pi\cdot 10^{-7} \text{ T m/A}\right)\left(0.40\cdot 300.\text{ W}\right)}{\pi\left(2\text{ m}\right)^2}} = 100 \text{ V/m}.$$

31.67. **THINK:** The peak magnetic field can be determined from the speed of light and the peak electric field. The power of the bulb can be determined from its intensity, which can be determined from the electric and magnetic fields. Use the values $r = 2.25$ m and $E = 21.2$ V/m.

SKETCH:

RESEARCH: $\dfrac{E}{B} = c,\ I = \dfrac{P}{A},\ I = \dfrac{E^2}{2c\mu_0}$

SIMPLIFY:

(a) $B = E/c$

(b) $P = IA = \dfrac{E^2}{2c\mu_0}A$. Light from a light-bulb is emitted isotropically, that is equally in all directions. To determine the power a distance, d, away from the light-bulb, the intensity at all points a distance, d, from the light-bulb must be summed. Hence, A should be the surface area of a sphere of radius, r:

$$A = 4\pi r^2 \Rightarrow P = \frac{4\pi r^2 E^2}{2c\mu_0}.$$

CALCULATE:

(a) $B = \dfrac{21.2 \text{ V/m}}{3\cdot 10^8 \text{ m/s}} = 7.067\cdot 10^{-8}$ T

(b) $P = \dfrac{4\pi\left(2.25\text{ m}\right)^2\left(21.2\text{ V/m}\right)^2}{2\left(3.00\cdot 10^8 \text{ m/s}\right)\left(4\pi\cdot 10^{-7}\text{ T m/A}\right)} = 37.92$ W

ROUND:

(a) $B = 7.07\cdot 10^{-8}$ T

(b) $P = 37.9$ W

DOUBLE-CHECK: These values are consistent with the power output for a regular household light bulb.

31.71. **THINK:** The laser will apply a force to the particle. Assume the particle starts from rest and 2.00% of the laser light is absorbed. The laser applies a force to a known mass for a time interval Δt from which we can calculate the impulse applied by the laser. Use the values: $P = 500./192$ TW, $d = 2.00$ mm, $\rho = 2.00 \text{ g/cm}^3 = 2.00\cdot 10^3 \text{ kg/m}^3$, $\Delta t = 1.00\cdot 10^{-9}$ s, and $\varepsilon = 0.0200$.

SKETCH:

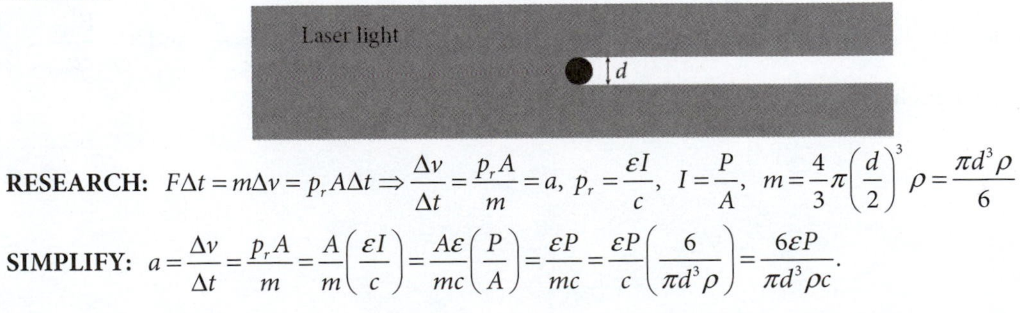

RESEARCH: $F\Delta t = m\Delta v = p_r A\Delta t \Rightarrow \dfrac{\Delta v}{\Delta t} = \dfrac{p_r A}{m} = a,\ p_r = \dfrac{\varepsilon I}{c},\ I = \dfrac{P}{A},\ m = \dfrac{4}{3}\pi\left(\dfrac{d}{2}\right)^3\rho = \dfrac{\pi d^3\rho}{6}$

SIMPLIFY: $a = \dfrac{\Delta v}{\Delta t} = \dfrac{p_r A}{m} = \dfrac{A}{m}\left(\dfrac{\varepsilon I}{c}\right) = \dfrac{A\varepsilon}{mc}\left(\dfrac{P}{A}\right) = \dfrac{\varepsilon P}{mc} = \dfrac{\varepsilon P}{c}\left(\dfrac{6}{\pi d^3\rho}\right) = \dfrac{6\varepsilon P}{\pi d^3\rho c}.$

CALCULATE: $a = \dfrac{6(0.0200)(500.\cdot 10^{12}\text{ W}/192)}{\pi(2.00\cdot 10^{-3}\text{ m})^3(2.00\cdot 10^3\text{ kg/m}^3)(3.00\cdot 10^8\text{ m/s})} = 2.0723\cdot 10^7\text{ m/s}^2.$

ROUND: To three significant figures, $a = 2.07\cdot 10^7\text{ m/s}^2$.

DOUBLE-CHECK: This is a reasonable result for 2% of the power of one very powerful laser.

31.75. **THINK:** To determine how long it takes the ice to melt, first determine how much total energy is required to melt the ice cube, and then determine the intensity of the microwaves at the location of the ice cube. To determine the number of photons hitting the ice per second, the energy of one photon must be calculated and compared to the total radiation power incident on the ice. $P_0 = 250.$ W, l = 2.00 cm, d = 10.0 cm, $\rho = 0.960\text{ g/cm}^3$, $\lambda = 10.0$ cm. The fraction of incident light absorbed by ice is $\varepsilon = 0.100$.

SKETCH:

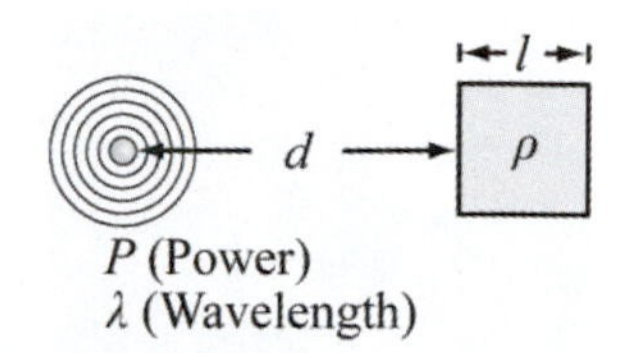

RESEARCH: The energy required to melt the ice is $c_f = 334$ J/g. The intensity of light at the cube is $I = P_0 / 4\pi d^2$. The radiation power incident on the cube is Il^2. The power absorbed by the cube is $P = \varepsilon I l^2 = E / t$. The mass of the ice is given by $m = \rho l^3$. The energy of one photon is given by $E_{ph} = hf = hc/\lambda$.

SIMPLIFY: The energy required to melt the ice is given by $E_m = mc_f = \rho l^3 c_f$. The power absorbed by the cube is given by:

$$P = \varepsilon I l^2 = \varepsilon l^2\left(\frac{P_0}{4\pi d^2}\right) = \frac{E}{t}.$$

The time required to melt the cube can be determined as follows:

$$\frac{E}{t} = \frac{\varepsilon l^2 P_0}{4\pi d^2} \Rightarrow t = \frac{4\pi E d^2}{\varepsilon l^2 P_0},\quad E = E_m = \rho l^3 c_f \quad \Rightarrow t = \frac{4\pi \rho l^3 c_f d^2}{\varepsilon l^2 P_0} = \frac{4\pi \rho l d^2 c_f}{\varepsilon P_0}.$$

The total power incident on the cube is given by: $Il^2 = P_0 l^2 / 4\pi d^2 = x$ J/s. x J/s is supplied by N photons of energy E_{ph} every second:

$$NE_{ph} = \frac{Nhc}{\lambda} \Rightarrow \frac{Nhc/\lambda}{\text{s}} = x\text{ J/s} \Rightarrow \frac{Nhc}{\lambda} = x\text{ J} \Rightarrow N = (x\text{ J})\frac{\lambda}{hc},$$

$$x\text{ J} = \frac{P_0 l^2}{4\pi d^2}\text{ s} \Rightarrow N = \frac{P_0 l^2 \lambda}{4\pi h d^2 c}\text{ s}.$$

CALCULATE: $t = \dfrac{4\pi(0.960\text{ g/cm}^3)(2.00\text{ cm})(10.0\text{ cm})^2(334\text{ J/g})}{(0.100)250.\text{ J/s}} = 3.223\cdot 10^4\text{ s} = 8.954\text{ h}$

$$N = \frac{(250.\text{ J/s})(2.00\text{ cm})^2(10.0\text{ cm})\text{ s}}{4\pi(6.626\cdot 10^{-34}\text{ J s})(10.0\text{ cm})^2(3.00\cdot 10^{10}\text{ cm/s})} = 4.003\cdot 10^{23}$$

ROUND: t = 8.95 h (or 8 hours 57 minutes), $N = 4.00\cdot 10^{23}$

DOUBLE-CHECK: The number of photons per second, N, is reasonable. The time is correct, although a real microwave will work much faster. This is because a real microwave is not a single point source. Also, a microwave has shielding which serves to reflect all waves hitting the walls, which keeps the intensity of the radiation high.

Multi-Version Exercises

31.77. $I = \dfrac{E_{\text{rms}}^2}{c\mu_0} \Rightarrow E_{\text{rms}} = \sqrt{Ic\mu_0} = \sqrt{(182.9 \text{ W/m}^2)(2.998 \cdot 10^8 \text{ m/s})(4\pi \cdot 10^{-7} \text{ Tm/A})} = 262.5 \text{ V/m}$

31.81. If I_0 is the intensity of the incoming sunlight, then the light passing through the first polarizer has intensity $I_1 = \frac{1}{2} I_0$. The intensity of the light passing through the second polarizer is given by $I_2 = I_1 \cos^2(\theta_2 - \theta_1)$, so that $I_2 = \frac{1}{2} I_0 \cos^2(\theta_2 - \theta_1)$. The reduction in intensity, then, is

$$R = \frac{I_0 - I_2}{I_0} = 1 - \tfrac{1}{2}\cos^2(\theta_2 - \theta_1) = 1 - \tfrac{1}{2}\cos^2(88.6° - 28.1°) = 87.9\%.$$

Chapter 32: Geometric Optics

Exercises

32.27. For plane mirrors, the object distance, $d_o = 1.00$ m, is always equal to the image distance, but the image is located behind the mirror. Therefore, the location of the image is $d_i = -d_o = -1.00$ m.

32.31. The focal length is given by $f = \frac{R}{2}$. For $R = -25.0$ cm, the focal length is: $f = \frac{R}{2} = \frac{(-25.0\text{ cm})}{2} \approx -12.5$ cm.

32.35. For an object a distance of $d_o = 2.0$ m in front of a convex mirror with magnification $m = 0.60$, the image distance is

$$m = -\frac{d_i}{d_o} \Rightarrow d_i = -md_o.$$

The focal length is:

$$\frac{1}{f} = \frac{1}{d_o} + \frac{1}{d_i} \Rightarrow \frac{1}{f} = \frac{1}{d_o} - \frac{1}{md_o} = \frac{m-1}{md_o} \Rightarrow f = \frac{md_o}{m-1} = \frac{(0.60)(2.0\text{ m})}{(0.60)-1} = -3.0\text{ m}.$$

32.39. The critical angle is given by $\sin\theta_c = n_2 / n_1$. The critical angles of the optical fiber in air, water and oil are:

$$\theta_{c,air} = \sin^{-1}\left(\frac{1.000}{1.50}\right) = 41.8°,\ \theta_{c,water} = \sin^{-1}\left(\frac{1.333}{1.50}\right) = 62.7° \text{ and } \theta_{c,oil} = \sin^{-1}\left(\frac{1.50}{1.50}\right) = 90.0°.$$

32.43. **THINK:** Since the normal line of the first surface bisects the opposite angle, the refracted ray must hit the other angled surface. Simple geometry must be utilized to determine all the angles involved. The index of refraction of air and the prism are $n_a = 1.00$ and $n_p = 1.23$, respectively.

SKETCH:

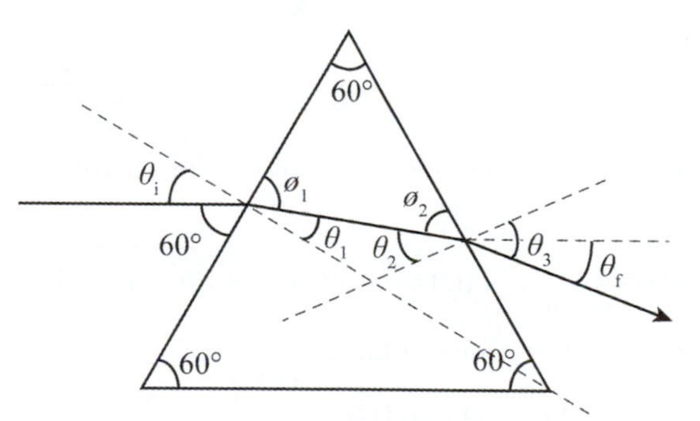

RESEARCH: Since the incident beam is parallel to the base, the incident angle is $\theta_i = 30°$. Snell's Law is used to determine refracted angles: $n_i \sin\theta_i = n_j \sin\theta_j$.

SIMPLIFY: At the first interface:

$$n_a \sin\theta_i = n_p \sin\theta_1 \Rightarrow \theta_1 = \sin^{-1}\left(\frac{n_a}{n_p}\sin\theta_i\right).$$

Based on the geometry shown in the figure above, $\phi_1 = 90° - \theta_1$ and $\phi_2 = 180° - (60° + \phi_1)$. Therefore,

$$\phi_2 = 120° - \phi_1 = 120° - (90° - \theta_1) = 30° + \theta_1.$$

Also, $\theta_2 = 90° - \phi_2$. Therefore,

$$\theta_2 = 90° - (30° + \theta_1) = 60° - \sin^{-1}\left(\frac{n_a}{n_p}\sin\theta_i\right).$$

At the second interface, Snell's Law is reapplied as the light exits the prism:

$$n_p \sin\theta_2 = n_a \sin\theta_3 \Rightarrow \theta_3 = \sin^{-1}\left(\frac{n_p}{n_a}\sin\theta_2\right) = \sin^{-1}\left[\frac{n_p}{n_a}\sin\left(60° - \sin^{-1}\left(\frac{n_a}{n_p}\sin\theta_i\right)\right)\right].$$

The change in direction is equal to the sum of the changes in angle at each interface:

$$\theta_f = (\theta_i - \theta_1) + (\theta_3 - \theta_2) = \theta_i - \theta_1 + \theta_3 - (60° - \theta_1) = \theta_i - 60° + \theta_3,$$

$$\theta_f = \theta_i - 60° + \sin^{-1}\left[\frac{n_p}{n_a}\sin\left(60° - \sin^{-1}\left(\frac{n_a}{n_p}\sin\theta_i\right)\right)\right].$$

CALCULATE: $\theta_f = (30°) - 60° + \sin^{-1}\left[\frac{(1.23)}{(1.00)}\sin\left(60° - \sin^{-1}\left(\frac{(1.00)}{(1.23)}\sin(30°)\right)\right)\right] = 16.322°$

ROUND: Rounding to three significant figures, $\theta_f = 16.3°$.

DOUBLE-CHECK: The change in direction depends on the initial incident angle, the refractive index of air and the refractive index of the prism, as expected. This is a reasonable angle for the ray of light to be deflected after going through a prism.

32.47. **THINK:** Fermat's Principle states that the path taken by a ray between two points in space is the path that takes the least amount of time. The law of reflection can be found by using this principle. To accomplish this, determine the time it takes for a ray to travel from one point to another by hitting the mirror. Using calculus, this time can be minimized and the law of reflection is recovered.

SKETCH:

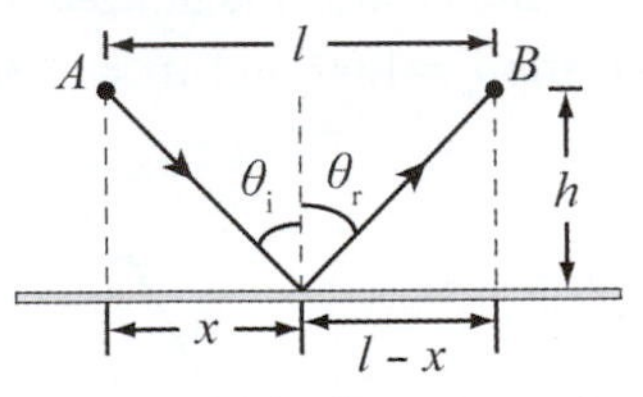

RESEARCH: The time it takes the ray to reach the mirror is $t = d/v$. To minimize the time, set $dt/dx = 0$.

SIMPLIFY: $t = \frac{d_1}{v} + \frac{d_2}{v} = \frac{1}{v}(d_1 + d_2) = \frac{n}{c}\left(\sqrt{h^2 + x^2} + \sqrt{h^2 + (l - x)^2}\right)$

The path of least time is determined from:

$$\frac{dt}{dx} = 0 = \frac{n}{c}\left[\frac{(1/2)2x}{\sqrt{h^2 + x^2}} - \frac{(1/2)2(l - x)}{\sqrt{h^2 + (l - x)^2}}\right] = \frac{n}{c}\left[\frac{x}{\sqrt{h^2 + x^2}} - \frac{(l - x)}{\sqrt{h^2 + (l - x)^2}}\right] = \frac{n}{c}(\sin\theta_i - \sin\theta_r)$$

$$\sin\theta_i - \sin\theta_r = 0 \Rightarrow \theta_i = \theta_r$$

CALCULATE: Not applicable.

ROUND: Not applicable.

DOUBLE-CHECK: The law of reflection was recovered using Fermat's Principle.

32.51. For the image to be twice the size of the object, the magnification is:

$$m = 2 = \left|\frac{d_i}{d_o}\right| \Rightarrow d_i = \pm 2d_o.$$

The spherical mirror equation is:

$$\frac{1}{d_o}+\frac{1}{d_i}=\frac{1}{f}=\frac{2}{R} \Rightarrow \frac{1}{d_o}\pm\frac{1}{2d_o}=\frac{2}{R}$$

$$\frac{(2\pm 1)}{2d_o}=\frac{2}{R} \Rightarrow d_o=\frac{(2\pm 1)R}{4}$$

The object can be placed at:

$$d_o=\frac{3}{4}R=\frac{3}{4}(20.0 \text{ cm})=15.0 \text{ cm or } d_o=\frac{R}{4}=\frac{(20.0 \text{ cm})}{4}=5.00 \text{ cm},$$

to produce an image that is twice the size of the object. If the object is placed at 15.0 cm, the image distance will be $d_i=2(15.0 \text{ cm})=30.0 \text{ cm}$. Since $d_i>0$, this image will be real. If the object is placed at 5.00 cm, the image distance will be $d_i=-2(5.00 \text{ cm})=-10.0 \text{ cm}$. Since $d_i<0$, this image will be virtual.

32.55. The critical angle is given by:

$$\sin\theta_c=\frac{n_2}{n_1} \Rightarrow \theta_c=\sin^{-1}\left(\frac{n_2}{n_1}\right).$$

The critical angle for the diamond-air interface is:

$$\theta_{c,a}=\sin^{-1}\left(\frac{1.000}{2.417}\right)=24.44°.$$

The critical angle for the diamond-water interface is:

$$\theta_{c,w}=\sin^{-1}\left(\frac{1.333}{2.417}\right)=33.47°.$$

Therefore, the critical angle in water is $9.03°$ greater than the critical angle in air.

32.59. **THINK:** The rays of light from the point are refracted before they reach the person, according to Snell's Law. Because the index of refraction of air is less than that of water, the image appears shallower. The point is $d=3.00 \text{ m}$ from the surface and $w=2.00 \text{ m}$ from the edge of the pool.

SKETCH:

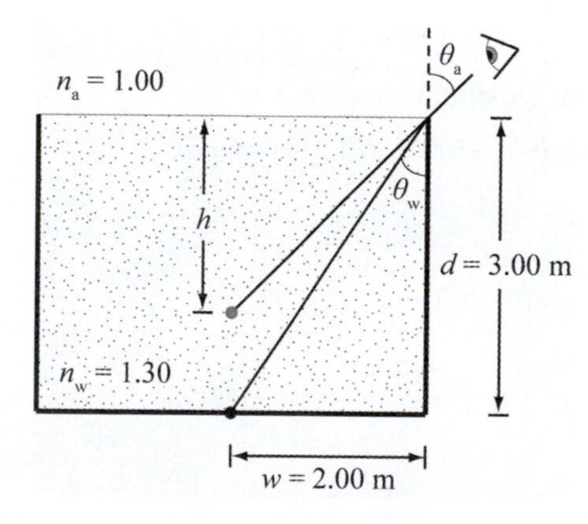

RESEARCH: The angle of the ray is given by Snell's Law: $n_w\sin\theta_w=n_a\sin\theta_a$. The triangles also relate the angles to the lengths:

$$\sin\theta_w=\frac{w}{\sqrt{w^2+d^2}} \text{ and } \sin\theta_a=\frac{w}{\sqrt{w^2+h^2}}.$$

SIMPLIFY: Combining the above equations gives:

$$n_w\sin\theta_w=\frac{n_w w}{\sqrt{w^2+d^2}}=n_a\sin\theta_a=\frac{n_a w}{\sqrt{w^2+h^2}}.$$

Solving for the apparent depth gives:

$$n_{\text{w}}\sqrt{w^2+h^2}=n_{\text{a}}\sqrt{w^2+d^2} \Rightarrow n_{\text{w}}^2\left(w^2+h^2\right)=n_{\text{a}}^2\left(w^2+d^2\right)$$

$$\Rightarrow n_{\text{w}}^2h^2=\left(n_{\text{a}}^2-n_{\text{w}}^2\right)w^2+n_{\text{a}}^2d^2 \Rightarrow h=\frac{1}{n_{\text{w}}}\sqrt{\left(n_{\text{a}}^2-n_{\text{w}}^2\right)w^2+n_{\text{a}}^2d^2}.$$

CALCULATE: $h=\dfrac{1}{(1.30)}\sqrt{\left((1.00)^2-(1.30)^2\right)(2.00\text{ m})^2+(1.00)^2(3.00\text{ m})^2}=1.92\text{ m}$

ROUND: Remaining at 3 significant figures, the apparent depth of the pool is $h=1.92$ m.

DOUBLE-CHECK: The apparent depth is less than the true depth of the pool, as expected.

32.63. The focal length of a liquid mirror is $f=\dfrac{g}{2\omega^2}$, where ω is the angular velocity of the rotating mirror. The angular velocity is:

$$\omega=\sqrt{\frac{g}{2f}}=\sqrt{\frac{\left(9.81\text{ m/s}^2\right)}{2(2.50\text{ m})}}=1.40\text{ rad/s}.$$

Multi-Version Exercises

32.65. $\dfrac{1}{f}=\dfrac{1}{d_{\text{i}}}+\dfrac{1}{d_{\text{o}}}\Rightarrow d_{\text{i}}=\dfrac{d_{\text{o}}f}{d_{\text{o}}-f}$

$$f=R/2 \text{ and } d_{\text{o}}=R+x_{\text{o}}\Rightarrow d_{\text{i}}=\frac{(R+x_{\text{o}})R/2}{R+x_{\text{o}}-R/2}=\frac{R^2+Rx_{\text{o}}}{R+2x_{\text{o}}}$$

$$d_{\text{i}}=R+x_{\text{i}}\Rightarrow x_{\text{i}}=d_{\text{i}}-R=\frac{R^2+Rx_{\text{o}}}{R+2x_{\text{o}}}-R=-\frac{Rx_{\text{o}}}{R+2x_{\text{o}}}=-11.7\text{ cm}$$

32.69. This is simply Snell's Law, but with a slight twist that the angles are measured relative to the interface between the two media, not relative to the normal.

$$\theta_1=90°-\varphi_1$$

$$\theta_2=\sin^{-1}(n_1\sin(\theta_1)/n_2)$$

$$\varphi_2=90°-\theta_2=90°-\sin^{-1}(n_1\sin(90°-\varphi_1)/n_2)$$

$$=90°-\sin^{-1}(1.329\sin(90°-61.07°)/1.310)=60.61°$$

Chapter 33: Lenses and Optical Instruments

Exercises

33.35. The distance to the image d_i is:

$$\frac{1}{d_o}+\frac{1}{d_i}=\frac{1}{f} \Rightarrow \frac{1}{d_i}=\frac{1}{f}-\frac{1}{d_o} \Rightarrow d_i=\frac{fd_o}{d_o-f}$$

Therefore, the magnification is

$$m=-\frac{\left(\dfrac{fd_o}{d_o-f}\right)}{d_o}=-\frac{f}{d_o-f}=-\frac{9.0 \text{ cm}}{6.0 \text{ cm}-9.0 \text{ cm}}=3.0.$$

33.39. **THINK:** The object height is $h_o=2.5$ cm, and is $d_o=5.0$ cm from a converging lens of focal length $f=3.0$ cm. The thin lens equation can be used to find the image distance and the magnification can be found from this.

SKETCH:

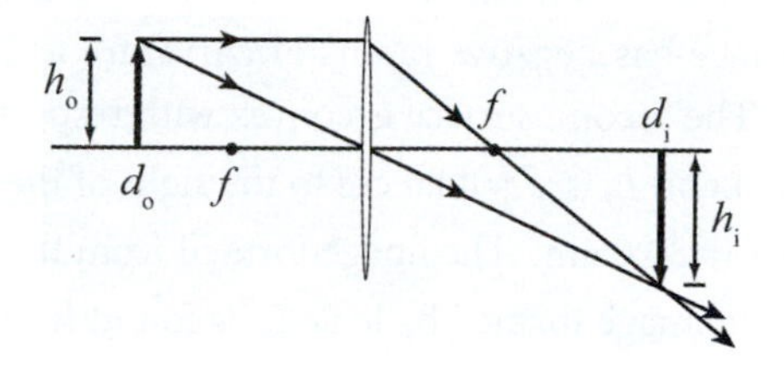

RESEARCH: The magnification m is: $m=-d_i/d_o$. The thin lens equation is: $1/d_o+1/d_i=1/f$.

SIMPLIFY: The image distance is: $\frac{1}{d_o}+\frac{1}{d_i}=\frac{1}{f} \Rightarrow \frac{1}{d_i}=\frac{1}{f}-\frac{1}{d_o} \Rightarrow d_i=\frac{fd_o}{d_o-f}$. Therefore, the magnification is: $m=-\frac{f}{d_o-f}$.

CALCULATE: The magnification is:] $m=-\frac{3.0 \text{ cm}}{5.0 \text{ cm}-3.0 \text{ cm}}=-1.5$. Since the magnification is negative, the image is inverted and since $|m|>1$, the image is enlarged.

ROUND: To two significant figures, the magnification of the image is $m=-1.5$.

DOUBLE-CHECK: The ray tracing shown above confirms that the image is inverted and enlarged. As seen in Table 33.1, this is what it is expected for $f<d_o<2f$.

33.43. The angular magnification of a magnifying glass is approximately $m_\theta \approx d_{\text{near}}/f$. With a given focal length of $f=5.0$ cm, and assuming a near point of $d_{\text{near}}=25$ cm, the magnifying power of this lens with the object placed at the near point is $m_\theta=\frac{25 \text{ cm}}{5.0 \text{ cm}}=5.0$.

33.47. The total magnification is the product of the magnification after passing through the first lens, m_1, and the magnification of the second lens, m_2. Magnification is $m=-d_i/d_o=h_i/h_o$. The focal length of each lens is $f=5.0$ cm, and the distance that the insect is from the first lens is $d_{o,1}=10.0$ cm. Using the thin lens equation the image distance from the first lens is:

$$d_{i,1}=\frac{fd_{o,1}}{d_{o,1}-f}=\frac{(5.0 \text{ cm})(10.0 \text{ cm})}{(10.0 \text{ cm}-5.0 \text{ cm})}=10. \text{ cm}.$$

Then $m_1 = -d_{i,1}/d_{o,1} = -(10.0 \text{ cm})/(10.0 \text{ cm}) = -1.00$. This image is inverted, but the size does not change. This image acts as an object for the second lens, and is a distance $d_{o,2} = L - d_{i,1}$ from the second lens, where L is the separation distance of the two lenses, $L = 12$ cm. Using the thin lens equation, the image distance from the second lens is:

$$d_{i,2} = \frac{fd_{o,2}}{d_{o,2} - f} = \frac{(5.0 \text{ cm})(12.0 \text{ cm} - 10.0 \text{ cm})}{((12.0 \text{ cm} - 10.0 \text{ cm}) - 5.0 \text{ cm})} = -3.333 \text{ cm}.$$

Then $m_2 = -d_{i,2}/d_{o,2} = -(-3.333 \text{ cm})/(12.0 \text{ cm} - 10.0 \text{ cm}) = 1.667$. This image is oriented the same way as the object (inverted). The final magnification of the insect is $m = m_1 m_2 = (-1.0)(1.667) = -1.667$. Therefore, the final image of the insect has a size of

$$h_i = mh_o = -1.667(5.0 \text{ mm}) = -8.333 \text{ mm} \approx -8.3 \text{ mm}.$$

With respect to the original insect, the final image is enlarged, inverted (since magnification is negative) and virtual (since $d_{i,2}$ is negative).

33.51. The object is $h_{o,1} = 10.0$ cm tall and is located $d_{o,1} = 30.0$ cm to the left of the first lens. Lens L_1 is a biconcave lens with index of refraction $n = 1.55$ and has a radius of curvature of 20.0 cm for both surfaces. The first surface has negative radius of curvature as its surfaces is concave with respect to the object: $R_1 = -20.0$ cm. The second surface is convex with respect to the object, so its radius of curvature is positive: $R_2 = 20.0$ cm. Lens L_2 is $d = 40.0$ cm to the right of the first lens L_1. Lens L_2 is a converging lens with a focal length of $f_2 = 30.0$ cm. The image formed from the first lens acts as the object for the second lens. The position of the image formed by lens L_1 is found from the Lens Maker's Formula with the thin lens approximation: $\frac{1}{d_o} + \frac{1}{d_i} = (n-1)\left(\frac{1}{R_1} - \frac{1}{R_2}\right)$. Then the image distance is:

$$d_{i,1} = \left[(1.55 - 1)\left(\frac{1}{-20.0 \text{ cm}} - \frac{1}{20.0 \text{ cm}}\right) - \frac{1}{30.0 \text{ cm}}\right]^{-1} = -11.32 \text{ cm}.$$

This image is on the left side of lens L_1 and it acts as the object for lens L_2. The object distance for lens L_2 is $d_{o,2} = d + |d_{i,1}| = 40.0 \text{ cm} + 11.32 \text{ cm} = 51.32 \text{ cm}$ from lens L_2. From the thin lens equation, the image distance of lens L_2 is: $\frac{1}{d_{i,2}} = \frac{1}{f_2} - \frac{1}{d_{o,2}} \Rightarrow d_{i,2} = \left(\frac{1}{30.0 \text{ cm}} - \frac{1}{51.32 \text{ cm}}\right)^{-1} = 72.2$ cm. Since this distance is positive, the final image is real and is 72.2 cm to the right of lens L_2, or $30.0 + 40.0 + 72.2 = 142$ cm to the right of the original object. The focal length of lens L_1 is required for a ray diagram. The Lens Maker's Formula gives:

$$f_1 = \left[(n-1)\left(\frac{1}{R_1} - \frac{1}{R_2}\right)\right]^{-1} = \left[(1.55 - 1)\left(\frac{1}{-20.0 \text{ cm}} - \frac{1}{20.0 \text{ cm}}\right)\right]^{-1} \Rightarrow f_1 = -18.2 \text{ cm}.$$

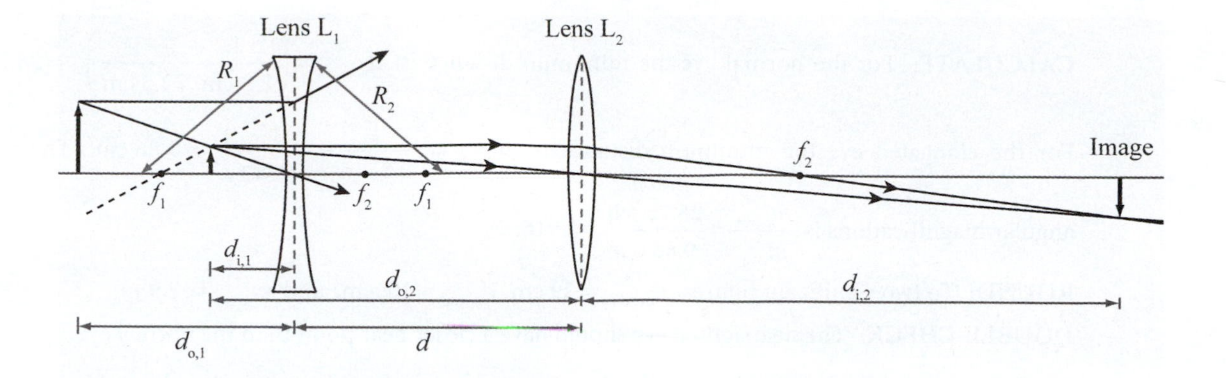

33.55. Jane's near point is $d_{\text{near}} = 125$ cm and the computer screen is $d_o = 40.$ cm from her eye. Use the thin lens equation: $\frac{1}{d_o}+\frac{1}{d_i}=\frac{1}{f}$. Also, the power of a lens (in diopters) is $P=\frac{1}{f}$ where f is in meters.

(a) The object distance is just the distance to the computer screen: $d_o = 40.$ cm.

(b) The image distance is Jane's near point: $d_i = -d_{\text{near}} = -125$ cm. It is negative because the image appears on the same side of the eye as the object (the image is virtual).

(c) The focal length is $f = \left[\frac{1}{40.\text{ cm}} + \frac{1}{-125\text{ cm}}\right]^{-1} = 59$ cm.

(d) Jane's near point is 1.25 m; to read the computer screen at $d_o = 0.40$ m, the image must be located at the near point, $d_i = -d_{\text{near}}$. The power of this corrective lens would be:

$$P = \frac{1}{f} = \frac{1}{d_o} + \frac{1}{d_i} = \frac{1}{0.40\text{ m}} + \frac{1}{-1.25\text{ m}} = +1.7\text{ Diopter.}$$

(e) Since the focal length is positive, the corrective lens is converging.

33.59. **THINK:** As objects are moved closer to the human eye the focal length of the lens decreases. The shortest focal length is $f_{\text{min}} = 2.3$ cm. The thin lens equation can be used to determine the closest one can bring an object to a normal human eye, $d_{\text{o,norm}}$, and still have the image of the object projected sharply onto the retina, which is $d_{\text{i,norm}} = 2.5$ cm. behind the lens. A near sighted human eye has the same f_{min} but has a retina that is 3.0 cm behind the lens. The thin lens equation can be used to determine the closest one can bring an object to this nearsighted human eye, $d_{\text{o,near}}$, and still have the image of the object projected sharply on the retina at $d_{\text{i,near}} = 3.0$ cm.

SKETCH: Provided with the problem.

RESEARCH: In each case the object is in front of the lens, and the image is formed behind the lens, so both d_o and d_i are positive. The thin lens equation is: $1/d_o + 1/d_i = 1/f$. The angular magnification is given by $m_\theta \approx d_{\text{near}}/f$.

SIMPLIFY: $\frac{1}{d_o} = \frac{1}{f} - \frac{1}{d_i} \Rightarrow d_o = \left(\frac{1}{f} - \frac{1}{d_i}\right)^{-1}$. The ratio of angular magnifications is:

$$\frac{m_{\text{norm}}}{m_{\text{near}}} = \left(\frac{d_{\text{near, norm}}}{f_{\text{norm}}}\right)\left(\frac{f_{\text{near}}}{d_{\text{near, near}}}\right).$$

Since the object is placed at the near point for the image to form on the retina and $f_{\text{near}} = f_{\text{norm}} = f_{\text{min}}$, this becomes

$$\frac{m_{\text{norm}}}{m_{\text{near}}} = \left(\frac{d_{\text{o, norm}}}{d_{\text{o, near}}}\right)\left(\frac{f_{\text{near}}}{f_{\text{norm}}}\right) = \left(\frac{d_{\text{o, norm}}}{d_{\text{o, near}}}\right).$$

CALCULATE: For the normal eye the minimum distance is: $d_{\text{o, norm}} = \left(\frac{1}{2.3 \text{ cm}} - \frac{1}{2.5 \text{ cm}} \right)^{-1} = 28.75 \text{ cm}.$

For the elongated eye the minimum distance is: $d_{\text{o, near}} = \left(\frac{1}{2.3 \text{ cm}} - \frac{1}{3.0 \text{ cm}} \right)^{-1} = 9.86 \text{ cm}.$ The ratio of angular magnifications is $\frac{m_{\text{norm}}}{m_{\text{near}}} = \frac{28.75 \text{ cm}}{9.86 \text{ cm}} = 2.916.$

ROUND: To two significant figures, $d_{\text{o, norm}} = 29 \text{ cm}$, $d_{\text{o, near}} = 10. \text{ cm}$, and $m_{\text{norm}} = 2.9\, m_{\text{near}}$.

DOUBLE-CHECK: The nearsighted eye should have a closer near point than the normal eye.

33.63. The focal length of the original lens is fixed at $f = 60. \text{ mm}$ and the zoom lens has a variable focal length. The object is a distance $d_{\text{o}} = \infty$ from the lens. Using the thin lens equation for the original lens shows $\frac{1}{d_0} + \frac{1}{d_i} = \frac{1}{f} \Rightarrow \frac{1}{\infty} + \frac{1}{d_i} = \frac{1}{f} \Rightarrow f = d_i$, the image appears at $d_{\text{i}} = f = 60. \text{ mm}$. With the zoom lens set to a focal length of $f' = 240. \text{ mm}$, the image appears at $d_{\text{i}}' = f' = 240. \text{ mm}$. The ratio of magnifications of each lens is:

$$\frac{m_{\text{original}}}{m_{\text{zoom}}} = \frac{-d_{\text{i}} / d_{\text{o}}}{-d_{\text{i}}' / d_{\text{o}}} = \frac{d_{\text{i}}}{d_{\text{i}}'} = \frac{60. \text{ mm}}{240. \text{ mm}} = \frac{1}{4.0}.$$

The zoom lens (at $f' = 240. \text{ mm}$) produces an image that is 4.0 times the size of the image produced by the original $f = 60. \text{ mm}$ lens.

33.67. The focal length of the objective lens is $f_{\text{o}} = 7.00 \text{ mm}$. The distance between the objective lens and the eyepiece lens is $L = 20.0 \text{ cm}$. The magnitude of the magnification is $|m| = 200$. The viewing distance to the image is $d_{\text{i, 2}} = 25.0 \text{ cm}$. The focal length of the eyepiece, f_{e}, can be found from the equation for the magnification of a microscope: $|m| = \frac{d_{\text{i, 1}} d_{\text{i, 2}}}{d_{\text{o, 1}} d_{\text{o, 2}}} = \frac{(25.0 \text{ cm})L}{f_{\text{o}} f_{\text{e}}}$. The focal length of the eyepiece is:

$$f_{\text{e}} = \frac{(0.250 \text{ m})L}{f_{\text{o}} |m|} = \frac{(25.0 \text{ cm})(20.0 \text{ cm})}{(0.700 \text{ cm})(200.)} = 3.57 \text{ cm}.$$

The best choice is the lens marked with a 4.00 cm focal length.

33.71. The angular magnification of a refracting telescope is $m_\theta = -f_{\text{o}} / f_{\text{e}}$. With an objective lens of focal length $f_{\text{o}} = 100. \text{ cm}$, and an eyepiece of focal length $f_{\text{e}} = 5.00 \text{ cm}$, the magnification of this telescope is: $m_\theta = -100. \text{ cm} / 5.00 \text{ cm} = -20.0$, where the negative sign indicates that the image is inverted.

33.75. **THINK:** The telescope is a refracting telescope with a magnification of $|m| = 180$. It is adjusted for a relaxed eye when the two lenses are $L = 1.30 \text{ m}$ apart. The telescope is designed such that the image formed by the objective lens (which appears at its focal length f_{o}) lies at the focal length of the eyepiece. Then the distance L between the two lenses is the sum of the two focal lengths: $L = f_{\text{o}} + f_{\text{e}}$. The magnification equation for a telescope can be used to find the focal length of each the objective lens, f_{o}, and the eyepiece lens, f_{e}.

SKETCH:

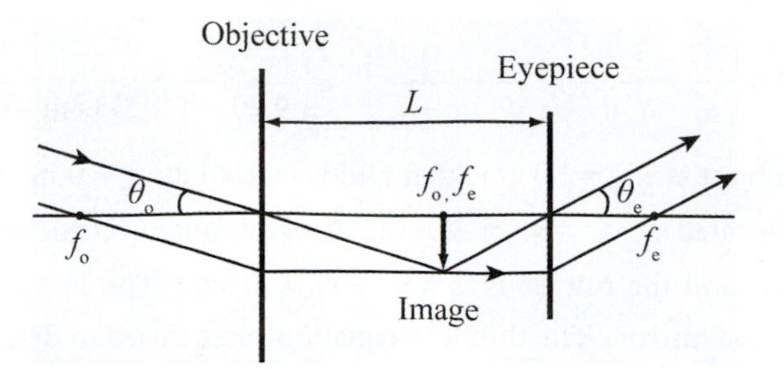

RESEARCH: The angular magnification of a refracting telescope is $m_\theta = -f_o / f_e$. With two equations and two unknowns, the two focal lengths f_o and f_e can be determined.

SIMPLIFY: $|m_\theta| = f_o / f_e \Rightarrow f_o = f_e|m_\theta|$, $L = f_o + f_e = f_e|m_\theta| + f_e \Rightarrow f_e = \dfrac{L}{1+|m_\theta|}$.

CALCULATE: $f_e = \dfrac{1.30 \text{ m}}{1+(180)} = 7.182 \text{ mm}$, $f_o = (7.182 \text{ mm})(180) = 1.293 \text{ m}$.

ROUND: Rounding to two significant figures, the focal length of the eyepiece is 7.2 mm. The focal length of the objective lens is 1.3 m.

DOUBLE-CHECK: The focal point of the objective should be much greater the focal length of the eyepiece for a refracting telescope.

33.79. The eyeglasses of a near sighted person use diverging lenses and create virtual images of objects for the near sighted wearer. When a normal person wears these eyeglasses, the person with normal vision will only be able to focus on these virtual images if they fall within the focusable distances of a normal eye, which is from 25 cm out to infinity. Since only the most distant objects can be focused on, the objects at infinity must be making virtual images at the normal near point of 25 cm. This will happen when: $\dfrac{1}{f} = \dfrac{1}{d_o} + \dfrac{1}{d_i} = \dfrac{1}{\infty} + \dfrac{1}{-0.25 \text{ m}} = -4.0 \text{ m}^{-1}$. Note that d_i is negative because the image is virtual. The prescription strength of the eyeglasses is about -4.0 diopter.

33.83. The object is $d_o = 6.0$ cm away from a thin lens of focal length $f = 9.0$ cm. The image distance d_i is determined from the thin lens equation: $\dfrac{1}{d_o} + \dfrac{1}{d_i} = \dfrac{1}{f}$. Therefore,

$$d_i = \left(\frac{1}{f} - \frac{1}{d_o}\right)^{-1} = \left(\frac{1}{9.0 \text{ cm}} - \frac{1}{6.0 \text{ cm}}\right)^{-1} = -18 \text{ cm}.$$

The image is 18 cm from the lens, and on the same side of the lens as the object (the negative sign indicates that it is a virtual image).

33.87. The magnifying glass is a converging lens. If you hold the magnifying glass at $d_i = 9.20$ cm above your desk you can form a real image on the desk of a light directly overhead. The distance from the light to the table is $h = 235$ cm.

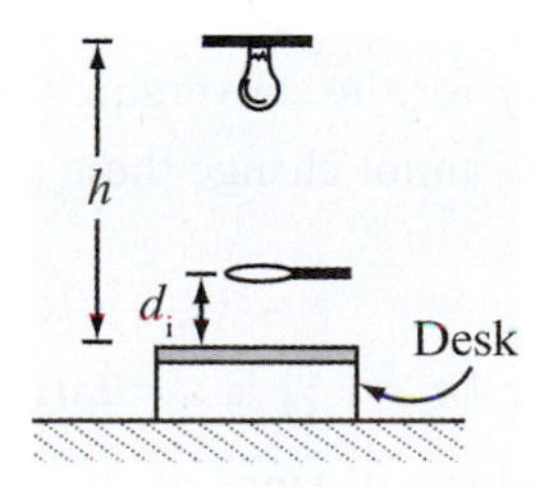

Using the thin lens equation, where $d_o = h - d_i$, the focal length of the magnifying glass is:

$$f=\left(\frac{1}{d_{\mathrm{i}}}+\frac{1}{d_{\mathrm{o}}}\right)^{-1}=\left(\frac{1}{d_{\mathrm{i}}}+\frac{1}{h-d_{\mathrm{i}}}\right)^{-1}=\left(\frac{1}{9.20\text{ cm}}+\frac{1}{235\text{ cm}-9.20\text{ cm}}\right)^{-1}=8.84\text{ cm}.$$

33.91. **THINK:** The object is $h_{\mathrm{o},1}=2.0$ cm high and is located at $x_0=0$ m. A converging lens with focal length $f=50.$ cm is located at $x_{\mathrm{L}}=d_{\mathrm{o},1}=30.$ cm. A plane mirror is located at $x_{\mathrm{m}}=70.$ cm, so the distance between the lens and the mirror is $L=x_{\mathrm{m}}-x_{\mathrm{L}}=40.$ cm. The image formed by the lens will act as the object for the plane mirror. The thin lens equation can be used to determine the position $x_{\mathrm{i},2}$ and the size $h_{\mathrm{i},2}$ of the final image.

SKETCH:

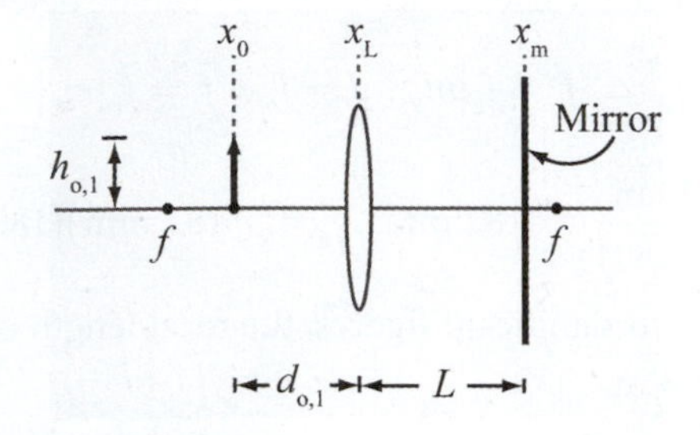

RESEARCH: The thin lens equation is $\frac{1}{d_{\mathrm{o}}}+\frac{1}{d_{\mathrm{i}}}=\frac{1}{f}$. The magnification of a lens is $m=h_{\mathrm{i}}/h_{\mathrm{o}}=-d_{\mathrm{i}}/d_{\mathrm{o}}$. For plane mirrors, $|d_{\mathrm{i}}|=|d_{\mathrm{o}}|$ and $h_{\mathrm{i}}=h_{\mathrm{o}}$.

SIMPLIFY: When the thin lens equation is rearranged to solve for the image distance, it becomes $d_{\mathrm{i}}=\frac{fd_{\mathrm{o}}}{d_{\mathrm{o}}-f}$. The image produced by the lens is located a distance of $d_{\mathrm{i},1}=\frac{fd_{\mathrm{o},1}}{d_{\mathrm{o},1}-f}$ from the lens. Since $f>d_{\mathrm{o},1}$, $d_{\mathrm{i},1}$ will be negative, and therefore, on the same side of the lens as the object. This image acts as the object for the mirror, and is a distance $d_{\mathrm{o},2}=L+|d_{\mathrm{i},1}|$ from the plane mirror. The final image is the image created by the plane mirror, and will appear $d_{\mathrm{i},2}=|d_{\mathrm{o},2}|$ to the right of the mirror. The final image position is given by $x_{\mathrm{i},2}=x_{\mathrm{m}}+d_{\mathrm{i},2}$. Since the mirror does not change the height of the image, the magnification is due to the lens, and the final height of the image is $h_{\mathrm{i},2}=-\frac{d_{\mathrm{i},1}}{d_{\mathrm{o},1}}h_{\mathrm{o},1}$.

CALCULATE: The image distance for the lens is $d_{\mathrm{i},1}=\frac{(30.\text{ cm})(50.\text{ cm})}{(30.\text{ cm})-(50.\text{ cm})}=-75$ cm. The object distance for the plane mirror is $d_{\mathrm{o},2}=40.\text{ cm}+|-75\text{ cm}|=115$ cm. Therefore, the position of the final image is $x_{\mathrm{i},2}=70.\text{ cm}+115\text{ cm}=185$ cm. The size of the final image is $h_{\mathrm{i},2}=-\frac{(-75\text{ cm})(2.0\text{ cm})}{(30.\text{ cm})}=5.0$ cm.

ROUND: To two significant figures, the final image is $x_{\mathrm{i},2}=190$ cm to the right of the object and the size of the final image is $h_{\mathrm{i},2}=5.0$ cm.

DOUBLE-CHECK: Since $d_{\mathrm{o}}<f$ for the converging lens, the image of the lens must be virtual, enlarged and upright. The plane mirror cannot change these attributes, so the calculated results agree with these expectations $(h_{\mathrm{i},2}>h_{\mathrm{o},1}>0)$.

33.95. The diameter of the glass marble $(n_{\mathrm{g}}=1.5)$ is $d=2.0\text{ in}=5.1\text{ cm}$. The radius of curvature of the marble is then $R=d/2$. Holding the marble a distance of $d_{\mathrm{o},1}=1.0\text{ ft}=30.\text{ cm}$ from your face, the distance of the image formed by the first side of the marble is:

$$\frac{1}{d_{o,1}}+\frac{n_g}{d_{i,1}}=\frac{2(n_g-1)}{d} \Rightarrow d_{i,1}=\frac{n_g d d_{o,1}}{2d_{o,1}(n_g-1)-d}=\frac{(1.5)(5.1\text{ cm})(30.\text{ cm})}{2(30.\text{ cm})(1.5-1)-(5.1\text{ cm})}=9.217\text{ cm}.$$

This image acts as the object for the second surface, for which the radius of curvature is negative (concave), $d=-5.1$ cm. Since $d_{i,1}>d$, the image for the second surface appears past it, so $d_{o,2}=d_{i,1}-d$. Therefore, the final image distance can be computed as follows.

$$\frac{n_g}{d_{o,2}}+\frac{1}{d_{i,2}}=\frac{2(1-n_g)}{d} \Rightarrow d_{i,2}=\frac{dd_{o,2}}{2d_{o,2}(1-n_g)+dn_g}=\frac{d(d_{i,1}-d)}{2(d_{i,1}-d)(1-n_g)+dn_g},$$

$$d_{i,2}=\frac{(-5.1\text{ cm})(9.217\text{ cm}+5.1\text{ cm})}{2(9.217\text{ cm}+5.1\text{ cm})(1-1.5)+(-5.1\text{ cm})(1.5)}=3.324\text{ cm}=1.3\text{ in}.$$

The magnification is $m=-\frac{d_{i,1}d_{i,2}}{d_{o,1}d_{o,2}}=-\frac{d_{i,1}d_{i,2}}{d_{o,1}(d_{i,1}-d)}=-\frac{(9.217\text{ cm})(3.324\text{ cm})}{(30.48\text{ cm})(9.217\text{ cm}+5.1\text{ cm})}=-0.070,$ where the negative sign indicates that the image is inverted.

33.99. **THINK:** The converging lens has a focal length $f_L=50.0$ cm. It is $L=175$ cm to the left of a metallic sphere. This metallic sphere acts as a convex mirror of radius $R=-100.$ cm (the radius of curvature of a diverging mirror is negative) and focal length $f_m=R/2=-50.0$ cm. The object of height, $h=20.0$ cm, is a distance $d_{o,1}=30.0$ cm to the left of the lens. The thin lens equation, the mirror equation, and the magnification for a system of optical elements can be used to find the height of the image formed by the metallic sphere, $h_{i,2}$. The image formed by the lens acts as the object for the mirror.

SKETCH:

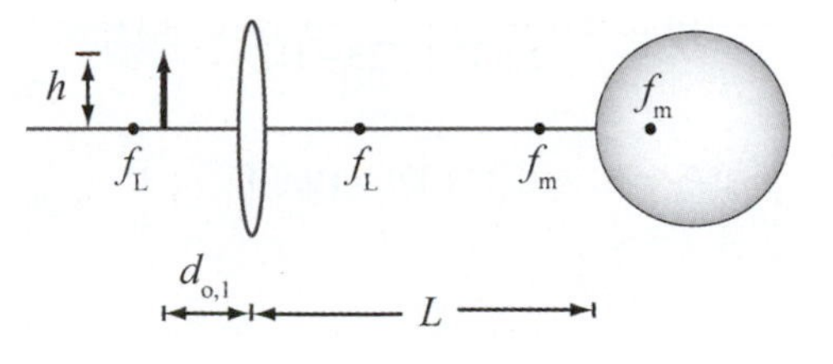

RESEARCH: The thin lens equation is $\frac{1}{f}=\frac{1}{d_i}+\frac{1}{d_o}$, The magnification (for lenses and mirrors) is $m=h_i/h_o=-d_i/d_o$. The total magnification m is the product of the magnification of the lens and the mirror: $m=m_L m_m$.

SIMPLIFY: For the lens, the thin lens equation can be rearranged as: $d_{i,1}=\frac{d_{o,1}f_L}{d_{o,1}-f_L}$. Since $f_L>d_{o,1}$, $d_{i,1}$ is negative, so the image is on the same side as the object (the image is virtual). This image acts as the object for the mirror at a distance of $d_{o,2}=L+|d_{i,1}|$ from the metallic sphere. The location of the image produced from the sphere is $d_{i,2}=\frac{d_{o,2}f_m}{d_{o,2}-f_m}$. The final image height is

$$h_{i,2}=mh=(m_L)(m_m)h=\left(\frac{d_{i,1}}{d_{o,1}}\right)\left(\frac{d_{i,2}}{d_{o,2}}\right)h=\left(\frac{\frac{d_{o,1}f_L}{d_{o,1}-f_L}}{d_{o,1}}\right)\left(\frac{\frac{d_{o,2}f_m}{d_{o,2}-f_m}}{d_{o,2}}\right)h=\frac{f_L f_m h}{(d_{o,1}-f_L)\left(L+\left|\frac{d_{o,1}f_L}{d_{o,1}-f_L}\right|-f_m\right)}.$$

CALCULATE: $h_{i,2} = \dfrac{(50.0\text{ cm})(-50.0\text{ cm})(20.0\text{ cm})}{(30.0\text{ cm}-50.0\text{ cm})\left(175\text{ cm}+\left|\dfrac{(30.0\text{ cm})(50.0\text{ cm})}{30.0\text{ cm}-50.0\text{ cm}}\right|-(-50.0\text{ cm})\right)} = 8.3333\text{ cm}$

ROUND: To three significant figures, the height of the image formed by the metallic sphere is $h_{i,2} = 8.33$ cm.

DOUBLE-CHECK: It is expected that $h_{i,2} < h$. For a converging lens, an image produced by an object placed within the focal length of the lens is enlarged, virtual and upright. For a diverging mirror, the image is always virtual, upright and reduced. Therefore, the height of the final image should be less than the height of the object since both the lens and the mirror act to reduce it.

Multi-Version Exercises

33.103. $P_{\text{water}} = \dfrac{1}{f} = \dfrac{n_{\text{lens}} - n_{\text{water}}}{n_{\text{water}}}\left(\dfrac{1}{R_1} - \dfrac{1}{R_2}\right)$

$P_{\text{air}} = (n_{\text{lens}} - 1)\left(\dfrac{1}{R_1} - \dfrac{1}{R_2}\right)$

Take the ratio:

$\dfrac{P_{\text{water}}}{P_{\text{air}}} = \dfrac{n_{\text{lens}} - n_{\text{water}}}{n_{\text{water}}(n_{\text{lens}} - 1)}$

Solve for power in water:

$P_{\text{water}} = P_{\text{air}}\dfrac{n_{\text{lens}} - n_{\text{water}}}{n_{\text{water}}(n_{\text{lens}} - 1)} = (4.29\text{ D})\dfrac{1.723-1.333}{1.333(1.723-1)} = 1.74\text{ D}$

33.106. $|m_\theta| = \dfrac{f_o}{f_e} = \dfrac{P_e}{P_o} \Rightarrow P_e = |m_\theta| P_o = 81.4(0.234\text{ D}) = 19.0\text{ D}$

Chapter 34: Wave Optics

Exercises

34.23. The wavelength of EM radiation in a medium with a refractive index n is $\lambda = \lambda_o / n$ where λ_o is the wavelength of light in a vacuum. Similarly the speed of light in the medium is $v = c / n$.

(a) The wavelength of a helium-neon laser in Lucite is $\lambda = \frac{\lambda_o}{n} = \frac{632.8 \text{ nm}}{1.500} = 421.9 \text{ nm}$

(b) The speed of light in the Lucite is $v = \frac{c}{n} = \frac{2.998 \cdot 10^8 \text{ m/s}}{1.500} = 1.999 \cdot 10^8 \text{ m/s}.$

34.27. For a Young's interference experiment, the maxima of the interference pattern is located at $y = m\lambda L / d.$ Substituting $m = 1$ for the first maximum intensity yields $y = \lambda L / d.$ Therefore, the distance between the slits and the screen is $L = \frac{yd}{\lambda} = \frac{(5.40 \cdot 10^{-3} \text{ m})(0.100 \cdot 10^{-3} \text{ m})}{540 \cdot 10^{-9} \text{ m}} = 1.0 \text{ m}.$

34.31. The minima of the interference pattern produced by a thin film is related to its thickness by $2t = m\lambda / n.$ The first dark band which corresponds to the thinnest and is when $m = D$ or when the thickness is much less than $\lambda.$ The next dark bands are for $m = 1$ and $m = 2.$ Therefore, the thicknesses that produces the dark bands are $t_1 = \frac{1}{2}\frac{\lambda}{n} = \frac{550 \text{ nm}}{2(1.32)} = 208 \text{ nm} \approx 210 \text{ nm}$ and $t_2 = \frac{2}{2}\frac{\lambda}{n} = \frac{\lambda}{n} = \frac{550 \text{ nm}}{1.32} = 417 \text{ nm} \approx 420 \text{ nm}.$

34.35. **THINK:** It is assumed that the refractive index of mica is independent of wavelength. In order to solve the problem, the condition for destructive interference of the reflected light is required. The film has thickness $t = 1.30$ μm. The wavelengths of interest are 433.3 nm, 487.5 nm, 557.1 nm, 650.0 nm, and 780.0 nm.

SKETCH:

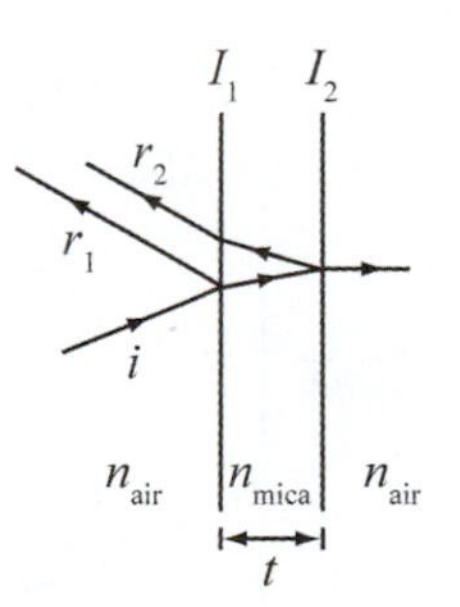

RESEARCH: Since $n_{air} < n_{mica}$, the light reflected by the first interface I_1 has a phase change of 180°. The light reflected by the second interface (I_2) has no phase change. The condition for destructive interference in the reflected light is

$$m\frac{\lambda_{air}}{n} = 2t \quad (m = 0, 1, 2, \ldots).$$

For two adjacent wavelengths with $\lambda_2 > \lambda_1$, $m_2 = m_1 - 1.$ Therefore,

$$m_1 = \frac{2nt}{\lambda_1} \text{ and } m_1 - 1 = \frac{2nt}{\lambda_2}.$$

SIMPLIFY: Solving these two equations for the refractive index n gives:

$$\frac{2nt}{\lambda_1} = \frac{2nt}{\lambda_2} + 1 \Rightarrow 2nt\left(\frac{1}{\lambda_1} - \frac{1}{\lambda_2}\right) = 1 \Rightarrow n = \frac{\lambda_1\lambda_2}{2t(\lambda_2 - \lambda_1)}.$$

CALCULATE: Choosing two adjacent wavelengths, $\lambda_1 = 433.3$ nm and $\lambda_2 = 487.5$ nm and substituting into the above equation yields $n = \dfrac{(433.3\cdot 10^{-9}\text{ m})(487.5\cdot 10^{-9}\text{ m})}{2(1.30\cdot 10^{-6}\text{ m})((487.5-433.3)\cdot 10^{-9}\text{ m})} = 1.499.$

ROUND: To three significant figures, the refractive index of the mica is $n = 1.50$.

DOUBLE-CHECK: Choosing another two adjacent wavelengths, $\lambda_1 = 650.0$ nm and $\lambda_2 = 780.0$ nm, the refractive index is found to be

$$n = \frac{(650.0\cdot 10^{-9}\text{ m})(780.0\cdot 10^{-9}\text{ m})}{2(1.30\cdot 10^{-6}\text{ m})((780.0-650.0)\cdot 10^{-9}\text{ m})} = 1.50.$$

This is in agreement with the previous result.

34.39. The number of fringes is given by the ratio of the path difference and the wavelength, that is, $N = \Delta x/\lambda = 2d/\lambda = 2(0.381\cdot 10^{-3}\text{ m})/449\cdot 10^{-9}\text{ m} = 1697 \approx 17.0\cdot 10^2$.

34.43. The minima of a single slit width are given by: $a\sin\theta = m\lambda$. The first minimum corresponds to $m = 1$, $a\sin\theta = \lambda$. Minima do not appear for $\theta = 90°$ or larger angles. Solving for a gives: $a = \lambda/\sin\theta \Rightarrow a = \lambda = 600.$ nm. If a is any larger θ would be less than $90°$, since $\sin\theta = \lambda/a$.

34.47. The angular resolution is given by Rayleigh's Criterion $\theta_R = \sin^{-1}(1.22\lambda/d)$. For the Hubble Space Telescope the value is $\theta_R = \sin^{-1}(1.22(450.\cdot 10^{-9}\text{ m})/2.40\text{ m}) = 1.31\cdot 10^{-5}$ degrees. For the Keck Telescope the value is $\theta_R = \sin^{-1}(1.22(450.\cdot 10^{-9}\text{ m})/10.0\text{ m}) = 3.15\cdot 10^{-6}$ degrees. For the Arecibo radio telescope, the value is $\theta_R = \sin^{-1}(1.22(0.210\text{ m})/305\text{ m}) = 0.0481$ degrees. The radio telescope is clearly worse than the other telescope in terms of angular resolution. The Keck Telescope is better than the Hubble Space Telescope due to its larger diameter.

34.51. **THINK:** Light of wavelength $\lambda = 600.$ nm illuminates two slits. The slits are separated by a distance $d = 24\ \mu\text{m}$ and the width of each slit is $a = 7.2\ \mu\text{m}$. A screen $w = 1.8$ m wide is $L = 2.0$ m from the slits. The problem can be approached by determining the number of fringes that appear due to the double slit and eliminate those removed by the minima due to single-slit diffraction.

SKETCH:

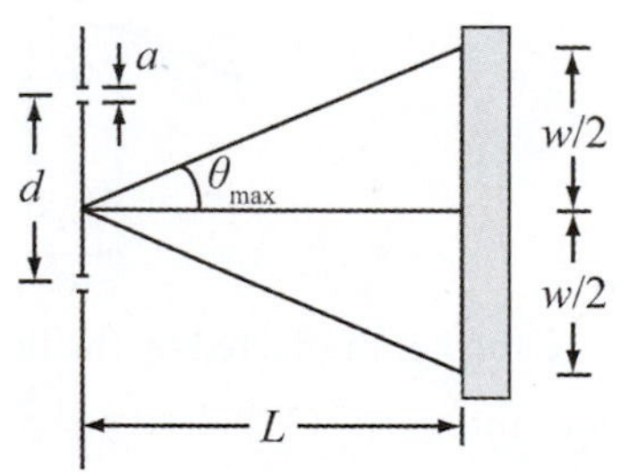

RESEARCH: The maximum angle θ_{max} is given by $\tan\theta_{\text{max}} = w/2L$. The bright fringes occur when $\sin\theta = m\lambda/d$. The disallowed fringes occur when $\sin\theta = n\lambda/a$.

SIMPLIFY: The maximum number of bright fringes that can appear on the screen is

$$m_{\text{max}} = \frac{d\sin\theta_{\text{max}}}{\lambda} = \frac{d\sin\left(\tan^{-1}\left(\frac{w}{2L}\right)\right)}{\lambda}.$$

The disallowed fringes occur when

$$\frac{m\lambda}{d} = \frac{n\lambda}{a} \Rightarrow \frac{m}{n} = \frac{d}{a}.$$

CALCULATE: The number of bright fringes is:

$$m_{\text{max}} = \frac{d\sin\theta_{\text{max}}}{\lambda} = \frac{(24\ \mu\text{m})\sin\left(\tan^{-1}\left(\frac{1.8\ \text{m}}{2(2.0\ \text{m})}\right)\right)}{(600.\ \text{nm})} = 16.4.$$

The disallowed fringes occur when:

$$\frac{m}{n} = \frac{d}{a} = \frac{24\ \mu\text{m}}{7.2\ \mu\text{m}} = \frac{10}{3}.$$

The only scenario this can occur for (since $m_{\text{max}} = 16$) is $m = 10$ and $n = 3$. Therefore, the only disallowed value of m is 10, so there are 15 bright fringes on either side of the central maximum.

ROUND: To the nearest integer, there are 31 fringes on the screen.

DOUBLE-CHECK: Without the effects from single-slit diffraction there would be 33. It is expected that there would be fewer fringes due to the effects of single-slit diffraction.

34.55.

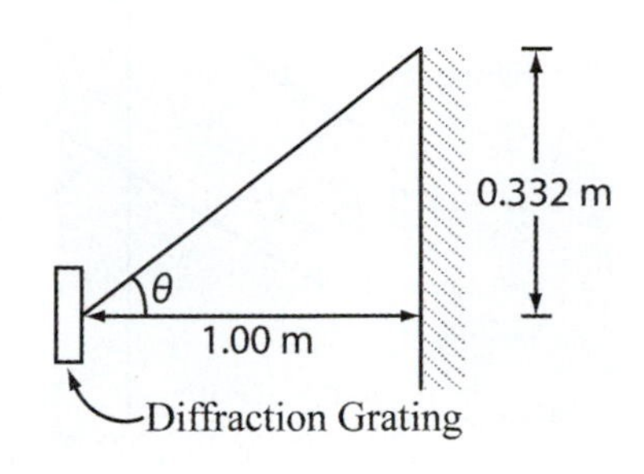

From the above diagram, $\tan\theta = \frac{0.332\ \text{m}}{1.00\ \text{m}}$. For a diffraction grating with $m = 1$, the wavelength of light is

$$\lambda = d\sin\theta \Rightarrow \lambda = \frac{1}{(7.02\cdot 10^{5}\ /\ \text{m})}\sin\left(\tan^{-1}\left(\frac{0.332\ \text{m}}{1.00\ \text{m}}\right)\right) = 4.49\cdot 10^{-7}\ \text{m} = 449\ \text{nm}.$$

34.59. The number of lines per centimeter is related to the slit separation d: $d\sin\theta = mx$. No second order spectrum occurs if for the smallest wavelength $\theta = 90°$,

$$d\sin 90° = 2(400.\ \text{nm}) \Rightarrow d = 800.\ \text{nm} \Rightarrow 1/d = 1.25\cdot 10^{6}\ \text{lines/m} = 1.25\cdot 10^{4}\ \text{lines/cm}.$$

34.63. The distance moved in an interferometer is given by $2d = N\lambda_{\text{water}}$,

$$n = 1.33 = \frac{c}{v_{\text{water}}} = \frac{c}{f_{\text{water}}\lambda_{\text{water}}} = \frac{f_{\text{air}}\lambda_{\text{air}}}{f_{\text{water}}\lambda_{\text{water}}},$$

since $f_{\text{air}} = f_{\text{water}}$, $\lambda_{\text{water}} = \lambda_{\text{air}}\ /\ n$.

$$2d = \frac{N\lambda_{\text{air}}}{n} \Rightarrow \lambda_{\text{air}} = \frac{2nd}{N} = \frac{2(1.33)(0.200\cdot 10^{-3}\ \text{m})}{800} = 6.65\cdot 10^{-7}\ \text{m} = 665\ \text{nm}.$$

34.67. Constructive interference for a thin film is given by $\frac{(m+1/2)\lambda_{\text{air}}}{n} = 2t$. For the minimum thickness, $m = 0$: $t = \frac{1}{4}\frac{\lambda_{\text{air}}}{n_{\text{coating}}} = \frac{1}{4}\frac{550.\ \text{nm}}{1.32} = 104\ \text{nm}.$

34.71. The Rayleigh criterion is given by:

$$\sin\theta_{\text{R}} = \frac{1.22\lambda}{d} \Rightarrow \frac{\Delta y}{L} = \frac{1.22\lambda}{d},$$

where $L = 384{,}000$ km is the distance to the Moon.

$$\Delta y = \frac{1.22\lambda}{d}L = \frac{1.22\left(550.\cdot 10^{-9}\text{ m}\right)}{12.0\cdot 10^{-2}\text{ m}}\left(384\cdot 10^{6}\text{ m}\right) = 2147.2\text{ m} \approx 2.15\text{ km}$$

34.75. **THINK:** Upon reflection, light undergoes a phase change of half a wavelength at the first interface, but not at the second interface. Since maxima are seen for two adjacent wavelengths, the layer thickness can be found by using the conditions for constructive interference.

SKETCH:

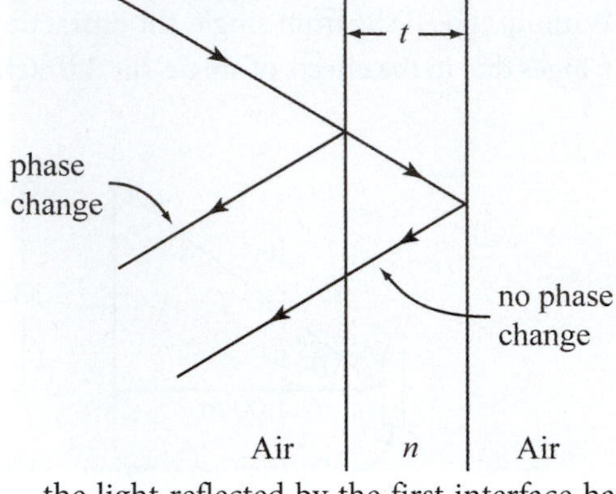

RESEARCH: Since $n_{\text{air}} < n_{\text{mica}}$, the light reflected by the first interface has a phase change of 180°. The light reflected by the second interface has no phase change. The condition for constructive interference in the reflected light is

$$\left(m+\frac{1}{2}\right)\frac{\lambda_{\text{air}}}{n} = 2t \quad (m = 0, 1, 2, \ldots).$$

For two adjacent wavelengths with $\lambda_2 > \lambda_1$, $m_2 = m_1 - 1$. Therefore,

$$m_1 = \frac{2nt}{\lambda_1} - \frac{1}{2} \text{ and } m_2 = m_1 - 1 = \frac{2nt}{\lambda_2} - \frac{1}{2}.$$

SIMPLIFY: Solving these two equations for the thickness *t* gives:

$$\frac{2nt}{\lambda_1} - \frac{1}{2} = \frac{2nt}{\lambda_2} + \frac{1}{2} \Rightarrow 2nt\left(\frac{1}{\lambda_1} - \frac{1}{\lambda_2}\right) = 1 \Rightarrow t = \frac{\lambda_1\lambda_2}{2n\left(\lambda_2 - \lambda_1\right)}.$$

CALCULATE: $t = \dfrac{(516.9\text{ nm})(610.9\text{ nm})}{2(1.57)(610.9\text{ nm} - 516.9\text{ nm})} = 1070.8\text{ nm} = 1.07\ \mu\text{m}.$

ROUND: To three significant figures, the thickness of the mica layer is $t = 1.07\ \mu$m.

DOUBLE-CHECK: As expected, the layer thickness is much larger than the observed wavelengths.

Multi-Version Exercises

34.77. $\Delta y = \dfrac{\lambda L \Delta m}{d} = \dfrac{(477\cdot 10^{-9}\text{m})(1.23\text{ m})\cdot 1}{(2.49\cdot 10^{-5}\text{m})} = 2.36\text{ cm}$

34.81. $w = \dfrac{2\lambda L}{a} = \dfrac{2(495\cdot 10^{-9}\text{m})(2.77\text{ m})}{(0.487\cdot 10^{-3}\text{m})} = 5.63\text{ mm}$

Chapter 35: Relativity

Exercises

35.27. The speed of light converted from SI to ft/ns is:

$$c = 2.9979 \cdot 10^8 \text{ m/s} = 2.9979 \cdot 10^8 \text{ m/s}\left(\frac{1 \text{ s}}{10^9 \text{ ns}}\right)\left(\frac{3.2808 \text{ ft}}{1 \text{ m}}\right) = 0.984 \text{ ft/ns}.$$

You can see that our result is quite close to 1 foot per nanosecond, which makes this a great way to visualize the speed of light: light moves about a foot in a time interval of a billionth of a second!

35.31. (a) Another astronaut on the ship sees the meter stick in the same (rest) frame as the astronaut holding the stick and so its length remains unchanged at one meter.

(b) For a ship moving at $v = 0.50c$, the length of the meter stick as measured by an observer on Earth is

$$L = \frac{L_0}{\gamma} = L_0\sqrt{1-(v/c)^2} = (1.00 \text{ m})\sqrt{1-(0.50c/c)^2} = 0.87 \text{ m}.$$

35.35. The fire truck of length $L_0 = 10.0$ m is traveling fast enough so a stationary observer sees its length contracted to $L = 8.00$ m. Therefore,

$$L = \frac{L_0}{\gamma} = L_0\sqrt{1-(v/c)^2} \Rightarrow \left(\frac{L}{L_0}\right)^2 = 1-(v/c)^2 \Rightarrow v = \left(1-\left(\frac{L}{L_0}\right)^2\right)^{1/2} c = \left(1-\left(\frac{8.00 \text{ m}}{10.0 \text{ m}}\right)^2\right)^{1/2} c = 0.600c.$$

(a) The time taken from the garage's point of view is

$$t_g = \frac{L}{v} = \frac{(8.00 \text{ m})}{0.600(3.00 \cdot 10^8 \text{ m/s})} = 4.44 \cdot 10^{-8} \text{ s}.$$

(b) From the fire truck's perspective the length of the garage will be contracted to

$$L = \frac{L_0}{\gamma} = L_0\sqrt{1-(v/c)^2} = (8.00 \text{ m})\sqrt{1-(0.600c/c)^2} = 6.40 \text{ m}.$$

Therefore, the truck will not fit inside the garage from the fire truck's point of view since the length of the truck from its rest frame is 10.0 m.

35.39. **THINK:** The tip of the triangle is the direction of the speed, $v = 0.400c$, so that only the length, $L = 50.0$ m, will be contracted and the width, $w = 20.0$ m, is not affected. The length of the ship L is not the same as the length of a side of the ship l. Relate the observed angle θ' to the speed of the ship.

SKETCH:

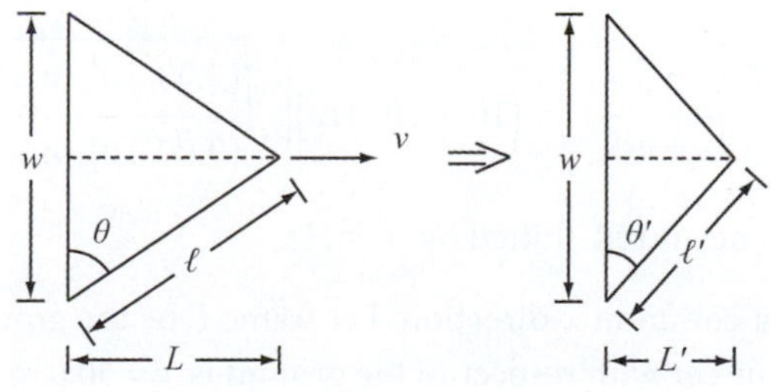

RESEARCH: The lengths are related to the angles, in both frames, by $l\cos\theta = w/2$, $l'\cos\theta' = w/2$, $L = l\sin\theta$, $L' = l'\sin\theta'$, and $\tan\theta = 2L/w$. The length of the ship contracts by $L' = L/\gamma$.

SIMPLIFY: Determine l' in terms of l:

$$\frac{w}{2} = l'\cos\theta' = l\cos\theta \Rightarrow l' = \frac{\cos\theta}{\cos\theta'}l.$$

The contracted length is then

$$L' = l'\sin\theta' = l\cos\theta\tan\theta' = \frac{L}{\gamma} = \frac{l\sin\theta}{\gamma} \Rightarrow \tan\theta' = \frac{\tan\theta}{\gamma} = \frac{2L}{w}\sqrt{1-(v/c)^2}.$$

Therefore, $\theta'(v) = \tan^{-1}\left(\frac{2L}{w}\sqrt{1-(v/c)^2}\right)$.

The plot of the angle between the base and side of the ship as a function of the speed of the ship as measured by a stationary observer is shown below.

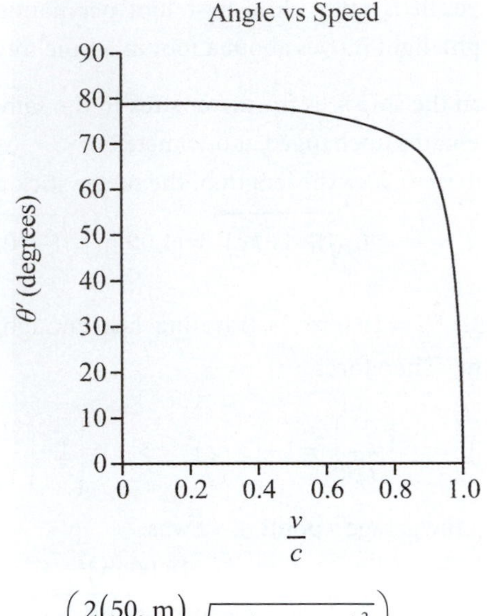

CALCULATE: $\theta'(v = 0.40c) = \tan^{-1}\left(\frac{2(50.\text{ m})}{(20.\text{ m})}\sqrt{1-(0.40c/c)^2}\right) = 77.69°$

ROUND: To three significant figures, $\theta'(v = 0.400c) = 77.7°$.

DOUBLE-CHECK: As v approaches c, the expression under the square root approaches zero and hence the angle will also approach zero. This agrees with the graph where the angle is smaller at higher velocities. When $v = c$, the side of the ship would effectively contract to zero, thus making an angle of zero with the width.

35.43. Since the car, moving with a speed $v = 32.0\text{ km/h}\left(\frac{1000\text{ m}}{1\text{ km}}\right)\left(\frac{1\text{ h}}{3600\text{ s}}\right) = 8.889\text{ m/s}$, is moving away from the radar of frequency $f_0 = 10.6$ GHz, the shift in frequency is,

$$\Delta f = f - f_0 = f_0\left(\sqrt{\frac{c-v}{c+v}} - 1\right) = (10.6\cdot 10^9\text{ Hz})\left(\sqrt{\frac{(3.00\cdot 10^8\text{ m/s})-(8.889\text{ m/s})}{(3.00\cdot 10^8\text{ m/s})+(8.889\text{ m/s})}} - 1\right) = -314.078\text{ Hz}.$$

Therefore, the frequency is red-shifted by 314 Hz.

35.47. Let all speeds be in a common *x*-direction. Let frame *F* be the ground and frame F' be the frame of your car. The speed of your car with respect to the ground is $v = 50.0$ m/s and the speed of the oncoming car is $u = -50.0$ m/s in frame *F*. Using the relativistic velocity transformation, the relative speed of the oncoming car is

$$u' = \frac{u-v}{1+uv/c^2} = \frac{(-50.0\text{ m/s})-(50.0\text{ m/s})}{\sqrt{1+(-50.0\text{ m/s})(50.0\text{ m/s})/(2.9979\cdot 10^8\text{ m/s})^2}} = -99.99999999999862\text{ m/s} \approx -100.\text{ m/s}.$$

The relative velocity is about the same as a Galilean velocity transformation $u' = u - v = 2u = -100$ m/s, since the speed of the cars is so small compared to the speed of light. In order to detect a difference, fourteen significant figures would need to be kept. This shows how close the values are.

35.51. **THINK:** The arrow has a velocity of $u' = 0.300c$ in Robert's reference frame. The railroad car has a length of $L_0 = 100.\text{ m}$ and travels at a speed of $v = 0.750c$. The velocity transformation equations and the equation for length contraction can be used to determine the values observed by Jenny.

SKETCH:

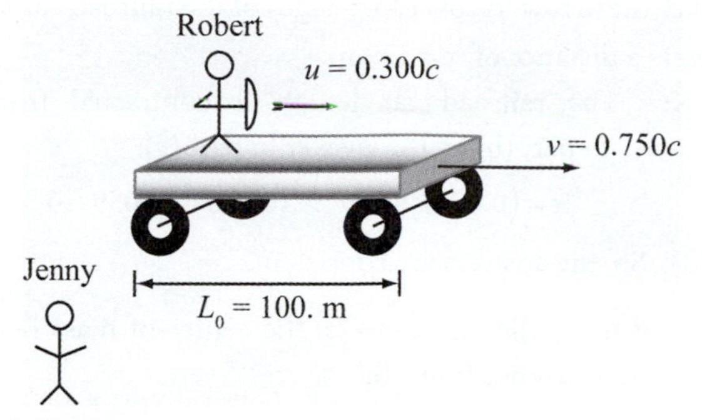

RESEARCH: As observed by Jenny,

(a) the railroad car is length contracted: $L = \dfrac{L_0}{\gamma}$,

(b) the velocity of the arrow is given by the inverse relativistic velocity transformation: $u = \dfrac{u' + v}{1 + vu'/c^2}$,

(c) the time of the arrow's flight is given by the inverse Lorentz transformation: $t = \gamma\left(t' + \dfrac{vx'}{c^2}\right)$, and

(d) the distance traveled by the arrow is given by the inverse Lorentz transformation: $x = \gamma(x' + vt')$.

SIMPLIFY: Here $x' = L_0$ is the length of the railroad car and $t' = L_0 / u'$ is the time of the arrow`s flight in Robert`s frame of reference. As observed by Jenny,

(a) $L = L_0\sqrt{1 - (v/c)^2}$,

(c) the time taken by the arrow to cover the length of the car is $t = \dfrac{L_0}{\sqrt{1-(v/c)^2}}\left(\dfrac{1}{u'} + \dfrac{v}{c^2}\right)$, , so if we take v as $v = kc$ and $u' = jc$, we have $t = \dfrac{L_0}{\sqrt{1-(v/c)^2}}\left(\dfrac{1}{jc} + \dfrac{kc}{c^2}\right) = \dfrac{L_0}{c} \cdot \dfrac{1}{\sqrt{1-(v/c)^2}}\left(\dfrac{1}{j} + k\right)$, and

(d) the distance covered by the arrow is $x = \dfrac{L_0}{\sqrt{1-(v/c)^2}}\left(1 + \dfrac{v}{u'}\right)$.

CALCULATE:

(a) $L = (100.\text{ m})\sqrt{1 - (0.750c/c)^2} = 66.14\text{ m}$

(b) $u = \dfrac{(0.300c) + (0.750c)}{1 + (0.750c)(0.300c)/c^2} = 0.85714c$

(c) $t = \dfrac{(100.\text{ m})}{(2.9979 \cdot 10^8\text{ m/s})\sqrt{1-(0.750c/c)^2}}\left(\dfrac{1}{(0.300)} + (0.750)\right) = 2.059 \cdot 10^{-6}\text{ s}$

(d) $x = \frac{(100.\text{ m})}{\sqrt{1-(0.750c/c)^2}}\left(1+\frac{(0.750c)}{(0.300c)}\right) = 529.2\text{ m}$

ROUND: The answers should be given to three significant figures. As observed by Jenny,

(a) the railroad car is $L = 66.1\text{ m}$ long,

(b) the velocity of the arrow is $u = 0.857c$,

(c) the time it takes the arrow to cover the length of the railroad car is $t = 2.06\ \mu\text{s}$, and

(d) the arrow covers a distance of $x = 529\text{ m}$.

DOUBLE-CHECK: The railroad car length is contracted from Jenny's viewpoint, as expected. Multiplying the answer to part (b) by the answer to part (c):

$$x = (0.8571)(2.9979\cdot 10^8\text{ m/s})(2.059\cdot 10^{-6}\text{ s}) = 529\text{ m},$$

as found in part (d). So, the answers are consistent.

35.55. The kinetic energy of the colliding beams in the center-of-mass reference frame is related to the fixed-target equivalent, or lab reference frame by

$$K^{\text{lab}} = 4K^{\text{cm}} + \frac{2\left(K^{\text{cm}}\right)^2}{m_p c^2} = 4(197)(100.\text{ GeV}) + \frac{2\left((197)(100.\text{ GeV})\right)^2}{(197)(1.00\text{ GeV})} = 4.02\cdot 10^6\text{ GeV}.$$

This is an incredibly large energy.

35.59. **THINK:** Electrons acquire kinetic energy as they accelerate through the potential difference. The speed acquired by the electron after moving through this potential can be found and then the appropriate classical and relativistic formulae can be used to find the total energy and momentum. Many of the answers only make sense if they are given to three significant figures, so rounding will be nonstandard.

SKETCH:

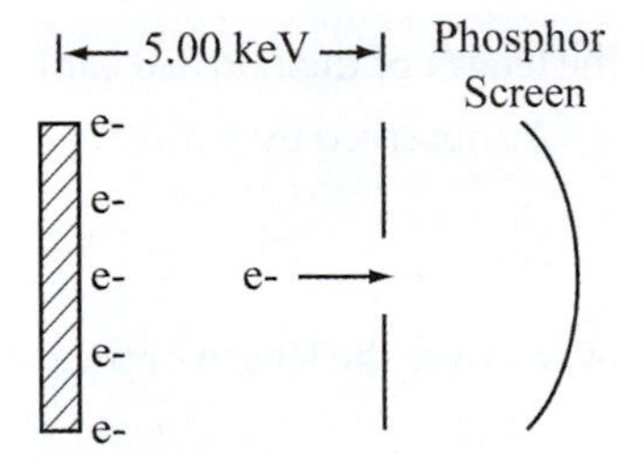

RESEARCH:

(a) The kinetic energy gained by the electron in moving through the potential difference V is equal to the work done by the potential difference: $W = K = qV$.

(b) The kinetic energy of a relativistic particle is $K = (\gamma - 1)E_0$.

(c) The relativistic values for the total energy and momentum are $E_R = \gamma E_0$ and $p_R = \gamma m v$. Classically, these values are given by $E_C = K = \frac{1}{2}mv_C^2$ and $p_C = mv_C$.

The rest mass energy of an electron is $E_0 = 511\text{ keV}$.

SIMPLIFY:

(a) $K = eV$

(b) The speed of the particle is found using the relativistic formula $K = (\gamma - 1)E_0$:

$$\gamma = \frac{1}{\sqrt{1-(v/c)^2}} = \frac{K+E_0}{E_0} \Rightarrow v = \sqrt{1-\left(\frac{E_0}{K+E_0}\right)^2}\,c = \sqrt{\frac{(K+E_0)^2 - E_0^2}{(K+E_0)^2}}\,c = \frac{\sqrt{K^2+2KE_0}}{K+E_0}c.$$

(c) The relativistic values for the total energy and momentum are

$$E_R = \gamma E_0 = K + E_0,\text{ and}$$

$$p_{\text{R}} = \gamma mv = \left(\frac{K+E_0}{E_0}\right)\left(\frac{E_0}{c^2}\right)\frac{\sqrt{K^2+2KE_0}}{K+E_0}c = \sqrt{K^2+2KE_0}\,/\,c.$$

Classically, the total energy and momentum are

$$E_{\text{C}} = K, \text{ and}$$

$$p_{\text{C}} = mv_{\text{C}} = m\sqrt{2K/m} = \sqrt{2Km} = \sqrt{2KE_0}\,/\,c.$$

CALCULATE:

(a) $K = e(5.00 \text{ kV}) = 5.00 \text{ keV}$

(b) $v = \dfrac{\sqrt{(5.00 \text{ keV})^2 + 2(5.00 \text{ keV})(511 \text{ keV})}}{(5.00 \text{ keV})+(511 \text{ keV})}c = 0.1389c$

(c) $E_{\text{R}} = (5.00 \text{ keV})+(511 \text{ keV}) = 516 \text{ keV}$

$p_{\text{R}} = \sqrt{(5.00 \text{ keV})^2 + 2(5.00 \text{ keV})(511 \text{ keV})}\,/\,c = 71.659 \text{ keV}/c$

$E_{\text{C}} = 5.00 \text{ keV}$

$p_{\text{C}} = \sqrt{2(5.00 \text{ keV})(511 \text{ keV})}\,/\,c = 71.484 \text{ keV}/c$

ROUND:

(a) The kinetic energy that the electron acquires is $K = 5.00$ keV.

(b) The electron has a speed of $v = 0.139c$, thus the electron will have only a small difference between its classical and relativistic values, but this can still be considered a relativistic speed.

(c) The relativistic and classical energies are $E_{\text{R}} = 516$ keV and 5.00 keV, respectively. (The difference is due to the fact that the relativistic energy includes the rest energy). The relativistic and classical momenta are $p_{\text{R}} = 71.7 \text{ keV}/c$ and $p_{\text{C}} = 71.5 \text{ keV}/c$, respectively.

DOUBLE-CHECK: The classical and relativistic momenta are similar, as expected for such a low speed.

35.63. THINK: The Lorentz transformations for energy and momentum in the frame F' can be used to write the quantity $E'^2 - p'^2c^2$ in terms of the values in the unprimed frame *F*.

SKETCH: Not required.

RESEARCH: The Lorentz transformations are

$$E' = \gamma(E - vp_x),\quad p'_x = \gamma(p_x - vE/c^2),\quad p'_y = p_y \text{ and } p'_z = p_z.$$

SIMPLIFY: Apply the transformations:

$$\begin{aligned}E'^2 - p'^2c^2 &= E'^2 - p'^2_xc^2 - p'^2_yc^2 - p'^2_zc^2 = \gamma^2(E - vp_x)^2 - \gamma^2(p_x - vE/c^2)^2c^2 - p_y^2c^2 - p_z^2c^2 \\ &= \gamma^2E^2 - 2\gamma^2Evp_x + \gamma^2v^2p_x^2 - \gamma^2p_x^2c^2 + 2\gamma^2p_xvE - v^2E^2\gamma^2/c^2 - p_y^2c^2 - p_z^2c^2 \\ &= \gamma^2\left(1-(v/c)^2\right)E^2 + \gamma^2(v^2 - c^2)p_x^2 - p_y^2c^2 - p_z^2c^2 \\ &= \gamma^2\left(1-(v/c)^2\right)E^2 - \gamma^2\left(1-(v/c)^2\right)p_x^2c^2 - p_y^2c^2 - p_z^2c^2 = E^2 - p^2c^2.\end{aligned}$$

CALCULATE: Not required.

ROUND: Not required.

DOUBLE-CHECK: The statement in the problem has been proved using only the Lorentz transformation equations. One could check the result for special and limiting cases. For example, if $v = 0$ then $\gamma = 1$ and the Lorentz transformations reduce to $E' = E$ and $p'_x = p_x$, so the result holds. When $p = 0$ in the frame *F*,

$$\begin{aligned}E'^2 - p'^2c^2 &= E'^2 - p'^2_xc^2 - p'^2_yc^2 - p'^2_zc^2 = \gamma^2E^2 - \gamma^2(-vE/c^2)^2c^2 \\ &= \gamma^2E^2 - v^2E^2\gamma^2/c^2 = \gamma^2\left(1-(v/c)^2\right)E^2 = E^2.\end{aligned}$$

35.67. The Schwarzschild radius of a black hole is $R_{\text{S}} = \dfrac{2GM}{c^2}$. The black hole at the center of the Milky Way in Example 12.4 was found to be $3.72 \cdot 10^6$ solar masses. The mass of the Sun is $1.989 \cdot 10^{30}$ kg. The Schwarzschild radius of this black hole is

$$R_{\text{S}} = \frac{2\left(6.674 \cdot 10^{-11} \text{ N m}^2/\text{kg}^2\right)\left(3.72 \cdot 10^6\right)\left(1.989 \cdot 10^{30} \text{ kg}\right)}{\left(2.9979 \cdot 10^8 \text{ m/s}\right)^2} = 10.99 \cdot 10^9 \text{ m}\left(\frac{\text{AU}}{149.60 \cdot 10^9 \text{ m}}\right)$$
$$= 0.0735 \text{ AU}.$$

35.71. The Newtonian and relativistic kinetic energies of a particle are $K_{\text{N}} = (1/2)mv^2$ and $K_{\text{R}} = (\gamma - 1)mc^2$, respectively. In Newtonian mechanics, the difference in their kinetic energy is

$$\Delta K_{\text{N}} = \frac{1}{2}mv_1^2 - \frac{1}{2}mv_2^2 = \frac{1}{2}m\left(v_1^2 - v_2^2\right) = \frac{1}{2}\left(0.9999^2 - 0.9900^2\right)mc^2$$
$$= \frac{1}{2}\left(0.9999^2 - 0.9900^2\right)\left(0.511 \text{ MeV}/c^2\right)c^2 = 5.03 \text{ keV}.$$

The difference using special relativity is

$$\Delta K_{\text{R}} = (\gamma_1 - 1)mc^2 - (\gamma_2 - 1)mc^2 = (\gamma_1 - \gamma_2)mc^2 = \left(\frac{1}{\sqrt{1-(v_1/c)^2}} - \frac{1}{\sqrt{1-(v_2/c)^2}}\right)mc^2$$
$$= \left(\frac{1}{\sqrt{1-0.9999^2}} - \frac{1}{\sqrt{1-0.9900^2}}\right)(0.511 \text{ MeV}) = 32.5 \text{ MeV}.$$

Therefore, we see that at velocities near *c*, the Newtonian approximation of Kinetic energy diverges from the relativistic Kinetic energy by several orders of magnitude.

35.75. Using the relativistic velocity transformation, the speed of object A relative to object B as measured by an observer on object B is

$$u' = \frac{v_{\text{A}} - v_{\text{B}}}{1 - v_{\text{A}}v_{\text{B}}/c^2} = \frac{(0.600c) - (-0.600c)}{1 - (0.600c)(-0.600c)/c^2} = 0.882c.$$

35.79. The distance of 100. ly was measured by someone on one of the space stations. Someone on the spaceship will measure a different distance, one that is shorter according to the formula for length contraction, $L = L_0/\gamma$. The time it takes to travel from one space station to the next as measured by someone on the spaceship is

$$t_1 = \frac{L}{v} = \frac{L_0}{\gamma v} = \frac{L_0}{v}\sqrt{1-(v/c)^2} = \frac{(100. \text{ ly})}{(0.950c)}\sqrt{1-(0.950c/c)^2} = 32.8684 \text{ years} \approx 33 \text{ years}.$$

As seen by someone on the space station, the time will be

$$t_2 = \frac{L}{v} = \frac{(100. \text{ ly})}{(0.950c)} = 105 \text{ years}.$$

35.83. **THINK:** The running back is travelling at 55.0% the speed of light relative to the field. He throws the ball to a receiver running at 65.0% the speed of light relative to the field in the same direction. The speed of the ball relative to the running back is 80.0% the speed of light. The relativistic velocity transformation can be used to find the speed that the receiver perceives the ball to be travelling at. Recall that the speed of light is the same in all reference frames.

SKETCH:

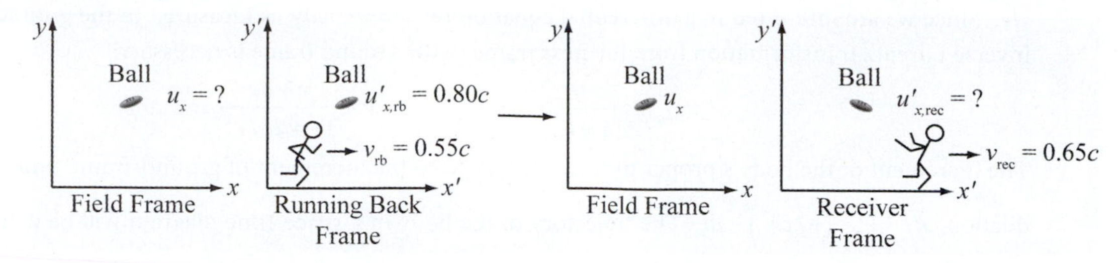

RESEARCH: The velocity of the ball with respect to the running back is $u_x' = 0.800c$. The velocity of the running back with respect to the field is $v_{\text{rb}} = 0.550c$. The inverse Lorentz transformation can be used to find the velocity u_x of the ball in the field frame:

$$u_x = \frac{u'_{x,\text{rb}} + v_{\text{rb}}}{1 + u'_{x,\text{rb}} v_{\text{rb}} / c^2}.$$

Using a Lorentz transform gives the speed of the ball relative to the receiver:

$$u'_{x,\text{rec}} = \frac{u_x - v_{\text{rec}}}{1 - u_x v_{\text{rec}} / c^2},$$

where $v_{\text{rec}} = 0.650c$ is the velocity of the receiver relative to the field.

SIMPLIFY: Not required.

CALCULATE:

(a) $u_x = \dfrac{(0.800c) + (0.550c)}{1 + (0.800c)(0.550c)/c^2} = 0.9375c \Rightarrow u'_{x,\text{rec}} = \dfrac{(0.9375c) - (0.650c)}{1 - (0.9375c)(0.650c)/c^2} = 0.7360c$

(b) Photons travel at the speed of light and the speed of light is the same in any reference frame; therefore, the photons would appear to be travelling at the speed of light to the receiver.

ROUND:

(a) To three significant figures, the speed of the ball perceived by the receiver is $u'_{x,\text{rec}} = 0.736c = 2.21 \cdot 10^8$ m/s.

DOUBLE-CHECK: The calculated value of the football's relative speed was less than the speed of light as it must be, since no massive object can travel at the speed of light.

35.87. **THINK:** In considering accelerating bodies with special relativity, the acceleration experienced by the moving body is constant; that is, in each increment of the body's own proper time, $d\tau$, the body acquires velocity increment $dv = g d\tau$ as measured in the body's frame (the inertial frame in which the body is momentarily at rest). Given this interpretation,

(a) Write a differential equation for the velocity v of the body, moving in one spatial dimension, as measured in the inertial frame in which the body was initially at rest (the "ground frame").

(b) Solve this equation for $v(t)$, where both v and t are measured in the ground frame.

(c) Verify that the solution behaves appropriately for small and large values of t.

(d) Calculate the position of the body $x(t)$, as measured in the ground frame.

(e) Identify the trajectory of the body on a Minkowski diagram with coordinates x and ct, as measured in the ground frame.

(f) For $g = 9.81$ m/s^2, calculate how much time t it takes the body to accelerate from rest to 70.7% of c, as measured in the ground frame, and how much ground-frame distance, Δx, the body covers in this time.

SKETCH: Not required.

RESEARCH: In moving from the ground frame to the next frame, the body's velocity was incremented by dv. Since we are interested in a differential equation for the velocity as measured in the ground frame, an inverse Lorentz transformation from the next frame to the ground frame is necessary:

$$u_{\text{ground}} = \frac{u_{\text{next}} + v}{1 + u_{\text{next}} v / c^2} \Rightarrow v + dv = \frac{v + dv}{1 + v dv / c^2}.$$

The increment of the body's proper time $d\tau$ is related to the increment of ground-frame time dt by time dilation, $d\tau = \left(1 - (v/c)^2\right)^{1/2} dt$. The trajectory of the body in a space-time diagram will be determined by examining the position as a function of time, which is determined in part (d).

SIMPLIFY:

(a) Ignoring squares and higher powers of differentials,

$$v + dv = \frac{v + g d\tau}{1 + v g d\tau / c^2} = (v + g d\tau)\left(1 - \frac{v g d\tau}{c^2} + \ldots\right) = v + g\left(1 - (v/c)^2\right) d\tau + \ldots, \text{ or } dv = g\left(1 - (v/c)^2\right) d\tau.$$

But the increment of proper time $d\tau$ is related to the increment of ground-frame time dt by time dilation so the differential equation, in terms of ground frame quantities, becomes

$$dv = g\left(1 - (v/c)^2\right) d\tau = g\left(1 - (v/c)^2\right)\left(1 - (v/c)^2\right)^{1/2} dt$$

$$\frac{dv}{dt} = g\left(1 - (v/c)^2\right)^{3/2}$$

(b) The above differential equation separates, yielding

$$g \int_0^t dt' = \int_0^{v(t)} \frac{dv'}{\left(1 - (v'/c)^2\right)^{3/2}} \Rightarrow gt = \frac{v(t)}{\left(1 - (v(t)/c)^2\right)^{1/2}}.$$

This is readily solved, giving $v(t) = \dfrac{gt}{\left(1 + (gt/c)^2\right)^{1/2}}$ for the ground-frame velocity of the accelerating body as a function of ground-frame time.

(c) For $gt << c$, i.e., the Newtonian limit, the above result takes the form $v(t) \cong gt$, exactly as expected. The relativistic limit, as time approaches infinity is:

$$\lim_{t \to \infty} v(t) = \lim_{t \to \infty} \frac{gt}{\sqrt{1 + (gt/c)^2}} = \frac{gt}{\sqrt{(gt/c)^2}} = c.$$

That is, the velocity of the accelerating body asymptotically approaches c, as expected.

(d) The position follows from the velocity through integration:

$$x(t) = \frac{c^2}{g} + \int_0^t v(t') dt' = \frac{c^2}{g} + \int_0^t \frac{g t' dt'}{\left[1 + (g t'/c)^2\right]^{1/2}} = \frac{c^2}{g} + \frac{c^2}{g}\left[1 + \left(\frac{g t'}{c}\right)^2\right]^{1/2}\Bigg|_0^t = \frac{c^2}{g}\left[1 + \left(\frac{gt}{c}\right)^2\right]^{1/2}$$

(e) The above result implies the simple relation $x^2 - (ct)^2 = c^4 / g^2$. The right-hand side is constant. Hence, the trajectory is a branch of a hyperbola on a Minkowski diagram.

(f) Consider the ground-frame speed as a function of ground-frame time from part (b),

$$v(t) = \frac{gt}{\sqrt{1 + (gt/c)^2}}.$$

The time t required for the body to accelerate from rest to $v = 0.707c$ is given by:

$$v = \frac{gt}{\sqrt{1 + (gt/c)^2}} \Rightarrow v^2\left(1 + (gt/c)^2\right) = (gt)^2 \Rightarrow t = \frac{v}{g\sqrt{1 - (v/c)^2}}.$$

The ground-frame distance travelled in this time is $\Delta x = x(t) - x(0)$. As stated in the problem, the ground-frame position at ground-frame time $t = 0$ is $x(0) = c^2 / g$. Then

$$\Delta x = \frac{c^2}{g}\left[1+\left(\frac{gt}{c}\right)^2\right]^{1/2} - \frac{c^2}{g} = \frac{c^2}{g}\left(\sqrt{1+\frac{(v/c)^2}{1-(v/c)^2}}-1\right).$$

CALCULATE:

(f) $t = \dfrac{(0.707)(2.998\cdot 10^8 \text{ m/s})}{(9.81 \text{ m/s}^2)\sqrt{1-((0.707c)/c)^2}} = 3.055\cdot 10^7 \text{ s} = 353.6 \text{ days},$

$$\Delta x = \frac{(2.998\cdot 10^8 \text{ m/s})^2}{(9.81 \text{ m/s}^2)}\left(\sqrt{1+\frac{((0.707c)/c)^2}{1-((0.707c)/c)^2}}-1\right) = 3.793\cdot 10^{15} \text{ m} = 0.4009 \text{ ly}$$

ROUND: The answers should be quoted to three significant figures:

(f) $t = 354$ days, and $\Delta x = 0.401$ ly.

DOUBLE-CHECK: The motion of an object with constant proper acceleration in special relativity should be described by a hyperbola, as found in parts (d) and (e). The values found in part (f) are reasonable considering the relatively slow acceleration of 9.81 m/s^2.

Multi-Version Exercises

35.88. $W = \Delta E = (\gamma_2 - \gamma_1)mc^2 = \left(\dfrac{1}{\sqrt{1-(v_2/c)^2}} - \dfrac{1}{\sqrt{1-(v_1/c)^2}}\right)mc^2$

$$= \left(\frac{1}{\sqrt{1-0.8433^2}} - \frac{1}{\sqrt{1-0.5785^2}}\right)(183.473 \text{ GeV}) = 116.4493263 \text{ GeV} = 116 \text{ GeV}$$

Note that because of the subtraction rule the answer has only three significant figures.

35.91. $K_{\text{lab}} = 4K_{\text{cm}} + 2K^2_{\text{cm}} / mc^2 = 4(503.01 \text{ GeV}) + 2(503.01 \text{ GeV})^2 / (50.30 \text{ GeV}) = 12.072 \text{ TeV}$

Chapter 36: Quantum Physics

Exercises

36.19. Assuming the surface temperature of the Sun is $T_{\text{Sun}} = 5800.\text{ K}$ and the surface temperature of the Earth is $T_{\text{earth}} = 300.\text{ K}$, Wien's displacement law can be used to calculate the respective peak wavelengths.

(a) $\lambda_{\text{m,Sun}} = \dfrac{2.90 \cdot 10^{-3}\text{ K m}}{T_{\text{Sun}}} = \dfrac{2.90 \cdot 10^{-3}\text{ K m}}{5800.\text{ K}} = 5.00 \cdot 10^{-7}\text{ m}$

(b) $\lambda_{\text{m,Earth}} = \dfrac{2.90 \cdot 10^{-3}\text{ K m}}{T_{\text{Earth}}} = \dfrac{2.90 \cdot 10^{-3}\text{ K m}}{300.\text{ K}} \approx 9.67 \cdot 10^{-6}\text{ m}$

36.23. **THINK:** The temperature of your skin is approximately $T = 35.0\text{ °C} = 308.15\text{ K}$. Assume that it is a blackbody. Consider a total surface area of $A = 2.00\text{ m}^2$. (a) The Wien displacement law can be used to determine the peak wavelength λ_{m} of the radiation emitted by the skin, (b) the Stefan-Boltzmann radiation law can be used to determine the total power P emitted by your skin, and (c) the wavelength of the radiation needs to be considered.

SKETCH:

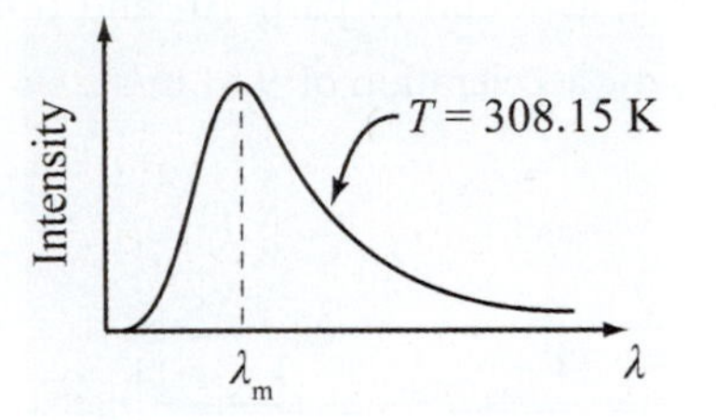

RESEARCH:

(a) The Wien displacement law is $\lambda_{\text{m}} T = 2.90 \cdot 10^{-3}\text{ K m}$.

(b) The total power is $P = IA$, where I, the intensity, is given by the Stefan-Boltzmann radiation law:

$$I = \sigma T^4, \text{ using } \sigma = 5.6704 \cdot 10^{-8}\text{ W}/\left(\text{m}^2\text{ K}^4\right).$$

(c) Power is energy per unit time. The relationship between energy and wavelength for photons is: $E = hc/\lambda$.

SIMPLIFY:

(a) $\lambda_{\text{m}} = \dfrac{2.90 \cdot 10^{-3}\text{ K m}}{T}$

(b) $P = IA$, substituting $P = \sigma T^4$ gives: $P = \sigma T^4 A$.

CALCULATE:

(a) $\lambda_{\text{m}} = \dfrac{2.90 \cdot 10^{-3}\text{ K m}}{308.15\text{ K}} = 9.4110 \cdot 10^{-6}\text{ m}$

(b) $P = \left(5.6704 \cdot 10^{-8}\text{ W}/\left(\text{m}^2\text{ K}^4\right)\right)\left(308.15\text{ K}\right)^4\left(2.00\text{ m}^2\right) = 1022.57\text{ W}$

(c) Considering that a typical light bulb has a power of 100 W, why is it that a person does not glow with a power output of about 1000 W? The reason is because in order for you to "glow" your wavelength must be in the visible spectrum. However, the peak wavelength calculated in part (a) is $\lambda_{\text{m}} = 9.4111\text{ µm}$. This wavelength is in the infrared part of the spectrum, not the visible part. Your wavelength is not in the visible spectrum.

ROUND:

(a) To three significant figures, the peak wavelength is $\lambda_{\text{m}} = 9.41\text{ µm}$.

(b) To three significant figures, the total power emitted by your skin is $P = 1.02\text{ kW}$.

DOUBLE-CHECK: The calculated values seem reasonable considering the given values. Comparing the calculated peak wavelength to the values in Figure 31.10 of the textbook shows that the wavelength of the emitted radiation is out of the visible spectrum.

36.27. The light that is incident on the sodium surface is $\lambda = 470\text{ nm} = 470\cdot 10^{-9}\text{ nm}$. The work function for sodium is $\phi = 2.3\text{ eV}$ (see Table 36.1 in textbook). The maximum kinetic energy of the electrons ejected from the sodium surface is $K_{\text{max}} = eV_0 = hf - \phi$. For photons $f = c/\lambda$,

$$f = \frac{3.00\cdot 10^{8}\text{ m/s}}{470\cdot 10^{-9}\text{ m}} = 6.38\cdot 10^{14}\text{ s}^{-1}.$$

Inserting this value into the equation for K_{max} gives:

$$K_{\text{max}} = \left(4.136\cdot 10^{-15}\text{ eV s}\right)\left(6.38\cdot 10^{14}\text{ s}^{-1}\right) - 2.3\text{ eV} = 0.34\text{ eV}.$$

36.31. White light is made up of photons with wavelengths ranging from $\lambda = 4.00\cdot 10^{2}\text{ nm}$ to $7.50\cdot 10^{2}\text{ nm}$ ($4.00\cdot 10^{-7}\text{ m}$ to $7.50\cdot 10^{-7}\text{ m}$). The work function of barium is given as $\phi = 2.48\text{ eV}$.

(a) The maximum kinetic energy of an electron ejected from the barium surface will correspond to a photon with the minimum wavelength.

$$K_{\text{max}} = \frac{hc}{\lambda_{\text{min}}} - \phi = \frac{\left(4.136\cdot 10^{-15}\text{ eV s}\right)\left(3.00\cdot 10^{8}\text{ m/s}\right)}{4.00\cdot 10^{-7}\text{ m}} - 2.48\text{ eV} = 0.622\text{ eV}$$

(b) The longest wavelength of light that could eject electrons is given by $\lambda = \frac{hc}{\phi} = \frac{\left(4.136\cdot 10^{-15}\text{ eV s}\right)\left(3.00\cdot 10^{8}\text{ m/s}\right)}{2.48\text{ eV}} = 5.00\cdot 10^{-7}\text{ m} = 5.00\cdot 10^{2}\text{ nm}.$ This means that the $7.50\cdot 10^{2}\text{ nm}$ wavelength light would not eject electrons from the barium surface.

(c) The wavelength of light that would eject electrons with zero kinetic energy is given by: $\lambda = hc/\phi$ which was solved in part (b). The wavelength was $\lambda = 5.00\cdot 10^{2}\text{ nm}$.

36.35. The wavelength of the incoming photon is $\lambda = 0.30\text{ nm}$; its original energy was: $E = hf = \frac{hc}{\lambda} = \frac{\left(4.13567\cdot 10^{-15}\text{ eV s}\right)\left(2.998\cdot 10^{8}\text{ m/s}\right)}{3.0\cdot 10^{-10}\text{ m}} = 4133\text{ eV}$. It rebounds at angle of $\theta = 160°$. Its new wavelength can be found using the Compton scattering formula.

$$\lambda' = \lambda + \frac{h}{m_e c}(1-\cos\theta) = \left(3.0\cdot 10^{-10}\text{ m}\right) + \frac{\left(6.626\cdot 10^{-34}\text{ J s}\right)\left(1-\cos 160°\right)}{\left(9.109\cdot 10^{-31}\text{ kg}\right)\left(2.998\cdot 10^{8}\text{ m/s}\right)} = 3.047\cdot 10^{-10}\text{ m}$$

Its new energy is: $E' = \frac{hc}{\lambda'} = \frac{\left(4.13567\cdot 10^{-15}\text{ eV s}\right)\left(2.998\cdot 10^{8}\text{ m/s}\right)}{3.047\cdot 10^{-10}\text{ m}} = 4069\text{ eV}$. The amount of energy lost is $\Delta E = E - E' = 4133\text{ eV} - 4069\text{ eV} = 64\text{ eV}$.

36.39. (a) The wavelength of a photon is $\lambda = hc/E$. For a photon of energy $E = 2.00\text{ eV}$, the wavelength is: $\lambda = \left(4.13567\cdot 10^{-15}\text{ eV s}\right)\left(2.998\cdot 10^{8}\text{ m/s}\right)/\left(2.00\text{ eV}\right) = 6.1994\cdot 10^{-7}\text{ m} \approx 620.\text{ nm}$.

(b) The wavelength of an electron is $\lambda = h/p = h/\left(m_e v\right)$, and its kinetic energy is $K = m_e v^2/2$. In terms of K, the velocity v is $v = \sqrt{\frac{2K}{m_e}}$. Then the wavelength of the electron is $\lambda = \frac{h}{m_e}\sqrt{\frac{m_e}{2K}} = \sqrt{\frac{h^2}{2Km_e}}$.

For an electron of kinetic energy $K = (2.00 \text{ eV})\cdot(1.602\cdot 10^{-19} \text{ J})/(1 \text{ eV}) = 3.204\cdot 10^{-19} \text{ J}$, the wavelength is:

$$\lambda = \sqrt{\frac{(6.626\cdot 10^{-34} \text{ J s})^2}{2(3.204\cdot 10^{-19} \text{ J})(9.109\cdot 10^{-31} \text{ kg})}} = 8.673\cdot 10^{-10} \text{ m} \approx 0.867 \text{ nm}.$$

36.43. The electron has a de Broglie wavelength of $\lambda = 550$ nm.

(a) The de Broglie wavelength is $\lambda = h/p = h/(mv)$. The speed of the electron is

$$v = \frac{h}{m_e\lambda} = \frac{6.626\cdot 10^{-34} \text{ J s}}{(9.109\cdot 10^{-31} \text{ kg})(5.5\cdot 10^{-7} \text{ m})} = 1323 \text{ m/s} \approx 1300 \text{ m/s}.$$

(b) This speed is much less than the speed of light, so the non-relativistic approximation is sufficient.

(c) In non-relativistic terms, the electron's kinetic energy is

$$K = \frac{mv^2}{2} = \frac{(9.109\cdot 10^{-31} \text{ kg})(1323 \text{ m/s})^2}{2} = 7.967\cdot 10^{-25} \text{ J}.$$

In eV, this becomes $K = (7.967\cdot 10^{-25} \text{ J})(1 \text{ eV})/(1.602\cdot 10^{-19} \text{ J}) = 4.973\cdot 10^{-6} \text{ eV} \approx 5.0 \text{ μeV}$.

36.47. The mass of the particle is $m = 50.0$ kg. It has a de Broglie wavelength of $\lambda = 20.0$ cm.

(a) The de Broglie wavelength is $\lambda = h/p = h/(mv)$. The speed is therefore

$$v = \frac{h}{m\lambda} = \frac{6.626\cdot 10^{-34} \text{ J s}}{(50.0 \text{ kg})(0.200 \text{ m})} = 6.626\cdot 10^{-35} \text{ m/s} \approx 6.63\cdot 10^{-35} \text{ m/s}.$$

(b) From the uncertainty relation $\Delta x\cdot\Delta p_x \geq (1/2)\hbar \Rightarrow \Delta x\cdot m\Delta v_x \geq (1/2)\hbar$, the uncertainty in the speed must be $\Delta v_x \geq (1/2)\hbar/(\Delta x\cdot m)$. The minimum uncertainty is

$$\Delta v_x = \frac{\hbar}{2\cdot\Delta x\cdot m} = \frac{1.0546\cdot 10^{-34} \text{ J s}}{2(0.200 \text{ m})(50.0 \text{ kg})} = 5.273\cdot 10^{-36} \text{ m/s} \approx 5.27\cdot 10^{-36} \text{ m/s}.$$

36.51. The uncertainty relation between position and momentum (in one dimension) is $\Delta x\cdot\Delta p \geq (1/2)\hbar$. In terms of speed, $\Delta p = m\Delta v$, and so the uncertainty relation becomes $\Delta x\cdot\Delta v \geq (1/2)\hbar/m$. The electron is confined to a box of dimensions $L = 20.0$ μm. The maximum uncertainty in the (one-dimensional) position of the electron is the dimension of the box, that is $\Delta x_{\text{max}} = L = 20.0$ μm. The minimum uncertainty in the speed of the electron is

$$\Delta v_{\text{min}} = (1/2)\hbar/(m\Delta x_{\text{max}}) = (1/2)(1.05457\cdot 10^{-34} \text{ J s})/\left[(9.109\cdot 10^{-31} \text{ kg})(20.0 \text{ μm})\right] = 2.894 \text{ m/s}.$$

The minimum speed the electron can have is 2.89 m/s.

36.55. **THINK:** The system is made up of N particles. The average energy per particle is given by

$$\langle E\rangle = \frac{\sum E_i e^{-E_i/k_B T}}{Z},$$

where Z is the partition function,

$$Z = \sum_i g_i e^{-E_i/k_B T},$$

and g_i is the degeneracy of the state with energy E_i. This system is a 2-state system with $E_1 = 0$ and $E_2 = E$ and $g_1 = g_2 = 1$. Calculate the heat capacity of the system, $C = N\left(d\langle E\rangle/dT\right)$, and approximate its behavior at very high and very low temperatures (i.e. $k_B T \gg 1$ and $k_B T \ll 1$).

SKETCH: Not applicable.

RESEARCH: Not applicable as the necessary equations were all given in the problem.

SIMPLIFY: The average energy per particle is for $E_1 = 0$ and $E_2 = E$ is:

$$\langle E\rangle = \frac{(0)\exp\left(-\frac{(0)}{k_B T}\right)+(E)\exp\left(-\left(\frac{(E)}{k_B T}\right)\right)}{(1)\exp\left(-\frac{(0)}{k_B T}\right)+(1)\exp\left(-\frac{(E)}{k_B T}\right)} = \frac{E\exp\left(-\frac{E}{k_B T}\right)}{1+\exp\left(-\frac{E}{k_B T}\right)}$$

Therefore,

$$\langle E\rangle = \frac{E}{1+\exp\left(\frac{E}{k_B T}\right)} \Rightarrow N\langle E\rangle = \frac{NE}{1+\exp\left(\frac{E}{k_B T}\right)}.$$

The heat capacity of the system is,

$$C = N\frac{d\langle E\rangle}{dT} = \frac{d(N\langle E\rangle)}{dT} = Nk_B\left(\frac{E}{k_B T}\right)^2 \frac{\exp\left(\frac{E}{k_B T}\right)}{\left(\exp\left(\frac{E}{k_B T}\right)+1\right)^2}.$$

For $k_B T >> 1$, $\exp\left(\frac{E}{k_B T}\right) \approx 1$:

$$C \approx \frac{Nk_B}{4}\left(\frac{E}{k_B T}\right)^2$$

For $0 < k_B T << 1$, $\exp(E/k_B T) >> 1$:

$$C \approx Nk_B\left(\frac{E}{k_B T}\right)^2 \frac{\exp\left(\frac{E}{k_B T}\right)}{\left(\exp\left(\frac{E}{k_B T}\right)\right)^2} \Rightarrow C \approx Nk_B\left(\frac{E}{k_B T}\right)^2 \exp\left(-\frac{E}{k_B T}\right)$$

For each temperature extreme, the heat capacity approaches zero.

CALCULATE: Not applicable.

ROUND: Not applicable.

DOUBLE-CHECK: In general, in the extremely low temperature limit, the heat capacity must approach zero to be consistent with the third law of thermodynamics.

36.59. The de Broglie wavelength is $\lambda = h/p$. The momentum of the baseball is:

$$p = mv = (0.100 \text{ kg})(100. \text{ mi/h})(1609 \text{ m/mi})(1 \text{ h/3600 s}) = 4.469 \text{ kg m/s}.$$

The de Broglie wavelength of the baseball is:

$$\lambda = h/p = (6.626\cdot 10^{-34} \text{ J s})/(4.469 \text{ kg m/s}) = 1.48\cdot 10^{-34} \text{ m}.$$

The momentum of the spacecraft is:

$$p = mv = (250. \text{ kg})(125000 \text{ km/h})(1000 \text{ m/km})(1 \text{ h/3600 s}) = 8.681\cdot 10^{6} \text{ kg m/s}.$$

The de Broglie wavelength of the spacecraft is:

$$\lambda = h/p = (6.626\cdot 10^{-34} \text{ J s})/(8.681\cdot 10^{6} \text{ kg m/s}) = 7.63\cdot 10^{-41} \text{ m}.$$

36.63. The plates have a potential difference of $V = 5.0$ V between them. The magnitude of the stopping potential is therefore $V_0 = 5.0$ V. The work function of silver is $\phi = 4.7$ eV. The largest wavelength (lowest frequency and energy) of light λ_{max} that can be shined on the cathode to produce a current through the anode is found from the equation, $eV_0 = hf - \phi$. The wavelength of light is

$$f_{\min} = \frac{(eV_0 + \phi)}{h} \Rightarrow \frac{c}{\lambda_{\max}} = \frac{(eV_0 + \phi)}{h} \Rightarrow \lambda_{\max} = \frac{hc}{(eV_0 + \phi)}$$

$$\lambda_{\max} = \frac{(4.136 \cdot 10^{-15} \text{ eV s})(3.00 \cdot 10^8 \text{ m/s})}{e(5.0 \text{ V}) + (4.7 \text{ eV})} = \frac{(4.136 \cdot 10^{-15} \text{ eV s})(3.00 \cdot 10^8 \text{ m/s})}{(5.0 \text{ eV}) + (4.7 \text{ eV})} = 1.279 \cdot 10^{-7} \text{ m}$$

$$\lambda_{\max} \approx 130 \text{ nm}.$$

36.67. A nocturnal bird's eye can detect monochromatic light of frequency $f = 5.8 \cdot 10^{14}$ Hz with a power as small as $P = 2.333 \cdot 10^{-17}$ W. The energy of each detected photon is

$$E_{\text{ph}} = hf = (6.626 \cdot 10^{-34} \text{ J s})(5.8 \cdot 10^{14} \text{ Hz}) = 3.843 \cdot 10^{-19} \text{ J}.$$

The number of photons, n, detected by the bird per second is:

$$n/s = P/E_{\text{ph}}$$

$$n/s = (2.333 \cdot 10^{-17} \text{ W})/(3.843 \cdot 10^{-19} \text{ J}) \approx 61 \text{ photons/s}.$$

That is, the minimum number of photons that this bird can detect in one second is about 61 photons.

36.71. **THINK:** To estimate the number of photons that impact the Earth, it is useful to know that the intensity of the Sun's radiation on the Earth is $I = 1370 \text{ W/m}^2$. Use the peak wavelength of the light emitted by the Sun, $\lambda = 500.$ nm, as stated in section 36.2. Note that the Earth's upper atmosphere, the ionosphere, is $d = 300.$ km above the Earth's surface. The radius of the Earth is $R = 6378$ km. Finally, keep in mind that only half of the Earth's surface can face the Sun at any given time. Note that one year has approximately

$$t = 1 \text{ year}(365.24 \text{ days/yr})(24 \text{ hr/day})(3600 \text{ s/hr}) = 31{,}556{,}736 \text{ s}.$$

SKETCH:

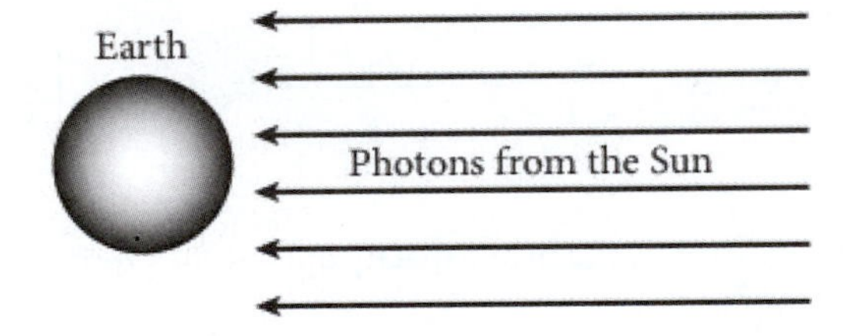

RESEARCH: The energy of a photon is $E_{\text{ph}} = hf = hc/\lambda$. The photon flux rate Φ (the number of photons per unit area per unit time) is found from $\Phi = I/E_{\text{ph}}$. The number of photons N that strike the Earth's upper atmosphere per year is $N = \Phi \cdot \frac{A_{\text{atm}}}{2} \cdot t_{\text{year}}$. The area that the Earth presents to the flux of photons from the Sun is $A = \pi r^2$.

SIMPLIFY: $\Phi = \frac{I}{E_{\text{ph}}} = \frac{I\lambda}{hc}$, $N = \frac{1}{2}\Phi A_{\text{atm}} \cdot t_{\text{year}} = \frac{1}{2}\left(\frac{I\lambda}{hc}\right)\left(\pi(R+d)^2\right)t_{\text{year}} = \frac{\pi I\lambda(R+d)^2 t_{\text{year}}}{2hc}$.

CALCULATE: $N = \frac{\pi(1370 \text{ W/m}^2)(5.00 \cdot 10^{-7} \text{ m})(6.678 \cdot 10^6 \text{ m})^2(1 \text{ yr})(31556736 \text{ s/yr})}{2(6.626 \cdot 10^{-34} \text{ J s})(3.00 \cdot 10^8 \text{ m/s})} = 7.618 \cdot 10^{42}$.

ROUND: To three significant figures, the number of photons received by Earth's upper atmosphere in one year is $N = 7.62 \cdot 10^{42}$.

DOUBLE-CHECK: This is a huge number, but is expected for the number of photons from the Sun to hit the Earth in one full year. Dimensional analysis confirms that the calculation yields a dimensionless result.

Multi-Version Exercises

36.73. **THINK:** An X-ray has an initial wavelength of $\lambda = 6.37$ nm. Its wavelength is increased by $\Delta\lambda = 1.13$ pm in a collision with an electron. Some of the energy of the photon will be imparted to the electron, giving it a velocity. To solve this problem, the conservation of energy is used. It is assumed that the electron is initially at rest.

SKETCH:

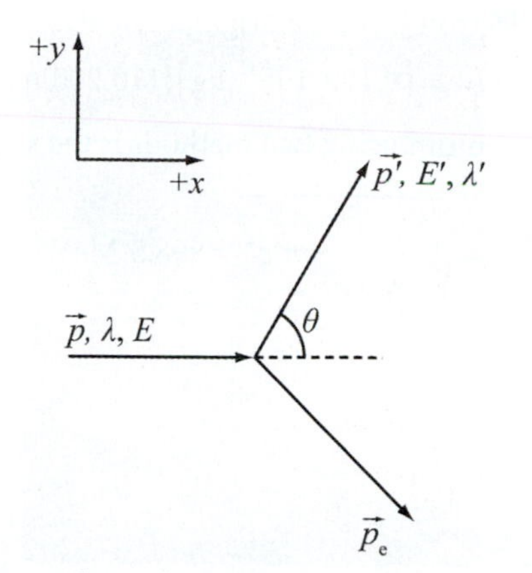

RESEARCH: Since energy is conserved in this scattering event, the kinetic energy that the electron receives is simply equal to the photon's energy loss:

$$K_e = E - E'.$$

Before the collision, the energy of the photon is $E = hc / \lambda$. After the collision, the energy of the X-ray is

$$E' = \frac{hc}{\lambda'} = \frac{hc}{\lambda + \Delta\lambda}.$$

SIMPLIFY: The kinetic energy of the electron after the collision is:

$$K_e = \frac{hc}{\lambda} - \frac{hc}{\lambda + \Delta\lambda} = hc\left(\frac{(\lambda + \Delta\lambda) - \lambda}{\lambda^2 + \lambda\Delta\lambda}\right) = \frac{hc\Delta\lambda}{\lambda^2 + \lambda\Delta\lambda} = \frac{1}{2}m_e v^2$$

$$v = \sqrt{\frac{2hc\Delta\lambda}{m_e\left(\lambda^2 + \lambda\Delta\lambda\right)}}$$

CALCULATE: $v = \sqrt{\frac{2\left(6.626 \cdot 10^{-34} \text{ J s}\right)\left(2.998 \cdot 10^{8} \text{ m/s}\right)\left(1.13 \cdot 10^{-12} \text{ m}\right)}{\left(9.109 \cdot 10^{-31} \text{ kg}\right)\left[\left(6.37 \cdot 10^{-9} \text{ m}\right)^2 + \left(6.37 \cdot 10^{-9} \text{ m}\right)\left(1.13 \cdot 10^{-12} \text{ m}\right)\right]}} = 110,200 \text{ m/s}$

ROUND: Rounding the result to two significant figures gives $v = 110.2$ km/s.

DOUBLE-CHECK: Momentum must also be conserved: $p = p' + p_e$. The initial momentum of the photon is $p = h / \lambda$. Since the *x*-direction is chosen to be the initial direction of the photon,

$$p = p_x \text{ and } p_y = 0.$$

The final direction of the photon is given by the Compton scattering formula,

$$\lambda' = \lambda + \frac{h}{m_e c}[1 - \cos\theta] \Rightarrow \theta = \cos^{-1}\left(1 - \frac{m_e c\Delta\lambda}{h}\right) = 57.70°.$$

The components of the final momentum of the photon are

$$p_x' = \frac{h}{\lambda'}\cos\theta = \frac{h}{\lambda + \Delta\lambda}\cos\theta \text{ and } p_y' = \frac{h}{\lambda'}\sin\theta = \frac{h}{\lambda + \Delta\lambda}\sin\theta.$$

The difference between the final and initial momentum of the photon must be equal to the final momentum of the electron.

$$p_{e,x} = p_x - p_x' = \frac{h}{\lambda} - \frac{h}{\lambda + \Delta\lambda}\cos\theta = \frac{6.626 \cdot 10^{-34} \text{ J s}}{6.37 \cdot 10^{-9} \text{ m}} - \frac{6.626 \cdot 10^{-34} \text{ J s}}{6.37113 \cdot 10^{-9} \text{ m}}\cos\left(57.70^\circ\right) = 4.8452 \cdot 10^{-26} \text{ kg m/s}$$

$$p_{e,y} = p_y - p_y' = 0 - \frac{h}{\lambda + \Delta\lambda}\sin\theta = -\frac{6.626 \cdot 10^{-34} \text{ J s}}{6.37113 \cdot 10^{-9} \text{ m}}\sin\left(57.70^\circ\right) = -8.7912 \cdot 10^{-26} \text{ kg m/s}$$

$$p_e = \sqrt{\left(p_{e,x}\right)^2 + \left(p_{e,y}\right)^2} = \sqrt{\left(4.8452 \cdot 10^{-26} \text{ J s/m}\right)^2 + \left(-8.7912 \cdot 10^{-26} \text{ J s/m}\right)^2} = 1.0038 \cdot 10^{-25} \text{ kg m/s}$$

The momentum of the electron from the original calculation is

$$p_e = m_e v = \left(9.109 \cdot 10^{-31} \text{ kg}\right)\left(110{,}200 \text{ m/s}\right) = 1.0038 \cdot 10^{-25} \text{ kg m/s}.$$

Since the calculated momentum using two methods is the same, the speed of the electron found is correct.

36.78. $p = \frac{h}{\lambda};\; E = \sqrt{p^2c^2 + m^2c^4} = \sqrt{\left(\frac{hc}{\lambda}\right)^2 + m^2c^4} = 998.5 \text{ MeV}$

Chapter 37: Quantum Mechanics

Exercises

37.23. The kinetic energy of a neutron is $10.0\text{ MeV} = 1.60 \cdot 10^{-12}$ J. The size of an object that is necessary to observe diffraction effects is on the order of the de Broglie wavelength of the neutron. The (relativistic) de Broglie wavelength is given by

$$\lambda = \frac{h}{p} = \frac{hc}{\sqrt{K^2 + 2Kmc^2}}.$$

$$\lambda = \frac{\left(6.63 \cdot 10^{-34}\text{ J s}\right)\left(3.00 \cdot 10^{8}\text{ m/s}\right)}{\sqrt{\left(1.60 \cdot 10^{-12}\text{ J}\right)^2 + 2\left(1.60 \cdot 10^{-12}\text{ J}\right)\left(1.67 \cdot 10^{-27}\text{ kg}\right)\left(3.00 \cdot 10^{8}\text{ m/s}\right)^2}} = 9.0454 \cdot 10^{-15}\text{ m} = 9.05\text{ fm}.$$

Since protons and neutrons have a diameter of about 1.00 fm, they would be useful targets to demonstrate the wave nature of 10.0-MeV neutrons.

37.27. The energies for a particle in an infinite square well are $E_n = \frac{\hbar^2\pi^2}{2mL^2}n^2$ for a square well of length L and $E_n = \frac{\hbar^2\pi^2}{2m(2L)^2}n^2$ for a square well of length $2L$. Therefore, $\frac{(E_2 - E_1)L}{(E_2 - E_1)2L} = \frac{\hbar^2\pi^2\left(2^2 - 1^2\right)/\left(2mL^2\right)}{\hbar^2\pi^2\left(2^2 - 1^2\right)/\left(8mL^2\right)} = 4.$

37.31. The potential energy for the well is given by:

$$U(x) = \begin{cases} \infty & \text{for } x < 0 \\ 0 & \text{for } 0 \le x \le a \\ U_1 & \text{for } x > a \end{cases}$$

This is illustrated in the diagram:

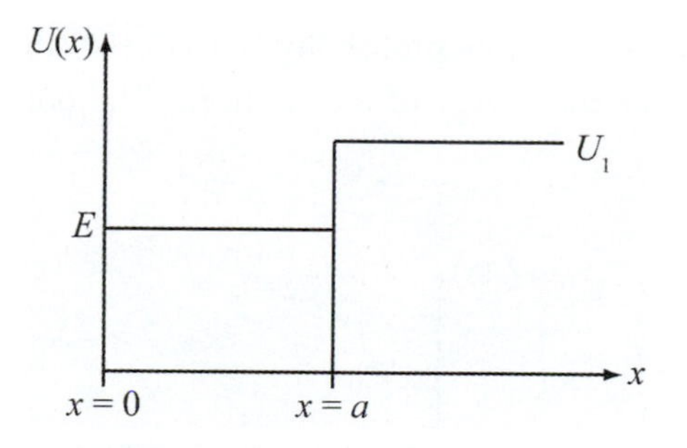

Since the question states that the electron is confined to the potential well, $E < U_1$. As shown in the text, the wave function for this finite potential well can be written as:

$$\psi(x) = \begin{cases} 0 & \text{for } x < 0 \\ A\sin(\kappa x) & \text{for } 0 \le x \le a \\ Be^{-\gamma x} & \text{for } x > a \end{cases}$$

Where $\kappa=\sqrt{\frac{2mE}{\hbar^2}}$ and $\gamma=\sqrt{\frac{2m(U_1-E)}{\hbar^2}}$. The wave function $\psi(x)$ must satisfy the boundary conditions at $x=a$:

$$(1)\ A\sin(\kappa a)=Be^{-\gamma a}$$
$$(2)\ \kappa A\cos(\kappa a)=-\gamma Be^{-\gamma a}.$$

Dividing (1) and (2) yields $\frac{\tan(\kappa a)}{\kappa}=-\frac{1}{\gamma}$ or $\tan(\kappa a)=-\frac{\kappa}{\gamma}$. Since κ and γ are positive, $\tan(\kappa a)$ must be negative. This is satisfied when

$$(2n-1)\frac{\pi}{2}<\kappa a<n\pi,\ n=1,2,3...$$

For the third state $(n=3)$:

$$\frac{5\pi}{2}<\kappa a<3\pi \Rightarrow \frac{25\pi^2}{4}<\kappa^2a^2<9\pi^2 \Rightarrow \frac{25\pi^2}{4}<\frac{2mEa^2}{\hbar^2}<9\pi^2.$$

Therefore,

$$\frac{25}{4}\frac{\hbar^2\pi^2}{2ma^2}<E_3<9\frac{\hbar^2\pi^2}{2ma^2} \Rightarrow \frac{25}{4}E_1<E_3<9E_1,$$

where E_1 is the ground state energy for the infinite square well:

$$E_1=\frac{\left(1.055\cdot 10^{-34}\text{ J s}\right)^2\pi^2}{2\left(9.11\cdot 10^{-31}\text{ kg}\right)\left(1.0\cdot 10^{-9}\text{ m}\right)^2}=6.03\cdot 10^{-20}\text{ J}=0.376\text{ eV}.$$

Therefore,

$$\frac{25}{4}(0.376\text{ eV})<E_3<9(0.376\text{ eV}) \Rightarrow 2.4\text{ eV}<E_3<3.4\text{ eV}.$$

Since $U_1=2.0\text{ eV}<E_3$, the third state is not a bound state.

37.35. **THINK:** Given that the tunneling probability is $T=0.100$, the equation for the transmission coefficient can be used to calculate the energy of the electron. The potential barrier is $b-a=2.00\text{ nm}$ wide and $U_1=7.00\text{ eV}$ high.

SKETCH:

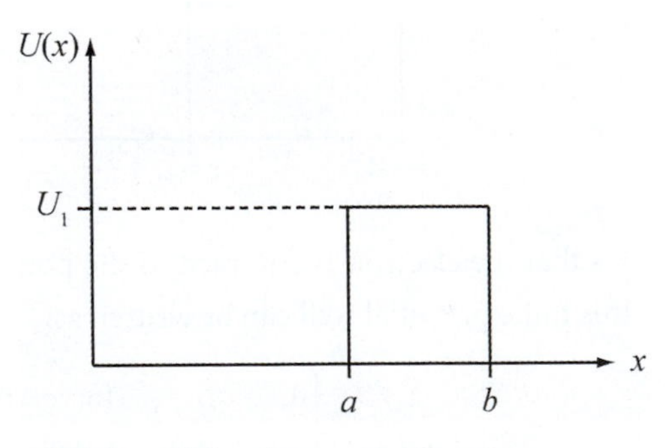

RESEARCH: The tunneling probability of the electron is given by:

$$T=e^{-2\gamma(b-a)}\text{ where }\gamma=\sqrt{\frac{2m(U_1-E)}{\hbar^2}}.$$

SIMPLIFY: Solving for E gives:

$$\ln(T) = -2\sqrt{\frac{2m(U_1 - E)}{\hbar^2}}(b-a)$$

$$\frac{2m(U_1 - E)}{\hbar^2} = \left(-\frac{\ln(T)}{2(b-a)}\right)^2$$

$$U_1 - E = \frac{\hbar^2}{2m}\left(-\frac{\ln(T)}{2(b-a)}\right)^2$$

$$E = U_1 - \frac{\hbar^2}{2m}\left(-\frac{\ln(T)}{2(b-a)}\right)^2.$$

CALCULATE:

$$E = (7.00 \text{ eV})(1.602 \cdot 10^{-19} \text{ J/eV}) - \frac{(1.055 \cdot 10^{-34} \text{ J s})^2}{2(9.11 \cdot 10^{-31} \text{ kg})}\left(-\frac{\ln(0.100)}{2(2.00 \cdot 10^{-9} \text{ m})}\right)^2 = 1.119 \cdot 10^{-18} \text{ J} = 6.987 \text{ eV}$$

ROUND: To three significant figures, the energy of the electron is $E = 6.99$ eV.

DOUBLE-CHECK: The electron energy comes out as less than the potential barrier, as expected.

37.39. **THINK:**

(a) The Schrödinger equation and the relevant boundary conditions can be used to find the wave function and the energy levels.

(b) The solution to the Schrödinger equation can be used to find the penetration distance η for a decrease in the wave function by a factor of $1/e$.

(c) This particular quantum well has width 1 nm and depth 0.300 eV with energy 0.125 eV.

The finite well potential is given by the function:

$$U(x) = \begin{cases} U_0 & \text{for } x \le -a/2 \\ 0 & \text{for } -a/2 \le x \le a/2 \\ U_0 & \text{for } x \ge a/2 \end{cases}$$

SKETCH:

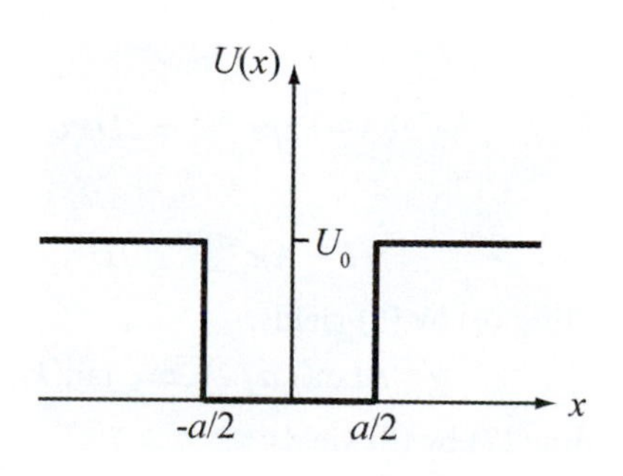

RESEARCH:

(a) The solution to the Schrödinger equation for each region is given by:

$$\psi(x) = \begin{cases} Ae^{\gamma x} + Be^{-\gamma x} & \text{for } x \le -a/2 \\ C\cos(\kappa x) + D\sin(\kappa x) & \text{for } -a/2 \le x \le a/2 \\ Ge^{\gamma x} + Fe^{-\gamma x} & \text{for } x \ge a/2 \end{cases}$$

where,

$$\kappa^2 = \frac{2mE}{\hbar^2}, \; \gamma^2 = \frac{2m(U_0 - E)}{\hbar^2}.$$

Combining these expressions gives

$$\gamma^2 = \frac{2m(U_0 - E)}{\hbar^2} = \frac{2mU_0}{\hbar^2} - \kappa^2 \Rightarrow \gamma^2 + \kappa^2 = \frac{2mU_0}{\hbar^2},$$

which represents circles in the $\kappa\gamma$-plane of radius $\frac{2mU_0}{\hbar^2}$. However, in the region $x < -a/2$, as $x \to -\infty$ the B term blows up and in the region $x > a/2$, as $x \to \infty$ the G term blows up. Therefore, the physical solution is given by:

$$\psi(x) = \begin{cases} Ae^{\gamma x} & \text{for } x \le -a/2 \\ C\cos(\kappa x) + D\sin(\kappa x) & \text{for } -a/2 \le x \le a/2 \\ Fe^{-\gamma x} & \text{for } x \ge a/2 \end{cases}$$

b) For $x \ge a/2$ the solution requires that

$$\psi(x)\big|_{x=a/2+\eta} = \frac{1}{e}\psi(x)\big|_{x=a/2}$$

SIMPLIFY:

(a) At $x = -a/2$, continuity of the function and its derivative requires:

$$Ae^{-\gamma a/2} = C\cos(-\kappa a/2) + D\sin(-\kappa a/2) = C\cos(\kappa a/2) - D\sin(\kappa a/2) \quad (1)$$

$$-A\gamma e^{-\gamma a/2} = -C\kappa\sin(\kappa a/2) - D\kappa\cos(\kappa a/2) = -(C\kappa\sin(\kappa a/2) + D\kappa\cos(\kappa a/2))$$

$$A\gamma e^{-\gamma a/2} = C\kappa\sin(\kappa a/2) + D\kappa\cos(\kappa a/2) \quad (2)$$

At $x = a/2$, continuity of the function and its derivative requires:

$$Fe^{-\gamma a/2} = C\cos(\kappa a/2) + D\sin(\kappa a/2) \quad (3)$$

$$-F\gamma e^{-\gamma a/2} = -C\kappa\sin(\kappa a/2) + D\kappa\cos(\kappa a/2) \quad (4)$$

These four equations can be simplified:

Adding (1) and (3):

$$(A+F)e^{-\gamma a/2} = 2C\cos(\kappa a/2) \quad (5)$$

Subtracting (4) from (2):

$$(A+F)\gamma e^{-\gamma a/2} = 2C\kappa\sin(\kappa a/2) \quad (6)$$

Adding (2) and (4):

$$(A-F)\gamma e^{-\gamma a/2} = 2D\kappa\cos(\kappa a/2) \quad (7)$$

Subtracting (1) from (3):

$$(F-A)e^{-\gamma a/2} = 2D\sin(\kappa a/2) \quad (8)$$

If $C \neq 0$ and $A \neq -F$, dividing (6) by (5) yields:

$$\gamma = \kappa\tan(\kappa a/2) \Rightarrow \tan(\kappa a/2) = \gamma/\kappa$$

If $D \neq 0$ and $A \neq F$, dividing (7) by (8) yields:

$$-\gamma = \kappa\cot(\kappa a/2) \Rightarrow \tan(\kappa a/2) = -\kappa/\gamma$$

If these two equations are simultaneously valid then they imply that $\tan^2(\kappa a/2) = -1$ which cannot be true for real values of the energy (i.e. κ must be real). This means that solutions can be divided into two separate classes. The wave functions split into even and odd parity solutions are given by:

(i) For even parity solutions where $\Psi(x) = C\cos(\kappa x)$ in the well, $D = 0$ and $A = F$. The wave function is given by:

$$\psi(x) = \begin{cases} Ae^{\gamma x} & \text{for } x \le -a/2 \\ C\cos(\kappa x) & \text{for } -a/2 \le x \le a/2 \\ Ae^{-\gamma x} & \text{for } x \ge a/2 \end{cases}$$

This leads to the solution $\kappa\tan(\kappa a/2)=\gamma$.

(ii) For odd parity solutions where $\Psi(x)=D\sin(\kappa x)$ in the well, $C=0$ and $A=-F$. The wave function is given by:

$$\psi(x)=\begin{cases} Ae^{\gamma x} & \text{for } x\le -a/2 \\ D\sin(\kappa x) & \text{for } -a/2\le x\le a/2 \\ -Ae^{-\gamma x} & \text{for } x\ge a/2 \end{cases}$$

This leads to the solution $\kappa\cot(\kappa a/2)=-\gamma$. The energy levels can be found by solving numerically or graphically each of these solutions with the required relation between κ and γ: $\gamma^2=\dfrac{2mU_0}{\hbar^2}-\kappa^2$. Solving $\kappa\tan(\kappa a/2)=\gamma$ and $\kappa\cot(\kappa a/2)=-\gamma$ graphically (intersection points) gives discrete values for κ and γ and hence the allowed energy levels are obtained from the κ values at the intersection points and $E=\dfrac{\hbar^2\kappa^2}{2m}$. A sketch of such a graph is shown:

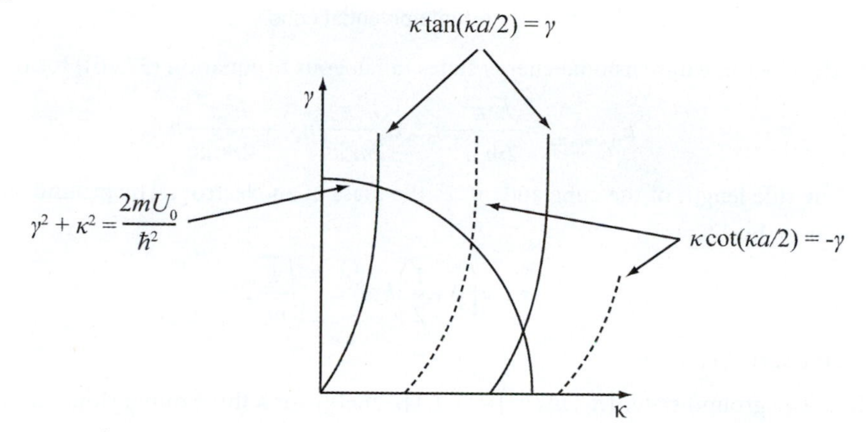

(b) $Fe^{-\gamma(a/2+\eta)}=\frac{1}{e}\left(Fe^{-\gamma(a/2)}\right)\Rightarrow e^{-\gamma\eta}=e^{-1}\Rightarrow \gamma\eta=1$

The penetration distance is given by $\eta=\dfrac{1}{\gamma}=\sqrt{\dfrac{\hbar^2}{2m(U_0-E)}}=\dfrac{\hbar}{\sqrt{2m(U_0-E)}}$.

CALCULATE:

(c) $\eta_{\text{GaAs-GaAlAs}}=\dfrac{(1.055\cdot 10^{-34}\text{ J s})}{\sqrt{2(9.109\cdot 10^{-31}\text{ kg})(0.300\text{ eV}-0.125\text{ eV})(1.602\cdot 10^{-19}\text{ J/eV})}}=4.668\cdot 10^{-10}\text{ m}$

ROUND:

(c) To three significant figures, the penetration distance is $\eta_{\text{GaAs-GaAlAs}}=467$ pm.

DOUBLE-CHECK: It is reasonable that the penetration depth is independent of the width of the well. A unit analysis of the units for the penetration depth provides the correct unit of length:

$$\frac{\text{J s}}{\sqrt{\text{J kg}}}=\sqrt{\frac{\text{J s}^2}{\text{kg}}}=\sqrt{\frac{(\text{kg m}^2/\text{s}^2)\text{ s}^2}{\text{kg}}}=\sqrt{\text{m}^2}=\text{m}.$$

37.43. **THINK:** Since the electron is confined to a cube, the electron can be treated as if it was inside a three-dimensional infinite potential well. In the text, the equation for the energy states for a two dimensional infinite potential is derived. An analogous form for the three dimensional case can be used to determine the ground state energy of the electron in the cube of side length $2R$, where $R=0.0529$ nm. The spring constant can be found by setting the ground state energy for a potential well equal to the ground state energy for a harmonic oscillator.

SKETCH:

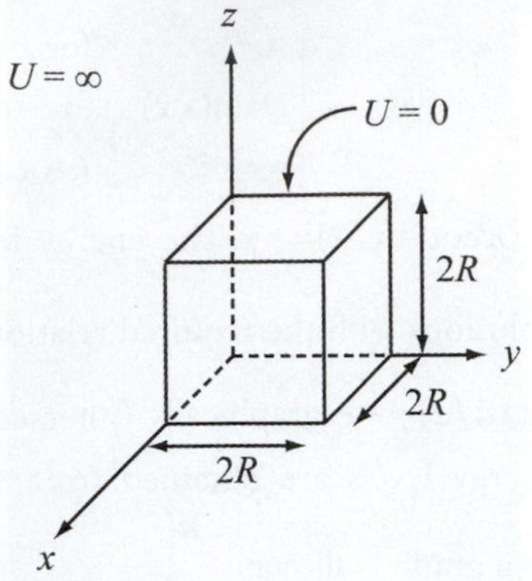

3D infinite potential cube

RESEARCH: The three dimensional energy states (analogous to equation (37.16)) for the electron are:

$$E_{n_x,n_y,n_z} = \frac{\hbar^2\pi^2}{2m_e a^2}n_x^{\ 2} + \frac{\hbar^2\pi^2}{2m_e a^2}n_y^{\ 2} + \frac{\hbar^2\pi^2}{2m_e a^2}n_z^{\ 2},$$

where a is the side length of the cube and m_e is the mass of an electron. The ground state of a harmonic oscillator is given by:

$$E_0 = \left(0+\frac{1}{2}\right)\hbar\omega_0 = \frac{\hbar}{2}\sqrt{\frac{k}{m_e}},$$

where k is the spring constant.

SIMPLIFY: The ground state, $(n_x, n_y, n_z) = (1,1,1)$, energy for a three dimensional infinite potential well of side length $a = 2R$ is:

$$E_{1,1,1} = \frac{3\hbar^2\pi^2}{8m_e R^2}.$$

For the case $E_{1,1,1} = E_0$: $E_{1,1,1} = \frac{3\hbar^2\pi^2}{8m_e R^2} = \frac{\hbar}{2}\sqrt{\frac{k}{m_e}} \Rightarrow \frac{3\hbar\pi^2}{4m_e R^2} = \sqrt{\frac{k}{m_e}} \Rightarrow k = \frac{9\hbar^2\pi^4}{16m_e R^4}.$

CALCULATE: $E_{1,1,1} = \frac{3\hbar^2\pi^2}{8m_e R^2} = \frac{3\left(1.055\cdot 10^{-34}\text{ J s}\right)^2\pi^2}{8\left(9.11\cdot 10^{-31}\text{ kg}\right)\left(0.0529\cdot 10^{-9}\text{ m}\right)^2} = 1.6159\cdot 10^{-17}\text{ J} = 100.87\text{ eV}$

$$k = \frac{9\left(1.055\cdot 10^{-34}\text{ J s}\right)^2\pi^4}{16\left(9.11\cdot 10^{-31}\text{ kg}\right)\left(0.0529\cdot 10^{-9}\text{ m}\right)^4} = 8.5484\cdot 10^4\text{ N/m}$$

ROUND: To three significant figures, the ground state energy for an electron confined to a cube of twice the Bohr radius is $E = 101\text{ eV}$ and the spring constant that would give the same ground state energy for a harmonic oscillator is $k = 85.5\text{ kN/m}$.

DOUBLE-CHECK: The ionization energy of an electron in a hydrogen atom is 13.6 eV and is comparable to the energy calculated.

37.47. (a) Normalization requires that $\int_{-\infty}^{\infty}|\psi(x)|^2\,dx=1$. Given that the wave function of the electron in the region $0<x<L$ is $\psi(x)=A\sin(2\pi x/L)$,

$$\int_0^L A^2\sin^2\left(\frac{2\pi x}{L}\right)dx=1$$

The identity $2\sin^2\theta=1-\cos2\theta$ can be used to simplify the integrand:

$$1=\frac{A^2}{2}\int_0^L\left[1-\cos\left(\frac{4\pi x}{L}\right)\right]dx$$
$$=\frac{A^2}{2}\left[x-\frac{L}{4\pi}\sin\left(\frac{4\pi x}{L}\right)\right]_0^L$$
$$=\frac{A^2}{2}L \Rightarrow A=\sqrt{\frac{2}{L}}.$$

(b) The probability of finding the electron in the region $0<x<L/3$ is:

$$\Pi=\int_0^{L/3}|\psi(x)|^2\,dx=\int_0^{L/3}\left|\sqrt{\frac{2}{L}}\sin\left(\frac{2\pi x}{L}\right)\right|^2dx=\frac{2}{L}\int_0^{L/3}\sin^2\left(\frac{2\pi x}{L}\right)dx$$
$$=\frac{1}{L}\left[x-\frac{L}{4\pi}\sin\left(\frac{4\pi x}{L}\right)\right]_0^{L/3}=\frac{1}{3}-\frac{1}{4\pi}\sin\left(\frac{4\pi}{3}\right)=0.402.$$

37.51. **THINK:** A one dimensional plane-wave wave function can be generalized for three dimensions to find $\Psi(\vec{r},t)$ for a non relativistic particle of mass m and momentum p. For a free particle, $U(\vec{r})=0$ identically. It is constantly zero.

SKETCH:

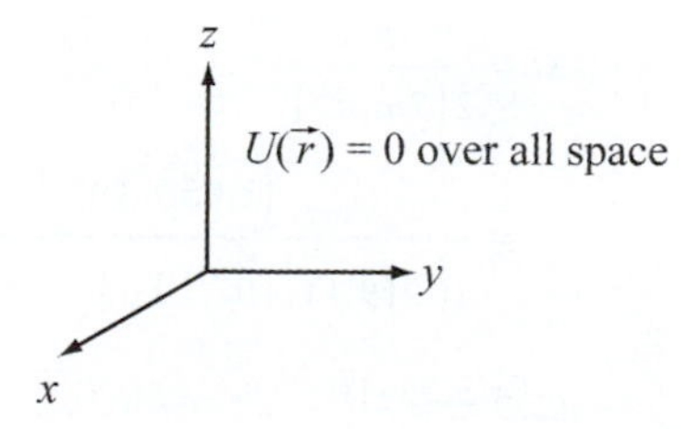

RESEARCH: A plane-wave wave function in one dimension is given by:

$$\Psi(\vec{x},t)=Ae^{i\kappa x}e^{-i\omega t},\text{ where }\kappa=p/\hbar\text{ and }\omega=E/\hbar.$$

The wave function can be assumed separable into spatial and time dependent parts. Here p is the momentum of the particle and E is the energy. The probability density function is $|\Psi(\vec{r},t)|^2$.

SIMPLIFY: The spatial wave function for such a particle can be written as the product of three plane waves. Hence, the wave function takes the form

$$\Psi(\vec{r},t)=Ae^{i(\vec{p}\cdot\vec{r})/\hbar}e^{-iEt/\hbar},$$

where $\vec{r}=\vec{x}+\vec{y}+\vec{z}$. κ and ω have been rewritten as $\kappa=p/\hbar$ and $\omega=E/\hbar$. Since $E=p^2/(2m)$ is the energy of a non relativistic particle, the full wave function can also be written as

$$\Psi(\vec{r},t)=Ae^{i(\vec{p}\cdot\vec{r})/\hbar}e^{-ip^2t/2m\hbar}.$$

The probability density is $|\Psi(\vec{r},t)|^2=\left(Ae^{-i(\vec{p}\cdot\vec{r})/\hbar}e^{ip^2t/2m\hbar}\right)\left(Ae^{i(\vec{p}\cdot\vec{r})/\hbar}e^{-ip^2t/2m\hbar}\right)=A^2.$

CALCULATE: Not applicable.
ROUND: Not applicable.

DOUBLE-CHECK: The spatial part of this wave function clearly represents a plane wave as $\vec{k}\cdot\vec{r}=c$, where c is a constant, is the general form of a plane perpendicular to $\vec{k}$. The wave function can also be substituted into the time dependent Schrödinger Equation satisfying the equation:

$$(-\hbar^2/2m)\partial^2\left(\Psi(\vec{r},t)\right)/\partial\vec{r}^2+U(r)\Psi(\vec{r},t)=i\hbar\partial\left(\Psi(\vec{r},t)\right)/\partial t$$

Since U = 0 for a free particle Schrödinger's Equation becomes:

$$(-\hbar^2/2m)\partial^2\left(\Psi(\vec{r},t)\right)/\partial\vec{r}^2=i\hbar\partial\left(\Psi(\vec{r},t)\right)/\partial t$$

Substituting for $\Psi(\vec{r},t)=\Psi(\vec{r})\Psi(t)=Ae^{i(\vec{p}\cdot\vec{r})/\hbar}e^{-ip^2t/2m\hbar}$ and differentiating, the left side is:

$$\frac{-\hbar^2(-p^2)Ae^{i(\vec{p}\cdot\vec{r})/\hbar}e^{-ip^2t/2m\hbar}}{2\hbar^2m}$$

and the right side is:

$$\frac{-i\hbar Aip^2e^{i(\vec{p}\cdot\vec{r})/\hbar}e^{-ip^2t/2m\hbar}}{2m\hbar}$$

After cancelling like terms and recalling that $i^2=-1$ these are equal and so the wave function does satisfy the time dependent Schrödinger Equation.

37.55. The energy time uncertainty relation is given by:

$$\Delta E\Delta t\geq\frac{\hbar}{2}.$$

A violation of classical energy conservation by an amount ΔE is then possible as long as the time interval over which this happens is at most Δt , the value of which is given by the minimal value obtained from the uncertainty relation. ΔE in this case is at least the sum of the rest energies of the particle and the antiparticle. This means that the maximum lifetime of the particle-antiparticle pair is given by:

(a) For an electron/positron pair:

$$\begin{aligned}\Delta t&=\frac{\hbar}{2\left(2m_ec^2\right)}\\&=\frac{\left(1.055\cdot10^{-34}\text{ J s}\right)}{2\left(2\left(9.11\cdot10^{-31}\text{ kg}\right)\left(3.00\cdot10^{8}\text{ m/s}\right)^2\right)}\\&=3.22\cdot10^{-22}\text{ s.}\end{aligned}$$

(b) For a proton/anti-proton pair:

$$\Delta t=\frac{\hbar}{2\left(2m_pc^2\right)}=\frac{\left(1.055\cdot10^{-34}\text{ J s}\right)}{2\left(2\left(1.67\cdot10^{-27}\text{ kg}\right)\left(3.00\cdot10^{8}\text{ m/s}\right)^2\right)}=1.75\cdot10^{-25}\text{ s.}$$

37.59. The normalized solution of the wave function in the ground state $(n=1)$ for an electron in an infinite cubic potential well of side length L is given by:

(a) $\psi=\psi_x(x)\psi_y(y)\psi_z(z)=\left(\sqrt{\frac{2}{L}}\right)^3\sin\frac{\pi x}{L}\sin\frac{\pi y}{L}\sin\frac{\pi z}{L};\ 0<x,y,z<L$

(b) Since the energies are given by

$$E=\frac{\hbar^2\pi^2}{2mL^2}\left(n_x^2+n_y^2+n_z^2\right),$$

the different energies depend on the energy state, $\left(n_x^2+n_y^2+n_z^2\right)$. The ground state is for $\left(n_x,n_y,n_z\right)=(1,1,1)$, the first excited state is for $\left(n_x,n_y,n_z\right)=(1,2,1)$, $(2,1,1)$, $(1,1,2)$, and the second excited state is for $(n_x, n_y, n_z) = (1,2,2),\ (2,1,2),\ (2,2,1)$. Since an electron has two spin states (up or down), there are a total of 14 possible energy states.

37.61. The distance between fringes (central maximum and first order peak) for a double slit setup is given by $\Delta y = \frac{\lambda L}{d}$. The wavelength is given by:

$$\lambda = \frac{h}{p} = \frac{h}{\sqrt{2mK}} = \frac{(6.626 \cdot 10^{-34}\ \text{J s})}{\sqrt{2(1.67 \cdot 10^{-27}\ \text{kg})(4.005 \cdot 10^{-21}\ \text{J})}} = 1.812 \cdot 10^{-10}\ \text{m}$$

The distance between interference peaks is:

$$\Delta y = \frac{(1.812 \cdot 10^{-10}\ \text{m})(1.5\ \text{m})}{(0.50 \cdot 10^{-3}\ \text{m})} = 5.44 \cdot 10^{-7}\ \text{m} = 0.54\ \mu\text{m}.$$

37.65. The fundamental state $(n=1)$ energy of a neutron between rigid walls (a one dimensional infinite potential well) $L = 8.4$ fm apart is:

$$E_1 = \frac{\hbar^2 \pi^2}{2mL^2} = \frac{(1.055 \cdot 10^{-34}\ \text{J s})^2 \pi^2}{2(1.67 \cdot 10^{-27}\ \text{kg})(8.4 \cdot 10^{-15}\ \text{m})^2} = 4.7 \cdot 10^{-13}\ \text{J} = 2.9\ \text{MeV}.$$

37.69. **THINK:** In the text, the equations for the energy states for a one and two dimensional infinite potential are derived. An analogous form for the three dimensional case can be used to determine the ground state energy of the electron in the potential cube of side length $a = 0.10$ nm.

SKETCH:

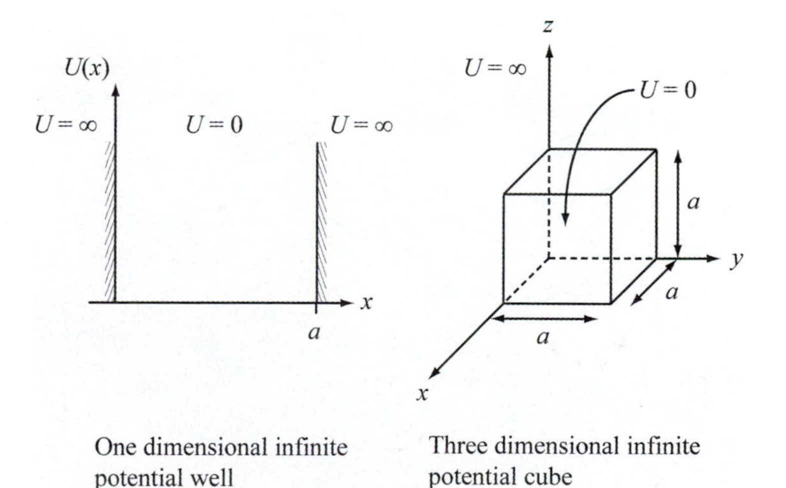

One dimensional infinite potential well

Three dimensional infinite potential cube

RESEARCH: The allowed energies for the one dimensional infinite potential well are given by:

$$E_{n,\text{1D}} = \frac{\hbar^2 \pi^2}{2ma^2} n^2.$$

The allowed energies for the three dimensional infinite potential cube in its ground state are given by:

$$E_{n,\text{3D}} = \frac{\hbar^2 \pi^2}{2ma^2}\left(n_x^2 + n_y^2 + n_z^2\right).$$

SIMPLIFY: The electron confined to the cube is in its ground state, so:

$$E_{1,\text{3D}} = \frac{3\hbar^2 \pi^2}{2ma^2}.$$

$E_{n,\,1\text{D}}$ is closest to $E_{1,\,3\text{D}}$ for $n=2$ (the first excited state), so the smallest energy difference is given by:

$$E_{\text{min}} = E_{2,1\text{D}} - E_{1,3\text{D}} = \frac{4\hbar^2\pi^2}{2ma^2} - \frac{3\hbar^2\pi^2}{2ma^2} = \frac{\hbar^2\pi^2}{2ma^2}.$$

CALCULATE: $E_{\text{min}} = \dfrac{\left(1.055\cdot 10^{-34}\text{ J s}\right)^2 \pi^2}{2\left(9.11\cdot 10^{-31}\text{ kg}\right)\left(0.10\cdot 10^{-9}\text{ m}\right)^2} = 6.029\cdot 10^{-18}\text{ J} = 37.6\text{ eV}$

ROUND: To two significant figures, the minimum energy difference is 38 eV.

DOUBLE-CHECK: The energy is of the same order of magnitude with the ionization energy of an electron (13.6 eV) in a hydrogen atom. Therefore, the answer is reasonable.

Multi-Version Exercises

37.72. $T = e^{-2\gamma w}$; width $= w$

decay constant $\gamma = \sqrt{2(mc^2)(U-E)/(\hbar c)^2}$

$= \sqrt{2(3727.4\text{ MeV})(15.5\text{ MeV} - 5.15\text{ MeV})/(197.327\text{ MeV fm})^2} = 1.40767\text{ fm}^{-1}$

$T = e^{-2(1.40767)(11.7)} = 4.95\cdot 10^{-15}$

37.76. $\Delta E = E_\text{f} - E_\text{i} = (n_\text{f}^2 - n_\text{i}^2)\dfrac{\hbar^2\pi^2}{2ma^2} = (25-1)\dfrac{(1.0546\cdot 10^{-34}\text{ J s})^2\pi^2}{2(9.109\cdot 10^{-31}\text{ kg})(13.5\cdot 10^{-9}\text{ m})^2} = 7.93\cdot 10^{-21}\text{ J}$

Chapter 38: Atomic Physics

Exercises

38.19. The shortest wavelength of light that a hydrogen atom can emit occurs when an electron is captured by the hydrogen atom and jumps directly to the ground state, i.e. a transition from $n_2 = \infty$ to $n_1 = 1$. The Rydberg formula gives a wavelength of:

$$\lambda = \left[R_{\mathrm{H}} \left(\frac{1}{n_1^2} - \frac{1}{n_2^2} \right) \right]^{-1} = \left[\left(1.097 \cdot 10^7 \text{ m}^{-1} \right) \left(\frac{1}{1} - \frac{1}{\infty} \right) \right]^{-1} = 91.16 \text{ nm}$$

or 91.2 nm to the usual three significant figures.

38.23. The quantum number of the fifth excited state is $n = 6$. The energy of the fifth excited state is:

$$E_n = -E_0 \frac{1}{n^2} \Rightarrow E_6 = -(13.6 \text{ eV}) \frac{1}{(6)^2} = -0.378 \text{ eV}.$$

38.27. **THINK:** An excited hydrogen atom emits a photon with an energy of $E_{\mathrm{ph}} = 1.133$ eV. When emitting a photon, the hydrogen atom loses the energy of the photon. To determine the initial and final states of the hydrogen atom, the energy levels must be analyzed to determine which set of two can be separated by the given photon energy. Start with the final energy level being the ground state, and then progress to higher states as the final state.

SKETCH:

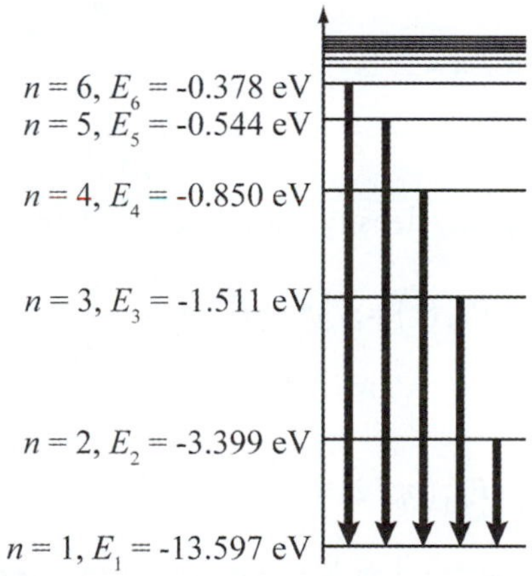

RESEARCH: The energy of the nth energy level in a hydrogen atom is $E_n = -13.6 \text{ eV}/n^2$. Since the hydrogen atom loses the energy that is gained by the emitted photon, $E_{\mathrm{ph}} = -\Delta E = E_{\mathrm{i}} - E_{\mathrm{f}}$.

SIMPLIFY: $E_{\mathrm{i}} = E_{\mathrm{f}} + E_{\mathrm{ph}}$

CALCULATE: The lowest energy levels of the hydrogen atom are: $E_1 = -13.597$ eV, $E_2 = -3.399$ eV, $E_3 = -1.511$ eV, $E_4 = -0.850$ eV, $E_5 = -0.544$ eV, $E_6 = -0.378$ eV. For $E_{\mathrm{f}} = E_1$, $E_{\mathrm{i}} = (-13.597 \text{ eV}) + (1.133 \text{ eV}) = -12.464 \text{ eV}$, which is not an allowed level. For $E_{\mathrm{f}} = E_2$, $E_{\mathrm{i}} = (-3.399 \text{ eV}) + (1.133 \text{ eV}) = -2.266 \text{ eV}$, which is not an allowed level. For $E_{\mathrm{f}} = E_3$, $E_{\mathrm{i}} = (-1.511 \text{ eV}) + (1.133 \text{ eV}) = -0.378 \text{ eV}$, which is the energy level E_6. The photon was emitted when an excited atom in the $n = 6$ state transitioned to the $n = 3$ state. This is the only allowed transition.

ROUND: Not applicable.

DOUBLE-CHECK: For a transition from the $n = 6$ state to the $n = 3$ state, the emitted photon has energy $E = E_6 - E_3 = (-0.378 \text{ eV}) - (-1.511 \text{ eV}) = 1.133 \text{ eV}$, as required.

38.31. For an electron in the $n=5$ shell, the orbital angular momentum quantum number can range from 0 to $n-1$; that is, $\ell=0, 1, 2, 3, 4$. The magnitude of the total orbital angular momentum is $L=\sqrt{\ell(\ell+1)}\hbar$. The largest possible value for the angular momentum occurs with $\ell=4$: $L=\sqrt{4(4+1)}\hbar=\sqrt{20}\hbar\approx 4.716\cdot 10^{-34}$ J s. The smallest possible value for the orbital angular momentum occurs with $\ell=0$: $L=\sqrt{0(0+1)}\hbar=0$.

38.35. **THINK:** The radial wave function for hydrogen in the $n=1$ state is given by $\psi_1(r)=A_1e^{-r/a_0}$, where the normalization constant is calculated in Example 38.2 to be $A_1=1/\left(\sqrt{\pi a_0^{3/2}}\right)$. The radial part of the hydrogen wave function must normalize to 1.

SKETCH:

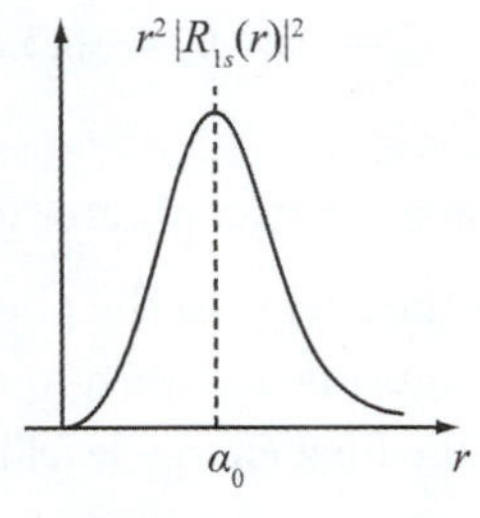

RESEARCH: The probability density is given by

$$P_n(r)=r^2\left|\psi_n(r)\right|^2.$$

The radial wave function for $n=1$ is

$$\psi_1(r)=\frac{1}{\sqrt{\pi a_0^{3/2}}}e^{-r/a_0}.$$

SIMPLIFY: (a) The probability density is given for $n=1$ by

$$P_{n=1}(r)=r^2\left|\psi_1(r)\right|^2=r^2\left|\frac{1}{\sqrt{\pi a_0^{3/2}}}e^{-r/a_0}\right|^2=\frac{r^2}{\pi a_0^3}e^{-2r/a_0}.$$

For $r=a_0/2$,

$$P_{n=1}(a_0/2)=\frac{(a_0/2)^2}{\pi a_0^3}e^{-2(a_0/2)/a_0}=\frac{1}{4\pi a_0}e^{-1}=\frac{1}{4\pi a_0 e}.$$

(b) The radial function, $\psi_1(r)$ has a maximum at the origin $(r=0)$. To find the maximum in the probability density we need to take its derivative with respect to r, set it equal to zero, and solve for r.

$$\frac{dP_1(r)}{dr}=\frac{d}{dr}\left(\frac{r^2}{\pi a_0^3}e^{-2r/a_0}\right)=\frac{1}{\pi a_0^3}\left(2re^{-2r/a_0}-\frac{2r^2}{a_0}e^{-2r/a_0}\right)=0.$$

$$2re^{-2r/a_0}=\frac{2r^2}{a_0}e^{-2r/a_0}\Rightarrow 1=\frac{r}{a_0}\Rightarrow r=a_0.$$

Thus, because the probability density contains an r^2 factor instead of the r in the wavefunction, it has a maximum at a_0 rather than the origin.

CALCULATE:
Not applicable

ROUND: Not applicable.

DOUBLE-CHECK: The probability density for $n=1$ peaks at $r=a_0$. We can see that this is true in Figure 38.9b. However, the probability density does not peak at $r=4a_0$ for $n=2$ and it does not peak at $r=9a_0$ for $n=3$.

38.39. By following the derivation of the Bohr radius for hydrogen from the textbook, but using $2e$ for the charge of the nucleus, one obtains:

$$\frac{k(2e)e}{r^2}=\frac{\mu v^2}{r} \Rightarrow r\mu k2e^2=\mu^2v^2r^2=n^2\hbar^2 \Rightarrow$$

$$r=\frac{n^2\hbar^2}{2\mu ke^2}=a_0'n^2 \Rightarrow a_0'=\frac{\hbar^2}{2\mu ke^2}=\frac{1}{2}\left(\frac{\hbar^2}{\mu ke^2}\right)=\frac{a_0}{2}=\frac{0.05295\text{ nm}}{2}=0.0265\text{ nm}.$$

The Bohr radius of He^+ is half that of hydrogen.

38.43. **THINK:** The derivations for the desired expressions are the same as the derivation in the textbook for hydrogen, except instead of the Coulomb force being proportional to e^2, it is now proportional to Ze^2.

SKETCH: Not applicable.

RESEARCH: The Coulomb force and the centripetal forces on the electron are equal:

$$k\frac{Ze^2}{r'^2}=\mu\frac{v'^2}{r'}.$$

The ground state energy is given by: $E_0'=\frac{1}{2}\mu v'^2-k\frac{Ze^2}{r'}$.

SIMPLIFY:

(a) Solving for r' in terms of Z gives:

$$k\frac{Ze^2}{r'^2}=\mu\frac{v'^2}{r'} \Rightarrow \mu r'kZe^2=\mu^2v'^2r'^2.$$

Recall $\mu^2v'^2r'^2=n^2\hbar^2$, so

$$\mu r'kZe^2=n^2\hbar^2 \Rightarrow r'=\left(\frac{\hbar^2}{\mu ke^2}\right)\frac{n^2}{Z}=\frac{a_0n^2}{Z}.$$

(b) Solving for v' in terms of Z gives:

$$k\frac{Ze^2}{r'^2}=\mu\frac{v'^2}{r'} \Rightarrow v'^2=\frac{Zke^2}{\mu r'}=\frac{Zke^2}{\mu\left(a_0n^2/Z\right)}=\frac{Z^2ke^2}{\mu a_0n^2} \Rightarrow v'=\frac{Z}{n}\sqrt{\frac{ke^2}{\mu a_0}}.$$

(c) Solving for E' in terms of Z gives:

$$E'=\frac{1}{2}\mu v'^2-k\frac{Ze^2}{r'}=\frac{1}{2}\left(k\frac{Ze^2}{r'}\right)-k\frac{Ze^2}{r'}=-k\frac{Ze^2}{2r'}=-\frac{kZe^2}{2\left(a_0n^2/Z\right)}=-Z^2\left(\frac{ke^2}{2a_0n^2}\right)=\frac{-Z^2}{n^2}\left(\frac{ke^2}{2a_0}\right)=Z^2\left(-\frac{E_0}{n^2}\right).$$

CALCULATE: There is nothing to calculate.

ROUND: Not applicable.

DOUBLE-CHECK: As Z increases, there are more protons attracting the electron. The radius decreases and the speed increases with Z, which is expected for a stronger attractive force. Likewise, as the electron gets closer and Z gets larger it will be more bound to the nucleus, so $E'\propto Z^2$ is also reasonable.

38.47. The laser has a power of $P=3.00\text{ kW}$ and a wavelength of $\lambda=694\text{ nm}$. It emits a light pulse of duration $\Delta t=10.0\text{ ns}$.

(a) The energy of each of the photons in the pulse is

$$E_0=hf=h\frac{c}{\lambda}=\left(6.626\cdot10^{-34}\text{ J s}\right)\frac{3.00\cdot10^8\text{ m/s}}{694\cdot10^{-9}}=2.86\cdot10^{-19}\text{ J}.$$

(b) The total energy in each laser pulse is

$$E=P\Delta t=\left(3.00\cdot10^3\text{ W}\right)\left(10.0\cdot10^{-9}\text{ s}\right)=30.0\ \mu\text{J}.$$

The number of chromium atoms undergoing stimulated emission is:

$$N=\frac{E}{E_0}=\frac{30.0\cdot10^{-6}\text{ J}}{2.86\cdot10^{-19}\text{ J}}=1.05\cdot10^{14}.$$

38.51. The mass of an electron is $m = 9.109 \cdot 10^{-31}$ kg. The mass of a proton is $M = 1.673 \cdot 10^{-27}$ kg. By using the reduced mass, $\mu = mM/(m+M)$, the percent change in mass of the electron is:

$$\left(\frac{m-\mu}{m}\right) = \left(1 - \frac{M}{m+M}\right)100\% = \left(1 - \frac{\left(1.673 \cdot 10^{-27}\text{ kg}\right)}{\left(9.109 \cdot 10^{-31}\text{ kg}\right) + \left(1.673 \cdot 10^{-27}\text{ kg}\right)}\right)100\% = 0.05442\%.$$

If $M = m,$ the reduced mass would be:

$$\mu = \frac{mM}{m+M} = \frac{m^2}{2m} = \frac{m}{2} = \frac{\left(9.109 \cdot 10^{-31}\text{ kg}\right)}{2} = 4.555 \cdot 10^{-31}\text{ kg}.$$

38.55. For the muonic hydrogen atom the energy is calculated in the same way as that of the usual hydrogen atom with an electron, but with a new reduced mass, μ_μ. The energy is

$$E_{n,\mu} = -\frac{1}{n^2}\frac{\mu_\mu k^2 e^4}{2\hbar^2} = -\frac{\mu_\mu}{\mu}\left(\frac{1}{n^2}\frac{\mu k^2 e^4}{2\hbar^2}\right) = -\frac{\mu_\mu}{\mu}\frac{1}{n^2}E_0,$$

where E_0 is the ionization energy of hydrogen. The ratio of reduced masses is

$$\frac{\mu_\mu}{\mu} = \left(\frac{m_\mu m_\text{p}}{m_\mu + m_\text{p}}\right)\left(\frac{m_\text{e} + m_\text{p}}{m_\text{e} m_\text{p}}\right) = \frac{m_\mu}{m_\text{e}}\left(\frac{m_\text{e} + m_\text{p}}{m_\mu + m_\text{p}}\right)$$

$$= \frac{\left(105.66\text{ MeV}/c^2\right)}{\left(0.511\text{ MeV}/c^2\right)}\left(\frac{\left(0.511\text{ MeV}/c^2\right) + \left(938.27\text{ MeV}/c^2\right)}{\left(105.66\text{ MeV}/c^2\right) + \left(938.27\text{ MeV}/c^2\right)}\right) = 185.94\text{ MeV}/c^2.$$

Therefore, the first three energy levels are:

$$E_{1,\mu} = -(185.94)\frac{1}{(1)^2}(13.6\text{ eV}) = -2530\text{ eV}$$

$$E_{2,\mu} = -(185.94)\frac{1}{(2)^2}(13.6\text{ eV}) = -632\text{ eV}$$

$$E_{3,\mu} = -(185.94)\frac{1}{(3)^2}(13.6\text{ eV}) = -281\text{ eV}$$

38.59. The energy of an electron in the nth orbital of a hydrogen atoms is given by:

$$E_n = -\frac{1}{n^2}E_0.$$

The energy of the orbiting electron in a hydrogen atom with quantum number $n = 45$ is:

$$E_{45} = -\frac{1}{(45)^2}(13.6\text{ eV}) = -6.72\text{ meV}.$$

38.61. **THINK:** The electron emits a photon in going from the $n = 4$ state to the ground state, so the atom will recoil since the photon carries momentum.

SKETCH:

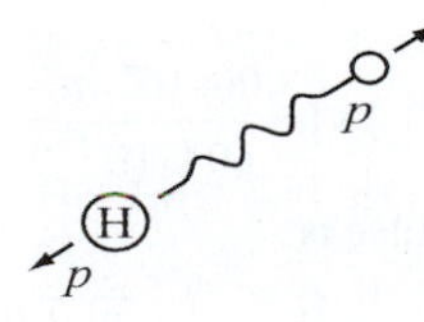

RESEARCH: The momentum of the photon is given by $p = E/c$. The energy of the electron is:

$$E_n = -\frac{1}{n^2}E_0$$

The momentum that the hydrogen atom gains is p in the opposite direction. The speed is given by $v = p / m_{\text{H}}$. The ground state energy for a hydrogen atom is $E_0 = 13.6$ eV.

SIMPLIFY: For a transition from the $n = 4$ state to the $n = 1$ state, the energy of the emitted photon is:

$$E = -E_0\left(\frac{1}{(4)^2} - \frac{1}{(1)^2}\right) = \frac{15}{16}E_0$$

Therefore, the speed of the hydrogen atom is

$$v = \frac{p}{m_{\text{H}}} = \frac{E}{m_{\text{H}}c} = \frac{15}{16}\frac{E_0}{m_{\text{H}}c}.$$

CALCULATE: $v = \dfrac{15}{16}\dfrac{(13.6 \text{ eV})(1.602 \cdot 10^{-19} \text{ J/eV})}{(1.674 \cdot 10^{-27} \text{ kg})(3.00 \cdot 10^8 \text{ m/s})} = 4.067 \text{ m/s}$

ROUND: To three significant figures, the speed of the recoiling hydrogen atom is $v = 4.07$ m/s.

DOUBLE-CHECK: This recoil speed is reasonable for a low energy photon emitted from a small mass.

Multi-Version Exercises

38.64. **THINK:** The kinetic energy of the electrons must provide enough energy for an electron in the hydrogen atom to move from the $n_1 = 1$ state to the $n_2 = 2$ state. Then the emission of light from the $n_2 = 2$ to $n_1 = 1$ can occur. Note that in this collision between the electron and the atom the total momentum is also conserved, and the hydrogen atom will recoil from the collision. However, the effect of this is a very small correction to the overall energy, because the mass of the hydrogen atom is ~2,000 that of the electron. This is why the problem statement instructs us to neglect the recoil.

SKETCH: No sketch is needed.

RESEARCH: The wavelength is given by: $\dfrac{1}{\lambda} = R_{\text{H}}\left(\dfrac{1}{n_1^2} - \dfrac{1}{n_2^2}\right)$. The kinetic energy must be equal to the energy of the photon $K = \dfrac{1}{2}m_e v^2 = \dfrac{hc}{\lambda}$.

SIMPLIFY: $\dfrac{1}{2}m_e v^2 = hcR_{\text{H}}\left(\dfrac{1}{n_1^2} - \dfrac{1}{n_2^2}\right) \Rightarrow v = \sqrt{\dfrac{2hcR_{\text{H}}}{m_e}\left(\dfrac{1}{n_1^2} - \dfrac{1}{n_2^2}\right)}$

CALCULATE: $v = \sqrt{\dfrac{2(6.626 \cdot 10^{-34} \text{ J s})(2.998 \cdot 10^8 \text{ m/s})(1.097 \cdot 10^7 \text{ m}^{-1})}{9.109 \cdot 10^{-31} \text{ kg}}\left(\dfrac{1}{1} - \dfrac{1}{4}\right)} = 1.894327754 \cdot 10^6 \text{ m/s}$

ROUND: Rounding to four significant figures, $v = 1.894 \cdot 10^6$ m/s. If we wanted to be conservative, due to our neglect of the recoil correction, we could round to three figures and give the result as $v = 1.89 \cdot 10^6$ m/s.

DOUBLE-CHECK: In the final expression, for larger values of m, v is smaller. This is reasonable. Also comforting is that the speed required increases with higher value of n_2 and decreases with higher value of n_1, both of which are expected.

38.67. $\dfrac{n_{\text{lower}}}{n_{\text{higher}}} = \exp\left[(-13.6 \text{ eV})\left(1/n_2^2 - 1/n_1^2\right)/(k_{\text{B}}T)\right]$

$$\frac{n_{\text{lower}}}{n_{\text{higher}}} = \exp\left[(-13.6 \text{ eV})\left(1/7^2 - 1/3^2\right)/\left((8.61733 \cdot 10^{-5} \text{eV/K})(528.3 \text{ K})\right)\right] = 5.85722 \cdot 10^{11} = 5.86 \cdot 10^{11}$$

Chapter 39: Elementary Particle Physics

Exercises

39.23. r_{min} occurs when the initial kinetic energy is equal to the potential energy at r_{min}.

$$K = 4.50 \text{ MeV} = U(r) = \frac{k(Z_{\text{p}}e)(Z_{\text{t}}e)}{r_{\text{min}}}$$

$$r_{\text{min}} = \frac{k(Z_{\text{p}}e)(Z_{\text{t}}e)}{K} = \frac{(ke^2)Z_{\text{p}}Z_{\text{t}}}{K} = \frac{(1.44 \text{ MeV fm})(2)(78)}{4.50 \text{ MeV}} = 49.9 \text{ fm}$$

39.27. **THINK:** The alpha particle has a de Broglie wavelength of $\lambda = 6.40 \text{ fm} = 6.40 \cdot 10^{-15}$ m and kinetic energy $K = 5.00 \text{ MeV} = 5.00 \cdot 10^6$ eV. The closest distance this alpha particle can get to the gold nucleus is $r_{\text{min}} = 45.5 \text{ fm} = 45.5 \cdot 10^{-15}$ m. Determine how the ratio r_{min} / λ varies with the kinetic energy of the alpha particle. Note that r_{min} / λ is unitless.

SKETCH:

alpha particle

gold nucleus — Au — r_{min} — He^{2+}

RESEARCH: At the point of closest approach all of the alpha particle's kinetic energy has been converted into potential energy: $U(r_{\text{min}}) = K$, where K is the initial kinetic energy of the alpha particle and $U(r_{\text{min}})$ is given by the Coulomb potential: $U(r_{\text{min}}) = kZ_{\alpha}Z_{\text{Au}}e^2 / r_{\text{min}}$. The de Broglie wavelength of the alpha particle is $\lambda = h/p = h/(m_{\alpha}v_{\alpha})$. The equation for K is $K = (1/2)m_{\alpha}v_{\alpha}^2$. The charge number for an alpha particle is $Z_{\alpha} = 2$, the charge number for gold is $Z_{\text{Au}} = 79$.

SIMPLIFY: $U(r_{\text{min}}) = K = \dfrac{kZ_{\alpha}Z_{\text{Au}}e^2}{r_{\text{min}}} \Rightarrow r_{\text{min}} = \dfrac{kZ_{\alpha}Z_{\text{Au}}e^2}{K} \Rightarrow \dfrac{r_{\text{min}}}{\lambda} = \dfrac{kZ_{\alpha}Z_{\text{Au}}e^2 p}{Kh} = \dfrac{kZ_{\alpha}Z_{\text{Au}}e^2 m_{\alpha}v_{\alpha}}{Kh}$,

but $m_{\alpha}v_{\alpha} = \sqrt{2m_{\alpha}K}$. Substituting this into the equation gives:

$$\frac{r_{\text{min}}}{\lambda} = \frac{kZ_{\alpha}Z_{\text{Au}}e^2\sqrt{2m_{\alpha}K}}{Kh}$$

$$= \frac{kZ_{\alpha}Z_{\text{Au}}e^2\sqrt{2m_{\alpha}}}{h\sqrt{K}} = \frac{Z_{\alpha}Z_{\text{Au}}ke^2\sqrt{2m_{\alpha}}}{h\sqrt{K}} = \frac{158ke^2\sqrt{2m_{\alpha}}}{h\sqrt{K}}.$$

Therefore, the ratio of r_{min} / λ is proportional to $1/\sqrt{K}$.

CALCULATE:

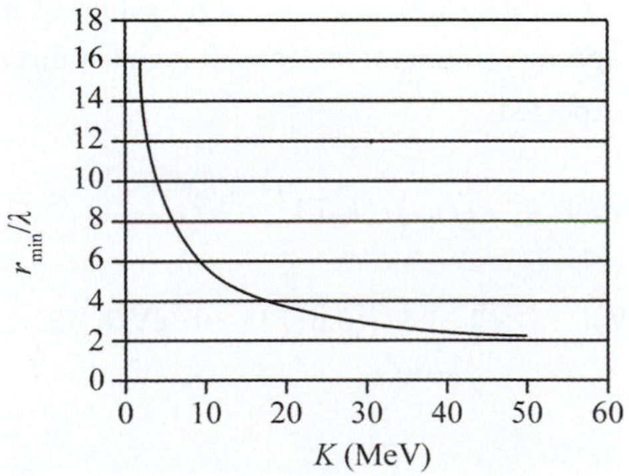

ROUND: Not applicable.

DOUBLE-CHECK: It makes sense that the ratio r_{min} / λ is affected by kinetic energy but not rapidly; that is, it is not inversely proportional to the square or cube of kinetic energy. Greater kinetic energy means one can probe smaller distances, and at the same time it means smaller wavelength. So r_{min} and λ both get smaller, but r_{min} does so a little faster.

39.31. **THINK:** The form factor is $F^2(\Delta p)$ and the Coulomb scattering differential cross section is $d\sigma / d\Omega$. Evaluate these quantities for an electron beam that is scattering off a uniform density sphere. The sphere has total charge Ze and radius R. Describe the scattering pattern.

SKETCH: See Figure 39.14 in the text.

RESEARCH: The differential cross section is given by equation 39.8 in the text: $\frac{d\sigma}{d\Omega} = \left(\frac{d\sigma}{d\Omega}\right)_{\text{point}} F^2(\Delta p)$, where $\left(\frac{d\sigma}{d\Omega}\right)_{\text{point}} = \frac{\left(2kZ_{\text{p}}Z_{\text{t}}e^2 m_{\text{p}}\right)^2}{(\Delta p)^4}$ (equation 39.4 in the text) and the form factor is given by equation 39.7: $F^2(\Delta p) = \left| \frac{1}{Ze} \int \rho(\vec{r}) e^{i\Delta\vec{p}\cdot\vec{r}/\hbar} dV \right|^2$. Note that in equation 39.4, m_{p} is the mass of the projectile. The total charge of the sphere is $Ze = \rho V = \rho \frac{4}{3}\pi R^3$. Because of the symmetry of the problem it will be most convenient to use spherical coordinates when evaluating the form factor. Note that by the definition of the dot product $\Delta\vec{p}\cdot\vec{r} = \Delta p r \cos\theta$.

SIMPLIFY: First determine the form factor. In spherical coordinates

$$F^2(\Delta p) = \left| \frac{\rho}{Ze} \int_0^{2\pi} \int_0^{\pi} \int_0^{R} e^{i\Delta pr\cos\theta/\hbar} r^2 \sin\theta dr d\theta d\phi \right|^2 = \left| \frac{2\pi\rho}{Ze} \int_0^{\pi} \int_0^{R} e^{i\Delta pr\cos\theta/\hbar} r^2 \sin\theta dr d\theta \right|^2.$$

Now, consider just the integral $\int_0^{\pi} e^{i\Delta pr\cos\theta/\hbar} \sin\theta d\theta$. This integral can be evaluated using substitution: let $i\Delta pr / \hbar = \alpha$ and let $\cos\theta = u$, then $du = -\sin\theta d\theta$. Substituting these values into the integral gives:

$$-\int e^{\alpha u} \frac{\sin\theta du}{\sin\theta} = -\int e^{\alpha u} du = -\frac{1}{\alpha} e^{\alpha u}.$$

Evaluating gives: $-\frac{\hbar e^{i\Delta pr\cos\theta/\hbar}}{i\Delta pr}\bigg|_0^{\pi} = \frac{-\hbar}{i\Delta pr}\left(e^{-i\Delta pr/\hbar} - e^{i\Delta pr/\hbar}\right)$, note that $\frac{e^{i\Delta pr/\hbar} - e^{-i\Delta pr/\hbar}}{2i} = \sin\left(\frac{\Delta pr}{\hbar}\right)$, so the evaluated value is $\frac{2\hbar}{\Delta pr}\sin\left(\frac{\Delta pr}{\hbar}\right)$. Substituting this back into the integral equation for $F^2(\Delta p)$ gives: $F^2(\Delta p) = \left| \frac{4\pi\rho\hbar}{Ze\Delta p} \int_0^R r \sin\left(\frac{\Delta pr}{\hbar}\right) dr \right|^2$. This integral can be solved using integration by parts. Let $\Delta p / \hbar = \beta$, let $u = r$, $du = dr$, $dv = \sin(\beta r)$ and $v = (-1/\beta)\cos(\beta r)$.

$$\frac{4\pi\rho\hbar}{Ze\Delta p} \int_0^R r\sin(\beta r) dr = \frac{4\pi\rho\hbar}{Ze\Delta p}\left(\frac{-r\cos(\beta r)}{\beta}\bigg|_0^R + \int_0^R \frac{1}{\beta}\cos(\beta r) dr \right) = \frac{4\pi\rho\hbar}{Ze\Delta p}\left(\frac{-R\cos(\beta R)}{\beta} + \frac{1}{\beta^2}\sin(\beta R) \right)$$

Substituting for β gives the desired version of the form factor:

$$F^2(\Delta p) = \left| \frac{4\pi\rho\hbar^3}{Ze(\Delta p)^3} \left(\sin\left(\frac{\Delta pR}{\hbar}\right) - \left(\frac{\Delta pR}{\hbar}\right)\cos\left(\frac{\Delta pR}{\hbar}\right) \right) \right|^2.$$

Substituting $F^2(\Delta p)$ and $\left(\frac{d\sigma}{d\Omega}\right)_{\text{point}}$ into the differential cross section gives:

$$\frac{d\sigma}{d\Omega}=\frac{(2kZ_pZ_te^2m_p)^2}{(\Delta p)^4}\left(\frac{4\pi\rho\hbar^3}{Ze(\Delta p)^3}\left[\sin\left(\frac{\Delta pR}{\hbar}\right)-\left(\frac{\Delta pR}{\hbar}\right)\cos\left(\frac{\Delta pR}{\hbar}\right)\right]\right)^2.$$

Recall that $Ze=\rho V=\rho\frac{4}{3}\pi R^3$ so this can be substituted into the equation to get:

$$\frac{d\sigma}{d\Omega}=\frac{(2kZ_pZ_te^2m_p)^2}{(\Delta p)^4}\left(\frac{3\hbar^3}{(\Delta p)^3R^3}\left[\sin\left(\frac{\Delta pR}{\hbar}\right)-\left(\frac{\Delta pR}{\hbar}\right)\cos\left(\frac{\Delta pR}{\hbar}\right)\right]\right)^2.$$

For $\frac{\Delta pR}{\hbar}\approx 1$ this equation gives the expected result for Rutherford scattering, but at large values of momentum transfer the differential cross section decreases much faster.

CALCULATE: Does not apply.

ROUND: Does not apply.

DOUBLE-CHECK: The derived equation implies that the differential cross section drops off quickly for larger values of Δp, this is consistent with the discussion of quantum limitations in the text and explains why at high kinetic energies the Rutherford model breaks down.

39.35. The Feynman diagram for an election–proton scattering event, $e^- + p \to e^- + p$, that is mediated by photon (γ) exchange can be drawn as follows:

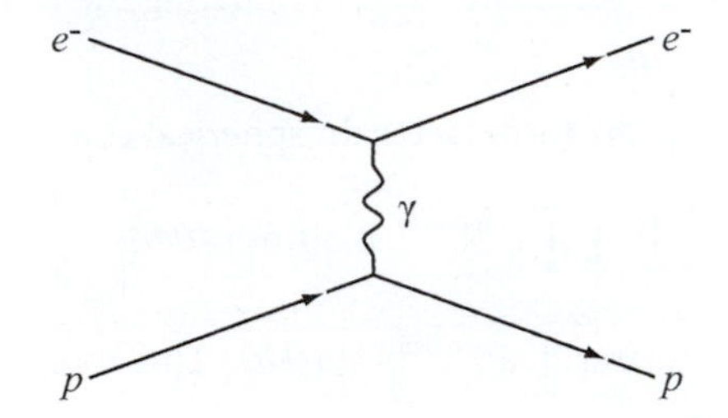

39.39. The Carbon dioxide molecule (CO_2) is made up of 1 carbon atom and 2 oxygen atoms. The atomic number for carbon is $Z=6$, so there are 6 protons, 6 neutrons and 6 electrons in this atom. Each proton is made of 3 quarks (uud) and each neutron is made of 3 quarks (udd), so the carbon atom has 6 electrons + 3(6) quarks + 3(6) quarks = 42 fermions. The atomic number for oxygen is $Z=8$, so there are 8 electrons + 3(8) quarks + 3(8) quarks = 56 fermions in each oxygen atom. The total number of fermions in the carbon dioxide molecule = 42+56 +56 = 154 fermions.

39.43. Equation 39.17 from the text gives the time dependence of the temperature during the radiation-dominated era. $T(t)=1.5\cdot 10^{10}\text{ K s}^{1/2}/\sqrt{t}$. In the text, in the section discussing Quark-Gluon plasma, it is mentioned that color singlets began to form at a temperature of approximately $2.1\cdot 10^{12}$ K. This can be used as the temperature when proton and neutrons began to form. Solving equation 39.17 in terms of time gives:

$$t=\frac{\left(1.5\cdot 10^{10}\text{ K s}^{1/2}\right)^2}{T^2}=\frac{\left(1.5\cdot 10^{10}\text{ K s}^{1/2}\right)^2}{\left(2.1\cdot 10^{12}\text{ K}\right)^2}=5.1\cdot 10^{-5}\text{ s}.$$

This would be the estimated age of the universe when protons and neutrons began to form.

39.47. **THINK:** The ratio of the wavelength of light received, λ_{rec} from a galaxy to its wavelength at emission, λ_{emit} is equal to the ratio of the scale factor (e.g., radius of curvature) of the Universe at reception to its value at emission ($a_{\text{rec}}/a_{\text{emit}}$). The redshift, z of the light is defined by $1+z=\lambda_{\text{rec}}/\lambda_{\text{emit}}=a_{\text{rec}}/a_{\text{emit}}$.

(a) Hubble's Law states that the redshift z of light from a galaxy is proportional to the galaxy's distance from us (for reasonably nearby galaxies). Derive this law from the first relationships above, and determine the Hubble constant in terms of the scale-factor function $a(t)$.
(b) If the present Hubble constant has the value $H_0 = 72$ (km/s)/Mpc determine the distance Δs from us to a galaxy which has light with a redshift of $z = 0.10$.
SKETCH: Not applicable.
RESEARCH:
(a) Hubble's law states $z \approx c^{-1}H\Delta s$, where c is the speed of light in vacuum, H is the Hubble constant, and Δs is the distance to the galaxy. The scale factor function $a_{\text{emit}}(t)$ can be expanded backwards in time from the present using a Taylor expansion. For a reasonably close source, expanding the series to first order should be a good approximation: $a_{\text{emit}}(t) \approx a_{\text{rec}} - (da/dt)_{\text{rec}}\Delta t$
(b) Although the numerical value is not required in the calculation, the megaparsec (Mpc) is a unit of length equal to $3.26 \cdot 10^6$ light years. Hubble's law can be used to calculate the distance.
SIMPLIFY:
(a) $a_{\text{emit}} \cong a_{\text{rec}} - (da/dt)_{\text{rec}}\Delta t$, substituting this into the equation that defines the redshift gives:

$$1+z \cong \frac{a_{\text{rec}}}{a_{\text{rec}} - \left(\frac{da}{dt}\right)_{\text{rec}}\Delta t} = \frac{a_{\text{rec}}}{a_{\text{rec}}\left(1 - \frac{1}{a_{\text{rec}}}\left(\frac{da}{dt}\right)_{\text{rec}}\Delta t\right)} = \left(1 - \left(\frac{1}{a}\frac{da}{dt}\right)_{\text{rec}}\Delta t\right)^{-1} \Rightarrow 1+z \cong 1 + \left(\frac{1}{a}\frac{da}{dt}\right)_{\text{rec}}\Delta t.$$ We

know that $\Delta t = \Delta s / c$. Therefore, $1+z \cong 1 + \frac{1}{c}\left(\frac{1}{a}\frac{da}{dt}\right)_{\text{rec}}\Delta s \Rightarrow z \cong \frac{1}{c}\left(\frac{1}{a}\frac{da}{dt}\right)_{\text{rec}}\Delta s$. Comparison of this

equation and the equation for Hubble's law shows that $H = \left(\frac{1}{a}\frac{da}{dt}\right)_{\text{rec}}$.

(b) $\Delta s \approx zc/H_0$
CALCULATE:
(a) Not applicable.

(b) $\Delta s \cong \dfrac{(0.10)(3.00 \cdot 10^5 \text{ km/s})}{72 \text{ (km/s)/Mpc}} = 416.66 \text{ Mpc}$

ROUND: The answer should be reported to two significant figures, therefore (b) $\Delta s = 420$ Mpc.
DOUBLE-CHECK: The calculated distance to the galaxy has units of length, which is expected. This distance is approximately 1.4 billion light years. This distance corresponds to a time that falls within the age of the universe.

39.51. The resolution provided by the electrons with kinetic energy $K = p^2/2m = 100.$ eV is dependent on their de Broglie wavelength ($\lambda = h/p$), which is inversely proportional to their momentum. With the rest mass of an electron being $m_e = 0.511$ MeV/c^2, the momentum of the electrons is:

$$p = (2m_e K)^{1/2} = \left(2 \cdot (0.511 \text{ MeV/c}^2)(100. \text{ eV})\right)^{1/2} = 10.1 \text{ keV/c} = 5.40 \cdot 10^{-24} \text{ kg m/s}$$

To obtain the same resolution, the neutrons must have the same momentum (wavelength) as the electrons. Since the mass of the neutron is $m_n = 1.675 \cdot 10^{-27}$ kg, this corresponds to a kinetic energy of:

$$K = p^2/2m_n = \left(5.40 \cdot 10^{-24} \text{ kg m/s}\right)^2 / \left(2 \cdot 1.675 \cdot 10^{-27} \text{ kg}\right) = 8.71 \cdot 10^{-21} \text{ J} = 0.0544 \text{ eV}.$$

The neutrons would require a kinetic energy of 54.4 meV.

39.55. The cross section of the interaction is $\sigma_\Lambda = (\Delta x)^2$ where Δx is the interaction range. The range is related to the decay time by $\Delta x \approx \tau c$. Therefore $\sigma_\Lambda = (\Delta x)^2 \approx (\tau c)^2 = \left((10^{-10} \text{ s}) \cdot (3.00 \cdot 10^8 \text{ m/s})\right)^2 = 9 \cdot 10^{-4} \text{ m}^2$.

39.58. THINK: The two photons created by the annihilation of the electron-positron pair must have the same energy as the combined energy of the particles. To conserve momentum, the photons must have equal and opposite momentum. The electron and positron have the same rest mass $E_0 = 0.511$ MeV and are each traveling at $v = 0.99c$ with respect to their center of mass.

SKETCH:

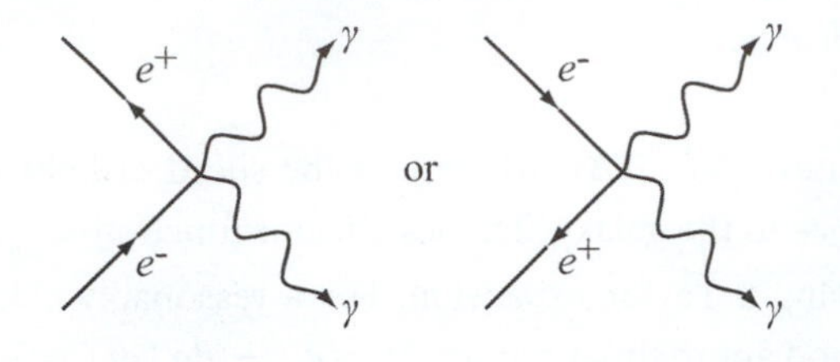

RESEARCH: The energy of a relativistic particle is $E = \gamma mc^2 = \gamma E_0 = E_0 / \sqrt{1-(v/c)^2}$. The wavelength of a photon of energy E_γ is $\lambda = hc / E_\gamma$.

SIMPLIFY: The energy of one photon is half that of the total energy of the electron-positron pair:

$$E_\gamma = \frac{1}{2}(E_{e^-} + E_{e^+}) = \frac{1}{2}(\gamma E_0 + \gamma E_0) = \gamma E_0 = \frac{E_0}{\sqrt{1-(v/c)^2}}.$$

The wavelength of the photon is then $\lambda = \frac{hc}{E_\gamma} = \frac{hc\sqrt{1-(v/c)^2}}{E_0}$.

CALCULATE: $\lambda = \frac{1240 \text{ eV nm}}{0.511 \cdot 10^6 \text{ eV}} \sqrt{1-0.99^2} = 342.3 \text{ fm}$

ROUND: The speed is given to two significant figures, so the wavelength is accurate to two significant figures. The annihilation produces photons of wavelength $\lambda = 3.4 \cdot 10^{-13}$ m = 340 fm.

DOUBLE-CHECK: Because the electrons are traveling at a speed close to that of light, the annihilation should produce very energetic photons. Energetic photons have a high frequency, but a small wavelength as these photons do.

39.61. THINK: What is the differential cross section for a beam of electrons Coulomb-scattering off a thin spherical shell of total charge *Ze* and radius *a*? Can this experiment distinguish between the thin-shell and a solid-sphere charge distribution?

SKETCH:

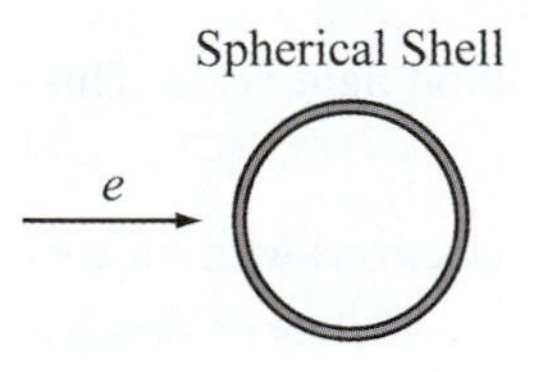

RESEARCH: The differential cross section is given by $\frac{d\sigma}{d\Omega} = \left(\frac{2Ze^2 m_e}{4\pi\varepsilon_0}\right)^2 \frac{1}{(\Delta p)^4} F^2(\Delta p)$, with form factor

$F^2(\Delta p) = \left| \frac{1}{Ze} \int \rho(\vec{r}) \exp\left(\frac{i}{\hbar}\Delta\vec{p}\cdot\vec{r}\right) dV \right|^2$. But in this case the target charge density ρ is concentrated in a thin spherical shell of radius *a*. The integral over *r* extends only over the thickness of the shell, with all the functions in the integrand evaluated at $r = a$. The charge density takes the form $\rho(r) = \frac{Ze}{4\pi a^2}\delta(r-a)$, using the Dirac delta function. Equivalently, the integral can be taken to be a surface integral, with the charge density ρ replaced by a surface charge density $Ze/(4\pi a^2)$. The cross section for a solid sphere of radius *a* is (see Solution 39.31)

$$\frac{d\sigma}{d\Omega}=\left(\frac{2Ze^2m_e}{4\pi\varepsilon_0}\right)^2\frac{1}{(\Delta p)^4}\left\{\frac{3\hbar}{(\Delta p)^3a^3}\left[\sin\left(\frac{(\Delta p)a}{\hbar}\right)-\frac{(\Delta p)a}{\hbar}\cos\left(\frac{(\Delta p)a}{\hbar}\right)\right]\right\}^2$$

SIMPLIFY: The form factor is given by

$$\begin{aligned}F(\Delta p)&=\frac{1}{4\pi a^2}\int_0^{2\pi}\int_0^{\pi}\exp\left(\frac{i(\Delta p)a}{\hbar}\cos\theta\right)a^2\sin\theta d\theta d\phi\\&=\frac{1}{4\pi a^2}a^2\int_0^{\pi}\exp\left(\frac{i(\Delta p)a}{\hbar}\cos\theta\right)\sin\theta\, d\theta\int_0^{2\pi}d\phi\\&=\frac{1}{2}\int_0^{\pi}\exp\left(\frac{i(\Delta p)a}{\hbar}\cos\theta\right)\sin\theta\, d\theta\\&=\frac{\hbar}{2(\Delta p)ai}\left[-\exp\left(\frac{i(\Delta p)a}{\hbar}\cos\theta\right)\right]\Bigg|_0^{\pi}\\&=\frac{\hbar}{2(\Delta p)ai}\left(\exp[i(\Delta p)a/\hbar]-\exp[-i(\Delta p)a/\hbar]\right)\\&=\frac{\sin[(\Delta p)a/\hbar]}{(\Delta p)a/\hbar}\end{aligned}$$

The cross section is therefore $\frac{d\sigma}{d\Omega}=\left(\frac{2Ze^2m_e}{4\pi\varepsilon_0}\right)^2\frac{1}{(\Delta p)^4}\left(\frac{\sin[(\Delta p)a/\hbar]}{(\Delta p)a/\hbar}\right)^2.$

CALCULATE: There is no need to calculate.

ROUND: There is no need to round. Like the cross section for the solid sphere, this matches the point-target result in the $\Delta p\to 0$ limit, but falls off much more rapidly for large momentum transfer. It also has zero for (in this case, periodic) values of the momentum transfer, so this scattering pattern too can show a central maximum surrounded by bright and dark rings. However, this pattern is distinguishable from that for the solid sphere target. It falls off more quickly with increasing Δp for small values, but for large values it falls off much more slowly: as $(\Delta p)^{-6}$, rather than $(\Delta p)^{-8}$ though still faster than the $(\Delta p)^{-4}$ fall off of the point-target cross section. So, yes, a scattering experiment with sufficient data could distinguish between a solid and a hollow spherical target.

DOUBLE-CHECK: The technique should be able to distinguish between the two charge distributions. One might expect that both spheres can be approximated by a point charge at the center of the sphere. The charge distribution, however, that the electron 'sees' is a function of its momentum. The faster the electron goes the closer it gets to the real particles and the more it can resolve the real charge distribution.

Multi-Version Exercises

39.62. $\Pi=\rho N_A t\sigma/M$

$$=\frac{(2.77\text{ g/cm}^3)(6.022\cdot 10^{23}/\text{mole})(68.5\text{ cm})(0.68\cdot 10^{-38}\text{ cm}^2/\text{GeV})(337\text{ GeV})}{27.0\text{ g/mole}}$$

$$=9.7\cdot 10^{-12}$$

39.65. $I(\theta)\propto\sin^{-4}(\tfrac{1}{2}\theta)\Rightarrow$

$$I(\theta_2)=I(\theta_1)\sin^4(\tfrac{1}{2}\theta_1)/\sin^4(\tfrac{1}{2}\theta_2)$$

$$=(853\text{ /s})\sin^4(47.45°)/\sin^4(30.25°)=3900.671\text{ /s}=3.90\cdot 10^3\text{ /s}$$

Chapter 40: Nuclear Physics

Exercises

40.23. In general, binding energy is given by

$$B(N,Z)=\left[Zm(0,1)+Nm_{\text{n}}-m(N,Z)\right]c^2,$$

where $m(0,1)=1.007825032$ u, $m_{\text{n}}=1.008664916$ u and $1\text{ u}=931.4940\text{ MeV}/c^2$. The choice of how to round the values is arbitrary. Round to the nearest integer.

(a) ^7Li has $N=4$, $Z=3$ and $m(4,3)=7.0160045$ u, so:

$$B(4,3)=\left[3(1.007825032)+4(1.008664916)-(7.0160045)\right]c^2\left(931.4940\text{ MeV}/c^2\right)$$
$$=39\text{ MeV}.$$

(b) ^{12}C has $N=6$, $Z=6$ and $m(6,6)=12.000000$ u, so:

$$B(6,6)=\left[6(1.007825032)+6(1.008664916)-(12.000000)\right]c^2\left(931.4940\text{ MeV}/c^2\right)$$
$$=92\text{ MeV}.$$

(c) ^{56}Fe has $N=30$, $Z=26$ and $m(30,26)=55.93493748$ u, so:

$$B(30,26)=\left[30(1.007825032)+26(1.008664916)-(55.93493748)\right]c^2\left(931.4940\text{ MeV}/c^2\right)$$
$$=489\text{ MeV}.$$

(d) ^{85}Rb has $N=48$, $Z=37$ and $m(48,37)=84.91178974$ u, so:

$$B(48,37)=\left[48(1.007825032)+37(1.008664916)-(84.91178974)\right]c^2\left(931.4940\text{ MeV}/c^2\right)$$
$$=731\text{ MeV}.$$

40.27. In β^- decay, the atomic number, Z, increases by one, and an electron and electron anti-neutrino is emitted.

(a) $^{60}_{27}\text{Co}\rightarrow{}^{60}_{28}\text{Ni}+e^-+\overline{\nu}_e$

(b) $^{3}_{1}\text{H}\rightarrow{}^{3}_{2}\text{He}+e^-+\overline{\nu}_e$

(c) $^{14}_{6}\text{C}\rightarrow{}^{14}_{7}\text{N}+e^-+\overline{\nu}_e$

40.31. Since the isotope decays to 1/8 its original amount in $t'=5.00$ h, it goes through three half-lives. Therefore the equation describing the amount of isotope remaining after t hours is

$$N(t)=N_0e^{-\lambda t}=N_0e^{-\ln(2)t/t_{1/2}}=N_0\left(\frac{1}{2}\right)^{t/t_{1/2}}$$

$$N(t)=0.900N_0=N_0\left(\frac{1}{2}\right)^{t/t_{1/2}}\Rightarrow 0.900=\left(\frac{1}{2}\right)^{t/t_{1/2}}\Rightarrow t=\frac{\ln(0.900)}{\ln(1/2)}t_{1/2}=\frac{\ln(0.900)}{\ln(1/2)}(5.00\text{ h}/3)=0.253\text{ h}$$

40.35. **THINK:** Carbon-14 has a half-life of $t_{1/2}=5730$ yr. For a piece of wood, the radioactive decay of ^{14}C follows an exponential decay law, while the number of ^{12}C isotopes stays constant in time because this isotope is stable. Since the ratio of the number of ^{14}C atoms to the number of ^{12}C stays constant until the intake of ^{14}C ceases (when the tree died), the initial amount of ^{14}C can be found. It can be assumed that ^{12}C and ^{14}C comprise all of the mass of the wood, $m=5.00$ g. Even though the wood was cut on January 1, 1700, the actual date today is not important to determine the activity today. It can be easily approximated by simply using the year associated with the date of the measurement (i.e. 2010). This is because the half-life is large (on a year scale), so being off by a few months (or even a full year) would only result in an error of less than 1%.

SKETCH:

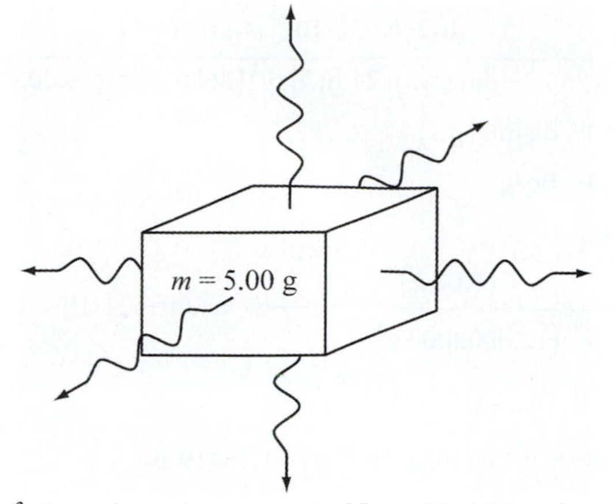

RESEARCH: The number of atoms in a given mass is $N = mN_A / M$, where M is the molar mass. The activity of a material is $A = \lambda N$, where the decay constant is $\lambda = \ln 2 / t_{1/2}$. The specific activity is $S_A = A / m\left({}^{14}\text{C}\right)$. The radioactive decay law is given by $N(t) = N_0 e^{-\lambda t}$. As stated in the text, the initial ratio of ${}^{14}\text{C}$ to ${}^{12}\text{C}$ atoms is $r = N_0\left({}^{14}\text{C}\right)/N\left({}^{12}\text{C}\right) = 1.20 \cdot 10^{-12}$.

SIMPLIFY:

(a) The number of ${}^{14}\text{C}$ atoms per gram is given by:

$$N\left({}^{14}\text{C}\right) = \frac{m\left({}^{14}\text{C}\right)N_A}{M\left({}^{14}\text{C}\right)} \Rightarrow \frac{N\left({}^{14}\text{C}\right)}{m\left({}^{14}\text{C}\right)} = \frac{N_A}{M\left({}^{14}\text{C}\right)}.$$

The specific activity of ${}^{14}\text{C}$ is then:

$$S_A = \frac{A}{m\left({}^{14}\text{C}\right)} = \frac{\lambda N\left({}^{14}\text{C}\right)}{m\left({}^{14}\text{C}\right)} = \frac{\ln 2}{t_{1/2}}\frac{N\left({}^{14}\text{C}\right)}{m\left({}^{14}\text{C}\right)} = \frac{\ln 2 N_A}{t_{1/2}M\left({}^{14}\text{C}\right)}.$$

(b) To find the initial activity of a piece of wood with $m = 5.00$ g, the mass of ${}^{14}\text{C}$ present in the piece of wood needs to be found:

$$m\left({}^{12}\text{C}\right) = \left(\frac{N\left({}^{12}\text{C}\right)M\left({}^{12}\text{C}\right)}{N_0\left({}^{14}\text{C}\right)M\left({}^{14}\text{C}\right)}\right)m\left({}^{14}\text{C}\right) \text{ and } m\left({}^{12}\text{C}\right) + m\left({}^{14}\text{C}\right) = m$$

$$m\left({}^{14}\text{C}\right) = \frac{m}{\left(\frac{N\left({}^{12}\text{C}\right)M\left({}^{12}\text{C}\right)}{N_0\left({}^{14}\text{C}\right)M\left({}^{14}\text{C}\right)}\right) + 1}$$

Then, the initial activity is given by:

$$A = S_A m\left({}^{14}\text{C}\right).$$

(c) The time passed in years is $\Delta t = 2010 \text{ yr} - 1700 \text{ yr} = 310.$ yr. Treating the year 1700 as $t = 0$, the change in the number of atoms, i.e. disintegrations is:

$$\Delta N\left({}^{14}\text{C}\right) = N_0\left({}^{14}\text{C}\right) - N\left({}^{14}\text{C}\right) = N_0\left({}^{14}\text{C}\right)\left(1 - e^{-\lambda t}\right) = \frac{m\left({}^{14}\text{C}\right)N_A}{M\left({}^{14}\text{C}\right)}\left(1 - e^{-t \ln 2 / t_{1/2}}\right).$$

CALCULATE:

(a) $$S_A = \frac{\ln 2\left(6.022\cdot 10^{23}\text{ atoms/mol}\right)}{(5730\text{ yr})(365.25\text{ days/yr})(24\text{ hr/day})(3600\text{ s/hr})(14.0032420\text{ g/mol})}$$
$$= 1.64846\cdot 10^{11}\text{ disint/(g s)}$$
$$= 1.64846\cdot 10^{11}\text{ Bq/g}$$
$$= 4.4553\text{ Ci/g}$$

(b) $$m\left({}^{14}\text{C}\right) = \frac{(5.00\text{ g})}{\left(\dfrac{(12.0000000\text{ g})}{\left(1.20\cdot 10^{-12}\right)(14.0032420\text{ g})}\right)+1} = 7.001621\cdot 10^{-12}\text{ g}$$

$$A = \left(1.64846\cdot 10^{11}\text{ Bq/g}\right)\left(7.001621\cdot 10^{-12}\text{ g}\right) = 1.15419\text{ Bq}$$

(c) $$\Delta N\left({}^{14}\text{C}\right) = \frac{\left(7.001621\cdot 10^{-12}\text{ g}\right)\left(6.022\cdot 10^{23}\text{ atoms/mol}\right)}{(14.0032420\text{ g/mol})}\left(1-e^{-(310.\text{ yr})\ln 2/5730\text{ yr}}\right) = 1.108219\cdot 10^{10}\text{ disint}$$

ROUND:

(a) $S_A = 1.65\cdot 10^{11}\text{ disint/(g s)} = 1.65\cdot 10^{11}\text{ Bq/g} = 4.46\text{ Ci/g}$

(b) $A = 1.15\text{ Bq}$

(c) $\Delta N\left({}^{14}\text{C}\right) = 1.11\cdot 10^{10}\text{ disint}$

DOUBLE-CHECK: These are reasonable results for radioactive decay over a long time period.

40.39. **THINK:** $N_B(t)$ decreases via decay $\text{B}\to\text{C}$, but is then replenished via $\text{A}\to\text{B}$. Thus, the total activity of B is the difference between the activity of A becoming B, and the activity of B becoming C. Initially, both A and B have some nuclei.

SKETCH:

$$\begin{array}{cccc} & \text{A} \xrightarrow{\lambda_A} & \text{B} \xrightarrow{\lambda_B} & \text{C} \\ t=0: & N_{A0} & N_{B0} & 0 \end{array}$$

RESEARCH: The activity of each nuclei at any time is $\left|dN_i(t)/dt\right| = -\lambda_i N_i(t)$. The number of atoms present at any time is $N_i(t) = N_{i0}e^{-\lambda_i t}$, where subscript *i* denotes either A, B or C.

SIMPLIFY: The total activity of atom B is:

$$\frac{dN_B(t)}{dt} = -\lambda_B N_B(t) + \lambda_A N_A(t) = -\lambda_B N_B(t) + \lambda_A N_{A0}e^{-\lambda_A t} \Rightarrow dN_B(t) = \lambda_A N_{A0}e^{-\lambda_A t}dt - \lambda_B N_B(t)dt$$
$$\Rightarrow dN_B(t)e^{\lambda_B t} + \lambda_B N_B(t)e^{\lambda_B t}dt = \lambda_A N_{A0}e^{(\lambda_B-\lambda_A)t}dt.$$

Consider $N_B(t)e^{\lambda_B t}$:

$$\frac{d}{dt}\left(N_B(t)e^{\lambda_B t}\right) = \frac{dN_B(t)}{dt}e^{\lambda_B t} + N_B(t)\lambda_B e^{\lambda_B t} \Rightarrow d\left(N_B(t)e^{\lambda_B t}\right) = dN_B(t)e^{\lambda_B t} + \lambda_B N_B(t)e^{\lambda_B t}dt.$$

Therefore,

$$\int_{N_{B,0}}^{N_B(t)} d\left(N_B(t)e^{\lambda_B t}\right) = \int_0^t \lambda_A N_{A0}e^{(\lambda_B-\lambda_A)t}dt$$
$$N_B(t)e^{\lambda_B t} - N_{B0} = \frac{\lambda_A}{\lambda_B-\lambda_A}N_{A0}e^{(\lambda_B-\lambda_A)t} - \frac{\lambda_A}{\lambda_B-\lambda_A}N_{A0}$$
$$N_B(t) = \frac{\lambda_A}{\lambda_B-\lambda_A}N_{A0}e^{-\lambda_A t} + \left(N_{B0} - \frac{\lambda_A}{\lambda_B-\lambda_A}N_{A0}\right)e^{-\lambda_B t}.$$

CALCULATE: Not applicable.
ROUND: Not applicable.
DOUBLE-CHECK: In general, the population of atom B would decrease. However, if $\lambda_A >> \lambda_B$, then B is replenished by A faster than B can decay to C. The solution when $\lambda_A >> \lambda_B$, simplifies to:

$$N_B(t) \approx -N_{A0}e^{-\lambda_A t} + N_{B0}e^{-\lambda_B t} + N_{A0}e^{-\lambda_B t} \approx (N_{B0} + N_{A0})e^{-\lambda_B t} - N_{A0}e^{-\lambda_A t}.$$

The first term is the regular decay of B if all of A was instantly converted to B (which is true when $\lambda_A >> \lambda_B$). The second term is the number subtracting the actual number of atoms still in A. Since this approximation validates what is known should happen in the limit, the solution is reasonable.

40.43. Given that the power plant with $P = 1.50$ GW has an efficiency of $\varepsilon = 0.350$, the total power that is created is $P_0 = P/\varepsilon$. The energy that the plant produces in $\Delta t = 1$ day is given by $E = P_0\Delta t$. Since each ^{235}U reaction is $\Delta E = 200.$ MeV, the number of ^{235}U consumed per day is given by $N = E/\Delta E$. Since ^{235}U has a molar mass of $m_M = 235.0439299$ g/mol, the mass of ^{235}U consumed in one day is:

$$m = \frac{N}{N_A}m_M = \frac{Em_M}{\Delta EN_A} = \frac{P\Delta tm_M}{\varepsilon\Delta EN_A} = \frac{(1.50\cdot 10^9 \text{ W})(86400 \text{ s/day})(235.0439299 \text{ g/mol})}{(0.350)(200.\cdot 10^6 \text{ eV})(1.602\cdot 10^{-19} \text{ J/eV})(6.022\cdot 10^{23} \text{ atoms/mol})}$$
$$= 4.51 \text{ kg}.$$

40.47. **THINK:** In order for the two ^3_2He atoms to bind, they must come close enough for the strong force to overcome the Coulomb repulsion. The closest the two can get is when their centers are separated by the sum of the radii of the atoms, i.e. the diameter of one atom. Assuming that one is at rest, the kinetic energy of the other atom must be greater than the potential barrier due to the repulsion force. The kinetic energy is directly proportional to the temperature of the surroundings.
SKETCH:

RESEARCH: Thermal energy of a particle is given by $K = \frac{3}{2}k_B T$. The Coulomb potential is given by $U_C = kq^2/d$, where $q = 2e$ for ^3_2He. The diameter of ^3_2He is $d = 2\,R(A)$.

SIMPLIFY: The temperature is given by: $K = U_C \Rightarrow \frac{3}{2}k_B T = \frac{k(2e)^2}{2R(A)} \Rightarrow T = \frac{4ke^2}{3k_B R_0 A^{1/3}}$.

CALCULATE: $T = \frac{4(8.99\cdot 10^9 \text{ N m}^2/\text{C}^2)(1.602\cdot 10^{-19} \text{ C})^2}{3(1.12\cdot 10^{-15} \text{ m})(3)^{1/3}(1.381\cdot 10^{-23} \text{ J/K})} = 13.790\cdot 10^9 \text{ K}$

ROUND: To three significant figures, the temperature required to make the fusion occur is $T = 13.8$ GK.
DOUBLE-CHECK: This result is about 1000 times hotter than the core of the Sun. However, the temperature really only needs to be a fraction of this because there will be nuclei in the high energy "tail" of the energy distribution of a lower temperature.

40.51. The average kinetic energy is related to the temperature by $K_{ave} = 3kT/2$. Substituting the numerical values gives:

$$K_{ave} = \frac{3}{2}(1.38\cdot 10^{-23} \text{ J/K})(1.00\cdot 10^7 \text{ K}) = 2.07\cdot 10^{-16} \text{ J}.$$

This corresponds to an average velocity of:

$$K_{\text{ave}}=\frac{1}{2}mv_{\text{ave}}^2 \Rightarrow v_{\text{ave}}=\sqrt{\frac{2K_{\text{ave}}}{m}}=\sqrt{\frac{2\left(2.07\cdot 10^{-16}\text{ J}\right)}{\left(1.67\cdot 10^{-27}\text{ kg}\right)}}=4.98\cdot 10^5\text{ m/s}.$$

40.55. The exponential decay law can be expressed as $N=N_0\left(1/2\right)^{t/t_{1/2}}$. It is given that $N/N_0=0.1000$. The time required is determined using:

$$\ln\left(\frac{N}{N_0}\right)=\ln\left(\frac{1}{2}\right)\frac{t}{t_{1/2}} \Rightarrow t=t_{1/2}\frac{\ln\left(N/N_0\right)}{\ln\left(1/2\right)}=\left(3.825\text{ days}\right)\frac{\ln\left(0.1000\right)}{\ln\left(1/2\right)}=12.71\text{ days}.$$

40.59. Initially, all of the energy is in the form of kinetic energy (assume the potential is zero at a large distance). When the particles are at their closest approach, all of the energy is potential energy. Therefore,

$$K_{\text{i}}=U_{\text{f}} \Rightarrow K_{\text{i}}=\frac{kZ_1Z_2q_{\text{p}}^2}{r} \Rightarrow r=\frac{kZ_1Z_2q_{\text{p}}^2}{K_{\text{i}}}$$

$$r=\frac{\left(8.99\cdot 10^9\text{ N m}^2/\text{C}^2\right)(2)(92)\left(1.602\cdot 10^{-19}\text{ C}\right)^2}{\left(5.00\cdot 10^6\text{ eV}\right)\left(1.602\cdot 10^{-19}\text{ J}\right)}=53.0\text{ fm}.$$

40.63. The binding energy of a nucleus is given by

$$B(N,Z)=\left[Zm(0,1)+Nm_{\text{n}}-m(N,Z)\right]c^2.$$

Substituting $Z=3$, $N=5$, $m(0,1)=1.007825032\text{ u}$, $m_{\text{n}}=1.008664916\text{ u}$, $m(5,3)=8.022485\text{ u}$ and $\text{u}=931.494\text{ MeV}/c^2$ gives:

$$B(5,3)=\left[3(1.007825032)+5(1.008664916)-(8.022485)\right]c^2\left(931.494\text{ MeV}/c^2\right)=41.279\text{ MeV}.$$

40.67. The binding energy per nucleon is:

$$\frac{B(N,Z)}{A}=\frac{1}{A}\left[Zm(0,1)+Nm_{\text{n}}-m(N,Z)\right]c^2.$$

(a) For ${}^4_2\text{He}$,

$$\frac{B(2,2)}{4}=\frac{1}{4}\left[2(1.007825032)+2(1.008664916)-(4.002603)\right]c^2\left(931.494\text{ MeV}/c^2\right)$$
$$=7.074\text{ MeV}$$

(b) For ${}^3_2\text{He}$,

$$\frac{B(1,2)}{3}=\frac{1}{3}\left[2(1.007825032)+(1.008664916)-(3.016030)\right]c^2\left(931.494\text{ MeV}/c^2\right)$$
$$=2.572\text{ MeV}$$

(c) For ${}^3_1\text{H}$,

$$\frac{B(2,1)}{3}=\frac{1}{3}\left[(1.007825032)+2(1.008664916)-(3.016050)\right]c^2\left(931.494\text{ MeV}/c^2\right)$$
$$=2.827\text{ MeV}$$

(d) For ${}^2_1\text{H}$,

$$\frac{B(1,1)}{2}=\frac{1}{2}\left[(1.007825032)+(1.008664916)-(2.014102)\right]c^2\left(931.494\text{ MeV}/c^2\right)$$
$$=1.112\text{ MeV}$$

40.71. **THINK:** The radioactive decay of ^{14}C follows an exponential decay law, while the number of ^{12}C isotopes stays constant in time because this isotope is stable. It can be assumed that ^{12}C comprises all of the mass of the ash; that is, $m = m\left(^{12}\text{C}\right) = 50.0\text{ g}$. The activity, $A = 20.0\text{ decays/hr}$, can be used along with the half-life, $t_{1/2} = 5730\text{ yr}$, to determine the current number of ^{14}C atoms. Using all of this information will provide an approximate age for the tree.

SKETCH:

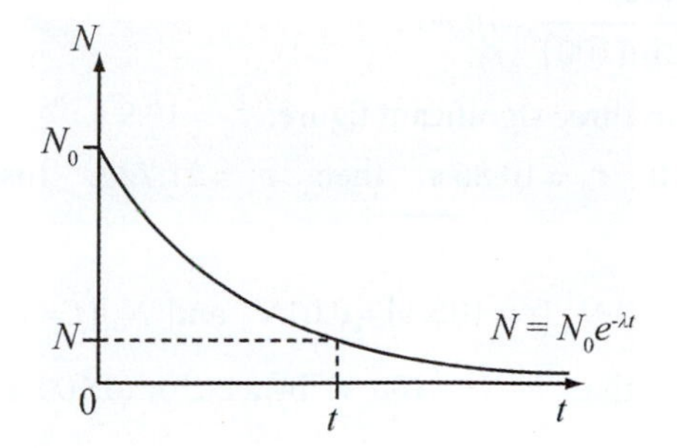

RESEARCH: The exponential decay law for the number of atoms remaining as a function of time is given by

$$N(t) = N_0 e^{-\lambda t},$$

where $\lambda = \ln 2 / t_{1/2}$. The activity of ^{14}C is given by $A = \lambda N\left(^{14}\text{C}\right)$. As stated in the question, the initial ratio of ^{14}C to ^{12}C atoms is $r = N_0\left(^{14}\text{C}\right)/N\left(^{12}\text{C}\right) = 1.300 \cdot 10^{-12}$. Therefore, the number of initial ^{14}C atoms is $N_0\left(^{14}\text{C}\right) = rN\left(^{12}\text{C}\right)$. The number of ^{12}C atoms is given by

$$N\left(^{12}\text{C}\right) = \frac{m\left(^{12}\text{C}\right)}{M\left(^{12}\text{C}\right)} N_\text{A},$$

where $M\left(^{12}\text{C}\right)$ is the molar mass of ^{12}C.

SIMPLIFY: The decay law for ^{14}C is

$$N\left(^{14}\text{C}\right) = N_0\left(^{14}\text{C}\right)e^{-\lambda t} = rN\left(^{12}\text{C}\right)e^{-\lambda t}.$$

Simplifying and solving for t gives:

$$\frac{A}{\lambda} = \frac{rm\left(^{12}\text{C}\right)N_\text{A}e^{-\lambda t}}{M\left(^{12}\text{C}\right)} \Rightarrow \frac{At_{1/2}}{\ln 2} = \frac{rm\left(^{12}\text{C}\right)N_\text{A}e^{-\ln 2t/t_{1/2}}}{M\left(^{12}\text{C}\right)} \Rightarrow e^{-\ln 2t/t_{1/2}} = \frac{AM\left(^{12}\text{C}\right)t_{1/2}}{rm\left(^{12}\text{C}\right)N_\text{A}\ln 2}$$

$$t = -\frac{t_{1/2}}{\ln 2}\ln\left(\frac{AM\left(^{12}\text{C}\right)t_{1/2}}{rm\left(^{12}\text{C}\right)N_\text{A}\ln 2}\right).$$

CALCULATE: $t = -\frac{(5730\text{ yr})}{\ln 2}\ln\left(\frac{(20.0\text{ decays/hr})(8760\text{ hr/yr})(12.000000\text{ g/mol})(5730\text{ yr})}{\left(1.300 \cdot 10^{-12}\right)(50.0\text{ g})\left(6.02 \cdot 10^{23}\text{ mol}^{-1}\right)\ln 2}\right) = 63813\text{ yr}.$

ROUND: Rounding to three significant figure yields $t = 63800\text{ yr}$.

DOUBLE-CHECK: This is a reasonable age for a campfire, considering that fossil evidence indicates that modern humans originated in Africa 200,000 years ago.

40.75. **THINK:** Radioactive decay follows an exponential decay law. In this problem, two species of radioactive nuclei, A and B, are compared. After a time of 100. s, it is observed that $N_\text{A} = 100N_\text{B}$ with $\tau_\text{A} = 2\tau_\text{B}$. The initial number of nuclei for both species is N_0.

SKETCH: A sketch is not necessary.

RESEARCH: The exponential decay law is given by $N = N_0 e^{-t/\tau}$. After an interval of time, t, the number of nuclei A and B are $N_A = N_0 e^{-t/\tau_A}$ and $N_B = N_0 e^{-t/\tau_B}$.

SIMPLIFY: Taking the ratio of N_A to N_B and using $\tau_A = 2\tau_B$ gives:

$$\frac{N_A}{N_B} = e^{-t(1/\tau_A - 1/\tau_B)} = e^{-(t/\tau_B)[(1/2)-1]} \Rightarrow \ln\left(\frac{N_A}{N_B}\right) = \frac{t}{2\tau_B} \Rightarrow \tau_B = \frac{t}{2\ln(N_A/N_B)}.$$

CALCULATE: $\tau_B = \dfrac{(100.\text{ s})}{2\ln(100)} = 10.86\text{ s}$

ROUND: Rounding to three significant figure, $\tau_B = 10.9$ s.

DOUBLE-CHECK: If $\tau_B = 10.86$ s, then $\tau_A = 21.72$ s. Inserting these results into the equation $N(t) = N_0 e^{-t/\tau}$ gives

$$N_A(t = 100.\text{ s}) = 0.01N_0 \text{ and } N_B(t = 100.\text{ s}) = 0.0001N_0.$$

$N_A(t = 100.\text{ s})$ is larger than $N_B(t = 100.\text{ s})$ by a factor of 100, as required.

Multi-Version Exercises

40.76. $\lambda = \dfrac{\ln 2}{t_{1/2}} = \dfrac{\ln 2}{5730\text{ y}} = 3.833 \cdot 10^{-12}\text{ s}^{-1}$

$$t = -\ln\left(\frac{A \cdot M(^{12}\text{C})}{r \cdot m(^{12}\text{C}) N_A \lambda}\right) / \lambda$$

$$= -\ln\left[\frac{(105.\text{ decays/min})(1\text{ min/60 s})(12\text{ g/mol})}{(1.20 \cdot 10^{-12})(12.43\text{ g})(6.022 \cdot 10^{23}\text{ mol}^{-1})(3.833 \cdot 10^{-12}\text{ s}^{-1})}\right] / (3.833 \cdot 10^{-12}\text{ s}^{-1})$$

$= 4100$ yr.